Genetic takeover

Genetic takeover

and the mineral origins of life

A. G. CAIRNS-SMITH

SENIOR LECTURER IN CHEMISTRY AT THE
UNIVERSITY OF GLASGOW

CAMBRIDGE UNIVERSITY PRESS

CAMBRIDGE

LONDON   NEW YORK   NEW ROCHELLE

MELBOURNE   SYDNEY

Published by the Press Syndicate of the University of Cambridge
The Pitt Building, Trumpington Street, Cambridge CB2 1RP
32 East 57th Street, New York, NY 10022, USA
296 Beaconsfield Parade, Middle Park, Melbourne 3206, Australia

First published 1982

Printed in Great Britain at the University Press, Cambridge

Library of Congress catalogue card number: 81–17070

British Library cataloguing in publication data
Cairns-Smith, A. G.
Genetic takeover and the mineral origins of life.
1. Evolution   2. Genetics
I. Title
575.1   QH430
ISBN 0 521 23312 7

Contents

Preface *p.* vii

Preview *p.* 1

I The problem

1 Current doctrine *p.* 9
2 Three doubts *p.* 61

} *On 'chemical evolution' and its difficulties*

3 Questions of evolution *p.* 78

On why first organisms would have been of a different kind from organisms now

II A change of view

4 Genetic takeover *p.* 121
5 On the nature of primary genetic materials *p.* 136
6 The first biochemicals *p.* 164

On minerals and mineral organisms

III A new story

7 First life *p.* 261
8 The entry of carbon *p.* 300
9 Revolution *p.* 372

On the origin of our present kind of biochemistry

Coda *p.* 415

References *p.* 427

v

To Dorothy Anne

Preface

The main idea in this book is that the first organisms on Earth had an altogether different biochemistry from ours – that they had a solid-state biochemistry. This idea was already in the opening sentence of the abstract of my first paper on the origin of life (1966): 'It is proposed that life on Earth evolved through natural selection from inorganic crystals'. It sounds an odd idea, but I meant it literally. I still do.

Other ideas have changed somewhat, especially on the question of where the first organic molecules came from. I still held to the usual view in my earlier book (1971) that the original source of organic molecules in organisms was a 'primordial soup' built up through non-biological processes. In writings since then (see References) I have moved increasingly away from this towards a more old fashioned notion that photosynthesis from atmospheric carbon dioxide has always been the source of organic molecules in organisms. In this book I suggest furthermore that the relevant photosynthesis was always biological; that organic molecules *in* organisms were never made other than *by* organisms – that in this respect anyway, life on Earth has always been as now.

A log-jam of objections has to be cleared before such a point of view can begin to make sense. I have to persuade you that the very first organisms need not have contained any organic molecules. I have to reinterpret results of experiments purporting to demonstrate that some of our present-day biochemicals would have been made by non-biological processes on the primitive Earth. I have to provide an account, at least in principle and in as much detail as possible, as to how early evolutionary processes could have transformed organisms with one kind of central control machinery into organisms based on control structures of an altogether different kind. (This is genetic takeover.) And I should explain too *why* evolution should

vii

have had such devious beginnings, and what kinds of experiments and observations can best help us to understand the likely nature of that deviousness.

I do not pretend to have written at a particular level. I doubt if this would be feasible for a book of moderate length on a subject that touches on so many fields and which often requires detailed discussion of particular topics. Any reader is bound to find bits of it difficult (as I did). But if you have a broad interest in science; if you enjoy reading, say, *Scientific American*, then I think that the main ideas and arguments should be accessible. To help in this I have included a fair amount of review material and textbook material.

I have many people to thank. First of all my wife and family for putting up with my frequent abrupt absences ('he's writing his book') and also for help in reading drafts, etc. Many colleagues, especially in the Chemistry Department here, have contributed to ideas developed in this book – with their enthusiasm or scepticism: I could hardly have asked for a wider range of chemical expertise immediately around me. I should mention particularly research students who have worked with me on topics related in some way to the problem of the origin of life – Donald Mackenzie, George Walker, Douglas Snell and Chris Davis. They helped me to keep in view the practicalities of chemistry – an important recurring theme in this book. For example, Chris Davis' single-minded attempts to make a replicating organic polymer revealed, more clearly than any thought-experiment, the mountains of difficulty that lie in that approach to the origin of life.

I have been very fortunate in being able to find experts in different areas who were willing to read and comment on chapters of the manuscript – in several cases they read the whole book. I would like to thank particularly Paul Braterman, Colin Brown, John Carnduff, Alan Cooper, Charles Fewson, Jim Lawless, Karl Overton, Neil Spurway and Jeff Wilson. I usually took their advice, but not quite always. In any case only I am to blame for the mistakes and misconceptions that you may find.

The whole task of writing was greatly helped by two other experts, Janet McIntyre who word-processed the words (again and again), and Dougal McIntyre who provided the necessary hardware and software.

The electron micrographs are an important part of this book and I am particularly grateful for the prints that were provided by Mr Alan Craig and by Drs Baird, Dixon, Fryer, Gildawie, Güven, Horiuchi, Keller, Kirkman, Mackenzie, Morimoto, Mumpton, Posner, Rautureau, Weir, Wilson and Yoshinaga.

Permission by authors to use material for line drawings (as indicated in figure legends) is gratefully acknowledged as are the permissions granted by the following publishers to redraw, reproduce or otherwise use material for the figures and tables indicated:

Academic Press Inc. (figures 1.8, 6.1, 8.13).
Akademische Verlagsgesellschaft (figures 6.17, 6.20, 6.25, 6.35, 6.45*a*).
The American Chemical Society (figure 1.11).
The American Institute of Physics (figures 6.5, 6.7, 6.9).
Butterworth and Co. Ltd (figure 6.14).
Canadian Journal of Chemistry (figure 6.12).
Chapman & Hall Ltd (figure 9.4).
The Clay Minerals Society (figures 6.26, 6.29, 6.32, 6.34, 6.48, 6.52, 6.55, 8.5, 8.6).
Cornell University Press (figures 6.21, 6.23).
Elsevier Scientific Publishing Co. (figures 6.46, 8.4, 8.9).
Elsevier Sequoia S.A. (figure 8.14).
The International Union of Crystallography (figure 7.2*b*).
Japan Scientific Societies Press (Table 1.5).
Johann Ambrosius Barth (figure 6.13).
Longman Group Ltd (figures 6.5, 6.7, 6.22, 6.44).
Macmillan Journals Ltd (figures 6.24*a*, 6.50).
The Mineralogical Society of America (figures 6.31, 6.45*b*, 6.51, 7.2*d*, 7.8).
The Mineralogical Society of London (figures 6.28, 6.30, 6.33, 6.38, 6.57, 7.2*c*, 8.7).
North-Holland Publishing Co. (figure 1.2).
Oxford University Press (figure 6.16).
Pergamon Press Ltd (figures 5.3, 5.8, 6.26).
D. Reidel Publishing Co. (figure 1.5, Table 1.1).
The Royal Society of Chemistry (figure 6.24).
The Royal Swedish Academy of Sciences (figure 7.2*a*).
The Society of Economic Paleontologists and Mineralogists (figure 6.54).
Springer-Verlag (figure 6.47, Table 1.2, 1.4).
Wiley Interscience (Table 6.1).
John Wiley and Sons Inc. (figure 6.15).

Graham Cairns-Smith
Glasgow, Summer 1981

Preview

Evolution, organism, life Let us start from Dobzhansky's (1973) assertion that 'nothing in biology makes sense except in the light of evolution'. **Evolution** is then the central idea and other ideas in biology are to be related to it.

Here evolution is not simply to be taken as any long-term trend or change (as we might talk of the evolution of the Universe or of the Earth's atmosphere). Biological evolution is a much more specific idea. It takes place through successions of physicochemical systems of a particular sort – systems that can reproduce and pass on specific characteristics to offspring. Such systems are **organisms**. What makes them special, what explains them, is precisely that they participate in evolution. A living organism is the latest link in a chain stretching far back into the past. Successions of organisms, and only successions of organisms, can be subject to natural selection and hence become adjusted to their environments in the kind of way that Darwin described.

Other ideas follow. What has to be passed on between generations of organisms, to maintain the line, cannot in the long term be a material: it can only be information if it is to be transmissible indefinitely. And what evolves is not the organisms exactly, but rather the information which they transmit and which determines their specific characteristics – **genetic information** it is called. Evolution depends absolutely on the existence of the means to transmit genetic information so that characteristics of parents can reappear in offspring: evolution depends on **heredity**.

All the same there must be material vehicles for genetic information. As we will discuss in Chapter 2, it looks as if the only chemically feasible way of transmitting more than trivial amounts of information would be through structures that, like DNA, replicate by some kind of templating mechanism. We can call such things **genes** without implying that they necessarily have to be made of DNA or anything at all chemically similar to this, our present **genetic material**.

Mutations are also needed for a Darwinian evolution although they represent little difficulty in principle: they fall into the category of *mistakes*. Genetic information must sometimes change to produce altered characteristics, and these alterations must be inheritable. Such changes can be effectively random. Usually they will be deleterious. But not quite always: there is the possibility that an arbitrary design modification may improve the adaptation of an organism to its environment. Such improvements will tend to catch on because individuals incorporating them are more likely on average to survive and have offspring. Eventually, and more easily in sexually reproducing populations, the improved modification is likely to become universal. Mutations should not be too common or their destructive effects will outweigh their advantages: evolution over the long term depends on the replication of genetic information being usually very accurate.

If organisms can be said to be prerequisites for evolution, **life** can be taken as the rather intangible *product* of that process. Evolved organisms of the sort that we are familiar with conform to but are not wholly explained by the laws of physics and chemistry. In this they are like manmade machines. Indeed it is a useful informal definition of life that it is naturally occurring machinery.

Certainly if you look closely enough at a supposedly humble bacterium like *Escherichia coli*, you will find it as packed as an aeroengine with systems and subsystems that co-operate – that *must* co-operate for the bacterium to survive and be able to pass on its complicated message of how to survive. Surely *E. coli* is 'life' by our informal definition – and presumably the cleverness of its design is a product of evolution.

The origin of life Because 'life' is a rather vague idea it would not be possible to locate the origin of life precisely in time. Life would have emerged gradually during the evolution of organisms that would not actually have been alive to begin with. One might draw an analogy between the term 'life' and the term 'high ground'. If you go for a walk in the hills you will reach high ground sooner or later. You might give various reasons for saying that you had reached high ground at some stage – because of a change in the vegetation, perhaps, or an improvement in the view, or the need for another pullover. But although your companions might agree that a day's walk had reached high ground there would be much less agreement as to exactly when. There would be arguments as to which were the definitive symptoms. In any case there would be no sharp line to be drawn.

Similarly with 'life' there are symptomatic characteristics that you can list if you like (metabolism, irritability, homeostasis, and such things), but the list, however erudite, will never amount to an unambiguous definition. 'Life', like 'high ground', is not a sufficiently sharp idea to be sharply defined. If there is no difficulty in practice in distinguishing now the living from the non-living, that is because evolution has by now climbed far beyond the regions of ambiguity. It is only because of this accident of viewpoint that life may seem to us to be a sharp idea.

'The origin of life', then, should be seen as describing a field of study rather than an event. Even the more particular question of the origin of life on the Earth has several parts to it. For example:

1 *Prevital conditions.* What was the Earth like when the first organisms arose?

2 *The origin of organisms.* What were those first systems like that were able to evolve but had not yet evolved? What were they made of? How did they arise?

3 *The emergence of life.* By what stage had early evolution given rise to seemingly designed, co-operative systems? What were organisms made of by then?

4 *The origin of our biochemical system.* How and when did evolution come to depend, as it now does, on nucleic acids and proteins?

Behind such questions there is the assumption that life arose spontaneously on the Earth – it was not brought here by spacemen or specially contrived by God. Here I am accepting a common view among scientists of the late twentieth century.

Another common view is that before there was life on Earth there had to be a build-up of the 'right' component molecules – amino acids, sugars, and such things. This seems to me to be a quite mistaken idea. The set of molecules that is now fixed at the centre of our biochemistry is bound in with a whole approach to molecular control that is highly sophisticated. As we shall see, it is doubtful whether our present 'biochemicals' would have accumulated in primordial waters; but even if they had, the first evolving systems could not in any sense have foreseen which of the molecules in their surroundings would have been particularly appropriate for the kind of machinery that was to emerge in the distant future. Natural selection is indeed an amazing engineer, but it works strictly without a drawing board. It does not think ahead. It only knows a good thing when it sees it in operation. It is not at all clear, for example, that amino acids would have been particularly significant to begin with, long before there were accurately made proteins. It is not clear indeed that *any* of the now universal 'biochemicals' would have been appropriate at the very start of evolution, even if they had been there on the Earth. For evolvable systems it is not amino acids, sugars, or lipids that are critical, but somewhat higher order structures such as catalysts and membranes – and above all hereditary control machinery. I will try to show that there are easier ways of making such things than with the ultimate units on which today's biochemical high technology is based. My specific suggestion will be that crystalline inorganic minerals of colloidal dimensions – clays – would have provided the main materials out of which organisms of a first kind were made, and that organic molecules had little if any part to play right at the beginning. Such minerals would have been forming on the primitive Earth as they do on the Earth now. And the units from which they are made are simple, symmetrical inorganic species such as silicic acid and hydrated cations. These are effectively indestructible and their endless supply would be maintained through rock weathering processes. Often the clay minerals that are produced from weathering solutions seem to organise themselves fortuitously, in a rough and ready way, into the kinds of things that might be needed in a primitive organism.

For membranes, for example, there is no need to presuppose processes that might make amphiphobic organic molecules such as lipids: clay membranes are much more easily made – they are forming in the environment in vast quantities all the time. Again, inorganic materials are the readiest form of control devices such as specific adsorbants or catalysts. Even primitive genes can most easily be envisaged as arising through the continuous crystallisation of colloidal inorganic materials. Indeed the idea of a wholly inorganic (non-carbon) genetic material is at the heart and at the start of my thesis.

To avoid confusion I must stress the distinction between what is the generally held view of the role of minerals in the origin of life and the view being proposed here. It is usual to follow Bernal (1951) in seeing minerals as assisting in that putative prevital build-up of organic molecules suitable for life. Clays, for example, are seen as having concentrated organic molecules and catalysed reactions between them until these organic molecules formed themselves into systems able to reproduce and evolve under natural selection. I would agree that clays were important for the origin of life; but not in the way that Bernal said. On the view being proposed here clays were the materials, perhaps the sole materials, out of which the earliest organisms were *made*.

It will be a general theme, when thinking about primitive evolving systems, that what can start most easily is unlikely to be at all similar to what will be selected in the long run. Consider an analogy with human technological development. Primitive ways of doing things – with a clay tablet or a stone axe – depend on very little fabrication. These devices are made from materials to hand that just happen to have (more or less) appropriate properties. (The materials are literally rough and ready.) Later, as techniques of fabrication improve, much more sophisticated means to similar ends become possible. Often these are not simply elaborations of the original means. Perhaps the axe head became a spear, and the spear a bow and arrow, but the intercontinental missile has a quite different history, one that started more likely in the jewellery trade with the discovery of gold and metal working.

That sort of deviousness is quite like evolution too – at least it is like later episodes of evolution that are still visible to us. Our lungs are not improved gills, nor is our way of walking related to the locomotion of an amoeba. During evolution quite new ways of doing things become possible from time to time and these may displace older ways. Only in the short term can a succession of organisms be seen as a line: over the longer term it should be seen rather as a rope that is made up of fibres constituting the various subsystems in the organisms. The rope may be continuous that joins us to our ultimate ancestors, those makeshift systems that could first evolve under natural selection: but that is not to say that each or any fibre in that rope is continuous from the beginning to the end. Subsystems may come and go while an overlapping continuity is maintained over the long term. It is in this kind of way, as we shall discuss in Chapter 3, that evolution has been able to escape from original design approaches – through takeovers: first ways may not only be transformed, they can be *replaced* by later ways that are based on unrelated structures.

Our central biochemistry, so fixed now, was evolving at one time. I will suggest that during that evolution its subsystems were updated, perhaps some of them several times; that protein, in particular, is a comparatively new idea replacing earlier, more cumbersome approaches to biomolecular control – and that nucleic acid is only a little older. I will try to indicate that switches of function, so common in that part of evolution still visible to us, would have been expected too in that now invisible early biochemical evolution and would have made possible the takeover of the most central controller of all – the genetic material. From the considerations touched on above – that it is unlikely that what is possible to begin with will re-semble what is optimal in the end – there would be a tendency for radical change in the genetic material (as in other things) during very early bio-chemical evolution. Given also a mechanism – genetic takeover – such a change would seem almost inevitable.

Two of the main preoccupations in this book, then, will be with the questions: 'is genetic takeover possible?' and 'can we imagine genetic materials that could more easily have been generated on the primitive Earth than nucleic acid?' Genetic takeover is the key that allows us to contemplate altogether different kinds of materials for the most primitive genes.

Part I of this book will lead towards the idea of genetic takeover. Part II will use this key to allow us to shift our view of the material nature of the very first organisms. Part III is then a new story of how life originated on the Earth.

Here are the kinds of answers that will emerge to those four sets of questions put earlier.

1 The most favourable prevital conditions would have been on an Earth that had land and sea and weathering cycles and an atmosphere dominated by nitrogen and carbon dioxide.

2 The first organisms were a subclass of colloidal mineral crystal-lites forming continuously in open systems.

3 These mineral organisms evolved modes of survival and propaga-tion that would have seemed highly engineered or contrived. That is to say they became a form of life.

4 Some evolved primary organisms started to make organic mole-cules through photosynthesis. This led to organisms that had both inorganic and organic genes. Eventually the control of their own synthesis passed entirely to the organic genes (nucleic acid) which by now operated through the synthesis of protein.

PART I

The problem

1

Current doctrine

A question of definition The established view about how life first arose is embodied in the doctrine of chemical evolution. This is not easy to define as it is not a single idea, but it can be understood from accounts such as those by Bernal (1967), Calvin (1969), Kenyon & Steinman (1969), Lemmon (1970), Miller & Orgel (1974), Dickerson (1978) and others.

At a very general level the doctrine of chemical evolution is simply:

1 that life arose from systems that obeyed the normal laws of physics and chemistry (Calvin, 1969). I would not cavil with that; but there is a more restricted meaning for the term 'chemical evolution' that is more or less explicit in most contemporary accounts. Almost invariably there are two other notions present which seem to me to be much more dubious, namely:

2 that there was a prevital progression, a natural long-term trend analogous in a limited way to biological evolution, that proceeded from atoms to small molecules to larger molecules – and finally to systems able to reproduce and evolve under natural selection; and also:

3 that the relevant molecules in prevital processes were, broadly speaking, the kinds of molecules relevant to life now (see figure 1.1).

In this chapter we will be concerned with some of the main successes of the doctrine of chemical evolution and also with some of its more immediate difficulties. These will be mainly difficulties of a practical sort – although some are very serious. I will leave until Chapter 2 my more fundamental scepticism.

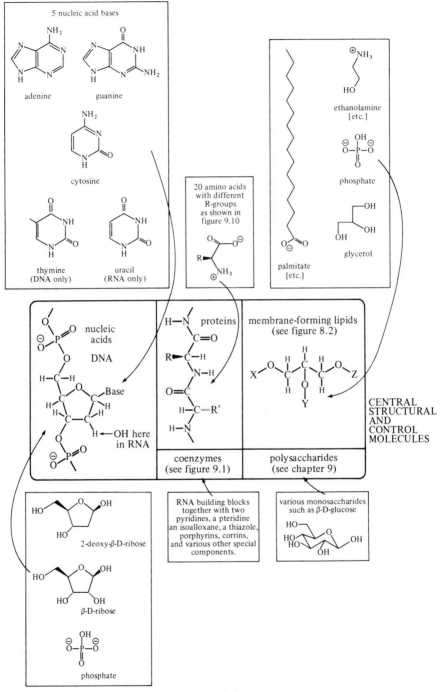

Figure 1.1.

The Oparin–Haldane hypothesis

A. I. Oparin in 1924, and J. B. S. Haldane in 1929, introduced the idea of a 'hot dilute soup' (in Haldane's words) from which life emerged and on which early life fed. Haldane saw carbon dioxide (CO_2) as the atmospheric carbon source and ultraviolet light as the means of generating reduced organic species from this source. In this he followed Allen (1899), Moore (1913), Moore & Webster (1913) and Becquerel (1924), and he was influenced by the studies of Baly (e.g. Baly, Heilbron & Barker, 1921; Baly, Heilbron & Hudson, 1922) on *in vitro* photosynthesis. Haldane took a genetic view of first life: 'The first living or half-living things were probably large molecules synthesised under the influence of the sun's radiation, and only capable of reproduction in the particularly favourable medium in which they originated.' This idea of some kind of genetic molecule at the origin of life has been a recurring one. Troland (1914) and C. B. Lipman (1924) suggested something of the sort as did Morgan (1926), Muller (1929), Dauvillier (1938), Beadle (1949), and Blum (1951). In more recent formulations of this idea life is generally seen to have started from molecules of nucleic acid (Horowitz, 1959; Maynard Smith, 1975).

Oparin (1938, 1957) was later to discuss these and numerous other ideas. For Oparin the main atmospheric carbon source was not CO_2 but methane (CH_4). According to Oparin the early atmosphere was somewhat like that of Jupiter, composed mainly of non-metal hydrides – all of which are volatile. This vision of a highly reduced early atmosphere was in line with Urey's ideas as to how the planets formed (Urey, 1952). The vast excess of hydrogen in the cloud of dust and gas from which the solar system formed would only gradually have dispersed, he said, leaving a reservoir of reduced materials that would dominate the atmosphere of the Earth for a considerable time.

Figure 1.1. All life now on Earth depends on a central set of large and rather complex molecules that can be made by joining together 100 or more 'building blocks' (shown in the outer boxes). **Nucleic acids** provide the genetic material for all present-day organisms. They also provide much of the machinery for the synthesis of proteins (see figures 9.6–9.9). **Proteins**, together with **coenzymes** provide catalysts and reagents for chemical reactions; while, with **lipids** and **polysaccharides**, proteins also provide the 'glassware' and other apparatus needed to perform organic chemical operations. Between them the central molecules create the conditions for their own resynthesis by controlling the synthesis (or acquisition) of their building blocks and the assembly of these components in an appropriate way.

Oparin also strongly rejected the idea of life originating from a for-tuitously forming genetic molecule. He regarded such an idea as unscien-tific on the grounds that it would be too improbable. 'How can one study a phenomenon which, at best, can only have occurred once in the whole lifetime of the Earth?' he asked. Oparin took a more 'metabolic' view of the nature and origin of life. Although reproduction was an essential pre-requisite for Darwinian evolution, for the appearance of life in the full sense, this property of reproduction was only gradually arrived at through systems that had a kind of metabolism and which could even evolve to some extent through a kind of selection. He saw these systems as co-acervate droplets – that is, aggregations of hydrophilic polymers such as those that had been studied in the laboratory by Bungenberg de Jong (1932). The small molecules made originally in the atmosphere and washed into the seas would have polymerised, he said, and formed discrete coacervate droplets which would have served to concentrate organic molecules gener-ally. Already these might have some of the characteristics of life: these systems are individuals, with reactions presumably taking place inside them. In some droplets the reactions cause more droplet material to be formed. Such droplets will grow at the expense of others. Even reproduc-tion of a sort is not very hard to imagine in a real ocean with breaking waves: big droplets will simply get broken into pieces which can then continue to grow. Thus slowly over a long period these coacervates become more life-like. Eventually they come to be able consistently to pass on characteristics to offspring; they can reproduce, that is, in the full bio-logical sense, and evolution through natural selection is under way.

The primitive atmosphere

On the two main points on which Oparin and Haldane differed – on primitive gases and primitive genes – it was Oparin's views that were to predominate. Not that the idea of a CO_2 carbon source was to disappear: that line continued particularly among those who saw sugars at the be-ginning of our biochemistry, for example Dauvillier (1938, 1965) and Lipmann (1965). And one of the earliest experiments simulating conditions on the primitive Earth was carried out by Calvin and his associates (Garrison *et al.*, 1951) – on the reduction of CO_2 by ionising radiation in the presence of hydrogen (H_2) and solutions containing iron (as Fe^{2+}). Subsequently, Getoff (1962, 1963a, b) was to show that ultraviolet radia-tion was effective too in making reduced species from CO_2 in water – in making, for example, carbon monoxide (CO), aldehydes and oxalic acid.

More recently, Hubbard, Hardy & Horowitz (1971) and Hubbard, Hardy, Voecks & Golub (1973), in studying simulated Martian conditions, have found that reduced organic species are formed when silicates are exposed to ultraviolet radiation in the presence of water and CO. (It is reasonable to suppose that there might have been some CO in a CO_2-containing atmosphere – through ultraviolet photolysis of CO_2 and from volcanoes: on the other hand, there is little CO in the Martian atmosphere.) More recently still, Bar-Nun & Hartman (1978) have found an efficient synthesis of alcohols, aldehydes and acids from ultraviolet photolysis of CO and H_2O.

So it is certainly plausible to suggest that reduced non-nitrogenous organic species could have been formed on the primitive Earth from CO_2 and sunlight, as both Allen and Moore had suggested so long ago. There had been doubts about this idea due to difficulties in repeating some of Baly's work (e.g. Mackinney, 1932), and this may have helped to suppress interest in the primordial fixation of CO_2.

Another consideration is that much less energy is needed to make organic molecules in a strongly reducing atmosphere (figure 1.2). And by the early nineteen fifties non-nitrogenous molecules, such as sugars, were no longer being seen at the origin of our biochemistry. Attention had shifted from metabolism itself to something still more central – to the control of metabolism, to proteins. Amino acids seemed more important than simple sugars. In 1953 Watson & Crick were to re-emphasise that life, or at least life now, is based on nitrogen-containing compounds. It became possible to imagine, perhaps, a minimal organism of nucleic acid or of nucleic acid and protein.

But it was Miller's experiment, also from 1953, that was to have the strongest effect in establishing that vision of Oparin's and Urey's of a strongly reducing early atmosphere for the Earth. Miller passed sparks through a mixture of CH_4, NH_3 (ammonia), H_2O and H_2 for a week at somewhat under 100 °C in an apparatus that allowed water-soluble products to ·be removed. The gas was the atmosphere, the water trap the oceans, and the sparking was a thunderstorm on the primitive Earth. Among the products identified were glycine and alanine. Later (Miller, 1955; Miller & Urey, 1959) it was found that 15 % of the original carbon added as CH_4 appeared in a fairly limited range of identifiable organic molecules. These are shown in figure 1.3. The rest of the carbon was mainly in the form of an unanalysed, probably largely polymeric, tar. Among the identified species there were four of the protein amino acids.

This was a sensational result: you might say a textbook example of

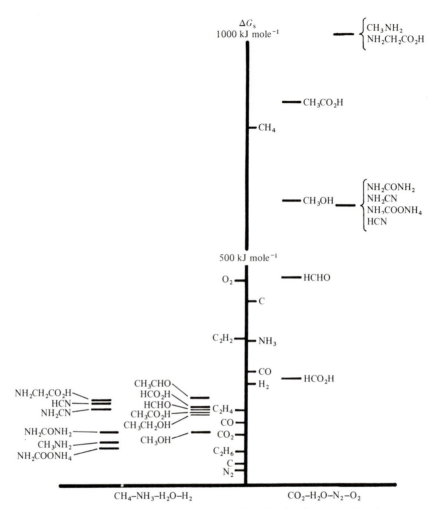

Figure 1.2. Standard free energies of synthesis of some molecules from a strongly reducing gaseous mixture (left) and from a strongly oxidising one (right). (After Toupance, Raulin & Buvet, 1971.)

scientific method at work. An hypothesis – in this case an hypothesis about important and remote events – had been tested by experiment, and not found wanting. Not only could organic molecules have been formed under a reducing primitive atmosphere, but there seemed to be a prejudice in favour of molecules at the centre of our biochemistry.

So the strongly reducing primitive atmosphere entered the general scientific imagination. Oparin's and Urey's vision was now generally

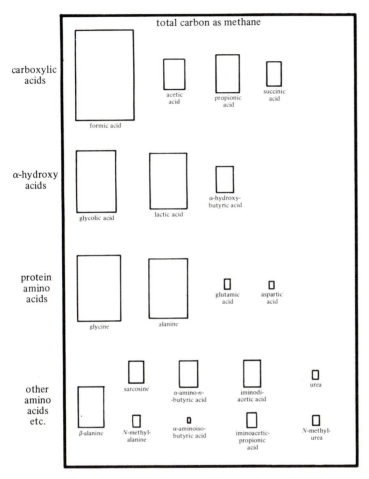

Figure 1.3. Yields from sparking a mixture of CH_4, NH_3, H_2O and H_2. The large outer box represents the total carbon available initially as methane. The other boxes represent the proportions of that carbon that became fixed in identified compounds. (Calculated from data in Miller & Orgel, 1974.)

believed to be true: even perhaps to be necessary for a rational explanation of the origin of life. At any rate the idea became almost obligatory for first pages of elementary textbooks on biology and geology.

Yet opinions based on the latest geological and astronomical evidence are increasingly against an early strongly reducing atmosphere for the Earth (e.g. Kerr, 1980). The trend in recent years has been towards the view that by the time the Earth had settled down sufficiently from its birth traumas to have sustained life, or to have started to accumulate a primordial soup, the general conditions would soon have been rather as now – but without the effects of life. The atmosphere would have been nitrogen (N_2) and CO_2, mainly, plus a small ration of noble gases with possibly some H_2 and CO as well.

Geologists have long favoured the idea that volcanoes have been the source of our atmosphere and hydrosphere (Rubey, 1951; Holland, 1962; Abelson, 1966). Even the earliest atmosphere, it would seem, must have been produced this way – it could not simply have been the residuum from the hydrogen-laden clouds from which the solar system formed. Had the first atmosphere been picked up in that way then the noble gases would have been caught at the same time – and the noble gases are not abundant enough on our Earth for that.

The question is: what was coming out of the volcanoes by the time the Earth had settled down from its birth traumas? The answer to that question depends on the oxidation state of the rocks in the crust and upper mantle of the early Earth. Holland (1962) had suggested that there might have been an early period when volcanic gases were quite strongly reducing – with plenty of hydrogen and carbon monoxide although not very much methane. This would have been possible if metallic iron had stayed near the surface for a time, only later sinking to the core.

Walker (1976a, b) has pointed to difficulties here. The $Fe^{3+}:Fe^{2+}$ ratio in present-day mantle-derived rocks is too high to suggest that these rocks were ever in equilibrium with metallic iron. Also there is too much nickel in surface rocks – if metallic iron had been present near the surface at one time, and then moved away, it should have taken the nickel with it (Clark, Turekian & Grossman, 1972). In any case recent ideas on planet formation suggest that the later accreting material, that would have formed the crust and upper mantle, would have been quite highly oxidised – anyway not metallic iron. Finally there is a fundamental problem in that any major change in the oxidation state of the mantle would have had cataclysmic effects at the surface – enormous amounts of gravitational energy would have been released as iron sank to the core. As Walker puts it: 'a highly reducing atmosphere, if it ever existed, would have had a short life and a violent end'. On the other hand, even present volcanoes produce a little hydrogen so that in Walker's view the primitive atmosphere might have contained a few per cent of hydrogen.

Admittedly such evidence is indirect. There is, however, direct evidence against there having been a strongly reducing atmosphere as long ago as 3.8 billion years. In Greenland there are metamorphosed sedimentary rocks of that age (Moorbath, O'Nions & Pankhurst, 1973). These contain abundant carbonate (Allaart, 1976) showing that at least by that time carbon dioxide was in the atmosphere (Schidlowski, 1978). Indeed Moorbath (1977) suggests that since the Greenland rocks are not so very abnormal, conditions when they were laid down could not have been so very different from now.

As I said, there is only the indirect evidence against some still earlier strongly reducing phase. But the time available for that hypothetical phase is getting rather tight. There is evidence, from moon rocks and elsewhere, for heavy impact cratering throughout the inner solar system that finished at about 4.0 billion years ago (Goodwin, 1976; Hartmann, 1975; Kirsten, 1978). That, seemingly, was the final stage of the accretion of the inner planets, and it gives us an earliest likely starting date for the origin of life (or for the consistent build-up of a primordial soup if there ever was such a thing). If a strongly reducing atmosphere was needed for the origin of life there was only about 0.2 billion years when it might have been there.

In addition to these considerations there are arguments against very much NH_3 having been present in the primitive atmosphere – even in a strongly reducing one – it is too soluble in water and too quickly photolysed (Bada & Miller, 1968; Ferris & Nicodem, 1974; Schwartz, 1981).

As doubts arose about the constitution of the primitive atmosphere, experiments were performed under a wider range of conditions, with different mixtures of gases and with different energy sources. The overall result as reviewed by Kenyon & Steinman (1969) was remarkable: neither the composition of the starting gas mixture (provided it is free of oxygen) nor the energy source is particularly critical. Electrical discharges could cause amino acids to be formed in any of the following mixtures for example: $CH_4 + NH_3 + H_2O$; $CO + N_2 + H_2O$; $CO + NH_3 + H_2$; $CO + N_2 + H_2$ (Abelson, 1956, 1957); or $CH_4 + N_2 + H_2O +$ traces of NH_3 (Ring, Wolman, Friedman & Miller, 1972; Wolman, Haverland & Miller, 1972). Also, as was pointed out by Miller (1955), it had been known for a long time that glycine can be synthesised by the action of a silent discharge on $CO + NH_3 + H_2O$ (Loeb, 1913).

Other energy sources that can generate amino acids from strongly reduced mixtures are ultraviolet radiation when H_2S is present as sensitiser (Sagan & Khare, 1971); ultraviolet radiation or sunlight in the presence of platinised TiO_2 (Reiche & Bard, 1979); ionising radiation (Palm & Calvin, 1962); heating at 1000 °C with silica catalysts (Harada & Fox, 1964, 1965); heating $CO + H_2 + NH_3$ with Ni/Al or clay catalysts (Hayatsu, Studier & Anders, 1971); by means of shock waves (Bar-Nun, Bar-Nun, Bauer &

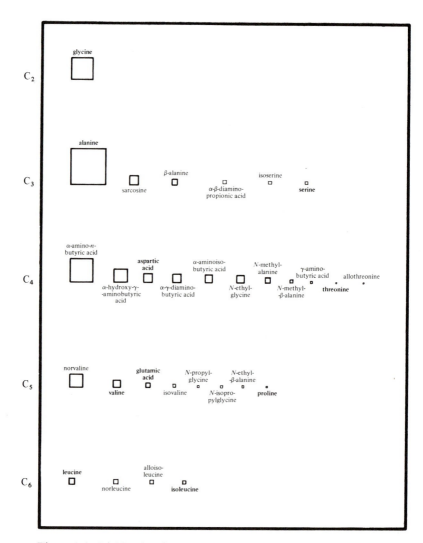

Figure 1.4. Yields of amino acids from sparking a mixture of CH$_4$, N$_2$, H$_2$O and a trace of NH$_3$. The large outer box represents the total carbon available initially as methane. The other boxes represent proportions of that carbon that became fixed in identified amino acids. (Calculated from data in Miller & Orgel, 1974, from Ring, Wolman, Friedman & Miller, 1972, and Wolman, Haverland & Miller, 1972.)

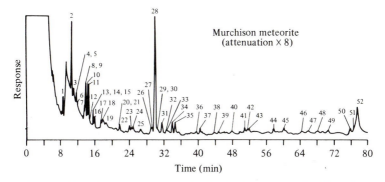

Figure 1.5. Gas chromatogram of the *N*-trifluoroacetyl-D-2-butyl esters of amino acids in an acid-hydrolysed extract of the Murchison meteorite. The numbers refer to Table 1.1. (From Lawless & Peterson, 1975.)

Sagan, 1970, 1971); by heating aqueous $NH_3 + CH_2O$ (Fox & Windsor, 1970) or aqueous hexamethylene tetramine with hydrochloric acid (Wolman, Miller, Ibanez & Oró, 1971).

Even the kinds and relative amounts of amino acids and other products are not so strikingly affected by these widely different conditions: glycine and alanine are generally still at the head of the list of the protein amino acids formed (although alanine sometimes wins); there may be small amounts of other small hydrophobic amino acids such as valine; very often aspartic and glutamic acids are there with serine not uncommon.

Unfortunately absolute yields are not quoted in many cases and quite often it seems that only protein amino acids were looked for. (There is also the possibility that if protein amino acids were not found then work was not published.) These objections do not apply to the results summarised in figure 1.4 which is probably a fair example of the kinds of yields and product distributions obtained from more successful syntheses. This set of amino acids was formed in a sparking experiment similar to Miller's original one. The main differences were in the absence of H_2, the presence of N_2 and the near absence of NH_3. These changes did not greatly affect the outcome although the yields were smaller (cf. figure 1.3). This set of amino acid products is strongly similar also to the products of another presumably abiotic synthesis: the amino acids in the Murchison meteorite. (See figure 1.5 and Table 1.1.) The comparison with the amino acid distribution in *E. coli* is much less striking (Lawless & Peterson, 1975).

I will leave until Chapter 2 the question of why any modern biochemicals should turn up in meteorites, etc. In the meantime we may ask why it should

Table 1.1. *Amino acids in the Murchison meteorite (from Lawless &*
Peterson, 1975)

1 Isovaline	27 L-Pipecolic acid
2 α-Aminoisobutyric acid	28 Glycine
3 D-Valine	29 Neutral cyclic
4 Linear neutral (C_6)	30 Neutral cyclic
5 L-Valine	31 β-Alanine
6 *N*-Methylalanine	32 Neutral cyclic
7 D-α-Amino-*n*-butyric acid	33 Polyfunctional linear aliphatic
8 D-Alanine	34 D-Proline
9 Linear neutral (C_5)	35 L-Proline
10 L-α-Amino-*n*-butyric acid	36 Linear neutral (C_5)
11 L-Alanine	37 Unknown[b]
12 Linear neutral (C_5)	38 Unknown[b]
13 NH_3[a]	39 Linear neutral (C_5)
14 Linear neutral (C_5)	40 Unknown[b]
15 Linear neutral (C_6)	41 γ-Aminobutyric acid
16 *N*-Methylglycine	42 D-Aspartic acid
17 *N*-Ethylglycine	43 L-Aspartic acid
18 D-Norvaline	44 Polyfunctional linear aliphatic
19 L-Norvaline	45 Polyfunctional linear aliphatic
20 Linear neutral (C_5)	46 Polyfunctional linear aliphatic
21 Linear neutral (C_6)	47 Polyfunctional linear aliphatic
22 D-β-Aminoisobutyric acid	48 Polyfunctional linear aliphatic
23 L-β-Aminoisobutyric acid	49 Unknown[b]
24 β-Amino-*n*-butyric acid	50 D-Glutamic acid
25 Unknown[b]	51 L-Glutamic acid
26 D-Pipecolic acid	52 Unknown[b]

[a] Present in blank.
[b] Peaks labelled 'unknown' do not appear to be amino acids. Their identification awaits the use of high-resolution mass spectrometry.

be that a fairly limited set of small molecules should keep on turning up under rather different conditions of synthesis. Alanine and glycine, for example, are particularly ubiquitous, with amino acids generally among favoured types. The answer is probably rather complicated. First we must not exaggerate: there are differences in the mixtures and most of the material is usually unanalysed polymeric tar. Second, the experiments are usually contrived so as to favour water-soluble species – if they can get into the water they escape the energy source and avoid being further transformed. And of course stable molecules will tend to be favoured in any case. But at least a large part of the explanation is probably kinetic:

because although the energy sources may be different these often generate the same small, high-energy, intermediate species. It is perhaps the limited number of these that leads to the relatively limited spread of products (Oró, 1965; Buvet & Le Port, 1973).

Abiotic routes to 'biochemicals'

Cyanide Cyanide (HCN) is one of those small, easily made high-energy species. Miller (1957) showed that in the early stages of his sparking experiments HCN was generated, along with aldehydes. He suggested that the mechanism for the formation of amino acids was a Strecker synthesis: for example glycine might be formed as follows:

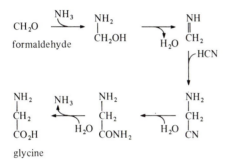

Other aldehydes should produce the corresponding amino acids – for example acetaldehyde would give alanine through similar reactions. The corresponding hydroxy acids (see figure 1.3) could be produced through the addition of HCN to aldehydes rather than, as in the above scheme, to imines. The relative amounts of amino and hydroxy acids that are produced depend on the concentration of free ammonia – the amounts are about equal when the concentration of free ammonia is about 10^{-2} M at 25 °C (Miller & Orgel, 1974).

Cyanoacetylene is another of the high-energy intermediates found in the sparking experiments, particularly with $CH_4 + N_2$ (Sanchez, Ferris & Orgel, 1966b). This hydrates to cyanoacetaldehyde:

$$
\begin{array}{ccc}
\text{CH} & & \text{CHO} \\
\| & \xrightarrow{\text{H}_2\text{O}} & | \\
\text{C} & & \text{CH}_2 \\
| & & | \\
\text{CN} & & \text{CN}
\end{array}
$$

which might be expected to yield asparagine and aspartic acid through a Strecker synthesis (the cyanide group in the initial aldehyde also hydrolysing, via primary amide, to acid). Glutamic acid, it has been suggested, could arise from acrolein obtained from the reaction of formaldehyde with acetaldehyde:

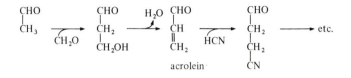

Serine might have arisen from glycolaldehyde – itself formed by condensing two formaldehyde molecules (the first step in the formose reaction discussed on page 28). These and other mechanisms are discussed by Miller & Orgel (1974), who emphasise that in the simulation experiments the yields of non-protein amino acids may be comparable to those of the common amino acids. Also the yields are always low by normal synthetic standards, and sometimes very low indeed.

Miller & Orgel suggest some (rather long) synthetic routes to the aromatic amino acids phenylalanine, tyrosine and tryptophan, although it is doubtful whether more than traces of these amino acids have ever been formed from very simple starting conditions. The basic amino acids lysine, arginine and histidine are still more difficult and lack any plausible prevital synthetic pathway. But these are by no means fatal difficulties for the doctrine of chemical evolution: it would be naive to suppose that all the present amino acids were needed for the very first organisms. Some 'primordial' set might have been enough to begin with – say glycine, alanine, valine, leucine, aspartic acid, glutamic acid and serine – the others being added during early biochemical evolution.

An additional source of aldehydes might have been available through the action of ultraviolet light on water vapour in an atmosphere that contained CH_4 or CO (Ferris & Chen, 1975a). If enough NH_3 had been present, HCN too could have been generated photochemically, as it probably still is in the atmosphere of Jupiter (Ferris & Chen, 1975b).

Already in 1955 Miller had noted that 'polymers' of hydrogen cyanide were formed during his experiments on the synthesis of amino acids by electric discharges. Stimulated partly by this and by a report from Abelson that such polymers gave amino acids on acid hydrolysis, Oró in the early sixties started investigating cyanide as a precursor for amino acids (Oró & Kamat, 1961). Interestingly cyanide also turned out to be a precursor for adenine (Oró, 1960; Oro & Kimball, 1961), which can be regarded as a pentamer of HCN:

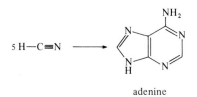

adenine

Ammonium cyanide was used in these early experiments with temperatures of 27–70 °C and concentrations in the range 1–15 M. Yields were of the order of half a per cent or so. More efficient syntheses of adenine, under admittedly very artificial conditions, were to follow: a 15–20 % yield from cyanide in liquid ammonia (Wakamatsu *et al.*, 1966), and a 40–50 % yield by heating formamide in a sealed vessel at 120 °C with $POCl_3$ (Morita, Ochiai & Marumoto, 1968). In this last case, although formamide can be regarded as the hydrate of HCN, HCN was *not* an intermediate in the reaction.

In Oró's experiments there was arguably too much cyanide to be realistic, but subsequent experiments under conditions nearer to those plausible for the primitive Earth have confirmed that amino acids, adenine and several other 'biochemicals' are formed from HCN (Lowe, Rees & Markham, 1963; Sanchez, Ferris & Orgel, 1967; Mathews & Moser, 1967; Ferris, Donner & Lobo, 1973; Ferris, Wos, Nooner & Oró, 1974; Ferris, Joshi, Edelson & Lawless, 1978; Ferris & Edelson, 1978).

A typical procedure is given by Ferris, Joshi, Edelson & Lawless (1978). A 0.1 M solution of HCN at pH 9.2 was left in a dark stoppered bottle at ordinary temperatures. After six months about half the cyanide had been converted to a complex mixture of HCN oligomers. The oligomeric material was then hydrolysed at 110 °C for 24 h in either 6 M HCl or at pH 8.5 to give a mix containing the 'biochemicals' shown in Table 1.2.

Of the oligomers of HCN the tetramers diaminomaleonitrile and diaminofumaronitrile:

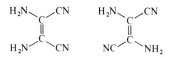

are seen as important intermediates on the way to higher oligomers and to the amino acids, purines and pyrimidines, etc. shown in Table 1.2.

Table 1.2. *Biomolecules from hydrolysis of HCN oligomers* (*from Ferris, Joshi, Edelson & Lawless, 1978*)[a]

	6 N HCl	pH 8.5[b]
4,5-Dihydroxypyrimidine	0.7–0.9 %	−
Orotic acid	−	0.009 %
5-Hydroxyuracil	0.003 %	nd[d]
Pyrimidine (structure unknown)[b]	−	+
Adenine	0.03–0.04 %	+
4-Aminoimidazole-5-carboxamide	+	+[e]
Glycine	0.6 %	0.1 %
Diaminosuccinic acid	0.1 %[c]	+
Aspartic acid	+	+
Alanine	+	+
β-Alanine	+	+
α-Aminoisobutyric acid	+	+
Guanidinoacetic acid	0.03 %	+
Guanidine	0.2 %	0.1 %

[a] HCN oligomers were prepared and hydrolysed as described in Ferris *et al.* (1978). The yields shown are based on starting HCN. Conversions are approximately $2\times$ as great since only about 50 % of the cyanide is consumed. It is assumed that 5 moles of HCN are required to form each mole of heterocyclic compound and 4 moles are required to form each mole of glycine. The identity of each compound was established by comparison of the mass spectrum of a volatile derivative with that of an authentic sample except for the following which were purified by chromatography and identified by their ultraviolet spectra and/or specific colour tests: adenine, 4-aminoimidazole-5-carboxamide, guanidinoacetic acid and guanidine. The presence of the compound is indicated by $(+)$ and absence by $(-)$. When quantitative analyses were not performed no analysis is indicated by (nd). The presence of guanine, xanthine and hypoxanthine is suggested by ultraviolet data but these were not reported because their identity has not been confirmed by mass spectral data.

[b] This compound gives a blue colour with diazotised sulphanilic acid and it is converted to 4,5-dihydroxypyrimidine by hydrolysis with 6 N HCl.

[c] Sum of the yields of *meso-* and racemic diaminosuccinic acid. They are formed in approximately a 2:1 ratio respectively.

[d] Control experiments demonstrated that this compound is destroyed by the hydrolysis conditions used in this study.

[e] 4-Aminoimidazole-5-carboxamide decomposes under acid hydrolysis conditions and when hydrolysed at pH 8.5 for 24 h. It can be detected in the hydrolysate of the HCN oligomers formed in the absence of oxygen when a 1.5 h hydrolysis time is used.

Whatever the mechanism, it is impressive that so many of the crucial types of nitrogen-containing molecules in our biochemistry can be made from a single very simple starting material. Yet there are difficulties. The yields are very small. And again there is a problem about initial concentrations: if the HCN is below 0.01 M there is no oligomerisation and the hydrolysis of cyanide to formate gets the upper hand:

$$H-C\equiv N \xrightarrow{H_2O} H-\overset{\displaystyle O}{\overset{\|}{C}}-NH_2 \xrightarrow[NH_4^{\oplus}]{H_2O} H-\overset{\displaystyle O}{\overset{\|}{C}}-O^{\ominus}$$

The most efficient synthetic sources of cyanide would probably have been electric discharges or shock waves in a strongly reducing atmosphere. As we have seen, it is questionable whether there ever was a methane-laden atmosphere. Even then it is hard to imagine thunderstorms making enough cyanide. Even if the raindrops in the thunderstorm had been 0.1 M cyanide, by the time this had reached the oceans it would have been very dilute – the more so from the tendency of cyanide to hydrolyse to formic acid. Some efficient and rapid concentration mechanism would seem to be an essential part of this story.

Sanchez, Ferris & Orgel (1966a) suggested that freezing might have provided the necessary effect. On being cooled below 0 °C a dilute cyanide solution would become more concentrated as the ice separated from it. (Under ideal circumstances, a 75 % solution of cyanide could be created in this way – by cooling to −21 °C which is the eutectic temperature of HCN/H$_2$O mixtures.) The oligomerisation reactions thus become possible (even if they would be rather slow) from initially very dilute solutions. At the same time the hydrolysis reaction would be strongly inhibited.

Guanine, xanthine and hypoxanthine were tentatively identified by Ferris *et al.* (1978). As the other purine of the nucleic acids, guanine is particularly interesting although it does not seem to be so easily come by as adenine.

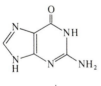

guanine

Syntheses of guanine have been reported by Ponnamperuma (1965) on irradiation of HCN solutions, and from 4-aminoimidazole-5-carbonitrile by reaction with cyanate, urea or cyanogen (Miller & Orgel, 1974).

Of the three nucleic acid pyrimidines, cytosine, uracil and thymine:

cytosine   uracil   thymine

cytosine has been successfully made in 20 % yield from cyanate and cyano-acetylene (Sanchez *et al.* 1966b; Ferris, Sanchez & Orgel, 1968). Both cyanoacetylene and cyanate might be regarded as belonging to that class of molecules that are small, reactive and relatively easy to make from available energy sources: the first is a substantial product of sparking mixtures of methane and nitrogen, the second of ultraviolet irradiation of CO and NH_3 (Hubbard *et al.*, 1975). So in view of the high yield of the reaction between cyanoacetylene and cyanate it is perhaps legitimate to regard cyto-sine as a possible product of primitive Earth chemistry. But we should be cautious with reactions that presuppose starting materials that have to be themselves the products of other reactions. Could cyanoacetylene and cyanate have been made under the same general conditions, and if not, could they nevertheless have come together?

Given a synthesis for cytosine, then uracil looks easy since it forms in very good yield from cytosine by hydrolysis. Indeed the reaction is rather too easy: with a half life of 200 years at 30 °C there is the question of how enough of the cystosine would survive (Miller & Orgel, 1974). Guanidine (an HCN product; see Table 1.2) and cyanoacetaldehyde provide alterna-tive possible routes to the synthesis of both of these pyrimidines (Ferris, Zamek, Altbuch & Freiman, 1974):

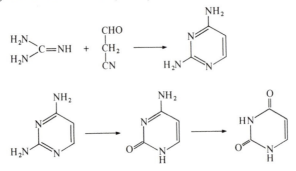

Yet another route to uracil that has been suggested starts from (the HCN products) urea and β-alanine giving, first, dihydrouracil:

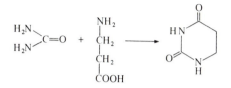

Such dihydropyrimidines can be dehydrogenated by irradiation with ultraviolet light in the presence of water vapour: this reaction is catalysed by clays – particularly montmorillonite (Chittenden & Schwartz, 1976).

The main interest in this last mechanism is perhaps that it can be modified to produce thymine – by adding acetate salts (Schwartz & Chittenden, 1977). Another synthesis of thymine – from uracil with formaldehyde and formic acid – has been reported recently (Choughuley, Subbaraman, Kazi & Chada, 1977).

We can see, then, that the entire set of nucleic acid bases *might* have been made on the primitive Earth. Routes from plausible intermediates can be formulated. If the 'wrong' pyrimidines predominate in the simpler kinds of synthesis this is not necessarily so important: as Ferris and his associates (1978) point out, the first kinds of nucleic acids might not have been quite the same as ours. These authors suggest that hydroxy and dihydroxy uracils might have been used and might have been more easily built into the necessary nucleosides. The main problems are in the yields: products are usually very minor constituents of tars.

Formaldehyde Cyanide and molecules related to it have so far dominated our discussions; but for the synthesis of sugars under loosely controlled conditions we must turn to formaldehyde. As long ago as 1861 Butlerov had discovered that formaldehyde generates a mixture of sugars in solutions of calcium hydroxide. Since then the formose reaction, as it has come to be called, has been demonstrated with a variety of catalysts, including common minerals, over a pH range from just acid to strongly alkaline (Reid & Orgel, 1967; Gabel & Ponnamperuma, 1967; Cairns-Smith, Ingram & Walker, 1972).

In principle the formose reaction is an oligomerisation of formaldehyde. You can regard glucose as $(CH_2O)_6$ rather as you can regard adenine as $(HCN)_5$. Like the oligomerisation of HCN the formose reaction is very

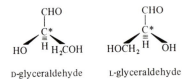

D-glyceraldehyde   L-glyceraldehyde

Figure 1.6. Glyceraldehyde has one **asymmetric carbon atom**
(starred), so called because this atom has four different things
attached to it. As a result glyceraldehyde has two mirror-related
forms designated D and L as shown. These are called **enantiomers.**
(In general, objects that are not superposable on their mirror images
are said to be **chiral**.) With most sugars there are two or more
asymmetric carbon atoms so that, in addition to enantiomers, other
kinds of **stereoisomers** become possible that are not mirror-related.
(As a left-hand-plus-right-hand is not mirror-related to a left-hand-
plus-left-hand.) Such isomers are called **diastereomers.** Usually if a
molecule has n asymmetric carbon atoms it will have 2^n
stereoisomers consisting of 2^{n-1} pairs of enantiomers.

complex, and in practice it generates products other than sugars – particularly polyalcohols. It can be discussed in terms of seven main classes of
reaction.

1. Formaldehyde self-reaction:

$$2CH_2O \rightarrow \begin{array}{l} CHO \\ | \\ CH_2OH \end{array}$$

glycolaldehyde

This is a slow reaction the mechanism of which is uncertain.

2. Aldol condensation of formaldehyde with a sugar: for example the
simplest sugar, glycolaldehyde, condenses to give glyceraldehyde:

This reaction depends on the sugar having at least one 'active hydrogen',
that is, at least one hydrogen atom attached to a carbon atom adjacent to
a carbonyl group. (It is because there is no such active hydrogen in formaldehyde that its self-condensation is difficult.)

In making glyceraldehyde, as in the making of most amino acids from
smaller molecules, an asymmetric carbon atom is generated so that there
is the additional complication that two kinds of glyceraldehyde are produced (see figure 1.6 for a further explanation and definition of terms).

3. Aldol and retro-aldol reactions of sugars: for example between glycolaldehyde and glyceraldehyde:

$$
\begin{array}{ccc}
\text{CHO} & & \text{CHO} \\
| & & | \\
\text{CH}_2\text{OH} & & \text{*CHOH} \\
| & & | \\
\text{CHO} & \rightleftharpoons & \text{*CHOH} \\
| & & | \\
\text{*CHOH} & & \text{*CHOH} \\
| & & | \\
\text{CH}_2\text{OH} & & \text{CH}_2\text{OH}
\end{array}
$$

There are three asymmetric centres now, in the aldopentose, so that really
this corresponds to eight different structures: namely the 'right-handed'
(D-) and 'left-handed' (L-) enantiomers of ribose, arabinose, xylose and
lyxose. There is good reason to suppose that each of these sugars is formed
to some extent and, in the case of right-handed and left-handed versions,
in about equal amounts (Mizuno & Weiss, 1974). (A mixture containing
an equal number of 'right-handed' and 'left-handed' enantiomers is described as a **racemic** mixture.)

4. Ring formation: ring structures become possible when there are more
than three carbon atoms in a sugar. For D-ribose, for example, there are
four possibilities:

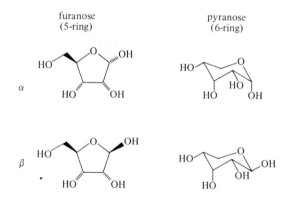

As we will discuss later, this adds complications when it comes to using
sugars as building blocks for larger molecules such as nucleotides or polysaccharides. Ring structures such as these are the predominent forms for
most natural sugars, but the free sugars interconvert fairly easily via the
open-chain structures – so we will keep to the simpler open-chain representations for the present.

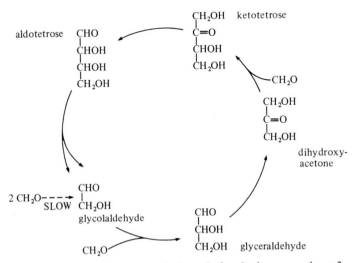

Figure 1.7. Breslow's autocatalytic cycle for the incorporation of formaldehyde into carbohydrate. The induction period for this reaction is explained by the need first to make some glycolaldehyde through a (slow) reaction between two formaldehyde molecules.

5. *Carbonyl shifts:* for example between aldopentoses and ketopentoses:

$$
\begin{array}{cc}
\text{CHO} & \text{CH}_2\text{OH} \\
| & | \\
\text{*CHOH} & \text{C}{=}\text{O} \\
| & | \\
\text{*CHOH} \rightleftarrows & \text{*CHOH} \\
| & | \\
\text{*CHOH} & \text{*CHOH} \\
| & | \\
\text{CH}_2\text{OH} & \text{CH}_2\text{OH}
\end{array}
$$

6. *α-Carbon epimerisation:* for example between D-ribose and D-arabinose:

$$
\begin{array}{ccc}
\text{CHO} & \text{CHO} & \text{CHO} \\
| & | & | \\
\text{H}{-}\text{C}{-}\text{OH} & ^{\ominus}\text{C(OH)} & \text{HO}{-}\text{C}{-}\text{H} \\
| & | & | \\
\text{H}{-}\text{C}{-}\text{OH} & \text{H}{-}\text{C}{-}\text{OH} & \text{H}{-}\text{C}{-}\text{OH} \\
| & | & | \\
\text{H}{-}\text{C}{-}\text{OH} & \text{H}{-}\text{C}{-}\text{OH} & \text{H}{-}\text{C}{-}\text{OH} \\
| & | & | \\
\text{CH}_2\text{OH} & \text{CH}_2\text{OH} & \text{CH}_2\text{OH}
\end{array}
$$

$\xrightarrow[+\text{H}^{\oplus}]{-\text{H}^{\oplus}}$ $\xrightarrow[-\text{H}^{\oplus}]{+\text{H}^{\oplus}}$

An induction period is typical of the formose reaction: a solution of formaldehyde and lime may sit for an hour with apparently nothing happening and then quite suddenly start to turn brown, which indicates that sugars (more strictly products of their further reaction to 'caramel') are being formed.

This induction period is explained by an autocatalytic mechanism such as that proposed by Breslow (1959) and shown in figure 1.7. This uses the reaction types 1, 2, 3 and 5 given above, but is very much a minimum mechanism: the true situation is certainly much more complex (Pfeil & Ruckert, 1961; Walker, 1971; Cairns-Smith & Walker, 1974). The various reaction types can, and probably do, produce all the possible sugars including branched sugars.

Indeed from more recent work (Mizuno & Weiss, 1974; Shigemasa, Sakazawa, Nakashima & Matsuura, 1978) it is clear that branched sugars are very often the most important products and it is easier to be selective in favour of branched-chain structures. Mizuno & Weiss describe the formose reaction as 'a unique method for producing branched-chain carbohydrates'. They report that while small proportions of many sugars are produced by a formose reaction catalysed by calcium hydroxide, the major products in the C_4–C_6 group are branched-chain aldoses and ketoses. Such structures readily arise either from type-2 or type-3 reactions, for example:

$$CH_2O + \begin{matrix} CHO \\ | \\ CHOH \\ | \\ CH_2OH \end{matrix} \rightarrow HOCH_2 - \begin{matrix} CHO \\ | \\ C-OH \\ | \\ CH_2OH \end{matrix}$$

In addition to all this Cannizzaro reactions take place such as:

7. Disproportionation between formaldehyde and a sugar:

$$CH_2O + \begin{matrix} CHO \\ | \\ CHOH \\ | \\ CHOH \\ | \\ CH_2OH \end{matrix} \rightarrow \begin{matrix} CH_2OH \\ | \\ CHOH \\ | \\ CHOH \\ | \\ CH_2OH \end{matrix} + HCO_2H$$

this adds complex polyalcohols – very often branched polyalcohols – to the formose mixture and these are often the predominant products (Shigemasa *et al.*, 1978). 'Caramelisation' reactions are a further complication; but even if you count only sugars the products are very complex indeed. For example Ruckert, Pfeil & Scharf (1965) identified 27 species by paper chromatography. A gas chromatogram of a trimethylsilylated formose mixture is illustrated in figure 1.8.

The oligomerisation of CH_2O, like that of HCN, produces, tantalisingly, some of the key molecules needed for our kind of biochemistry; but the products are very much more complicated mixtures than had previously been thought, containing only small amounts of natural sugars. Again with CH_2O, as with HCN, there are problems of concentration: the

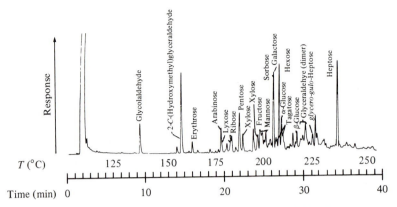

Figure 1.8. Gas chromatogram of a trimethylsilylated formose.
(From Mizuno & Weiss, 1974.)

formose reaction has not been demonstrated with solutions more dilute
than 0.01 M (Miller & Orgel, 1974). Because formaldehyde is so generally
reactive it is hard to see how large volumes of sufficiently concentrated
solutions could have accumulated anywhere on the primitive Earth.

Carboxylic acids Formic acid is perhaps the easiest of all organic
molecules to synthesise under conditions simulating those on the primitive
Earth. It was the major product in Miller's original experiment (figure 1.3):
it was produced in 22 % yield by reduction of CO_2 by ionising radiation
(Garrison *et al.*, 1951). Formic acid was produced also on ultraviolet
irradiation of CO and H_2O in the presence of silicates (Hubbard *et al.*,
1971, 1973).

Getoff (1962, 1963a, b) suggested that in the reduction of CO_2 in aqueous
Fe^{2+} solutions by ultraviolet radiation the effective reducing agent was the
solvated electron and he proposed the following scheme for the production
of formic and oxalic acids:

$$H^+ + e^-_{aq} \rightarrow H\cdot + H_2O$$
$$CO_2 + e^-_{aq} \rightarrow \cdot CO_2^- + H_2O \text{ (with } Fe^{2+} \text{ present)}$$
$$\cdot CO_2^- + H\cdot \rightarrow CO_2H^- \text{ (or } CO + OH^-)$$
$$\cdot CO_2^- + H^+ \rightarrow \cdot CO_2H$$
$$CO_2H^- + H^+ \rightarrow HCO_2H$$
$$2\cdot CO_2H \rightarrow \begin{matrix} CO_2H \\ | \\ CO_2H \end{matrix}$$

The intermediate carboxyl radical is generated also by ultraviolet irradiation of CO, CO_2, or both, adsorbed on silica gel (Tseng & Chang, 1975).

Formic acid is also produced by hydrolysis of cyanide (as discussed already) or CO:

$$CO + H_2O \rightarrow HCO_2H$$

In any complex system where redox processes are taking place – for example in the formose–Cannizzaro reactions – formic acid will often tend to be made and then persist since formate is not easily reduced.

There are fewer routes to acetic acid but it, and somewhat higher acids, are reasonably accessible. Propionic acid, for example, is ten times as abundant as glycine in the Murchison meteorite. On the other hand, the amount of any carboxylic acid in this meteorite falls away fairly rapidly as the molecular weight increases (Lawless & Yuen, 1979).

Long straight-chain fatty acids have proved to be very difficult to make under plausibly prevital conditions. Allen & Ponnamperuma (1967) obtained some monocarboxylic acids in the C_2–C_{12} range by exposing a mixture of methane and water to a semicorona discharge: but branched chains seem to be produced this way. Leach, Nooner & Oró (1978) obtained straight-chain fatty acids under the rather extreme conditions of a 'Fischer–Tropsch-type' synthesis, that is, catalytic reaction of CO and H_2 at high temperatures (here in the range 375–485 °C). Yields in terms of initial CO were in the range of 0.002–0.05 %.

The small hydroxy acids, glycolic and lactic acid, were major products from Miller's experiments (figure 1.3) and occur in the Murchison meteorite in similar amounts to the corresponding amino acids (Lawless, 1980).

Radiolysis and ultraviolet photolysis of aqueous solutions of acetic acid have been shown to give polycarboxylic acids including succinic, tricarballylic and malonic acid (Negrón-Mendoza & Ponnamperuma, 1978). Yields, although not quoted, appear to be small. Fourteen dicarboxylic acids have been identified in the Murchison meteorite (Lawless *et al.*, 1974).

Production of glycerol by uncontrolled reaction does not appear to have attracted much attention: it is a minor product of the formose–Cannizzaro reaction, but that is not in itself very promising. Perhaps early membranes could have been made with less elaborate molecules than our lipids: but anything at all like our membranes would presumably depend on the consistent synthesis of long alkyl chains, or at least of large regular hydrocarbon moieties of some sort. In the event, on the basis of current evidence, even the simplest kinds of soapy micelles would be difficult to envisage on the prevital Earth.

Chirality

If the production of relatively small molecules on the primitive Earth is to be seen in the light of chemical evolution as part of a progression through polymers to higher order systems, it is high time to pause to consider more seriously the question of chirality ('handedness', see figure 1.6). Unless the chirality of monomers, such as amino acids and sugars, is defined no well organised polymers can be made from them.

Through the looking glass there is that other world of objects that are obviously similar to the objects of our world – but very few are exactly the same. Most objects, that is, are chiral: you can not superimpose them on their mirror images. Small molecules are often sufficiently symmetrical to have superposable mirror images, and human artefacts are often approximately mirror symmetrical; but most things in Nature – pebbles, mountains, even moderately sized molecules – are very unlikely to be identical to their mirror images.

There is nothing particular, then, in an organism not being superposable on its mirror image: nor is it in the least odd to find chiral molecules in organisms – since organisms contain many very large molecules. What is odd is that the chirality of a given molecule in an organism is, on the whole, always the same. You can imagine the mirror image of a haemoglobin molecule, but you will never find a real molecule corresponding to it. Even among the smallest chiral molecules you will find at least a very strong predominance in organisms of one of the enantiomers – that is, for the 'left-handed' or 'right-handed' version of a particular chiral molecule. Sugars are like this – they are with few exceptions D-, that is, related to D-glyceraldehyde (figure 1.6). So are the amino acids (except glycine, which is achiral). Almost all amino acids, and all protein amino acids, are L- with their groups attached to the central carbon atom as in L-alanine:

We have to dig a little to get to the essential oddness of this situation. The teleological question 'why are organisms like this?' is quite easy to answer. Organisms are machines with molecules as their smallest working parts. In so far as these parts make contact interactions it is important that the chirality of the parts is specified, just as it is important that other features – materials and dimensions for example – are specified.

Even to make such a simple machine as a pair of scissors you have to be

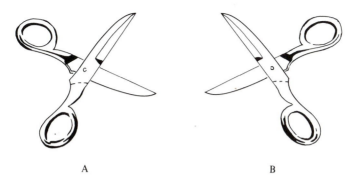

A   B

Figure 1.9. Each of the blades of a pair of scissors like these has to be either 'right-handed' or 'left-handed'. Either enantiomer, A or B, will do, but the diastereomer consisting of one 'right-handed' and one 'left-handed' blade would not work (cf. caption to figure 1.6).

careful – because the normal design of scissors is chiral. The scissors in your hand and the scissors in the mirror would both work: but a pair made up of one blade from each would not (see figure 1.9).

You could design scissors with achiral blades if you wanted to; you might even be able to design a typewriter whose components (apart from the typeface) were achiral. But as the machinery becomes more complex the restrictions of having to use achiral components would be increasingly troublesome. Apart from anything else you would not be able to use nuts and bolts or screws.

It would be even more difficult to design a complex machine – say a typewriter – which contained chiral components but for which it did not matter whether these components were 'left-handed' or 'right-handed'. I do not know how you would design such a 'racemic' machine when even D, L-scissors don't work.

The situation, then, seems to be this: (i) If a complex machine is to be assembled from premade components, then it is likely that, at least for an optimal design, some of these components will be chiral. We only have to think about the complexity of modern biomolecular machinery to be unsurprised that Nature uses chiral components. (ii) We should not be surprised that, using chiral components, Nature has tight conventions about which enantiomer of each molecule to use.

But you might ask why, given that some set of chiral conventions is a good idea, we have life as it is rather than its mirror image (or both). This can be seen as part of a bigger puzzle: not only does all life on Earth agree about the chirality of the most central small molecules, it shares many other quite intricate details of structure as well. The metabolic pathways are similar, and the choice of protein amino acids is identical throughout life on Earth – and the genetic code is very nearly universal. This points to an explanation in terms of descent of all life now extant from a common ancestor that was already highly evolved biochemically. As part of the complex package that we have all inherited from this last common ancestor there happened to be a particular set of chiral choices. It may well be that what gave our common ancestor the edge over organisms using other biochemical systems (that happened perhaps to have other chiral choices) had nothing to do with chirality: perhaps our ancestor discovered a neater way of fixing carbon dioxide, or happened first on the ribosome.

In Chapter 3 we will come to consider the interesting question of why all life should share such a late biochemical ancestor; but for the present the fact is enough: being a highly evolved intricate piece of machinery this ancestor would have had to have made some consistent set of chiral choices. And it is really no surprise that mirror life is not with us. Even if you were to imagine two groups of organisms that were identical except that one was biochemically the mirror image of the other, they would not stay identical for long. Each would be evolving separately, experiencing a different mutational history. One or the other would win: only one or the other would produce that particular design that was (much later still) to become the common design for all life.

The origin of chiral discrimination

So far we have been mainly concerned with questions of chiral choice. Given that an efficient biochemistry must use chiral molecules, we have been thinking about how our biochemistry might have arrived at some single set of chiral choices. We may have little idea as to why it settled on the particular set that it did. But in principle, at least, all such questions of chiral choice can be understood in terms of evolution, as arising from some mix of historical accident and biochemical efficiency. These questions are difficult and interesting, but they are not the real nub of the matter. The most central of the nest of puzzles about biochemical chirality is not one of choices – of why early evolving systems chose D- this or L- that: it is how evolving systems came to see any difference between D's and L's in the first place. Before you can choose you must discriminate. The nub question is the origin of chiral discrimination.

Ordinarily, chemical reactions are indifferent to chirality. A reaction, such as the formose reaction, that produces chiral molecules is expected to make racemic mixtures with 'left-handed' and 'right-handed' forms of the products equally represented. If biochemistry is different in this respect, if cells can treat enantiomers differently, then this is because the molecules in cells, especially the enzymes, are already chiral.

Bonner (1972) has reviewed the long history of this fundamental question of origin. In terms of the doctrine of chemical evolution there are two sorts of explanation that have been given: an external 'abiotic' explanation and an internal 'biotic' explanation, that is:

1 The environment created an asymmetric bias: in particular, at least somewhere, it provided especially L-amino acids and D-sugars.

and 2 The environment always produced racemic mixtures of chiral molecules, but at some stage in their evolution organisms came to select molecules of uniform chirality from these mixtures.

Bernal (1951) suggested that quartz crystals in the primitive environment might have biased a mixture of molecules locally. Quartz crystals come in 'left-handed' and 'right-handed' versions which are self-seeding; so quite extensive regions might well have contained only one of the kinds of quartz, and hence tended to accumulate or destroy enantiomers preferentially. The units out of which quartz crystals are built – silicic acid – are achiral, so in making a quartz crystal chirality is literally being generated. Presumably in any particular case this is from an initial chance that might have gone either way: an initial seed happened to start, say, left-handed quartz and it simply went on that way. Examples of such spontaneous generation of chirality have been found in the laboratory and are discussed by Bonner (1972) and Wald (1957). Wald was very sceptical of this idea, though, doubting whether more than a small bias could thus have been transferred from crystalline minerals to molecules in the environment. Modest effects have been reported since (Bonner, Kavasmaneck, Martin & Flores, ·1974, 1975; Bonner & Kavasmaneck, 1976, 1977; Morimoto, Kawashiro & Yoshida, 1978) with 30 % enrichment in the most favourable case (about 1 % is more usual). Wald was sceptical too of the idea that D- and L-isomers might separate through crystallisation in the way that, say, Pasteur (1848) discovered for sodium ammonium tartrate. Wald lists the following problems: there would have to be supersaturated solutions; moderately sized areas would be bound to contain both kinds of crystals (because, unlike the quartz case, D would become more concentrated in solution as L crystallised out, and Pasteur would not have been there to separate the crystals); finally the two enantiomers would have to form separate crystals, which enantiomers only rarely do. Perhaps Wald over-estimated this rarity, but we might add that even if all three of these conditions were satisfied, the products of simulation experiments are so grossly contaminated that it looks as if prevital organic molecules would very seldom have crystallised at all.

Other asymmetric influences have been considered. In laboratory experiments circularly polarised light can show small discriminations in the photochemical decomposition of chiral organic molecules. But it is doubtful if there was much circular polarisation of light in primordial skies: the effect in any case could only have been very feeble. Other circularly polarised photons that have been considered are Bremsstrahlung γ-rays which are generated when longitudinally polarised β-decay electrons interact with matter. Thus, it has been suggested, a fundamental asymmetry of matter – expressed in the non-conservation of parity – might have left its mark at the chemical level: Bremsstrahlung photons are always polarised the same way. It has been shown recently (Bonner, Van Dort & Yearian, 1975) that longitudinally polarised electrons can have a different effect on enantiomers. But again the effect found was small – a 1 % enantiomeric bias in the destruction of D, L-leucine.

A recent suggestion that a soup might have been significantly chirally biased comes from the report by Bondy & Harrington (1979) that L-leucine, L-aspartic acid and D-glucose are preferentially adsorbed on bentonite. The total amount of binding was small, but it seemed that the 'correct' enantiomers were bound about ten times as much as the 'incorrect' ones.

Friebele, Shimoyama, Hare & Ponnamperuma (1981) found a very much smaller effect in the adsorption of amino acids by sodium montmorillonite – a 0.5–2 % preference for L-amino acids, while Youatt & Brown (1981), reporting negative results, suggest that the original effects may have arisen from an artefact of the technique being used.

Clays such as these would have adsorbed organic molecules on the primitive Earth as Bernal (1951) was the first to point out. It is not clear how such clays could have shown net asymmetric effects. One would have expected that there would have been a balanced mix of D-binding and L-binding features in clays on the primitive Earth – unless there had been a chiral bias in some particular environment within which crystallites were forming. Even if it turns out that some modern clays do show chiral preferences in certain cases, it will not necessarily follow that clays formed on the primitive Earth would have been chirally selective. Modern (chiral) biochemicals and their degradation products are very commonly present in the environment within which modern clays form. And we know that chiral organic molecules can impose a chiral bias on crystals forming in their presence (Pincock, Bradshaw & Perkins, 1974).

The most radical cause that has been suggested for a possible asymmetry in the early environment is that enantiomers are not, after all, chemically equivalent – only very nearly so (Yamagata, 1966). The idea here is that the fundamental asymmetry between matter and antimatter might show up directly, creating differences in the energies of enantiomers (the true enantiomer of, say, L-alanine being D-alanine made of antimatter). For example reaction rates, melting points and solubilities might be slightly different. (They cannot be very different or the phenomenon would be well known.) Evidence for such effects has been presented recently and amplification devices – for example through polymerisation or precipitation – have been discussed (see papers in the *Journal of Molecular Evolution*, 1974, and *Origins of Life*, 1981, in particular, papers by Thiemann and by Wagener; also Thiemann & Darge, 1974).

So where does this leave us: supposing that there was a primordial soup, was it chirally biased? 'I don't know' is clearly the best answer; but next to that we might venture that it might have been very slightly asymmetric in a global sense – because of possible minor effects arising from either Bremsstrahlung or non-equivalence of enantiomers or both; and since these effects would act consistently in one direction. Somewhat stronger biases might have existed locally, but these would probably not have persisted for very long by geological standards. In any case, anything more than a very moderate bias can not reasonably be hoped for within any moderately sized persisting environment.

The trouble is that a moderate bias in favour of particular enantiomers in the primitive environment would not have been nearly good enough. It might indeed be used, in 'biotic' theories, to explain why life was to end up with L-amino acids and D-sugars: life evolving under conditions in which one set of enantiomers are easier to come by than another might well eventually be pushed one way. But, as we saw, that is not the real problem: even without an imposed bias evolving organisms would flip one way or the other. The real problem is right at the start: it lies in the seemingly hopeless incompetence of mixed D/L machines on the one hand and in the need, seemingly, for working machinery before natural selection can get started. This problem afflicts too those 'biotic' theories where chiral discrimination is acquired gradually during evolution.

What would we mean by saying that evolving systems had acquired chiral discrimination? We would mean (i) that they were able to distinguish, if only partially, the enantiomers of some molecules. Operationally this might show up in a particular appetite that the systems were showing for, say, L-alanine rather than D-alanine. Then (ii) we would mean that this peculiarity was indefinitely transmissable to offspring. There have to be, in some sense, receptors in the systems that are stereospecific to some degree to some molecules; and then these receptors must be reproducible, or the means by which they are produced must be reproducible. Condition (i) says what chiral discrimination is, while condition (ii) defines 'acquire'. It is because of condition (ii) that simple fluctuations within the systems are no good: the adventitious acquisition of a quartz crystallite, say, or a polymer tangle that momentarily happened to have a crevice in it that bound L-alanine. To reappear in subsequent generations, something much more distinct is required: some sort of hereditary machinery. The problem is how to make such machinery in the first place from parts that have only a slightly better than evens chance of having a given chirality (on the 'abiotic' model) or an evens chance (on the 'biotic' model).

This is a serious problem: for molecular machinery to work at all precisely – and indefinite heritability of characteristics would seem to need precisely working machinery – it looks as if there must be multiple contact interactions between components. But in that case, if the components are chiral, then their chiralities would have to be specified. I cannot be dogmatic, but I can no more imagine an effective racemic molecular biology than I can imagine an effective racemic typewriter (and the same goes for biased D, L molecular biologies or biased D, L typewriters). There is another possibility though. Let us stay with the typewriters.

People who assemble typewriters in factories may not have realised it, but there is an ingenious device of the management in such factories for cutting down errors in assembly: where components like screws or little bent levers are chiral only one of the enantiomers is provided. The situation is analogous to that presumed in the purely 'abiotic' theory of the origin of chiral discrimination. Mistakes that might have arisen through the lack of chiral discrimination initially were simply prevented by there being no choice. 'The Management' in the form of prior abiotic processes had it rigged.

Suppose that you had a typewriter factory in which the unfortunate operatives were provided with racemic boxes of nuts and bolts, racemic heaps of little bent levers and so on with which to make the machines. Even if the operatives could not tell right from left, provided they got the relationships between the parts correct, the result could be a production of competent typewriters – with, incidentally, half of them right-handed and half of them left-handed. But it would be a slow business. Half the times when trying to put in a chiral component the operative would find that its relationship to the preassembled parts was wrong and he would have to try again. And, of course, this would only work if the operatives knew when a relationship was correct: we are assuming knowledge on their part that could not be ascribed to that operative, kinetic motion, that had to put together the first evolving systems. But let us go a little further and suppose that an over-ingenious management, instead of providing chirally pure components, had designed the typewriter in such a way that each piece as it was added would only fit if it was of the correct chirality in relation to the already assembled parts.

If we are to avoid the need either for a well resolved soup to begin with, or alternatively, for chirally mixed if not racemic initial hereditary systems, then I think that something like this sort of foolproof jigsaw puzzle of a hereditary machine is what we must be thinking about. Once an initial system has started with one chirality it must have to go on with the same chirality. And I mean *have* to, as near as matters. Anything less than that would not create a long-term persisting effect: there would just be trends that would fluctuate to and fro.

If this is correct, then it places very severe restrictions on plausible initial hereditary mechanisms. Under what circumstance can molecules self-assemble into a structure that is determined (a) by the form of the components and (b) by the way its assembly happened to start?

Do protein or nucleic acid molecules fall into this restricted class? Are there conditions under which, say, D, L mixtures of amino acids will polymerise into a mixture of D- and L-polypeptides rather than D, L-polypeptides? Or, if you seed such a reaction mixture with, say, L-polypeptide, does this tend to induce specifically the polymerisation of the L-amino acids? Wald (1957) discusses experiments that were designed to answer such questions. These experiments used 4 % solutions in anhydrous dioxan of the activated glutamic acid derivative γ-benzyl glutamic acid anhydride (Blout & Idelson, 1956; Doty & Lundberg, 1956). The results of these experiments could be taken to support the idea that life would be much better off making a decision one way or the other about D or L: the pure enantiomers of the monomer polymerised much more quickly than D, L mixtures. The reason was, apparently, that the pure enantiomers gave a more stable α-helix and that the formation of such an α-helix had an accelerating effect on polymer growth. There was little selectivity, however, between D and L: racemic monomers gave D, L-polymers rather than a mixture of D-polymers and L-polymers.

There are two reasons why selective seeding would be expected to be much less effective for such polymer growth processes than for crystal growth processes. First there are more ways of going wrong. The configuration of the previous unit may influence the relative rates of incorporation of new D and L units, but polymers generally are too flexible, there is too much room around them, for this influence to be decisive: even the relatively rigid α-helix was not able to insist that a new amino acid adding was to be of the same chirality as the units already there.

Then secondly, as Wald points out, the irreversible character of the reaction mitigates against selectivity. Mistakes once made cannot be put right. With crystal growth processes on the other hand, the conditions of packing are very restrictive: it may simply be impossible to fit the wrong kind of molecule or if it is possible the energy difference due to the distortions created is likely to be substantial. Since the forces holding molecules in crystals are generally reversible, the molecules can come and go repeatedly as the crystal is built up: they can feel their way to an ideal lowest energy arrangement. There is a fundamental difficulty here because the main biopolymers are all energetically uphill from their monomers in solution: thus 'error correction', so easy for crystal growth, is difficult and energetically expensive for biopolymers. We will be returning to this point in Chapter 5.

If we imagine replication of a nucleic acid molecule through the attachment of monomers to a separated strand as in Watson & Crick's original vivid picture, then it would be reasonable to suppose, as Wald did, that an initial chirally uniform strand would exert a selectivity on nucleotides in the surroundings: in particular that the template strand would select for the correct chirality (that is, it would select D-ribonucleotides if the template strand is a poly-D-ribonucleotide). As it has turned out, DNA does not replicate like this in modern cells. Instead the process, as it now operates, depends on very elaborate assistance of enzymes. (We will discuss this in detail in Chapter 5.) Nor is there any indication from experiment that a nucleic acid molecule is, on its own, sufficiently discriminating to replicate without some rather elaborate assistance. There has to be an 'operative', it seems, who knows something about how things are to be arranged: the selectivity is partly in the enzymes. That at least is a reasonable view on the basis of current knowledge, although it might be overturned by an experiment tomorrow that might show that there are fairly simple conditions, for example in the presence of some mineral surface, where activated nucleotides would polymerise spontaneously on a nucleic acid template. But no such experiment has yet been successful (Orgel & Lohrmann, 1974).

Part of the problem of the origin of chiral discrimination is to see that there is a problem at all. Always at the back of one's mind is the feeling that there must be some simple physical effect that will work because crystals are often chirally discriminating with such high efficiency. The crystal is the ideal simple example of the sort of foolproof jigsaw puzzle object that we had been thinking about: a crystal structure is implicit in the structure of its units under given conditions plus, in the case of chiral crystals, one extra bit of information that can be provided by a seed – whether to choose right or left. But having started one way it goes on that way.

The difficulties arise when you start to think out in detail how in practice that effect could be used. Its only obvious application is to the 'abiotic' mechanism: but there are severe difficulties with this mechanism as we have seen. To be used in 'biotic' mechanisms there would have to be crystals in the organisms that were of a particular chirality and which grew and broke up to seed subsequent generations and so maintain an original handedness. That sounds all right in principle, and indeed it is an old idea, but in practice it is very difficult to see generations of coacervates keeping with, say, L-glutamic acid, and leaving D-glutamic acid in the soup, through such a trick. (The coacervates would have to create supersaturated conditions; the solutions would have to be reasonably pure to crystallise at all; as time passed D-glutamic acid would accumulate in the surroundings and make adventitious seeding of this other enantiomer more likely. And so on.)

And yet efficient resolution of enantiomers through crystal seeding is at least possible under idealised laboratory conditions. That is more than can be said about resolution through polymerisation processes, where even in the laboratory under the most contrived conditions there are no good examples of polymers that, made from one enantiomer of a chiral monomer, can seed specifically further growth through selective addition of that enantiomer. Yet for Wald's mechanism to work, nothing much less will do.

We might recall that the modern biochemical technique of keeping control of chirality is largely through the specification of 'sockets' in proteins. And it is worth emphasising how unprimordial that technique seems – it looks like a way of maintaining chiral discrimination rather than of originating it. Consider: the control of specific sockets in proteins depends in turn, and in an intricate way, on the control of protein folding. Suppose that somehow a polypeptide sequence had been specified with 100 chiral amino acids in it. This sequence would only specify the folding if the chirality of the amino acids could be relied upon – otherwise there would be 2^{100} (i.e. approximately 10^{30}) different structures possible, with side chains pointing different ways. This would be bound to affect the folding which depends so critically on side-chain packing. A protein molecule that hardly ever folds the same way twice would be a poor basis for specifying a groove that could feel the difference between L-alanine and D-alanine.

Problems of molecular discrimination in early evolving systems go far beyond those of chirality. (As they do in making machines generally: before such a nice question arises as to whether one has the correct enantiomer of a component there is the cruder question of whether one has the correct component.) To imagine a DNA molecule replicating in the kind of way that Wald supposed, the parent strand must pick out nucleotides in the first place from a presumably complex mixture of molecules that, if it contained nucleotides in any quantity, would contain many nucleotide-like molecules. In a primitive unevolved environment chiral discrimination might seem the least of the problems; yet the discrimination would seemingly have to be that good: only a chirally uniform nucleic acid looks as if it could work (Wald, 1957; Miller & Orgel, 1974). (And these are not the only problems associated with the idea that there were nucleic acid molecules in the very first organisms, as we shall see later.)

Attitudes to the origin of chirality in organisms have varied from surprise that anyone should still think that there is a problem (Frank, 1953) through a refusal to be baffled (Miller & Orgel, 1974) all the way to talk of enigma (Kenyon & Steinman, 1969).

I would agree with Frank to the extent that I do not see the problem of the origin of chirality as raising fundamental philosophical issues. The difficulties are of a practical sort. But they are very great difficulties all the same. Perhaps we might settle for the comment by Briggs with which Bonner concludes his review of the origin of chirality that 'it presents problems to the hypothesis of chemical evolution that are at present insoluble'.

In Part III of this book I will give a possible solution to the problem of the origin of chirality that depends on the idea that the first systems able to evolve under natural selection were hereditary machines made from achiral components. Chiral discrimination could thus be a comparatively late evolutionary achievement since it was not needed to begin with. Without being drawn into details at the moment I can perhaps best indicate what I mean by adapting an analogy of Thiemann & Darge (1974). They see the origin of chirality in organisms as analogous to the spontaneous crystallisation of one enantiomer from a supersaturated racemic solution. On the view that I will be presenting the appropriate analogy becomes rather the spontaneous appearance of enantiomeric crystals (like quartz) from an achiral mother solution. Chirality does not have to be built into units, it can appear at a higher level of organisation. Thus organisms that start with achiral units can come eventually to be able to handle chiral molecules (which later still take over). Of course such an idea only becomes possible if you are prepared to contemplate altogether different initial hereditary machinery. But if you are prepared to follow through this somewhat more complicated story you will find another advantage: crystallisation, with its error correction mechanisms intact, can become the means through which the initial hereditary machinery worked.

Previtar polymers

Let us nevertheless pursue the story of the origin of life according to the doctrine of chemical evolution and consider in more detail the question of how biopolymers might have arisen. Which of these would have been the most important to begin with is a matter of some dispute; but it is commonly believed that proteins of a sort, or nucleic acids of a sort (or both) would have been necessary for the making of those first systems that could evolve under natural selection and so take off from the launching platform provided by prevital chemical processes.

We have already come to a major difficulty here: much of the point of protein and the whole point of nucleic acid would seem to be lost unless these molecules have appropriate secondary/tertiary structures; and that is only possible with chirally defined units. As we saw, the 'abiotic' way of circumventing this problem (by prevital resolution of enantiomers) seems hopelessly inadequate, and 'biotic' mechanisms depend on efficient machinery already in action. At all points these difficulties can be seen as technical – some effect, such as a chirally discriminating polymerisation, might be discovered to sweep these difficulties away. If you are a believer in chemical evolution then that is the attitude to take. But there are further severe difficulties ahead. A clutch of them appear when we come to consider how biopolymers might have been made consistently without biology.

Purification

First of all there is a problem which is seldom discussed. The starting monomers would have been grossly impure. On the basis of simulation experiments they would have been present in complex mixtures that contained a great variety of variously reactive molecules.

No sensible organic chemist would hope to get much out of a reaction from starting materials that were tars containing the reactants as minor constituents. Perhaps because they are sensible organic chemists most experimenters, in trying to establish some prevital path to biopolymers, do not start with such complex mixtures. Instead they say something like this: 'monomer A has been shown to be formed under prebiotic conditions and so has reagent B: so we treated A and B (obtained from Maxipure Chemical Corporation) under prebiotic conditions such and such and made the biochemically significant molecule C'. Suggestions as to how A and B might have been purified under prevital conditions are seldom made.

In organic chemistry it is often the work-up rather than the reaction that causes most of the trouble. Think about the techniques that are used: pH adjustments, solvent extractions, chromatography, evaporations to dryness, recrystallisations, filtrations and so on. Now you can say that such things might have taken place fortuitously under primitive geological conditions. Each individual operation can be imagined – a transfer of a solution, a washing of a precipitate, an evaporation, and so on. But very many such operations would have had to take place consistently and in the right order. In a typical work-up procedure there are subtle things that can make the difference between success and mess – how long to wait, say, after the pH adjustment before filtering. Practical organic chemistry is not easy. Very much has to be engineered. It is not sensible to suppose that an uninformed geochemistry would fortuitously be expert in such things.

Concentration

Next there is the problem of the concentrations of the monomers in primordial waters. It has been emphasised repeatedly that the idea of an oceanic primordial soup is difficult to sustain on thermodynamic and kinetic grounds. For example Hull (1960) says: 'First, thermodynamic calculations predict vanishingly small concentrations of even the smallest organic compounds. Second, the reactions invoked to synthesise such compounds are seen to be much more effective in decomposition.' Hull was discussing particularly the effects of ultraviolet radiation which he calculated would have destroyed 97 % of amino acids produced in the atmosphere before they reached the oceans. Glycine, he reckoned, would have formed at best a 10^{-12} M solution. Sillen (1965) was similarly pessimistic and talked of the 'myth of the probiotic soup', and Abelson (1966) pointed to the lack of any geological evidence for a thick oceanic soup in the past. The matter was pursued again by Hulett (1969) in an influential paper in which he considered various energy sources but stressed particularly photochemical effects. (The main point is that short-wavelength photons are needed for the synthesis of molecules such as aldehydes and amino acids, while the much more abundant longer wavelength photons are effective in decomposing them.) The importance of ultraviolet radiation in destroying organic molecules was again emphasised by Rein, Nir & Stamadiadou (1971). Dose (1974, 1975) has calculated that primitive oceanic concentrations of amino acids would have been around 10^{-7} M if photochemical effects are taken into account. Thus Dose points out that the concentration of amino acids in the 'soup' would have been about the same as their concentrations in the oceans now. A similar conclusion has been reached by Nissenbaum (1976) on the basis of other (non-biological) geochemical processes that scavenge organic molecules from the oceans (for example adsorption on sinking minerals). As a result the mean residence time for organic molecules is 1000–3000 years. A further effect that can be seen in action now on Mars would have tended to keep surface regions clean of organic molecules. Ultraviolet radiation that can penetrate an anoxic atmosphere can have an oxidising effect on inorganic surface materials creating species that are ready to destroy any organic molecules that are formed (Ponnamperuma *et al.*, 1977; Klein, 1978).

It can be reasonably argued that an oceanic soup was not needed, that what was required was that stocks of organic molecules of the right kind should have been able to build up somewhere to a sufficient concentration, and consistently enough, to have provided the materials to make the first evolving systems – and to have provided them with food. Perhaps at the margins of oceans there would have been evaporating rock pools that would have served to concentrate materials. It would take a lot of evaporating, though, from 10^{-7} M to the kinds of concentrations that might be needed. And the ultraviolet sunlight would have been a nuisance for any concentration process that uses evaporation. Here Bernal's idea seems better. He pointed out that organic molecules would have been adsorbed on solid particles, particularly clays, so that potential monomers would have been brought closer together. Hence, perhaps, amino acids might have been able to form polymers on clays even if the general solutions surrounding them were quite dilute.

Lahav & Chang (1976) have recently taken up this idea in an analysis of published data on the adsorption and condensation on clays of amino acids, purine and pyrimidine bases, sugars, nucleosides and nucleotides. None of these types of molecule binds very strongly to ordinary clays. At equilibrium, between ocean and clay sediments, they would be very thinly spread over the available clay surface. Such estimates are necessarily rough, but even supposing that Dose (1975) was pessimistic by two orders of magnitude in his estimate of oceanic amino acid concentrations, the molecules would have been on average at least 10 nm apart. Additional factors would still be needed to concentrate the molecules sufficiently, for example evaporation or freezing.

Condensation

There is a third difficulty in prevital synthesis of biopolymers, and this is the most generally recognised: all the major biopolymers are metastable in aqueous solution in relation to their (deactivated) monomers. Left to itself in water, a polypeptide will hydrolyse to its constituent amino acids: Miller & Orgel (1974) estimate that the half life of alanylalanine is about 8×10^7 years at 0 °C and about 6×10^5 years at 25 °C.

Most experimental work on prevital simulation of polymer synthesis has been concerned with this third problem. There have been two main approaches.

1. Devices for shifting the equilibrium. The idea here is to separate the products, water and polymer, as they are formed so that the reaction is driven to the right, for example:

$$R—CO_2H + H_2N—R' \rightleftarrows R—\overset{\overset{\displaystyle O}{\|}}{C}—\underset{\underset{\displaystyle H}{|}}{N}—R' + H_2O$$

As this equilibrium is only reached very slowly at ordinary temperatures and near neutral pH, heat, catalyst or both are also needed.

The simplest technique is to heat the amino acids together so that the water is driven off as steam. This tends to decompose the amino acids (Katchalski, 1951), but Fox found that if a relatively high proportion of aspartic acid, glutamic acid or lysine were present in a dry mixture of amino acids, then a clean protein-like material, 'proteinoid', could be formed. The temperature required was usually around 150–180 °C (Fox, 1956, 1969; Fox & Harada, 1960; Fox & Dose, 1972). More recently amino acids have been polymerised at 65–85 °C (Rohlfing, 1976). Protenoid shows catalytic properties (see also Dose, 1971) and on contact with water forms very striking cell-like objects ('microspheres') which Fox sees as models for systems that are on the way to becoming cells.

Lahav, White & Chang (1978) have made oligomers (dimers and trimers mainly) of amino acids. In a typical experiment 1 ml of 0.023 M glycine

Table 1.3. *Oligomerisation of glycine on clays* (*after Lahav, White &* *Chang, 1978*)

Expt no.	Glycine (nmoles/ mg clay)	No. of cycles	Net heating time (days)	Approx. yields of oligomers (nmoles/mg clay)				Total yield of glycine in oligomers	
				Di	Tri	Tetra	Penta	nmoles/ mg clay	approx. %
Kaolinite									
1	374	11	33.7	2.3	0.5	—	—	5.9 ± 0.9	1.6
3	374	27	67.4	2.3	1.0	0.3	Trace	8.9 ± 0.6	2.4
5	123	27	67.4	1.0	0.4	0.1	Trace	3.5 ± 0.3	2.8
7	791	27	67.4	3.5	1.6	0.6	Trace	14.14	1.8
Bentonite									
15	1070	11	32.8	6.4	0.2	—	—	13.3 ± 1.7	1.2
17	1070	27	67.4	4.9	0.6	Trace	—	11.7 ± 1.7	1.1
20	32000	11	57.0	37	8.2	—	2.5	111 ± 28	0.3
23	93000	11	57.0	40	7.9	1.2	0.8	113 ± 10	0.1

was added to 60 mg of kaolinite or 20 mg bentonite (Na^+ form) and put through cycles of wetting and drying, and temperature changes, as follows:

(i) dehydrate at 60 °C for 1–2 days
(ii) heat at 94 °C for 2–3 days
(iii) rehydrate with 1 ml of water
(iv) repeat (i)–(iii) n times

Some results are given in Table 1.3. One might imagine inland or coastal regions on the primitive Earth that were being alternately flooded and baked dry in the sun.

In similar experiments with bentonite Lawless & Levi (1979) have shown that yields may be improved considerably by transition metal ions in the exchange positions of the clays – Cu^{2+} in particular, but also Zn^{2+} and Ni^{2+} (see Table 1.4).

As with the formation of proteinoid, the tyranny of thermodynamics is avoided in the above experiments by using open systems. In each case water is being removed and energy used to do that. The clays are also, presumably, exerting catalytic effects – particularly those clays which have exchangeable transition metal cations.

2. Raising the energy of the monomer. One problem with trying to poly-merise amino acids at around neutral pH in water is that under these

Table 1.4. *Influence of exchangeable cations on the oligomerisation of glycine on bentonite*

Exchangeable cation	Yields of oligomers (nmoles glycine incorporated/mg clay)				Glycine incorporated into oligomer (%)
	Di	Tri	Tetra	Penta	
Na$^+$	17.2	—	—	—	0.9
Ni^{2+}	29.5	6.3	—	—	1.8
Cu^{2+}	69.9	41.4	10.4	2.8	6.2
Zn^{2+}	25.6	3.3	—	—	1.4

2 ml of 2×10^{-2} M glycine at neutral pH were added to 20 mg of bentonite in each case and put through eight wet–dry cycles (after Lawless & Levi, 1979).

conditions neither the amino nor the acid group is very reactive – because the amino acid molecules spend most of their time as zwitterions:

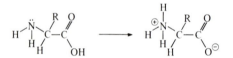

When the molecule is in the zwitterion form the negative charge on the carboxyl group destroys the electrophilic character of its carbon atom. At the same time the nucleophilic character of the nitrogen is destroyed by protonation of its lone pair of electrons.

The reactivity of the carboxyl can be increased by changing its hydroxyl for a group that does not have a proton to lose – a methyl ester, for example, is a mildly activated form of the carboxyl group. An acid chloride is much more strongly activated since in this case the hydroxyl has been replaced with an electron-withdrawing group that can easily leave as an anion to push the reaction in the required direction:

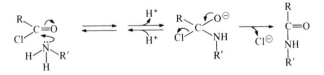

Indeed acid chlorides are rather too reactive to be useful in peptide synthesis – they react too easily with alcohols and with water for example. Somewhat milder forms of carboxyl activation are more selective for reaction with amines, and numerous alternative leaving groups have been used in peptide syntheses (Bodansky, Klausner & Ondetti, 1976). Two examples are *p*-nitrophenyl esters and imidazole amides:

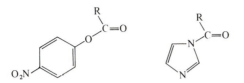

There are several problems, though, in attempting to apply this idea to primitive chemistry. Firstly, energy must come from somewhere to activate an amino acid. Secondly, even mildly activated species are more or less susceptible to hydrolysis. Thirdly, the amino group of an amino acid will usually have to be protected during activation if only to prevent premature polymerisation. This means that separate deprotecting steps will also be required. In addition side groups of the more reactive amino acids will have to be protected and deprotected.

Amino acyl adenylates are activated forms of amino acids involved in protein biosynthesis (figure 9.9). For example alanyl adenylate is:

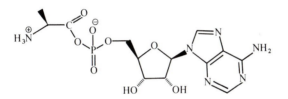

As we will discuss in the next section, such molecules would be very implausible prevital products but, interestingly, they can be made to polymerise in aqueous solutions in the presence of the clay montmorillonite (Paecht-Horowitz, Berger & Katchalsky, 1970; Paecht-Horowitz, 1974; although, see also Brack, 1976).

Instead of using isolable activated intermediates, amide links can be formed by means of coupling or condensing agents such as carbodiimides. These coupling agents first join to the carboxyl to create a good leaving group which is then expelled following attack by the amino nitrogen (see figure 1.10). Carbodiimides can also be used in making internucleotide phosphodiester bonds. Thus, with a coupling agent, activation and condensation can be performed in one operation.

Cyanamide is a tautomer of carbodiimide itself:

$$H_2N—C{\equiv}N \rightleftarrows HN{=}C{=}NH$$

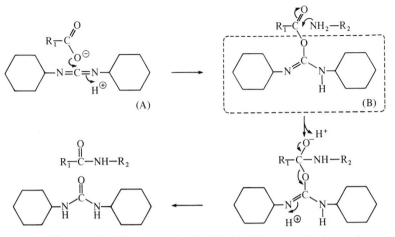

Figure 1.10. Dicyclohexylcarbodiimide (A) is a typical coupling agent. In joining an acid to an amine it reacts first with the acid to make a leaving group (B).

and has been suggested as a primordial coupling agent. Small yields of triglycine, leucylglycine and glycylleucine have been obtained with this reagent (Ponnamperuma & Peterson, 1965), but under conditions more acid than would have been expected in primordial oceans. Ponnamperuma (1978) has reviewed the possible prevital condensing agents shown in Table 1.5. Of the first five cyanide-derived agents he considers that only hydrogen cyanide tetramer operates at a sufficiently high pH to be realistic. This agent on the other hand is rather unstable. Ponnamperuma concludes that linear or cyclic polyphosphates are the best idea. On the other hand, Sherwood *et al.* (1978) suggest that the regions between clay layers might sometimes be sufficiently acidic to allow condensation reactions mediated by cyanamide.

It seems to me that the idea of coupling agents putting together polypeptides on a lifeless Earth adds another dimension of unreality to an already unreal line of thought. Remember that primordial simulations generally give only low yields of amino acids. Remember that the products are tars and that suggestions for prevital work-up procedures are usually absent. Remember the difficulties anyway in building up concentrations of solutions of amino acids or of the cyanide or phosphate to make a coupling agent. Remember that even from laboratory bottles the agents in question do not work very well. Remembering all that, now add the thought that coupling agents are rather unspecific. If a well chosen coupling agent under

Table 1.5. *Some possible prevital condensing agents (after Ponnamperuma, 1978)*

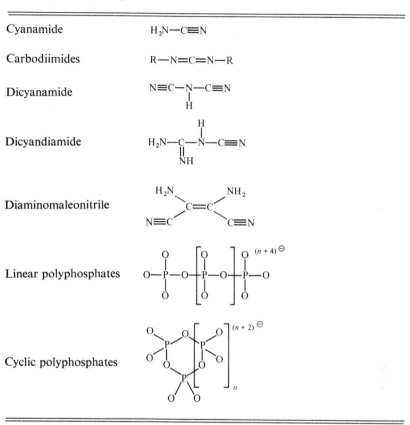

Cyanamide	$H_2N-C\equiv N$
Carbodiimides	$R-N=C=N-R$
Dicyanamide	$N\equiv C-\underset{H}{N}-C\equiv N$
Dicyandiamide	$H_2N-\underset{NH}{C}-\underset{H}{N}-C\equiv N$
Diaminomaleonitrile	
Linear polyphosphates	
Cyclic polyphosphates	

well chosen laboratory conditions can effectively join the acyl group A to the nucleophile B that is because among the choices exercised by the experimenter was the crucial one of only putting A and B into a flask for the coupling agent to couple. Compared with such carefully arranged marriages the affairs of a primordial soup would have been grossly promiscuous.

One can get an impression of what is needed in practice for the synthesis of peptides by considering the machinery that is used in automated procedures. One such piece of equipment is shown in figure 1.11. Merrifield, Stewart & Jernberg (1966) describe its construction and operation in nine close pages of diagrams and descriptions. I quote (more or less at random) from the middle of their paper: '... the rear disk contains a center port

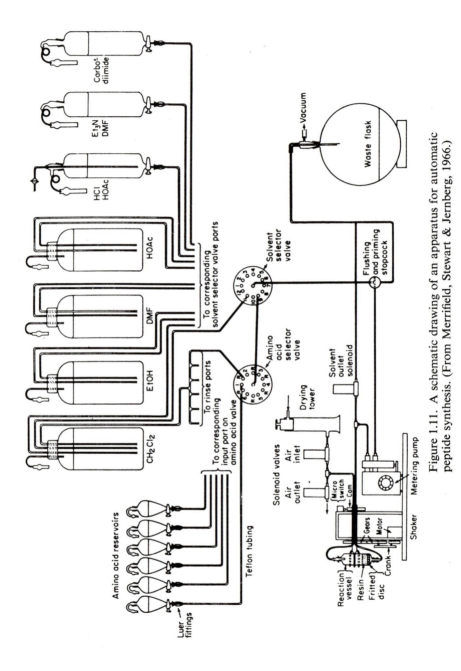

Figure 1.11. A schematic drawing of an apparatus for automatic peptide synthesis. (From Merrifield, Stewart & Jernberg, 1966.)

and one circumferential port which are joined by a 1.5 mm hole within the disk. As this disk is turned it connects one at a time the 12 inlet ports to the central outlet port. A leak-free seal between the two teflon disks of the valve . . .' And that is one of the less terse passages. Not shown in figure 1.11 is a programmer, like a musical box drum, that puts appropriate operations (mixings, rinsings, shakings, etc.) in sequence. There have to be many pegs on the drum because one cycle of the automatic synthetic procedure that extends the peptide chain by one unit requires nearly 90 steps.

Now I am not saying that for peptide synthesis without human intervention there has to be something physically like Merrifield's machine. There does not have to be that particular piece of engineering. But I think there has to be engineering.

Another example of automatic peptide synthesis is the synthesis by the ribosome in the modern cell. (We will be considering this in more detail in Chapter 9.) There are no tubes or valves or metering pumps here: but in the design of the ribosome, the adaptor RNA molecules and their activating enzymes; in the whole system, with its message tapes and its code, there is surely at least as much engineering as in Merrifield's machine. (See figures 9.6–9.10.)

Perhaps there is some other way of making peptides with more or less specified amino acid sequences; and perhaps this way does not need detailed control. Perhaps it could then have operated before there was life on Earth, before that engineer, natural selection, appeared on the scene. But it is difficult to see how this could have been so. I think we would know by now if there was some much easier way.

It is similarly difficult to imagine anything like polysaccharide being accumulated in primordial waters. As we saw, the monosaccharides could only have been made easily from formaldehyde, as far as anyone knows, and there is doubt if there could have been sufficient concentrations of that. In any case, as we saw, the product of the formose reaction is a very complex mixture that easily leads to higher polymers and to caramel.

More realistic is the suggestion by Kenyon & Nissenbaum (1976) that more random kinds of polymeric material 'melanoidin' and 'aldocyanoin' were implicated in the earliest precursors of life. Melanoidin was made by heating solutions of a sugar (e.g. 0.1 M glucose) and an amino acid (also 0.1 M) at around 100 °C for 2–4 days at around neutral pH. Aldocyanoin was made by keeping a solution of NaCN, NH_4Cl, CH_2O and NH_4SCN (each 0.2 M) at pH 9.3 in a stoppered flask at room temperature for 2–4 weeks. Each of these kinds of polymer gave microspheres – the melanoidin made two distinct sizes (about 3 μm and 8 μm) while in the turbid dark-brown aldocyanoin reaction mixtures the microspheres were more uniform (at about 2 μm).

For such polymers to form, concentration mechanisms would still be needed; but presumably impure starting materials would suffice for these much less ambitious products. Then again the energy problems are lighter – at least if we can assume a supply of cyanide and formaldehyde since these products are energetically downhill from there.

The implausibility of prevital nucleic acid

If it is hard to imagine polypeptides or polysaccharides in primordial waters it is harder still to imagine polynucleotides. But so powerful has been the effect of Miller's experiment on the scientific imagination that to read some of the literature on the origin of life (including many elementary texts) you might think that it had been well demonstrated that nucleotides were probable constituents of a primordial soup and hence that prevital nucleic acid replication was a plausible speculation based on the results of experiments. There have indeed been many interesting and detailed experiments in this area. But the importance of this work lies, to my mind, not in demonstrating how nucleotides could have formed on the primitive Earth, but in precisely the opposite: these experiments allow us to see, in much greater detail than would otherwise have been possible, just why prevital nucleic acids are highly implausible.

Let us consider some of the difficulties. *First*, as we have seen, it is not even clear that the primitive Earth would have generated and maintained organic molecules. All that we can say is that there might have been prevital organic chemistry going on, at least in special locations. *Second*, high-energy precursors of purines and pyrimidines had to be produced in a sufficiently concentrated form (for example at least 0.01 M HCN). *Third*, the conditions must now have been right for reactions to give perceptible yields of at least two bases that could pair with each other. *Fourth*, these bases must then have been separated from the confusing jumble of similar molecules that would also have been made, and the solutions must have been sufficiently concentrated. *Fifth*, in some other location a formaldehyde concentration of above 0.01 M must have built up. *Sixth*, this accumulated formaldehyde had to oligomerise to sugars. *Seventh*, somehow the sugars must have been separated and resolved, so as to give a moderately good concentration of, for example, D-ribose. *Eighth*, bases and sugars must now have come together. *Ninth*, they must have been induced to react to make nucleosides. (There are no known ways of bringing about this thermodynamically uphill reaction in aqueous solution: purine nucleosides have been made by dry-phase synthesis, but not even this method has been successful for condensing pyrimidine bases and ribose to give nucleosides (Orgel & Lohrmann, 1974).) *Tenth,* whatever the mode of joining base and sugar it had to be between the correct nitrogen atom of the base and the correct carbon atom of the sugar. This junction will fix the pentose sugar as either the α- or β-anomer of either the furanose or pyranose forms (see page 29). For nucleic acids it has to be the β-furanose. (In the dry-phase purine nucleoside syntheses referred to above, all four of these isomers were present with never more than 8 % of the correct structure.) *Eleventh,* phosphate must have been, or must now come to have been, present at reasonable concentrations. (The concentrations in the oceans would have been very low, so we must think about special situations – evaporating lagoons and such things (Ponnamperuma, 1978).) *Twelfth,* the phosphate must be activated in some way – for example as a linear or cyclic poly-phosphate – so that (energetically uphill) phosphorylation of the nucleo-side is possible. *Thirteenth,* to make standard nucleotides only the 5'-hydroxyl of the ribose should be phosphorylated. (In solid-state reactions with urea and inorganic phosphates as a phosphorylating agent, this was the dominant species to begin with (Lohrmann & Orgel, 1971). Longer heating gave the nucleoside cyclic 2',3'-phosphate as the major product although various dinucleotide derivatives and nucleoside polyphosphates are also formed (Österberg, Orgel & Lohrmann, 1973).) *Fourteenth,* if not already activated – for example as the cyclic 2',3'-phosphate – the nucleo-tides must now be activated (for example with polyphosphate; Lohrmann, 1976) and a reasonably pure solution of these species created of reasonable concentration. Alternatively, a suitable coupling agent must now have been fed into the system. *Fifteenth,* the activated nucleotides (or the nucleotides with coupling agent) must now have polymerised. Initially this must have happened without a pre-existing polynucleotide template (this has proved very difficult to simulate (Orgel & Lohrmann, 1974)); but more important, it must have come to take place on pre-existing polynucleotides if the key function of transmitting information to daughter molecules was to be achieved by abiotic means. This has proved difficult too. Orgel & Lohr-mann give three main classes of problem. (i) While it has been shown that adenosine derivatives form stable helical structures with poly(U) – they are in fact triple helixes – and while this enhances the condensation of adenylic acid with either adenosine or another adenylic acid – mainly to di(A) – stable helical structures were not formed when either poly(A) or poly(G) were used as templates. (ii) It was difficult to find a suitable means of making the internucleotide bonds. Specially designed water-soluble carbodiimides were used in the experiments described above, but the obvious pre-activated nucleotides – ATP or cyclic 2′,3′-phosphates – were unsatisfactory. Nucleoside 5′-phosphorimidazolides, for example:

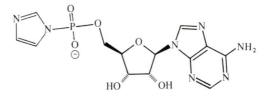

were more successful, but these now involve further steps and a supply of imidazole, for their synthesis (Lohrmann & Orgel, 1978). (iii) Internucleotide bonds formed on a template are usually a mixture of 2′–5′ and the normal 3′–5′ types. Often the 2′–5′ bonds predominate although it has been found that Zn^{2+}, as well as acting as an efficient catalyst for the template-directed oligomerisation of guanosine 5′-phosphorimidazolide also leads to a preference for the 3′–5′ bonds (Lohrmann, Bridson & Orgel, 1980). *Sixteenth*, the physical and chemical environment must at all times have been suitable – for example the pH, the temperature, the M^{2+} concentrations. *Seventeenth*, all reactions must have taken place well out of the ultraviolet sunlight; that is, not only away from its direct, highly destructive effects on nucleic acid-like molecules, but away too from the radicals produced by the sunlight, and from the various longer lived reactive species produced by these radicals. *Eighteenth*, unlike polypeptides, where you can easily imagine functions for imprecisely made products (for capsules, ion-exchange materials, etc.), a genetic material must work rather well to be any use at all – otherwise it will quickly let slip any information that it has managed to accumulate. *Nineteenth*, what is required here is not some wild one-off freak of an event: it is not true to say 'it only had to happen once'. A whole set-up had to be maintained for perhaps millions of years: a reliable means of production of activated nucleotides at the least.

Now you may say that there are alternative ways of building up nucleotides, and perhaps there was some geochemical way on the early Earth. But what we know of the experimental difficulties in nucleotide synthesis speaks strongly against any such supposition. However it is to be put together, a nucleotide is too complex and metastable a molecule for there to be any reason to expect an easy synthesis. You might want to argue about the nineteen problems that I chose: and I agree that there is a certain arbitrariness in the sequence of operations chosen. But if in the compounding of improbabilities nineteen is wrong as a number that would be mainly because it is much too small a number. If you were to consider in more detail a process such as the purification of an intermediate you would find many subsidiary operations – washings, pH changes and so on. (Remember Merrifield's machine: for one overall reaction, making one peptide bond, there were about 90 distinct operations required.)

$$A \longrightarrow B \longrightarrow C \longrightarrow D$$

Figure 1.12. According to Horowitz (1945), a metabolic pathway would have evolved backwards. D was at first a vital molecule available in the environment. D gradually ran out, giving organisms time to evolve an internal source – by converting C, some simpler precursor, that was still in the environment. As C ran out there would then be selection pressures to find some other environmental molecule, B, and the means to convert it to C. Hence complex molecules that were originally provided by a primordial soup came to be made instead from simple commonly available molecules such as CO_2 and N_2.

Problems for primitive heterotrophs

Let us suppose that all the difficulties that we have been discussing were somehow overcome, and let us now consider how the very first organisms might have fared. According to the doctrine of chemical evolution these organisms were heterotrophs, that is to say they depended on organic foods. The diet of primordial soup was so adequate, it is said, that these organisms had no need for metabolic pathways to begin with. Such pathways could evolve gradually as the foods ran out (by the mechanism proposed by Horowitz in 1945; see figure 1.12).

To have one's food provided sounds like an easy sort of life, but in reality there would be great difficulties with such an idea. There are problems of assimilation. To be a heterotroph implies an ability to recognise molecules, or at the very least to distinguish between classes of them. For the eventual evolution of metabolic pathways, specific recognition devices would be required. Thinking along the lines of current means of biomolecular control, some kind of structure would seem to be needed that could form specific sockets corresponding to the molecules in the environment. But until you have the ability to recognise at least some molecular units, how do you reach the point of being able to manufacture such specific devices? Organisms now can presuppose protein-synthesising machinery. And a great variety of transport proteins located in the cell membranes can actively and selectively pull in particular molecules from the environment (Lin, 1970; Wilson, 1978; Rosen, 1978).

The trouble is that a socket (such as that in an enzyme or a transport protein) that can recognise another molecule is much more difficult to engineer than the molecule itself. Organic chemists are only just coming round to this means of molecular control, and the structures being made are very complex (Cram & Cram, 1978). So what were the control techniques? How was tarry chaos avoided?

If the enzymes in today's cells can cope so well this is partly because the molecules that they come across belong to a quite limited set. An enzyme may distinguish between D-glucose and D-fructose, because these are among the relatively few kinds of molecules that it encounters: but it can easily be confused by molecules from a larger range. *E. coli*'s enzymes work because, among other things, *E. coli*'s protein-loaded cell membrane is highly selective about the molecules it lets in. A primitive organism, lacking such customs control and living in a tarry 'broth' that contained for every 'correct' molecule a myriad of similar 'incorrect' ones would have to have far more accurate enzymes to bring about any particular sequence of reactions.

So that is the problem: how to evolve accurate recognising structures from a molecular technology that probably could not tell glycine from alanine, let alone D from L. Until you know one molecule from another how do you start to do the kind of sophisticated chemistry needed to make the membranes, the active centres and so on, on which molecular discrimination depends?

Another problem is heredity. In modern organisms, at least, the holding and printing of genetic information depends on an ability to do some very sophisticated chemistry – beyond the abilities of organic chemists, never mind primordial thunderstorms. Perhaps there are easier ways than ours of performing this most crucial of all functions; but attempts to replicate nucleic acids without enzymes have been unsuccessful and, as we saw, nucleotides are not among the expected components of a prevital soup. As mentioned earlier, one of the ways of attempting to overcome this last difficulty is by postulating a preliminary kind of semi-Darwinian evolution, where microsystems are not able to reproduce but they are already subject to selection. Hence organic chemical expertise could have evolved, it is said, to the point at which the synthesis of nucleic acids becomes feasible. In the next chapter we will take up again this very important question of whether a useful kind of evolution is possible without some means of storing and replicating information. I do not think it is.

2

Three doubts

In this chapter we will move from the practical problems about the origin of life that mainly concerned us in Chapter 1 to more philosophical ones. We will be concerned here with three fundamental doubts about the doctrine of chemical evolution: doubts about the significance of Miller's experiment; doubts about the relevance of a probiotic evolution; and doubts about the whole idea that there have been any 'molecules of life' invariant through evolution from the start.

Puzzles or anomalies?

Few would deny that there are difficulties in the doctrine of chemical evolution. The question at issue is whether these are to be taken as puzzles or anomalies in the sense defined by Thomas Kuhn (1970) in his well known account of how changes of view can take place in science. If the difficulties are puzzles then they are soluble within the framework of chemical evolution; the cracks can safely be ignored in the meantime (although they should not be papered over); something will turn up to account for the difficulties. But if these difficulties are anomalies they are to be seen rather as signs that the doctrine of chemical evolution is an inadequate frame within which to discuss the origin and early evolution of life; the cracks are structural and further building work will make them worse. So are the difficulties that we have been coming across puzzles or anomalies?

So far we have been discussing mainly acknowledged difficulties and treating them, as they usually are treated, as puzzles. Some seem to be very hard puzzles indeed – the origin of chiral discrimination, for example, or the problems of prevital work-up procedures: but never is it inconceivable that a solution could be found tomorrow that will fit the established doctrine. Indeed, if you believe in the established doctrine, if you hold to the idea that the difficulties are puzzles, then they are positively to be welcomed as fining down the possibilities. For example, the difficulties about having to have sufficiently concentrated solutions of HCN and of preventing HCN from hydrolysing to formate can both be seen as indicators that the Earth was a cold place when life originated. What is seen as yet another *ad hoc* assumption by the sceptic is seen rather as clarification of details by the believer.

If you hold to the doctrine of chemical evolution, how do you explain the gigantic implausibility of prevital nucleic acids? There are several ways. In line with, say, Oparin or Bernal you may say that the question is irrelevant; that chemical systems can evolve, in the sense of becoming more competent at doing organic chemistry, before evolving in the full Darwinian sense – that to begin with hereditary machinery was not needed. Or you may say that hereditary processes can be efficient enough to begin with without replicating molecules. I will be trying to show later in this chapter, and in Chapter 3, that neither of these statements will do: that a genetic view of the origin of life is the only tenable one. But even so, you might accept a genetic view but deny the implausibility. Prevital nucleic acid only seems to be implausible you may say, but the very fact that it is so puzzling on present knowledge will make the explanation, when it turns up, the more convincing. If we could find one or a few unexpected effects that can account for all nineteen of those problem areas that we considered towards the end of Chapter 1, then we would not only have confirmed the doctrine of chemical evolution in a convincing way, we would know in some detail how life on Earth must have originated.

So there is nothing irrational in not giving in to difficulties. But as the difficulties accumulate the stakes get higher: success would be all the more resounding, but it becomes less likely. Sooner or later it becomes wiser to put your money elsewhere. The puzzles may be, after all, anomalies.

A pattern of success and failure has been emerging through the experimental work on the origin of life that started in the early nineteen fifties. The small molecules of our biochemistry have proved to be chemically much more accessible than one might have hoped, with amino acids and purines as perhaps the best examples. That fits the doctrine of chemical evolution: we can say that the doctrine is confirmed (in the weak scientific sense) by these findings. But, equally consistently, other things that are needed for the doctrine to work out have not been so readily forthcoming. Problems of concentrations are now well recognised: they are a recurring theme. Then there are the much less well recognised problems of purification. And as one goes on to consider the larger molecules, such as lipids and nucleotides, chemical accessibility dissolves. These molecules are not at all easy to make. Part of the trouble is how to get up energy gradients; but that is not the only trouble. Because these molecules are bigger, and hence have richer possibilities for isomerism, the problems of discrimination – of which chiral discrimination is only one – become so difficult that there almost seems to be a conspiracy not to discuss them, but to design experiments as if they had been solved, by using purified starting materials. And then there are the biopolymers . . . In all there is an impression of a field that started well and then fell away.

Some analogies One of the troubles with starting well is that you may lock onto a point of view and then stay with it long after it has failed to maintain its initial promise. (This kind of thing is well described by catastrophe theory: cf. Zeeman, 1976.) There are many examples from the history of science. The phlogiston theory was one. This was the idea that an inflammable material contained a 'principle of fire' – phlogiston – which was given off when the material was burnt. This explains the obvious facts of combustion so well that it was quite natural to accept it and then interpret new data in terms of it. Suppose that as a respectable eighteenth century phlogistonist, you had been asked to explain why a candle goes out when a jar is put over it. Would you have taken this as a puzzle or an anomaly? Neither, you would have taken it as a confirmation of the phlogiston idea because it is so easily explained: the air in the jar has become saturated with phlogiston and cannot take up any more. What about the gain in weight when a metal is converted to a calx (i.e. oxide)? Here you would have put on a more patient expression. This is not so much a puzzle, you would explain, as a misunderstanding: phlogiston is the principle of fire, you see, not an isolable substance; metals all contain this principle (which is why incidentally they have so many properties in common). When the corpuscles of metals are imbued with phlogiston the effect of gravity is lessened . . . You would, I daresay, have been able to ward off all comers with such explanations. Nothing would force you to change sides. (Nothing ever forced Priestly.)

Then again the alchemists found plenty to confirm the idea that the synthesis of gold was a feasible project. The problem was seen as that of combining properties – heaviness, shininess, yellowness and so on – in one material. It was found that certain solutions (of polysulphides) could tinge metals yellow (for the recipe see Liecester, 1956). The more perceptive of the alchemists knew, I daresay, that they had not yet made gold. But on one point all were surely agreed: this was a step in the right direction.

You can see the dangers of supposing that an entire doctrine is confirmed when experiments that test only a part of it come up with expected results. Much depends on whether there are other explanations for the positive results. What might be the signs of some other, yet unthought-of explanation? One sign might be where, in spite of tests of many predictions in different areas of the doctrine, success is restricted to only one or a few areas. As we have seen, this is the case for chemical evolution.

Another symptom might be where success (in some limited area) is too easy. Here is a story that caricatures what I mean:

Lady G has lost the diamond from her ring. She summons the Brownies to look for it. Late in the afternoon, sitting at the French windows, Lady G sees in the distance a small brown figure running towards her. Through her opera glasses she notes a look of delight on the Brownie's face and the glint of a small object held by excited fingers. Being of a scientific turn of mind, Lady G frames an hypothesis: 'the diamond has been found'. Then she notices another Brownie appearing from a similar direction also clutching a shining object. Then several more, some with several shining objects in their hands. Is Lady G's conjecture further confirmed? Why not? Surely the more glinting objects there are in view the better is the chance that one will be the diamond? But long before the first Brownie reaches Lady G she has abandoned her hypothesis in favour of 'There Must Be Some Other Explanation'. And of course there was: a car with a shattered windscreen had spread pseudo-diamonds on the road outside . . .

I think that the history of the prevital simulation experiments has something of that story in them. In the limited zone of the synthesis of small biochemical molecules the experiments have been embarrassingly successful – with 'blanks' often coming out positive, that is, biochemicals being favoured by conditions that would *not* have been present on the primitive Earth. Even among the plausibly prevital syntheses we are left with too many possibilities for their significance to be very clear. Was it from the sparks of lightning or the shocks of thunder or the glare of ultraviolet light or the infall of meteorites or the baking of volcanoes; or was it through a sweeping up of cosmic dust that we acquired the first biochemicals? Or was it a bit of each? With so many possibilities to choose from one might be as well not to settle on any. There is the thought that other possibilities will turn up; that the true explanation is perhaps still to be found.

First doubt; the significance of Miller's experiment

It is really too naive simply to assert that the prevital simulation experiments confirm the doctrine of chemical evolution. Well, yes they do (in the weak scientific sense) – but with crippling provisos. The whole subject of confirmation in science is notoriously tricky, and here we have an example, I think, of a hypothesis that has become *less* plausible the more it has been 'confirmed' (cf. Gardner, 1976).

It is those 'Other Explanations' that cause the trouble. Proposition A may be confirmed by observation L in the scientific sense, that is L is consistent with A. But suppose that observation L is also consistent with proposition B? Then L does not confirm A as against B, it only confirms A *or* B as against (some) other possibilities. Any observation inevitably confirms a whole set of propositions A, B, C . . . and only some of these you will have thought of. The normal way round this difficulty is to test the proposition in as many diverse ways as possible. Observation M, let us say, also confirms proposition A (including other unthought-of possibilities) but not proposition B. Then you have confirmed A as against B. But you have not eliminated the possibility of some unthought-of proposition that is common to the sets that are confirmed by both observations L and M. So you carry on with observations N, O, P, and so on – all as diverse as possible. If proposition A continues to hold out against these diverse tests, then it is reasonable to hope that there is not some unthought-of, quite different explanation that would also fit all these observations.

But, as we have seen, the really successful tests of chemical evolution have been restricted to one aspect – chemical accessibility of a number of the smallest molecular units. These molecules can be said to be chemically accessible because they are so often present in complex mixtures formed by uncontrolled reactions. The other aspects are mainly a catalogue of difficulties that have been getting worse: the early environment now looks less clement than was once thought; hopeful possibilities for resolution of enantiomers have not worked out; purification, concentration, activation, when any kind of solution has been proposed, all call for rather special situations, and they are anyway inefficient. And the significance of those seemingly easy routes to biochemicals from cyanide and formaldehyde becomes less clear when examined in more detail. As we saw, neither cyanide nor formaldehyde would easily have built up to sufficient concentrations. Yields in these uncontrolled syntheses are very low, in general, and the products always very complex. Many of these products – especially sugars – would not have survived long enough to have accumulated in a 'primordial soup'.

Is there, then, some Other Explanation? Is there an hypothesis that will account for the (rather general) accessibility of (some) of our biochemicals, and yet accommodate the difficulties?

Biochemical economy

Biochemical economy is a label that I will attach to the idea that on the whole you do not expect organisms to use molecules that are much more elaborate or unstable than they have to be. Whatever else, it is a likely *outcome* of evolution that a biochemistry be based on chemically accessible molecules.

Other things being equal, it would be an advantage to an organism to be able to use molecules that were relatively easy to make – because then there would be fewer catalysts needed, less genetic information, fewer metabolic products to interfere with each other, and so on. During the early evolution of our biochemistry there would be a continual pressure to be more efficient in this sense. Provided early evolutionary processes were flexible enough to allow experimentation with different subsystems, then the final outcome would predictably contain numerous fairly easily made molecules using metabolic pathways of least resistance.

One can make the expectation a little more precise. You might expect three rough classes of small molecules to have emerged as components of an evolved biochemistry – let us say classes A, B and C. The A-class are accessible molecules such as formate, glycine, alanine or adenine. They are accessible because thermodynamically favoured: they are simple and rather unreactive molecules. For that reason they naturally tend to emerge as products of abiotic processes of various sorts in which atoms are being more or less violently shuffled about. (Once you have made glycine, for example, it tends to persist because it is a zwitterion – page 50.) But one can easily see advantages for a biochemical system in having some molecules that are not only easily made but fairly stable. Such units would be less liable to misreact. A-Class molecules are simple basic nuts and bolts.

The B-class would again be accessible types but this time only because kinetically favoured. They would be easily made from available starting materials but would not be particularly stable. There are plenty of such comparatively short-lived molecules in our biochemistry. Sugars are evidently in this category. That sugars appear temporarily in the condensation of formaldehyde may have some relevance to how they were first biosynthesised: but you do not find sugars in mature products of non-biological semi-chaotic reactions – they are not in meteorites for example. B-class molecules are simple active metabolites.

The C-class biochemicals are not easy to make. They are common types in our biochemistry nevertheless because they perform critical functions that make their manufacture worthwhile. Nucleotides, coenzymes, chloro-

phyll and lipids are in this category. So are many of the amino acids –
especially the aromatic and the basic amino acids. Higher animals can get
the best of both worlds, avoiding the trouble of making some of their
C-class molecules by depending instead on dietary intake – vitamins and
essential amino acids are very often C-class biochemicals. (Dietary requirements vary of course, but for the white rat the essential amino acids are
phenylalanine, tryptophan, lysine, histidine, arginine, threonine, methionine, leucine, isoleucine and valine; Mahler & Cordes, 1971.) If the A-class
molecules are like nuts and bolts, those in the C-class are like more specialised, expensive pieces of a machine. C-class molecules are specially made
for particular purposes.

Now only the first of these rough categories, only the A-class, might
accumulate through probiotic processes to allow the possibility of a probiotic explanation for present biochemicals. If our biochemical system was
really put together from components of some oceanic or localised soup
then you would expect some minimum version of our system to be makeable entirely from A-class molecules. Yet that seems very far from the case
– since sugars, lipids and especially nucleotides are absent from this
supposedly primordial class.

Why are nucleotides difficult to make? That nucleotides are very
difficult to make whereas several of the amino acids are quite easy can be
understood in terms of biochemical economy, along the following lines.

Our whole biochemical system, including the choice of the molecular
units out of which it is now made, is an outcome of fully Darwinian evolutionary processes. Always in making these choices there were questions of
cost effectiveness; of taking the most easily made molecule that could do
the job. Several easily made amino acids could be chosen for the protein
set because it did not matter very much which were chosen – so long as
there was some variety of amino acid types. If you doubt this just think of
the variety of functions that protein can perform and how many very
effective functions of protein, for example making feathers or antibodies,
were only discovered long after the twenty amino acid set was decided on.
The cleverness of protein lies only somewhat in the choice of the twenty
amino acids. Most of the cleverness is in the stitching of these units into
specific sequences. Functions in proteins are predominantly sequence-
dependent rather than unit-dependent.

By contrast, the process of the replication of a nucleic acid molecule is
independent of the sequence of monomers in that nucleic acid molecule.
The whole point is that the sequence is replicated whatever it is. Hence the

cleverness has to be far more in the units themselves. Units for replicating organic molecules have far more design restraints on them: it would be lucky if something that happened to be easily made would do. (I will be going into this in greater detail in Chapter 5.)

'Probiotic pathways' In any case biochemical economy seems the most plausible general explanation for the tendency for reaction pathways to be similar within and without organisms. For example, Degani & Halmann (1967) found that the spontaneous decomposition of glucose takes place along routes that are similar to the glycolytic pathway. We might say, then, that if there had been a primordial soup and if sugars had formed in it, something like the glycolytic pathway might have come before organisms. This would be a misleading way of talking however. It might seem to imply that the glycolytic pathway had to be *rehearsed* in a pro-biotic soup before it could be taken up by organisms. To talk about a 'probiotic pathway' at all is misleading; it biases the imagination in favour of an historical connection which may not have been there.

Is it not more likely that early evolving systems simply discovered a bit of chemistry: not something out there in an actual soup, but something inherent in the way molecules behave? If the organic chemistry of meteor-ites; or of Fischer–Tropsch reactions; or of spark plasmas or shock waves in gases; or of reactions of formaldehyde in base; or of cyanide solution (and so on) bear some resemblances sometimes to present-day biochem-istry (and to each other) that is not to say that there was ever any causal connection in the sense that any of these processes caused or copied any of the others. The connection, I think, is universal and parallel, not historical and serial. All these processes are confined by rules of chemistry.

Biochemical economy is sufficient According to the view to be elaborated in this book, biochemical economy provides virtually the whole explanation for any overlap there may be between the set of easily made molecules and the set of our biochemicals. We will have to take a more complex view then of very early evolution: we will have to find other approaches to the nature of the very first organisms; and we will have to explain the situation within which an evolving biochemistry would have been able to make choices about certain molecules when now there can be no more choice. We will see that these are not insuperable paradoxes, and that furthermore biochemical economy allows us to dispense with a pri-mordial soup. But one advantage we can already see: biochemical economy explains most easily why, although some biochemicals are stable and easily made, other very central ones are either not very stable or not at all easily made.

Second doubt; the relevance of a probiotic evolution

The two kinds of evolution

The Universe evolves and with it the stars, the solar system, the Earth, and environments on the Earth. That is one kind of evolution. The other kind is biological evolution (what I mean in this book by the unqualified term 'evolution').

Environmental evolution was under way, presumably, before life started to evolve on Earth. So there is a serial connection in that the evolution of the Earth created the conditions for the evolution of life. But this is not to say that the two processes are the same or even similar, except that each is a kind of change. And it is hard to see *why* one kind of evolution should have led to the other.

Was 'chemical evolution' the connection? I do not think so. The building up of a primordial soup, if such a thing ever happened, would have been part of environmental evolution. The oceans would have accumulated organic molecules in much the same way as any other geochemical process would have taken place. Unless you take a religious or mystical view there was no guiding hand to contrive an outcome suitable for the origin of life. Mountains were made and worn down, the wind blew, the sun shone – and a soup did or did not form: all such processes were on an equal footing; it would only have been with an eye to the future that some of these processes might have been given a special label and called 'chemical evolution'.

Biological evolution, on the other hand, *is* special, as discussed in the opening pages of this book. Above all what makes it special is heredity. This is the great divide: either there is a long-term hereditary mechanism working or there is not. If there is not then there is no accumulation of 'know-how' as Kuhn (1976) put it: the survival or non-survival of some putative half-organism will not be 'remembered' in the distant future to have any effect. Things would change, systems such as coacervates would come and go, but you could not expect them to become more *efficient*: you would not expect them to become more efficient at organic chemical operations, for example. Only evolving organisms can progress in that sort of way.

Suppose that by chance some particular coacervate droplet in a primordial ocean happened to have a set of catalysts, etc. that could convert carbon dioxide into D-glucose. Would this have been a major step forward towards life? Probably not. Sooner or later the droplet would have sunk to the bottom of the ocean and never have been heard of again. It would not have mattered how ingenious or life-like some early system was; if it lacked the ability to pass on to offspring the secret of its success then it might as well never have existed.

So I do not see life as emerging as a matter of course from the general evolution of the cosmos, via chemical evolution, in one grand gradual process of complexification. Instead, following Muller (1929) and others, I would take a genetic view and see the origin of life as hinging on a rather precise technical puzzle. What would have been the easiest way that hereditary machinery could have formed on the primitive Earth?

Model for a biological evolution

Are genes necessary? What is the minimum sort of hereditary system? Can we say what kinds of structures would be necessary? The reproduction of organisms now on the Earth depends on the replicability of molecules – DNA molecules: is it a necessary minimum requirement for a system to be subject to natural selection that it should contain replicable molecules of some sort?

Consider an ocean that is coacervating spontaneously, giving rise to droplets of a certain general composition. We are interested in one rather odd droplet that happens to contain n particles of a catalyst X that increases the rate of growth of the droplet. In all other respects our droplet is typical. The trouble is that even if n is 10^{18} – an unreasonably large number – the particles would have been diluted out after about 60 generations. Clearly to continue to have an effect indefinitely the number of catalyst particles in a droplet must, on average, double between droplet divisions. Anything less than that and the catalyst will sooner or later be 'forgotten'.

There are two ways in which the catalyst particles could increase in numbers between droplet divisions:

(*1*) *by assimilation* – they could be picked up from the environment – or

(*2*) *by synthesis* – they could be put together from other units (d, e, etc.) themselves ultimately acquired from the environment. In either case there is an additional requirement that X must exert some more or less specific effect in bringing about the increase in its numbers. Otherwise it might simply be that a droplet would contain X, not because its parent did, but because the environment was such that droplets coacervating from it contained X.

In the case of assimilation there must be some kind of X-receptor in the droplets: not necessarily a specific socket, but some effect of X must predispose the droplet to acquire more X. In the case of synthesis, X must either predispose the droplet to acquire suitable units (say d and e) or catalyse the synthesis of X from units that are there anyway, or both. This gives us seven cases to consider.

A1 (direct X-assimilation)

X is itself an X-receptor, or part of an X-receptor.

A2 (indirect X-assimilation)

X helps to make some other X-receptor: that is, X is a control structure (for example a catalyst) in the making of the X-receptor.

S1 (direct d, e-assimilation)

X is a receptor for components d and e and hence contributes to the making of more X, because elsewhere in the droplets d and e are converted into X (without the need for further control).

S2 (indirect d, e-assimilation)

X helps to make receptors for d and e by acting as a control structure in their synthesis. Hence indirectly it contrives to pull d and e from the environment: again these form spontaneously into more X without the need for further control.

S3 (X-synthesis)

The droplet is a receptor for d and e in any case (that is, to be a d, e-receptor is a general property of coacervates forming under the given conditions). X converts d and e into more X.

S4 (direct d, e-assimilation + X-synthesis)

As S3 but X is also a d, e-receptor.

S5 (indirect d, e-assimilation + X-synthesis)

X helps to make d, e-receptors and also converts d and e into more X.

At least formally, then, hereditary processes can be imagined that do not explicitly involve replicating molecules: only in the schemes S3, S4 and S5 do the growth- and/or fission-promoting structures X also control their own synthesis – and even here this control might not involve template-directed synthesis. The acid hydrolysis of an ester is a reaction in which protons can be said to be 'self-reproducing' species. This reaction is both catalysed by hydrogen ions and it produces them. Such autocatalyses are indeed well known in chemistry and no templating processes need be involved; no process, that is, in which the form of a parent structure is imposed on a daughter structure. In the schemes S3, S4 and S5, X is simply an autocatalyst. In other cases X makes its reappearance more likely in some other way, by picking up X from the surroundings, by helping to collect the bits to make more X, and so on. In general and in principle what X has to do is to look after its own reappearance in the next generations of coacervate droplets. It must reproduce in the ordinary meaning of that word – it must be produced again – but there is seemingly nothing to insist that that reproduction should involve template copying (what I will from now on call **replication**).

Evolution through natural selection, though, requires more than just the inheritance of characteristics. It requires an indefinite future potential for those characteristics to be modified. It must be possible for X to change to X′, a modification of X, and then to X″, and so on: and then these modifications must be inheritable.

Horowitz (1959) stressed the insufficiency of simple autocatalytic processes as a basis for heredity. The key point is that autocatalytic processes do not in general have the property of reproducing modifications. If you feed protons to an aqueous solution of an ester you will get back more protons, but if you feed deuterons you will not get deuterons reproduced instead. Simple autocatalytic processes are not only limited in themselves, but they have no future, they are isolated. The species that is being reproduced does not belong to a set of similar species that could alternatively be reproduced. In simple autocatalysis only the information corresponding to the presence or absence of a particular species is reproduced – nothing about its detailed form.

Now one might imagine the evolution of a system taking place through the setting up of several autocatalytic processes. The catalyst set X, let us say, consists initially of an autocatalyst m: X then evolves, through X′, to X″ which consists of three autocatalysts l, m and n. The trouble here is that each reproducing molecular species only contributes a limited amount of genetic information (about one bit – the species is either there or it is not). Also it is very hard to imagine in practice several autocatalytic processes going on together without interfering with each other. (Here is a challenge to a chemist's ingenuity: a vessel contains a mixture such that if A is added, more A forms in the vessel; if B is added, more B, and if C, more C. It might just be possible to construct such a three-bit genome using non-template autocatalytic processes, but I think that if you were to try to design a multi-bit system you would find that the difficulties rapidly got out of hand.)

There would be a similarly low ceiling for the evolution of systems using hereditary mechanisms A1 or A2. Here what is inherited is the ability to pick up X from the environment rather than the ability to make X: X is 'autoacquisitive' rather than autocatalytic. Even if you could imagine a mechanism for this, such an evolution would be limited by the number of (non-interfering) 'autoacquisitive' catalysts in the environment. The whole idea seems absurd.

There is nothing wrong with the idea that the early environment might have provided catalysts for early organisms. The environment is still a source of metal ions, such as Zn^{2+}, that are important parts of enzymes. The trouble arises when one tries to imagine the environment as the source of the *specificity* of catalysts, or indeed of any other kind of specificity. If it is the environment that entirely controls the sequences of reactions in some microsystem – because, say, the environment contains catalysts that the microsystem passively acquires – then the microsystem is simply part of the environment. To evolve within the environment the microsystems must acquire specific characters of their own so that a million years later the microsystems could be different even if the environment had remained similar. That is not to say that some aspects of the microsystems' behaviour cannot be controlled by the environment – by catalysts and other things – what is important if a microsystem could be said to have evolved is that at least some aspects of its behaviour are not so controlled.

Expect replicating structures

Let us summarise our conclusions about what is necessary for an extended evolution of microsystems, and then take the argument a little further.

Simple reproduction is no good: for example, coacervates growing and dividing would not be evolving if that is all they were doing; in that case their characteristics would be determined by the environment. (If you came back to look in a million years you would find no difference if the environment was no different.) Particular structures that happened to be possessed by some droplets – for example catalytic particles – might be passed on when the droplets divided and might provide some individuality for different lines of droplets, but these effects would soon be diluted out. Not the goods but their means of production must be inheritable. (Their means of acquisition was another idea but this turned out to be too limited.) And the means of production has to be in the form of specificity or information. Even this, though, is not enough. For evolution, the information must be modifiable and the modifications must also be inheritable. This eliminates the kind of simple autocatalysis of which ester hydrolysis is a model. Indeed it implies copying of some sort.

Taking a more complex example, suppose that X catalyses the formation of Y and that likewise Y catalyses the formation of X. That would be a reproducing system with the structures of X and Y not necessarily related. But to be evolvable X′ must catalyse the formation of Y′ *and Y′ must catalyse the formation of X′*. It is this last consideration that demands some structural correspondence between the Xs and the Ys: X and Y could be + and − strands of a DNA molecule for example. It is not necessary that X, X′, X″, etc. are structurally identical to Y, Y′, Y″, etc. or even at all similar in composition, but there must be a detailed correspondence between them – and for an extended evolution there would have to be some vast set of such corresponding member states. Templating is still not forced on us as the means of ensuring such a correspondence. For example, a text can be reproduced by being translated first into a temporal sequence of blips, sent along a telephone wire and reconstructed a thousand miles away. With sufficiently sophisticated equipment, translation and back-translation could be effected by other means than the direct testing out of units for size and shape as happens, say, in DNA replication. All we can say is that, for a molecular system, templating looks the easiest way.

Our rejection of a set of independent autocatalytic processes was similarly based on chemical plausibility: in principle genetic information could be carried in the form of a check list of presences and absences, but this would be impracticable, and in any case inefficient from the point of view of information capacity. Suppose that there were four possible items on the check list. This would give, in all, sixteen possible lists (none; A; B; C; D; AB; AC; ... ABCD) which is to say that the system could have an information capacity of four bits. By contrast, if it is not merely the existence or absence of items that constitutes the information but their arrangement – for example the arrangement of the four bases in DNA – the information capacity is limited only by the number (not by the number of kinds) of the units. The chemistry can thus remain simple while the information capacity tends to infinity.

Although other hereditary systems are possible in principle, it would seem that in practice any extended evolution of microsystems would depend on the copying of stable arrangements in some sort of genetic material. Such a genetic view of life is now widely held; but if you add to it the conclusion from Chapter 1 that prevital nucleic acid replication is wholly implausible, then you arrive at a much less popular view: *there must have been at least one other kind of genetic material before nucleic acid.*

Third doubt; original biochemical similarity

Perhaps the main barrier to accepting the above conclusion is the idea that the central biochemicals have always been more or less a fixed set. It is as if the origin of life was being seen as a kind of chess problem to be solved, if at all, by using only the pieces now visible on the board. I will conclude this chapter with comments on some of the rather superficial arguments that are sometimes put up in favour of original biochemical similarity.

Perhaps life can only work in one way Pirie (1957, 1959) has pointed out the emptiness of the assertion that life could only work in one way. This assertion becomes less plausible the more we know about the details of how our kind of life works. There are surely other catalysts than proteins, other membrane materials than lipids, other genetic materials than nucleic acid. Life is primarily about systems of a certain kind, not substances as such.

Present biochemical similarity The close biochemical similarities between all organisms now on Earth is not a good argument for the idea that organisms have always been biochemically similar to present-day organisms. When one looks more closely at the common features of all life now on Earth one sees machinery that could not have been picked out of a soup; that is evidently the product of prolonged evolution. All life now is evidently descended from a common ancestor that was high up the evolutionary tree – a considerable distance from the start of the evolutionary process that gave rise to that common ancestor. The crucial question is what happened over that distance.

Miller's experiment Then again, as discussed at length in this chapter, the simulation experiments, although promising to begin with, have turned out to be ambiguous: if anything they are better understood in terms of the idea that our present set of central biochemicals was a product of evolution rather than a precondition.

Surely to make a machine you first need the components? Then again it is a naive view of evolution, even if it sounds like logic, that to make a machine you have to start with the components. That is the way we make machines, but evolution does not work like this. I will devote much of the next chapter to the question of how natural selection brings machinery into existence; how subsystems come and go in the trying out of designs and combinations of them; how, eventually, mechanisms become frozen in; how one subsystem may become universal – not because it was the first but because it became the best. And we will think in detail about just why the first is unlikely to be at all similar in design to what eventually turns out to be the best.

Should we not choose the simplest hypothesis? Others would appeal to Occam's Razor in defence of the principle of original biochemical simi-larity. They would say that this is the simplest hypothesis and should be adopted in the absence of clear evidence one way or the other. Simplicity has its place in science: where there is nothing else to be said between two or more views – then choose the simplest by all means. But there is very much to be said about early evolution, and there is no alternative but to think through the arguments in detail. In the circumstances to hold to the simplest view may show less a leaning to elegance than to laziness. There is no indication that the very early evolution that made our biochemistry was a simple process. Rather the reverse, if later evolution is anything to go by.

In any case the place for Occam's Razor is in natural philosophy, not in natural history. Of historical questions we may ask simply 'is it true?' (is it true, say, that Harold got an arrow in his eye at the battle of Hastings?). Maybe we do not know, but the question is not one of formal elegance, it is one of fact: either it happened or it did not. Similarly, questions of early evolution are questions of fact, however remote these facts may seem. This is not at all like, say, physics where the choice between two ways of describing known facts may rest on which lets you write the simpler equations.

Where simplicity is most to be sought in biology is in the construction of the very first systems capable of indefinite evolution. They must have been simple, in the sense of uncontrived, if chance and chemistry put them to-gether. It is surely at least as plausible to suggest that the first organisms were simple but dissimilar to us and that our biochemistry was the outcome of a subsequent (complicated) evolution, as to suppose that the first organisms were based on our complicated way of doing things, followed by a subsequent (simple) evolution.

Would not the genetic material be (virtually) invariant? Then again it is sometimes stated more or less explicitly that (even if the rest of evolution was devious) there could be no place for radical change at the very centre – that the genetic material itself must be invariant or nearly so. I will try to show in the following two chapters that, on the contrary, you would expect there to have been at least one radical switch in genetic materials during early evolution.

3

Questions of evolution

I define life . . . as a whole that is pre-supposed by all its parts.
S. T. Coleridge (ca *1820*)

. . . organisms – that is to say, systems whose parts co-operate.
J. B. S. Haldane (*1929*)

We shall regard as alive any population of entities which has the properties of multiplication, heredity and variation.
J. Maynard Smith (*1975*)

I suggest that these three properties – mutability, self-duplication and heterocatalysis – comprise a necessary and sufficient definition of living matter.
N. H. Horowitz (*1959*)

. . . it might be claimed that the most important fact about them [living things] is that they take part in the long term processes of evolution.
C. H. Waddington (*1968*)

Definitions of life commonly fall into either of two classes. First, and most immediately understandable, there are definitions like those given by Coleridge and Haldane. Here organisms are to be distinguished from other physicochemical systems in being seemingly purpose-built. Organisms are to be seen as machines analogous in some detail to man-made machines which are made up of components and subcomponents acting together. According to modern science the machinery of life was not put together with forethought by an intelligent being. The only engineer was natural selection. Definitions of this sort are **teleonomic**, that is, they use the idea of apparent purpose.

The other kind are the **genetic** definitions of life which concentrate not on products but on prerequisites for evolution. The definitions according to Maynard Smith and Horowitz are of this kind. We discussed in the last chapter why replicating structures of some sort – genes in the most general sense – would seem to be necessary for heredity and hence for evolution through natural selection. (The entities referred to by both Maynard Smith and Horowitz, for example, could be DNA molecules.)

Waddington's description neatly connects both kinds of definition of life and avoids some of the quibbles that can be put up against each. Even organisms that do not reproduce are part of the long-term processes of evolution – they are an outcome of these processes – although they no longer directly further them. Then again the very first organisms on Earth must fail to conform to the teleonomic definition if you bar miracles: these were not overtly machines in the sense of being highly contrived and seemingly improbable in their mode of survival and propagation. Yet they too can be claimed to have been part of life – in retrospect at least – as having been at the other end of those same long-term processes of evolution. Life in the teleonomic sense could only have emerged gradually during the very early evolution of systems that from the start conformed to the genetic definition.

As implied in the Preview, I am using the teleonomic definition of 'life' but the genetic definition of 'organism'. Hence I will never talk about unevolved life – which would be a contradiction – but I will have much to say about putative unevolved organisms. Indeed to search for the origin of life is to search, first, for such systems.

Much of the difficulty in the problem of the origin of life and much of the confusion in the definition of life has arisen, I think, from a confusion between prerequisites for evolution and products of evolution. Part of the trouble is that products have become prerequisites – that is, products of early stages of evolution have become prerequisites for evolution now.

In this chapter we will be concerned very largely with how evolution ties such knots and in doing so conceals the means by which it creates that most characteristic property of life – co-operation between components. But to begin at the beginning: what sort of system is it that we are looking for, that is not an improbable collaboration of parts and yet can evolve to become one?

What is a phenotype for?

An organism is said to consist of a **genotype** and a **phenotype** in an environment. We discussed in the Preview and in the last chapter how any system if it is to evolve within an environment must be able to hold and replicate information of some sort – instructions about how to make its characteristic features. It is only such instructions that can be transmitted indefinitely between generations to provide the basis for a long-term evolution. The replicable instructions constitute the genotype of an organism, while the characteristics are the phenotype. The environment has to be included in the description because in the end it is the environment that the genetic instructions instruct. The phenotype of a modern organism can be seen as a kind of go-between: itself a bit of the environment locally modified by the genetic information.

Following Sherrington (1940) you can think of the environment for life as a stream, with organisms as eddies in the stream that are locally moving against the entropy trend. On the surface of the Earth the main motive stream is the stream of photons from the Sun. Sooner or later the energy that arrives on the Earth will be reradiated into space in a more degraded form. But between this arrival and this departure the gigantic eddies of the weather and the water cycle are driven continuously. Much more intricate are the convolutions inserted by life on Earth between the absorption of sunlight in green leaves and the reradiation of that energy, eventually, into space.

But really it is only the phenotypes of modern organisms that should be viewed in this highly dynamic way, as being a dynamic part of the environment with 'the motion of the eddy ... drawn from the stream', as Sherrington put it. By contrast, for an individual organism, the genotype is static. We might extend Sherrington's analogy by thinking of a stone of a certain shape placed in a stream and causing a specific kind of turbulence. The (static) shape of the stone is a genotype which when placed in a particular environmental stream informs that stream to generate a localised, persistent, but dynamic turbulence – a phenotype.

There are two counter-intuitive aspects here. Using higher animals as models we would be much more inclined to see the organism as dynamic and the environment as static. But the only bit of an organism that is unambiguously not part of the environment is the bit that is static – the genotype.

The other counter-intuitive idea is that, in computer jargon, it is software in organisms that lasts, while hardware is being perpetually replaced. Consider, for example, the instructions about how to make cytochrome *c* molecules: that software has remained little altered in essentials while mountain ranges have risen and been worn away many times. Yet the hardware, the actual individual protein molecules, individual DNA molecules, and so on, have been quite evanescent, flickering in and out of existence on a geological time scale. And this is very close to the heart of the problem: following our discussions in the last chapter we might say that life can begin to appear when mechanisms exist for retaining and propagating a kind of software – genetic information – indefinitely.

Perhaps the simplest kinds of organisms would be hardly more than pieces of unencumbered information-printing machinery – 'naked genes' as they have been called (Muller, 1929). To have the potential for indefinite evolution into the future, the potential information capacity of these naked genes would have to be very high.

As discussed in Chapter 1, the idea of a 'naked gene', as the simplest and first kind of organism, has a long history. It is somewhat out of favour now mainly on account of two kinds of argument that are put up against it.

First, there is a practical argument. Even if it could evolve in principle, it is said, such a structure would be too improbable in practice: it would be exceedingly unlikely to form, and the Earth would be exceedingly unlikely to continue to provide the highly specialised components needed to keep it replicating. If we think about a naked nucleic acid molecule such an attitude seems justified.

Second, there is a formal argument. To evolve, a system must have both a genotype and a phenotype. Pure information is no use: it is the phenotype on which selection operates to give genetic information a meaning. Formally this argument is impeccable, but it is largely irrelevant. A 'naked gene' would not be – could not be – pure genotype. Clearly what is meant by a gene, in this context at least, is some sort of structure that is holding information – something analogous to a DNA molecule or a punched card. Such a thing is not pure software as it includes the structure that is holding the information, and that is hardware. And at least some aspects of hardware could very well be phenotype.

There are two ways in which this could be so. First, we may note in passing that in a DNA molecule only the base sequence (specified by a copying process) is genotype: everything else about it is phenotype because the processes that give rise to all those other aspects of the DNA molecule – processes such as making deoxyribose, collecting the phosphate, and so on – are under the control of information disseminated throughout the genome. They arise by following rather than by copying instructions. Of more immediate interest, nucleic acid molecules can often be both genotype and phenotype in the sense that a sequence of bases directly specifies some functional object. The transfer RNAs and the ribosomal RNAs are like this: here a (potentially replicable) message tape can become a machine by twisting up on itself in a way that is determined by the message.

That this duality in RNA can allow RNA molecules to evolve has been beautifully illustrated by Spiegelman and his school (Spiegelman, 1970; Spiegelman, Mills & Kramer, 1975; Mills, Kramer & Spiegelman, 1973) in their studies of RNA molecules replicating with the aid of an enzyme system *in vitro*: natural selection was found to operate so as to favour particular RNA molecules according to their size and to how they twist. For more recent discussions see Orgel (1979) and Eigen, Gardiner, Schuster & Winkler-Oswatitsch (1981).

So the formal argument against evolving 'naked genes' is too formal. All that is necessary is that, in addition to holding and printing some sort of pattern (analogous to nucleic acid base sequences), a primitive gene could have at least some of its properties affected by the pattern that it happened to be holding. Simple properties such as shape and size might do. Suitably shaped particles might tend to adhere to each other slightly better. In a particular environment this might have survival value – for example in a stream, by reducing the chances of a mass of such particles being washed away. To my mind it is only the practical arguments that hold up, but then these arguments are against naked nucleic acid genes, not against 'naked genes' as such.

The term 'naked gene' means 'consisting only of genetic material' and not 'lacking any phenotype'. The key point is that in so far as a gene has physical and chemical properties that are affected by the information that it contains; and in so far as those properties affect the survival chances of the gene, then that gene itself has phenotypic characteristics. Phenotype and genotype are separate ideas, but they are not necessarily separate structures.

In any case we must try to disengage from the idea that a phenotype has to be something elaborate. True, all the phenotypes that we have ever come across in real organisms are highly elaborate. Yet if we ask 'what is a phenotype for?' the answer does not seem to demand that it be particularly ingenious. A phenotype should provide conditions that encourage the preservation and/or the replication and/or the propagation of genetic information. In the most general case these conditions are strictly unspecified. It does not matter in what way the phenotype helps, or even if its assistance is only marginal. So long as it is some sort of consequence of genetic information and so long as it tends to promote that information then a phenotypic characteristic can catch on.

It is because nucleic acid is so difficult to make that the phenotypic machinery serving it has to be so complicated. We might say that the necessary complexity of the go-between machinery – machinery between a genetic material and its environment – depends on how mismatched a genetic material is with its environment. The modern cell is like a spacecraft providing life support systems for the fragile DNA molecules that contain its essential information. A naked DNA gene would be mismatched on our Earth as a spaceman without his space suit would be mismatched on Mars. As Dawkins (1976) puts it, organisms are survival machines for their genes.

One possible way of explaining the present relationship between DNA and the external environment is to say that there was a time in the remote past on the Earth when the external environment was matched to DNA and that the go-between apparatus evolved gradually as the environment became less clement. That is logical, but from our discussions in Chapter 1 it does not seem to work. It really does not look as if any purely physico-chemical environment would have had the expertise to carry through the necessary cycles of operations for a nucleic acid genetic material.

Another way of imagining a first match between a genetic material and a primitive environment is to suppose that it was the genetic material, rather than the environment, that was different to begin with. We have already concluded something of the sort (page 75). But would such a change be evolutionarily possible? Would it fit with more general ideas on the nature of evolutionary change?

What does evolution do?

Four aspects of organisation

Most people would agree that evolution has created highly organised systems. Indeed we tend to equate, in a rough sort of way, the level of evolution with the level of organisation in an organism. Unfortunately the question 'how organised is this system?' is ambiguous. A system might be said to be highly organised because it is very orderly, like a crystal; or because it is complicated like a television set; or because it is arranged into distinct organs like a bureaucracy; or because the parts out of which it is made are strongly dependent on each other, as in machines of many sorts. Or you might say that efficiency is the only measure of organisation. Organisation is difficult to measure just because it is a compound idea – only its elements are measurable in suitable circumstances. But we can say, I think, that an organised system is some arrangement of units that are at least to some extent orderly, efficient, complex and made of parts that co-operate. The question is which of these aspects is the most typical product of evolution.

Orderliness I will try to show now that orderliness is not a particularly significant attribute of organisms.

First let us define orderliness in a way that is in line with its meaning in statistical thermodynamics. We will say in a general way that to be well

ordered is to belong to a small subset of possibilities. Suppose that you
have put some coffee cups, that had been lying about in the kitchen, into
the kitchen cupboard. You would have been ordering the cups because
their positions in the cupboard, although not completely defined, are more
closely defined than they were when the cups were just somewhere in the
kitchen. There are fewer possibilities for the state 'stowed away in the
cupboard' than there are for the state 'somewhere in the kitchen'. If you
could count the respective numbers of possibilities then you would have a
measure of the amount of tidying up that you had done.

Molecular ordering processes can often be measured in this fashion,
because there is a way of estimating molecular possibilities – the number
of particular ways that a set of molecules could be arranged, including the
distribution of their energy between them. Such a number, W, is vast, even
for moderately sized physicochemical systems. But W for a system can be
estimated because it is related to the entropy (S) through the Boltzmann
equation, $S = k \ln W$ (k is a constant, *ca* 1.4×10^{-23} JK^{-1}).

Now let us try to make a rough but very conservative estimate of the
physicochemical orderliness of a bacterium such as *E. coli*. Let us think
only about the water in *E. coli* – about 10^{-13} moles. The molar entropy of
water under ordinary conditions of temperature and pressure is about
70 JK^{-1}: so the entropy of an *E. coli*-sized blob of water is about 7×10^{-12}
JK^{-1}. That gives us a value for W of about $10^{200\,000\,000\,000}$. For the same
mass of water vapour the entropy would be about 19×10^{-12} JK^{-1}: W in
that case would be about $10^{600\,000\,000\,000}$. The ratio W (vapour)/W (liquid)
represents a measure of the ordering that takes place when this tiny
amount of water condenses from the vapour. It represents the 'improbability' of the condensation – or rather how improbable that condensation
would seem if one did not take into account intermolecular forces. It is a
big 'improbability' – about $10^{400\,000\,000\,000}$.

Biological systems have another kind of 'improbability'. Even if you
take into account intermolecular forces, an organism still seems to be an
improbable system if you do not also take into account its evolutionary
history. To measure what we might call the biological orderliness of *E. coli*
we should attempt to get some estimate of the constraints placed by evolution on molecular arrangements within *E. coli*. How much of the order in
E. coli has been generated by evolution?

It is not difficult to make a maximum estimate for this. The product of
evolution, the only thing that is passed on between organisms over the long
term, is genetic information. We can estimate the maximum amount of
that information in *E. coli* by counting the number of base pairs in its

DNA. This number is about 4.5×10^6. With four possibilities for each base pair the total number of ways in which *E. coli*'s DNA might have been arranged is $4^{4500\,000}$, i.e. $10^{2700\,000}$. Since no doubt there are many slightly different DNA sequences that would do to make an *E. coli*, this is very much a maximum measure of the 'improbability' of *E. coli*. It is a measure of the maximum amount of ordering that evolution could have been said to have achieved in making *E. coli*. Looked at one way $10^{2700\,000}$ is a staggering number, far exceeding, say, the number of electrons in the known Universe (around 10^{80}). But compared with the coralling of possibilities achieved when a speck of dew condenses, it seems rather a poor performance. The generation of order as such cannot be what evolution is especially good at.

Efficiency Components of organisms are not just tidily made. They have functions. They are machines. As such it is appropriate to ask how efficient they are.

Natural selection can be seen as a kind of optimising procedure that leads to increasingly efficient machines; and we can discuss evolution, as we can discuss optimising procedures generally, in terms of movements across a diagram – what is sometimes called a 'functional landscape'.

A very simple example is shown in figure 3.1. This refers to some imagined machine with two variable features, say a carburettor with two adjusting screws on it. The contour lines represent screw settings of equal efficiency. To find the optimum setting for the carburettor you might first adjust screw X in the direction that improved performance, until there was no further improvement, then similarly adjust Y, and then X again, and so on – following a path like (*a*) in figure 3.1. Or you might make alternate small changes to X and Y sensing the gradient, as it were, and following a more direct path – like (*b*) in figure 3.1. If you imagine the contours defining an altitude so that more efficient positions are lower, then the path taken through an optimising procedure could be compared with that of a ball rolling in a basin. It too can find the lowest point *without testing every point*. That is the key idea: if there are smooth relationships between the efficiency of an object and some aspects of its structure, then the structure can be optimised without having to try out every possibility.

You can think of a protein molecule, say an enzyme, as a machine containing as many adjusting screws as there are amino acids in the molecule. Each screw has twenty possible settings corresponding to the twenty amino acids. Some of the screws – near the active centre particularly – are coarse adjusters. Others, more remote from the active centre, have much less

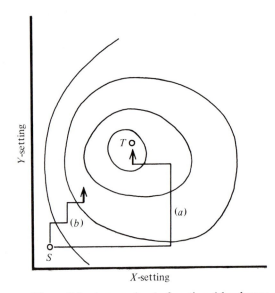

Figure 3.1. A very simple functional landscape corresponding to a
machine with two adjusting screws that have some single optimum
setting (*T*). The contours represent settings of equal efficiency. (*a*)
and (*b*) correspond to alternative tuning procedures starting from
some inefficient setting (*S*). If we regard more efficient settings as
downhill this diagram becomes a contour map for a basin, and the
optimisation of screw settings is analogous to a ball starting at *S* and
finding the lowest point *T*.

effect – they are fine adjusters. Occasionally a mutation takes place that, in
effect, flicks one of the screws to a new position. If as a result the enzyme
works better, then the gene specifying it, and within which the mutation
took place, will tend to become more prevalent and eventually displace the
earlier 'setting' for that amino acid. (Because on average organisms that
contain the improved modification are more likely to have offspring and
pass on the gene specifying it. Even if the benefit is only marginal the long-
term effect will be inexorable because always tending in the same direction.
Eventually the improved modification will displace the other.)

For the evolution of a protein the relevant functional landscape contains
not two but perhaps 200 structural dimensions, with the number of possible
combinations of settings so vast that only a minute fraction of them could
ever have been tried out in the whole history of the Earth. But in spite of
the vast multidimensional volume of this protein hyperspace, distances in
it are quite short. There are indeed 20^{200} sets of settings for the screws on a
protein made up of 200 amino acids, but any two of these can be inter-

converted in no more than 200 moves. It is a question of knowing which direction to take.

It is sometimes said that since mutations are chance events the exploration of the space of protein sequences must be a random walk and hence the chances of any particularly effective sequence being stumbled on is too remote to be worth thinking about. But this fails to understand the nature of evolutionary exploration and the role of chance in it. The exploration of the protein space is not like a random walk, it is like a succession of random steps checked at each step to see if it seems to be in the right direction. A single step that improves function will be amplified and become the most likely basis for the next step. The most important effect of mutations is thus far from throwing up new random amino acid sequences on the off-chance that they might be useful. It is to make small random modifications to sequences that already are useful. Having hit on an idea that works, the immediate surroundings are explored. When a better one is found this in turn becomes established by the automatic process of selection and then the new surroundings are again explored. Then the best modification of the original modification is selected . . . Eventually some optimal sequence is discovered. Such a procedure is a progression from a member of some large set of possibilities – for example the set of all amino acid sequences that have at least some sort of activity of some very general kind – to members of smaller and smaller subsets. Eventually very particular structures can be arrived at that have efficient and specific action.

Now chance has a part to play in all this, as it has a part to play in chemical reactions or in the diffusion of a gas through a hole. But that is not to say that the course of evolution is 'pure chance' any more than chemical reactions are or the behaviour of gases is. Natural selection is an optimising procedure analogous to the procedure for tuning a carburettor, or to the way in which a ball in a basin discovers the lowest point. All points do not have to be visited if there is a reasonably smooth relationship between structures and functions. Optimisation procedures work on this basis: when you have hit on a good idea a better idea is most likely to be next door. Natural selection is no exception.

Natural selection could arrive at seemingly wildly improbable structures quite quickly if the circumstances were right. Once natural selection in moving from set to subset to sub-subset, etc. had made the equivalent of a series of 200 choices between equally probable alternatives, that is to say once 200 bits of genetic information had been accumulated, the *a priori* probability of any one system so selected would be one in 2^{200} (i.e. 10^{-60}). Two hundred selections may not seem very many, but if they can be accumulated in series, if the outcome of one selection can consistently become the basis for the next (and this is possible in Nature only for replicating structures) then these two hundred selections each between two alternatives becomes equivalent to a single selection from *ca* 10^{60} alternatives. When you consider that the number of atoms in the Earth is about 10^{50} it is clear that natural selection, because it operates on replicating structures, is capable of choosing quite quickly from a notional bag of possibilities that is far too big ever to have been actual.

One might discuss in similar terms the optimisation of other parts of organisms, say the shape, size, flexibility, etc. of a fly's wing, where the lowest point on some multidimensional diagram would be determined by a compromise between such things as efficiency in flight, ease of manufacture and invisibility to predators. The evolution of whole organisms can be described in terms of a huge imaginary diagram – a vast landscape with the net tendency to survive and leave offspring (biological fitness) as the only measure of efficiency.

Landscapes of this sort were introduced by Wright (1932) and are frequently used in the discussion of evolutionary processes (for example by Dobzhansky, Ayala, Stebbins & Valentine, 1977; Lewontin, 1978). The usual convention is to regard optimisation through natural selection as a climbing of mountains and to talk of 'adaptive peaks' in the evolutionary landscape. I prefer to think in terms of basins – downhill as fitter – because that is more like a potential energy diagram describing simple physical systems. Also one can imagine adaptation as being like a ball rolling in a basin.

This improving effect of natural selection (adaptation) is not the whole explanation for evolution by any means. Most of the time, indeed, natural selection operates to keep things as they are, with all subsystems near-optimised. It is one thing to be able to adjust a pre-existing machine, it is another to invent the machine in the first place. In terms of the functional landscape: once you are in a basin natural selection will find its lowest point, but how is the basin found? Then there is the problem of multiple minima. A difficulty with optimisation procedures is that they only seek out a best local solution, the local lowest point. But there may be some much lower point separated by intervening high ground.

The form of a functional landscape depends on the environment in which the organism finds itself (think of those flies: the optimised wing design could depend on, say, average wind speeds or on the predators that were around). So if the environment changes, the functional landscape may undergo 'earth movements': what was a basin may become a slope allowing movement to a nearby lower point. Indeed the functional landscape should be regarded as variable and mobile (Waddington, 1957). One can see the possible effects of this by imagining a collection of ball bearings placed in one hollow of a dynamically undulating and complex surface. They might well become spread out in time into separate collections occupying different hollows.

I will return to the question of how natural selection can become an inventor. In the meantime let us note that evolution is not completely to be described in terms of optimisation procedures. A man is not an optimised fish, or a fish an optimised bacterium.

What seems more serious still is that a man is not overtly more efficient than a fish, or a fish than a bacterium. The overall function of an organism – the function to which in the end all its most elaborate machinery is subservient – is to be fruitful and multiply. What is optimised overall is fitness – the tendency to leave offspring in the long term. Lower organisms are often very good at doing that.

So where does this leave us? Orderliness, we decided, was not sufficiently characteristic of biology to be the crucial product of evolution. Natural selection indeed generates order to the extent that it seeks out efficient subsets of possibilities, but the amount of ordering thus achieved is, as we saw, tiny compared to the ordering that goes on when, say, a teaspoonful of water freezes. Efficiency is more to the point in that it introduces the idea of function. The overall function of genetic information turns out to be something rather simple – to persist or at least to be continuous (although possibly changing) in time.

Rocks are good at persisting too. And the 'function' of a crystal structure might be said to be persistence, a stable polymorph being the most persistent arrangement of component molecules in the circumstances. To avoid these quibbles we must add that in biology what persists in the long term is not an object, such as a rock; or a thermodynamic fact, such as the stability of a crystal structure; but a 'plan'. In biology it is software that persists in the long term. But there is still a difficulty in seeing biological efficiency – fitness – as the key feature that increases with evolution. On that score man would be less highly evolved than many bacteria.

Complexity One has to be careful too with complexity as the measure of evolution. After all, any heap of rubbish can be complex (that is, needing a lot of information to specify it). Yet, organisms that are in our eyes highly evolved are more complex than primitive organisms. Why does evolution go with increasing complexity?

I think there are three main reasons. First, there is the point made by Maynard Smith (1969): the first organisms must have been simple, to become more complex was then the only way to go. This is not sufficient to explain why organisms became as complex as they did, however. Even the space of protein molecules of up to 200 amino acids long should have been roomy enough for a virtually endless exploration. Shortage of possibilities cannot have been the only consideration.

Second, we might very well argue, in answer to the last point of the previous section, that what evolution really does is not to increase efficiency, to create organisms of increasing fitness, but to come to produce organisms that can fill niches of increasing difficulty. To be a land animal is, in many ways, more difficult than to be a sea animal, and, on the whole, land animals are perhaps more interesting, more highly evolved we would say. To come to live up in the trees, our ancestors had to develop some pretty advanced technology in the way of eyesight, hand grip, and so on. As a result of all this they were no more biologically efficient than the soil bacteria beneath them; but they were a good bit more interesting.

Now any sort of advanced technology usually calls for complicated answers. Engineers may, for aesthetic or economic reasons, seek simple designs for machines, and natural selection will tend to favour simplicity in organisms. But in the event, in both cases, the most efficient machine is likely to be complex: to get something just right for some difficult task usually calls for nests of subsystems. The latest kind of aeroplane may be simple in conception, but infernally complicated when it comes down to the nuts and bolts. The much greater number of possibilities inherent in more complex structures improves the chance that among these possibilities something suitable will be found. Thus organisms, if they are to occupy difficult niches, may be pushed on to complex answers. And it is the organisms that have been so pushed that we choose to call 'highly evolved'.

Huxley (1974) defined evolutionary progress in terms of an increasing ability to control the environment and to be able to cope with changes in the environment. The means to those ends – ways of controlling water loss, or of exerting force, or of maintaining body temperature – are bound to call for sophisticated engineering.

A third factor would be effects of irreversibility in evolution. Following Maynard Smith (1970), Saunders & Ho (1976) argue that in optimising fitness reductions of complexity are less likely to be effective than increases (roughly, because usually anything an organism has got it needs).

We will discuss later, and on a somewhat different tack, how features might tend to become locked-in when they have been present for a long time in a line of organisms. If, for whatever reason, features become difficult to remove (even when their removal might have been appropriate in adapting to new circumstances) then the only way of changing is to add new features on top.

Co-operation between parts Some would say that to be organised at all a system must be made up of co-operating parts. Some overall function is divided into subsidiary functions that are carried out by distinct 'organs' working in collaboration. (The system is literally organ-ised.) Certainly this is what machines are very often like – and it is what organisms are very strikingly like. Both Coleridge and Haldane used this characteristic to define living systems. Descartes had a similar view, by implication, when he likened a living thing to a clock. Let us consider this example.

At the highest level a clock is a device for telling the time. This function arises from a collaboration between a number of subsidiary devices – a chassis, an energy store, a gear train, a governor, a display, and so on. Each of these has further subsystems. For example, the energy store may consist of a ratchet and a spring; the governor of an oscillator and escapement; the display of a dial and hands. At the next level down there are further divisions – the ratchet contains a pawl and notched wheel perhaps. Even this is not the end: the materials must be suitably chosen, the atoms and the way they hold together is important too.

At every level there is room for design modifications, for adjustments to shapes and sizes. Some of these will have little effect on function; others will affect function strongly. (For example, the spoke structure of a wheel in the gear train will be less important than the form of the cogs.) The clock on your mantelpiece could be represented as a point on the surface of a multidimensional functional landscape. (One dimension would represent efficiency in time-keeping while each of the other n dimensions would be concerned with specifying some size, shape, material and so on: together some set of specifications should be sufficient in principle to define the clock on your mantelpiece.) One can even imagine this landscape being explored by successive arbitrary modifications to its components – these either being retained, if they improve the overall function, or being rejected. Thus a better clock might be 'evolved'.

But an imaginative designer is not restricted to such minor adjustments in discovering new and more effective machines. The clockmakers have come up with quite different design approaches to various subsystems. It was found that a pendulum could be replaced with a spring-loaded wheel in the oscillator, for example. Such different design approaches would be represented by distinct basins in the clock landscape. Either you do it one way or the other: intermediate structures are no use. There seems, then, to be no possibility of going from one to the other through successive small design changes. Yet if it is only a subsystem that has been radically re-designed the new basin will still be in a similar part of clock space.

Of course there are also quite different clock designs – sand clocks, water clocks, sundials, piezoelectric clocks have little in common structurally. Intermediates between these are even less likely to be satisfactory.

Turning to organisms, we can consider the ultimate co-operating parts to be small molecules, such as amino acids, etc. There are three features worth noting here.

1. Co-operation is (to some extent) modular. The small molecules do not all interact with each other on an equal footing: very often groups of molecules interact strongly with each other and then the group as a whole has some functional relationship with other molecules or groups. For example, the amino acids that make up a protein molecule interact strongly and in a complicated way with each other in making a folded structure. This machine may then have some rather simply defined function – for example it catalyses reaction A → B. The situation might be compared to a hi-fi system which consists of modules, such as amplifiers and speakers, of the most advanced design which any fool can plug together. A consequence of modular design is that it allows sections to be interchanged with others whose detailed working might be somewhat different. A module is interchangeable with another that has the same function.

Modular construction of organisms makes sexual reproduction possible. Here, in effect, half the components of one pre-existing machine are mixed in with half from another. This is only possible because gene products are interchangeable modules – at least within a given species.

2. Co-operation is (to some extent) hierarchical. To be modular is already to be hierarchical to at least one level, but there may be many levels for some parts of the organisation of an organism. For example, enzymes are often made from several subunits: ribosomes are made from many proteins and RNA molecules. More complex organelles, such as mitochondria, are made up of large numbers of proteins, lipids, and other components. For multicellular organisms a structural hierarchy extends upwards through cell, tissue and organ. It seems to be a general principle of efficient and complex organisations that they tend to be structurally somewhat hier-archical (Simon, 1962). Also the module idea appears at all levels – the kidney, for example, is practically a plug-in unit.

3. Co-operation is between standard micro-parts. Life on Earth now embodies another idea that is familiar to engineers: it keeps to a standard kit of (small molecule) components. The real design work only starts at the next level up: not in the choice of micro-components – because there is very little choice – but in the way in which these components are arranged.

Why co-operation between parts is characteristic of life Efficient machines (clocks, organisms) are often organised in a strongly co-operative way. It is a feature, and a snag, about such organisations that their components have to be critically made and put together exactly. A deck chair is a more co-operative organisation than a heap of stones – and more comfortable: but the latter has the advantage that its components can be put together almost any way and still be something to sit on.

So far, in considering evolutionary landscapes, we have been using the convention that downhill is fitter. But for this part of the discussion the normal 'uphill' convention is, I admit, better. One then imagines the highly co-operative machines that organisms are, as being perched precariously on high, narrow ridges in the functional landscape.

The question of how organisms could ever get to such places will be the subject of the next section; but we can see already how it is that only organisms among naturally occurring systems could negotiate such hazardous regions of object space. Where one false move can be fatal all will be lost sooner or later; so you do not expect to find among non-living objects very many whose survival depends on being highly co-operative. If organisms can survive and indeed continue to evolve in spite of this it is because for them one false move does not matter. If there are many copies around moves of all sorts can be made without the plan being lost – provided the moves are not too frequent and provided the viable systems continue to reproduce. It is life's way of surviving in the long term, not by just being there but by continuously making copies, that allows that most characteristic feature of the evanescent systems through which it persists in the long term. Haldane's comment that organisms are systems whose parts co-operate is not strictly a definition, but it points to the most characteristic aspect of the organisation that is generated by evolution. But how is it generated?

The evolution of functional interdependence

The question 'which came first, nucleic acid or protein?' is only one of the chicken-and-egg questions that can be asked about central

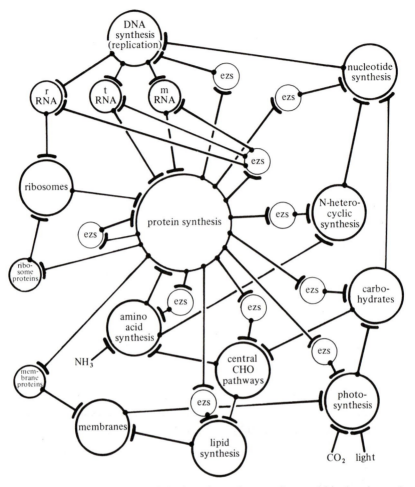

Figure 3.2. Some of the interdependences of central biochemistry. A dependence is represented by a 'crutch' symbol. ezs, enzymes.

biochemistry. Quite simply, in central biochemistry everything depends on everything. A sketch of just some of these interdependences is given in figure 3.2. How could such a situation have evolved? This question is the more pertinent when we consider that co-operation between parts seems to be the most characteristic feature of the kind of organisation generated by evolution. Yet it is not immediately obvious how that main engine of evolution – natural selection – could have produced it. One can see how, once it is there, such an organisation can be locally optimised, and kept like that, through natural selection. But how are these interacting machines

Figure 3.3. How could a structure like this be built if you were only allowed to touch one stone at a time?

set up in the first place when the significance of each depends on the existence, already, of all the others – when 'the whole is pre-supposed by all the parts', as Coleridge put it? The ball may be able to find the lowest point in its basin, but, with so many parameters having to be right, the mouth of the basin in the relevant evolutionary landscape seems too tiny ever to have been hit on. So many things have to be there before a machine built along the lines of modern organisms could start to work at all. How can you arrive at a situation in which everything depends on everything, through a succession of small modifications to some initial system that was presumably not so critically made?

Getting stones to co-operate Let us take an arch of stones, as in figure 3.3, as a model. The integrity of this structure depends on everything being simultaneously in place. It is a highly interdependent structure, like the modern biochemical system (cf. figure 3.2). Would it be possible to arrive at such a structure through a series of small steps? Suppose that we were to build an arch such as that in figure 3.3 using only stones that we found lying about and using only one hand; and we were restricted in our building work to touching only one stone at a time. At first sight this task might seem impossible – and it would be impossible if the stones in the final product were the only ones allowed to be used. But there was nothing in the rules to say that: some add-and-subtract procedure such as that shown in figure 3.4 could provide a solution.

One might imagine a situation in which highly arched structures like this were generated preferentially. Imagine a community of primitive people who worship heaps of stones. They see an accidental heap as the work of a god architect worthy of being faithfully copied. Occasionally among the sets of identical cairns that were being made one is a bit different: a stone has been changed, added or removed. This was, of course, carelessness on

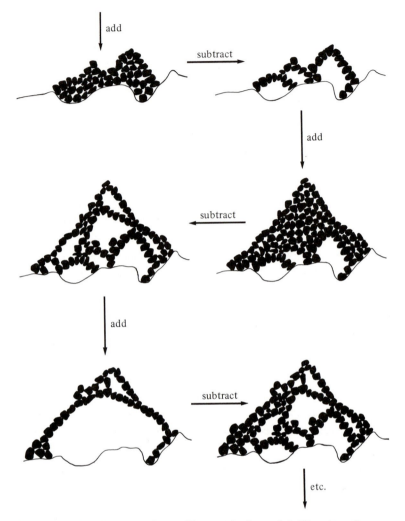

Figure 3.4. A solution to the problem set in figure 3.3. The strongly interdependent structure could have been made one stone at a time if stones had been removed as well as added.

the part of the builders; but the high priests of the society see in it the hand of their god. They ordain that in such circumstances two considerations will determine whether the novel design is worthy of being copied in preference to the old design. First if it is taller and second if it uses fewer stones. Come back in a few thousand years and you would find great arched structures, like cathedrals, all over the place. Two selection pressures on the replicating structures – to perform some function (i.e. to be tall), and yet to do it as economically as possible – would favour designs in which a relatively small number of components were strongly interdependent. Such seemingly paradoxical designs become accessible through evolutionary processes that involve subtractions as well as additions.

Add-and-subtract procedures are common in manufacturing. It may be difficult to see how an object was made because something necessary to hold the parts together during manufacture is no longer there in the final product. How did the ancient Britons put in place those huge lintels in Stonehenge? You can be sure there was some sort of 'scaffolding' which was later dismantled – earth ramps most likely.

The elephant's trunk is a paradoxical structure. How could it evolve from a short snout when structures of intermediate length would seemingly be no good? (Too floppy for nuzzling, not long enough to be properly prehensile.) Maynard Smith (1975) has discussed this evolutionary conjuring trick. It appears to have depended on a feature no longer present – a lengthened lower jaw that allowed a short forerunner of the trunk to be functional (Watson, 1946).

We know that features, subsystems, are subtracted as well as added during evolution – we can see that in later parts of the process that are still visible to us. It would be strange if early biochemical evolution had been different in this respect; it would be an improbable assumption even if there was no evidence for earlier subtractions in the outcome. But there is such evidence: the multiple interdependences of central biochemistry surely call for some kind of 'scaffolding', for other subsystems at one time (no longer present because no longer needed), on which our now interdependent subsystems leaned during an earlier era while they were evolving. Of course this 'scaffolding' was inadvertent, it was not erected *so that* a subsequent elaboration would be possible. Natural selection has no forethought, but it can be efficiently opportunist. If a stage is reached when subsystem X can be dropped to the advantage of the system as a whole, you can be fairly sure that subsystem X will be dropped. But when that happens the remaining subsystems may very well be more strongly interdependent. By analogy with the stone building game, later stones depend on the earlier ones but not *vice versa*. But when some of the earlier stones can be removed because the later stones can now lean on each other, a stage may be reached at which a co-operative set of later subsystems works better than earlier, less critically interdependent, subsystems.

One very critically interdependent set of subsystems is the set of machines used in protein synthesis – the transfer RNAs, the enzymes for attaching amino acids to them, the ribosome, etc. (These will be discussed further in Chapter 9.) For the reason given at the end of the last section, a strongly co-operative organisation only becomes plausible in Nature in fully reproducing systems. However it happened, the invention of protein (that is, the ability to put amino acids together into long sequences specified by genetic information) was an achievement that only fully reproducing systems could have brought off. However protein synthesis evolved, it did so within organisms that already had highly competent hereditary machinery and that could work well enough without protein. Think of what would have happened, though, when in some line of organisms machinery first appeared that could consistently generate particular amino acid sequences. New ways of making catalysts, new ways of making structures of all sorts would become accessible. With the invention of protein nothing in biochemistry would ever be the same again. Then doubtless earlier ways of catalysing reactions and building structures would have been displaced. But only then: up to that point, while a mark-I protein synthesiser was being perfected, protein could not have had the central place that it now has. What is so evidently a product of evolution could not have been a requirement for evolution to begin with. That only happened when those subsystems went missing that had inadvertently provided the 'scaffolding' to support an otherwise impossible-looking enterprise.

We can perhaps best think of long-term evolution as a combination of many (always small) increases in ordered complexity combined with (sometimes large) decreases (a 'forward creep–back leap' progression; Zuckerkandl, 1975). Paradoxically, perhaps, it is the character of the decreases that imposes a trend to greater complexity in the long term. Think again about the stone building game. You might dismantle a heap of stones in the exact reverse order of its construction – through a succession of stable smaller heaps. In a similar way you might imagine an evolutionary process going into reverse. (Although it is hard to see what selection pressures could bring about such an exact reversal, it is at least possible in principle in the sense that the process would be through a succession of viable intermediates.) The trouble comes when evolutionary pressures favour the deletion, not of the last item that had been added, but of some earlier item(s): when one or several stones lower in the pile can be removed; when the dismantling is not in the exact reverse order of construction; when what had previously been done by an earlier subsystem can be done better by some more sophisticated combination of later subsystems – when an arch of some sort is formed. Further simplification is then blocked. You can not dismantle an arch any more than you can build it by only touching the stones in the arch itself. On the other hand, on the stone building model there is no such restriction to increases in complexity at any time; to adding a new stone on top.

Clearly a heap of stones, even a partly arched heap is too simple a model; but it is probably fair to say that the well known tendency for ancient design features in organisms to become fixed arises in part from the continual pressure under natural selection to simplify in the short term. One highly rigidifying economy would be where a structure, for example a molecule, is put to many diverse uses (think of glucose or adenine or any of the protein amino acids). Such a structure becomes unchangeable, not because some modification might not be beneficial for one of its functions under some circumstances, but because the same modification could hardly ever fail to be disastrous for at least one of its other functions. Note, though, that multifunctionality could only be the gradually acquired outcome of evolution. A structure that is indispensable now would have been less so at a much earlier time. Earlier still it may not have been there at all.

In the most general terms we might say that a subsystem will tend to become fixed during evolution when other subsystems come to depend on it. At first sight you might suppose that later subsystems will consistently tend to fix earlier ones. This might be true if the organisation of organisms had a simple 'heap' structure – if later additions always depended on earlier ones but not *vice versa*. But that is not quite so. Organisms have an 'arched' organisation to a very considerable extent, in that very many subsystems depend mutually on each other. As we have already seen, this implies that, as well as there being subsystems present now that were not there in the past, there were subsystems in the past that are not there now.

We might think of a feature that has recently appeared in an evolving line as having the status of an 'optional extra' (and likely to be variable, as Darwin noted). If it remains for long it may become a necessity (and less variable) because other subsystems have come to presuppose it – or perhaps because the organism changes its way of life.

Sometimes, subsystems become redundant. This is less likely to happen if it has acquired many diverse functions, if many other subsystems lean on it. But it will never be quite safe. An altogether more sophisticated 'technology' may oust it in each of its roles, as the invention of protein no doubt displaced previously established means of carrying out organic reactions (Chapter 9). On the stone building model, even an arch can be dismantled through a new arch formed *above* it (figure 3.4).

Evolutionary trees Darwin pointed out that the familiar fairly ready classification of organisms into groups and subgroups was evidence for evolution – that the present structure of relationships could be under-

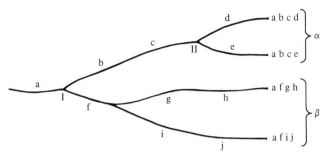

Figure 3.5. Each of the four species on the right-hand side of this evolutionary tree shares a characteristic, a, that became fixed before the last common ancestor, I. The subgroup α has two characteristic features, b and c, that were fixed between I and II. For the subgroup β only f is a defining characteristic.

stood in terms of descent from common ancestors (figure 3.5). As a rule the characteristics used to define some group of organisms should have been present in the common ancestor of that group, and yet would probably not have been in the line that led to the common ancestor for too long – otherwise it is unlikely to be characteristic. On the other hand, it should have been an old enough idea, by the time the common ancestor appeared, to have been already fixed – otherwise the feature would probably not be maintained in subsequent independently evolving lines (see figure 3.5 and its legend).

Symbiotic relationships may complicate things. In the origin of eukaryotic cells, for example, mitochondria look like degenerate forms of previously symbiotic bacteria, which before that were independent (Margulis, 1970; Richmond & Smith, 1979). Clearly we cannot rule out the possibility that early biochemical evolution involved such collaborations. Figure 3.6 illustrates an evolutionary tree that contains a fusion of previously divergent lines. There is an ambiguity here as to whether A or B is the proper last common ancestor of, say, D and E. But the ambiguity does not seriously affect the idea that all life on Earth (because of its biochemical near-identity) must be derived from some common ancestor. If D and E have features in common that are rather detailed and seemingly arbitrary so that convergent effects can be discounted (for example if they have settled on exactly the same set of amino acids), then B is the last common ancestor at least as regards these features. All these features must have been in B, whether or not they were in A.

Sexual reproduction also mixes genes – to allow beneficial design modi-

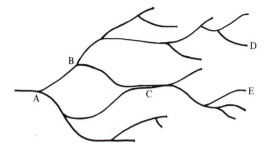

Figure 3.6. Compare figure 3.5. Because of the symbiosis, and subsequent fusion at C, E may have inherited features that became fixed between A and C, and that were in neither B nor D. But any features that are shared by D and E (and which do not have a convergent explanation) must have been present in B. B is then the effective last common ancestor of D and E at least with respect to these features. It is in this sense that we may say that the universality of so much of present-day biochemistry indicates that all life now on Earth is descended from a common ancestor, whether or not symbioses were involved.

fications to come together in the same individual. Such fusions are, however, between very close branches of the tree – usually between members of the same species. If we take a long enough view, then, and see the branches of the tree as representing species, this kind of gene mixing does not change the general picture.

In the construction of evolutionary trees there are various levels of anthropomorphism to be guarded against. First there is the gross anthropomorphism of thinking of ourselves at the end of a line that started with some ultimate ancestor, some first ancestor able to evolve under natural selection. Of course, in a rough sort of way, and forgetting about symbiotic effects, there is a line – and it raises a marvellous image. There you are standing beside your father, your grandfather, your great grandfather . . . your (great)^n grandfather. By the time the line is about 200 km long your (great)^n grandfather's knuckles are touching the ground. By 7000 km the appearance of the company would be less like a military parade than a fishmonger's slab. Seen in this way, as a line, evolution would not only be comic but incomprehensible. Yet sometimes it seems almost to be taken for granted that early biochemical evolution ought to be seen as a line; that the tree of evolution was like a standard apple tree (figure 3.7), straight to begin with (knowing where it was going perhaps?) and then bursting into a flourish of alternatives.

The tree in figure 3.8 is of the sort in which species that no longer have

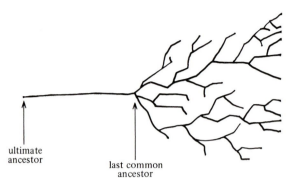

ultimate
ancestor

last common
ancestor

Figure 3.7. It is often implied that before the last common ancestor of all life now on Earth, evolution followed an unbranched course, it being felt, perhaps, that this is the most economical hypothesis. But branching – experimentation with different design modifications – is an essential part of the mechanism of evolution as we now see it. The most economical hypothesis should be that the tree has always been branched. This picture is too much like a standard apple tree – another human artefact.

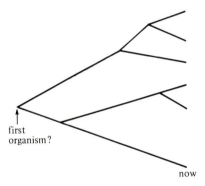

first
organism?

now

Figure 3.8. Thinking only about organisms that are alive today one may be persuaded that their common ancestor must have been the first organism. This is not so.

living descendants are left out. Such trees are very common in discussions of the evolution of proteins, and understandably so since only the proteins of living species are generally accessible. Although useful, such trees are clearly incomplete and they can be misleading. From such trees it is easy to get the impression that the last common ancestor is to be identified with the ultimate ancestor of life on Earth.

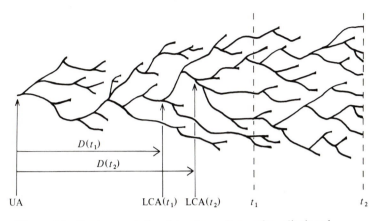

Figure 3.9. During evolution later forms have often displaced earlier ones entirely: the vast majority of species that have ever lived are extinct. So the evolutionary tree was something like this – heavily pruned. At a given time (say t_1) all living species could be referred to some last common ancestor, LCA(t_1). But at a later time, t_2, there is a new common ancestor, LCA(t_2). Any such change can only increase the distance between the ultimate ancestor, UA, and the LCA (from $D(t_1)$ to $D(t_2)$). That our last common ancestor was evidently highly sophisticated biochemically suggests that, for $t =$ now, D is very considerable.

Darwin saw clearly the effect of extinctions on the form of the evolutionary tree. Towards the end of the chapter on natural selection in *The Origin of Species* he says: 'Of the many twigs which flourished when the tree was a mere bush, only two or three, now grown into great branches, yet survive and bear all the other branches; so with the species that lived during long past geological periods, *very few now have living and modified descendants*' (my italics).

One might add that of the species living at a given time the number that will have living and modified descendants at increasingly distant times in the future can only become less. The last common ancestor of all life at any given time can only increase its distance from the ultimate ancestor (see figure 3.9). That we all share such a complicated biochemistry indicates that in Darwin's tree of life on Earth this distance is now very considerable. What is decidedly not indicated is that the last common ancestor had the only kind of biochemistry there ever was. This would be like supposing, on the basis of an examination of a few mammals, that all life must have hair. Hair was fixed in the common ancestor of what is now a large and successful group of organisms. Other features, such as a backbone, were present and fixed in earlier progenitors and so are common to still larger groups of organisms still living. The more fundamental the design idea, the longer ago it was fixed and the fewer alternative ideas there now are. It is quite in keeping with this that at the most fundamental levels the number of design approaches should have fallen to just one. The unity of biochemistry is an artefact generated by our point of view: the unity to be seen at the base of life at any given time reflects those parts of the organisation that happened to be frozen into the last common ancestor of life at that time (see again figure 3.9 and its legend).

To conclude this section: it is typical that features become invariant in an evolving line. There is no reason to suppose that our now central biochemistry was any different in this respect – that it was born invariant. Surely, like other kinds of invariance that have appeared at all levels, our biochemistry had invariance thrust upon it. Part of that invariance was the set of small molecules that we ambiguously call 'the molecules of life'. If the invariance was a product of evolution so too, presumably, was the choice of those molecules. They *became* 'the molecules of life'.

How radical was very early evolution?

If it is true that the multiple interdependences of our biochemical subsystems is an outcome of subtractive episodes in early evolution, what was it that was subtracted? Before there were enzymes and lipid membranes, for example, how were organic reactions controlled? By simpler kinds of peptides and simpler kinds of amphiphobic molecules? Not necessarily at all. The evolutionary mechanisms that we have been thinking about do not operate wholly through gradual transformations of subsystems through continuous successions of intermediate forms. These mechanisms have been takeovers. Here there is no necessary implication even of similarity between the new structures and the structures they replace. It was rather functions that were continuous: the new structures had to perform some similar task to the old ones – and presumably rather better – but they did not need to be made of the same stuff, or for that matter to operate at all similarly. You do not have to use stones as the scaffolding for a stone arch. Later in this book we will try to identify more closely the precursors of our now central biochemicals. In this present section we will be concerned only with general expectations – that if later evolution and the properties of machines generally are anything to go by, the basis of the first biochemistry would have been not only simpler than ours but of a different kind.

We are so used to the idea that evolution must proceed through continuous successions of small modifications to organisms – which is broadly true – that it may seem a contradiction to suggest that the evolution of the subsystems of organisms could be different. In so many discussions of the origin of life it seems to be taken as a necessary feature of the Darwinian idea that if, say, a complex protein is now used for some vital function, then the original way that function was carried out must have used some very simple kind of 'protein' (a couple of amino acids joined together perhaps, or a semi-chaotic polymer). Yet evolution at the higher levels is so evidently not just the elaboration of a pre-existing set of subsystems – many new things have been invented – that it is naive to suppose that during the evolution of our biochemical subsystems radically new inventions were not allowed.

There is no inconsistency in the idea that a succession of systems may be continuously related to each other through small modifications while the subsystems are not. In the preview we considered a rope as a model for this: a continuous rope may be made up of discontinuous fibres. A rope might even have different kinds of fibres in it, so that at one end it was hemp and at the other nylon, with no sharp changeover anywhere.

To be a more realistic model for an evolving line of organisms, we should think of a rope in which the fibres making up more central strands are longer than others. The fibres that together make up the most central strand of all stretch back as far as we have any sight of the rope – to some last common ancestor. Only the more peripheral fibres are easily modified – with sometimes new ones appearing while others peter out. That is to say, the locus of active evolution is mainly in the more peripheral zones. Darwin makes this point in chapter 5 of *The Origin of Species*: that the features in organisms that are most variable are typically those that have been acquired most recently.

Depending on that now most unvarying strand of central biochemistry, there are other layers of strands which are also well hardened in. If you think of higher organisms, say the land vertebrates, there is not so much about them that is variable. Not only is the biochemical ground plan common, as it is for all life on Earth, but at much higher levels conservatism is more remarkable than variety – for example in tissues, organs, in the way these make up still higher order systems such as the digestive system or the circulatory system. Even numbers of bones often seem hard to change.

Of course there are differences between a mouse and a tortoise; but these differences are not all that far in. And the differences between a mouse and an elephant are still more superficial. If this is not immediately evident it is because we tend to be over-impressed by the latest design modifications. Again, as Darwin pointed out, it is on (often elaborately developed) recent features that an organism may depend for its way of life. And that is because, in adapting to circumstances, it is the more recent features that are more easily changed. (On the stone building analogy, stones that have just been added, such as the top stones in figure 3.3, can easily be adjusted since the integrity of the organisation as a whole does not depend on their exact arrangement.)

Keratin was probably invented by early land vertebrates as a material for scales to cut down the evaporation of water (Fraser, 1969). In any case keratin is a comparatively recent idea – and it has a large effect on our impression of variety in land vertebrates. Different groups have put keratin to different uses. The reptiles have stayed with scales, the birds prefer feathers, the mammals hair. Different subgroups use these features in different ways – the scales of the tortoise are not just for keeping the water in; feathers are not always for flying; hair may be a secondary sexual characteristic, and so on. In hooves, claws, fingernails or horns, we see some of the multitude of individual uses to which this amazing material has been put. At a superficial level the uses of keratin are still evidently somewhat variable under evolutionary pressures. But at the level of actually changing, say, feathers into something else, birds already seem to have passed a critical point. They can seemingly only modify the form of keratin they have got – not change it in a fundamental way.

So it is, on the whole, only the more superficial strands of the rope, corresponding to comparatively recent features, that are the loci of active evolution. Deeper strands go further back to be common to larger groups of organisms – and the deeper they are the larger the groups of living organisms that share them. It is not only because organisms are descended from common ancestors but also because such strands become fixed that organisms can fairly readily be classified into a hierarchy of groups and subgroups (cf. figure 3.5). But the strands *become* fixed. Of course they were not fixed to begin with: they were not even necessary to begin with. Even such a widely well established idea as a backbone must have been hardly more than an optional extra when its first rudimentary forms made their appearance. On the most plausible extrapolation the now universally fixed strands of (now) central biochemistry were not always fixed. What was biochemistry like when protein was an optional extra? To begin to answer that sort of question we should look at mechanisms of radical change in evolution as they operate still – if mainly now at high levels of organisation.

Preadaptation and functional ambivalence

Most of the time natural selection is a conserver, keeping organisms in some local basin in the evolutionary landscape; keeping them adapted to their environment. If environmental pressures change, natural selection may operate as an improver in relation to the new conditions – appropriately retuning sizes, shapes and other parameters to optimal performance: adjusting the pre-existing strands.

One of the most interesting questions in evolution is how natural selection could be an inventor: how new strands originate. The main part of the answer seems to be this: through the discovery of alternative functions for pre-existing structures. Such structures are said to be **preadapted** to the alternative functions. Preadaptation is strictly fortuitous, but by no means implausible when we come to consider particular examples. It arises from a general property of machines of all kinds – what we might call **functional ambivalence**.

This book, you will find, is a tolerable flower press. (Not heavy enough to be really good, but the slightly water absorbent paper is helpful, and so are the numbers on the pages for recording the location of specimens.) The amateur handyman is familiar with the idea that a use can often be found for an object that was designed with no thought for that use. Indeed the idea of functional ambivalence is too familiar to need much elaboration: the chair is a rudimentary step-ladder, the rolling pin an offensive weapon, the pocket calculator a paper weight, and so on. Indeed the difficulty might be to think of objects that did not have endless alternative uses.

At all levels organisms can be seen to cash in on the functional ambivalence of structures. The glucose molecule might seem an ideal sort of water-soluble fuel; but then it is very well adapted too to be the unit for a highly water-insoluble polymer for making plant cell walls. And cellulose is not only used for that: in the cotton plant, for example, it puts wings on seeds.

The functions of protein are too many to enumerate, and most of these must have been accidentally hit on long after protein synthesis had been perfected. The immunoglobulins, that are so characteristic of vertebrates, were probably nevertheless derived from a class of invertebrate proteins (Barker, McLaughlin & Dayhoff, 1972). It seems to be common, indeed, for protein classes to be older than particular protein functions (Zuckerkandl, 1975).

Even a given protein may turn out to have diverse uses – keratin for example. Enzymes almost always show some activity over a range of substrates in the laboratory, and this may be important *in vivo*. Jensen (1976) has discussed the likely importance of 'substrate ambiguity' for the evolution of the enzymes for new metabolic pathways. Apparently this happens through 'recruitment' of enzymes that are already present in the organisms, catalysing analogous reactions. Provided an existing enzyme can catalyse a reaction, even if only incompetently to begin with, if that catalysis confers an advantage on the organism then it is likely that subsequent mutations will be selected that will improve competence so that, in effect, a new enzyme has been discovered.

Jensen lists a number of families of proteins with similar amino acid sequences and yet more or less divergent functions; for example, the serine proteases (Stroud, 1974) and the enzymes that work with pyridine nucleotides (Bennett, 1974).

At a somewhat higher structural level, the microtubule provides another example of a structure that has been put to many uses (Watson, 1976; Dustin, 1980). Microtubules are cylinders that may be many micrometers long and have a hollow core of about 14 nm diameter, the wall being about 5 nm thick. They are made up of proteins with molecular weights of 55000 and 57000. One function for microtubules is as a sort of internal skeleton for cells, allowing them to assume highly non-spherical shapes – as for example in nerve cells. Restricted to eukaryotic cells the microtubules are also involved in that most characteristic eukaryotic function of mitosis, where they make up part of the machinery that draws the chromosomes apart. They are also the main structural elements of cilia which contain twenty microtubules apiece and provide engines of locomotion for many single-celled organisms (Satir, 1974). But then, as you might have guessed, that is not all that cilia are used for. Our lungs, for example, are kept clean through a continual upward flow of mucus powered by the concerted beat of cilia attached to the linings of the air passages.

Then again tissues have diverse functions. The main function of bone may be as a load-bearing material, but latterly mammals found that bone is useful too to transmit sound – in the middle ear. And it is hard to think of an organ that does not have several functions. The lungs, you might say, are gas exchangers – but in the opera house they also have a noisier function. Toothed jaws may be weapons or tongs as well as food processors, and so on and on.

Why is functional ambivalence so common? It is perhaps easier to comprehend the functional landscape for an organism than for a subsystem of an organism. For an organism there is one overall function that matters – fitness – and hence only one functional dimension to be added to the multitude of structural dimensions. But, in contributing to fitness, a subsystem may, as we have seen, perform several functions. The ideal set of structural specifications for one of these functions is unlikely to be ideal also for the others. The design of a multi-purpose object is almost bound to be a compromise. There is not just one hypersurface but many, and a local lowest point on one is unlikely to correspond exactly with lowest points on the others. Indeed, thinking in wholly abstract terms, one might be inclined to say that any correspondence at all would be pure chance; and that for functions that depend on highly specified objects the probability of such a correspondence is too low to be worth considering – because the basins for highly specified objects are, relatively speaking, very tiny. But if we think more realistically this pessimism is evidently not justified: quite often a highly specified object – that appears to have been designed in some detail for some particular function – turns out to be pre-adapted, and often in some detail, for something else. (A book really is not a bad flower press, and the crocodile's mouth is a rather good playpen for its children.) When we say that functional ambivalence is a factor to be reckoned with in evolution, we mean that there is, after all, a tendency for basins on seemingly disparate landscapes to overlap; and that this principle applies to the machines in organisms as to machines in general.

The reason seems to be something like this. Much of the detailed specification needed to make a machine (i.e. functional object) are in its subsidiary components. A tube of constant diameter, for example, might be what is required to make an ideal X – although a makeshift X might be possible with something less perfect. So under selection pressures generated by X's usefulness the means of manufacturing the required tube is perfected. But tubes are very generally useful things, they are major or minor components of other machines that now become possible – some perhaps for the first time because, for them, a makeshift, inaccurate tube would be no use at all. Speculating on the origin of microtubules one might perhaps guess that the skeletal function came first, because so long as they had some rigidity they would do. But if they were of constant diameter they would be better because they could pack and align better. Then cilia, for which packing and alignment is more critical, become possible.

That particular story may very well be wrong, but the general idea seems understandable – that the perfection of a structure for one function may make another function possible for the first time. The first function may correspond to a basin with a wide mouth that is easy to find; but near its lowest point it overlaps with another basin, on an interpenetrating landscape, that would have been too small to have been found *de novo*. Then one of two things may happen. The original structure is further modified to be an optimal compromise for the new and old function(s), or two (or more) structures diverge, each specialising in one of the functions (for example in the appearance of new proteins through gene doubling, or in the evolution of the lung as separate from the oesophagus).

Referring to protein evolution, Zuckerkandl (1975) says: 'Any new functional development in an old molecule will benefit from the accumulated "capital" of optimisation of the old molecule with respect to the various general functions that any protein is bound to carry out.' Simply to have a definite tertiary structure is quite an achievement already; and not to interfere with other processes – not to misbehave – is another. Much the same could be said about books: a number of alternative uses is opened up by their simply being flat and chemically stable.

Technological analogies are often used in discussions of evolution (for example by Pirie, 1951, 1957, 1959; Rensch, 1959; Maynard Smith, 1975), and they seem particularly appropriate to this part of the discussion. In evolution, as in the development of our present civilisation, new 'technologies' are discovered from time to time. Some technique of fabrication is developed first in one area, perhaps for quite trivial reasons, and then turns out to have a very general usefulness. Was fire first invented to keep warm, or to fend off large animals, or to tenderise their meat? Perhaps it was at first a toy; but in any case its invention let in other technologies – adding metal and glass to the range of available materials. Perhaps in turn these materials were used for beads and trinkets to begin with. It is hard to say; as it is hard to say exactly when and for what purpose bone-like materials were first used in animals. But, for large classes of animals, once the technology of being able to control the crystallisation of apatite into hard, rigid masses had been perfected, nothing was ever to be the same again. Of course we are inclined to say that bone was invented for load-bearing functions. That was doubtless the reason for its success; but we cannot be sure that the first partly controlled apatite precipitations would have been strong enough to be useful in that way. Perhaps they were simply a means of storing calcium and phosphate – still one of the functions of bone.

Beginnings are often special. If some new technology, some new manufacturing process, is to catch on then there must first be a use for poorly made products. That is to say, on one of the many interpenetrating landscapes corresponding to different conceivable functions for these new kinds of products there must be a catchment area wide enough to be found. As evolution proceeds other basins are discovered which may be much narrower. Some of these will also be shallower than the original basin. But there is no necessary correlation here. Narrow basins may very well be deep.

It is the discovery of deep but narrow basins that constitutes the kind of technological breakthrough that is such a familiar part of the recent history of our civilisation – in the discovery of the means of making mild steel, or PVC, or transparent glass; or, at higher levels, insulated wires, polymer films, photographic emulsions; or at higher levels still, printed circuits or internal combustion engines. Such techniques are highly sophisticated, depending on detailed know-how. Yet they are very generally useful and our whole civilisation has come to depend on them.

In organisms there are equivalent pieces of high technology – in the techniques of manufacture of, say, lipid membranes or microtubules; or of bone or keratin; or of neurone-based data processing systems. But beneath all these, at the base of everything, there is a particularly intricate piece of engineering – that hi-fi equipment on which protein synthesis depends. As I have already remarked, we cannot suppose that the early development of this technology was with any sort of view to the functions that were eventually to become possible.

Functional overlap

If a given structure can often have several functions, a given function, at least of a general sort, can often be achieved through quite different structures. We might call this **functional overlap**. Examples from everyday life are obvious and numerous – such as the various kinds of clocks that we talked about. Among organisms too there is a great variety of structural means to similar ends. Such general animal functions as breathing, locomotion or defence may be carried out by subsystems that are different in detail or very different indeed. At a deeper level structurally dissimilar enzymes may bring about the same reaction; for example, Campbell, Lengyel & Langridge (1973) found a β-galactosidase different from the usual one in a strain of *E. coli* that lacked the normal gene.

Within organisms too there are often areas of overlap in the functions of subsystems. The most obvious are again at the highest levels – a bird may walk or fly to catch the worm. Such an internal functional overlap between strictly alternative modes would be pointless if it was perfect. Many birds have the option of walking or flying to get about; but that only makes sense because, although there may be times when either would do, there are other times when one or the other is preferable or obligatory. The overlap here is incidental to a means of extending the range of the general function of locomotion – it is the regions of non-overlap that matter.

More usually functional overlaps are not strictly between alternatives. To keep warm a cat may move to the fire, or fluff up its fur, or shiver; but when it is really cold it may very well do all three.

These 'and' and 'or' aspects of functional overlap can be seen too at the level of protein molecules. The multiplicity of transport proteins in *E. coli*, for example, provides this organism not only with the ability to pick up a wide range of useful molecules, but it allows it to cope with the same molecule present under a wide range of external concentrations (Jensen, 1976; Lin, 1970). There may be more general as well as more specific transport proteins: for example, there are two aspartate transport systems in *E. coli*, one of which operates on a wide variety of C_4-dicarboxylic acids while the other is more fussy (Kay, 1971). Similarly with enzymes: the digestive enzymes trypsin and chymotrypsin hydrolyse overlapping but different sets of peptide bonds, while at the same time the trypsin-susceptible set is smaller. Then again, *Aspergillus nidulans* has an enzyme that will hydrolyse amide links in general as well as having one that is specific for acetamides and another for formamides (Hynes, 1975).

There is no need to continue this catalogue: it is both general and understandable that it may be advantageous for an organism to be able to carry out an important general function in a number of different ways. This may allow a wider range of situations to be coped with. At all organisational levels one can think of alternative or back-up subsystems that can be switched in; and some of these latter are effective precisely because they work differently from the systems whose functions they are supplementing.

Takeovers in evolution

Functional ambivalence, functional overlap and, a third element, redundancy, can together create the most radical transformations in evolution. They can bring about **takeovers**. Here some general function that had previously been carried out by one set of structures comes later to be carried out instead by some different set of structures.

A takeover is to be inferred from the evolution of the lungs (as described by, for example, Rensch, 1959; Schmalhausen, 1968). The stage was most probably set here by a functional ambivalence of the food canal of early fish – it can hold air as well as food. This means that the food canal might act as a buoyancy device and also as an alternative means of absorbing oxygen. Selection could then operate to improve either or both of these functions without interfering too seriously with the primary function of the food canal. It seems that a swelling or sac formed off the upper oesophagus which gradually ballooned out to form, in some lines, swim bladders, as in many modern fish; and in other lines lungs, as in the modern lungfish. It would be difficult to disentangle the exact sequence of events or the early roles of buoyancy and respiration in providing selection pressures – or indeed to be sure that there were not other functions involved in the earliest stages. But the overall story seems clear enough: part of the upper food canal of early fish evolved into an elaborate organ of respiration through successive small modifications set in train through an initial functional ambivalence. This led to bimodal organisms with a strong functional over-lap between gills and lungs. Finally, for later land animals, the gill mode became redundant, the gills themselves disappearing – at least in the fully developed animals.

It is typical that this takeover was only of a general function – acquiring oxygen and disposing of carbon dioxide. In detail the function changed in that the external medium changed. This was a genuine takeover nevertheless: the general function was always vital and always maintained; and the transformation, although through a continuous succession of organisms, was not through a continuous succession of subsystems. One strand of the rope gradually petered out while another became stronger.

A takeover can be seen too between the amphibian and reptilian way of preventing the loss of water from the skin – in the change from secreting mucus to keratinised scales. And there have been many other examples of radical design changes in evolution: in, say, defence mechanisms, in modes of communication, or of protection of the growing embryo. Again it is not necessarily the case that the new means evolved from the old.

In keeping with our earlier discussions it is now mainly at the higher levels of organisation that radical takeovers are most easily effected. It is here that evolutionary experimentation is most active, where subsystems are provisional enough to come and go. It is here that the strategies of active evolution are now most easily to be seen – and it is with such strategies in mind that we should try to extrapolate back to that much earlier era when our central biochemistry was truly in the making.

Not that biochemical innovation is finished by any means. Secondary metabolism can be highly individual. For example, new enzymes are being invented still, and being discarded still, in the pathways leading to such things as flower colouring materials. And there must be some biochemical basis for such higher order evolutionary transformations as the change in the shape of a limb, even if little is known at present of the detailed chemistry that controls such things.

Even in primary pathways there are alternatives, often, to suggest that some redesigning of the details of central biochemistry is still possible. And there can be redundancies in primary metabolism. That there are 'essential' amino acids, and vitamins, illustrates our incapacity now to synthesise vital components that were made by our remoter ancestors.

Efficiency range

Let us move somewhat from 'how' questions to 'why' questions. If you have a way of making machines that work perfectly well, why change?

Because, of course, they do not work perfectly well – nothing does. In particular you do not expect that the first general way (design approach) that becomes possible for carrying out a given function will remain the best way. In any case the design approach with the greatest **efficiency range** (that is, potential for improvement in the long term) cannot be recognised as such by natural selection – not to begin with.

In evolution new opportunities for new design approaches arise. There are both external and internal opportunities to be considered. For example, the evolution of the lung was pushed, latterly, by the opportunities inherent in the conquest of the land. But this evolution depended on internal things too. And these were more than a convenient oesophagus. One can think of any subsystem within an organism existing in a 'niche' created by the rest of the organism. The subsystem and its 'niche' co-evolve. The gills, for example, co-evolved with a heart and circulation – a prerequisite for effective lungs. Thus the lungs might be said to have taken over a 'niche' that had been created in part by the gills. The deviousness of the evolution of the lung, then, was not some haphazard changing of plan (there was no plan of course) but a necessary general way of going.

On chance alone you would expect that, sooner or later through evolution, some improvement to an initial design approach would turn up – you are more likely to draw an ace the more cards you take. But the expectation of improved opportunities is stronger than this. As discussed earlier, as 'technologies' in organisms evolve – for example the ability to make tubes precisely – opportunities may appear of a kind that could not have been available to begin with, that depend absolutely on techniques of fabrication that have to be closely controlled. As any technology evolves machines of higher efficiency range come into view.

For example, there are more ways than one to kill a rabbit. A stone is a

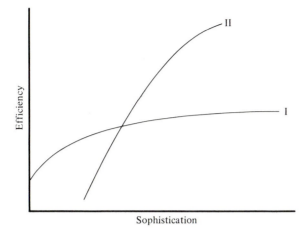

Figure 3.10. The law of diminishing returns usually applies to a particular approach to an engineering problem and can be expressed as some sort of concave-down curve of efficiency *vs* sophistication. As technology as a whole advances new design approaches to engineering problems emerge and usually some of these have a greater efficiency range than initial, 'zero-technology', systems. For example, I might be the curve for stones and II for shot guns as rabbit killers. A central theme of this book is that our nucleic acid–protein based biochemistry has a curve like II: there is no 'zero-technology' version of it that would work. Rather our system took over, beyond some crossover point, from a system with a curve like I. This earlier system was structurally as unlike our biochemistry as stones are from shot guns. Its virtue was that it could start from scratch (that is, its curve cut the *y*-axis).

zero-technology rabbit killer, an object to be found in the environment, that is fortuitously preadapted. This approach can be improved on: for example a spear as a sharpened stone attached to a stick. But the shot-gun is much more efficient although its invention depended on the development of a surrounding technology that was not always or primarily pursued with weapons in mind. There is hardly a field of engineering where there has not been a similar leap-frogging progress; where later devices were often more than improved versions of earlier ones; where alternative design approaches of increasing efficiency range have superseded each other (see figure 3.10).

It may be more difficult for that sort of thing to happen in evolution, and eventually the more central subsystems may become fixed; but we should expect such things to tend to happen – and indeed where evolution is still active they do.

Life on Earth started with zero-technology devices, some set of microsystems fortuitously preadapted to evolve under natural selection. We know from the sophistication of the eventual outcome that there must have been an extended period of active evolution between those first systems that could start to evolve and those nucleic acid–protein–lipid machines that were eventually to inherit the Earth. From chemical considerations, given in Chapters 1 and 2; from the highly co-operative character of present biochemistry; by analogy with active evolution as it is still visible at higher organisational levels; and from the consideration of the efficiency range of machines in general, there is no reason to suppose that those zero-technology devices that we seek were made in the way that organisms now are made.

Looking at our central biochemistry; at its 2000-odd reactions remaining unconfused, each controlled by a separate protein; at the interlocking allosteric circuits within that system; at the synthesis of these proteins which depends on a technology that seems to presuppose them; at a technology all the more ingenious in being based on a small group of small chemically accessible molecules, it is perhaps no wonder if we are inclined to identify 'life' with protein and then, by extension, to imbue amino acids with a touch of vitality – at least to the extent of labelling them as 'molecules of life' and seeing in their abiotic production a step towards life.

Yet this reaction is precisely the opposite of what it should be. It might have been appropriate rather if our biochemistry was simple or simply an elaboration of a demonstrably zero-technology machine. But it is not. This is no heap built only by one-at-a-time additions and adjustments. This is a cathedral of arches built now on each other, depending now on each other with almost every stone a key-stone in a dozen dimensions. To build these arches there were presumably earlier forms, less rigid, less ingeniously interdependent, that were to be the inadvertent discardable scaffolding for the later design.

Even Occam might have agreed that a proliferation of hypotheses beyond that of a mere adjusting and heaping up of subsystems is needed to understand the making of our biochemistry – particularly when takeovers cry out to be considered. And this is hardly a proliferation. Takeovers and the elements of takeovers – preadaptation, functional overlap and redundancy – can be seen operating in later evolution as the means of radical invention. We should doubt whether any of the molecular structures in our present biochemistry were there at the start, whether any first subsystems would have had the efficiency range to have stayed the course.

Conclusion

What is there to hang onto then: what has been invariant through evolution from the start?

The search for some constant underlying substance through change is natural perhaps to our way of thinking – it is an inheritance from Aristotle who distinguished *substance*, the root of constancy in the universe, from *form* which is changeable. If a form persists – for example the form of a piece of sculpture – that is thanks to the persistence of the material on which that form is impressed (so the story went).

Clearly, though, the persistence over a billion years of, say, cytochrome *c* is not like this. Even on the much shorter time scales of an individual life, molecules do not persist as such – they are being continually remade. It is not cytochrome *c* that persists within organisms and between generations of organisms; it is the ability to make cytochrome *c* molecules. That depends on other molecules – in the end on DNA molecules. But neither do these persist in the long term between generations in the kind of way that the Venus de Milo has persisted through the centuries. DNA molecules too are remade. What is inherited in the long term – indeed perhaps all that is now inherited in the long term – are particular *forms* of DNA molecules, particular base sequences. Only genetic information provides the long-term lines of continuity through evolution, to allow the repeated reappearance of phenotype structures – such as cytochrome *c*. Here it is form that lasts, not substance.

If we ask why certain phenotype forms now recur so reliably, the answer is to be found in the highly co-operative character of the machinery of which these forms are a part. As we discussed, such co-operativeness, and hence the invariances that it imposes, can only be a *product* of evolution. There is no reason to expect that this form-invariance was original.

But perhaps DNA itself is an exception? One might suppose that if it is only genetic information that provides the long lines of continuity in evolution then at least the kind of material that holds that information must be invariant. (RNA, say, might have changed to DNA because these can read each other's messages, but you might suppose that really radical changes would be impossible.) That might be so if the evolution of a genetic material was restricted to successive small modifications to that material – if takeovers could be discounted. Then indeed the evolution of a genetic material would be particularly hemmed in. But if, as we suspect, those now fixed central strands of biochemistry were not always fixed (and not always central) then takeovers could have been a part of the story of the invention of our biochemistry as they have been part of the story of much later evolution. In particular takeovers of the genetic material, through organisms using more than one kind of genetic material, might have provided mechanisms for change that would be no more restricted in principle than takeovers of other kinds: a **genetic takeover** might be between wholly different kinds of materials. In Part II we will use this idea to shift our view of the problem of the origin of life.

PART II

A change of view

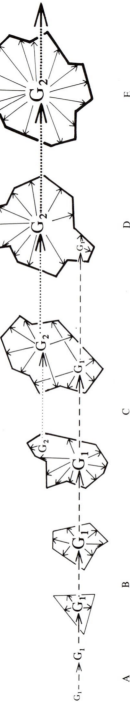
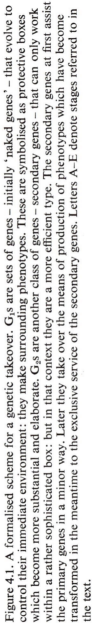

Figure 4.1. A formalised scheme for a genetic takeover. G_1s are sets of genes – initially 'naked genes' – that evolve to control their immediate environment: they make surrounding phenotypes. These are symbolised as protective boxes which become more substantial and elaborate. G_2s are another class of genes – secondary genes – that can only work within a rather sophisticated box: but in that context they are a more efficient type. The secondary genes at first assist the primary genes in a minor way. Later they take over the means of production of phenotypes which have become transformed in the meantime to the exclusive service of the secondary genes. Letters A–E denote stages referred to in the text.

4

Genetic takeover

In the last chapter we came to the conclusion that, most probably, the components of our central biochemistry are 'high-technology' replacements of earlier designs – ultimately of 'zero-technology' designs with which evolution must have started. Takeovers, analogous to those visible in more recent evolution, could have provided the means for such re-designing; and, as we saw, the multiple interdependence of the present set of subsystems is itself evidence that here we are looking at the outcome of an evolution in which takeovers played a part. This is consistent with the conclusion that we had arrived at already in Chapter 2, that while a genetic material is a prerequisite for evolution through natural selection, nucleic acid could not have been the first genetic material. The question now is whether takeovers could have operated also at the very centre as a means of radically changing genetic materials. Is a genetic takeover possible?

An overall mechanism for a genetic takeover is suggested in figure 4.1. Referring to ideas developed in the last chapter I have here represented the earliest systems, A, as 'naked genes'. The evolution of such organisms would be expected to proceed nevertheless through the elaboration of phenotypes (as in systems B) in branching lines occupying various niches. In particular, in more difficult niches more elaborate phenotypes would be necessary go-between structures.

A sketch I will leave until later in the book speculations as to workings of primary organisms – that is, organisms based on primary ('self-starting') genetic materials. These speculations will depend on inorganic crystal chemistry to be developed in Chapter 6. And I will leave also till later some of the more detailed speculations as to how our biochemical system might have caught on. In the meantime here is a rough speculative sketch, based on the scheme in figure 4.1.

We must suppose that within the phenotypes of some of the evolving primary organisms (by stage B) metabolic pathways were being gradually built up that created consistent supplies of new molecules of various sorts with functions relevant to these primary organisms. In particular, new kinds of organic polymers were invented by some of the more highly evolved species. The original functions for such molecules would not have depended on particularly accurate specification – they were used for such things as glues, gels or fibres perhaps. Nevertheless these functions were often improved by improved specification. From time to time such polymers turned out to be preadapted to other functions; sometimes to more sophisticated functions so that new selection pressures arose for more precisely made, and often more complex, materials. To be a good structural polymer, for example, it would often be important for the monomers to be uniform and regularly arranged – as in, say, cellulose – and if these monomers were chiral then the chirality too should be controlled.

Now we might suppose that among the polymers being used for rather sophisticated structural purposes there was a class that contained a hydrophilic backbone to which were attached more hydrophobic groups. The structure-forming technique depended on these more hydrophobic side chains tending to cluster together. Under selection pressures to define these structures more accurately, more specific associations of the side groups were encouraged through modifications that allowed hydrogen bonding between them. Eventually something like RNA had been made because of useful effects, gradually discovered, resulting simply from its ability to twist into defined objects. The base sequences in these RNA-like molecules were perhaps rather simple – say stretches of poly(A) and poly(U) – and the information for specifying them was contained in the primary genetic material.

The genetic function of this RNA-like polymer was a fortuitous preadaptation: because it so happens that one way to make a structure-forming polymer is to have side groups on the polymer such that they key with each other – and that is also part of the design for a replicating molecule. But this functional ambivalence could only have appeared at a late stage in the evolution of organisms that had solved all sorts of much more fundamental problems: how to make ribose and distinguish it from its isomers; how to bring about energetically uphill condensations, and much more besides. Such a concatenation of expertise would be inconceivable except in an evolved organism. But to arrive at this new kind of replicating molecule the primary organisms need in no sense have had that goal in view. Each of the 'technologies' required would have been of a general usefulness – providing graded selection pressures for their evolution.

Even the replication of a molecule might have had a purely phenotypic role to begin with. For example, for some structural purpose it might have been important that stretches of different RNA strands locked with each other. It would not matter then what the sequences were so long as the sequence in one strand was the complement of whatever sequence happened to be in the other: and the way to ensure that would be by forming one strand on the other.

By supposing that at first replication had a phenotypic role, we can perhaps overcome one of the great problems of imagining the genesis of a genetic material: that unless replication is pretty accurate it is no use, information being quickly lost between generations. Continuing our speculations, if the phenotypic structures being made were fairly crude – for example for some sort of capsule – the replicative technique for matching up stretches could still have been useful even if it was somewhat haphazard. And there would be no information to lose. (The matched stretches, you remember, only had to be complementary – they did not have to be anything in particular.) But now there would be evolutionary pressures for improving the fidelity of replication – to make less crude products. Sooner or later, following the evolution of efficient catalysts for assisting replication, particular sequence information would be being passed on indefinitely between generations. This would allow much more intricate intermolecular and intramolecular twist structures to reappear in successive generations. (This idea will be elaborated further in Chapter 9.)

By now the organisms would be genetically bimodal (i.e. at stage C in figure 4.1) since part of the information needed to specify the common phenotype – that is, the information specifying the twisting of RNA molecules – was now being transmitted directly between these molecules. Given that these RNA structures remained useful to the organisms, then mutations would be selected – in either of the genetic materials – that tended to improve the conditions for this new technique of RNA polymerisation. In particular the 'replicase' catalysts would be improved.

If, perhaps, nucleic acid first got a toe-hold in some evolved primary organism through its structure-forming prowess, I doubt whether that would have been sufficient in itself to complete a takeover. Continuing our story, we might now suppose that twisted RNA structures came to interact with amino acids. The amino acids did not come from an external soup – which would have been far too messy – rather they were metabolites of the organisms and their synthesis had been perfected previously over long periods. They were there because they were useful as such – they were metal chelators perhaps. Dipeptides might also have been similarly useful and hence any technique for their synthesis. Among the techniques chanced on was one that depended to some extent on twisted structures made by RNA. The RNA molecules only had to help here, to provide some specificity. They formed structures, perhaps, in which more or less specific pairs of activated amino acids were orientated for reaction. Under new selection pressures created by this and by the usefulness of dipeptides (and then, later, higher peptides) the RNA-controlled technique of peptide synthesis was perfected, until eventually a Mark-I protein synthesiser emerged.

However it happened, the invention of the means of specifying long polypeptide sequences would have been a long haul, and surely only possible within an already quite highly evolved organism. Even when this was complete, the primary genetic machinery must still have been there. But now the breakthrough had been made, perhaps, to a new technology, to an altogether faster and more accurate way of making structures and controlling organic reactions. The protein revolution was imminent (i.e. $C \rightarrow D \rightarrow E$ in figure 4.1). One by one reactions could now be taken over by catalysts made from this new material – until eventually all the reactions needed to make the parts of the new system were under the control of this system itself. Then, only then, was the necessarily highly evolved 'scaffolding' kicked away to leave an 'arch' of subsystems to surprise us.

Two points in particular Extracting two salient points from the foregoing discussions we may note that G_1 and G_2 genes were both preadapted to their genetic functions – but under very different circumstances. G_1 was a mineral of some sort – it was necessarily a geochemical product. Its preadaptation was wholly fortuitous. G_2 was a biological product: its preadaptation depended on their being common design elements in polymers with genetic and other simpler, but biologically relevant functions. In view of these very different circumstances of birth you might have expected G_1 and G_2 to have been structurally very different. In any case the mechanism of genetic takeover allows them to be different *because G_2 appears, not from a modification of G_1, but from molecules that G_1 is controlling.*

Whatever the details, G_1 (or for more complex cases *some* pre-nucleic acid genetic material(s)) had to support an extended evolution during which the 'technologies' were established that were eventually to allow the synthesis of (at least) nucleic acid molecules. Genetic takeover thus allows specifically what in any case seems to make sense that *our central biochemical machinery was invented by quite highly evolved organisms,* by systems that would have clearly conformed to both the genetic and the teleonomic definitions of life discussed at the start of the last chapter. Whatever else, we know this much about G_1: it must have been capable (eventually) of holding and replicating considerable amounts of information.

Objections to genetic takeover

I will devote the rest of this chapter to considering objections to the idea of genetic takeover as they might be raised by someone who has picked up the general suggestion that evolution started with some kind of genetic material radically different from DNA and then changed over. We might imagine a coffee room discussion between this critic and an advocate of genetic takeover.

Dr Kritic: I can't say I think much of this idea that our kind of life took over from some earlier kind. It seems to me to have all the qualities of a bad theory – it is unnecessary, speculative, vague, complicated, defensive, not fitting the known facts, and, anyway, impossible.

Dr Advo: But apart from that it's O.K.? May I pick you up first on the *unnecessary* charge? You have to be very satisfied with current ideas on anything to say that alternatives are unnecessary. Can we agree that we are looking for the most probable general sequence of events in the light of the evidence available?

Dr Kritic: Sure. But genetic takeover cannot be part of such a sequence: it is *impossible*. You see, evolution is concerned with the elaboration of genetic information – in the end that is what it consists of: DNA molecules containing messages that contrive, by specifying phenotypes, to bring about their own propagation. But the whole thing keeps going only because a message in one DNA molecule can get copied into a new one. That would not be possible between radically different kinds of genetic material. Also the reading equipment, by which genetic messages are translated into an effective phenotype, would only work with one or a very restricted range of genetic materials. With a change to a new genetic material all the information in the old one would become illegible – it would be lost. That's not really a takeover: it's simply a waste of time.

Dr Advo: Nature can be enormously wasteful: she has certainly wasted time during evolution if you like to think of it that way – laboriously evolving one structure only to have it displaced by others. Yet often this is an essential part of invention. (Darwin's central idea should really have been called Natural Rejection since mostly this is what happens. And the 'amount of organisation' in an organism (whatever that means) is anyway related to the size of the pile of design modifications rejected on the way.) We might ask whether a life based on a non-nucleic acid genetic system could have provided some sort of booster rocket to get nucleic acid life off the ground. Then, even if the booster later falls away, it may have been part of the evolutionary history of the winning system without which that history cannot properly be understood.

Dr Kritic: Mmmmmm.

Dr Advo: Now you have been assuming that the only help that one kind of life could be for another would be to provide it with genetic information. That is not so; at least not in the sense of the first kind of life handing over messages in its own admittedly illegible script.

Think about it this way. The optimal design for an organism depends on what the environment is like. That would have been true at the most fundamental levels to begin with. The environment must provide the units out of which the first genetic material is made. First life is thus limited by what primitive geochemistry can do: it starts the best way it can with some genetic material that can put itself together. Once this life evolves, however, it changes the available environments. Its evolved phenotypes – possibly a variety of them in different niches – now provide a range of new opportunities for new kinds of life; if only by providing consistent supplies of particular complex molecules not previously available, or only available in very impure form. It is no novelty for one set of organisms to create the possibility for another set of organisms (plants made animals possible, for example). A first life system might well increase the probability of a completely new system simply by providing new kinds of foods. Suppose that it made monomers for an alternative genetic material . . .

Dr Kritic: Are you trying to say that a primary life would have evolved the ability to make nucleotides? What selection pressures, may I ask, would have brought about that remarkable circumstance?

Dr Advo: I will avoid asking you in return what selection pressures would have caused the primitive Earth to make nucleotides, since that might sound like bickering: but really it is you, not me, that is falling into the anthropomorphic trap. Primary life, as you so aptly call it, would not have been in any sense looking for new genetic units, it would simply have been evolving the ability to do organic chemistry because that is a generally useful ability. As it evolved this competence it would happen on a great variety of molecules – for making phenotypic polymers and so on. This would increase the chances, over the chances that had been there for the unevolved Earth, that some alternative units would become available for some alternative genetic material. Maybe several were produced in this way but, among them, one species of the more primitive life discovered nucleotides. It was subsequent events that were to make that discovery so significant for us.

Dr Kritic: Let me see then, so primary life made nucleotides and there was this secondary life waiting in the sidelines to gobble them up?

Dr Advo: Not exactly. I am asking you in the meantime to think along analogous lines to those who believe that nucleic acid molecules evolved on the primitive Earth: the only difference is in the source of the units.

Dr Kritic: But the trouble with ideas that depend on nucleic acid molecules forming on the primitive Earth is not only that the nucleotides might be hard to come by, but that a polymerase substitute would have been needed to make them join up and copy pre-existing sequences. That problem would still be there?

Dr Advo: Yes – according to this line of thought.

Dr Kritic: But there are now new problems. This source is not a general environmental pool, but a set of micro-environments provided by the set of individuals of some particular species or group. That would drastically reduce the available places where a second kind of life could catch on; also it would be very dependent on primary life. And how would the nucleotides get out of the primary organisms? How would primary life hand them over, and how would secondary life pick them up?

Dr Advo: Secondary life would need nucleotides before it could *begin*. (It might be molecules like nucleotides, but let's say nucleotides for the sake of argument.)

Dr Kritic: I see; this is the 'naked nucleic acid gene' idea; but that was never a very good one. Even given activated nucleotides and a pre-existing nucleic acid strand, this strand would not by itself select the appropriate new monomers – single nucleotides do not bind at all well to nucleic acid strands. Even if they were emplaced and polymerised before coming away again as a completed complementary strand, how would this sequence of events be controlled? Even if somehow it was, all you would have would be replicating nucleic acid molecules unable to do anything except replicate.

Dr Advo: Much of what you say is true, and it suggests that the role of primary life was not simply as a provider of suitable molecules. But you make two mistakes. First when you complain that secondary life would depend on primary life. Well, it would to begin with, and perhaps the amounts of the key molecules being made by primary life would be small. But that wouldn't matter so long as the supplies were consistent and clean. It can be said in favour of a two-stage take-off that primary life does not build up stock-piles of essential nutrients for secondary life – it creates a means of production. The problems inherent in stock-piling molecules are thus largely removed. Your second mistake is in supposing that replicating nucleic acid molecules could not evolve: in the right circumstances they could – as Spiegelman has shown. But I would agree that there is difficulty in imagining the circumstances under which 'naked' nucleic acid molecules would replicate.

Dr Kritic: Not the only difficulty by any means: to become independent eventually, secondary life has to learn to make those molecular bits that primary life had provided for it. Now how is it going to do that? It can't read the genetic messages in primary life.

Dr Advo: Something like Horowitz's mechanism might work (figure 1.12): not only some vital product like a nucleotide is provided, let us say, but also the immediate precursor metabolite. A secondary organism that can use that too will have an advantage, and it will be able to use it if it evolves the means of catalysing the last step. Similarly all the way back, step by step, it copies the metabolic route in the primitive organism. It never has to read the genetic messages: the metabolic intermediates have told it how to go.

Dr Kritic: Very neat. But the trouble with this sort of mechanism is that many metabolic intermediates are unstable: the question of how the goods got across (which you never answered incidentally) would be very acute here. Still worse, this whole scheme presupposes an ability in the secondary organisms to evolve a variety of catalysts at will (as it were). How is it going to do that until it has invented protein? And what selection pressures are going to lead to the invention of protein when the key function of specific catalysis could not be expected to emerge until a late stage? In any case the invention of protein is going to call for some very well made catalysts. I see only two conceivable ways of getting out of this: either nucleic acids themselves can be catalysts – RNA molecules, say, twisting up into objects with catalytic grooves in them – or an obliging primary organism makes and hands over the required catalysts. I suppose that twisted tRNA-like molecules might do, but I cannot see them having the kind of universal ability to evolve a given catalytic function 'on demand' that the Horowitz mechanism requires. And as for primitive organisms making and handing over such things as tRNA activating catalysts – well, why should they? Anyway you admitted that the whole idea had fallen through with the need to start from replicating nucleic acid molecules. By making a consistent supply of nucleotides, a primary life might have made this less impossible, but nucleotide supply was not the only problem: the idea is still impossible.

Dr Advo: I agree.

Dr Kritic: Then I rest my case.

Dr Advo: What case?

Dr Kritic: The case against genetic takeover.

Dr Advo: But we haven't been talking about genetic takeover.

Dr Kritic: What?

Dr Advo: No, all I was wanting to do was to show that by providing supplies of molecules not previously available a primary life could at least make the origin of our system 'less impossible'. I was also wanting you to clarify some of the remaining problems: it helps one to see how to adjust the hypothesis towards something that might actually work. The necessary adjustment, I think, is to go from the kind of saprophytic idea that we have been discussing to something more like an endosymbiotic one: to suppose that the secondary genetic systems started *within* primary organisms. (*Dr Advo draws figure 4.1 on the back of an envelope and explains the idea in more detail.*)

Dr Kritic: Well I can see now how the primary system could transfer metabolites to the secondary system, but your account of the evolution of protein synthesis was sketchy to say the least; and you are still supposing that accurate nucleic acid replication was achieved long before protein enzymes would have been possible. Is this really much of an advance?

Dr Advo: It is not just that with secondary genes evolving within the same organism as primary genes there would be an easy means of transferring nucleotides. There would also be a reason for maintaining supplies. Even before the new genetic material started working, nucleotides, and also nucleic acid-like polymers, were useful to the primary organisms, and the means of production of these molecules was maintained only for that reason. Similarly, this newly discovered trick of direct information transfer between the nucleic acid-like molecules would be continued only if it was useful to the organism as a whole. That is the *sine qua non* of any further progress: but, given that, there would be no limit in principle to the assistance that the evolved and evolving 'technologies' of (a still mainly primary) organism might provide for the younger partner – or should I say younger part. There is no need to imagine any sort of genetic altruism: these primary genes would have been as selfish as any other kind; but it would have been in their short term interest to help the secondary genes if the secondary genes in any way helped them. You could say the same about any two genes cohabiting any organism. Of course nowadays all genes are made of the same stuff and work the same way – organisms now are 'homogenetic': but I see no reason why there should not be 'heterogenetic' organisms. Indeed the account I gave of the evolution of protein synthesis was sketchy – I do not know, for example, how the genetic code evolved. I could speculate on that if you like or refer you to several interesting speculations on this question in the literature. (To be discussed in Chapter 9.) But that is not the main issue here. All such speculations that I have come across are evolutionary – they talk of the gradual perfection of this and that subsystem. But there is only one engine for the evolution of ingenious competence that I know of and that is natural selection. To evolve, the subsystems have to be part of an organism of some sort. Now there might be no need to postulate an earlier kind of life if some minimum nucleic acid–protein system could be conceived of as having formed spontaneously on the primitive Earth. But I do not see such a system as conceivable. You say yourself that naked nucleic acid genes are no good, and anything else would be more complicated – nucleic acid plus something else. I see no alternative to postulating some other kind of starter life to provide the milieu within which our kind of life system began its evolution.

Dr Kritic: But do we not have the same problems all over again in trying to understand how starter life got off the ground?

Dr Advo: Oh, we have a whole lot of problems: but they are not the *same* problems. There is nothing to say that starter life relied on an interdependent set of complex subsystems, as our life does. Indeed one of the things that we can definitely say about primary organisms is that they were not built that way.

Dr Kritic: If they were that simple how did they come to be so good at doing organic chemistry (you say in effect that they became nucleotide chemists, for goodness sake).

Dr Advo: There is a distinction to be made between having to be complicated before you can start to evolve, and being able to become complicated through evolution. But I agree there is a whole nest of problems here. There must have been at least one pre-nucleic acid genetic material with considerable information capacity, and there must have been some way in which this information could specify organic reactions. I would see these questions as constituting the main part of the problem of the origin of life: what was the primary genetic material and how did it work?

Dr Kritic: I'm not so sure. Even if there were such 'self-starting' organisms I doubt whether 'heterogenetic' organisms would ever catch on. They sound ungainly beasts to me. Two different genetic systems in the same organism? A recipe for confusion.

Dr Advo: People usually try to think of an alternative genetic material as being something like the one that they know; but the takeover mechanism does not require primary and secondary genes to be similar, and indeed it might be very much better if they weren't, for the reason you mention. If the genetic control systems were very different chemically there might be no confusion. But I daresay a genetically bimodal organism is not the ideal type.

Dr Kritic: Why then would they have appeared?

Dr Advo: Because a primary organism is even less likely to be the last word in biochemical competence. The primary genetic material was surely a nasty compromise between what would replicate well enough and what units were around. New genetic materials might well have been a good idea when they became possible, because of limitations on what the primary material could do. Working through protein our system is very versatile, but before that Jack-of-all-trades appeared there might have been a case for having different means for making different kinds of phenotype structures.

Dr Kritic: O.K. I will concede that genetic takeover is possible logically; even that some earlier different kind of life might have been relevant to the early evolution of our life. But this whole idea is so *speculative*. You seem to think that the world is full of genetic materials – actual ones on the primitive Earth, and then potential ones for subsequent evolution to stumble on. Genetic materials don't grow on trees you know. Within bio-chemistry there are only two and they are very alike. And there are no known replicating molecules outside biochemistry – no one has ever made one.

Dr Advo: Well actually I think that secondary genetic materials *did* grow on trees – evolutionary ones. I think that primitive life acted in effect as a *search device* for secondary genetic materials – by setting up the means of production of various different sets of small molecules in various evolved lines; by developing generalised techniques for organic synthesis; by creat-ing uses for various polymers and the means of producing them; and then later by making higher order structures that depended on precise lining up between polymers. All such evolutionary advances would have edged to-wards the discovery of a new genetic material. As for speculative, surely this whole subject is speculative?

Dr Kritic: Yes, but genetic takeover is much more so: and you admit it is *vague*?

Dr Advo: What I have said so far is. But genetic takeover is meant to be a background hypothesis. It is meant to be vague – in the sense that it is a general frame within which to make more specific speculations. Chemical evolution is like that as well: it says that matter formed into atoms, then small molecules and then bigger ones; that these then came together to make higher order structures; that some of these could reproduce, and so on. Put at its baldest it's pretty vague. But I would not criticise it on that count. The question is whether the frame is correct – can it hold more specific speculations that are at the same time plausible? Now I do not think that there is any picture of the origin of life that fits the frame of chemical evolution. On the other hand, the genetic . . .

Dr Kritic: But the genetic takeover frame is more *complicated*, and that's always bad for a hypothesis. If you make a hypothesis complicated enough it can be made to fit anything.

Dr Advo: If it is more complicated it is less vague – but let that pass. Life is complicated – our life anyway: a simple explanation for its evolution would be bound to be wrong. Not that the frame itself is much more complicated (is figure 4.1 very complicated?). It is only the picture that is bound to be complicated. My criticism of the chemical evolution frame is that it does not easily let in a certain kind of complexity that should be in the picture. I mean those evolutionary processes that gave rise to the functional interdependence of the subsystems, and led to the choice of such a non-primitive-looking genetic material as DNA.

Dr Kritic: Maybe; but there is an openness about chemical evolution and it *fits the facts*: maybe nucleotides were not in the primitive oceans but Miller's experiment, and others like it, have shown that many of our biochemicals could have been there. That cannot be a coincidence.

Dr Advo: No, I'm sure it is not a coincidence that our biochemistry uses many easily made molecules. But if you remember that biochemistry is a kind of chemistry you should not be so surprised at that. The question is *when* the choice of the particular set of molecules now common to life was made. If it was made wholly before life started then all the present basic parts should be easy to make. If on the other hand the choice of biochemicals was made at some later stage of evolution, as part of the invention of a secondary genetic system and its means of control, then you might expect only some of the component molecules to be easy to make: because by that time the already well evolved organisms had the means to make complex molecules if need be. But there would always be a preference, nevertheless, for easily made molecules if these would do. Well, what are the facts? Our biochemistry is made from some basic molecules that are easy to make (amino acids, sugars) and others that are not (nucleotides, lipids). Genetic takeover fits the facts better.

Dr Kritic: Yes, if you insist on either a 'wholly before' or a 'wholly after' origin for our biochemicals. But no chemical evolutionist would insist that every biochemical was premade. Some, I daresay, did come in later.

Dr Advo: But the late starters include two of the most important types. And, anyhow, who is it that is making *ad hoc* hypotheses now?

Dr Kritic: You are, for goodness sake! I have never heard of a greater bit of ad hoccery than this primary life that you talk of. Even that I might not mind if the whole idea was not so obscure. *Defensive* is perhaps the best word: failing signally to put its neck out. The explanation for our bio-chemistry is to be put out of sight, is it? Neither where it can be checked by experiments simulating primitive Earth conditions nor by extrapolation backwards of the life-system that we have direct knowledge of. The whole secret is hidden in some long lost 'self-starting' ancestor. And all you can tell me about it is that it might have been 'very different'. Your idea of an initial booster rocket that fell away is appealing but useless. Dimly we see the preparations for the launching of the rocket: but the launch itself takes place in a mist that gets thicker, if anything, as the rocket rises. When the mist clears we find that the rocket is powered by a motor that we are familiar with, but you tell me that this was not the original motor – which fell away before the clouds dispersed?

Dr Advo: Roughly speaking yes, but the rocket that emerged was not there at all on the launch pad: it was actually assembled in the clouds (I think this analogy is feeling the strain).

Dr Kritic: But you see what I mean by obscure and defensive: all the interesting stuff is out of sight – so as not to hamper speculation. Can you think of an experiment that could conceivably disprove genetic takeover? That is the real test: if there is no conceivable way in which a hypo . . .

Dr Advo: Oh come off it. You know how hard it is to disprove background hypotheses of any sort. 'Pictures' can easily be disproved perhaps, but it is more difficult to disprove a 'frame' – to show that there is no 'picture' that could fit it. The best you can hope for are considerations that will either increase or decrease their plausibility. Taken together such con-siderations can amount to virtual disproof. But you will be lucky to find one neat experiment. So my answer to your question is, no, I cannot think of a single experiment that could disprove genetic takeover any more than I can think of one that would disprove chemical evolution. But I think chemical evolution has become less likely as a result of experiments, observations and ideas from a number of fields. (Discussed in detail in Chapters 1 and 2.) The same could very well happen to genetic takeover: for example, if life that was closely similar to ours was detected elsewhere in the Universe, or if a one-step simulation experiment generated nucleo-tides, or if in spite of determined searching no simpler kind of genetic material than nucleic acid was found experimentally; or if it could be shown theoretically that nothing simpler than nucleic acid could work.

As for the charge of deliberate obscurity, it is the subject that is obscure

– or rather remote (although I confess to finding remote subjects interesting). Certainly if you shy away from remote subjects they will remain that way. The corpuscular theory of matter of Newton's time was far from sterile even if, at that time, it was hard to get experimental confirmation of the idea that matter is made up of particles too tiny to see. That was an obscure idea, about remote entities, that I have no doubt raised sarcastic comment at the time. But if you are prepared to live with a plausible but unproved idea, formulating experiments and further speculations on the provisional supposition that it is true, then its plausibility may grow or shrink. Of course you do need data, as directly relevant as possible. But even the most indirect data and the most tentative inferences can add up.

Now the evolution of our biochemistry is not by any means lacking in relevant data, even if direct evidence is lacking. We know much about what atoms and molecules can do and this places limits on conceivable or likely hardware for any kind of life, given also what we can deduce as to minimum conditions for evolution. We may still be very uncertain about conditions on the early Earth, but we can already place limits, and these should get tighter with the present activity in planetary and Precambrian research. We should then be able to draw up a general specification for a first organism on the basis that it had to appear spontaneously and evolve on the early Earth. And the abiotic syntheses of biochemicals are relevant here too. One would like as complete a list as possible of all the kinds of molecule that are easy to make from materials and energy sources that might have been available to early organisms. Such lists are being built up. Nor is the hi-fi system that finally emerged irrelevant to that starter life that we seek. The original genetic control apparatus had to come to control organic reactions, as I have said, but of course in particular it had to be able to handle at least some of our now central biochemicals. And then the general way in which our biochemistry is put together – what leans on what – can help us, perhaps, to see a sequence for its original construction; and in the details of its interdependences let us see, perhaps, just where things went missing and what sorts of things they must have been.

No, there are plenty of data: if genetic takeover seems more obscure than 'straight' chemical evolution that is because it has not been thought about enough, and experiments have not been designed with that possibility in mind. You scorn the thought that the secret of the origin of life is in a first link that is now missing and was made differently from us. Perhaps you feel that the rules of the game demand that we stick to the pieces that we can see. But that rule is not in the parsimony of Nature, it is in the paucity of the imagination.

Dr Kritic: All right, suppose that you have put two and two and two together and from these various considerations you have managed to come up with what you think is the answer. How are you going to know? What sort of experimental test would be possible? How do you recognise an unevolved primary organism if you found one – or made one? If you don't have a million years to spare how do you know that it can evolve at all, never mind that it is going to invent nucleic acid? It's not the paucity of the imagination, it's the paucity of experimental approaches that worries me.

Dr Advo: That sounds so like the sceptics of atomism who could see no way of verifying the totally invisible; or the ancients for whom the stuff of the stars was unknowable; or Mendel himself who did not believe that the material basis of heredity would ever be discovered. Not being able to see experimental approaches is part of the paucity of the imagination. First we would have to put two and two and two together. Then we might see.

Dr Kritic: Go ahead then.

Dr Advo: It might take half a book . . .

5

On the nature of primary genetic materials

> There seems to be little in the last resort that a gene can be but a fragment of a specific macromolecular pattern, capable of being copied in growth by the laying down of fresh molecules in conformity with the same design.　*C. N. Hinshelwood (1951)*

Introduction

In the last chapter, we distinguished between primary genetic materials, which can arise without a previous biological evolution, and secondary genetic materials which can only appear within already evolved organisms. It seemed clear enough that nucleic acid was a secondary genetic material, in which case there must have been at least one truly primary genetic material in our evolutionary history – and at least one such material in the world to be found. In this chapter, we start looking.

If genetic takeover obscures the view of first life from the standpoint of present-day biochemistry, it by no means sets us adrift. The design specifications for any system able to evolve indefinitely under natural selection are quite restrictive even if these specifications do not tell us immediately what kinds of materials must be involved. Indeed, the specifications for a primary organism – one that can evolve from scratch – may seem at first to be so restrictive as to exclude everything that is physically plausible. Our approach here will be to take full advantage of the freedom in choosing initial materials that genetic takeover provides, and try to find *anything* that might have been able to start to evolve under natural selection.

The nature of primary genetic material(s) is at the centre of this problem. We will consider first the possibility that the material(s) we seek might have been something like DNA, and then pursue a line of thought that leads away from that, away from linear encoding of information, away from the organic chemist's idea of a polymer – and indeed, away from organic chemistry altogether. Even through such a journey, the essential requirements for a genetic material can be retained. In Chapters 6 and 7, we will become concerned mainly with inorganic crystal chemistry and mineralogy, since, as I will try to show, it is during mineral precipitation processes that primary genetic and other control structures would be most likely to appear. In Chapter 8, we will discuss how an evolving mineral-based life could 'learn to do organic chemistry', evolving to the point at which metabolic pathways developed – some of these perhaps similar to our present central metabolic systems. Chapter 9 will be concerned with the appearance of organic genetic polymers within some advanced primary organisms; with the subsequent evolution of our nucleic acid–protein system; and with the eventual establishment of this system as the basis of the only life form on Earth.

Genes and chemostats

Following our discussions in Chapters 2 and 3, we can guess quite a bit about our ultimate ancestor. It was able to reproduce in a hereditary manner. It belonged to a very large set of possible structures related to each other by minor modifications – there was evolutionary landscape open to exploration. It was made from units that were available in the primitive environment. Unlike us, it could be explained without reference to pre-existing genetic information. Its hereditary machinery was based on structures that could template-replicate fairly accurately from the start, or which could rapidly evolve accuracy of replication. This primary genetic material, although containing initially no genetic information, had a large potential capacity: it could contain any of a very large number of possible replicable patterns that could exert some control on their immediate environment so as directly or indirectly to affect their future prevalence – by increasing the efficiency either of replication or of the distribution of the products of replication. Also, we know something about where our ultimate ancestor lived. Like all organisms, it must have been an open system and hence, to have been maintained indefinitely, it must have lived in another open system – in a matter–energy stream of some sort.

Indeed, the easiest environment for life is the microbiologist's chemostat (figure 5.1) in which identical conditions are maintained, or at least recur reliably. It is this, not the petri dish, that should be our model for the ideal environment in which a delicate early life might thrive. An oceanic soup of organic molecules, built up in some previous era, looks too much like a

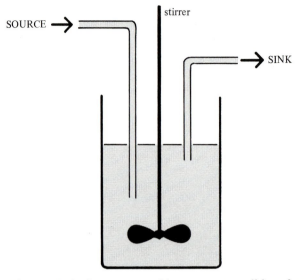

Figure 5.1. A chemostat provides constant conditions for the culture of micro-organisms. These are removed in suspension at the same rate as fresh nutrient solution is supplied.

gigantic petri dish; too liable to be altered by, and to the detriment of, any life that might appear in it. Organisms now are not set up like that: instead they are locked into geochemical and biological cycles which ensure the long-term maintenance of their environments. Indeed, every stable biological niche is a chemostat if you like to look at it in that way: there must be an endless supply of essential feedstock and an endless provision for disposing of waste products and corpses (see Bull, 1974).

Without carrying this discussion any further for the moment, let us formalise the ideal situation for those simplest kinds of evolving entities that we discussed in Chapter 4 – 'naked genes'. Such genes are represented by the black boxes in figure 5.2a. To operate, a black box must be provided with suitable component molecules – 'genetic units' – premade by the environment and out of which the box makes another like itself. There must be many ways in which the units can, in principle, be put together to make different boxes and, occasionally, in the manufacture of a new one, there must be a mistake in copying that allows for the exploration of this field of possibilities. Furthermore, some boxes must be, for various reasons, more efficient at having offspring than others. Such a system will evolve, with more efficient boxes emerging as time goes on. In practice, the environment must also provide other materials – if only solvent molecules – and to go

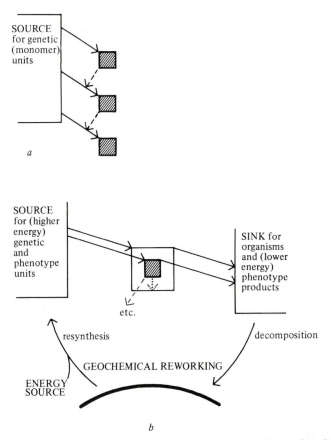

Figure 5.2. (*a*) The simplest evolving system would consist of 'naked genes' (hatched boxes) in an environment providing its monomer units (solid arrows). The synthesis of a new gene also requires a transfer of information from a pre-existing gene (dashed arrows). (*b*) Early organisms most probably lived in open systems, the environment maintaining sources of nutrient and sinks for waste products. These organisms would have consisted of genes whose immediate environment had been more or less modified as a result of information in the genes (dotted arrow). Such a modified environment would protect or assist the replication of the genes, that is, it would be a phenotype (outer white box).

very far, and in any case to set up a situation in which a genetic takeover could occur, the black box must transform its immediate environment as part of its technique of efficient survival and replication. The evolution of such 'clothed genes', i.e. genes + phenotype, will almost certainly call for other environmental molecular supplies. A minimum 'chemostat' that would be necessary for the long-term evolution of such an organism, is illustrated in figure 5.2*b*.

This chapter, and the next two chapters, will be an attempt to explicate figure 5.2.

Some design principles for replicating molecules

First, let us try to prise open that black box to see something of the machine inside. 'What is it made of?' may seem the most pressing question, but we must start nearer the known functional requirements with 'how did it work?' In particular, let us concentrate on its replicative function through the question: 'among physiochemical systems, what are the easiest imaginable microscopic information-replicating machines?'

We might seek a *self*-replicating polymer molecule; by which I mean a molecule that could hold information and replicate it without pre-evolved machinery. It should be necessary only to add copolymer with a given sequence of monomer units to a mixture of free monomers in solution for these monomers to link up, forming copies of the original copolymer sequence. This should go on with little or no help. There is no indication that DNA is like this – it seems to need a great deal of catalytic machinery, as we will be discussing in a few pages. But let us start with the assumption that the primitive genes we seek were molecules something like DNA.

The simplest DNA-like replicating molecule would be a binary ladder copolymer, for example:

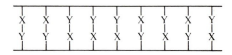

made from trifunctional monomers:

such that groups A and B can form irreversible bonds with each other, while groups X and Y can associate reversibly. Figure 5.3 illustrates one possible scheme for the self-replication of such a molecule. The problem is to find a monomer pair with physical and chemical properties that conform to a set of design specifications. Let us consider some of the features that would be needed.

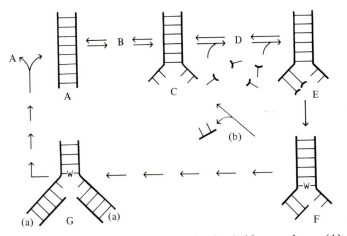

Figure 5.3. A scheme for a self-replicating ladder copolymer (A). Rung bonds must be sufficiently reversible for A occasionally to 'fray' at one end (B, C), allowing monomers to bind (D, E) and to polymerise irreversibly (E → F → G) so that the chains are forced apart (G → 2 × A). W represents a weakened rung, (a) partly made ladders and (b) a short single chain. (After Cairns-Smith & Davis, 1977.)

1. Solubility

Monomers, ladder and intermediates must be soluble: if the ladder is to be a primitive genetic material, then, presumably, soluble in water. It may be necessary to have solubilising groups (e.g. charged groups) for this purpose.

2. Stereochemical feasibility

(a) The three functional groups in each monomer must be so arranged that, with the A and B groups joined up to form a polymer chain, the side groups X and Y can point in such a direction that they can mate with the side groups of another similar chain to make a ladder. This constraint eliminates, among others, any very small molecules for the monomers – there must be at least three atoms on the main chain between the side groups if the latter are to be arranged parallel to each other on the same side of the chain. (In a double helical ladder, successive side groups are at an angle to each other but, in that case, the length of chain between them is longer.)

(b) The various intermediate structures, such as those in figure 5.3, must be stereochemically possible.

(c) All transition states should be (at least) stereochemically possible.

3. Specificity

(a) X and Y must either cross-pair or self-pair, but not both.

(b) A and B must be able to react quantitatively to form main chain bonds, but there must be no other irreversible reactions between A, B, X or Y, or any other (e.g. solubilising) groups.

(c) A and B must be relatively unreactive in the free monomers which should neither polymerise spontaneously in solution nor tend to extend existing chains.

4. X–Y association equilibria

Considering the particular scheme in figure 5.3, the various X–Y association constants must have suitable values, such that 'fraying' at one end of a ladder occurs occasionally to give structures such as C. Also, the concentration of free monomers must be high enough for structures such as E to form (monomers might tend to pair with each other). Such initiation steps must not happen too readily, however, or partly made ladders (e.g. (a)) would start to replicate prematurely leading, often, to incomplete new ladders. Also, if the rung-bonding is too weak, short single chains would be generated (b).

Ideally, the two end-rungs of A, and any monomer pairs, should have low association constants: those attaching the monomers in E should be higher. It would be useful if after the irreversible main bond-forming reaction, E → F, the next rung was weakened, due to stereochemical or electronic strain in the product F, so that the subsequent propagation processes, F → G → 2 × A, could be relatively fast, discouraging side reactions that would lead to partly formed chains.

5. A–B reaction rates

These, too, must be fast in the propagation stages to prevent side reactions (although inhibited in other circumstances – 3(c) above).

Alternatively, one may imagine self-replication schemes in which chains separate completely, as in formalised diagrams of DNA replication. But here an essential problem of self-replication is even more acute: that under the same conditions, rungs of ladders must tend to break while rungs between monomers and single strands tend to form. This is the opposite of what you would expect on entropy grounds. Also, the sequencing problem is still there: one stage must be inhibited until the previous stage is complete. Figure 5.4 illustrates a conceivable scheme of this kind.

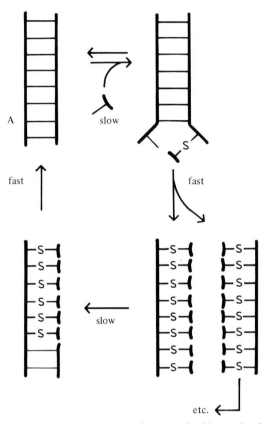

Figure 5.4. An alternative conceivable mechanism for a self-replicating ladder copolymer (cf. figure 5.3). Here the monomers themselves pull the ladder (A) apart. The monomers are present in excess and form particularly strong bonds (indicated by the letters S) with single strands. Slow and fast processes alternate as shown to maintain a correct sequence of stages and so prevent the formation of incomplete chains.

6. Stereochemical controls

Under *2*, we discussed stereochemical features needed to allow appropriate processes: there are others that would be required to prevent inappropriate ones, even assuming the functional group specificities given in *3*. Mainly this new class of problems arises because soluble organic polymers are generally flexible. There would always be a danger of insoluble tangles:

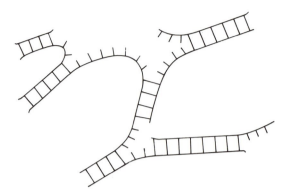

And some of the attached monomers might be left out during the polymerisation stage; for example:

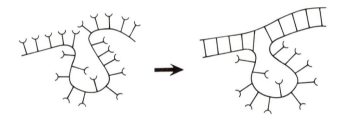

Then, again, the direction of polymerisation has to be constrained in some
way. Otherwise, the formation of daughter chains might become blocked;
for example:

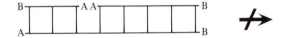

Chance and engineering

We have discussed in all about twenty largely independent design
considerations for a self-replicating molecule of an apparently rather simple
type. None of these considerations is, by itself, particularly taxing: one
can find water-soluble molecules, potential A–B pairs that link quantitatively and irreversibly with each other, even perhaps X–Y pairs that
associate sufficiently specifically. And one could find an X that would not
react with an A, or a Y that would not react with a B, or itself, and so on.
The problem is to find a monomer pair that combines all twenty or more
desirable features. The trouble with such molecular engineering is that one
cannot adjust or machine molecules: if a bond angle or a van der Waals

radius or an activation energy barrier is inappropriate, that is just too bad – one must choose another molecule. The scope for engineering is thus firmly tied to the number of molecules available in principle to choose from.

Here is an analogous situation. Suppose you want to buy some object that can only be had ready-made – let us call it a snark. You go into a shop that sells snarks and ask the assistant if he has a green one. The shop has about 1000 of these things in stock and, generally speaking, about one in ten snarks that come in from the suppliers are green. The assistant is confident: 'Certainly sir, I will go and fetch one.' 'But, wait a minute,' you say, 'it must be the three-wheeled type.' Again, as it happens, one in ten, or so, snarks have three wheels. There should be no problem: the chances are that of the 1000 snarks in stock, about ten will be *both* green and have three wheels. But then you remember that it has to fit into the cupboard under the stairs – the snark must be no longer than 2 metres. Again, 10 % or so are the short kind – and you might be lucky. But you only have to add a few more conditions like these (each, perhaps, easy in itself) for the assistant to start pursing his lips and saying that no one has ever asked for *that* before.

To return to the question of monomers for self-replicating polymers. Suppose that you have a very large box of miscellaneous organic molecules. You sift through them and put aside all those that are soluble in water. You then select, from these, molecules that have a possible A-group. From these, you now select those with *also* a B-group that will allow polymers to form: and you proceed like this to smaller and smaller subsets by applying successively the various design constraints that we discussed. Given that some of these constraints depend on such features as activation energy barriers, bond angles, etc. being rather accurately suitable, you would be lucky if, on average, these siftings would leave you with more than one-tenth of your initial set at each stage. Even on this rather optimistic reckoning, to have a reasonable chance of being left with one molecule worth testing at the end of your search you would need about 10^{20} different ones to choose from to begin with. That is more than there are in *Chemical Abstracts* (by a factor of around 10^{14}).

Such numbers can be out by factors of billions without affecting the main argument: a self-replicating polymer of the kind discussed would be difficult enough for an organic chemist to design with the whole field of organic molecules to choose from. As a primary genetic material, it is wholly implausible. Apart from tending to prefer stable and perhaps water-soluble molecules, I can see no reason why prevital organic reactions should have tended to converge on some particular combination of func-

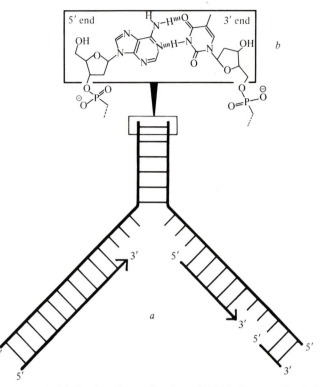

Figure 5.5. (*a*) During the replication of DNA the strands of the parent molecule are peeled apart. The single strands act as templates for new synthesis. But the strands run in opposite directions, and synthesis on the 5'-ending strand takes place discontinuously. (*b*) Closer view of 5'- and 3'-ends, showing in this case an adenine side group paired with a thymine.

tions required for such a molecular machine. To make amino acids is one thing, to make self-replicating organic ladder polymers – whether nucleic acid or anything else – is quite another.

Assisted replication of organic polymers: (a) catalytic assistance

A case in point As I said earlier, there is no indication that DNA is a self-replicating molecule in the strict sense that I have been using that term. Organisms today use complex enzyme systems to catalyse DNA replication. For the light that it may throw on our immediate discussion and because I will be returning to the question of nucleic acid replication in Chapter 9, let us consider this matter in some detail.

The simplicity of Watson & Crick's original scheme has given way to an amazing complexity of the still incompletely understood reality (see, for example, Watson, 1976; Alberts & Sternglanz, 1977; Cozzarelli, 1980). There is not just one enzyme involved but a multi-enzyme complex or **replication apparatus**. In *E. coli*, at least thirteen proteins operate in the replication fork. And the mode of operation of this machinery makes figure 5.3 look childish. In *E. coli* there is a 'gyrase' that uses ATP energy to put negative twists into the DNA double helix: this makes it easier for another enzyme to unwind the DNA into single strands which are then stabilised by binding of yet other proteins. The single strands then act as templates for synthesis of new double helix. These unwound strands necessarily run in opposite directions: synthesis of new DNA however, by a DNA polymerase, only takes place in one direction – towards the 3'-end. As a result, one of the DNA molecules has to be made in a series of short stretches (figure 5.5): this is the more complicated because the polymerase apparently cannot start a new chain of DNA on a single strand. Another primer enzyme is needed for this and it puts on a piece of RNA. With a (hybrid) double strand thus started, the DNA polymerase can get to work extending (as proper DNA) towards the previously placed segment. The RNA piece that had been on the end of this segment has to be removed (with another enzyme) before the final sealing of the gap with yet another enzyme – DNA ligase (figure 5.6).

Much of this seemingly cumbersome manufacturing procedure can be understood in terms of the overriding need for fidelity in DNA replication. (It has been estimated that *E. coli* makes, on average, one mistake in replication for every 10^9–10^{10} base pairs (Drake, 1969): that is about one mistake for every *metre* of DNA processed.) It would seem that starting a new strand on an existing strand is particularly liable to errors. Presumably, there is insufficient stereochemical control until a fair sized piece of double helix has been formed. Using RNA for starting is a clever way of labelling the dubious stretches for subsequent identification and removal. But why make so many starts, by insisting on a $5' \rightarrow 3'$ direction of new strand synthesis? Because, among other things, this direction makes easier an error-correcting mechanism that is built into the DNA polymerase. A DNA polymerase is really two enzymes in tandem – a polymerase proper and an exonuclease. Faced with a mismatched base pair at the growing end of a new strand, the polymerase part delays the addition of another deoxynucleotide, allowing the exonuclease part to hydrolyse off the misplaced deoxynucleotide, and continues in this manner until a correct base pair is reached. It then stops, or at least slows down to a lower rate than that at

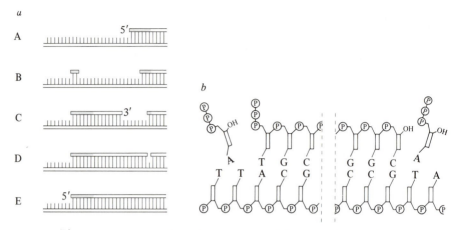

Figure 5.6. (*a*) Overall synthesis of a new DNA strand in the 5′ direction (between A and E) actually takes place with the strands growing in the 3′ direction (as indicated in figure 5.5). Since DNA synthesis can only take place by adding to an existing strand, the many new starts implied by this back-to-front mechanism each requires initiation through emplacement of a section of RNA (hatched) that is subsequently removed. Many enzymes are involved in these processes. (*b*) On the right we have a closer view of a daughter chain growing in the (normal) 3′ direction. Growth in the 5′ direction (left) would be precarious – because liable to be blocked by accidental hydrolysis of the activating pyrophosphate group attached to the growing chain.

which the polymerase part can now proceed to add new deoxynucleotides – the exonuclease always poised to undo future errors.

One can see, from figure 5.6*b*, why such a device would not work so easily in the 3′ → 5′ direction. In that direction, the growing chain would hold the activating pyrophosphate needed to make the main chain phosphoester linkages. This would be a nuisance in any case since accidental hydrolysis of the pyrophosphate group would prevent further synthesis until the 5′-phosphate had been reactivated. Removal of a misplaced nucleotide by the exonuclease would have a similar effect – the growing chain would have to be reactivated each time.

No doubt the present machinery for DNA replication has been the product of an extended evolution. No doubt there was much simpler machinery to begin with. And it is very plausible that RNA, whose replication seems to need less machinery than DNA, was the earlier genetic material – we will be assuming as much later. But RNA replication still needs evolved machinery, as far as anyone knows: neither DNA nor RNA have been

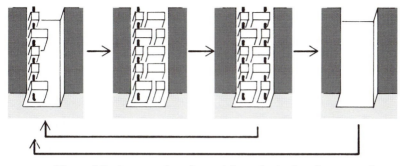

Figure 5.7. A grooved surface might assist in the alignment of monomer units during the replication of a copolymer consisting of units of different sizes.

shown to replicate at all effectively without enzyme assistance (Orgel, 1979; Orgel & Lohrmann, 1974). It is an extrapolation that does not yet fit the experimental evidence that nucleic acid could ever replicate without sophisticated catalysts. Nor does such an idea fit with theoretical expectations – remember how many constraints there are on the design of self-replicating polymers generally. It looks very much as if the nucleotide, which is far too big to be a plausible probiotic product (Chapter 1), is, at the same time, far too small to be a monomer for a *self*-replicating polymer: there are not enough possibilities to choose from with molecules of this sort of size.

An advantage of the idea of genetic takeover is that a secondary genetic material, such as DNA, need never have been a *self*-replicating molecule: it would have evolved in a context that was already biologically evolved. As touched on in the last chapter, and as we will discuss in more detail in Chapter 9, nucleic acid replication could have been a function gradually arrived at through a number of preadaptations. Part of that gradual evolution would have been the evolution of assisting catalysts (probably not protein to begin with).

But perhaps the environment might have provided a catalyst for the replication of some DNA-like replicating polymer. One might think of a grooved crystal surface orienting a large–small copolymer so as to create sites with a complementary specificity (see figure 5.7).

Such a mechanism dispenses with specific X and Y functional groups. On the other hand, a fortuitously adapted environment would be needed. The groove must be of the right shape: the polymer must tend to adsorb in it; then all the monomers must shuffle into place; then they must completely polymerise; then they must come away. A neat balancing of energies would still be required, and many of the problems discussed before are still there. For example, some effect must stop premature polymerisation or, alternatively, premature desorption of incomplete chains. Something must make double occupancy of a groove unstable. There must be no poisoning of the groove with strongly binding impurities. And so on.

Assisted replication of organic polymers: (b) *environmental programming*

A second form of assistance for replicating organic polymers would be through macroscopic manipulation of environmental conditions. This is the way that an organic chemist would approach the problem. Looking at any of the schemes in figures 5.3, 5.4 or 5.7, the problems of monomer design would be greatly relieved if one could take batches of molecules synchronously through different stages by suitably changing temperature, pH, etc. One might imagine a machine, rather like an automatic peptide synthesiser, that could take the molecules through blocking, activating, condensing, and deblocking stages to mitigate what is perhaps the worst of all the problems – inappropriate cross-reactions and side-reactions. (For discussions of the design of laboratory replicating polymers, see Cairns-Smith & Davis, 1977; and Stillinger & Wasserman, 1978.)

Could the primitive Earth have been in any sense such an automatic machine? Kuhn (1976) has suggested something of the sort: that small environments could have had characteristic programmed sequences of change, some of which might be suitable for assisting nucleic acid replication. An example which Kuhn gives is of a number of rocks in the sun: as the day proceeds, shadows will fall in different places and in complicated ways, creating different characteristic diurnal temperature programmes in different sites among the rocks. On a seashore, the tides might add another sequence of wetting and drying; or morning dew might have a similar effect. Whether such an idea is feasible or not depends on how reliable and how particular a programme would be needed.

But the basic problem remains. If a number of independent considerations have to coincide for a given circumstance to arise, then the probability of that circumstance rapidly approaches zero as the number of considerations increases. Even for individually plausible considerations, a few becomes too many and several becomes absurd. And that seems to be the trouble with imagining a circumstance that would allow the consistent replication of any organic polymer that is analogous to DNA. That sort of thing looks too ingenious, however we may try to distribute the ingenuity between the design of monomers, or of assisting catalysts, or of a fortuitously preadapted environment. Penrose (1959a and b) in his designs for macroscopic replicating devices, found that, here too, what appears to be quite a simple problem turns out to be rather complex when it comes down to the mechanical details.

Replication through crystallisation

The difficulties discussed in the last section arose because there are insufficient reasons to see why a combination of properties and conditions required for ladder replication should be generated purely by physico-chemical processes. Too much has to be put down to coincidence. What we seek are models of genetic replication in which such coincidences are reduced: where not only individual aspects of design are physicochemically possible, but where their appropriate combination is also somehow explainable in physicochemical terms. To illustrate what I mean, think again about the hypothetical grooved crystal acting as a jig for a copolymer replication (figure 5.7). One of the difficulties was that the jig had to be just the right size for the polymer. This need not be such a coincidence if the crystal that acts as the jig is the polymer itself. Crystals often have grooves on their surfaces that fit the molecules out of which they are made.

I will not take that idea any further because there is a better related idea illustrated in figure 5.8. Here, the pairing of chains is dispensed with altogether, and the grooves are templates as well as jigs – that is, they specify the sequence of incoming monomer units as well as holding them in a position suitable for polymerisation. I have chosen a short–long copolymer sequence in this illustration, but there are several other possibilities that might be imagined – for example, with wide–narrow, or anionic–cationic copolymer sequences that aligned in a complementary way in the crystal. In such models, crystal packing forces – repulsive as well as attractive – take the place of the mutually recognising X and Y groups in the ladder model. Instead of one polymer chain growing on another single chain, the polymers grow on crystalline bundles of chains. The chains in a given bundle would each have the same sequence (or alternatively, for some models, the bundle would contain an equal number of complementary sequences).

How do the chains eventually separate? They don't. These 'crystal genes' would not be separate molecules but crystallites of colloidal dimensions. The genetic information would be highly redundant, it is true, repeated perhaps thousands of times in the chains of a given bundle; but this

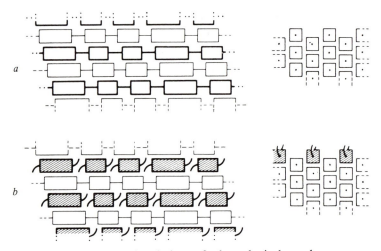

Figure 5.8. (*a*) Top and end views of a hypothetical copolymer crystal showing surface grooves. (*b*) Unpolymerised units (hatched) are located and aligned in the grooves in a manner similar to that in the polymer. (From Cairns-Smith & Davis, 1977.)

could make the process of replication more reliable: a given template groove is specified by at least three adjacent chains and more weakly by more distant chains that together make that groove. A mistake in one chain might thus be overlooked by a growing polymer which conforms, as it were, to a majority opinion. To complete a replication cycle, it would not be necessary that the individual chains separate, but the crystallites would have to break up from time to time and so maintain a population of colloidal dimensions.

Notice, particularly, that the sequencing of the various processes would not be so critical as in the corresponding processes for ladder replication. It would not matter if the monomers began to polymerise before a complete row was in place, nor would it matter if the crystallite broke up while there were still incomplete chains attached to it. The whole thing can be more casual just because it is, from the point of view of information density, less efficient. But the density of information that a genetic material can hold is much less important than the fidelity with which it can transmit that information to offspring.

There would still be plenty of things to go wrong though. There would be the possibility that spontaneous polymerisation in solution would compete with the surface reaction. And the surface reaction would have to be sterically possible. And how good would these grooves really be in recognising and aligning the monomers? Strictly the grooves will fit the polymer out of which they are made – not the corresponding monomer row: that would be misfitted to the extent that the monomers have a different shape and electronic structure from the polymer. In any case, would the binding forces really be strong enough for small monomers?

In general, the misfitting of the monomers in the grooves would be less for rather *large* monomers since there would be less proportionate change in structure on polymerisation. Also, the binding would be stronger. The surface polymerisation might even be specifically preferred, from the following consideration. Suppose that the monomers in figure 5.8 were really oligomers of two different kinds and lengths. If they were long enough they might bind quite well in the corresponding crystal grooves due to favourable interactions along most of their length – in spite of misfitting at their ends (designated by the bent sticks at the ends of the blocks in figure 5.8*b*).

One might compare this with the binding of the lysozyme substrate in its groove in the enzyme: the catalytic activity of lysozyme is thought to depend on its substrate fitting well in the enzyme groove except in one place where it is strained – a strain that is relieved by the substrate molecule passing along the reaction co-ordinate (Phillips, 1966). In the model in figure 5.8, the end-strains would be relieved by polymerisation which might thus be specifically catalysed by the grooves.

However interesting as a project for polymer chemistry, I doubt whether we are yet talking about a primary genetic material. We would seem to need rather large molecules as monomer units – so that they would 'crystallise' well on the polymer crystal surfaces. But (the old problem) primordial clean supplies of large molecules – even of molecules of the size of glucose – are particularly difficult to imagine since such molecules are likely to belong to a very large set of similar molecules that would form under similar conditions and interfere with crystallisation processes.

We seem to have arrived at another dead end. Not, I think, because we have been trying to rely too much on crystallisation processes, but because we have still not made enough use of them. The simple picture of units locating and then irreversibly locking into place fits our intuition about what macroscopic manufacturing processes should be like, rather than what most crystallisation processes really are like. The astounding overall accuracy of crystallisation depends on local reversibility: a site on the surface of a growing crystal does not have to be all that selective – a newly inserted molecule can usually come away again if it turns out later to have been misplaced. By making some of the bonds in our model irreversible, we threw away that key to the accuracy of crystallisation. We had to rely

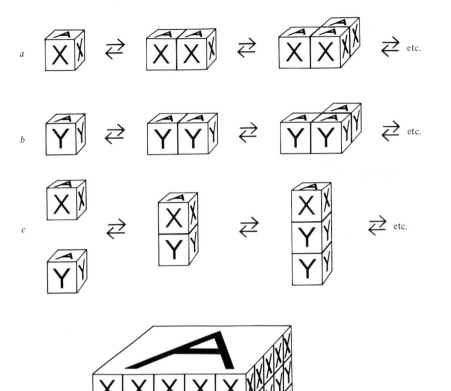

Figure 5.9. Suppose that there are two kinds of molecular cubes, each with a pair of opposite A-faces, the other faces being either all X or all Y. Suppose also that like faces associate and unlike faces do not. X–X and Y–Y associations would themselves tend to give two classes of platelets (*a* and *b*). A–A associations would give any of an immense number of possible linear copolymers (*c*). The combination of all three types of association would tend to lead to crystals like (*d*). This could be a genetic crystal provided that, in addition, (i) A-face growth was usually inhibited, and (ii) the crystals periodically cleaved, but only normal to the A-faces. Such would be a 'type-1' genetic crystal.

instead on accurate surface–monomer recognition and hence on implausibly large units. The question sharpens into this: how can you reliably retain genetic information in polymers that have reversible main bonding? For crystal-gene models this is not, as it may sound, a contradiction in terms.

Models for two types of genetic crystals

Consider the following pair of units:

Let us suppose that the faces of these cubes tend to stick together reversibly such that A–A, X–X and Y–Y interactions have lower energies than any mixed interaction. Furthermore, these energies are determined solely by the faces in contact. A supersaturated solution of a mixture of these units would give rise to disordered crystals with two opposite 'A-faces' and four 'X/Y-faces', as in figure 5.9. Each crystal would contain its own 'information' in the form of identical X–Y sequences being replicated through growth on any of the X/Y-faces. (Growth on the A-faces should be inhibited, since this would lead to new 'information' being added in a random manner.) In the model in figure 5.9, information is being held in one dimension and being replicated through crystal growth in the other two dimensions. (This was true also for the model shown in figure 5.8.) I will call such things **type-1 genetic crystals**. A **type-2 genetic crystal** would be one in which information is held in two dimensions and replicated in the third. A block model for a type-2 genetic crystal is illustrated in figure 5.10.

Figures 5.11 and 5.12 illustrate replication cycles for type-1 and type-2 genetic crystals respectively.

Crystal growth and error correction We have moved, it seems, very far from DNA. Yet, curiously, in one crucial respect we are closer to DNA in these last models than in any of the previous ones – because these models have a 'proof-reading' mechanism built into them. Think about the growth of a crystal face by successive surface nucleations. In figure 5.13*a*, we have the first difficult step of such a nucleation with a single unit precariously attached to a site. Under only slightly supersaturated conditions the chances are that that unit will come away again before others

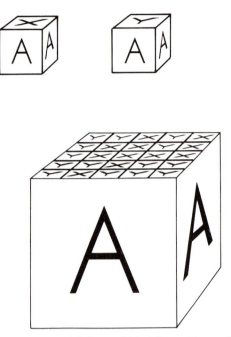

Figure 5.10. (cf. figure 5.9.) A 'type-2' genetic crystal might be based on cubes with four A-faces and either two opposite X-faces or two opposite Y-faces. Such a crystal would contain information in two dimensions.

come alongside. Eventually, by a chance fluctuation, a stable island forms (figure 5.13*b*). Even then, new units will come and go many times: the difference between crystal growth and crystal dissolution is not absolute under such circumstances: both happen, it is a question of which is marginally faster. And mistakes are frequent – perhaps usual. That is no matter since a crystal with a mistake in it is less stable – more soluble – than a more perfect crystal: the crystal will tend to redissolve until the mistake has been eliminated. Trial and error – and error correction – are the secrets of success. It is perhaps significant that the two most accurate processes that we know of in the universe – the faultless replication of a metre of DNA and the growth of a hand-sized crystal – each depend on a continuous product control. For crystal growth, this 'proof-reading' is probably much the most important element: only when a molecule is already snuggled into a lattice can the lattice 'feel' whether it is really the right molecule, rightly orientated. (Is it generally true, perhaps, that it is easier to know when something is wrong than to get it right in the first

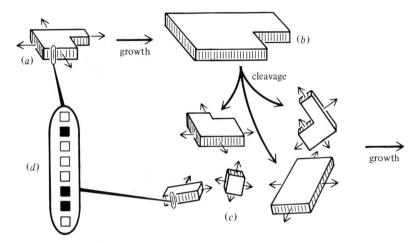

Figure 5.11. A replication cycle for a type-1 genetic crystal. (*a*) is a crystallite consisting of a bundle of identical sequences (*d*). Growth takes place on faces, as shown, by the incorporation of monomers such that polymer chains are not extended. This gives a platey morphology (*b*). (*b*) Shatters into smaller fragments by cleaving parallel to the growing faces (*b* → *c*). Each of the crystallites (*c*) still contains only the copolymer sequence that was in (*a*). Through this combination of crystal growth and crystal cleavage, replication continues indefinitely, maintaining a population of crystallites with a roughly constant distribution of sizes.

place?). With DNA replication too, it is likely that product control is at least as important for fidelity as initial accuracy in the placement of incoming nucleotides – perhaps more so (Hopfield, 1974; Ninio, 1975; Topal & Fresco, 1976). The exonuclease is one of the main keepers of our genetic inheritance through the element of 'reversibility' that it (and other repair enzymes) effectively introduce into the main bonding of DNA. But this 'reversibility' must be very specific: faced with a misfit, the enzyme must know which strand to believe, it must distinguish between parent and daughter. (It is hard to see how this could happen in a primitive ladder polymer without yet more design constraints to be added to an already daunting list. Yet after-the-event control is, perhaps, a necessary part of tolerably accurate molecular replication.)

In genetic crystals such as those shown in figures 5.9 and 5.10, the genetic information could be retained in spite of surface main bond reversibility, because most of the copies of the information are safely inside the crystal. If there is a misfit, it is easy to see which sequence is wrong – the one that is different from all the others. Usually, this will be near the

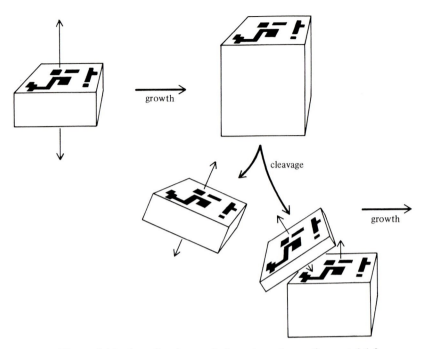

Figure 5.12. A replication cycle for a type-2 genetic crystal (cf. figure 5.11). Here information is held as a two-dimensional pattern that replicates indefinitely through exclusive growth on two opposite faces and cleavage only along planes parallel to the growing faces.

surface and so will tend to be removed automatically through local disso-lution resulting from lattice instability caused by the misfit.

An energy problem These models involve an assumption that is arguably unlikely, however: there is no significant energy difference be-tween different X–Y arrangements. Suppose there was a big difference – let us say that in figure 5.9 an alternating X–Y sequence is much more stable than anything else. Then misfits during growth that increased the extent of this alternation would be less disfavoured. Indeed, if the energy drop resulting from alternation within a polymer was greater than the energy rise due to misfitting between the polymer and its neighbours, then any initial information would be quickly wiped out. The initial 'imperfect' crystal would have learnt to be 'perfect'. In any case, even a small energy difference would lead to eventual rubbing out of initial information – here thermodynamics would be on the side of finding that unique information-ally boring solution.

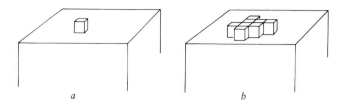

Figure 5.13. Surface nucleation of a new layer during crystal growth (see text).

A question of supersaturation This is the negative side of reversible main-bonding, but is by no means a fatal objection: there are many crystals that contain partial irregularities that have very little effect on stability. Provided the energy difference between a disordered and ordered structure is small enough, the disordered structure could still be preferred. The crucial question is one of balance between rate of erasure and rate of selection of a functionally significant disordered structure. I will return to this topic in Chapter 7. But, we can already see this much about what is required for the replication of crystal genes: the environment must provide a solution of suitable supersaturation. If the solution is too supersaturated, then error correction mechanisms will too often be overtaken and crystals will be formed that are heavily mutated with inappropriate disorders; if it is kept only barely supersaturated, then even small differences in energy between the informationally disordered and 'blank' regular structures could lead to the latter being preferred in the long term. Only an open system could maintain an appropriate level of supersaturation. Over the next two chapters, we will consider the question of how the primitive Earth could have set up open systems of the kind required.

Generalised genetic crystals

At the most general level, a genetic crystal would be a small crystal, probably of colloidal dimensions, that contained some irregularity of structure that was accurately copied in growth by the laying down of fresh molecules in conformity with the pre-existing irregularity. That is close to Hinshelwood's pre-Watson–Crick description of what a gene must be like (quoted at the beginning of this chapter). Our last models were both substitutional crystals, with genetic information held as arrangements of chemically distinct units, thus reflecting, still, an aspect of DNA's structure. But even this homage to DNA is not necessary. As we will discuss in Chapter 7, the irregularity in question might be purely physical and not chemical – a pattern of dislocations, perhaps, or a complex form of twinning.

Big writing to begin with? We should bear in mind the possibility of another kind of redundancy, apart from the multiplicity of copies inherent in the genetic crystal idea: primitive genetic information might have been 'writ large', the irregular features (the information units) being bigger than the units out of which those features were built (the genetic monomer units). Think of the reproduction of a printed page by photography: here, the information units are the letters on the page, but these are formed from silver grains that are several orders of magnitude smaller. Where fidelity of replication is a major problem – and this would be so right at the start of evolution – we should not assume that the genetic information was written as it is now, at the level of the genetic monomer units, even if subsequent evolution might lead to systems operating more intricately. In terms of the block model in figure 5.10, the information might consist of relatively large patches – domains – of X and Y co-operatively more stable and more easily copied. We will return to this idea in Chapter 7 when we have more particular examples to consider.

Inorganic genes

Under cover of abstractions, we have now, I think, moved well into inorganic chemistry. Our course was set in that direction as soon as we started to think about replication through crystallisation.

Mainly it is a question of bonding. The genetic units, we said, should be smallish, water-soluble, yet able to form colloidal crystallites, stable when dispersed in water. This suggests strong bonding throughout the crystallites. Yet all this bonding has to be quantitatively reversible at (some) crystal surfaces. A typical organic polymer crystal is held together by a combination of secondary forces, which are reversible but not strong, and covalent bonds that are strong but not reversible.

This would be too neat a dismissal of any conceivable organic polymer crystal gene: perhaps hydrogen bonding could provide strong enough template forces. And the main bonding is reversible in some organic polymers (e.g. polyoxymethylene). But it would seem that, in organic polymerisations, activation energies are usually too high, and/or reaction trajectories too sterically demanding, for controlled polymerisation on the surface of preformed polymer crystals to happen very often. Usually, monomers *first* polymerise into a chain which *then* crystallises. This is the wrong order of events for the crystal structure to have much chance of controlling the character of the polymerisation. Indeed, usually organic polymer crystallisation is only partial, zones of aligned chains being embedded in chaotic tangles. This can be seen as a likely consequence of polymerisation and crystallisation being separate events taking place in that order.

If we cannot dismiss completely the possibility of a primary organic crystal gene, I think it fair to say nevertheless that the main bonds of organic polymers are generally too difficult to form, and too stable when formed, for interchain forces to exert a decisive control on polymerisation.

Inorganic polymers present a far more promising field – in particular, those kinds of polymers that underly the majority of minerals. Most of these are essentially crystalline – they do not exist free in solution. Many are colloidal, their crystallites having formed at ordinary temperatures from dilute aqueous solutions barely supersaturated with small units such as silicic acid and hydrated cations. These crystalline polymers are far more complex than our idealised models: but often they contain features of the bonding arrangement required. There are often bonds of different kinds arrayed in different directions which may give rise to perfect cleavage in certain planes and preferential growth in certain directions. The bonds between the units are generally strong, ranging between electrovalent and polar covalent types. Inside, the crystals are thus stable at ordinary temperatures. In contact with water, however, at the surfaces of the crystals, these bonds are more labile – allowing perhaps the kind of error correction during crystal growth that we have talked about. In any case, the extensiveness of crystallinity typical of many colloidal minerals makes it clear that, however it is done, the crystal structures themselves must largely control the way in which their covalent bonds are made. Furthermore, mineral crystals are frequently substitutional, and possess indeed a variety of irregular features that might store genetic information.

Three objections

Later, we will come to the idea that not only primary genes but early phenotypes might have been purely inorganic – that minerals might have been the materials out of which an early primary life was made. Our journey will be via chemistry that may be unfamiliar to some readers. Perhaps we should check that we are not obviously off course. I can imagine three objections that Dr Kritic of Chapter 4 might make at this stage.

First, he might make a general point to doubt the validity of the journey so far: 'Long serial arguments are always unsatisfactory – where the conclusions from argument A are used as a basis for argument B, whose conclusions are then used for C and so on. Even if individually the arguments are plausible, the string of them taken as a whole may be quite implausible.'

Dr Kritic would be quite correct in saying this and, indeed, all of us interested in very early evolution should have his remark written on our wall. But the discussion so far has not been an extended argument series. It has been, as I said, a journey; a sightseeing tour through the landscape of possible primary genetic materials to show what a barren sort of place it is without attempting to visit everywhere. If there seems to be an oasis in the region of inorganic crystal chemistry, I cannot quite prove that there are no other oases elsewhere. And the sceptical arguments suggesting that the regions around DNA are indeed barren do not depend on considerations in series: they depend on considerations piled on top of each other. Unlike serial arguments, which become weaker as they become more numerous, parallel arguments become stronger when taken together. (Another thing to write up on the wall.) Having arrived at inorganic crystal chemistry, there is only one way to check that there is an oasis here and not a mirage: we must take a closer look. There is nothing for it but to continue the journey, to find specific examples in inorganic crystal chemistry and see if these can be worked out to give a detailed plausible story. Only after that, with detailed considerations in mind, can we hope to design clearly relevant experiments.

There is another kind of objection that I have sometimes come across – of the 'you're-not-going-to-tell-me' kind. For example: 'you're not going to tell me that life started as a kind of mud'.

A similar stone wall might have been put up in earlier times to refute the suggestion that moulds are alive, since they too are macroscopically unprepossessing. That kind of long view can be decidedly unhelpful. The critical questions concern structures in the size range visible under the electron microscope, and the most intricate details of crystalline architecture. At these levels, colloidal inorganic minerals – 'muds' – can be far richer in structure than mineral organic materials like raw petroleum. (The microstructures of 'muds' will be considered in detail in the next chapter.)

A more sophisticated objection is along these lines: 'life must be founded on organic molecules because only then is the necessary variety of molecular machinery possible – indeed, there must be strong faithful bonds between atoms to make proper molecular machines at all. That means that there must be hard covalent bonds of the sort that form typically between first-row elements. In particular, carbon is implicated because of the usual stability of the C–C bond under aqueous conditions and because of the high valency of carbon – and hence the unique richness of variety in its compounds.'

This argument cannot be dismissed so easily. Life is nothing if not naturally occurring machinery. But it makes the opposite mistake to the previous one: it takes too close a view – it assumes that an atomic grain size is essential for living machinery. This is not even true of all components of present-day life. (I doubt whether the exact position of every calcium atom in your teeth was genetically determined or really matters very much.) Some living machines can work very well without being specified down to the last atom in the way that, say, an enzyme is. There are structures specified at various levels in life now. Who is to say that primary life used any truly molecular machines? We have indeed started to answer this argument in pointing out that a primary genetic material might work better if its information was not written so finely as genetic information is written today. When we come to discuss primitive phenotypes, I will pursue this question of grain size further. We will then think of the ultimate machines of primary life as having been made on a slightly larger scale than ours.

The necessary and the efficient Returning to an earlier theme, we should always be careful to distinguish the necessary from the efficient. Indeed, it is efficient to be able to make the most intricate possible machinery. But, if this is a more difficult way, and if it is not necessary for making evolvable systems, then it is not to be expected for primary life. On the other hand, once clumsy first organisms had started to evolve, we can predict selection pressures tending to make them less clumsy: a continual drive towards finer and finer control through increasingly fine-grained machinery. Genetic takeover provides a mechanism for even the most central machinery to give in to such pressures through radical changes. The result should be predictable: machines will arise eventually of the finest possible grain size with atoms as nuts and bolts and molecules as working components. Arguments about the excellence of carbon are thus true but misdirected: they do not tell us what to expect of first life: they tell us what to expect of biochemically highly evolved life. And, of course, all life now on Earth (that we know of) is firmly in this latter category.

6

The first biochemicals

Introduction

By assuming that at least one genetic takeover was involved in the
origin of our biochemical system, we are released from any need to conform to present biochemical ground rules in attempting to formulate likely
structures and *modus operandi* of a primary life. In the last chapter, we took
full advantage of this freedom in trying to arrive at the kind of stuff that
primary genes were made of. What processes, occurring in the non-
biological world, might inadvertently conform to specifications for a self-
operating primary genetic material? That was the central question, and our
tentative answer was 'crystallisation'. This seems to be the only non-
biological process that would be accurate enough to sustain an extended
evolution. It was possible to be somewhat more specific about what kind
of crystallisation we should be thinking about and what kinds of crystals:
and we were pushed into contemplating inorganic materials (inorganic in
the chemist's sense).

The change of view

This change of view from organic to inorganic has to be radical, I
think, to be satisfactory. It is not just a question of imagining a mineral to
begin with instead of DNA. Primitive biochemistry should be seen as
having been quite different from present biochemistry in all but the most
abstract essentials. The change of view that I suggest is altogether away
from the idea that life started with some class of organic Earth products
(which the Earth is no longer making), to the idea that the materials of
first life, all of these materials, were inorganic minerals – presumably of

kinds still being produced by the Earth. To understand the beginnings of life, then, it is present geochemistry, not present biochemistry, that is most germane.

This chapter is a long pause for information. Some readers will become impatient with it, wondering why so many minerals are being brought in. The reason is partly that we are engaged in a search: first of all we should survey the field of possibilities; we should try to get some feel for the variety and variability of the materials that the Earth can most easily produce. And we will be at least touching on a substantial fraction of the possibilities. I will grant that it may not be important to know, say, all the names and formulae of the zeolites given on pages 189–91. But one should have some notion of the variety of zeolites, of their crystal structures and morphologies, and of how they are formed in Nature. Then, again, I do not imagine that at first reading all the subtleties of layer silicate chemistry that we will be discussing will be retained by the reader who is not already familiar with them. But it should at least be noted that there are such subtleties: that silicate layers contain, very often, fixed charges; that they are often flexible; that they are highly articulated structures that can transmit effects through them; that many layer silicate types can form from dilute solutions at around ordinary temperatures. It may be that the most important thing is to get a general sense of what these minerals are like; but even that can only be arrived at by considering some examples in detail.

In any case, the search is not for one mineral – the true stuff of first life. It will be part of the view to be developed in Chapters 7 and 8 that early organisms were assemblages of minerals of different sorts that had in common the ability to crystallise from solutions derived from the weathering of rocks.

Such materials are, broadly speaking, **clays**; and it will be convenient to define clay for our purposes as any microcrystalline material (particle size less than 10 μm or so) that can form from aqueous solution near the Earth's surface. On this definition, a clay is very often, but not necessarily, a layer silicate. Framework silicates, metal oxides, hydroxides and sulphides are among other types that can be 'clays' within the above definition.

Other minerals of interest In this chapter, we will consider a number of minerals that are not themselves clays but whose structures are similar to major clay types and are known in greater detail, for example muscovite. In other cases, the mineral may be of interest as a source of weathering solutions: olivines and pyroxenes are discussed briefly for this reason, and because they also help to illuminate general aspects of silicate structures.

Two main themes Two of our main concerns in this chapter will be with the structures and conditions of synthesis of clays. These themes will be returned to in later chapters to provide the basis for more detailed discussions of crystal genes, to be given in Chapter 7; and also for the development of ideas, in Chapter 8, about how minerals organisms might have started to photosynthesise and to evolve a competence in organic chemistry.

But, in this chapter, I will refrain from such speculations. First, let us study the materials: then we will be ready to think about a putative stone-age biochemistry at the beginning of our evolution.

Cation polymerisation

Metal cations in aqueous solution are invariably hydrated, the water molecules in their immediate vicinity being more or less regularly arranged. An inner shell of octahedrally co-ordinated water molecules is common:

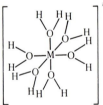

Such an object – an aquo ion – is a kind of molecule: the co-ordination bonding between oxygens and metal is very strong, even if activation energy barriers are low enough for water exchange to take place rapidly. $Cr(H_2O)_6^{3+}$ and $Rh(H_2O)_6^{3+}$ are exceptions here with exchange rates on the time scale of hours, but the usual time scales are from seconds to nanoseconds (Cotton & Wilkinson, 1972). These molecules are acids, they can donate a proton to a water molecule, for example:

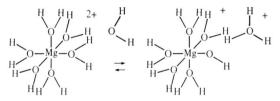

The conjugate base on the right hand side is called a hydroxo complex.

Almost all hydroxo complexes undergo **olation** to give, in the first place, dimers joined by one or two hydroxo bridges. Further deprotonation and olation can lead to higher oligomers. For example:

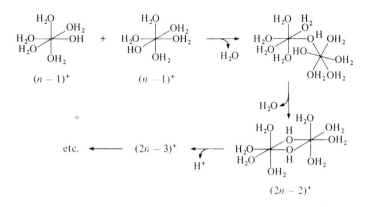

$(n - 1)^+$ $(n - 1)^+$

etc. ←——— $(2n - 3)^+$ ←——

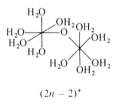

$(2n - 2)^+$

This may be further complicated by anions displacing water molecules, and perhaps also hydroxo groups, from co-ordinating positions, for example in oligomers such as:

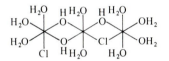

Another way in which positive charge on polymers may be reduced is by **oxolation** where oxo bonds are formed through loss of protons, for example:

$(2n - 2)^+$

As the pH of a solution containing a given cation is increased, so is the tendency to oligomerisation. A precipitate of high polymer, for example an hydroxide or oxide, usually appears within 1 or 2 pH units from the onset of oligomerisation.

Although the structures shown above are vaguely reminiscent of oligosaccharides, there is one important respect in which they are usually quite different from organic structures. These cation oligomers are seldom isolable, being in a dynamic equilibrium that shifts fairly quickly on dilution, change of pH, etc. Cr^{3+} is one interesting exception – a blue dimer and a green trimer can be separated from solutions of chromium (III) that have been refluxed and cooled (Laswick & Plane, 1959). But even here the kinetic stability is less than that of, say, a sucrose molecule in neutral aqueous solution.

According to Baes & Mesmer (1976), cations are generally monomeric at sufficiently low concentrations with only a small number of kinds of oligomers predominating as concentrations increase. Very often, from one to three of the following types will be found:

$$M_2OH^{(2n-1)+} \qquad M_2(OH)_2^{(2n-2)+} \qquad M_3(OH)_3^{(3n-3)+}$$
$$M_3(OH)_4^{(3n-4)+} \qquad M_3(OH)_5^{(3n-5)+} \qquad M_4(OH)_4^{(4n-4)+}$$

(Co-ordinated water molecules are ignored in these formulae.)

Between the generally small inorganic oligomer and the crystalline precipitate of high polymer, which becomes accessible to X-ray structural analysis, there is much ignorance. Particularly with highly charged cations, metastable materials may persist for a considerable time at ordinary temperatures. For example, Rausch & Bale (1964) made Al^{3+} solutions in which platelets (*ca* 1×30 nm) formed at first, and this gave rise to a gel. The gel reverted to a sol after a few days. In another study, Turner & Ross (1970) found that a precipitate was formed first which contained chloride. This then dissolved while the concentration of oligomers in solution increased: these in turn decreased as the stable crystalline form of $Al(OH)_3$ appeared after a few weeks.

The polymerisation of Fe^{3+} at ordinary temperatures provides a useful example to consider in more detail. Murphy, Posner & Quirk (1976a) investigated the reaction of partly neutralised ferric nitrate solutions using an electron microscope. The first products identified were small discrete spheres of apparently amorphous polymer (diameter 1.5–3.0 nm, mol. wt 3000–30000). On ageing, the spheres joined up into rods of two to three spheres. With solutions of higher ionic strength and/or Fe^{3+} concentration these rods remained the same length and aggregated into extensive raft-like arrays. The spheres slowly coalesced within the rods and the ferric oxide-hydroxide polymorph goethite was formed. With lower ionic strength and/or Fe^{3+} concentration the rods increased in length without forming rafts but again crystallised into goethite. After extended ageing, some of the rods had coalesced to give lath-like goethite microcrystals. Figure 6.1a is an electron micrograph taken after ageing for one year. Laths, and single and double mature rods are visible, but also, still, single

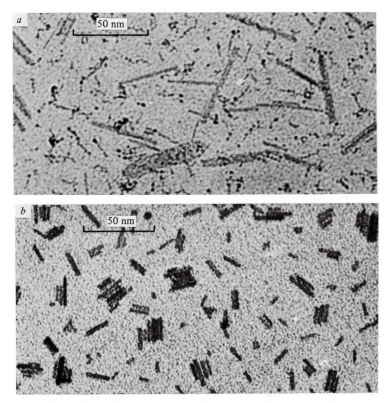

Figure 6.1. Transmission electron micrographs of products of polymerisation of ferric solutions for 1 year. (From Murphy, Posner & Quirk, 1976a and b.) (*a*) 0.0165 M Fe(NO₃)₃, OH:Fe ratio 2.08. *ca* 3 nm diameter spheres as well as rods and laths are visible. (*b*) 0.0165 M FeCl₃, OH:Fe ratio 2.10. Rods, and rafts made from these rods, are visible. The rods often seem to be made up of two *ca* 2 nm thick rods side-by-side.

spheres and bundles of joined up spheres. Note especially the size range of these objects.

Changing the anion to chloride (Murphy, Posner & Quirk, 1976b) did not affect the initial spheres but the subsequent development was different (figure 6.1*b*). Here β-FeO(OH) was the crystalline product. With perchlorate as anion (Murphy, Posner & Quirk, 1976c) laths of yet a third polymorph of ferric oxide-hydroxide were formed – lepidocrocite – as well as goethite rods. (The crystal structures of these polymorphs will be described shortly.)

Spiro *et al.* (1966) had previously noted polymer spheres of about 7 nm

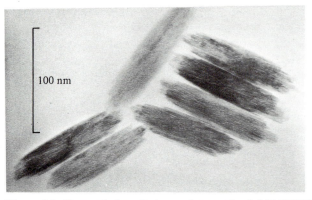

Figure 6.2. Transmission electron micrograph of β-FeO(OH) 'somatoids' formed at room temperature from 0.01 M $FeCl_3$ solution after 14 days. (From Gildawie, 1977.)

diameter (mol. wt about 240000) in similar experiments with ferric nitrate neutralised with bicarbonate. When separated from other solution components, these spheres were apparently indefinitely stable at ordinary temperatures. In a subsequent study (Brady *et al.*, 1968) the iron atoms in the polymer were found to be tetrahedrally co-ordinated with oxo as well as hydroxo bridges between them. In another series of studies, Dousma & de Bruyn (1976, 1978) describe the formation of initial particles (again of a remarkably definite size – here around 4 nm) and discuss mechanisms for their formation through olation and oxolation, their subsequent clustering and the linking of these clusters into chains.

Matijevic & Scheiner (1978) prepared a number of sols of haematite (α-Fe_2O_3) with variously shaped particles (*ca* 200 nm) by ageing ferric solutions for 1–3 days at 100 °C. Again, both shapes and sizes were very uniform within a given preparation. Preparations of 'monodisperse' sols from aluminium, titanium, chromium, iron, copper and thorium are described by Matijevic (1976).

Micromorphologies of β-FeO(OH) have attracted attention for some time (Mackay, 1964). Spindle-shaped crystals obtained from dilute ferric chloride solutions are particularly striking (figure 6.2). These 'somatoids', as they are called, vary from 100–500 nm long and 20–100 nm diameter according to the conditions of preparation, but within a given preparation they are again very uniform in size and shape. The somatoids tend to aggregate into sheets with their long axes perpendicular to the sheets.

Nucleation by crystalline phases can have a profound effect on the course of ferric ion polymerisation (Atkinson, Posner & Quirk, 1977). The number of initial nuclei can determine whether an indefinitely stable sol or a murky precipitate forms (Hsu, 1972). Silica can inhibit the formation of γ-FeO(OH) during oxidation of ferrous chloride solutions (Schwertmann & Thalmann, 1976). And organic molecules can have an important effect: oxalate can accelerate the crystallisation of haematite from 'amorphous' iron hydroxides (Fischer & Schwertmann, 1975). These authors suggest that chelated ferric ions act as templates for crystal growth. On the other hand, traces of inorganic cation impurities in minute early haematite crystals were found to inhibit further growth as a result of lattice distortions (Nalovic, Pedro & Janot, 1976).

What is true of ferric ions is true very generally in that the hydroxides, oxide-hydroxides and oxides that are the typical products of cation polymerisation in aqueous solution may seem often 'amorphous' or 'crypto-crystalline'. But such words are expressions of ignorance of what are complex but far from chaotic colloidal products.

Hydroxide crystal structures

Brucite (magnesium hydroxide, $Mg(OH)_2$) is an important archetype structure. Here, two-dimensional polymers are stacked on top of each other. A ball-and-spoke representation of the structure of one of these polymer layers is shown in figure 6.3a. One can see it as a final outcome of olation of $Mg(H_2O)_6^{2+}$ (cf. scheme on page 167), where all the hydroxyl oxygens are co-ordinated with three magnesium atoms. The hydroxyl bonds are normal to the plane of the layers which thus have knobbly surfaces. These key together in the stacked layers – the hydrogen atoms on adjacent surfaces are in contact in such a way that any hydrogen atom in any layer is nested into the surface niche created by a triplet of hydrogens on an adjacent layer. Alternatively, a brucite layer can be imagined as two close-packed sheets of spheres, representing the hydroxyl groups, with the much smaller magnesium atoms filling all the octahedral sites thus created (figure 6.3b). Hydroxides of Ca^{2+}, Mn^{2+}, Fe^{2+}, Co^{2+} and Ni^{2+} have essentially the same structure as brucite. Fe^{2+} and Mn^{2+} frequently replace Mg^{2+} in the lattice to the extent of about one or two per thousand. (Cation substitution in minerals is so general that I will not remark on it in the following unless it is particularly extensive.) Brucite most commonly occurs as a metamorphic alteration product of *periclase* (MgO) in contact-metamorphosed dolomites and limestones. It is also formed **hydrothermally** (i.e. with water at high temperatures and pressures) in serpentine veins.

Gibbsite, or hydrargillite (aluminium hydroxide, $Al(OH)_3$), has a similar

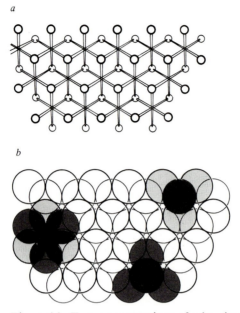

Figure 6.3. Two representations of a brucite layer. (*a*) Upper and lower planes of spheres represent hydroxyl groups. Magnesium ions are located at spoke intersections. (*b*) Two closest packed planes of spheres have **octahedral sites** between them, that is, niches with six surrounding spheres (like those shaded on the left). In brucite the spheres represent hydroxyls, and magnesium ions fill all the octahedral sites. Closest packing of spheres also generates two kinds of somewhat smaller sites surrounded by four oxygens (like those shaded on the right), but none of these **tetrahedral sites** is occupied in brucite.

structure to brucite except that only two of the three octahedral sites are occupied (figure 6.4). This can be seen as an end result of olation in which hydroxyls are only co-ordinated twice to metal atoms. Also, the stacking of the layers is different from that in brucite: the oxygen atoms at the base of one layer are directly above those on the top of the layer beneath. Only half the hydrogen atoms point towards the adjacent layers so that complete interlayer hydrogen bonding is possible. Chromic oxide-hydroxide has a somewhat similar structure except that the octahedral sites are here fully occupied and the hydrogens, of which there are only half as many, all form interlayer hydrogen bonds.

Along with boehmite and diaspore, discussed below, gibbsite is a major constituent of bauxites. It is often a late weathering product of alumino–silicate minerals and found particularly in zones that have been subject to

a

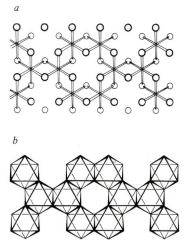

b

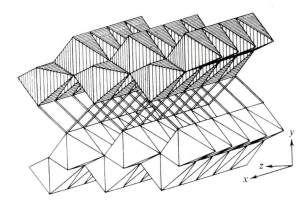

Figure 6.4. Two representations of the ideal gibbsite structure. (*a*) A ball-and-spoke model (cf. figure 6.3*a*). Here only two out of three of the octahedral positions are occupied by aluminium ions. Otherwise similar to brucite. (*b*) Alternatively a gibbsite layer can be represented as a flat network of octahedra joined through shared edges.

Figure 6.5. Representation of boehmite (γ-AlO(OH)) using solid octahedra (cf. figure 6.4*b*). (From Deer, Howie & Zussman, 1962, after Ewing, 1935a).

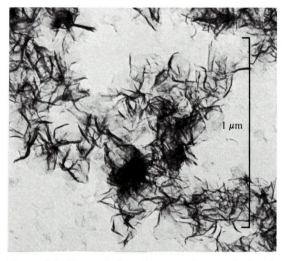

Figure 6.6. Transmission electron micrograph showing crumpled and rolled up sheets in boehmite that had been prepared by refluxing 0.002 M Al(NO$_3$)$_3$ solution for 24 days and then ageing for 11 days at room temperature. (From Gildawie, 1977.)

intense weathering. (Roughly speaking, the silica has been washed away.) On the other hand, gibbsite may be a transitory early weathering product too. Busenberg found that feldspar suspended in distilled water under 1 atmosphere of CO$_2$ gave a solution saturated with gibbsite within 6 minutes (see Clemency, 1976): also, Smith & Hem (1972) have shown that crypto-crystalline gibbsite can form fairly rapidly in certain solutions becoming microcrystalline in one or two months.

Boehmite (aluminium oxide-hydroxide, γ-AlO(OH)) has a different kind of layer structure. The aluminium atoms are located in a double sheet of distorted octahedral sites within the layers which are stacked on each other through oblique hydrogen bonds (figure 6.5). This structure can be seen as a product of both olation and oxolation. The layered structure of boehmite is vividly suggested by the crumpled and rolled-up crystals shown in figure 6.6.

Lepidocrocite (ferric oxide-hydroxide, γ-FeO(OH)) has a similar structure to boehmite. It occurs under oxidising conditions as a weathering product of iron-bearing minerals.

Diaspore (aluminium oxide-hydroxide, α-AlO(OH)) is different again. This is not a layer structure but consists of double strips of octahedrally co-ordinated aluminium atoms running in the z-direction (figure 6.7). All

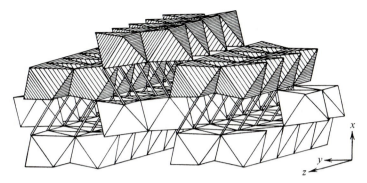

Figure 6.7. Representation of diaspore (α-AlO(OH)) using solid
octahedra. (From Deer, Howie & Zussman, 1962, after Ewing,
1935b.)

the oxygens are involved in hydrogen bonds which are in paired rows
between the strips.

Goethite (α-FeO(OH)) has a similar structure to diaspore. Goethite is
very common and occurs as a weathering product of iron-bearing minerals.
As already discussed, it can be obtained artificially from ferric salts.
Goethite appears to be stable in water between 0 °C and 100 °C.

Other minerals with a diaspore-like structure are *groutite* (MnO(OH)),
montrosite ((Fe, V)O(OH)), and *para-montrosite* (VO$_2$).

Akaganéite (β-FeO(OH)) is uncommon in Nature, having only comparatively recently been identified as a mineral (Mackay, 1964). Its main
interest lies in its frequent formation from ferric chloride solutions as
already discussed. Its crystal structure (Mackay, 1960) is similar to diaspore
in that it contains chains of double octahedra. The chains are arranged
differently, however, leaving channels containing chloride ions and water
molecules (figure 6.8.).

δ-FeO(OH) is yet another polymorph of ferric oxide-hydroxide. It is
obtainable by rapid oxidation of ferrous hydroxide (Francombe & Rooksby,
1959).

The structures and interrelationships of various iron oxides and hydroxides have been discussed in detail by Bernal, Dasgupta & Mackay (1959)
and by Misawa, Hashimoto & Shimodaira (1974).

Metal oxides

Almost all metal oxides are densely bonded three-dimensional
crystalline polymers. Many of the common mineral oxides can be repre-

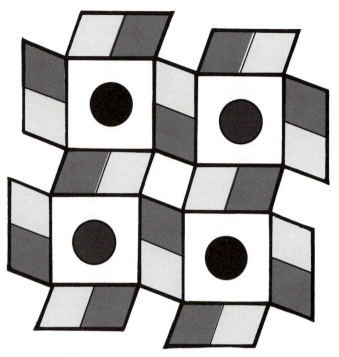

Figure 6.8. β-FeO(OH) has a similar structure to α-MnO$_2$ which is shown here. (Data from Bystrom & Bystrom, 1950.) Strings of paired octahedra (cf. figure 6.7) are represented as parallelograms. These form channels that hold chloride ions and water (large circles).

sented as a succession of close-packed layers of oxygen spheres with cations occupying some of the octahedral sites (but between *all* the layers, thus differing from true layer structures such as brucite). *Corundum* (α-Al$_2$O$_3$) and the similar *haematite* (α-Fe$_2$O$_3$) are like this. So are the complex and various families of *spinels* (e.g. spinel itself, MgAl$_2$O$_4$) except that here tetrahedral as well as octahedral sites are partly occupied by divalent and trivalent cations respectively. The arrangement is shown in figure 6.9. *Inverse spinels* have trivalent cations in the tetrahedral sites and an equal mix of divalent and trivalent ions in the octahedral positions. *Magnetite* (Fe^{2+}Fe$_2^{3+}$O$_4$) is an important member of this class.

One can formally imagine metal oxides as final oxolation products from the polymerisation and cross-linking of aquo cations. And, indeed, some of the above mineral metal oxides may sometimes form in this way, for example haematite and magnetite. And many metal oxides such as Cu$_2$O, CuO, Ag$_2$O, AgO, HgO, SnO, PbO and PbO$_2$ are obtainable as products

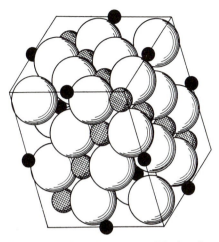

Figure 6.9. Spinel structure. Black spheres are in tetrahedral and hatched spheres in octahedral sites between the larger oxygen spheres. (From Verwey & Heilmann, 1947.)

of cation polymerisation from aqueous solutions in the laboratory (Feit-knecht & Schindler, 1963). Many such products are initially colloidal, metastable and 'amorphous' or 'cryptocrystalline' – which means that, from our point of view, they look interesting.

There is a great variety of manganese oxides, and many occur in soils and deep sea sediments, and are typical constituents of clays (Chukhrov, Gorshkov, Rudnitskaya, Beresovskaya & Sivtsov, 1980). Little is known about how they are formed in Nature. Schwertmann (1979) has discussed this question in a short review of the genesis of mineral oxides, oxy-hydroxides and hydroxides under ambient conditions.

Schwertmann considers also titanium oxides associated with clays – a topic of particular interest for our later discussions on photosynthesis. Among TiO_2 forms, *anatase* appears to be the main weathering product. The high-temperature polymorph *rutile* is derived mainly from meta-morphic rocks, although it does sometimes crystallise from solution (see also Force, 1980). The mixed oxide *pseudorutile* ($Fe_2Ti_3O_9$) occurs as a weathering product, and a whole range of mixed iron and titanium co-precipitates can be formed in the laboratory and matured under quite mild conditions (70 °C for 70 days at pH 5.5) to give iron-containing rutile and anatase phases (Fitzpatrick, le Roux & Schwertmann, 1978).

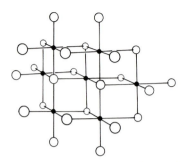

Figure 6.10. Ball-and-spoke model of the paramolybdate anion (open circles, oxygen).

Metal sulphides

Rickard (1969) discusses the formation of iron sulphides at low temperatures and pressures. All six of the known mineral forms can be synthesised via Fe^{2+} in aqueous solutions with sulphide, polysulphide, or sulphur. These six minerals are *pyrite* and *marcasite* (FeS_2); *pyrrhotite* ($Fe_{(1-x)}S$, where x lies between 0 and 0.25); *smythite* and *greigite* (Fe_3S_4); and *mackinawite* $FeS_{(1-x)}$. Pyrite and pyrrhotite are the stable phases, but metastable phases, especially mackinawite, are easily formed (Taylor, 1980). Iron sulphides, like many other sulphides, readily form colloidal suspensions.

In Chapter 8, we will take up the idea that sulphides might have taken part in electron-transfer processes in primitive forms of photosynthesis (Granick, 1957, 1965) and that perhaps colloidal iron sulphides were the ancestors of present day iron–sulphur redox proteins (Hall, Cammack & Rao, 1974).

Anion polymerisation

Whereas aquo cations tend to polymerise when the pH's of their solutions are raised, a number of oxyanions form polynuclear species in solution when the pH is lowered. Molybdenum and tungsten have a rich oligo-anion chemistry. MoO_4^{2-}, for example, gives, under a wide range of conditions, the paramolybdate anion $Mo_7O_{24}^{2-}$ with little sign of intermediate species. The structure of this ion is shown in figure 6.10. The paratungstate ion, $W_{12}O_{42}^{12-}$, has a similarly compact structure.

Germanium (IV) in aqueous solution can form oligo-oxyanions containing eight or so germanium atoms. On the other hand, tin (IV) is monomeric in alkaline solution but yields the very insoluble stannic oxide (SnO_2) in the neutral pH ranges (8.5–4.0) (Johnson & Kraus, 1959).

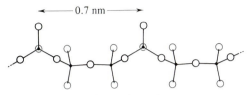

Figure 6.11. Conformation of the polyphosphate chain in Maddrell's salt. (From data in Thilo, 1962.)

Boron and aluminium are somewhat similarly related: borate oligomerises to a limited extent in aqueous solution while the aluminium oxyanion (probably $Al(OH)_4^-$) remains essentially as monomer in alkaline solution. Like tin (IV), aluminium precipitates insoluble products (this time hydroxides) in the middle pH range.

Phosphate

A number of salts contain infinite polyphosphate chains in the solid, for example Maddrell's salt II (figure 6.11). Others, for example Graham's salt, contain branched chains; still others cyclic structures (Van Wazer, 1958). Also phosphorus pentoxide exists in the solid either as the oligomer P_4O_{10} or as infinite three-dimensional polymer. These materials are formed under anhydrous conditions, however; none is thermodynamically stable in water. Here they all tend to depolymerise to orthophosphate. Branched phosphate chains are the most quickly degraded, but straight chains may, under suitable conditions, remain metastable for a considerable time in aqueous solution. (This property is made use of by all modern organisms in the triphosphate chain of ATP; and there are bacteria that store phosphorus as polyphosphate.)

Table 6.1 lists factors that influence the rate of depolymerisation of condensed phosphates. Figure 6.12 illustrates in particular the effect of pH on the hydrolysis of the linear trimer. Figure 6.13 gives an idea of the time scales involved in the degradation of high-polyphosphate polymers. Even after three weeks in this experiment, only about 50 % of the polymer was degraded as far as orthophosphate. That was at 60 °C, pH 5. With no catalysts, and at room temperature and neutral pH, the rate of hydrolysis of polyphosphate chains is usually to be measured in years.

This ability of polyphosphate to persist kinetically against the thermodynamic odds is quite exceptional for polyoxyanions (Van Wazer, 1958, p. 452); indeed, it is exceptional for aqueous solutions of inorganic polynuclear species generally. The polyphosphates are the nearest things to

Table 6.1. *Factors influencing the rate of hydrolysis of condensed phosphates (from Van Wazer, 1958, p. 453)*

Factor	Approximate effect on rate
Temperature	10^5–10^6 times faster from freezing to boiling
pH	10^3–10^4 times slower from strong acid to base
Enzymes	As much as 10^5–10^6 faster
Colloidal gels	As much as 10^4–10^5 faster
Complexing cations	Very many-fold faster in most cases
Concentration	Roughly proportional
Ionic environment in solution	Several-fold change

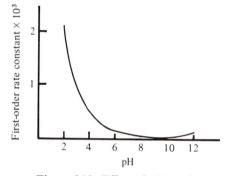

Figure 6.12. Effect of pH on the rate of hydrolysis of sodium triphosphate. (From Crowther & Westman, 1954.)

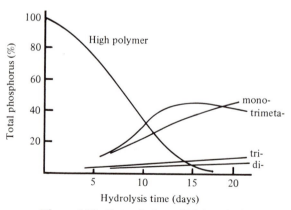

Figure 6.13. Degradation of high polyphosphate polymers at pH 5. (From Thilo & Wieker, 1957.)

organic biopolymers that we have yet come across in this chapter. Poly-arsenates do not share this property of indefinite metastability in water. (Nor do arsenate–phosphate copolymers – we have yet to come across a soluble inorganic polymer that might be able to hold much information.)

Silicate

Like the corresponding phosphates, sodium polysilicate $(Na_2SiO_3)_n$ is prepared under anhydrous conditions. It dissolves in water to give viscous concentrated alkaline solutions. On dilution, smaller oligomers are found, and below a concentration of about 1 mM the species present are essentially monomeric. On acidification, a solution of sodium silicate gives the hydrated oxide – as silica sol or silica gel. Since this process can be reversed, it would appear that such polymeric and oligomeric species as exist in solution are reversibly formed and degraded (Van Wazer, 1958, p. 449). Silicate is unlike phosphate both in precipitating an oxide from aqueous solution (one can hardly imagine P_2O_5 forming this way) and having a more dynamic population of polynuclear species in solution. Phosphate remains exceptional.

Again in a way that is becoming familiar, it is at the next structural levels that metastability appears. At room temperature, silica sols may remain metastable for a considerable time: often they develop into gels of $SiO_2 \cdot nH_2O$ that mature further losing some of the chemically bound water. But, at these temperatures, and on laboratory time scales, they never arrive at the thermodynamic ideal – quartz.

Although the extent of sodium silicate polymerisation increases with reducing pH, the rate has a maximum around pH 8–9 (figure 6.14). This is close to the pH (around 9.5) at which silicic acid and its mono anion are present in equal concentrations: it is therefore reasonable to postulate a bimolecular nucleophilic mechanism for the polymerisation (Gimblett, 1963). One possibility would be:

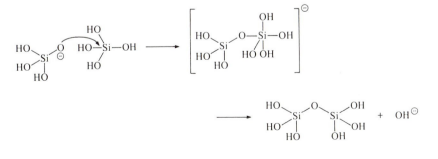

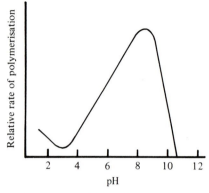

Figure 6.14. The influence of pH on the rate of silicate polymerisation. (From Gimblett, 1963.)

Unlike the polyphosphates, branched polysilicates are not specially disfavoured. This, and the tetravalence of the silicic acid unit, allows complex, densely cross-linked polymers. Presumably, the network becomes so rigid that reactions such as the one above can no longer take place because they are too sterically demanding. Thus unreactable Si–OH groups remain inside the structure.

Even at neutral pH and ordinary temperatures, silica is appreciably soluble in water. For example, Wey & Siffert (1962) found that different forms of amorphous silica in contact with water gave solutions of 120–140 parts per million (p.p.m.) of silica within a week or so. The silica was present in solution as monomeric silicic acid (Alexander, Heston & Iler, 1954). Silica is far more soluble at around neutral pH than most of the oxides, oxide-hydroxides and hydroxides that we discussed under cation polymerisation. This has geochemical implications that we will return to. Figure 6.15 shows how the solubility of silica and of alumina changes with pH.

We can perhaps understand why silicic acid monomer is the dominant solution species in quasi-equilibrium with silica. A mature silica particle will be highly cross-linked so that even surface silicic acid units will be multiply bonded and thus much more difficult to detach than they would be if singly bonded (one can think of analogies with the chelate effect or the relatively much greater stability of multiply base-paired oligonucleotides). Monomeric silicic acid units will nevertheless come away from time to time, but dimers or trimers would require greater numbers of simultaneous bond breakages to do likewise. Hence, at dynamic equilibrium, the

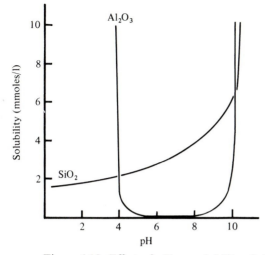

Figure 6.15. Effect of pH on solubility of silica and alumina at ordinary temperatures. (From Mason, 1966.)

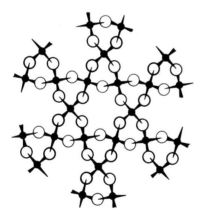

Figure 6.16. Plan of the structure of β-quartz. (From Wells, 1975.) Small black circles represent silicon atoms. The oxygen atoms lie at different heights above the plane of the paper, those nearest the reader being drawn with the heaviest lines. Each atom is repeated at a certain distance above (and below) the plane of the paper along the normal to that plane so that the Si_3O_3 rings in the plan represent helical chains.

supply of solution species will be almost exclusively of monomers. Some of these, it is true, will dimerise, trimerise and so on; and there will thus be a small concentration of small oligomeric species. But the bigger these become the more easily will they be mopped up by mature silica particles with which they can the more easily make multiple bonds. This argument would not apply to ideal linear polymers with no kind of cross-linking or folding since then, say, a 100-mer might be as likely to break into a 61-mer and a 39-mer as a 99-mer and a monomer. Here a smoother equilibrium distribution would be expected.

The crystalline forms of silica are considerably less soluble – about 27 p.p.m. for cristobalite (Fournier & Rowe, 1962) and about 6 p.p.m. for the more common stable form, quartz, at 25 °C (Morey, Fournier & Rowe, 1962). Each of these is a three-dimensional network of SiO_4 tetrahedra, linked through all the corners. In the quartz structure (figure 6.16) one can distinguish helices of linked tetrahedra: in a given crystal the helices of a given type are either right-handed or left-handed but not both. Crystals of quartz are thus chiral. An interesting two-dimensional crystalline polymer of silicic acid has been described by Kalt, Perati & Wey (1979).

Non-layer silicate minerals

Most of the minerals that make up the rocks in the Earth's crust are silicates. As a group they combine structural features that we have discussed under cationic and anionic polymerisation – although there are also many new features.

Silicates are generally classified according to the state of polymerisation of the formally anionic silicate component. This may extend in zero, one, two or three dimensions. We will consider all but the two-dimensional type in this section.

Orthosilicates

These are the simplest class from the above point of view: here the silicate units are unpolymerised. *Olivines* are examples. They have the general formula M_2SiO_4 where M is a divalent metal. For example, *forsterite* is Mg_2SiO_4, and *fayalite* Fe_2SiO_4. Between these there is a whole series $(Mg, Fe)_2 SiO_4$ with different proportions of Mg^{2+} and Fe^{2+}. Manganese, calcium and small amounts of many other metals are also found in olivines. The idealised structure is given in figure 6.17. The divalent ions occupy octahedral sites between silicate tetrahedra. Alternatively, the

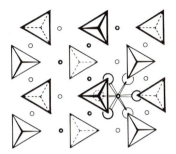

Figure 6.17. Structure of olivine as determined by Bragg & Brown (1926). SiO_4 groups are represented as tetrahedra. Here two sets are shown with their bases at levels 1/4 and 3/4. Mg^{2+}, Fe^{2+}, or other divalent cations are the small circles. Two sets of these ions are shown here at levels 0 and 1/2. The octahedral co-ordination of one of these ions is shown in more detail: here the large circles are oxygen atoms, and the dashed circle is an oxygen atom of a tetrahedron below the one shown.

structure may be seen as an infinite array of approximately hexagonally close packed oxygen spheres with silicon in one-eighth of the tetrahedral niches and metal in half of the octahedral niches.

Olivine is a major constituent of the volcanic basalt rocks. It is among the most easily weathered primary minerals. That the silicate units in it are not self-polymerised may have something to do with olivine's relative water solubility, but it is clearly not the whole story since other orthosilicates, for example the *garnets*, $M_3^{2+}, M_2^{3+}(SiO_4)_3$, are very stable to weathering. So is *zircon*, $ZrSiO_4$. In spite of the separation of the silicate tetrahedra these orthosilicates are really three-dimensional polymers if we remember their metal–oxygen bonding. We may note that the oxides of tri- and tetravalent metals are typically very much less soluble in water, much less inclined to depolymerise, than silica.

Chain silicates

The *pyroxenes* are the commonest iron–magnesium rock-forming silicates. Their structures contain linear siloxane chains with a two-repeat conformation (figure 6.18*a*). The chains have a truncated triangular cross-section (figure 6.18*b*) and are held together by cations between them (figure 6.18*c*). For example, in *diopside*, $CaMg[Si_2O_6]$ there are magnesium and calcium ions in six-fold and eight-fold co-ordination respectively.

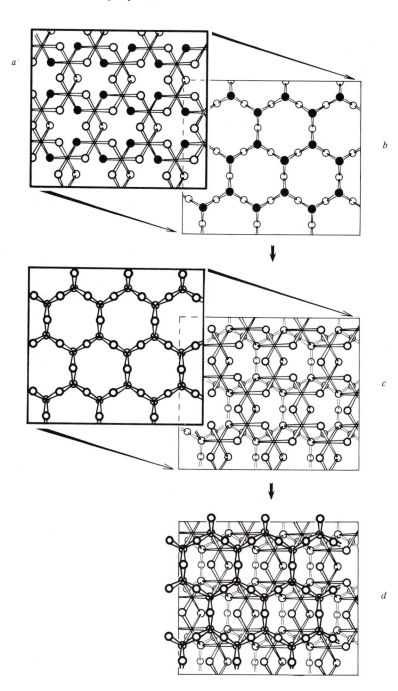

Figure 6.27.

and associated silicate minerals have been found in deposits of saline alkaline non-marine environments. He lists also 89 marine and fresh-water locations. Tuffs are common but not invariable host rocks. The youngest marine occurrence reported is in the gulf of Naples (Muller, 1961) in the top 10–15 cm of sediment thought to have been deposited over the last 500 years, and consisting of volcanic ash and its alteration products – which make up about 50 % of the material. Some of this consists of very fine-grained analcite, and (usually microscopic) authigenic quartz crystals, but the best crystallised of these products are not framework silicates at all but layer silicates, including kaolinite crystals of up to 1 mm long, and mica with short pseudo-hexagonal crystals as much as 1.5 mm maximum diameter. Which brings us to the next topic.

Layer silicates

Like chain and band silicates, the layer silicates are a fusion of two polymer types – here two-dimensional metal oxy-hydroxy and siloxane types. The mineral silicates that we have considered so far have been three-dimensional polymers with covalent bonds of one sort or another extending throughout the structure. In the layer silicates, the covalence is usually restricted to two dimensions. As a result, the layer silicates contain clearly defined, more or less separable two-dimensional macromolecules.

Two of the main structural types of silicate layers can be imagined as being built up, as shown in figure 6.27, by fusing an extended siloxane net either onto one side or onto each side of a gibbsite sheet. This would give the ideal (1 : 1) kaolinite and (2 : 1) pyrophyllite layer structures respectively. These are described as dioctahedral layer silicates because, as in gibbsite, only two of the three octahedral sites are occupied. (Side views, showing the stacking of kaolinite and pyrophyllite layers, are given in figure 6.28.)

Figure 6.27. Formal construction of silicate unit layers. Starting with a gibbsite sheet (*a*) remove two-thirds of the hydroxyls (shaded) from the lower side and plug in instead the apical oxygens of an extended amphibole-like siloxane network. Such a network is shown in (*b*) with the apical oxygens shaded. This would give the 1 : 1 dioctahedral (kaolinite) layer type (*c*). Performing a similar operation also on the upper surface of the gibbsite sheet would lead to the 2 : 1 dioctahedral (pyrophyllite) structure (*d*). Repeating these operations, but starting from brucite instead of gibbsite, would lead to the corresponding trioctahedral archetypes (serpentine and talc). In these models the balls are either oxygens or hydroxyls and cations are located at spoke intersections. The real structures are considerably distorted from these ideals.

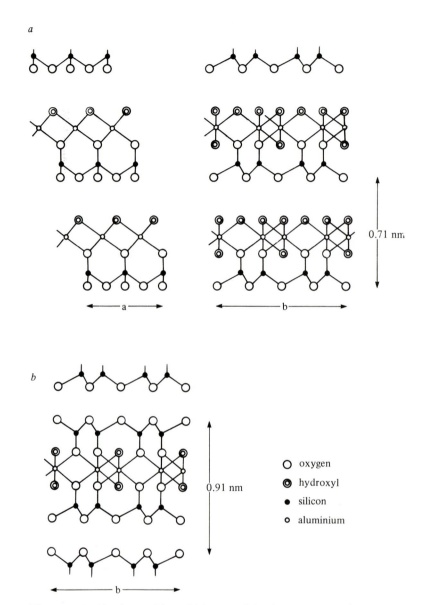

Figure 6.28. (*a*) The stacking of ideal kaolinite layers as seen along the *y*- and *x*-directions respectively. (After Brindley, 1951.) (*b*) The stacking of ideal pyrophyllite layers as seen along the *x*-direction.

Table 6.2. *A classification of layer silicates (see text)*

	DIOCTAHEDRAL		TRIOCTAHEDRAL	
1:1				
	A		B	
	Kaolinite clays:		*Serpentines:*	
	kaolinite		chrysotile	
	dickite		lizardite	
	halloysite		antigorite	
			amesite	
			greenalite	
			berthierine	
			cronstedtite	
2:1	C		D	
	pyrophyllite		talc	
	muscovite mica		phlogopite mica	
	margarite		biotite mica	
			E:	
			chlorite	
	F: Vermiculite, illite and smectite clays		vermiculite	
		most illites	some illites	
	Smectites:		*Smectites:*	
	montmorillonite		saponite	
	beidellite		hectorite	
	nontronite		sauconite	

The corresponding trioctahedral silicate layers, based on brucite, are those of serpentine and talc. These four major structural classes of layer silicates are:

> 1:1 dioctahedral
> 1:1 trioctahedral
> 2:1 dioctahedral
> 2:1 trioctahedral

Table 6.2 lists examples of these types. A typical layer silicate crystal consists of a stack of such units either directly in contact with each other (e.g. kaolinite itself or talc); or with intervening water molecules (halloysite); or with intervening cations (e.g. micas), or both (e.g. montmorillonite). Alternatively, a brucite-like layer may regularly interleave with 2:1 unit layers (chlorite). In addition, different kinds of silicate layers may

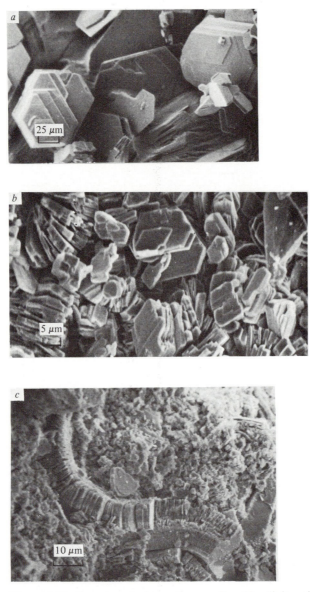

Figure 6.29. Scanning electron micrographs of kaolinite minerals. (From Keller & Hanson, 1975, and Keller, 1976, 1977.) (*a*) Dickite from Anglesey, Wales. (*b*) Kaolinite from Keokuk, Iowa. (*c*) S-shaped 'book' or 'vermiform' of kaolinite from Hodges mine, Georgia Kaolin Co. (*d*) Grooved vermiforms of kaolinite from Jacal, Mexico. (*e*) Kaolinite vermiforms and books showing complex grooving.

interleave, either regularly or irregularly. Even where the layers are identical, they may stack in different ways. And real layers are always distorted
out of the ideals of figure 6.27. Add to all this the possibilities inherent in
cation substitutions within the layers as well as between them – and the
possibilities, often, of organic molecules between the layers – and one has
an impression of a complex subject of study. This impression is correct.

The layer silicates in Table 6.2 are divided into the four main structural
classes as well as showing a cross-group (E) and a cross-group (F). We will
follow the sequence A, B, C (except F), D (except E and F), E, F.

A. *The kaolinite group of clays*, $Al_4[Si_4O_{10}](OH)_8$

Kaolinite itself is the major constituent of kaolin (china clay)
where it occurs as massive deposits in a limited number of localities. But
it is also widely dispersed in Nature and clays of the kaolinite group are
possibly the commonest of all clay minerals. Kaolinites form either hydrothermally or as a weathering product of other silicates, particularly feldspars. Typical examples are shown in figures 6.29a and b, illustrating well
made microcrystalline plates. 'Book' and 'vermiform' (worm-like) morphologies are often found in low-temperature products (Keller & Hanson,
1975) (see figures 6.29c, d, and e). These pictures vividly reflect a crystal
structure consisting of molecular layers stacked on top of each other. The
thinnest plates in figure 6.29 contain about 200 unit layers of the kind
shown in figures 6.27c and 6.28a. The unit layers are held together by
hydrogen bonding between hydroxyls on one side and siloxane oxygens on
the other.

The unit layers in kaolinite clays are considerably distorted from the
ideals of figures 6.27c and 6.28a. There are a number of reasons for this.
For one thing, the apical oxygen atoms in an unconstrained siloxane net
would be somewhat further apart than the corresponding array of oxygen
atoms in an ideal gibbsite sheet – that is to say, the two sets of shaded
atoms in figure 6.27a and b would not match up exactly for size. As a
result, the siloxane net is compressed. This can happen quite easily through
alternate contrary rotations of the tetrahedra (we will return to this idea).
But the array of connecting oxygen atoms on the gibbsite sheet would not
only be undersized, it would be distorted. The gibbsite structure shown is
an idealisation: in fact the aluminium-occupied sites are considerably
smaller than the vacant sites. As a result, some of the oxygen atoms are
actually too far apart. Pairs of tetrahedra connected to such pairs of
oxygen atoms are tilted towards each other. These effects can be seen
readily in the kaolinite mineral, *dickite* (figure 6.30).

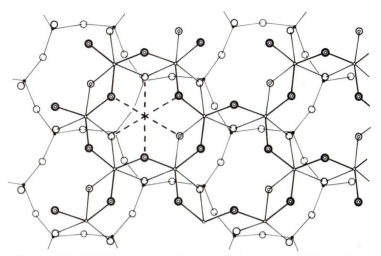

Figure 6.30. Dickite illustrates divergences from the ideal kaolinite layer shown in figure 6.27*c*. Note the distortions of the octahedral environments of the aluminium atoms. One of the (much larger and more symmetrical) octahedral vacancies is indicated by dashed lines. There are both rotations and tiltings of the SiO_4 tetrahedra. (After Newnham, 1961.) Oxygens and hydroxyls are the larger open and double circles; aluminium and silicon atoms are smaller open and black circles.

The distortions in dickite and in kaolinite itself are somewhat different, the tetrahedra being rotated by 7.3° and 11.3° respectively. Also, while in dickite the surface hydroxyls are about normal to the layer, in kaolinite they are more like the hydroxyls of gibbsite, that is, one in three lies almost in the layer plane (Giese & Datta, 1973; Rouxhet, Samudacheata, Jacobs & Anton, 1977). The hydroxyls countersunk in the other surface of the layer are tilted towards the vacant octahedral sites.

Kaolinite itself and dickite are described as polytypes; that is, they differ in the way in which the unit layers are stacked on top of each other: more precisely, it is a difference in the relationship between the positions of the octahedral vacancies between adjacent layers. The dioctahedral character of the kaolinite group produces, inevitably, directionality in the unit cell: the vacancy may be in any of the positions A, B or C:

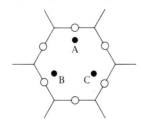

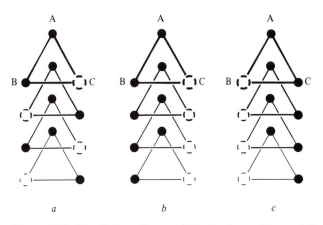

Figure 6.31. Kaolinite polytypes differ in the positions of the octahedral vacancies in successive layers. The vacancies alternate in dickite (*a*) whereas in kaolinite itself they are either all on the left (*b*) or all on the right (*c*) of the direction of the stacking off-set (see text). (Modified from Bailey, 1963.)

In kaolinite itself, the vacancies all point the same way; in dickite, they alternate. Although there are, in principle, very many stacking modes with different rotations and displacements of adjacent layers that would allow complete interlayer hydrogen bonding (Newnham, 1961), the stacking turns out to be the same in both kaolinite and dickite – successive layers being displaced along the *x*-axis by $-a/3$. By convention, they are said to be stacked in the direction of the A site (figure 6.31). This makes the three sites different – which raises new questions. Which site is vacant in kaolinite itself, and between which two do the vacancies alternate in dickite? For dickite at least there is a straight answer: between B and C (figure 6.31*a*). Of the three possibilities for kaolinite, two – all-B and all-C – are enantiomers, while the third – all-A – is achiral and would be expected to have a different energy (Newnham, 1961; Bailey, 1963; Wolfe & Giese, 1973). It is generally assumed that most samples of kaolinite consist of a mixture of enantiomeric crystals.

Unfortunately, it has not so far been possible to demonstrate the chirality or otherwise of single kaolinite crystals – crystals available are too small for present techniques (Wolfe & Giese, 1973). The investigation of a large crystal might still be inconclusive since an occasional changeover between B and C vacancy positions could make a large kaolinite crystal achiral although it contained chiral domains. 'Dickite stacking' has indeed been found in kaolinite crystals (Plançon & Tchoubar, 1977) and highly disordered kaolinites are common.

Ideally, polytypes would correspond to different ways of stacking other-wise identical layers. With kaolinite itself and dickite the situation is not as simple as this since the layers are distorted in different ways. Seemingly this difference is a cause as well as a consequence of the different stacking modes: Wolfe & Giese found that the energy was lower if a crystal that had started one way – as dickite, or as either enantiomer of kaolinite – con-tinued the same way, although the commonness of disordered structures indicates that this is not an overriding factor in many cases.

Halloysite is another mineral with a kaolinite layer. Here, the layers are less well ordered in relation to each other, less firmly stuck together: indeed there is very often a sheet of water molecules between the silicate layers. This is readily detected by X-ray diffraction through an increase in the interlayer spacing from the normal 0.7 nm to 1.0 nm. With the electron microscope halloysite is often seen to be tubular. Originally thought to be a simple consequence of the misfitting of siloxane and gibbsite sheets, the origin of the tubular morphology in halloysite no longer seems so clear since contrary rotations of siloxane tetrahedra provide such an easy way of relieving the misfit. It is clear, however, that unit layers can be flexible and hence are able to roll up.

Figure 6.32 (*a* and *b*) shows electron micrographs of a tubular halloysite from Wagon Wheel Gap, Colorado, investigated by Dixon & McKee (1974). The tubes are loosely swiss-rolled packets of unit layers stacked without water between them. Narrower tubes of circular cross-section may have about five layers per packet; the larger tubes with polyhedral cross-section have about 35. Spaces between the packets can be seen in figure 6.32*a*. The internal void diameters of nine tubes measured were between 7 nm and 38 nm.

These authors also studied a halloysite from Minnesota consisting of spheroids (figure 6.32*c*). These appear often to be hollow with an internal void of about 150 nm diameter and a patchy exterior of kaolinite plates.

Kirkman (1977, 1981) prefers the term 'multifaceted squat cylinder' to describe similar halloysites from New Zealand: the 'spheroids' appear to have a more or less definite axis. This became more obvious on treatment with alkali which produced objects resembling stuffed olives. (Indeed, figure 6.32*c* shows some spheroids like this.) Among the most striking objects in Kirkmam's papers are the spiral discs shown in figure 6.33. These are very flat (perhaps as little as 20 nm thick) and were found securely jammed between silica flakes where they had evidently grown (probably over a period of some thousands of years). Often several of the discs are intergrown (figure 6.33*b*).

Nagasawa & Miyazaki (1976) describe halloysites obtained from a

Figure 6.32. Transmission electron micrographs of tubular and spheroidal halloysite. (From Dixon & McKee, 1974.) (*a*) Tubular form from Wagon Wheel Gap. (*b*) Replicas of tubes. (*c*) Replicas of spheroidal halloysite from Minnesota showing flat kaolinite plates (P).

variety of Japanese sites: some having arisen from hydrothermal alteration of volcanic glass in pumice beds, others by weathering of feldspar in granitic rocks, still others by alteration of clayey matrix of sand by circulating ground waters. Various morphologies – tubes, laths, balls, scrolls, and plates with rolled edges – are among the forms described and illustrated. All of these had interlayer water. Whatever the kind of location, the water

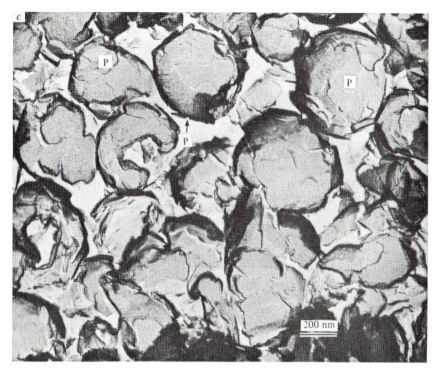

Figure 6.32 (*c*).

was more easily removed from the older samples, suggesting a tendency to transform towards kaolinite itself with age. In any case, kaolinite itself was commoner as an associated mineral in older locations.

Parham (1969) simulated tropical weathering of potassium feldspar with water (at about 78 °C) in a soxhlet extractor for 140 days. Surface replicas from the feldspar were then examined under the electron microscope. Already in this time, flame-like membranes were seen to be attached to the feldspar. These were taken to be 'primitive halloysite' on account of their morphological identity to features found in naturally weathered feldspars where halloysite is the product. These membranous projections often occurred in lines suggestive of underlying crystallographic features such as dislocations. In some cases, tubes appeared to be growing out of pits in the feldspar (see also Kirkman, 1981). Similar morphological features were found by Parham on feldspar crystals that had been exposed in a road cut in the warm rainy climate of Hong Kong for less than five years.

From these results and from a survey of localities of recent weathering in which halloysite has been identified, Parham concludes that halloysite

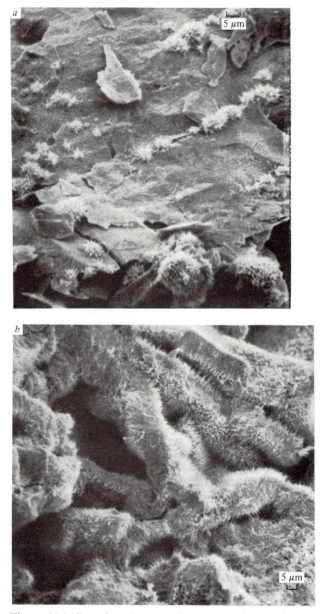

Figure 6.34. Scanning electron micrographs of clay minerals of the kaolinite group growing from solution. (From Keller, 1977.) (*a*) Clay beginning to form along microjoints in Sparta granite. (*b*) Feldspar fragments almost completely coated by elongated clay crystallites. (*c*) Closer view of (*b*). (*d*) Long clay crystals clumping like wet grass.

Figure 6.34.

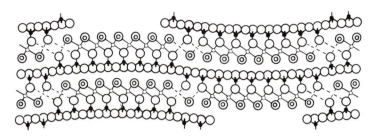

Figure 6.35. Structure of antigorite viewed along the *y*-axis. The curl imposed on the layers by an oversized octahedral sheet is accommodated by periodic inversions so that on a larger scale the sheets are more or less flat. (After Kunze, 1956.) Oxygens and hydroxyls are larger open and double circles; magnesium and silicon atoms are smaller open and black circles.

also for the existence of fibrous antigorites which are elongated in the *y*-direction. (Chrysotile, like halloysite, prefers to be rolled up along the *x*-axis.)

The whiteness of white asbestos underlines another similarity between it and the kaolinite group: there is seldom much substitution of cations in tetrahedral or octahedral positions – this in spite of the ultrabasic rocks from which serpentines are derived usually having plenty of coloured transition metal ions in them.

Exotic serpentines can be synthesised hydrothermally (at *ca* 500 °C) in the laboratory. Germanium may replace silicon for example. Interestingly, this variety is never tubular, the larger germanium atoms relieving the intersheet misfit. Non-tubular serpentine can arise also where magnesium in the octahedral sheets has been suitably partly substituted with (the smaller) aluminium (Roy & Roy, 1954).

Tubular chrysotiles have been made in which nickel, cobalt, or iron partly or completely replace magnesium. (See Jasmund, Sylla & Freund, 1976.) Serpentines with different amounts of nickel in them occur in Nature as a series between end members (Brindley & Maksimovic, 1974).

Amesite is a highly substituted 1:1 trioctahedral mineral with an ideal formula of $(Mg_4Al_2)[Si_2Al_2O_{10}](OH)_8$. The octahedral aluminium atoms generate positive charges that are balanced by the negative charges generated by the tetrahedral aluminium atoms. These opposite charges will tend to come near each other in the crystal structure, both within and between layers, while at the same time like charges keep their distance. There are a number of ways in which this might happen. In an amesite sample from Antarctica, Hall & Bailey (1979) found that aluminium-rich tetrahedral

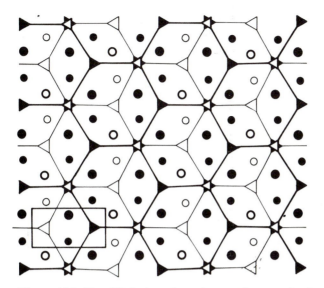

Figure 6.36. Simplified view of two layers of an amesite from Antarctica. The two alumino–siloxane nets are represented by heavy and light lines connecting triangles which are the tetrahedral positions (white, aluminium-rich and black, silicon-rich). The magnesium-rich octahedral sites are black circles with those at the lower level slightly smaller. The octahedral aluminium-rich sites at the two levels are heavy and light open circles. The rows of aluminium atoms here run in one direction. If we call this 'six o'clock' then there are five other possibilities – at 'eight o'clock', 'ten o'clock', 'twelve o'clock', etc. (From data in Hall & Bailey, 1979.) The rectangle in the lower left-hand corner indicates a class of octahedral and tetrahedral aluminium-rich sites in an amesite from the Urals described in the text.

and octahedral sites of adjacent layers were arranged in slightly zig-zag rows in one of six directions as shown in figure 6.36. These crystals were often hexagonal prisms made up of six twin sectors. It seems clear that these sectors correspond to the six different ways of arranging the rows of aluminium atoms. (We will be discussing twinning of crystals in some detail in the next chapter.)

Another kind of interlayer ordering of aluminium substitution is described by Anderson & Bailey (1981) who studied an untwinned portion of an amesite crystal from the Urals. The general structure was the same as in figure 6.36, except that here the aluminium-rich sites were arranged in left-handed helical columns parallel to the z-axis. In projection normal to the layer plane, the aluminium atoms lie at the corners of a diamond. The rectangle in the lower left corner of figure 6.36 shows four sites which, for the Urals amesite, would be occupied by aluminium – as would the other identically placed sites. But again, as with the amesite from Antarctica, there are alternative geometrically equivalent arrangements – this time there are three similar ordering patterns. (There are two other similar positions – at $\pm 120°$ – for rectangles like the one in figure 6.36.)

Anderson & Bailey (1981) see this sort of arrangement as consistent with the much more complex twinning typical of the Urals amesite as compared with the Antarctic sample: there would be less lateral distortion to the crystal structure, and it would thus be easier for there to be a good lateral fit between domains that had these more subtly different arrangements.

Deer, Howie & Zussman (1962) give typical analyses for three iron-bearing 1:1 trioctahedral layer silicates:

> *Greenalite:* $Fe^{2+}_{4.5}$, $Fe^{3+}_{1.0}[Si_4O_{10}](OH)_8$
>
> *Berthierine:* (previously '7 Å chamosite'):
> $Fe^{2+}_{3.6}Al_{1.6}(Mg, Fe^{3+}$ etc.$)_{0.8}[Si_{2.6}Al_{1.4}O_{10}](OH)_8$
>
> *Cronstedtite:* $(Fe^{2+}, Fe^{3+}, Al)_{4.6}[Fe^{3+}_{2.0}, Si_{2.0}, O_{10}](OH)_8$

Greenalite appears to be something like a ferrous analogue of chrysotile or lizardite, although in a study of synthetic hydrothermal products containing different proportions of magnesium and iron, Jasmund, Sylla & Freund (1976) found that (orthorhombic platey) greenalite was distinct from (monoclinic tubular) chrysotile in the sense that there was no single series of intermediate structures between them. Iron was unlike nickel or cobalt in this respect. Berthierine can be seen as somewhat like an amesite with Fe^{2+} in place of Mg^{2+}: whereas in cronstedtite, iron (Fe^{3+}) stands in for tetrahedral Al^{3+} as well.

Greenalite, cronstedtite and particularly berthierine, have characteristics that justify their being considered in a class somewhat apart from the 'classical' serpentines, chrysotile, lizardite and antigorite. Chemically, they are similar to chlorites; physically their very small particle size pushes them towards the category of clays. And they are more like clays too in the way in which they appear to be formed in Nature – often as low-temperature weathering products.

That at least is the conclusion arrived at by Harder (1978) from experiments on low-temperature synthesis (20 °C and 3 °C) of iron layer silicate minerals in which, among other things, greenalite-like and berthierine-like products were formed. Harder has found indeed that iron-bearing layer silicate minerals form rapidly from hydroxide precipitates and dilute silica solutions if the conditions are right. The following are favourable factors: (1) Reducing conditions. (Fe^{2+} must be present; reducing agents, dithionite or hydrazine were used.) (2) High pH, generally 8–9. (3) Presence of Mg^{2+}. (4) Low SiO_2 concentrations (10–20 p.p.m.). (5) A stoichiometric proportion of adsorbed silica in the hydroxide precipitates. In such circumstances, iron layer silicates became detectable by X-ray diffraction in a matter of days.

The formulae of these iron-bearing minerals are enough to indicate a considerable complexity – the partly dioctahedral character of greenalite, for example, and the necessarily strongly negatively charged tetrahedral sheets in berthierine and cronstedtite. Further insight into berthierine's structure has been obtained by Mossbauer spectroscopic studies by Yershova, Nikitina, Perfil'ev & Babeshkin (1976). They found iron in octahedral sites of two different kinds with different degrees of distortion from cubic symmetry: 65 % of the iron was in the less distorted set of sites. This iron was more readily oxidised on heating.

C. Dioctahedral micas

Muscovite, $KAl_2[Si_3AlO_{10}](OH, F)_2$, rivals feldspar and quartz in ubiquity, if not in abundance: it can form and survive under a wide range of conditions and is found in igneous and metamorphic rocks, and in sediments either as detrital particles or having crystallised *in situ*. We will discuss the structure of mica, particularly muscovite, in some detail; other 2:1 layer silicates can then be described largely through alternative or progressive moves from this 'archetype'.

Let us recall first that the kaolinite unit layer is derived formally by fusing a siloxane net onto one side of a gibbsite sheet (figure 6.27c). Many of the characteristic features of the kaolinite group of clays arise more or less directly from the asymmetry of this arrangement and through the possibility of extensive interlayer hydrogen bonding. By having a siloxane net on both sides of a gibbsite sheet, as in pyrophyllite (figure 6.27d) these effects are eliminated – the resulting material has (more or less) flat layers only weakly held together by van der Waals forces. Pyrophyllite has indeed a somewhat slippery feel reminiscent of graphite. It is not really typical of 2:1 layer silicates however. Usually, there are quite strong interlayer forces of a kind different from the direct hydrogen bonding interactions in the kaolinite group. Muscovite shows these (electrovalent) forces at work in a relatively straightforward way.

In muscovite, as in many micas, about one in four of the silicon atoms of

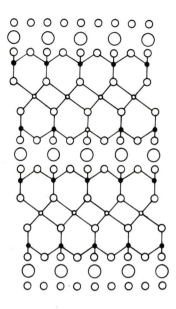

Figure 6.37. Side view of mica along the *y*-axis. Large, middle and small open circles are respectively potassium ions, oxygens, and and octahedral cations. Black circles are silicon *or* aluminium atoms in tetrahedral positions.

the siloxane nets are substituted by aluminium. As in the framework silicates each substitution creates one negative charge. The muscovite layers are thus strongly anionic. They also have arrays of indentations in the six-ring siloxane nets that cover their surfaces. As it happens, the radius of the potassium ion is such that it can nest fairly well between two such indentations in layers above it and below it. The muscovite crystal thus consists of 2:1 unit layers held together by potassium ions between them. The extent of substitution of aluminium for silicon is such that, with the interlayer sites fully occupied, the negative charge in the layers is exactly balanced. An idealised side view of mica layers is given in figure 6.37: in muscovite, two of the three octahedral positions are occupied by Al^{3+}.

As a result of the keying effect of the potassium ions, the upper siloxane net of one layer lies almost exactly opposite the lower net of the layer above it – there is no *inter*layer displacement in the ideal structure and little in real structures. There is, inevitably, an *intra*layer displacement however: it is not possible to put siloxane nets exactly opposite each other on each side of a gibbsite sheet. The intralayer displacement (of $a/3$) can be seen in figures 6.37 and 6.38.

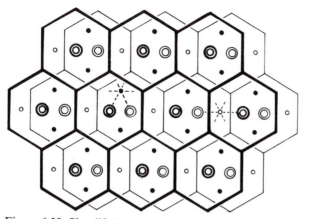

Figure 6.38. Simplified top view of a mica layer showing displacement, by $a/3$, of the upper from the lower alumino–siloxane nets (to the left in this picture – but there are six possibilities). Double circles are hydroxyls. The central plane of octahedral sites is represented by small circles of which there are two kinds, designated M(1) and M(2) (open and black circles respectively). (After Smith & Yoder, 1956.)

Polytypism in micas This intralayer displacement immediately provides possibilities for polytypism. The unit layer of a mica can be said to have an arrow in it pointing in the direction of the displacement of the upper siloxane net in relation to the lower net (see particularly figure 6.38). The question arises as to how the arrows in a stack of layers are related to each other. If the keying requirements of the potassium ions was the only factor to be optimised there would seem to be an immense number of possibilities for a multilayer crystal. In the stacking of ideal layers each can be placed equally well in one of six orientations, obtained by rotations of multiples of 60°. A crystallite with, say, 1000 layers in it would be one of 6^{999} (i.e. about 10^{777}) possibilities – all of equal energy. In the event, most muscovite crystals are found to be stacked in only one of a few ways – most often in only one way.

Whatever the explanation for this regularity of stacking, it is reasonable to look for rules. Smith & Yoder (1956) assumed that the angle between the 'arrows' would most likely remain the same in successive layers. On the further assumption that these angles are either *always* in the same sense or *always* alternate, Smith & Yoder showed that there would be six preferred stacking modes for a crystal such as muscovite, however many layers it might contain; that is, there are six particularly simple polytypes. Figure 6.39 displays these possibilities. It has turned out that regularly stacked 2:1 layer silicates very often fall within this set.

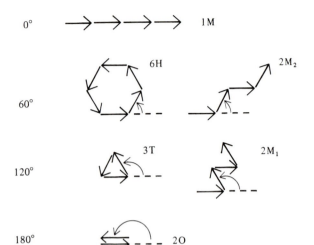

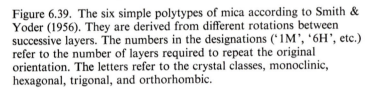

Figure 6.39. The six simple polytypes of mica according to Smith & Yoder (1956). They are derived from different rotations between successive layers. The numbers in the designations ('1M', '6H', etc.) refer to the number of layers required to repeat the original orientation. The letters refer to the crystal classes, monoclinic, hexagonal, trigonal, and orthorhombic.

The commonest polytype for muscovite is $2M_1$ (see figure 6.39) and this is the stable form (Velde, 1965). 3T and 1M are also found as are more randomly stacked crystals (Ross, Takeda & Wones, 1966). The 1M polytype may be kinetically favoured by crystal growth mechanisms: it appears first during laboratory synthesis of muscovites (Yoder & Eugster, 1955). Bailey (1966) in reviewing this topic concludes that, generally, stable polytypes of layer silicates occur nearly to the exclusion of others in natural environments in which sufficient thermal energy or pressure had been available. Metastable polytypes may persist otherwise. Thus when muscovites are found in sediments, crystals of the 1M polytype are more likely to be authigenic (formed *in situ*) and $2M_1$ crystals detrital (carried from elsewhere).

The mutual displacement of the siloxane nets within a single layer of micas generates two kinds of octahedral sites between them: those in the direction of the displacement and those on either side. This is illustrated in figure 6.38. The sites are designated M(1) and M(2). That they are different can perhaps be most easily seen by inspecting the arrangement of the hydroxyl groups in relation to them. The single M(1) site has two '*trans*' OH groups; the pair of M(2) sites have their OH groups '*cis*'. It is in the

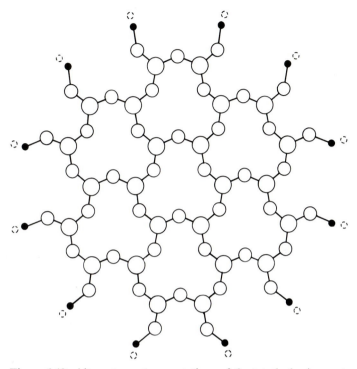

Figure 6.40. Alternate contrary rotations of the tetrahedra in a net such as that shown in figure 6.27*b*. Oxygens are shown as open circles with the nearer apical ones larger. Silicon and aluminium atoms are concealed except round the outside. The extent to which a tetrahedral sheet is contracted is indicated by the shift in the positions of the outermost (exposed) silicon and aluminium atoms.

latter that the octahedral aluminium atoms are located in muscovite. The vacant M(1) site is larger than the other two.

As in the kaolinite group, the siloxane sheets in muscovite are contracted through symmetrical contrary rotations of alternate tetrahedra. This has the effect of lowering the symmetry of the surface oxygens from hexagonal to ditrigonal (see figure 6.40) and it changes the co-ordination of the potassium ions from twelve-fold to six-fold – three of the oxygen atoms in each of the surface indentations will come closer and three will move away from the potassium ions. Furthermore, instead of six orientations for adjacent layers giving identical environments for the potassium ions, there will now be two sets of three. In the 1M polytype, for example, the potassium ion finds itself with triads of oxygen atoms above and below staggered in relation to each other, that is, octahedrally arranged. On rotating the

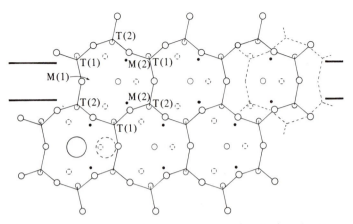

Figure 6.41. Muscovite $2M_1$ drawn from a unit cell given by Güven (1971). The alumino–siloxane net on the underside is omitted except at the top right of the picture where it illustrates a direction of intralayer off-set from right to left. Only one upper and lower potassium ion is shown (lower left). The tetrahedral rotations are such as to bring the bridging oxygens closer to the aluminium atoms. The tetrahedra are also tilted to create shallow grooves that lie horizontally in this picture. (The groove direction is indicated by the solid bars.) The tilting arises from the vacant octahedral sites (M(1)) being bigger than the occupied sites (M(2)). The two classes of tetrahedral sites are designated T(1) and T(2).

layers by 60°, the oxygen triads are now opposite each other at the corners of a prism. A further 60° rotation regenerates an octahedral environment, and so on. The octahedral arrangement is presumably the most stable – hence, presumably, the preference is muscovite for 1M, $2M_1$ or 3T polytypes (see figure 6.39).

As remarked earlier, the contrary rotations of the tetrahedra in a siloxane net can never lead to a perfect match between its apical oxygen atoms and a gibbsite sheet, since the latter is insufficiently symmetrical. In muscovite, as in dickite, the tetrahedra are tilted. Pairs of oxygen atoms on the same side of the gibbsite sheet that straddle vacant M(1) sites are further apart than the others and two out of the six Si–O–Si groups in each siloxane ring are actually stretched. As a result, the bridging oxygens are depressed by some 0.02 nm below the other basal oxygens so that the surfaces of muscovite's unit layers are corrugated. The corrugations run in the direction of the intralayer offset between the siloxane nets. The structure of a $2M_1$ muscovite layer including these features is shown in figure 6.41.

Figure 6.42 shows how this grooving must disturb the local symmetry of the K[+] sites. Instead of layer rotation generating three equivalent arrange-

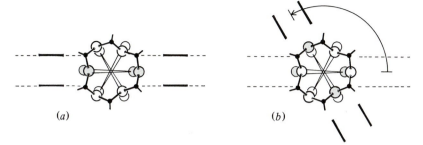

Figure 6.42. (*a*) The arrangement of oxygens around interlayer K^+ in 1M muscovite. Tetrahedral rotations cause six of the twelve oxygens to be nearer the K^+. This octahedral environment is not quite symmetrical, however, because the tetrahedra are also tilted so that two of the six co-ordinating oxygens are slightly further from the K^+ (that is, those in the grooves; see figure 6.41). These oxygens (shaded) are orientated '*trans*'. (*b*) By rotating the top layer by 120° an octahedral arrangement of oxygens is recreated, but with the more distant oxygens '*cis*'. This is the arrangement for the $2M_1$ or 3T polytypes.

ments, there are now two distinct possibilities according to whether the 'special' ligand oxygens, i.e. those in the grooves, are '*trans*' (0°) or '*cis*' (120°). Indeed, there are two possibilities anyway if one looks beyond the basal oxygens: the OH groups and especially the octahedral Al^{3+} ions are differently related between layers rotated by 0° on the one hand or $\pm 120°$ on the other (cf. figure 6.41).

It would have been more immediately understandable, perhaps, if the stable polytype of muscovite had been 1M rather than $2M_1$ because in 1M muscovite the grooves of successive layers are parallel. In the event, we are left with two questions: why is 120° rotation favoured and why do the rotations alternate + and − ?

It would appear that the answer to the first question lies in parallel grooves being energetically *dis*favoured. Parallel grooves and ridges cannot nest into each other (because the potassium ions prevent the necessary offset): rather the ridges clash all along their lengths (Güven, 1971). A cross-ridged arrangement, although still not perfect, is a better compromise in reducing the total energy of interaction between adjacent negatively charged surfaces. (As a model for this, one might imagine two ridged air beds on top of each other: they would stack slightly closer if the ridges were angled rather than directly opposed.) Without this effect, the 1M type might indeed have been favoured – as it generally is for the ungrooved trioctahedral micas.

We are left with the perplexing question of selectivity between + and −

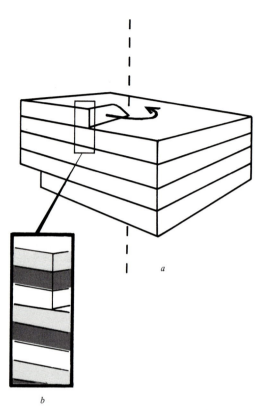

Figure 6.43. (*a*) A screw dislocation (cf. figure 7.1*d*) creates a
continuous ramp spiralling around a central line of maximum
disturbance – the screw dislocation line (dashed). Crystal growth at a
screw dislocation may be particularly easy because new units can add
to an advancing step (arrow) which is never eliminated. Hence the
difficult process of surface nucleation (cf. figure 5.13 on page 159)
is avoided once growth through a screw dislocation is under way.
(*b*) A step caused by a screw dislocation may have several layers in
it. If in addition there are different possible orientations for the
layers (polytypism) then a complex repeating pattern may be
generated.

rotations which seem to be structurally equivalent. To alternate, layer n
must know which way layer $n-2$ is lying. How can it feel that through
layer $n-1$? (These layers are a nanometer apart – too far probably for
direct electrostatic interactions to be effective.)

A multilayer spiral growth mechanism can explain the existence of
polytypes with complicated long-range repeats, as shown in figure 6.43.
Smith & Yoder (1956) suggested that such effects could account for polytypism in micas. (This might give a neat answer to our question: layer n

knows about layer $n-2$ because layer n *is* layer $n-2$ – reappearing at the next turn of the growth ramp.) One has to assume only that once started each layer retains its orientation. No transfer of information vertically between layers is needed. This cannot, however, be a complete explanation where a polytype is a thermodynamically favoured form. Whatever its mode of birth, a $2M_1$ crystal of muscovite is a special lowest energy arrangement. If you were to rotate layer n in the middle of such a crystal, the energy would rise and only return to the lowest energy state at 360°. Layer n still knows which way it should be. There must be a vertical transfer of information across a distance of at least 1 nm. We are left with the question: 'how does layer n know which way layer $n-2$ is lying?'

A mechanical model Megaw introduced an idea into the discussion of silicate structures that may be useful here. She thought of the structure of the feldspar anorthite in the sort of way that an engineer thinks about a framework of girders – in terms of stresses and strains which may be transmitted over considerable distances (Megaw, Kempster & Radoslovich, 1962). How and how far an effect can be transmitted like this will depend on the compliance of the 'building elements'. (Some are more compliant than others: for example, bond angles at oxygen are more compliant than angles at silicon or aluminium.)

Look again at the siloxane network in figure 6.40. Suppose that you were forcibly to rotate one of the tetrahedra. Then the whole set would embark on a dance of alternate contrary rotations. Such an effect would diminish with distance according to the compliance of the bonds in the structure.

More generally, any kind of stress applied to a silicate layer will produce strains spreading sideways to some extent within its nicely articulated structure. Such a transmission of effect can be imagined, for example, within a set of adjoining octahedral sites:

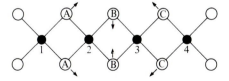

Suppose that a large cation is put into site 1 in place of a smaller cation. As indicated by the arrows, this would push the A-oxygens outward and tend to compress the other oxygens attached to the cation in site 2. Hence the B-oxygens would tend to move inward a little. The C-oxygens around the cation in site 3 would now have more space to move away from each

other – outwards. As a result of this site 4 would become better adapted
for a larger cation. All this because there was a larger cation three sites
away.

Vertical effects of this sort can be imagined that might account for the
transmission of information *through* silicate layers. For example, the disposition of the surface oxygens of a siloxane net would be directly affected
by the repeated forces exerted on these oxygens by the adjacent layer. But
the disposition of the apical oxygens that join the siloxane net to the octahedral sheet would be affected too. For example, if one siloxane net in a
2:1 layer was stressed as in our imaginary experiment on figure 6.40 so as
to counter-rotate the tetrahedra still further from the 'ideal' arrangement,
this would compress *one* of the hexagonal arrays of apical oxygens. The
octahedral cations would then be pushed closer together, and hence the
other array of apical oxygens would be compressed too. The other surface
oxygens would have to accommodate – they would tend to adjust through
contrary rotations in a manner similar to the directly stressed oxygens on
the opposite surface of the layer.

Radoslovich (1963) was thinking in 'mechanical' terms when he said of
potassium ions in muscovite that they occupy 'an equilibrium position
determined by a complex balanced system of interlocking strong bonds
reaching right through the adjacent layers to K's at the next level above
and below'.

Allostery There is a good analogy between this kind of transmission of effects in articulated silicate structures and **allosteric** effects in proteins. Haemoglobin provides the classical example of allostery. This molecule has four oxygen binding sites some 3–4 nm away from each other. The
binding constant of any site increases substantially as the other sites are
occupied. How do the sites know about each other's state of occupancy?
The answer seems to be through concerted pushings and pullings of an
articulated intervening structure – somewhat like a system of 'levers'
(Perutz, 1978). More precisely, pairs of identical subunits are related
through an axis of symmetry which tends to be preserved in spite of the
distortion produced on oxygen binding: thus the imposed distortion in one
site is copied in the others (rather as opening or closing one end of a pair
of lazy-tongs correspondingly opens or closes the other end).

A symmetrical structure was an essential part of allostery as originally
formulated (Monod, Wyman & Changeaux, 1965). Soon, it was realised
(Koshland, Némethy & Filmer, 1966) that symmetry is not essential for
long-range transmission of information through folded protein molecules:
a distortion in one site may produce a different or even an opposite effect

on a similar site elsewhere. It all depends on the detailed characteristics of the intervening articulated network.

For really complex networks – like proteins or layer silicates – one would expect a given structure to produce diverse distant effects according to detailed structural considerations.

Other examples of 'allosteric' effects in layer silicates We have already come across what seems to be an 'allosteric' effect, in the polytypism of the kaolinite group – where incidentally the layers are not symmetrical. Here, the layers are apparently distorted by the way they stack in such a way as to favour further stacking as before.

Takeda & Ross (1975) found that 1M and $2M_1$ polytypes of the trioctahedral mica biotite with very similar chemical constitutions had differently distorted layers, and they proposed that once a stacking sequence starts it tends to persist through a similar train of cause and effect.

What appears to be an example of 'negative allostery' in a silicate is provided by the altered micas discussed by Norrish (1973). Here, binding in one place inhibits a similar binding elsewhere. Na^+ may sometimes replace K^+ between *alternate* layers: it seems that a plane of sodium ions alters the orientations of the hydroxyl groups within the layers above and below so as to make it more difficult for potassium ions to be removed from the opposite sides of the layers above and below. Hence the regular alternation – with the K^+ planes 'knowing' that there are Na^+ planes next door.

Chemical effects Much of our discussion so far may give an over-simplified impression of the processes through which layer silicate crystals are formed. Almost certainly they do not form by simply stacking premade unit layers: these two-dimensional macromolecules are probably synthesised at the same time as they crystallise (at least the ones that we have been talking about so far). This lets in another possible mechanism for control of layer stacking – through chemical effects; in particular, through effects arising from different arrangements of cation substitutions. The stacking of a given layer, on the layer below it, may predispose it to a particular arrangement of cations which in turn predisposes the next layer. In that case, transmission of information might be partly chemical and partly physical.

A 3T muscovite whose structure was determined by Guven & Burnham (1967) may be a case in point. Its gross composition was:

$$K_{0.90}Na_{0.06}Ca_{0.01}Ba_{0.01}(Al_{1.83}Fe^{2+}_{0.04}Fe^{3+}_{0.04}Mg_{0.09}Ti_{0.01})$$
$$[Si_{3.11}Al_{0.89}O_{10}](OH_{1.98}F_{0.03})$$

This is very similar to a typical $2M_1$ muscovite analysed by Radoslovich (1960):

$$K_{0.94}Na_{0.06}(Al_{1.84}Fe^{3+}_{0.12}Mg_{0.06})[Si_{3.11}Al_{0.89}O_{10}](OH)_2$$

Yet the layers of the 3T polytype are substantially different from the $2M_1$ type, not only physically (having shallower grooves and a different space group), but chemically. For one thing, the aluminium substitutions were ordered into only one of the two classes of tetrahedral sites – unlike other muscovites where the substitutions are equally distributed. Also, the minor octahedral constituents were concentrated in one of the two aluminium sites. Such a chemical asymmetry might well help to transmit the one-handed style of the 3T polytype stacking. In any case, the chemical asymmetry within the layers is ordered (repeated) between the layers, presumably concomitantly with the crystal growth processes that determined the 3T stacking mode.

Ordering *within* layers is another important topic. As already discussed, dioctahedral layer silicates invariably show long-range ordering of octahedral vacancies. Also, as we have just seen, more elaborate octahedral ordering is possible as is the ordering of aluminium in tetrahedral sites. It is surprising that long-range tetrahedral ordering is not more common in 2:1 silicates since the effect of replacing a silicon atom with an aluminium atom must be particularly powerful: Al is bigger than Si and, most important, it introduces a negative charge. For example, for potassium ions to be locally neutralised, exactly three of the atoms in the twelve nearest tetrahedral sites above and below a potassium ion should be aluminium. That consideration alone makes a random distribution of tetrahedral substitutions most unlikely. Bailey (1975) has pointed out that seeming disorder in tetrahedral substitutions may be a result of assumptions made in crystal structure determinations. Tetrahedral ordering in domains was suggested by Gatineau (1964) from diffuse X-ray scattering evidence. This idea involves zig-zag chains of aluminium substitutions arrayed in one of three different directions in different domains. Most interestingly, in Gatineau's model, aluminium-rich domains face similar aluminium-vacant domains across the K^+ planes. Gatineau's specific suggestion has been viewed with caution by Bailey (1975) on the grounds that it violates a general principle that aluminium atoms avoid being beside each other as far as possible in silicate structures, and also because since 1964 other kinds of domaining have been found that could produce diffuse scattering without Si, Al ordering. This matter remains unsettled, but short-range ordering of some sort seems likely.

In the brittle micas, there is about twice as much aluminium substitution as in the normal micas, to balance out the calcium ions that mainly fill the interlayer spaces. Margarite is the closest type to muscovite: in this case there is nearly complete alternate ordering of the aluminium substitutions in a specimen of composition:

$$Ca_{0.81}Na_{0.19}K_{0.01}(Al_{1.99}Mg_{0.03}Fe_{0.01})[Si_{2.11}Al_{1.89}O_{10}](OH)_2$$

In addition, there was a slight difference between the two tetrahedral sheets of any layer, one containing somewhat more aluminium than the other (Guggenheim & Bailey, 1975).

D. *Trioctahedral micas*

The simplest trioctahedral 2:1 layer silicate is *talc* with the ideal formula $Mg_3[Si_4O_{10}](OH)_2$. The unit layer is a brucite sheet with siloxane nets fused on each side. Talc, like its dioctahedral counterpart pyrophyllite, is atypical of the group as a whole having only weak, van der Waals, interlayer bonding.

Phlogopite is the better archetype for trioctahedral micas corresponding to muscovite for the dioctahedral class. Its ideal formula is $K(Mg_3)$ $[Si_3AlO_{10}](OH, F)_2$, but this is seldom approached in natural samples where Fe^{2+}, Fe^{3+}, Al^{3+} and Ti^{4+} in particular may occupy a substantial proportion of octahedral positions. Also, Na^+ and, to a smaller extent, Ca^{2+}, are found standing in for K^+ in interlayer positions. While muscovite tends to be short on tetrahedral aluminium, in phlogopite micas there is usually more aluminium than the ideal.

Biotite is a very common mica of igneous and metamorphic rocks. It is somewhat arbitrarily distinguished from phlogopite in having a Fe:Mg ratio of more than 1:2. Figure 6.44 illustrates the wide range of compositions of these two micas in relation to four extreme structures. Even this underestimates the possibilities, though, as the formula of a particular biotite illustrates:

$$K_{0.9}Na_{0.07}Co_{0.01}(Fe^{2+}_{1.42}Mg_{0.85}Mn_{0.01}Al_{0.41}Ti_{0.13}Fe^{3+}_{0.09})$$
$$[Si_{2.8}Al_{1.2}O_{10}](OH_{1.7}F_{0.02})$$

This biotite, chosen from 32 examples given in Deer, Howie & Zussman (1962), is from a garnet mica schist from Morar in Scotland (Lambert, 1959). It is not untypical in being short on (OH, F) and having 0.1 octahedral positions out of three vacant.

The hydroxyl groups in micas are well placed to be spectroscopic probes

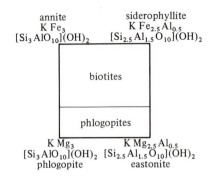

Figure 6.44. The phlogopite–biotite composition fields. Most phlogopites and biotites lie within these fields. The division between them is arbitrarily chosen to be where $Mg:Fe = 2:1$. (From Deer, Howie & Zussman, 1962.)

of octahedral cation distribution. The OH stretching frequency is sensitive to the cations (usually three in trioctahedral micas) to which OH is co-ordinated. A glance at the formula of the Morar biotite is enough to realise the large number of possible sets of nearest neighbours that any one OH group might have. The OH stretching bands are understandably complex (not helped by the influence, also, of interlayer cations). Vedder (1964) distinguished nevertheless three general classes of OH groups – 'Normal', 'Impurity' and 'Vacancy':

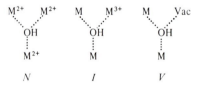

which gave distinctive composite bands. More detailed examination of the bands can give clues as to which ions are involved (Vedder, 1964; Wilkins, 1967; Farmer *et al.*, 1971; Rousseaux, Gomez Laverde, Nathan & Rouxhet, 1973; Rausell-Colom, Sanz, Fernandez & Serratosa, 1979). An adjacent vacancy has the effect of altering the orientation of the OH groups from approximately normal to the layer plane to approximately in the plane – the hydrogen atom then pointing towards the vacancy as in dioctahedral silicates. Using polarised infrared radiation, Chaussidon (1973) was thus able to distinguish OH groups lying '*trans*' from those lying '*cis*' across vacancies. Both types were present in comparable amounts indicating both M(1) and M(2) vacancies. This is in contrast to dioctahedral micas where the vacancies are ordered in the M(1) site.

In all, one has an impression of considerable freedom in the arrangement of octahedral components in the trioctahedral micas. A review by Hazen & Wones (1972) of synthetic products confirms also a considerable freedom in choice of kinds of ions – provided they are small enough and divalent they will pass as octahedral guests. Even small trivalent ions, the tetravalent Ti^{4+}, and the monovalent Li^+, are tolerated. Indeed, divalent ions may sometimes be quite absent, being replaced by a balance of Li^+ and Al^{3+} (in the *lepidolite* micas).

The variety of possible ways of arranging the octahedral cations in trioctahedral micas may be very great, yet clearly there are restraints over and above the requirements of small ionic size and overall charge balance. The number of vacancies, for example, seldom rises much above 0.3 for every three sites. If these vacancies are not so obviously ordered as in the dioctahedral micas, the cations are ordered in many cases, with larger cations generally preferring $M(1)$ sites (Bailey, 1966, 1975). This goes to emphasise that a layer silicate structure only partly resembles a rigid array of pigeon holes within which ions can be distributed. A cation will control the size of its pigeon hole and that will affect the preferred size of adjacent pigeon holes. Lateral 'allosteric' as well as electrostatic effects can be expected. Veitch & Radoslovich (1963) have discussed the theoretical grounds for octahedral ordering on the basis of both size and charge. The 'allosteric' effect in particular helps us to understand the fairly clear distinction that exists in layer silicates between dioctahedral and trioctahedral types: the structural restraints induced by one or other of these forms of organisation seem to be self-perpetuating. Dioctahedral domains within a trioctahedral structure, or vice versa, might be electrostatically balanced but are likely to be sterically strained and hence disfavoured unless some other compensating factor is at work.

Although trioctahedral sheets are bigger than dioctahedral sheets, they are still usually smaller (in 2:1 silicates) than the tetrahedral sheets – because the latter are also enlarged by substantial substitutions of aluminium for silicon. As a result, tetrahedra are alternately rotated as usual; but since $M(1)$ and $M(2)$ sites are more similar in size, there is little corrugation of the surfaces. Hence, 1M polytypes are favoured for trioctahedral micas.

E. *Chlorite and vermiculite*

Chlorites occur in igneous, metamorphic and sedimentary rocks – in the latter both as detrital and authigenic particles. There is a wide range of sizes for the crystals: they may be found as large blocks or as clay-sized particles.

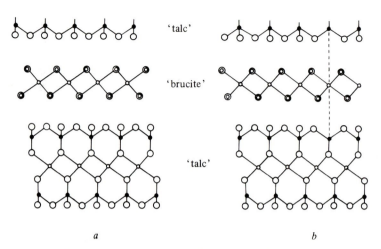

a   *b*

Figure 6.45. (*a*) and (*b*) A typical chlorite crystal is a stack of alternating talc-like and brucite-like units. Two of the numerous stacking modes are shown. In (*b*) one of the sites in the 'brucite' interlayer lies exactly between tetrahedral sites in the 'talc' layers above and below. (After McMurchy, 1934, and Bailey, 1975.) Larger open circles, oxygens; double circles, hydroxyls; small open circles, octahedral cations; and small black circles, tetrahedral cations.

An ideal chlorite is generally taken to consist of talc and brucite layers regularly interleaved (figure 6.45). Thus, a 2:1 structure is combined with the kind of interlayer hydrogen bonding typical of the kaolinites and serpentines. Real chlorites, however, are far from this ideal, the 2:1 unit being more like biotite than talc in the variety and extent of its cation substitutions. Aluminium for silicon substitutions occur within the range $(Si_{3.5}Al_{0.5})$–(Si_2Al_2). The negative charges thus created are compensated by octahedral substitutions – generally R^{3+} for R^{2+} – which may be either exclusively in the brucite layers or also in the talc layers. Al^{3+} is usually the main source of these positive charges although Fe^{3+} commonly contributes a substantial fraction. Iron is indeed an important component of most chlorites – mainly as Fe^{2+} standing in for Mg^{2+}. A typical formula would be:

$$(Mg, Al, Fe)_6[(Si, Al)_4O_{10}](OH)_8$$

with minor amounts of other ions such as Mn^{2+}, Cr^{2+}, Ni^{2+} and Ti^{4+} in octahedral positions.

As might be imagined, the greater complexity of chlorites as compared with micas multiplies the possibilities for polytypism. Brown & Bailey

(1962) have derived twelve possible one-layer polytypes and Lister & Bailey (1967) no less than 1134 two-layer polytypes on the basis of a classification that distinguishes different ways in which a brucite layer can be placed on a talc layer with maximum hydrogen bonding, different ways of superposing these, and different directions of intralayer offset in the talc subunits. Most chlorites have a semi-random stacking due to arbitrary displacements ($\pm b/3$) between adjacent chlorite units, but the relationship between 'talc' and 'brucite' subunits tends to remain fixed.

Steinfink (1958a) determined the crystal structure of a monoclinic chlorite of composition:

$$(Mg_{2.6}Fe_{1.7}Al_{1.2})[Si_{2.2}Al_{1.8}O_{10}](OH)_8$$

From differences in oxygen–tetrahedral cation distances, it was clear that the tetrahedral aluminium was concentrated in one of the two kinds of site. The distribution of octahedral cations was also far from arbitrary. Judging from bond lengths and electron densities, of the three octahedral 'talc' sites the first contained magnesium, the second magnesium + iron (about 3:1) and the third iron. The octahedral aluminium was in two of the three interlayer sites which contained also about 25 % iron. The third of these 'brucite' sites contained magnesium with about 25 % vacancies. A triclinic chlorite also studied by Steinfink (1958b) showed no ordering in the 'talc' part of the structure, but again aluminium was concentrated in the interlayer, with signs of ordering there.

Brown & Bailey (1963) determined the crystal structure of a chromium chlorite from Erzincan, Turkey, which had the following composition:

$$(Mg_{5.0}Fe^{2+}_{0.1}Cr^{3+}_{0.7}Al_{0.2})[Si_{3.0}Al_{1.0}O_{10}](OH)_8$$

In this particular polytype, the trivalent cations in the 'brucite' layers come opposite the set of tetrahedral sites of the 'talc' layers within which the aluminium ions are concentrated. Thus the centres of excess positive charge in the 'brucite' layers lie exactly between the centres of excess negative charge in the 'talc' layers above and below. The arrangement is illustrated in figure 6.45b.

Most chlorites are trioctahedral both in the 2:1 and interlayer components. Donbassite is the name given to doubly dioctahedral chlorites, that is, those based on pyrophyllite and gibbsite rather than talc and brucite. The vacancy in the gibbsite interlayer is partly occupied to make the required balance of charges. In a Ia donbassite:

$$(Al_{4.1}Fe^{3+}_{0.04}Fe^{2+}_{0.01}Mg_{0.08}Li)_{0.26}[Si_{3.14}Al_{0.86}O_{10}](OH)_8$$

Aleksandrova, Drits & Sokolova (1972) found both tetrahedral ordering of aluminium and an ordering between gibbsite and pyrophyllite components analogous to that in the Erzincan chlorite. Here, however, it is the least charged, that is, partly vacant site of the gibbsite interlayer that lies over the aluminium-rich tetrahedra.

Cookeite is a still more curious chlorite with a dioctahedral 2:1 layer and a trioctahedral interlayer. Bailey (1975) reports on a preliminary study of a Ia cookeite of approximate composition:

$$(Al_{4.02}Li_{0.86})[Si_{3.02}Al_{0.97}O_{10}](OH)_8$$

The lithium is ordered in one of the positions in the interlayer and the aluminium is apparently in only one of the tetrahedral sheets of the 2:1 layer.

It is a feature of the micas that their layers are firmly locked together. The chlorites are similar in this respect, and they are similar too in having a high level of tetrahedral substitution which locates negative charges close to the 2:1 layer surfaces. It is the interlayer junction that is quite different in the two cases, a simple plane of cations in the micas being replaced by a hydroxide sheet. As a result, the characteristic distance between 2:1 layers changes from about 1 nm in the micas to about 1.4 nm in chlorites.

Going from chlorites to *vermiculites*, we move two steps further from the micas. The negative layer charge is still dominated by aluminium for silicon substitutions, but the charge is generally somewhat less than that in either micas or chlorites due to excess positive charge in the central octahedral sheets of the 2:1 layers. These are still about 1.4 nm apart, but the junction between them is significantly different from the arrangement in chlorite. The hydroxide interlayer is replaced by two (incomplete) planes of water molecules with divalent cations, usually Mg^{2+}, octahedrally hydrated between them. The arrangement of octahedral sites is similar to that in brucite, but they are much less heavily populated: here (unlike brucite) each magnesium ion contributes two net positive charges to the interlayer – and occupation of between one in six to one in nine of the sites is generally enough to compensate the 2:1 layer charges – and there is no need for higher valence interlayer cations, although these are sometimes present. An analysis of vermiculite from Llano, Texas (Foster, 1963) illustrates how the charges may be balanced:

$Mg_{0.48}K_{0.01}$	$(Mg_{2.83}Fe^{3+}_{0.01}Al_{0.15})$	$[Si_{2.86}Al_{1.14}O_{10}]$	$(OH)_2$
interlayer	2:1 octahedral	tetrahedral	

Charges: $+0.97$ $+0.14$ -1.14

Mathieson & Walker (1954) had suggested that the interlayer cations in vermiculite would, as a result of mutual repulsion, tend to be distributed within a hexagonal network of sites. Shirozu & Bailey (1966) defined the situation more precisely in a crystal structure determination of a vermiculite from the Llano location. Here 0.41 magnesium atoms are concentrated in one of the three available octahedral interlayer sites. There is also an ordering of aluminium between aluminium-rich (T1) and aluminium-poor (T2) tetrahedra. The magnesium-rich interlayer sites lie exactly between the aluminium-rich tetrahedral site in a manner analogous to the Erzincan Cr-chlorite (see figure 6.45b). The (double) charges on the magnesium ions can thus be locally balanced.

The partial occupancy of the interlayer sites in vermiculite still leaves open the possibility of super-ordering. Where, as is common, only a third of the possible sites are occupied by magnesium, Bradley & Serratosa (1960) suggested that these might be arrayed in a regular manner over three unit cells – that is, as far away from each other as possible. An arrangement like this has been found by Alcover, Gatineau & Mering (1973) in a vermiculite from Kenya. Here, interlayer magnesium is present in domains. In each domain only one of three sets of sites is occupied, but different sets can be occupied in different domains. This ordering is two-dimensional, there being no relation between successive interlayers. These authors suggest that there may be a corresponding ordering of the aluminium tetrahedral substitutions to balance the charges locally, and Bailey (1975) has pointed to the possibility that the Llano vermiculite might combine a similar short-range domaining of aluminium substitutions within tetrahedral sheets with the long-range intersheet ordering implied by Shirozu & Bailey's study (1966).

Much, if not most, vermiculite in Nature is formed by alteration of trioctahedral micas. The main part of this process would appear to be simply the replacement of interlayer potassium by magnesium and water – a reaction that can indeed be simulated in the laboratory by leaching biotite or phlogopite with MgCl (Barshad, 1948; Caillère & Hénin, 1949). The Llano vermiculite referred to above was evidently a product of weathering of a phlogopite that was found unaltered just below the surface. There is a puzzle here, however, since the ordered arrangement of aluminium substitutions deduced as being in the vermiculite could not have been there in the supposed parent phlogopite (Shirozu & Bailey, 1966).

F. *2:1 clays*

The *Illites* are an abundant if not very well defined group of clays that are structurally quite close to the micas – generally to muscovite. Being clays, their crystallites are small. They are also less well crystallised than muscovite, a somewhat lower layer charge contributing, perhaps, to a less accurate stacking of layers. In addition to having less aluminium for silicon substitution there is some replacement of interlayer potassium by, for example, Ca^{2+}, Mg^{2+} and H^+. There is, however, little interlayer water. An idealised formula is:

$$K_{0.5-0.75}Al_2[Si_{3.5-3.25}Al_{0.5-0.75}O_{10}](OH)_2$$

Glauconite found in marine sediments is an iron-bearing illite in which Fe^{3+}, Fe^{2+} and Mg^{2+} occupy a substantial proportion of the octahedral sites.

The *smectites* are again related to the micas but now more remotely. The interlayer cations are here mainly Ca^{2+} and Na^+, rather than K^+, and they are readily exchangeable. The crystallites are smaller than those of the illites and usually have a very irregular outline. The layer charges are low enough to allow the most characteristic property of the group – that the unit layers readily intercalate water molecules. Smectites are the main class of *expanding clays*. Calcium smectites generally have two sheets of interlayer water but the sodium smectites often pull in many more. Indeed the unit layers of smectites can come more or less completely apart in aqueous suspension to give, in effect, a solution of two-dimensional macromolecules. These are quite flexible. Often the appearance under the electron microscope is less like a conventional crystal than a piece of crumpled linen (figure 6.46). Even when the layers of a smectite crystal are apparently stacked neatly on top of each other there may be little regularity in the interlayer relationships. Regular crystalline organisation may be largely confined to two dimensions – and even this may be quite short-range.

Often, too, a clay crystallite consists of layers of different kinds. Illite and montmorillonite, for example, are frequently interstratified – indeed kaolinite layers may be included too. Such interstratification may be either regular or irregular.

In spite of the potential and actual structural variability of smectites, it has been possible to obtain from certain localities clays of a more or less consistent type. *Montmorillonite* is one of these. It can be given the following idealised formula:

$$(\tfrac{1}{2}Ca, Na)_{0.33}(Al_{1.66}Mg_{0.33})[Si_4O_{10}](OH)_2 \cdot nH_2O$$

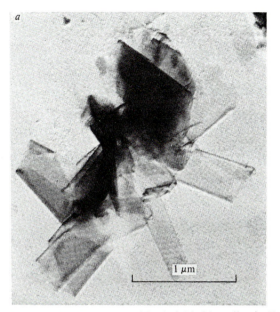

Figure 6.46. Aggregate of back-folded lamellae in Tatatilla
montmorillonite after dispersion. (From Grim & Guven, 1978.)

The ease of expansion of montmorillonite is sometimes attributed to the
layer charge arising mainly in the octahedral sheet – some distance from
the compensating interlayer cations.

In *beidellite* however, the layer charges arise mainly from tetrahedral
substitutions:

$$(\tfrac{1}{2}Ca, Na)_{0.33}(Al_2)[Si_{3.66}Al_{0.33}](OH)_2 \cdot nH_2O$$

Nontronite is similar except that Fe^{3+} replaces octahedral Al^{3+} in the
idealised formula.

Saponite is the trioctahedral equivalent of beidellite with a unit layer
thus similar to that of vermiculite:

$$(\tfrac{1}{2}Ca, Na)_{0.33}(Mg_3)[Si_{3.66}Al_{0.33}](OH)_2 \cdot nH_2O$$

There may be little if any distinction between a high-charge saponite and
a low-charge vermiculite (Suquet, Iiyama, Kodama & Pezerat, 1977).

Completing the set, *hectorite* is the trioctahedral equivalent of montmorillonite, the octahedral charge here being provided by substitution of
some Li^+ for octahedral Mg^{2+}:

$$(\tfrac{1}{2}Ca, Na)_{0.33}(Mg_{2.66}Li_{0.33})[Si_4O_{10}](OH, F)_2 \cdot nH_2O$$

A *bentonite* is a mixture of (mainly) smectite clays – in particular
montmorillonite and beidellite.

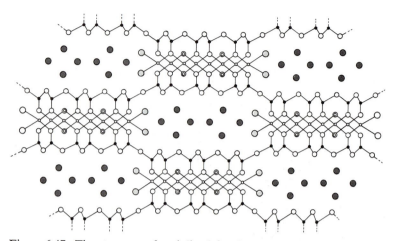

Figure 6.47. The structure of sepiolite (after Brauner & Preisinger, 1956) showing channels containing both mobile or 'zeolitic' water (dark shading) as well as more strongly fixed water (lighter shading). Small open circles, magnesium atoms; small black circles, silicon atoms.

Other clay minerals

Sepiolite has a crystal structure consisting of narrow talc-like strips joined at their corners, as in figure 6.47, so that channels are formed. The crystals are typically fibrous as shown in figure 6.48*a*. Note the groove that is about 10 nm wide running along the length of the larger of the crystals shown here. Figure 6.48*b* is a view across such a sepiolite crystal at higher magnification. The channels of the ideal crystal structure are clearly visible here – these are about 0.7×1.3 nm. There are also tubular pores of between 2 nm and 20 nm across.

Palygorskite is somewhat similar, but with narrower strips and a more complex pattern of octahedral cations.

Although structurally close to the layer silicates, these two are really framework minerals, similar to zeolites. A discussion of their surface properties has been given by Serratosa (1979). Sepiolite and palygorskite tend to form when magnesium-rich waters evaporate.

Imogolite and *allophane* are hydrous alumino–silicates. Imogolite can be seen under the electron microscope to consist of long thin tubes with an outer diameter of about 2 nm and an inner pore of about 1 nm – like the finest imaginable capillary tubing (figure 6.49). Cradwick *et al.* (1972) have proposed a structure that is like kaolinite in so far as silica units are fused on one side of a gibbsite sheet, but the units are inverted: here three of the

Figure 6.48. Electron micrographs of sepiolite (from Rautureau & Tchoubar, 1976). (*a*) Grooved sepiolite laths. (*b*) Cross-section at high magnification showing channels in the crystal structure (cf. figure 6.47) as well as larger holes running through the crystal.

oxygen atoms of each SiO_4 unit are shared with the gibbsite sheet leaving one as a hydroxyl group. There is no Si–O–Si bonding – the SiO_4 units are separate as in an orthosilicate (figure 6.50*a*). This kind of bonding imposes a powerful distortion on the gibbsite sheet which thus curls into a tube with ten to twelve –Si–OH units on the inside circumference (figure 6.50*b*).

Allophane particles are also visible in figure 6.49. Henmi & Wada (1976)

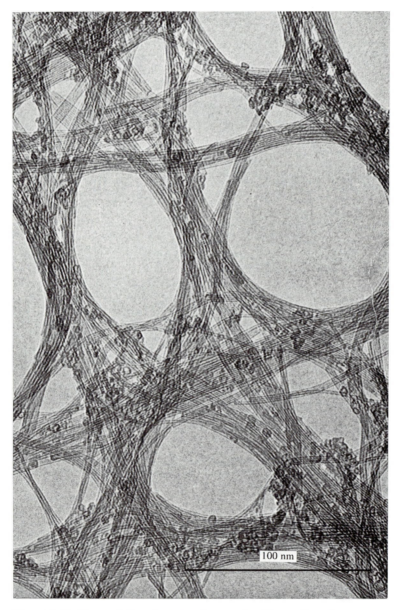

Figure 6.49. Transmission electron micrograph of imogolite tubes and allophane pods from Kurayoshi. (From Nakai & Yoshinaga, 1978.)

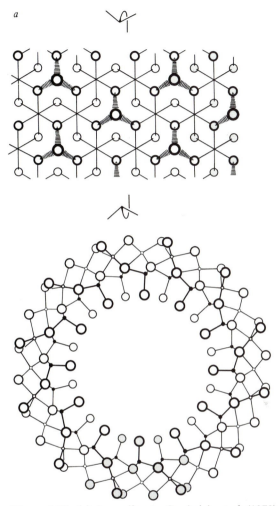

Figure 6.50. (*a*) According to Cradwick *et al.* (1972) imogolite consists of a gibbsite sheet with separate silicate units fused to one side (these are shown with 'spring' bonds). The flat structure shown would be strained – the 'springs' would be under tension – because silicate units are not big enough. So instead the structure rolls up as indicated by the arrows. (*b*) A view down an imogolite tube showing the 1 nm diameter pore. This view is as from the right-hand side in (*a*). Corresponding atoms visible in both (*a*) and (*b*) are shaded.

have proposed that such particles are hollow spheres with an outer diameter of 3.5–5.0 nm and a wall thickness of about 1 nm or less. The walls seem to have pores in them that can let water molecules through (Wada & Wada, 1977; Wada, 1979), and are sites for phosphate binding (Parfitt & Henmi, 1980).

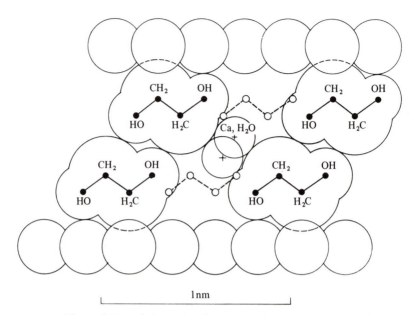

Figure 6.51. Ethylene glycol molecules between the layers of a calcium montmorillonite. (After Reynolds, 1965.)

Both imogolite and allophane can apparently form quite quickly under suitable conditions. A 'proto-imogolite' can be made by mixing dilute solutions of aluminium salts and monomeric silicic acid at pH 4.5; on heating (40–100 °C) this gave tubes very similar to natural imogolite (Farmer, Fraser & Tait, 1977; Farmer & Fraser, 1979). Allophane precipitates from the CO_2-charged waters of 'Silica Springs' in New Zealand (Wells, Childs & Downes, 1977). These minerals are particularly important constituents of soils derived from the weathering of volcanic ash, but their occurrence is widespread – they can form from plagioclase feldspar (Tazaki, 1979a). As they are difficult to detect by X-ray methods, both imogolite and allophane may have been underestimated as soil components.

Parfitt & Henmi (1980) conclude that the allophane membrane can often be regarded as a defective sort of imogolite; but they distinguish such 'proto-imogolite allophane' with an Al:Si ratio of *ca* 2 from other forms, such as the Silica Springs material which has an infrared spectrum more like felspathoid.

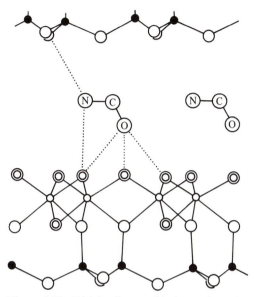

Figure 6.52. Dickite–formamide intercalate viewed along the *x*-axis. Dotted lines represent hydrogen bonds between the nitrogens and carbonyl oxygens of the formamide molecules, and the hydroxyls below and siloxane oxygens above. (After Adams, 1978.)

Clay–organic interactions

Smectites, particularly montmorillonite, have long been known to take up organic molecules of various kinds. Often this is between the silicate layers and in that case changes in the interlayer spacing can easily be detected by X-ray diffraction. For example, ethylene glycol can replace most of the interlayer water molecules in calcium montmorillonite to form highly organised complexes such as that shown in figure 6.51. Many other kinds of polar organic molecule can form such organo–clay complexes – for example, alcohols, ketones, ethers and amines may come in, instead of water, between the layers of smectites and vermiculites. To be adsorbed from dilute aqueous solutions polar molecules generally have to be fairly big ($> C_5$) if they are to compete well enough with water (Hoffmann & Brindley, 1960).

Organic cations may replace inorganic interlayer ions in many 2:1 layer silicates. *N*-Alkyl ammonium ions and ionised heterocyclic compounds such as pyridinium are particularly effective.

Small highly polar molecules are favoured types for intercalating kaolinite layers – for example, formamide intercalates in a regular way in dickite (figure 6.52). Urea, hydrazine, and dimethyl sulphoxide are other examples.

SILICATE LAYER

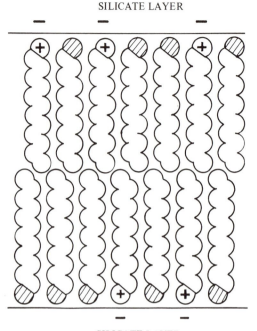

SILICATE LAYER

Figure 6.53. Packing of (positively charged) long-chain alkyl ammonium ions together with corresponding long-chain alcohols between mica-like silicate layers. (After Weiss, 1963.)

Rather high concentrations are needed, however, and the interlayer species are often rather easily removed in contact with water.

This whole topic of clay–organic interactions has been thoroughly reviewed in a book by Theng (1974). Here, particular attention is paid to molecules of current biological importance: not only 'biochemicals' but pesticides, antibiotics, alkaloids and other classes, illustrating the great variety of organic molecules that will bind to clays of some sort in some way or another. Kaolin has been used medically for a long time as a means of taking up various toxic materials – although 2:1 clays are generally the most effective in forming organo–clay complexes.

Generally, too, it is basic organic molecules that are most readily taken up – in the neutral to acid pH range. Nucleotides and their heterocyclic components have been studied particularly (Lailach, Thompson & Brindley, 1968a, b; Lailach & Brindley, 1969; Thompson & Brindley, 1969). Most clays do not readily adsorb the major biomonomers at the kind of pH (7.1 $\pm$ 1) thought to have existed in the primitive ocean, but the binding of nucleotides is enhanced in the presence of transition metal cations (Odom, Rao, Lawless & Oró, 1979). Porphyrins, too, are more readily bound along with such metal ions (Cady & Pinnavaia, 1978).

Fatty acids may be removed from aqueous solutions by clays (Meyers & Quinn, 1971), but complexes between fatty acids and montmorillonite are not formed very easily. Edges rather than interlayers are likely to be the adsorption sites (Sieskind & Ourisson, 1971). Triglycerides, too, tend to give rather indefinite complexes with 2:1 layer silicates unless these already have organic interlayer material in the form of suitable long-chain alkyl ammonium ions: then triglycerides may be taken up between the layers in a regular way (Weiss & Roloff, 1966). A similar effect had been noted by Weiss (1963) with the co-adsorption of long chain alcohols and corresponding alkyl ammonium ions (figure 6.53).

Simple sugars such as glucose do not bind well to clays. Apparently they are too highly hydrated; methylated sugars complex more easily (Greenland, 1956).

Not only layer silicates, but many other inorganic materials with layer crystals can intercalate organic molecules – for example niobates, titanates and molybdates (Lagaly, 1981).

The genesis of clay minerals

Conjurors make use of our tendency to take mental short cuts. The egg broken in the teacup cannot be the same egg as the perfect one found later in the lady's handbag; but the clever conjuror persuades us that it is. Your perception (if not your intellect) can be persuaded of the most preposterous notions on the basis of a principle that normally works very well in everyday life. We cannot be for ever wondering if what seems to be the same object really is: if the green book on the table this morning is the same green book that was left there last night, or if the train emerging from one side of a distant tunnel is the same train that we saw entering the other side a minute before. Such suppositions are common sense but they are not logically secure – and it is on this distinction that conjurors make their living.

Nature cannot be credited with the conjuror's guile, but it sometimes seems to play tricks on us by producing the same kinds of molecules, or the same or similar crystal structures in different circumstances; and we may be persuaded to draw causal lines between them like the spurious causal line between the egg in the teacup and the egg in the handbag. One of these tricks we have already discussed at length: amino acids and purines are found (for example) in meteorites. They are also found in us. Therefore, (short cut thinking tells us) our amino acids and purines came in the first place from meteorites (or alternatively from thunderstorms or whatever). The synthesis of clays in Nature provides another interesting example of an often false set of causal connections that have been quite hard to disentangle and that have likewise their origins in thermodynamics – here in the tendency for certain types of crystal structure to reappear under different circumstances.

Consider a piece of granite. It contains feldspars, quartz and some layer silicates – biotite probably and perhaps some muscovite. Now granite is strong stuff, but it does not last for ever; it is worn down by the weather and we can often see it disintegrating. We can see little pieces being broken off, being carried by rainwater to rivers to be further transported, further broken up and dumped in the sea. This grain of sand from the sea shore may well have originated as a quartz crystal in granite.

Much of the products of rock weathering that are carried by rivers is clay, very much finer than our quartz grain and structurally somewhat similar to the micas of such igneous and metamorphic rocks as granite. It is natural to suppose that here we are seeing another component of the original rock very finely ground down now, modified perhaps, but still essentially the same stuff. Examining a tiny illite crystal from a marine sediment, we might be persuaded that, with its structure so like that of muscovite, it must be a piece of an originally igneous or metamorphic muscovite crystal.

Such a conclusion would be far too hasty. Clays are not just bits knocked off hard rocks. For one thing, much clay is derived from parent minerals that are structurally dissimilar. Kaolinite is a frequent major weathering product of granite. Its structure is quite different from the feldspars that may provide most of its component atoms. As illustrated now so vividly with the scanning electron microscope, kaolinite clays grow from solution (figures 6.29 and 6.34). Often, this is through a series from less to more crystalline phases. In weathering there is not only decomposition but resynthesis to be reckoned with.

Again, in considering a succession of clay phases, we can easily be misled by short cut thinking. If kaolinite is like halloysite from which it often seems to be derived, one's first idea might be that the rolled up layers of halloysite straighten out, sweep off their water molecules and stack up into kaolinite. That might seem a sensible way of avoiding the trouble of making new layers. But Nature is not concerned with saving trouble: it is clear from the pictures that the kaolinite stacks are too neat to have been formed in that way. For silicate crystals forming at low temperatures it is not always, or even usually, a question of one crystal gradually transforming into another: it is a question of one crystal dissolving to provide the nutrient for another (similar or quite different) crystal. Where the structures are similar (for example where halloysite changes to kaolinite itself) the second crystal may not so much remember the structure of its immediate predecessor as rediscover that structure in more perfect form. Seeding by the old phase or epitaxial growth on it may perhaps transfer information to the new phase, but general similarity of crystal structure (for example when both are in the kaolinite group) is no good evidence for this. Millot (1973) comments on such series that they 'testify the transition from disorder to order and seem to necessitate successive recrystallisations rather than unlikely thermodynamic transformations'.

Or consider the abundant illites of marine sediments. Their structure is close to that of muscovite but not necessarily because they are derived from it by minor transformations. More usually, indeed, they are derived from smectites (Jonas, 1976) which are much less like muscovite. Again, though, we must not suppose that to be 'derived from smectite' implies that the smectite layers stacked up on each other, changed sodium and calcium to potassium, some silicon to aluminium, and so on. It would be better to think of the smectite as possibly creating initial seeds but being mainly a reservoir of ionic units that were to reform via solution into brand new illite crystal (Jonas suggests as fringes on the smectite seeds).

True, there are sometimes direct successions, where layers are altered without being remade. You expect the product in such cases to be less well organised than the starting material – as perhaps sometimes when montmorillonite forms from micas. In the making of vermiculites, too, there can be, at least to begin with, a direct inheritance of major structural elements from a parent mineral: laboratory studies suggest that biotite or phlogopite can have their K^+ exchanged for Mg^{2+} (aq.) without very seriously disturbing the silicate layers. But that may not be all that happens in Nature over long periods of time. Maybe a particular polymorph (with a particular arrangement of tetrahedral substitutions perhaps) that suits K^+ as the interlayer cation is no longer the ideal for Mg^{2+} (aq.). If that is so, the resulting crystal will be slightly metastable compared with some more ideal polymorph. The first formed vermiculite, that is, will be slightly more soluble than this putative ideal. Eventually, given a seed of the more perfect form recrystallisation to it will be inevitable, that is:

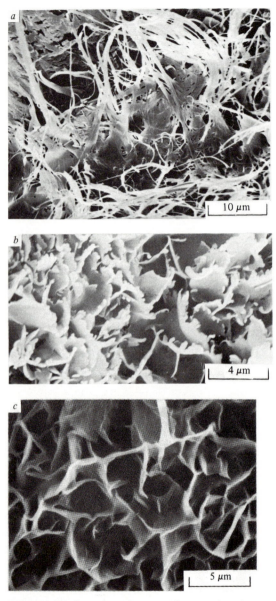

Figure 6.54. Scanning electron micrographs of some clays that have grown *in situ* in sandstones. (From Wilson & Pittman, 1977.) (*a*) Illite. (*b*) A mixed-layer smectite–illite. (*c*) Smectite. (*d*) Smectite with kaolinite. (*e*) Web-like smectite. (*f*) and (*g*) 'Honeycomb' and 'cabbagehead' chlorite. (*h*) Enlargement of (*g*). (*i*) Cliachite. (*j*) Cristobalite. (*k*) Iron oxide (?) grains with kaolinite.

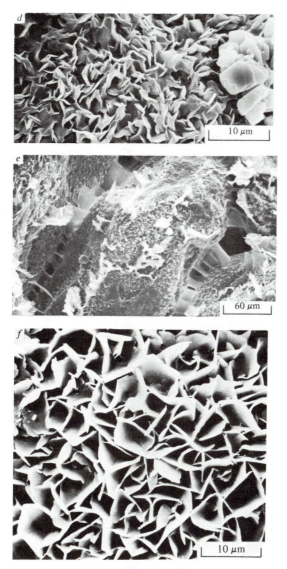

Figure 6.54.

Such a recrystallisation of an initial vermiculitised phlogopite might perhaps account for the paradoxical situation that Shirozu & Bailey (1966) noted in the Llano vermiculite and which we discussed earlier: this did not have the same detailed structure as the parent phlogopite – in particular, the arrangement of aluminium substitutions in the tetrahedral sheets was

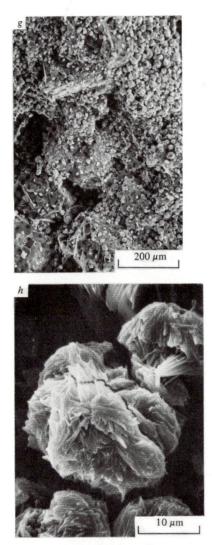

Figure 6.54 (*contd.*)

quite different. These authors remark that 'it is difficult to believe that sufficient energy would be available during surface weathering to order an original disordered Si, Al distribution'. Then again, Meunier & Velde (1979) have found that some vermiculitic mineral derived from the weathering of biotites in granites represents entirely new recrystallised material.

In sum, then, we may at least suspect that even a change in siloxane organisation that seems quite small, if it takes place at low temperatures,

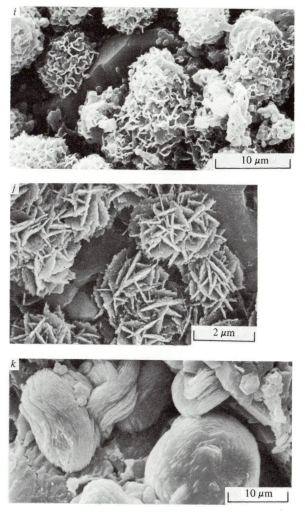

Figure 6.54 (*contd.*)

will call for a radical reconstruction because there is little if any scope for 'annealing' (e.g. through aluminium and silicon atoms hopping from one site to another). Only those units at surfaces and edges are accessible to reorganisation: 'Ostwalt ripening' seems to be the only way of creating new kinds of long-range order – that is to say, defective crystals or less stable crystal structures dissolve at the expense of more perfect crystals or more stable polymorphs. As with a piece of knitting, one can only put things right at the edges: to correct a faulty plain–purl arrangement in the middle, there is nothing for it but to undo the structure and remake from scratch.

From an extensive study of sandstones, Wilson & Pittman (1977) found that authigenic clays are very common, indeed usual constituents. Of 785 samples from 55 formations, 91% were judged to contain clay crystals that had grown in the pores of the sandstone after the sandstone had been formed. The electron micrographs shown in figure 6.54 illustrate a rich variety of microcrystalline forms. The lath-like illite shown in figure 6.54*a* is attached like seaweed to a detrital grain. Figure 6.54*b* is a clay composed of both smectite and illite layers. The membranous character of smectite, and the tendency for authigenic smectite to form a compartmentalised or cellular structure, is illustrated in figure 6.54*c*. Another typical growth habit of authigenic smectite is as a crinkly coating on detrital grains. This is illustrated in figure 6.54*d* which shows also some kaolinite crystals. The membranes of smectite bridging detrital grains in figure 6.54*e* are particularly convincing indicators of authigenic origin.

Like smectite, chlorite can create a highly compartmentalised structure – as with the chlorite crystals coating a sand grain in figure 6.54*f*. The crystallites here are more clearly individual than with smectite. Figure 6.54*g* shows a barnacle-like growth of complex chlorite structures ('cabbageheads') shown at higher magnification in figure 6.54*h*. In addition to stacked plates there were occasional vermicular forms of kaolinite (cf. figure 6.29).

Those ubiquitous framework minerals potassium feldspar and quartz were also found to crystallise *in situ*, here adding new material to existing detrital grains. Sometimes there were irregular spheroidal aggregates of curled flakey material identified as cliachite – a form of aluminium hydroxide (figure 6.54*i*) and there might be minute rosettes of cristobalite (figure 6.54*j*) or rounded flakey aggregates (figure 6.54*k*) (thought to be a form of iron oxide).

There can be little doubt that these objects formed *in situ* from solution, although exactly when, over what period, and at what temperatures and pressures may be more difficult to judge. As such clays are so common, it seems unlikely that the conditions for their formation are particularly critical.

Velde (1977) says of soils that 'most clay minerals are found to develop in one soil profile or another'. Wilson & Pittman (1977) make a similar comment in relation to sandstones: 'all the major varieties of clay minerals are known to form authigenically in sandstone either as a direct precipitate from formation waters (neoformation) or through reactions between precursor materials and the contained water (regeneration)'. This constant reappearance of a fairly limited group of crystal types within the porous matrices of the Earth's upper crust must have, in the end, a thermodynamic explanation. The typical minerals of the igneous and metamorphic rocks, particularly those formed at the highest temperatures such as olivine, are, on the whole, the least stable when subjected to conditions so different from their conditions of formation. They tend to dissolve in water and recrystallise into more stable phases. Among the most stable are micro-crystalline materials – clays – although exactly which clays appear where, may depend in complicated ways on conditions that change from place to place and with time.

Laboratory studies

I have referred already to a number of microcrystalline minerals that can be made fairly easily in the laboratory at ordinary temperatures, for example ferric oxide-hydroxides, gibbsite and iron-bearing 1:1 clays. Kaolinites, illites and smectites have proved more difficult in spite of being the dominant clays in Nature, formed mainly by weathering under surface conditions (Harder, 1977). A main reason for this difficulty would appear to be that, at ordinary temperatures, clay synthesis is only successful from very dilute solutions. Furthermore, the reactions are slow, often needing periods of months or even years. These studies were initiated mainly by French chemists (see, for example, Caillère & Hénin, 1948, and papers in Hocart, 1962; an account of recent developments is given by Siffert, 1979).

It was soon evident that clays containing aluminium – by far the most important class – were particularly tricky. Reviewing this topic, Siffert & Wey (1973) saw the main problems in the strong hydration of aluminium and in the insolubility of aluminium over the near-neutral pH range. Wey & Siffert (1962) had managed partly to overcome these problems in the synthesis of 'protokaolin' by decomposition of the aluminium oxalate complex – to supply suitably hexa-co-ordinated aluminium in solutions containing monomeric silica. Fulvic acid – a complex mixture of organic molecules obtained from peat – was also effective as a complexing agent for aluminium in low-temperature kaolinite synthesis (Linares & Huertas, 1971; La Iglesia & Martin-Vivaldi, 1973). Kaolinite was also obtained using ion-exchange resins (La Iglesia & Martin-Vivaldi, 1975) or feldspars (La Iglesia, Martin-Vivaldi & Lopez Aguayo, 1976) to generate a slowly changing pH as a means of maintaining a suitable and homogeneous level of supersaturation. In this last case, micas and traces of smectite were also obtained.

Harder (1977) discusses another way of creating conditions suitable for

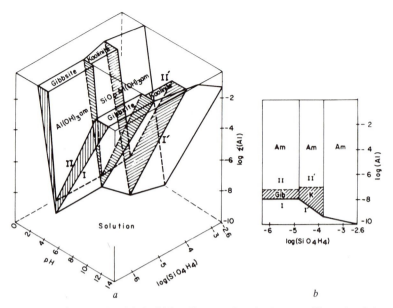

Figure 6.55. (*a*) Stability diagram for the system silica–aluminium–water at 25 °C and 1 atm. The solid phases considered are gibbsite (I), amorphous aluminium hydroxides (II), kaolinite (I′), and the corresponding amorphous silico–alumina gel (II′). (*b*) Cut at pH 5.6 from (*a*). Hatched zones correspond to the crystalline phases gibbsite (Gib) and kaolinite (K). Am corresponds to the different amorphous phases indicated in (*a*). (From La Iglesia & Van Oosterwyck-Gastuche, 1978.)

clay synthesis. When hydroxides are precipitated from solutions containing silica the latter is strongly chemisorbed and mixed silica-hydroxides are formed. Many of these start to generate clays after a few days ageing at room temperature. It is necessary that silica concentrations in the initial solutions are low – undersaturated with respect to amorphous silica (less than 100 p.p.m. at 20 °C). Also precipitation should be slow. This can be achieved by the kinds of homogeneous precipitation methods mentioned above – slow change of pH or decomposition of complex ions – or alternatively by oxidation (e.g. $Fe^{2+} \rightarrow Fe^{3+}$; cf. Harder, 1978). Also, the overall composition of the precipitate should be similar to that of the clay mineral being made. Not all insoluble hydroxides are effective. The best results are obtained where the radius of the cation is between 0.083 nm and 0.078 nm (as for Mg^{2+}, Ni^{2+}, Co^{2+}, Zn^{2+} and Fe^{2+}). With a radius of 0.057 nm, aluminium is seemingly too small, but even this can be incorporated in solid solutions with the more easily made clays. Harder lists about two

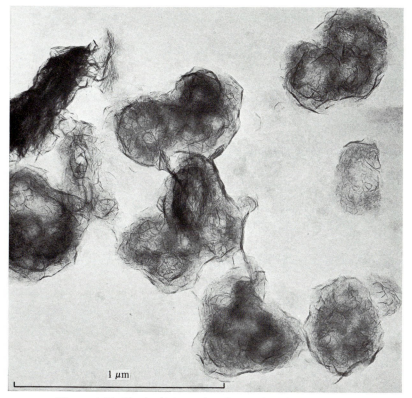

Figure 6.56. Synthetic hectorite. Transmission electron micrograph showing partly transformed brucite crystals, 'doughnuts', enclosed in hectorite 'bags'. (From Mackenzie, 1971.) This stage was reached after 6 days reflux of a slurry of composition $SiO_2:MgO:LiF = 0.01:1.0:0.25$ in pyrex glass.

dozen clays of all four layer types that have been made in this way. Kaolinite is the most notable absentee.

La Iglesia & Van Oosterwyck-Gastuche (1978) have discussed thermodynamic restraints on conditions for kaolinite synthesis. They stress the need for very dilute solutions. Figure 6.55 shows furthermore how narrow is the region of concentrations and pH for kaolinite synthesis. Even when thermodynamic conditions are satisfied it is clear that other factors can stand in the way of synthesis – in particular, suitable nuclei for crystal growth may have to be present. In another paper, Van Oosterwyck-Gastuche & La Iglesia (1978) review laboratory syntheses of kaolinite at various temperatures and stress that it is the laws of crystal growth rather than Arrhenius kinetics that must be considered.

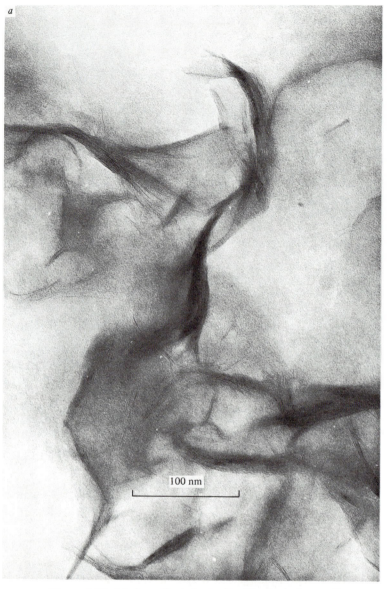

Figure 6.57. (*a*) A synthetic hectorite. The typical smectite appearance is similar to that in figure 6.54*c*. (From Baird, Cairns-Smith, Mackenzie & Snell, 1971.) Conditions: heating for 7 days at 90–95 °C in polypropylene: SiO_2:MgO:LiF = 2.0:1.0:2.5. (*b*) At a magnification of a million, 2:1 layers are clearly visible when they happen to lie edge-on. This is part of the picture (*a*) in which the typical smectite fabric can be seen as arising from flexible layers that have become connected together through regions in which they are stacked on top of each other.

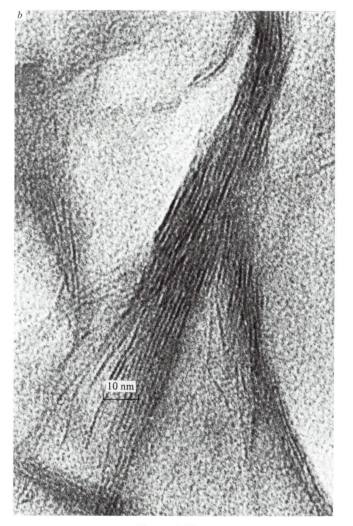

Figure 6.57.

Baird, Cairns-Smith & MacKenzie (1973) made an electron microscope study of the efficient and rapid synthesis of hectorite that had been discovered by Granquist & Pollack (1959) – by boiling slurries of $Mg(OH)_2$, SiO_2 and LiF. In this reaction, brucite crystals are quickly attacked by silica to give an intermediate material with some of the properties of smectite. The membranous hectorite that appeared after about 24 h was much more extensive than the initial brucite crystals (figure 6.56). It had evidently crystallised from solution, and seemingly at the expense of the 'pseudo-smectite' which had partly redissolved. A more mature product is shown in figure 6.57.

Such intermediate materials as 'pseudo-smectite', and the similar amorphous silicified hydroxides used by Harder, can be thought of as reservoirs of suitable units for subsequent crystallisation from solution. The units themselves may be monomeric, or possibly they are small oligomers particularly suited to the synthesis of clays – hence perhaps the importance of the conditions of formation of the intermediate materials. Siffert & Wey (1973) have suggested that aluminium clays are made from oligomers such as:

$$(HO)_3-Si-O-\{Al(OH)_2(H_2O)_3\}$$

or $\qquad (HO)_3-Si-O-\{Al(OH)(H_2O)_3\}-O-Si-(OH)_3$

to give respectively 1:1 and 2:1 clays. But whatever the nature of the units, the burden of the laboratory studies has been that clay synthesis from aqueous solutions takes place through unit-by-unit crystal growth mechanisms.

It is generally true that for accurate crystal growth a solution should be only barely supersaturated with respect to the perfect crystal, so that at the same time it remains undersaturated with respect to imperfectly formed crystal. This is the basis of the 'error correction' mechanism that we talked about in the last chapter. If levels of supersaturation are raised then the number of kinds of imperfect organisation that can persist increases: the resulting crystals are increasingly flawed – until eventually there is more flaw than crystal and we describe the precipitate as 'amorphous'.

On the whole, it is the more insoluble materials that are most inclined to give amorphous precipitates – and clays such as kaolinite are very insoluble at ordinary temperatures. For clay synthesis at around 20 °C, it is evidently quite a short step between solutions being too dilute for anything to come out and too concentrated to allow the appearance of anything recognisably crystalline.

But we should remember, too, that making a clay is not just a crystallisation – not just an orderly emplacement of units as is, say, the crystallisation of sucrose. It is also a polymerisation. Covalent bonds have to be made and broken in the process with several water molecules having to be eliminated as each unit adds. Although little is known here about the details, it is generally the case that to make and break covalent bonds rather particular orientations of groups are required and activation energy barriers have to be surmounted. In these circumstances, we can imagine metastable polymers forming and persisting – as in the polymerisation of silicic acid to silica gel (discussed on pp. 181–2). We can imagine the orderly unit-by-unit processes of crystal growth easily becoming blocked by tangles of polymer attached to the crystals. Only at concentrations low enough (at least) to prevent spontaneous net polymerisation in solution would you expect such tangles to be unstable – for the crystal surfaces to be cleared to expose those relatively few sites at which orderly growth can proceed.

Complex systems in Nature In view of the rather narrow range of conditions often needed for low-temperature clay synthesis, we might wonder how it is that Nature so often seems to get it right. There are perhaps two general answers to this question. First, the weathering solutions from which clays form are rather dilute, so the conditions are likely to be roughly right in that respect at least. Then, secondly, Nature only has to get it roughly right – at least in the first place. The most stable phases – for example well crystallised kaolinite – may only be arrived at after a succession of other minerals that are simpler or less fussy about the requirements of orderly growth. Millot (1973) gives allophane $\rightarrow$ halloysite $\rightarrow$ disordered kaolinite $\rightarrow$ ordered kaolinite as a typical sequence.

To a first approximation, then, we can perhaps consider the synthesis of clays in Nature as taking place as follows. The materials that come out of solution from freshly weathered rocks tend to be metastable and poorly organised; these eventually come into quasi-equilibrium with their surrounding solutions which are now at a lower level of supersaturation than before. New, more orderly phases crystallise from these solutions, so reducing their concentrations still further so that still more orderly phases become kinetically possible. Each new phase reduces the level of supersaturation of the surrounding solutions to just above that at which it can grow – a level at which the previous phase dissolves. More stable, more insoluble, more orderly phases can thus be arrived at. Thermodynamics eventually gets its way.

Indeed, that must be too simple a picture: solutions in contact with minerals in Nature are not necessarily in quasi-equilibrium with those minerals. In open systems, species in solution may be added or subtracted so that the new phase that appears may have a different chemical composition. To make matters more complicated, solid particles may be being transported. Also, there may be seasonal variations – alterations of high and low silica levels in solution, changes of pH, and so on. Thus, different minerals may be favoured in the same place at different times. In discussing conditions for kaolinite *versus* smectite, Johnson (1970) points out that, while a general environment might favour one, there could nevertheless be micro-environments that were favourable to the synthesis of the other. In any case, smectite and kaolinite are often found together in spite of supposedly different conditions needed for their synthesis. We should remember that while 'pure' clay deposits may catch the eye, clay minerals occur mainly in complex assemblages.

As yet a further complication, the growth of a new phase may very well affect the environment in which it is forming. For example, a clay may clog up the pores of a sandstone within which it is forming, altering the patterns of solution flows and hence the patterns of variation of species in the solutions: if they stop moving, pore solutions will come nearer to equilibrium with the surrounding rocks. Hence, and in other ways, long-term variation can be superimposed on shorter term fluctuations: a mineral assemblage may 'evolve' in a manner that is perhaps analogous to the 'evolution' of an ecological system – in so far as we can imagine a succession of species, with each predisposing the conditions for the next. This may go on indefinitely or alternatively the system may stabilise if, by chance, an assemblage is arrived at which happens to stabilise the environment for itself. A rather simple case in which that may have happened was found by Clemency (1976). He found two different kinds of clays alongside each other, a smectite and kaolinite, each apparently self-maintaining. Clemency suggests that the relatively impervious smectite would favour a slow percolation of water and so a higher concentration of silica, thus stabilising smectite, while the more porous kaolinitic rock would be in equilibrium with the more dilute solutions passing more rapidly through it. 'The pore water characteristics within each rock type thus impresses its own particular equilibrium constraints on the clay minerals forming within each rock type.'

Conclusion

With this reminder that the conditions for synthesis of clay minerals in Nature are often rather complicated, and that the products are often mixtures of many different crystal types, we come to the end of this chapter. My main objective has been to illustrate the variety and complexity of clay mineral types; to persuade those of my readers who might have needed persuading that *mud* can be interesting, highly structured stuff. In Part III, I will argue that several particular features of clay mineral structures correspond to what is really needed for the components of organisms that could generate spontaneously, evolve under primitive conditions and then, through photosynthesis, begin to invent an organic biochemistry.

PART III

A new story

7

First life

Introduction and review

In this chapter we will start to pick up some of the threads from earlier chapters to try to arrive at some more specific speculations about how life on Earth might have originated. This will be to illustrate the general idea that our first ancestors might have been inorganic minerals, rather than with much hope of the details being correct. Like the fictional details of an historical novel that help to make more real some long past way of life, it is perhaps only when one tries to see how a mineral life might have worked, in terms of what we happen to know now about the details of mineral structures and synthesis, that such a way of life becomes thinkable.

Even just to think about thoroughly non-nucleic acid–protein life styles may be a help – because it was the main conclusion of Part I of this book that first life would almost certainly have been of an altogether different kind from ours: that it would have been made from different basic units that were put together differently and worked differently. However unpalatable Dr Kritic and others may find this idea, it seems to me that an assertion of original biochemical *dis*similarity is an essential starting point from which to try to understand the origin of life. We must detach our fingers from one guide rail before we can grasp others that can lead us much closer to the answer.

As we saw, this 'unpalatable' proposition is in any case based on common sense. Whatever else, organisms are machines: life is natural engineering – and engineering principles apply. You do not expect first ways of doing things to be like more sophisticated ways – not in mechanical details anyway. First life would have been different from us, not so much because conditions on the early Earth would have been particularly different from now, but because the specifications for a first organism are quite different from the specifications for an advanced organism. The first organisms had to start up a new enterprise with no pre-existing technology. Later, evolved techniques of fabrication could be presupposed.

As we saw also in Part I, a too facile view of the way evolution works stands in the way of a general acceptance of this common sense. It is often supposed without argument that the unity of our central biochemistry makes a powerful case for the absolute invariance of certain small molecules for life, or at least for life on Earth. Alanine, glucose and so on are seen as invariants through evolution, analogous perhaps to the atoms that are invariant through chemical reactions. Amid all the changes that evolution can bring about, at least these basic small molecules can be relied upon – so it is said.

That may be true now, but there is no more an absolute law of the conservation of alanine in biochemistry than there is a law of the conservation of 1 cm right-handed star screws for automobiles. The invariance of our central biochemicals arises from their multiple and intricate interdependence in our present biochemical machine. Biochemical invariance is evidently a system invariance; it would have appeared as the system evolved. But there would have been no rules till that happened. And no foresight either – no way of knowing what was going to be just the thing in the remote future.

It is not as if mechanisms for radically updating original design approaches cannot be imagined through the early evolution of our biochemistry as they can be seen still in operation at higher organisational levels. Evolution is not always restricted to the optimisation of some initial set of subsystems. New subsystems have evidently been invented. As discussed in Chapter 3, this arises through switchings between structures and functions (as well as redundancies) in conjunction with the adaptation of structures to functions. That is the main inventive technique of evolution, the way quite new design approaches can be arrived at. Eventually some elaborately interdependent set of subsystems becomes frozen in, and the period of radical invention is over. If our central biochemistry may seem always to have been frozen, this is an artefact of our point of view: the last common ancestor of all life on Earth now happens to be more recent than that time at which the central machinery of biomolecular control became rigid.

As for the argument that the components of our life were all ready and waiting on the primitive Earth, it is not clear in the first place that any of our central biochemicals would have been present on the Earth in reasonable amounts. And in any case the interpretation of experiments supposed to show this is ambiguous. Evolving organisms would tend to follow biochemical pathways of least resistance, and these would tend to be like abiochemical pathways; and stable molecules would tend to be preferred biochemically and abiochemically. This can explain why some of our smaller biomolecular components keep on turning up all over the place (inside and outside organisms). On the other hand, some very critical structures that are central to our biochemistry – nucleotides and lipids, for example – are difficult to make; do not turn up in plausible simulations of conditions on the primitive Earth; would, as far as one can see, need an evolved biochemical system to produce them consistently or polymerise them in an organised way. Needless to say the evolved biochemical system that synthesised such molecules in the first place did not absolutely depend on these same molecules – they must have been useful to begin with rather than vital. The evolved biochemical systems that we must postulate were organisms of another kind, with takeover mechanisms, in particular genetic takeover, providing the continuity between them and us. The continuity of evolution can be compared with that of a long rope where no one fibre stretches from one end to the other. Or we might see it as resembling the continuity of a business enterprise where staff, premises, even the nature of the main product might have changed over the years without any sudden point at which everything changed.

Or we might use the ancient idea that life is a kind of fire, to illustrate the central theme of this book. How do you set fire to coal? Often by setting fire to paper. First you have to get *something* burning. Evolution has been a kind of spreading fire and the problem of the origin of life can be seen as a problem about how something started to evolve, to transform matter into 'survival machines', in Dawkins' phrase, and to build up a 'technology'. Then, inevitably, first ways would be improved on and displaced. Sooner or later the coal would be alight with no trace left of the paper.

There are many other examples of processes for which beginnings are special and may involve mechanisms that are then dispensed with (think of booster rockets, starter motors, detonators, eggs). Particularly for any sort of 'evolutionary' process – where one thing leads to another – the process itself can easily be transformed by its own effects (think of the progress of a game of chess or a piece of music). This all seems to be so much common sense that we should surely stop insisting in the name of Occam or anyone else that life would have started in anything like the way in which it was eventually to be most successful. It would not. It would not have known how to. Let us insist rather that there must have been *organisms* that invented our biochemistry: that the origin of our kind of molecular life is to be distinguished from a remoter origin on which that depended. Darwin's idea must be taken further back, I think, and away from present bio-molecular hardware to a different region of chemistry altogether. And there, somewhere there, is to be found the true origin of life.

That was the gist of Part I of this book. In Part II we started looking for the appropriate region of chemistry. That search was started in earnest in Chapter 5, although Chapter 1 had already suggested that organic minerals might have been inappropriate starting materials. We concentrated attention on the hardest part of the problem, on the one essential piece of machinery for any organism able to take part in long-term processes of evolution. Whatever else, if the 'fire' was to keep alight through millions of years, if 'know-how' was to be retained and improved on so that ultimately our nucleic acid–protein–lipid machine could be made, there had to be some sort of accurate information-printing machinery in those very early organisms. The main clue to the nature of this first machinery is that it must have been able to put itself together without any evolved machinery of any sort already there. From such considerations, and by thinking about bonding characteristics in organic and inorganic molecules and crystals, crystals with covalent bonds in them seemed a much better bet than organic polymers. Our course was thus set towards inorganic chemistry – towards colloidal minerals that could crystallise from aqueous solutions at around ordinary temperatures. We arrived at clays, defined rather broadly.

Chapter 6 provided more detailed information and ideas which will form the basis of a further discussion in this chapter on the nature of the primitive genetic material. The accounts of mineral structures given in Chapter 6, particularly the electron micrographs, also served to illustrate that even without genetic information the common colloidal minerals often have the forms of membranes, tubes and such things – that here we have, perhaps, potential phenotypic apparatus for first life, a clue to a rough but ready means of making some of the hardware of early organisms. We will also take this idea further – in this and in the next chapter. But since, to be relevant, the synthesis of any such hardware must have been to some extent under genetic control, it is with possible genetic structures that we must start.

Information storage

In looking for a primary genetic mineral we can apply first the constraint that, whatever it was, it must have been able to store information. An object can in principle store information if it can be in one of a large number of alternative, reasonably stable configurations. A book stores information because there are many ways of arranging letters, etc. on pages, and a particular book is just one such way. Similarly, a DNA molecule stores information through having one of a very large number of possible base sequences. On the other hand, a perfect crystal could not store information because there is only one way of being perfect; nor does a drop of water, because although there are many ways of arranging molecules in water none of them lasts.

The slightest feature can make the difference between a structure having zero information capacity, and its having a very large capacity. For example, you might suppose that there would be no way of storing information in a straight polyethene molecule:

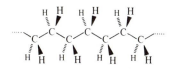

In practice that may be so, but in principle you could store more information in a polyethene molecule than in the base-pair sequence of a DNA molecule of the same weight. Since one in every 90 or so carbon atoms is ^{13}C, information could be stored as a particular arrangement of these atoms on a long polyethene chain. Similarly, polyvinyl chloride (PVC) could hold much information if the chlorine isotope sequence could somehow be specified and read. But here there is another way that does not depend on there being any two-letter or *n*-letter alphabet. A sequence of As could hold information if some of the As were printed upside down. Similarly, (atactic) PVC could hold much more information than the same weight of DNA – through a particular left–right sequence of chlorines along the chain:

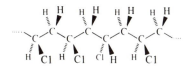

Crystals as information stores

There are very many ways in which real crystals (as opposed to the text book ideals) could hold information. Real crystals almost invariably contain a multitude of defects of various sorts. Over and above its regularly repeating **crystal structure** any given crystal will have various irregular

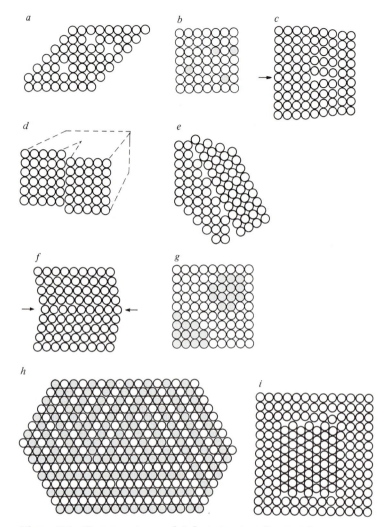

Figure 7.1. Common types of defect structure in crystals.
(*a*) Vacancies. (*b*) Substitutions. (*c*) An edge dislocation showing an extra plane of units. (*d*) A screw dislocation (cf. figure 6.43).
(*e*) A grain boundary where similar but differently oriented lattices meet. (*f*)–(*i*) show more intimate forms of crystal intergrowth.
(*f*) As in (*e*) except that the misorientation is such that there is a common plane of units at the junction (the composition plane, arrowed). Such crystals are twinned. (*g*) As with (*b*) except that the substitutions are in domains. (*h*) Another kind of domain structure: here the domains have the same overall composition but differ in the alignment of units. (*i*) Different crystal forms of the same material may be intergrown, or alternatively the intermingled phases may be chemically distinct. Here there is a domain of one phase within another.

features – it will have some sort of **defect structure**. It may contain point and line defects such as **vacancies, substitutions** and **dislocations** (figure 7.1*a–d*). The positioning of such features could in principle represent specific information analogous to a DNA base sequence or to the arrangement of letters on a page. Alternatively, information might be held on a somewhat larger scale. A crystal usually consists of a mosaic of at least slightly misaligned blocks; and very often crystals have regions, or **domains**, that are grossly misaligned (e.g. figure 7.1*e*). In **twinned crystals** there is at least one major misalignment, such that at least one plane of atoms – **the composition plane** – is shared by adjacent domains. Twinning is very common, particularly in minerals. Twinning and other forms of **crystal intergrowth** are illustrated in figure 7.1*f–i*.

Crystal defect structures may sometimes be visible to the naked eye – for example twinning is often visible. But such structuring goes down to a very small scale indeed, as the high-resolution electron microscope is revealing increasingly. It is possible to see minute heterogeneities in most sorts of crystals that have been examined – a row of atoms missing, a slight misalignment, and so on. Very often there is a profusion of such tiny imperfections. Such features can be described as **ultrastructures**. (For a review of the ultrastructure of minerals see Hutchison, Jefferson & Thomas, 1977.)

Domaining can be seen to be very common – and is often on a minute scale. Some of the very best pictures have so far come from high-temperature materials. In the 1 MV electron microscope image of tungsten–niobium oxide ($4Nb_2O_5 \cdot 9WO_3$) shown in figure 7.2*a*, rows of atoms can be seen, as can the changing directions of the rows in different minute domains (figure 7.2*a*).

In figure 7.2*b* one can see a misalignment of blocks in a β-FeO(OH) 'somatoid' that was formed at low temperatures. (Look also at the picture of a sepiolite crystal in figure 6.48*b*.) Another low-temperature mineral, beidellite, reveals a complicated mass of dislocations (figure 7.2*c*; compare figure 8.7). Very numerous microtwin boundaries can be seen in the pyrrhotite crystal shown in figure 7.2*d*. Zeolites too are beginning to reveal a rich ultrastructure (Bursill, Thomas & Rao, 1981).

Zeolites, and more especially feldspars, can have complicated twin structures. Feldspar crystals are generally highly twinned and various types of twinning have been classified according to the orientation of the composition plane and the symmetry relations between the parts of the crystal on either side of the plane. Empirical 'twin laws' for feldspars were given a crystal structural interpretation by W. H. Taylor: across a composition plane there is some change in the structure of the alumino–siloxane frame

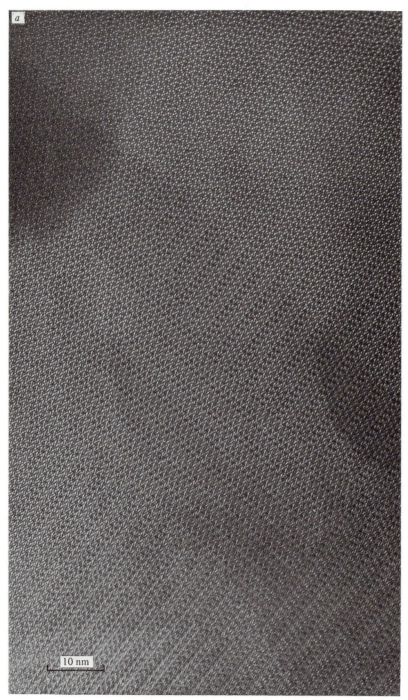

Figure 7.2*a. For explanation, see caption on p. 270.*

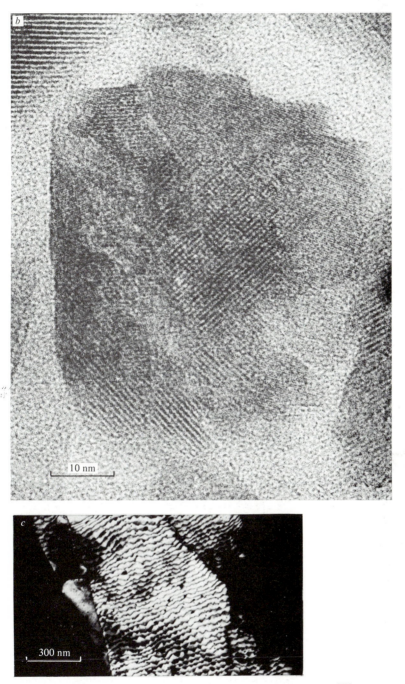

Figure 7.2*b* and *c. For explanation, see caption on p. 270.*

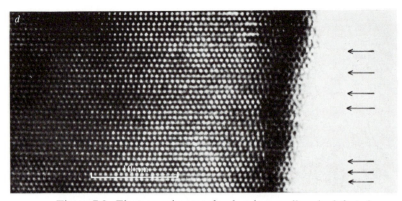

Figure 7.2. Electron micrographs showing small-scale defects in crystals (see text). (*a*) $4Nb_2O_5 \cdot 9WO_3$ (from Horiuchi, 1978). (*b*) Cross-section of β-FeO(OH) crystal (from Galbraith, Baird & Fryer, 1979). (*c*) Dislocations in a beidellite crystallite (from Guven, Pease & Murr, 1977). (*d*) Microtwin boundaries in pyrrhotite (arrowed) corresponding to changes in the arrangement of iron atom vacancies (from Nakazawa, Morimoto & Watanabe, 1975).

which, although a major change in the sense that a wholesale reconstruction of the lattice would be needed to eliminate it, nevertheless involves little distortion of covalent bond angles and lengths. There are several such subtle structural modulations that can be correlated with the classical 'twin laws'. Interpretations of the 'Manebach law' and the 'Baveno law' as given by Taylor, Darbyshire & Strunz (1934) are illustrated in figure 7.3.

We have already discussed the idea, in Chapter 5, that *chemical* defect structures might have provided the basis for a crystalline genetic material. As we saw in the last chapter, it is very common indeed for there to be some choice about which ion is present at some particular site in a mineral lattice. And for many minerals, especially silicates, such a chemical heterogeneity is an essential feature. Montmorillonite would not have negatively charged layers that can disperse in water if there were not a suitable degree of substitution of cations within the layers – in that case it would simply not be montmorillonite.

Information and metastability To store information the 'defects', whatever they are, must stay put: they must not move about inside the crystals as the vacancies and dislocations in metals easily do. (Metals are malleable for this reason.) There are signs, though, that the (much less malleable) silicates and metal oxides are not nearly so internally mobile as metals – that at ordinary temperatures the main Si–O and M–O bonding

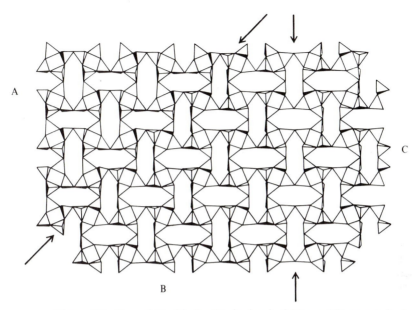

Figure 7.3. Two of the kinds of twinning in feldspar. 'Baveno twins' are related as between sections A and B; 'Manebach twins' are related as between B and C. Arrows indicate the composition planes.

can be maintained within crystals over very long periods of time. One sign is the common persistence of metastable silicate and oxide phases at ordinary temperatures. The technique of thermoluminescent dating also depends on metastable features in crystals persisting indefinitely at ordinary temperatures. This technique depends on the accumulation of radiation damage within silicate and oxide crystals. The damaged crystals cannot anneal except at somewhat elevated temperatures when they lose their stored energy in a burst of heat and light.

Tracks left by high-energy particles in micas and many other minerals represent minute metastable features which can persist for millions of years at ordinary temperatures (Fleischer, Price & Walker, 1975).

As we saw in the last chapter, mineral crystals can reform, via aqueous solution, at ordinary temperatures: but that is no bad thing since this kind of limited reformation could have provided a mechanism for correcting errors in the replication of information in crystals without necessarily losing the information itself (Chapter 5). What is important is that deep inside the crystals there is no wholesale mobility of ions – that at least some of them can stay put for long periods.

From considerations of the site energies of different ions in mineral

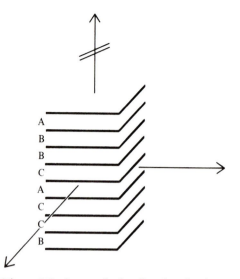

Figure 7.4. A crystal of a disordered polytype could be said to contain information in the form of a particular stacking sequence, that is, a particular sequence of relationships between adjacent layers. Such 'information' would be amplified by exclusive sideways crystal growth (and replicated through subsequent crystal cleavage as in figure 5.11).

genetic information, whether as a pattern of chemical substitutions, or of dislocations, or of crystal domains or whatever, would have to direct the formation of new material 'in conformity with the same design' to use Hinshelwood's phrase.

For a crystal gene the matching-up mechanism would make use of the orderly addition of new units that takes place as a crystal grows. As discussed in Chapter 5, the 'come and go' aspect of this operation is an essential part of it: there has to be a way of putting mistakes right, the bonding between units and crystal has to be reversible as long as it is at the surface.

One can see in all this a particularly tight constraint. The matching-up mechanism must be able to correct certain kinds of defects – mistakes in matching; but it must leave other kinds of defect alone – those defects whose patterning constitutes the genetic information. Indeed these kinds of defects it must copy.

Polytypism as a key phenomenon? A matching-up of one-dimensional information in the form of a particular stacking sequence would take place in a disordered polytype that grew exclusively sideways while

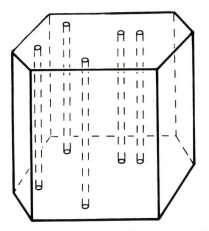

Figure 7.5. The disposition of a set of parallel screw dislocation lines in a crystal might be said to constitute information that would be amplified by crystal growth that was exclusively in the direction of the dislocation lines. (It would be replicated through subsequent crystal cleavage as in figure 5.12.) Such a genetic crystal would be analogous to a stack of punched cards each with the same pattern of holes.

retaining its initial stacking sequence (figure 7.4). If the thin laminar crystals thus produced were to break up suitably this could provide the basis for a genetic crystal of the kind given in figure 5.11.

The screw dislocation as a key phenomenon? A screw dislocation is an example of a defect that is often replicated through crystal growth (compare again figure 6.43). This kind of defect is faithfully copied, while other kinds, arising from initial misplacement of adding units, etc., can be put right by local reversals.

A crystal containing many parallel screw dislocation lines might conceivably replicate information in the form of a particular disposition of these lines (cf. figure 5.12). This would be an example of a two-dimensional information store, like a stack of identical punched cards with their lines of superimposed holes analogous to the dislocation lines (figure 7.5).

Interlayer ordering as a key phenomenon? Alternatively, coming closer to the formal model shown in figure 5.12, information might be held in the form of a particular arrangement of cations – a specific irregularity within a silicate layer – that was copied as the crystals grew through some layer-on-layer mechanism. In such a way very complicated two-dimensional information might conceivably be printed off. The resulting crystals would be 'semi-ordered' like the stack of punched cards referred to above: within the layers there would be some element of disorder but between the layers this disorder would be repeated in an orderly way.

In Chapter 6 we discussed several examples of interlayer ordering. There was the possible ordering of domains across K^+ planes as suggested by Gatineau for muscovite: there were indications of ordering of aluminium-rich sites between the layers of Ia chromium chlorite from Erzincan, and for vermiculites more generally. The orientation of substitutions was clearly ordered between layers in amesites. The ordered forms of poly-typism that we discussed in Chapter 6 are further examples of, often, quite subtle ordering between layers. It would not be surprising, and indeed it seems sometimes to be the case, that a substitutional feature of one layer can induce a similar feature in the layer that grows on top of it.

What kind of substitutional feature might be the most promising? One might think first of a very fine-grained information store where the in-formation units – the letters – were individual cations within layers. One might imagine, say, a 2:1 layer silicate with information written as a particular cation-by-cation arrangement of Al^{3+} and Si^{4+} in tetrahedral positions. And we might suppose that the negative charge pattern resulting from this would locate counterions, such as Mg^{2+}, which in turn deter-mined the locations of Al^{3+} and Si^{4+} in the lower tetrahedral sheet of the newly forming layer. Then we might suppose that there was a similar con-trol of the upper tetrahedral substitution pattern through electrostatic or distortion effects operating through the newly forming layer. This whole process would then repeat until there was a stack of layers each containing the identical very complex pattern of cation substitutions.

Such a device could have a marvellous information capacity – more than the same weight of DNA. But there are a number of difficulties. There is a purely geometrical problem in that the intralayer offset between the lower and upper tetrahedral sheets (figures 6.37 and 6.38) creates an ambiguity – any one tetrahedral site is equally close, in the ideal structure, to two sites in the opposite sheet of the same layer. We saw, when discussing poly-typism, that it must indeed be possible for information of some sort to be transmitted through 2:1 silicate layers, and the mechanism suggested – via mechanical distortions – might operate here. But in normal polytypism what is transmitted through a layer is not an individual piece of informa-tion about some particular atom, it is information about sets of atoms – for example about the direction in which grooves are aligned.

Information held as the complex irregular patterning of individual cations might seem too fragile in any case – too liable to be wiped out by mistakes in copying. And the tendency for intralayer ordering of cations – sideways ordering – would positively tend to wipe out the information: it would be like having a typesetter with an obsessive preference for regular repetitions of letters. (That would be worse than having a typesetter who was just careless sometimes.)

Another difficulty would be that the very irregularity that constituted the information would mitigate against its accurate replication because it would create a rather distorted kind of crystal with, probably, a weak discrimination for incoming units.

A good genetic crystal would have to be rather well organised, with tight discrimination – with far more order in it than disorder. Although some disorder is necessary to provide the capacity for information, capacity would not be the main problem: it would be much more important to begin with that information was transmitted accurately even if its amount was rather small.

We might ask then whether *domains* of a particular cation arrangement might be units of information so that what is copied between layers is not the positions of individual cations but the nature or alignments of more extensive two-dimensional arrays of cations. Such domains pre-existing in a growing crystal might exert a more decisive control so as to be copied more accurately.

Crystal intergrowth as a key phenomenon? More generally we might see a genetic crystal as intergrown – with genetic information held as some sort of patterning of domains. Twinning is a particularly pure and prevalent form of crystal intergrowth so let us start with this question. How would the composition planes be disposed in a genetic crystal that held its information in the form of a particular twin structure?

A twin in crystallography is Siamese. It consists of two or more lattices joined in an abnormal but understandable way (see figures 7.4f, 7.3 and 7.6). Buerger (1945) has considered the question of how twinned crystals can arise through the processes of crystal growth. During early stages of growth, when crystal nuclei are poorly organised and when solutions are likely to be more highly supersaturated, 'mistakes' in the growing crystals are more easily made and more likely to remain. This is particularly true of twin structures. Here the units that are misplaced are misplaced in relation to their next-but-one neighbours across a composition plane (see figure 7.1f); but the nearest neighbour relationships, which are energetically most important, are correct. Twinning is a particularly persistent kind of defect

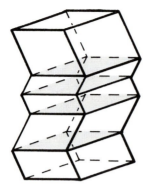

Figure 7.6. Multiple twinning often reveals itself in grooves and ridges parallel to the composition planes (here shaded). Compare figure 7.1*f*.

structure because once the mistake has been made growth beyond the composition plane (for example vertically in figure 7.6) soon becomes quite normal.

But if twinning, and the persistence of twin structure through crystal growth, are to be the basis of a replicable genetic information store, then the twinning must be multiple and microscopic; and the composition planes should not only be retained but extended during crystal growth. Furthermore, the number and mutual alignments of these composition planes must not change during crystal growth.

We have then some geometrical constraints on the morphologies of genetic crystals of this sort. If a patterning of composition planes is to remain constant while being extended, then the composition planes must be parallel to each other in the direction(s) in which they are being extended. There are only two general ways in which this could be so. The first is an example of a type-1 genetic crystal (page 155) in which the information is held in one dimension and amplified through growth in two (cf. figure 5.11). For example, if we look at the twinned crystal in figure 7.6, the disposition of composition planes is a vertical, one-dimensional 'message' that would be amplified through exclusive sideways growth.

In the second possibility, an example of a type-2 genetic crystal (cf. figure 5.12), growth would only be in one dimension. The composition planes would have to be parallel to that and hence, in this direction, to each other; but they need not otherwise be parallel to each other. The crystal that we are now thinking about is somewhat similar to the model in figure 7.5; only here it is planes rather than lines that intersect the layers within

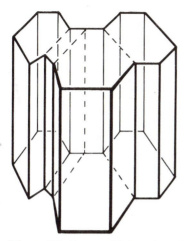

Figure 7.7. In a multiply twinned crystal like this the disposition of a set of vertical planes (the internal composition planes and the sides of the crystal) might constitute information that would be amplified through growth exclusively in the vertical direction (cf. figure 5.12).

which the information is held. In place of the pile of punched cards we might imagine instead a stack of cards on each of which there is an identical, complex, irregular, crazy-paving pattern (figure 7.7). Each multidomained layer would have the same shape (a more or less complex polyhedron, probably) giving an overall tabular or columnar morphology (possibly grooved or fluted as shown in figure 7.7). Since growth must not take place sideways in this case, a constant cross-section should be a characteristic feature.

The Antarctic amesite crystals studied by Hall & Bailey (1979) (see page 213), have the sort of twin structure required except that the twinning here was too regular and on too large a scale to be able to hold much information. The Urals amesite (Anderson & Bailey, 1981) is a better example, with its more complex twinning.

Although for the sake of discussion I have taken the simplest and most ideal form of intergrowth – twinning – the main arguments have been purely geometrical: any kind of intergrowth would do that conformed to the general requirements, and we might be able to identify a potential genetic crystal from its morphology – without knowing what sort of intergrowth was involved.

Kaolinite vermiforms of the kind shown in figure 6.29 (pages 200–1) are of interest in this connection. Some of the crystals, particularly in figure 6.29*e*, almost look as if they had been *extruded* from an irregularly shaped hole. Presumably this was not so and the constant cross-section arose from the processes of crystal growth. But in the complex environment in which they grew it is surprising that sideways growth should have been so uniform, if indeed these crystals were growing sideways. An alternative interpretation would be that, after some initial phase, growth was exclusively at the ends extending the vermiforms which then, from time to time, broke into shorter lengths (cf. figure 5.12).

Grooves in crystals are often a sign of twinning, or of some other form of crystal intergrowth (figures 7.6 and 7.7), and the grooves that characteristically run along kaolinite vermiforms suggest something like twinning here, with composition planes lying parallel to the main (or exclusive) direction of growth (cf. figure 7.7). (Some such internal defect structure is suggested particularly by the broken crystals in figure 6.29*d*.)

So we might take these electron micrographs as providing visible evidence for the replication of information through processes of crystal growth – information at least in the form of a complex cross-section but probably also as some kind of crazy-paving domaining of the individual layers. These pictures also provide visible evidence for another feature that would be necessary for a genetic crystal: cleavage preferentially in the planes in which the information is held and across the direction of growth (figure 5.12). This is not what you normally expect: normally twinned crystals tend to cleave along composition planes, not across them.

But then kaolinite is not at all a simple sort of crystal. A study by Mansfield & Bailey (1972) of large kaolinite vermiforms emphasises the subtlety of kaolinite crystallography and it may provide the clue as to the nature of this putative domaining in Keller's (much smaller) crystals. Mansfield & Bailey's crystals were divided into domains characterised by different positions, among three possibilities, for the orientation of the vacant octahedral sites (see page 204). The intersection between three such domains is illustrated in figure 7.8. From the sharpness of the diffraction spots, Mansfield & Bailey reckoned that the domains were at least a few nanometres across. This might be compared with the mosaic patterns in replicas of etched kaolinite surfaces that were observed by Williams & Garey (1974). From their published electron micrographs the surface mosaic units here appeared to be mainly in the 100–1000 nm range. (The grooves along vermiforms in figure 6.29*d* and *e* suggest that here domains are commonly about 1000 nm across.)

According to Mansfield & Bailey the 'pseudotwinning' of kaolinite would most likely have arisen in the first place through an early growing

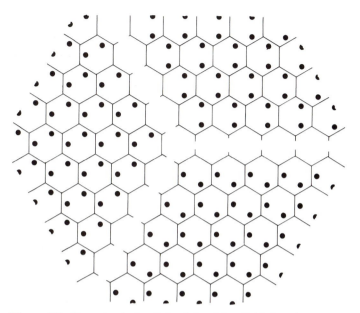

Figure 7.8. Domains in kaolinite (after Mansfield & Bailey, 1972): see text.

together of initially separate crystals that happened to have started with different vacancy orientations. The lattices of such crystals could never match up with each other perfectly because they would be distorted in different ways: the domain boundaries must be rather complicated – more than just planes across which octahedral vacancy positions change (hence the term 'pseudotwinning').

In view of the common tendency for mistakes in the stacking of layers in kaolinite crystals, one might suspect that in these vermiforms, vacancy orientations might not always be accurately copied. (They must *tend* to be, otherwise the crystals would not be kaolinite.)

Mansfield & Bailey found also another kind of stacking fault, although not in all of the crystals. This was a true twinning where successive packets of layers were turned upside down. (Somewhat like figure 7.6, making grooves parallel to the kaolinite layers.) Occasional switching of vacancy orientations (or layer inversions even) might not matter because the information is seen to be primarily in a pattern of domain boundaries. What is important is that these boundaries do not wander as the crystal grows. But as indicated already we have some direct evidence for fidelity here from the morphology of some of those microscopic vermiforms – from their 'extruded' appearance – whether the more detailed structural considerations are correct or not.

Crystals as replicators We can see, then, a variety of possible ways in which defect structures of various sorts might be copied through crystal growth processes. Each of these ideas depends on the notion that crystal growth mechanisms, although reversible, are only locally reversible under conditions of supersaturation. The mechanisms have no foresight, they cannot see the most perfect way of placing all the units even if they can, in a sense, 'feel' the best way of placing one or a small number of units. They can follow some easiest path. Often the easiest way of proceeding, even the most orderly way of proceeding in the circumstances, may be not to eliminate a defect but to *repeat* it: to extend a screw dislocation or a disordered layer stacking sequence; or to repeat a pattern of domains. That way there may be less local disturbance. Suppose, for example, that a pre-existing layer consists of domains each with one of three possible orientations for rows of units (as in figure 7.1*h*). What would be the most orderly way of making another layer? With all the rows, this time in the same direction? It would take foresight to see that that was the way to perfection because locally there would be a severe mismatch: two-thirds of the new rows would now be misaligned with respect to the rows underneath. Even to shift a domain boundary a little will involve some higher energy intermediate, especially if the interlayer forces are strong and depend on the matching of orientations for that strength.

Mansfield & Bailey suggested that in kaolinite the 1:1 layer type, and the adoption of the same vacant site in each layer, produces an asymmetric and rather strained structure – and that that is why kaolinite crystals usually remain very small. The pseudotwinning, and the inversions of packets of layers, they see as ways of 'alleviating the strain by redirecting the distortions on a domain scale'. We saw something broadly similar to this in antigorite with its curious periodic lateral inversions of 1:1 serpentine layers as a means, here, of alleviating the considerable strain inherent in the mismatching of tetrahedral and octahedral sheets in that structure.

It is an interesting thought that in layer silicate minerals the perfect structure may very often not be the structure of lowest energy: structures with various forms of domaining may be more stable with the energies of these various forms very similar. After all it would be sheer coincidence if there was no mismatch at all between the elements of an asymmetric layer, such as the 1:1 silicate layer. The chances are that on its own, *in vacuo*, it would tend to curl one way or the other. To that extent there must be some energy penalty, however small, in having the layers flat in the crystals. As a crude model you might imagine an old painting which has developed a tendency to curl because the paint film has contracted. If the painting is kept on a stretcher the paint film will craze.

We can take this analogy one step further. I daresay that for a painting there would be one craze pattern that was of minimum energy – some regular array of hexagons I suppose. But who ever saw that on a painting, or exactly the same such craze pattern twice? Similarly, a silicate layer could make a very good information store, following our earlier discussion (page 272), if some sort of 'crazing' is better than none at all, but if at the same time no particular craze structure – no special regular superlattice – is much more stable than countless others.

3. What was the locking mechanism?

For DNA replication the locking mechanism is quite distinct – it is in the formation of phosphodiester bonds. The security of this lock depends on the relatively high activation energy barriers that protect such bonds from hydrolysis. This bonding is not normally reversible.

As we saw in Chapter 6, linear silicate and metal-oxy polymers in aqueous solutions are not able on their own to maintain a complex meta-stable pattern – their bonding is too readily reversible. For crystal genes, however, the security of the information would not depend on the integrity of bonds exposed to water in the same way: it would depend rather on there being many copies of the information with most of the copies inside the crystal. In the growth of silicates or metal oxide crystals from aqueous solution there is local surface reversibility. That is important for the matching-up mechanism. But because water participates in those bond-making and bond-breaking processes which are necessarily involved in the growth of crystals of this sort, this reversibility will generally be restricted to the surfaces. Eventually the information in, say, a layer of a growing layer silicate becomes locked in through the growth of further layers on top.

For crystal genes, then, the matching and locking mechanisms would be less distinct than for DNA with its two kinds of bonding for these two functions – reversible hydrogen bonding (mainly) for the matching, and irreversible covalent bonds for locking. For crystal genes of the kind that we have been thinking about the strategy of holding and replicating information may be similar, but the tactics are here not the same at all. In particular the matching and locking mechanisms are not so distinct from each other, and the bond types used mainly are like neither of the bond types used mainly in DNA.

4. How did master and copy separate?

One rather unexpected difficulty in the idea of an organic polymer as a primary genetic material was with the question of how, having synthesised one chain on another, the chains could be made to come apart (Chapter 5). The trouble was that, if the matching forces were strong enough to hold the units reliably in place, then they would together be too strong to let the chains separate. There were various ways in which this problem might be overcome, but they all seemed to need too much contrivance to provide the basis for a primary genetic material. The corresponding final stage of the replication of information in growing crystals looks rather easier – it would be through the break up of the crystals in such a way as to expose the information on new surfaces.

Crystals very often break up as they grow. A flask containing a supersaturated solution may remain for days with nothing happening – and then quite suddenly be full of thousands and thousands of crystals. The initial delay showed that the rate of spontaneous nucleation was far too slow for thousands of independent nucleations. It is more likely that what happens in such circumstances is that one crystal nucleates spontaneously and then grows quickly and breaks off pieces to make new nuclei which in turn grow and break up. This kind of thing is called 'breeding' (with no implication of course that any very complex information is necessarily being propagated in the process). But we can at least say this about a working crystal genetic material: it must 'breed' in this unsophisticated sense as a necessary but insufficient condition. More precisely, it must have a strong preference for cleaving across the direction or directions of growth and in the planes or plane within which the information is held (*i.e.* as in figures 5.11 and 5.12).

These last considerations make a layered structure attractive since there is here a strong preference for cleavage in the plane of the layers and, as discussed earlier, there are ways in which information could be held within the plane of the layers. Nor is it asking too much that crystal growth must not take place in this same plane – growth must be exclusively layer-on-layer. Highly elongated habits are common enough in crystals, indicating that there can often be a strong preference for one direction of growth. The kaolinite vermiforms, indeed, can most easily be explained as having arisen from a crystallisation process that was mainly in one direction – in the correct direction for our purposes. And of course these long fragile vermiforms often break up as they should – in the plane of the layers, across the preferred direction of growth.

It seems too good to be true that kaolinite, the commonest clay on Earth perhaps, might also be a primary genetic material. But it combines nevertheless many of the characteristics that you would want. Maybe the first genetic materials were something like kaolinite.

Primitive genetic control

We come now to the third basic characteristic of genetic information: it must not only be held and replicated, it must have a meaning. We might say that the meaning of genetic information depends on its exerting some sort of control that tends to increase its own prevalence. To be genetic information it must tend more or less indirectly to be self-propagating.

The indirectness is important – one of the characteristics of highly evolved organisms. Physical scientists naturally tend to seek the explanation for some persistent or recurring patterning of atoms in terms of the relative stability of that pattern, or at least of the more or less immediate processes involved in its formation. But evolved genetic patterns are not to be explained away like that. If you found, let us say, that part of a mammal's genome read ATTGCGTAGCGTAAGTCG, you should not expect to find any direct physicochemical explanation for it. Your efforts to show that this sequence was especially thermodynamically stable or easily formed would be unsuccessful. More likely the sequence specifies the positions of a few amino acid side chains in a protein molecule – helping to make an enzyme, perhaps, that helps the animal to see better, that makes it aware more quickly when there is a cat in the vicinity. That would be a very indirect reason for ATTGCGTAGCGTAAGTCG.

We might say, then, that while life depends on replicating structures, it depends too on there being circumstances in which particular complicated structures are preferred. This in turn depends on there being possibilities for indirect and yet successful control of the environment by replicating patterns. Perhaps, indeed, evolution can be described most simply as being the appearance of increasingly indirect ways in which replicating patterns can show their mettle.

We will take it, then, that the first genetic materials evolved along such lines – that, although the patterns they contained exerted a rather direct kind of control to begin with, a major early evolutionary trend was towards more indirect modes.

Direct effects of millions and millions of gene copies

To begin with, the replicating defect structures in a crystal genetic material might have operated simply through the effects of such defects on the bulk properties of that material – via effects on crystallite shapes and sizes – and on the way the crystallites interacted with each other.

Van Olphen (1977) discusses in detail various factors that can influence the bulk properties of clay suspensions. Quite small differences in the composition of the solutions within which clay particles are suspended can make the difference between, say, a drilling mud being very runny or far too stiff; or it might make the difference between a badly drained and a well drained soil, or between a place where you might build a house or be very ill advised to. Such differences are to be explained in colloid chemical terms, in terms of effects operating at the microscopic level. If adding a relatively small amount of lime or gypsum to a soil greatly improves the drainage, this is probably because negatively charged clay crystallites have had their charges reduced or neutralised so that the particles now adhere to each other, via the calcium ions to form separate flocs. But, depending on the shapes of the crystallites present, on their sizes, on their electric charges – and often too on details of the distributions of these charges – a given flocculating agent, like Ca^{2+}, may operate more or less efficiently or it may even operate in reverse. Perhaps the particles are positively charged anyway, but they have special sites on them with a strong selective affinity for Ca^{2+}. Then they would become more highly charged and more likely to deflocculate when the calcium concentrations of the surrounding solutions increased.

For clays, anions can act as deflocculators by cancelling positive charges on the edges of clay platelets – where there are exposed intralayer cations. Polyanions are particularly good at this: indeed polymetaphosphate can even reverse the edge charge (see figure 7.9). This is thought to be effective in destroying a kind of 'card-house' association between the clay particles that can produce apparently very solid gels (see figure 7.10; compare figure 6.54*f* on page 247).

Not only can small changes in the environment of a given clay suspension make a big difference to porosity, ion exchange properties, etc., but between different clay samples small differences in structure may also be

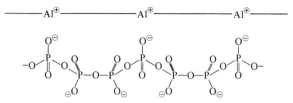

Figure 7.9. Partly exposed cations at clay edges can lead to positive charges there. Polyphosphate can reverse such charges, possibly, as suggested by Van Olphen (1977), through co-ordination of some of the anionic groups of the polyphosphate leaving others to build up a net negative edge charge.

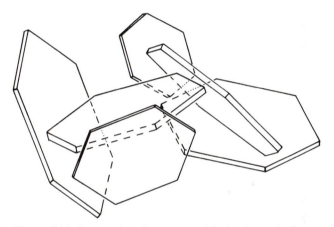

Figure 7.10. Interactions between positively charged edges and negatively charged faces of clay particles can give rise to 'card-house' associations.

important. In particular the shapes and sizes of the crystallites and their defect structures are likely to be important – the latter because they affect the distribution of permanent charges in the crystallites and the locations of special adsorption sites. On such features may depend, for example, the stability of a given 'card-house' to changes in the environmental solutions, or whether an impervious gel-like association is preferred or a very porous lumpy flocculation.

Some genetic control of some quite pedestrian bulk properties would be enough to allow us to imagine the beginnings of a Darwinian evolution. Figure 7.11 illustrates the kind of situation that we touched on in Chapter 3 when discussing how 'naked genes' might work. Here we suppose that

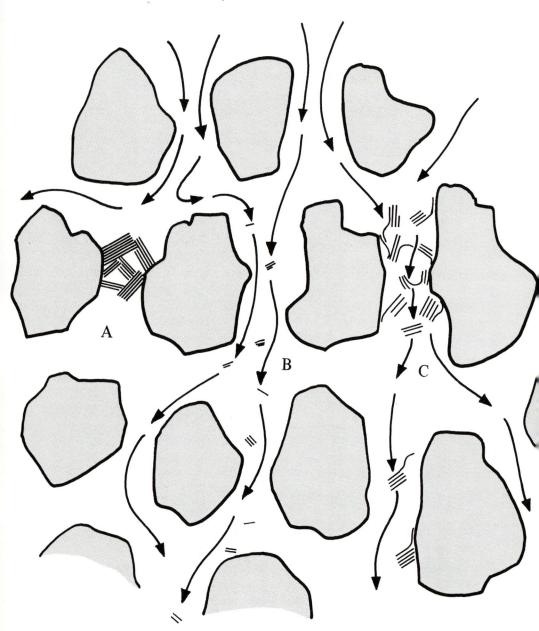

Figure 7.11. We imagine here replicating clay crystallites forming in a sandstone through which nutrient solutions are moving. In zone A the clay being formed is too impervious – it deflects the nutrient flow. In zone B the crystallites do not adhere sufficiently and they are being washed out of the sandstone. The clay type forming in zone C is the most successful of the three: it is porous and yet not easily dislodged wholesale. Sometimes, however, a piece breaks off to seed a similar colony downstream.

clays are growing in a sandstone (cf. figure 6.54) and we suppose that one clay is printing off copies of a defect pattern that makes crystals that associate with each other strongly to form a plug. This stops the local flow of nutrient solutions, and the clay crystallites stop growing there. The pattern of defects that makes that kind of crystal is a failure. On the other hand, it would not be a good idea either to be too loose. A defect structure that produced crystallites that did not stick at all to each other or to the walls of the pores would also be selected against because the crystallites containing such patterns would be washed away. The best compromise might be to stick to the walls but let the nutrient solutions flow through by having a rather open porous texture.

Because it is a consequence of interactions between millions and millions of gene copies, you might say that the consistency of such a clay would be a phenotype. Even this seemingly commonplace property, provided it could be influenced by (usually) reliably replicating patterns, could provide the basis for an evolution through natural selection. Inevitably the clays in a location for which consistency was important would evolve an appropriate one. No wildly improbable event would be needed to start such a process off because nothing very clever is required. Of all the crystallite shapes, etc. that might be being replicated perhaps half would be adequate as starters, because they stick (at least a little) and do not stop the flow (at least not completely). Of that large fraction of all the possibilities that were adequate, there would be a subfraction that were a bit better, and a sub-subfraction that were better still. In line with the discussion of optimisation procedures given in Chapter 3, evolution would be expected to proceed from patterns belonging to larger subsets into those belonging to smaller subsets (cf. figure 3.1), that is, from structures that could have been almost anything to those that were rather special. How far this would go, how small the subset that could eventually be arrived at, would depend partly on how intricate the control of consistency through defect structure could be. It would depend also on whether a very particular consistency made a perceptible difference to survival chances or replicative efficiency in the available environments. That is to say the evolutionary potential would depend on the characteristics of a relevant functional landscape: to get started there must be a large catchment area provided by some easily achieved function, but to go very far there must be fairly deep valleys somewhere.

How odd might the properties of genetic crystals become? If we had no experience of highly organised protein molecules made from perfectly specified amino acid sequences, if the only long polypeptides that we had ever come across had been copolymers of the sort that arise from normal chemical polymerisation procedures, we would have no conception of the subtle pieces of machinery that a well made protein can generate. For one thing we would not even have come across a sample of material in which all the polypeptide chains were identical. We would have had experience only of mixtures and hence of visible properties that were the statistical outcome of mixed assemblages. (We would be lucky, indeed, to have come across a sample of long polypeptide chains in which any one chain was the same as another.) Blends are likely to be boring, average things concealing special properties of rare, or even not-so-rare, individuals.

Straight away, then, you might expect the visible properties of a material composed of replicating crystallites to be quirky: at least now there is the possibility that characteristics arising from individual defect structures, rather than from mixtures of certain general classes of such structures, could show up. But of course the possibility of evolution through natural selection inherent in (nearly perfectly) replicating structures leads to a still more interesting thought: that exceedingly rare individual structures might be discovered in time, with properties that would have been very odd indeed for the normal average sort of material. Indeed in proceeding from set to subset to sub-subset, in exploring an evolutionary landscape, natural selection can quite quickly arrive at subsets that are so small that you can discount the possibility that that subset could have been hit on in any other way (cf. pages 87–8).

Let us not suppose, then, that microcrystalline minerals, such as clays, could never have very interesting properties because the stuff at the end of the garden spade does not look very interesting. (And let us not be too put off by the unfortunate term 'defect' – a 'defective' silicon crystal can be a microprocessor – it is a question of *what* 'defects' and where they are.)

The ultimate limit to a crystal's performance as a genetic material would be set by its maximum genetic information capacity. In an idealised case this would be given by:

$$I_{max} = \log_2 w \text{ (bits)}$$

where w is the number of *a priori* equally likely ways in which the replicable features could be disposed. As an example, suppose that there was complete freedom about the disposition of the screw dislocation lines in the hypothetical genetic crystal shown in figure 7.5, and suppose that there were 100 such lines in a crystal 1 μm across. If such a crystal was a layer silicate it would have about 10^6 unit cells in a layer, with about that number of positions for an intersecting dislocation line. There would then be about $10^6 \times 10^6$ ways of putting two dislocation lines – and about 10^{600} ways of putting 100 such lines, giving a maximum information capacity of about 2000 bits. By comparison, an average sort of globular protein with 200 amino acids is one of 20^{200} possibilities, which means that the information capacity here is about 860 bits.

There are dubious assumptions in calculations of this sort – the independences of the screw dislocations and the accuracy of replication of their disposition are the most serious here – but we can see all the same that quite a moderately complex defect structure could give rise to microcrystals – objects still of colloidal dimensions – that could in principle be as ingenious as protein molecules (that is to say selected from a similarly large set of possibilities). If the information was in the form of a pattern of 100 replicating domains the capacity could be very much greater than for a screw dislocation array (cf. figure 7.7). (Kaolinite vermiforms may combine each of these ideas – if each of the columnar domains implied by the electron micrographs contains one or more screw dislocation lines in the direction of growth.)

How to 'focus' information Although the information content of such genetic crystallites might be similar to that of a protein molecule, the information density would be very much less. Does this limit the effects of such information to effects on consistency and other such bulk properties? In particular, does it rule out anything really clever – like the intricate control of organic reactions? Would such information be altogether too coarse-fingered ever to be able to manipulate individual molecules – ever to arrive at that point that we know early evolution did arrive at where a competence in protein and nucleic acid chemistry had been achieved?

There are four things to be said here. *First,* there is nothing in the definition of life to say how large or small genetic information should be written. These are practical, not theoretical questions. *Second,* coarse fingers, human fingers, can in effect manipulate molecules: if 100 ml flasks and such things can control organic reactions when suitably used, I do not see why contrived (evolved) micrometre-scale apparatus might not be effective too. There is nothing in the definition of life either to insist that a protein-like technique of fine molecular discrimination should have been the basis of the first forms of biomolecular control. (I will come back to this topic in the next chapter.) *Third,* modern genetic information is not in any case all that fine-grained: it takes about 100–1000 nm of DNA to specify a protein molecule. There are two stages in the 'focussing' of this rather diffuse information so that some precise molecular activity can be specified. The information is translated into a somewhat more compact kind of molecule with a higher information density – a protein – and then this molecule folds up. The information in a protein is written still rather coarsely – as a sequence of choices of *ca* nanometre-sized objects; but, because of the folding, a globular protein can discriminate at a much finer level. As we discussed in Chapter 3, the positioning of the few side chains at the active centre is the consequence of many, if not most of the other side chains in the molecule, with the choice of remoter side chains allowing 'fine adjustment' of the positioning at the active centre. The effect of the choice of perhaps 100 amino acids is thus concentrated at the active centre. Each choice may seem rather coarse but 100 one-in-twenty choices represents enough information to allow a very detailed specification in principle.

This kind of focussing of information originally spread out in a linear sequence depends absolutely on the beautifully articulated quasi-crystalline structure of the folded protein molecule. Real crystals too can transmit effects over a distance through concerted pushings and pullings, as we discussed, in particular, for layer silicates (pages 223–5). And information of a sort can thus be focussed in real crystals in the sense that characteristics of a given site may be determined by the cumulative effects of atoms distributed within a quite extensive adjacent region. So the *fourth* point to make is that relatively diffuse information in a crystal could very well exert effects that were concentrated or expressed at particular sites. With reference to the suggestion made by Mansfield & Bailey (1972) that a large kaolinite domain becomes unstable because of the accumulation of distortions created by the unsymmetrical unit cell of kaolinite, we might speculate that the detailed character of a given domain boundary, the extent to which it represented a misfit between differently distorted lattices, might depend on the sizes and shapes of the adjacent domains. (The bigger they were, the more out of step they would be with each other.) In such a case a given crazy paving pattern of domain boundaries would be chemically a very interesting structure, it would be a particular array of high-energy sites with particular adsorptive and catalytic properties. The edges of layers, and the grooves created by stacks of layers, might also represent replicable active features.

Threats and opportunities

To imagine how more particular and intricate genetic patterns might catch on we should try to imagine more complicated situations than that illustrated in figure 7.11. Suppose for example that the flow rate was variable – and in the rainy season it increased while at the same time the ion concentrations dropped. The successful clays would be those whose crystallites would cohere in solutions of low ionic strength. Perhaps indeed the most successful would adhere to each other more strongly under those circumstances, so that they formed an impervious plug to stop the flow locally, to reduce the rate at which they were being redissolved by the now undersaturated solutions. When the ion concentrations increased again, in the dry season, the crystallites would 'know' to separate because conditions were no longer dangerous – it would be time to be printing off some more crystals.

Although abnormal, perhaps, such behaviour for a clay would not be absurd. But anyway we are not talking about normal clays. We are talking about clays of a sort that we have no direct experience of (well, probably not, but I will return to this at the end of the book). Suppose that we really could tailor-make a clay particle; suppose we could place substitutions precisely, we could arrange lattice tensions so that they were just so; we could contrive shapes and sizes exactly; suppose that we had the techniques to fabricate clay unit layers in the kind of way that we can fabricate micro-processors – only at a still much finer level. Then we would have little trouble, I think, in engineering much more ingeniously adapted clays than the one imagined above. You could make materials with properties unlike those of any natural clay, with specially engineered crevices, perhaps, that had an affinity for very particular ions or molecules. Or you might make a clay that self-assembled into a mansion of a card-house with rooms of just certain sizes interconnected with each other in certain definite ways. Can you doubt that if we could fabricate silicate layers with atomic precision we could produce materials that to the layman would have magical properties? Such a material might seem alive. It would not be alive because it would be a human artefact, its multiple ingenuities built in by hand. But suppose instead natural selection had been the engineer. Then it would be alive.

What for a normal physicochemical system would be an extraordinary coincidence becomes almost an expectation for a structural feature that is subject to evolution through natural selection. Among such features are those that help to propagate the genetic patterns more widely. (Because those genetic patterns that had in the past been adept at publishing themselves are bound to be the most prevalent now.) A simple way would be for a clay to be of such a consistency that flocs occasionally broke off and seeded further growth downstream (figure 7.11c). But in a finite open system there is a limit to the amount of spreading that is possible downstream. So perhaps clays might be selected that on drying made a friable powder that was easily spread by the wind to seed further growth of similarly adapted material elsewhere. Or perhaps, like the salmon, the clays would swim upstream to lay their eggs. Ridiculous? Not altogether: we might imagine clays forming in groundwaters that occasionally reversed the direction of flow (because of an overflowing river or a very high tide or something). Then the clever thing would be to break into flocs when the flow was 'upstream'. For a clay that had been engineered by natural selection such a feature would be quite understandable because it would be based on a suitable conditional property that would not in itself be un-claylike. For example when the concentration of Na^+ reached a certain level (a sign of back-flow) loose flocs would be formed.

Nor is there anything very extraordinary in a material having 'conditional properties' – all materials have them (for example, *if* I heat this piece of plastic *then* it will soften). What would seem extraordinary about an evolved genetic clay would be the detailed appropriateness of the conditional behaviour. It would be a teleological way of speaking to say that such a material had learned to cope with particular habits of the environment; as it is a teleological way of speaking to say that the crocuses know when the spring has arrived. But such teleological talk becomes justifiable (at least as kind of shorthand for longer-winded mechanistic explanations) when talking about products of natural selection – and sooner or later it would become justifiable when talking about the behaviour of replicating mineral crystallites. If the total behaviour of the crystallites was seemingly highly improbable; if it depended on complex individual features of the crystallites being just so; if these features were replicable or produced under the control of replicable structures – and anyway if their configuration that had generated the strange behaviour was indeed a product of natural selection – why then it would be plain silly not to call such minerals alive.

Early phenotypes

The main oversimplification in the foregoing discussion of modes of genetic control has been in thinking about pure genetic crystals. Pure mineral deposits are not at all common and nor is it necessarily so that primitive genetic crystals could catch on most easily if theirs was the only kind that was crystallising out in a given locality. Let us consider what might happen if both genetic and non-genetic crystals were being formed in the same place.

To begin with you could think of the non-genetic crystals as simply part of the environment of the genetic crystals – something the latter had to put up with, perhaps. But sometimes, by chance, the non-genetic crystals might be an advantage. For example, they might tend to offset the main snag about a genetic crystal – that to work most effectively it must be in a solution of just the right level of supersaturation (cf. page 159). We might imagine non-genetic crystals acting as a buffer for the units out of which the genetic crystals form: they might be some relatively quickly crystallising material – such as imogolite or allophane. These would hold concentrations down during periods of plenty preventing a too rapid growth of the genetic crystals (that would lead to too many mistakes in replication). On the other hand, when the solutions flowing into the regions became undersaturated the non-genetic crystals would dissolve to maintain supplies. Or perhaps non-genetic crystals could act as other kinds of stabilisers, as pH buffers or flow controllers.

The point at which such non-genetic minerals might be said to be part of the phenotype of the primary genetic crystals would be when the genetic crystals evolved characteristics that encouraged favourable non-genetic minerals to form. It is only one step away from helping yourself to help someone else who helps you. But how might the genetic crystals affect the formation of some other, perhaps quite different kind of material?

There are several ways. The genetic crystals will have effects on flow rates, for example, and this is bound to alter the compositions of the surrounding solutions and influence the sets of minerals crystallising from them. Not only the types, but the morphology of a given mineral type can be strongly influenced by conditions of synthesis. Newly formed halloysite appears to be particularly sensitive here, showing a great variety of micromorphologies that appear to result from differences in such things as local permeability, pH of waters and grain size of primary minerals (Tazaki, 1979b).

Control through epitaxy is another possibility. One crystal can nucleate another's growth and it is typical that such nucleation is at defect sites (see for example Parham (1969) on the growth of halloysite on feldspar). So there might be a marvellous scope here for a critical control by highly contrived defect structures at the surfaces of genetic crystals. Special defect sites on evolved genetic crystals might be able to nucleate the growth of mineral crystals that would not normally have formed – minerals for which nucleation tends to be the difficult step.

Even if only the morphologies of the newly forming minerals was being controlled by the genetic crystals this might still have a profound effect on the local physicochemical environment of the genetic crystals. Look again at the varied morphologies of zeolites (figure 6.26 on page 194) or at the various forms of ferric oxide-hydroxides (figure 6.1 on page 169) or at the variety of forms and types of clay minerals in sandstones (figure 6.54 on page 246). If genetic crystals could control the kinds and forms of such minerals, even if they could only exert a general kind of control by helping one kind of growth process or hindering another, they would be controlling indirectly such things as local porosity, local ion-exchange characteristics, local catalytic effects and so on. They would have a more indirect and complicated means of affecting factors that affected their own survival and propagation.

Advantages of specialist phenotype structures

Given reliably replicating mineral crystallites we can, then, begin to see simple mechanisms for indirect self-help. But we can see also a driving force – a reason for a long-term trend in that direction. There are positive advantages for genes to operate through other structures. The ways in which a genetic material can directly control its environment are bound to be limited by design constraints imposed by the overriding need for such a material to be able to replicate information accurately. Crystals that were good at pattern printing would probably have to be rather rigid with quite strong inter-unit forces so that crystal growth proceeded in a very orderly manner. On the other hand, although some control structures could be made from rigid materials, others might be more suitably made from flexible ones. The control of more intricate and heterogeneous micro-environments might become possible if, say, tubes and vesicles were present as well as plates and grooved rods. It is, I suppose, more difficult to template-replicate a tube than a plate. A tube might be reproduced without being template-replicated so long as the characteristics of the tube could be specified by information in, say, a plate – and so long as that information was replicable. Modern organisms use this device of dividing the roles of information replicator on the one hand and main control structure on the other between quite different kinds of molecules – between DNA and the much more versatile and flexible protein.

Embarking yet again on the hazards of more particular speculation we might imagine a primary organism with an analogous device. We might imagine that from time to time, when groundwaters were fairly concentrated, the layer-on-layer growth of domained genetic kaolinite crystals was interrupted by a few layers of halloysite instead, these layers becoming unstable eventually and rolling up to form independent tubes whose exact swiss-roll structure was affected to some extent by domain patterns which they had inherited. The maze of tubes thus made then helped to create an appropriate consistency perhaps, or was a pH buffer, or bound certain heavy metal ions that would otherwise interfere with kaolinite synthesis.

The general technique of gene action

To improve the general analogy between the way in which primary genes might have been effective and the way in which modern genes work it is worth emphasising what genes of any sort do and what they do not do.

In the first place genes can only operate passively because they are not themselves sources of energy – they can only redirect flows or potential flows in the environment. And genes do not specify a phenotype through specifying a brick-by-brick construction in the kind of way that you might physically build a house. The construction procedure is more like the way an architect builds a house – relying on the known tendency for others to work in certain ways. As remarked in Chapter 3, most of the explanation for the arrangement of the atoms in an organism is to be given in physico-chemical terms. Genes only cue the outcome. That is all they could possibly do since they contain far too little information for them to be able to specify the positions of every atom in the phenotype. Even in the making of a protein molecule the final functioning object is not intricately made by the genetic information: most of the explanation for why a protein folds the way it does is to be found in books on physical chemistry. There is only one factor that you will not find there – why the amino acid sequence is as it is. (If the primary structure of a protein really was the sole determinant of its tertiary structure you would in principle be able to make any adjustment to the position of any atom in the folded molecule by suitably altering the sequence. This is clearly not the case, only some adjustments to the tertiary structure are physicochemically possible.)

Although in the details of how they would work and what they were made of primary mineral genes would have been quite different from DNA genes, they would have been similar in that they too would have worked by 'cueing' the environment, by triggering events. The main difference would have been that the kinds of events triggered would have been closer to the ground in all senses – nearer to what tends to happen anyway.

Vital mud

As for the size of primary organisms, they were not the neatly packaged boxes of genes that micro-organisms are now – with roughly one gene of each kind needed to run and remake the localised 'survival machine' that contains them. The microscopic size of the modern bacterium should be seen as one of its most notable achievements rather than any sign of evolutionary infancy. The bacterium is a highly concentrated package of ingenuity. Before that sort of thing became possible there would have been clumsier, bulkier ways towards our biochemistry (as I will discuss in the next chapter).

The first primitive organisms that we have been trying to conjure up were not really microscopic objects at all. The control of the environment by primitive genes depended, not on the individual acts of individual genes, but on effects depending on millions and millions of copies of them. There was thus no need for anything as neat as a cell. If you are a gene very close to the ground, if your modes of preferential survival and propagation depend on deflecting somewhat, and to your advantage, processes that are going on in any case around you, there is no need to be so cordoned off. Indeed it is better not to be. It is only much later, when you are a gene made of an unnatural sort of stuff having no direct connection with the usual products of the Earth, that you must have your life support systems immediately around you. Then you *have* to be in the kind of cell that has now become the identifiable unit of life – but only then.

For a picture of first life do not think about cells, think instead about a kind of mud, an assemblage of clays (including perhaps, zeolites, iron sulphides, oxides, etc.) actively crystallising from solutions. It might not have seemed to be very prepossessing stuff, this vital mud, if you had gone back to the primitive Earth to look; but it would have had some strange properties if you had taken the trouble to examine it carefully. It is not just that it would have tended to alter its properties (consistency, ion-exchange characteristics, etc.) according to circumstances; this substance would have seemed to know about the world, because its behaviour would have been appropriate in quite complicated ways to its survival – and to its spreading.

A sequence of places to live

Where on the primitive Earth would you have looked for these earliest forms of life? You would have looked in open systems – you would have looked for conditions for continuous culture where food supplies were assured and where conditions for growth remained reasonably constant over the long term. For these mineral organisms that would have meant somewhere between weathering and sedimentation. You might have looked in the porous weathering crust of a piece of granite, or in a volcanic tuff, or in precipitates from groundwaters far from the origin of the ions in those waters. Or you might have looked in streams or rivers or lakes, or in the sea and its sediments.

It is well understood that the descendants of an organism that evolved in one place may now live somewhere else. We do not live in the trees any more, and yet the design of our hand and eye owes much to an arboreal way of life. Going further back, our circulatory system was basically invented by animals that lived in the sea, while cell components such as mitochondria were first put together by micro-organisms with some now unknown way of life. Scene changes have been an important part of evolution – it is almost as if the machine had to be sent to different workshops to have different components at different levels of organisation perfected.

I think that there would have been such scene changes at the start of evolution because, whether you believe a mineral story or not, there were very special problems for organisms at the beginning. Life would have started, I think, in a special workshop that happened to be set up to produce a first working genetic material. An unevolved first mineral life, for example, would have been restricted to particular regions, where conditions happened to be precisely right for orderly replicative crystal growth – a particular region of a particular sandstone perhaps. But then, with the evolution of indirect modes of control, with the cued collusion of other materials, the genetic information could have been preserved and propagated in a wider range of places.

So the first workshop would have been somewhere stable, deep down probably, below ground or near the bottom of the sea. To begin with, security would have been more important than opportunity. But adjacent to the easy places where evolution could start there would have been more difficult places to provide a clear field for some of those with more elaborate ways of surviving. As part of this adventure we might suppose that some kinds of mineral organisms found places to live at the surface, in rivers or lakes perhaps; that they found places in the sunlight – and another kind of chemistry.

8

The entry of carbon

Figure 8.1 gives an outline of the plot of our story – about how the last common ancestor of life on Earth came into being. In the last chapter we were concerned with the first three stages shown in this figure, arriving at the idea that the first organisms were, quite simply, made of clays; and that *life* would soon have become a wholly appropriate form of description for these systems, since they would only have been plausibly explicable in terms of a Darwinian evolutionary history.

Everything about this story rests on the supposition that there was some class of colloidal inorganic crystalline minerals forming on the primitive Earth that could replicate more than trivial amounts of information accurately. From our discussions of conditions on the primitive Earth; of how microcrystalline minerals form today; of what is needed for accurate replication – and from considerations of bonding characteristics in organic and inorganic materials – it seems altogether more plausible that growing mineral crystallites rather than organic polymers should have been the primitive genes. We cannot yet put our finger on one mineral and say that it combines all the required features, but we can say that clays show between them all the features required. We can say, even, that one kind of clay – kaolinite – is a conceivable candidate. In any case this question is open to further clarification through observation and experiment; we will come back to this in the coda.

The details of the life-styles of the earliest organisms are much less accessible to experiments and observations. But they are less important. In the latter part of Chapter 7 we saw a number of ways in which that engineer natural selection might have taken hold if reliably replicating structures were there. A more important point was the more general one – that a

STAGE 1: unevolved crystal genes	STAGE 2: evolved crystal genes	STAGE 3: complex 'vital muds'
Clay minerals on the primitive Earth include some with complex replicable defect structures	Some patterns of replicating defects become more common because they confer on assemblages of clay crystallites properties (including large-scale properties) that favour clay synthesis	Replicating defect patterns evolve more indirect modes of control: complex assemblages are formed in which genetic and nongenetic clays promote each other's synthesis and propagation

STAGE 4: photosynthesis	**STAGE 5: metabolic pathways**	**STAGE 6: structural organic polymers**
In a number of lines, evolved clays provide photochemical machinery for making a few metastable molecules, especially polyphosphates, very simple C, H, O compounds and ammonia	Elaborately specified clays make the apparatus for multi-step organic syntheses	The availability now through biosynthesis of particular chirally specified monomers provides the basis for the manufacture of precisely made organic polymers for various structural purposes

STAGE 7: molecular genes	STAGE 8: protein synthesis	STAGE 9: organic organic-synthetic machinery
In one line of organisms a class of interlocking structural polymers (protopolynucleotides) can replicate as an alternative mode of synthesis creating a secondary minor genetic material within the organisms	The secondary genetic material proves to be more versatile: it comes to control some steps in the synthesis of organic molecules, especially the joining of amino acids into peptides	With the ability to make protein, and the (late) discovery of enzymes and of lipid membranes, alternative, more efficient techniques of organic synthesis become available: all the clay machinery is dispensed with

Figure 8.1.

trend towards indirect action by a primary genetic material might be expected. Our questions for this chapter are on why and how organic molecules became involved in this trend. We will be concentrating on the middle three of the nine major stages suggested in figure 8.1.

From inorganic crystals to organic molecules

If an inorganic crystal life is easier to start up, no doubt an organic molecular life is more efficient in the end. So given a mechanism (genetic takeover) you would expect early evolution to change from one kind of life to the other. You might say that evolution started with inorganic materials not because it wanted to, but because it could not have started in any other way – practical organic chemistry is altogether too difficult for unevolved organisms. In organic synthesis there are not the error correction mechanisms of crystal growth processes. The main virtue of organic chemistry – the fidelity of intramolecular bonding – is thus also at the root of its difficulty: the outcome of a loosely controlled organic synthesis is usually a very complicated mixture of metastable molecules – a tar of some sort.

While semi-chaotic organic mixtures can be useful – and we will be thinking about some possible primitive biological uses for such mixtures later – it is hard to see very precise higher order structures being put together from organic molecules on the primitive Earth. From our discussions in Chapter 5 it is especially difficult to see how information could be replicated through semi-chaotic organic mixtures.

'Self-assembly'

Looking at our present biochemical subsystems we can see organic molecules specifying more or less precise higher order structures. A membrane of a particular thickness and fluidity will put itself together from a suitable collection of lipid molecules (figure 8.2). A groove with a particular catalytic effect may be specified by a folded polypeptide sequence. These and other forms of 'self-assembly' are a crucial part of the whole technique through which modern genetic information is expressed – through which a message is converted into action. Given the competence to specify a suitable set of covalent bonds, such objects as membranes and catalysts may need no further specification. In terms of an idea introduced towards the end of Chapter 3, organic chemistry has evidently an enormous efficiency range for several key biochemical functions. But that is not to say that organic molecules could have carried out these functions to begin with. I would suggest that for most biochemical functions devices made from organic molecules would have a curve like curve II in figure 3.10: great potential but no competence at low levels of preorganisation.

Even to make a lipid membrane, for example, the molecules should be amphiphilic and at least roughly of the right shape (Israelachvili, 1978; see

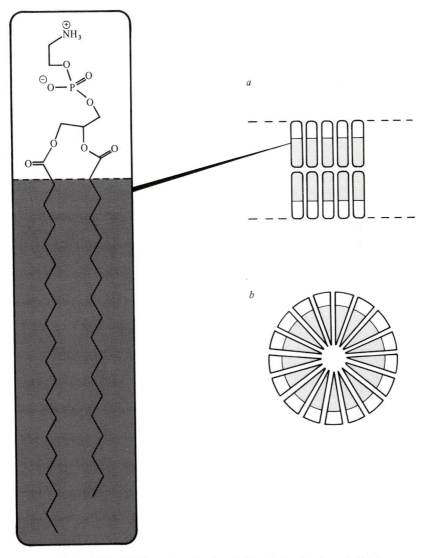

Figure 8.2. (*a*) The pair of hydrophobic chains in phosphatidylethanolamine has a similar cross-section to the (hydrated)
hydrophilic head group. Such lipid molecules may 'self-assemble'
into an extensive membrane structure. (*b*) Single-chained lipids are
effectively wedge-shaped and tend to form finite micelles instead.

legend to figure 8.2). And much more prearrangement is needed for a protein to fold up consistently so as to make, say, a catalytic groove. Generally speaking organic molecules have to be preorganised to a considerable extent before they will organise themselves at all precisely.

We can see this in the familiar phenomenon of crystallisation, which is an example of a 'self-assembly'. In order for an organic material to crystallise it is usually necessary that the material should be fairly pure.

Requirements for a 'self-assembly' We can list three very general requirements for a 'self-assembly' from molecular units:

1 There should not be too many kinds of molecular units present.
2 There should be a sufficient cohesion between the units.
3 The bonds between the units should be reversible.

Now it is difficult for a loosely controlled organic product to satisfy conditions 1 and 2 at the same time. And for this reason: the forces between organic molecules are intrinsically weak, so these units have to be fairly large if they are to hold together. But if they are fairly large, and products of poorly controlled reactions, they are likely to consist of a plethora of structural types. It is no coincidence that organic geochemical products are typically tars.

Why are inorganic minerals so often crystalline? The reasons are that here the units are held together by strong bonds (polar covalent mainly), and because this bonding is reversible under the conditions of crystal growth. Hence the effective units are atoms rather than molecules. The units thus belong to a far more restricted set of possibilities. All three of the above conditions can thus be satisfied with much less preorganisation.

It might very well be argued that if the units for crystal life are to be atoms there would be too little scope for variety. It is all very well, you might say, to have a very limited set of units that can thus organise themselves more easily; but if the units are that simple they could only give rise to a very limited number of kinds of higher order structures. With only the choice between a handful of kinds of atoms (rather than between the astronomical numbers of organic molecules) how could really specific objects ever be engineered?

To answer this we must note again the distinction between the X-ray crystallographer's 'crystal structure' and the real structure of a crystal – the electron microscopist's structure. It is only the idealised crystal structure that is truly inherent in the collection of units out of which the crystal forms: the defect structure of a real crystal – its arrangement of dislocations, twin boundaries, substitutions, and so on – is something over and above. And of course it was just such features that we saw as constituting the genetic information of crystal life, and through such features that its phenotypic characteristics could arise. A defect structure is a metastable patterning that is not itself makeable by a 'self-assembly'. *In crystal life the defect structures of crystals are the nearest analogues of the covalent structures of molecules in molecular life.*

A defect structure is one shade more abstract than a molecule – it is not so much a thing in itself as a distortion or patterning of another thing, a form of writing in an otherwise blank crystal structure. That more abstract form of organisation of matter is more easily replicated I think; and when it comes to the specification of control structures, such as membranes or specific adsorptive or catalytic sites, it has a head start over a collection of small molecules, which have a great deal of highly co-ordinated joinings and foldings ahead of them before they can be built into a precise coherent structure large enough to exert a control on others. The task is easier when it is through the modification of a structure that is at least large enough, and coherent, in the first place. One might say that crystal life is a modified perfection, while molecular life is a tamed chaos. The second is the more creative approach and has greater potential, but the first is easier to bring off.

The origin of chiral discrimination (continued from Chapter 1)

The essential sophistication of the molecular way of life can be seen at its sharpest in the problem of the origin of chiral discrimination. The gist of the problem, as we discussed it in Chapter 1, is this. Most organic molecules, beyond the very simplest, are chiral: so there would be severe limitations on any molecular life that had to stick with achiral units. But if you use chiral components to make specific sockets, or indeed machines of any sort, you must be able to distinguish between enantiomers. If you were to build up some molecular machine you would soon begin to see this: one molecule or side chain would not fit with another because the chiral relationship between them was wrong. And as more units were built in, the number of relationships and number of ways in which these could be wrong would increase very rapidly. More particularly, the functioning tertiary structure of a protein molecule could not be specified at all by an amino acid sequence unless the chirality of the amino acids was specified. This would be true for the specification of a socket of any sort, never mind a chiral one. It is not that you might get a mixture of D- and L-sockets, you would simply never get the same socket twice because each protein molecule would contain a different mix of right-handed and left-handed amino acid units. (All the protein molecules would be chemically distinct diastereomers with their own ways of folding.)

The problem of how to handle chiral molecules is pressing and immediate for *de novo* molecular life – or at least for anything like our kind of molecular life. Yet the problem is not there at all at the beginnings of an inorganic crystal life – because that kind of life would not use chiral units.

The information in a crystal life is built in at the next level up from its units – exclusively in the way the units are arranged. And this would often include chiral information. One kaolinite domain, for example, must be either 'left-handed' or 'right-handed', and generally this handedness is copied between layers. Indeed the transmission of chiral information in growing crystals is very well known – even quartz can do it. This is the one piece of information that the X-ray crystallographer's ideal crystal structure can convey – whether to be right-handed or left-handed. But the defect structure could also, independently, be chiral: defects such as domain boundaries would almost certainly create chiral surface features – replicable perhaps if the speculations of the last chapter have any truth in them. In any case, a screw dislocation cannot help being of one chirality and, in any case, often replicates as a crystal grows.

Organisms would have been needed It must be stressed that this way of approaching the problem of the origin of chiral discrimination depends on the idea that the crystals that we have been talking about – or rather their defect structures – were products of natural selection. Otherwise there are too many objections of the sort that we discussed in Chapter 1.

One of these objections to the notion that minerals first injected chirality into life on Earth was that for every 'right-handed' crystal structure or crystal defect there should be, by and large, another 'left-handed' one. A universal bias one way, although quite conceivable, has not been clearly demonstrated and would be weak anyway. Chiral discrimination by the environment would either be localised or very weak; and multiple chiral discrimination – where the environment was making some appropriate *set* of chiral choices – could only be hoped for in restricted, probably evanescent, environments. In any case multiple chiral discrimination would rapidly become quite implausible as the number of members of the set increased. But this objection only holds for non-biological systems.

The extraordinarily non-statistical character of the outcome of evolutionary processes is one of the main distinctions between biology and the rest of physics and chemistry. This arises from the repeated selections of selections of selections. It might not take very long, on a geological time scale, for an entire species to share a genetic modification that originated in one individual. There are admittedly a few examples in the physico-chemical world where, seemingly, a single molecular event triggers a macroscopic effect: the whole pattern of frost flowers on a window may have depended on just where and how one crystal nucleus became stable enough to grow. But on the whole the physicochemical world is more statistical than that. What one molecule does doesn't matter. But what is quirky and unreliable at best in the physicochemical world is part of normal procedure in the evolution of organisms. Here the effects of a few key molecular events can be retained and amplified indefinitely because there exists the hereditary machinery to do it. Just one mutation in one gene in one individual of one species can become established to represent, eventually, the main form of that gene throughout a group of organisms spread worldwide.

In the kind of crystal life that we were thinking about at the end of the last chapter the crystals that were forming continuously in the environment had, like other kinds of organisms, been evolving the means to stabilise their own production, to improve the rate and fidelity of the printing of the genetic information and to propagate that information more widely. In very easy environments the spontaneous generation of (unevolved) organisms might complicate matters; but in more difficult environments only organisms that had phenotypes that were evolved to some extent could survive. In such places there would be no more spontaneous generation and 'common ancestor effects' of the kind discussed above could be expected: you could explain the universal occurrence of, say, a particular left-handed socket within a certain species of the primary life in the same way as you might explain the (near) universality of our genetic code – because all the extant systems are remote progeny of a single ancestral system that happened to be like that.

The general explanation for the origin of chiral discrimination, then, would be as follows. First life, a crystal life, did not use chiral units and did not need to be chirally discriminating. As it evolved, some of the species of this crystal life came to use organic molecules as phenotype components. These molecules were at first 'optional extras', useful but not vital, not part of central biochemistry which was still wholly inorganic. To begin with, the organic components were probably ill organised mixtures with unsophisticated functions. But in some biological niches there were advantages to organisms that had slightly better organised mixtures. Then, later still, when polymers and multimolecular structures became important, there were advantages to organisms that could make more chirally uniform products. Gradually, with the invention of increasingly sophisticated organic phenotypes, chiral discrimination improved through surface crystal sites becoming more closely engineered to interact with particular enantiomers of particular molecules.

Chiral discrimination would have emerged gradually, then, and as part of the emergence of a more general competence in recognising and handling molecules. In so far as molecules in the primary biochemistry were recognised through interactions with sockets of some sort, the recognition of an enantiomer would not be qualitatively different from other acts of recognition: from the point of view of a socket the wrong enantiomer of the 'plug' may be a very poor fit – like a right hand in a left glove. Once an organism could distinguish between, say, isoleucine and valine it would probably have been able to see the difference between D and L. But I do not think that either of these functions was a very early evolutionary achievement.

First uses for organic molecules (*including polyphosphate*)

So far in this chapter we have been enlarging on arguments introduced earlier as to why life would have started with inorganic control machinery (because that was easier) and then would have moved to organic machinery (because that would have been better). In the very broadest terms we can see why. But first life would not have been able to foresee that organic chemistry was to be the coming thing. So we must ask a more short range 'why?' Why were organic molecules first brought into organisms?

One can only speculate. But the question is not particularly perplexing so long as one does not try to relate the uses of organic molecules then to the uses of similar molecules now. Early organisms would not only have been less competent at handling organic molecules but they would have had different needs from organisms now. Thinking about the kinds of clay organisms that we discussed at the end of the last chapter one can see how quite crude mixtures of organic molecules might have affected the flocculation behaviour of clay assemblages through their influence on inter-particle associations.

Organic anions, particularly oligo- or polyanions can have major effects on the rheological properties of clays by acting as peptising (deflocculating) agents. They stick to the edges of clay layers cancelling net positive charges there and hence breaking up 'card-house' structures. Such effects can be

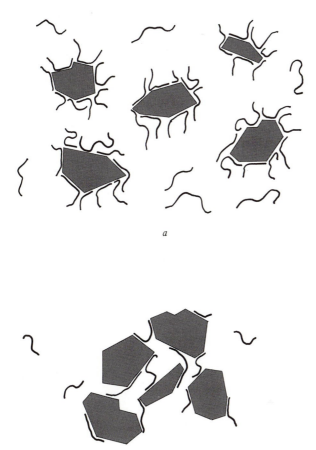

Figure 8.3. (*a*) Clay crystallites being kept apart by a charged hydrophilic polymer. (*b*) Smaller amounts of the same polymer may flocculate the crystallites.

quite complicated. Small amounts of the polyanion carboxymethyl cellulose can act as a peptiser for montmorillonite while *very* small amounts have the opposite effect, helping instead to join the particles together (Van Olphen, 1977; see also figure 8.3). Tannins – which are phenolic oligomers – are also effective peptisers and are used in the oil industry to control the rheology of drilling muds. Polyphosphates are potent peptising agents that probably act by co-ordinating to exposed cations along the edges of clay layers (figure 7.9). A firm lump of clay may liquify and flow if treated with 0.5 % of its weight of a polyphosphate with an average chain length of fifteen units (Corbridge, 1978).

Again physical characteristics of clays can be greatly affected by the intercalation of organic molecules between the layers – for example organic cations standing in for interlayer metal cations, or polar organic molecules for interlayer water.

Apart from effects on rheological properties, flocculation behaviour, and so on, organic molecules may be catalysts for clay synthesis. As discussed towards the end of Chapter 6, oxalate as an aluminium chelator may encourage kaolinite synthesis. Even a crude mixture such as fulvic acid may be effective. More generally one might see it as a major early function for organic molecules, that were serving a fundamentally inorganic kind of life, that they should be effective ligands for metal cations.

We can see, then, a number of reasons why it might have been useful for clay organisms to have been able to synthesise (or somehow acquire) certain classes of organic molecule. Speculating further one might guess that Krebs acids and amino acids were originally incorporated for their metal complexing prowess; that heterocyclic bases of various shapes and sizes were designed originally to intercalate clay layers; that oligophosphate chains like that on ATP were originally means of attaching organic molecules to clay edges. I will come back to more specific speculations of this sort later. But there are more serious questions, 'how' questions of a general sort, that we must tackle first. How did primitive organisms acquire such expertise in organic chemistry that they could reach the point at which molecules of the complexity of nucleotides were being made? How was organised organic chemistry possible without either organic chemists or enzymes?

Organic chemistry without enzymes

The first techniques for controlling organic reactions would not have been determined by any tendency to anticipate protein. Clays would have controlled organic molecules in ways that made appropriate use of inherent properties of this kind of material. There would have been similarities, I dare say, between some aspects of the first and later techniques. For example, chiral discrimination was perhaps achieved through specific binding sites for clay life as it is for ours. But we are probably being too optimistic if we imagine that any inorganic crystal surface, however contrived by specified defects, could achieve the selectivity of proteins, or seriously mimic the selectivity and catalytic ability of a protein enzyme.

Another way of controlling organic reactions is the organic chemist's way – with larger scale apparatus. Here you do not have specific catalytic sites that can pull particular molecules from a complex mixture so as to specify particular reactions and sequences of reactions. You do not manipulate molecules as individuals. Instead you manipulate the conditions to which groups of molecules are subjected. But operating that way you will not be able to carry out different operations at the same time in the same reaction vessel. In the absence of highly efficient selective catalysts, sequences of reactions must generally be controlled through sequences of operations: solutions must be mixed, conditions adjusted, side products removed, and so on. In those earlier organisms that we are trying to imagine there must have been, I think, in place of protein, some quite elaborate chemical automation. At its simplest, to keep the wrong molecules apart there had to be 'glassware', and to put the right molecules together there would have had to be 'tubes' of some sort – and 'pumps' as well, I dare say.

Not that the modern cell is entirely without such techniques – it is not just a bag of enzymes by any means. Organisms today do not rely solely on the specificity of their enzymes to keep control. Even the prokaryotic cell is quite highly structured. Referring to the internal morphology of cells generally, Mahler & Cordes (1971) remark: 'These complex structures, we are beginning to suspect, hold much of the secret of how cellular processes are controlled in both space and time: the secret may consist, at least in part, of isolating and maintaining the different cellular constituents – mainly enzymes, together with their substrates, products and modifiers (activators, inhibitors) – in different compartments; sometimes allowing, sometimes denying mutual accessibility.'

An indication that early approaches to organic chemistry might have had to put much more reliance on compartmentalisation comes also from the consideration, from Chapter 1, that modern enzymes are on the whole only discriminating enough to be able to work with the sorts of mixtures that they normally come across. Within cells these mixtures are of a fairly limited number of kinds of molecules. Much more complex mixtures would have existed if the control had been weaker in the first place – if the transport proteins and enzymes had been less competent. Where a system goes very well (in a certain way), provided it is going very well (in that way), you naturally look for some other way for it to have started. In this case you look for another way, I think, that relies much less on sockets, because not only is it hard to see such particular things as sockets being made from very complicated mixtures of organic molecules, but it is doubtful whether socket control would work on very complex mixtures. You look for something that is more like the organic chemist's kind of practical chemistry.

Elements for an organic synthesiser

So let us start to think about very early biochemistry in terms of what sorts of things are needed to bring about sequences of organic reactions automatically – if you are not allowed to use enzymes or other such high-fidelity microscopic control devices. Let us first draw up a list of necessary or desirable hardware. After that we can try to see which if any clay materials might be structurally 'preadapted' for such elements, and under what general circumstances evolutionary processes might have, in effect, assembled them, first into rudimentary and then into increasingly sophisticated automatic organic-synthesising machines.

1. Basic 'glassware' There would have to be functional equivalents of the chemist's flasks, tubes, T-junctions, and so on, for controlling solutions. There would have to be the means of 'sometimes allowing, sometimes denying mutual accessibility' as Mahler & Cordes put it when describing the membranous 'glassware' in modern cells. It is not essential that anti-mixing devices should be membranes. Different populations of molecules might be kept apart by being adsorbed on separated surfaces, or by being immobilised in a gel, with mutual accessibility being controlled in this case by diffusion processes in the gel. Also much of the 'glassware' might be in the environment – for example if the pores of a sandstone acted as containment devices for very early clay organisms so that they needed no boundary membranes of their own. But, all the same, if the 'glassware' is eventually to be highly contrived by genetic information, some kind of inert, manipulable, membranous material would seem to be at least highly desirable.

2. Energy transducers This term can cover a wide range of devices in which there is a change in a form of energy. For example, a furnace converts chemical energy into heat; a steam engine converts heat into mechanical motion, and so on. Within organisms there are many energy transductions. In photosynthesis, for example, light energy collected from the environment goes through several forms before being used or stored as a fuel. As we will discuss in more detail later, one of these forms is as a proton potential across a membrane. Other potentials are generated through various kinds of pumps in modern cells – cation pumps, anion pumps, water pumps, molecular pumps of many sorts. Such forms of active transport now depend on cleverly asymmetric lipid–protein assemblages (Lodish & Rothman, 1979) that are very far from primitive in their means of operation.

But pumps of one sort or another would have been needed, I think, long before the present pumping techniques were invented. What a serious design restriction it would be for an automatic organic synthesiser that it should always have to run down osmotic gradients. Sooner or later a solution would have to be concentrated or somehow 'unmixed', and you cannot reasonably expect the environment to provide evaporating pools just when needed except perhaps for some especially simple synthetic operations. One might imagine that chemical operations were sequenced, in the pre-protein era, through 'production lines' of some sort, in which case at the very least there would have to be the means to move solutions. Again one might imagine first ways of using flows in the environment (groundwater movements, tides, etc.) but again there would be great limitations in being so dependent: there would be strong selection pressures, if there were pressures to do organic chemistry at all, for more positive forms of the control of the means to do it. An automatic organic synthesiser without pumps may be conceivable, but I doubt whether it would work very well.

3. Reagents Condensation reactions are encouraged by conditions under which water is being continuously removed, and so one can imagine an automatic system in which a water pump helps to bring about a condensation reaction by creating a dehydrated region. Similarly, an electron pump can in principle bring about a reduction by providing electrons, or an oxidation by removing them. But 'dehydrating power' and 'reducing power' do not have to be tied to energy transducers in this way; they can be made portable in the form of soluble reagents. ATP and NADH are biochemical devices of this sort and of course the organic chemist has many dehydrating agents, reducing agents, and so on, on his shelves. There are also more specialised group-transferring agents – acetylating agents such as acetyl-coenzyme A or the chemist's acetic anhydride, or indeed ATP, again, as a phosphorylating agent. We could predict, I think, that quite soon in the evolution of a competence in organic synthesis, early organisms would have hit on the idea of using some kind of energy transducer to make some kind of high-energy reagent(s). There would not have to be many such reagents greatly to enlarge the field of accessible organic molecules. (There are not after all so very many kinds of reagents – coenzymes – in our biochemistry; see pages 376–8.)

4. Catalysts Homogeneous catalysts such as acids, bases and metal ions would be useful not only to accelerate reactions but to exert control: A and B react *when* catalyst C is present, for example. Similarly, catalytic sites on solid surfaces might exert control by determining just *where* some thermodynamically favourable reaction took place. And I dare say there would always be some choice about which molecules reacted. But, as we have already discussed, we should not expect anything approaching absolute specificity for any of the catalysts in a pre-protein life. It would be reasonable to suppose only that members of some (fairly large) class of molecules would react at a given type of site, and that within that class there would be differences of rate.

5. Binding sites Various kinds of binding sites are inevitable when solid surfaces are present, and these can be affected by defects, such as dislocations, that intersect the surfaces. We discussed in Chapter 7 how genetic information in crystals might have been 'focussed' to create special high-energy sites.

6. Collectors We discussed at the end of Chapter 1 the very great difficulties in the idea that primitive life forms could select molecules from a complicated environmental 'soup'. Today's heterotrophic micro-organisms select molecules by means of exceedingly sophisticated lipid–protein transport machinery. Even if one can imagine how a primitive organism could have coped with the tarry organic materials that might have been present in the primitive environment, it is not clear whether indeed such material would have been present except at best in rather restricted locations. But there must have been carbon feedstock(s) of some sorts for those organisms to evolve the techniques of organic synthesis. Where and what would these feedstocks have been? How would they have been collected?

7. Separators It is easy to mix things, but in most organic syntheses there are separations to be done as well. In organisms, excretion involves separating what is not wanted from what is, and similarly the working-up of an organic reaction involves the 'excretion' of unwanted side products. As remarked before the work-up is perhaps the most difficult part of practical organic chemistry. How do you 'demix' the components of solutions when you lack very specific binding sites? As I said, pumps of some sort are likely to be needed for automatic demixing operations. Conceivably, given sufficient 'glassware', early organisms could have used some kind of chromatography, that is, some technique based on relatively small differences in the affinities of static binding sites for components in moving solutions.

8. Sequencing mechanisms Even if early organisms had evolved the means of constructing all the types of apparatus that we have referred to already, there would still be problems of master control. Procedures would have to be specified. There would have to be somehow written into the assemblage of apparatus such instructions as: 'after this has happened, do that'. A production line is one way of building in such instructions, but you might need other sequencing techniques as well. It would be at least desirable to have timed procedures in some cases, as with Merrifield's automatic peptide synthesiser (figure 1.11), that is, to have a set-up that was more like a washing machine than a conveyor belt – in which different things happen in the same place at different times. Diurnal and other environmental cycles might be able to impose a programme on early organisms as Kuhn (1976) suggested (see page 150). The world might provide a set of time switches. But it would be much better all the same to have a more positive kind of control; not to have to depend on the outside world in this way. And it becomes less and less likely that more complicated procedures could be made to fit in with what the world happened to be doing. Built-in means of performing such operations as periodically reversing flows, or opening and closing control valves, would be very much better.

9. Feedback control Even programmed time sequences are not enough for the more sophisticated forms of automatic control. In a washing machine, for example, rather than have instructions of the sort 'pump water in for 30 seconds' or 'heat the water for 2 minutes' it may be better for the instructions to be along the lines of 'pump water in to such and such a level' or 'heat the water to such and such a temperature'. Where water pressures and ambient temperatures are variable there are obvious advantages in this more intelligent way of working – where sometimes goals are specified rather than all the details of the procedures needed to arrive at these goals. But to work like this a mechanism must contain some kind of sensor to see how things are going, and then some means of communicating this information to an effector that can initiate, or maintain, or adjust, or stop an appropriate action. Such mechanisms can be fairly simple. For example, in a ballcock valve the ball is the sensor which communicates information about the water level to an effector, an inlet tap, through a connecting rod. A self-controlling mechanism of this sort is said to contain a feedback loop – information about an output is fed back to control an input.

It is almost what you mean by an automatic device that it usurps the main function of the human operator in that it can see when things are

wrong and make appropriate adjustments. So we might guess that to replace the considerable skills needed to perform a multi-stage organic synthesis one would need automatic machinery with many feedback loops
built in. I doubt if the synthesis of, say, a nucleotide could be controlled
wholly by a blind sequencing of operations. Apart from inevitable variations in conditions – in the composition of the feedstock, in the ambient
temperature, and so on – the machinery itself would be time-variable:
often a tube would become slightly clogged or a catalyst slightly poisoned.
Sooner or later something would go out of synchronisation.

In any case feedback loops are part of industrial techniques of automatic
control of chemical operations, although there is nothing approaching the
automatic synthesis of a nucleotide here. The only chemical factory with
that sort of expertise is the modern living cell. It is also well endowed with
mechanisms for feedback control. For example, the concentrations of
certain molecules are held at appropriate levels through their acting as
inhibitors for enzymes involved in their production. This control is generally allosteric (see page 224). One site (or set of sites) acts as sensor – it
binds the inhibiting molecule. The information that the molecule is bound
is transmitted by allosteric means – through the articulated intervening
structure of the enzyme protein – to the effector which is the active centre
of the enzyme. Often the reaction which is being inhibited is several steps
back along the production sequence, so there may be very little structural
relationship between the molecule that is acting as inhibitor and the molecule whose synthesis is being directly inhibited.

It is typical of a clever control device that it makes a contrived causal
connection like this. Why should a rise in temperature cause an electric
current to flow in a motor winding? That happens in a refrigerator because
the thermostat makes the connection. In modern organisms proteins are
almost invariably the means of making contrived connections: they know
what to do and when to do it. They know to shut off the catalysis of this
reaction, or let in that sodium ion, or turn on the synthesis of the other
messenger RNA molecule. (Looked at one at a time such control techniques might seem to be mere refinements: but taken together I suspect
that they make our kind of biochemistry *possible* in much the same way
that the various feedback control devices in an airliner make that way of
flying possible.)

So we might add to our shopping list of components for pre-protein
organic-synthesising machinery, the elements needed for some sorts of
feedback control devices – sensors, communicating links, and effectors –
without for the moment being very clear as to exactly what these objects

might be like, except that they could not have been protein. Perhaps when we get into the shop we will see some things that might do.

Minerals and the primary synthetic apparatus

I am not suggesting that to start to do organic chemistry organisms would have needed all nine of the types of apparatus discussed in the previous section. But it seems to me to be a common sense, based on observations of how organic chemists go about things, that many if not most of such items would have to be included in practicable machinery that was to bring about automatically the consistent clean manufacture of particular moderately complex organic molecules. As discussed in Chapter 1, nucleotides and lipids are really very difficult molecules to make and even a particular set of a few amino acids of a particular chirality would not be easily contrived. And yet such supplies would be the minimum, it would seem, before our kind of biochemistry could even start to evolve. What is more, as discussed in Chapter 5, the general requirements for *any* kind of spontaneously replicating organic polymer are such that its monomer units could not be very simple. Then, as discussed in this chapter, the 'self-assembly' of organic molecules into any very definite kind of structure would call for an ability, already, to specify a set of covalent bonds, that is, a certain minimum competence at organic chemistry. Perhaps vesicles could be made from semi-chaotic organic mixtures, but that would be far from enough to direct the sequences of operations needed to do organic chemistry properly – to arrive at the stage at which molecules were being produced competently enough to provide the machinery on which competent molecular synthesis could be based. There is a clear vicious circle here that cannot be broken, I think, within the ambit of organic chemistry itself. Just how vicious the circle is can be seen from the following consideration. Even with explicit knowledge of organic chemistry and the ability to plan sequences of reactions; even with supplies of pure and often quite complex organic molecules; even with the ability to choose and set up appropriate apparatus, and to choose an appropriate solvent, temperature, pH, and reaction time; even with the ability to monitor reactions, to make adjustments if need be, and then eventually to separate out a desired product; even with all such things on his side the organic chemist is hard put to it to make molecules that would form themselves into the kinds of microscopic apparatus that can control the making of other organic molecules.

It is not, it seems, in the nature of *simple* organic molecules, or of semi-chaotic mixtures, to constitute the elements of organic-synthetic machinery: as I will try to make more explicit in the rest of this chapter, inorganic minerals seem much nearer the edges of the necessary competence.

Virtues of inertness The membranous character of so many clays, and their predisposition to form vesicles, tubes, etc. has already been commented on, and illustrated in many of the electron micrographs of Chapter 6. Apparatus made from clays would often be relatively inert with respect to organic molecules. These minerals do not readily react covalently *with* organic molecules under aqueous conditions. And the catalytic effects of clays in aqueous solutions and at ordinary temperatures are quite limited – indeed clays may often stabilise organic molecules (see for example Van Olphen, 1966).

In any case it is not usually the general clay structure that shows catalytic activity but special sites – typically Lewis or Bronsted acid sites, or oxidising or reducing sites, on the surfaces or edges of layer silicates (Solomon, Loft & Swift, 1968a, b; Theng, 1974; Van Olphen, 1977). In looking for the ideal kind of control material for an organism (cf. protein) one positively does not want a material with promiscuous catalytic effects: one wants a material which shows catalytic effects when suitably contrived by genetic information. For crystal life that means, probably, that the effect should depend on the defect structure of the (crystalline) control material. And that seems to be the case, very often, with clays.

Because clays do not readily form covalent bonds with organic molecules the controllers and the controlled could be more easily kept apart. By contrast, a semi-chaotic mixture of organic polymers and oligomers would be a poor sort of material for a reaction vessel if, as would seem likely, it became embroiled with the reactions that it was supposed to be containing. Organic chemists are not just being old fashioned in preferring, very often, glass to plastic for reaction vessels – even for low-temperature reactions. In making a complex chemical circuitry it would be as important to have chemically inert materials to hand as it is important to have insulators in making electronic circuitry.

On the other hand, the term 'living clay' might not be altogether inappropriate as a description of this microscopic chemical apparatus that we are trying to imagine. As we discussed already in the last chapter, an evolved clay, even a slightly evolved clay, might have some queer properties. The question now is this: from what we know about the properties of ordinary clays can we indeed imagine apparatus being specified by clays with defect structures that had been contrived by natural selection? Or put another way: suppose that you could engineer the sizes and shapes of clay crystallites, and suppose that you could build in more or less particular arrangements of cation substitutions, could you thus make crystallites that would be, or would self-assemble into, components of organic-synthetic apparatus?

Card-houses and bundles of sticks One possible construction system we have discussed already – the 'card-houses' that result from edge–face associations between clay platelets (figure 7.10). A collection of clay crystallites of defined shapes, sizes and charge distributions might come together to make a complex defined microporosity where again and again there were similar suites of 'rooms' similarly interconnected.

The electron micrographs of zeolites shown in figures 6.26 and the lath-like sepiolite of figure 6.48 suggest another way in which crystal habit might have been able to define compartments and connections between them. Not only do these minerals have channels in them as part of their crystal structures but the packing together of their crystallites, and especially the packing of grooved rods into bundles, could define pores – perhaps quite precisely.

Mixed layers and micro-origami Smectite layers that have been separated in solutions of low salt concentration will reform multilayer crystallites when the salt concentration is increased. Where mixtures of differently charged layers are present these may come together selectively to give two kinds of crystals, or there may be a segregation into vertical domains, or regular or irregular mixed layers may be formed (Frey & Lagaly, 1979). These possibilities are illustrated in figure 8.4a–d respectively.

This is a kind of 'self-assembly'. There is even an element of specificity about the mode of assembly in all but the disordered case. The layers must be recognising each other to some extent to arrange themselves as they do.

How this happens is not fully understood. But it is not altogether mysterious when one considers how subtle can be the dependence of interlayer forces on the structures of clay layers. Our discussions on polytypism in Chapter 6 illustrated this. To begin with, the strength of the interlayer force is greatly affected by the charges in the layers. Giese (1978) calculates that the energy required to separate (by 0.9 nm) the layers of the highly charged mica margerite is about 750 kJ mole^{-1}, whereas for muscovite, with half as many charges, the energy is about 125 kJ mole^{-1}. For (uncharged) pyrophyllite layers the separation energy is only about 25 kJ mole^{-1}.

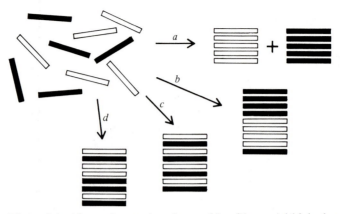

Figure 8.4. Alternative modes of assembly of low- and high-charged smectite layers. Interlayer cations are not shown. (After Frey & Lagaly, 1979.)

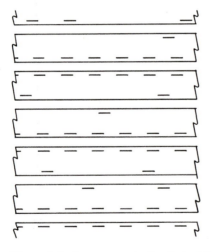

Figure 8.5. Arrangement of polarised layers in rectorite. Interlayer cations are not shown. (After Lagaly, 1979.)

The stronger forces depend on negatively charged silicate layers being held together by an intervening sheet of cations. This will only work as a cohesive mechanism if both the layers are charged – and presumably it is strongest when there is a similar charge density on each of the layers. Anyway there is often a tendency for highly charged layers to prefer each other's company and for less highly charged layers to do likewise (Frey & Lagaly, 1979; Lagaly, 1979).

Figure 8.6. Rectorite from Allevard, France. (From Weir, Nixon & Woods, 1962.)

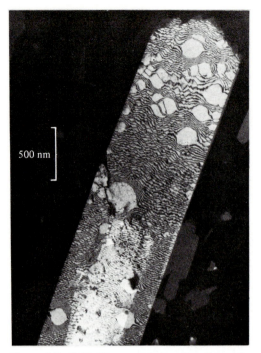

Figure 8.7. Beidellite crystal showing dislocations. (From Guven, Pease & Murr, 1977.)

More strictly it is the charge densities of facing tetrahedral sheets that are important. 2:1 layers are often quite highly polarised with aluminium substitutions concentrated in one of the two tetrahedral sheets. Rectorite is a case in point. In this structure mica-like and smectite-like interlayers alternate, but there is only one kind of (polarised) layer (figure 8.5). The mica-like junctions are very firmly glued, but the smectite-like junctions are not – with the result that rectorite may easily separate into double 2:1 layers. The thinner plates and ribbons in the electron micrograph of rectorite shown in figure 8.6 are such double layers.

Apart from the charge densities on adjacent tetrahedral sheets, the lateral distribution of charges within the sheets must have some effect. For example, if the charging is patchy there will only be strong forces between the layers where the patches on facing tetrahedral sheets at least roughly correspond. And then there is the question of being in register. Substitutions alter bond lengths as well as, often, introducing charges. Now mica-like cohesion depends on the interlayer cations being located in the hexagonally arrayed indentations of the layer surfaces, and these will only lie

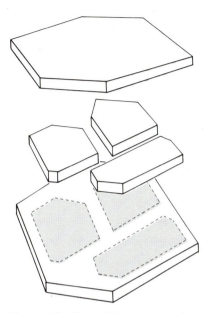

Figure 8.8. If small layers, or packets of them, are located at particular positions between more extensive layers then a complex maze of compartments might be defined (see text).

exactly opposite each other over an extensive area if the two opposing layers have exactly the same dimensions. If either layer is distorted then the other should be similarly distorted if they are to stick together well. The electron micrograph of a beidellite crystal in figure 8.7 shows how smectite layers may be in register in small patches but largely out of register, as indicated by the profusion of dislocation lines. That the layers are indeed firmly glued together where they are in register is suggested by the greater density of dislocation lines around the clear patches.

We can begin to see in all this how tetrahedral and other substitutions in smectite layers might affect the ways in which such layers would 'self-assemble'. Now think about what might be possible if substitutions, especially tetrahedral substitutions, could be contrived. The ways in which 2:1 layers come together could then also be contrived. For example, a 2:1 layer with a complex pattern of high-charge patches would have a specific affinity for another layer with a corresponding (mirror image) pattern of patches – particularly if the pattern was rather complicated and irregular; and particularly too if the patches were exactly in register. Layers with such prearranged mutual cohesion might be of different lateral extents and hence come together to specify very complex interconnected compart-

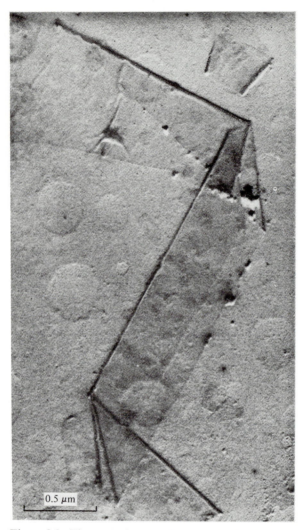

Figure 8.9. Electron micrograph of a folded beidellite ribbon from Black Jack Mine. (From Grim & Guven, 1978.)

ments. A simple example of such a 'self-assembling maze' is imagined in figure 8.8.

But perhaps the most interesting possibility arises when we come to think about a two-dimensional analogue of the globular protein's construction system – when we come to think about forms that might be defined by folding up a layer in a way that was specified by intralayer interactions. The toughness and flexibility of silicate layers or small packets

Figure 8.10. Defining a complex object through folding a flexible strip with self-cohesive patches (hatched in (*a*)). See text.

of them is vividly illustrated in many of the electron micrographs in Chapter 6, and especially in rectorite (figure 8.6) and in the crystal of beidellite in figure 8.9. These structures can fold profusely without breaking.

We seem to have the raw materials here for a sort of micro-origami. (Origami: the Japanese art of making objects by folding paper.) Imagine an extensive 2:1 layer, or a small packet of such layers, and imagine that, due to specific substitutions in the tetrahedral sheets, there are pairs of charged patches that key with each other. A specific folding mode might thus be prearranged. (It would be as if one had a piece of paper with areas marked out on it and with instructions such as 'stick area A to area A', B to B'', and so on – and then when you have blindly followed these instructions you find that you have a box or a toy aeroplane or something.) Figure 8.10 shows an example of the kind of thing that I mean. The patches on the strip shown in figure 8.10*a* can overlap each other perfectly in only one way – and that requires making a pair of intersecting rucks as well as a 180° fold. In such ways tubes of different dimensions might be specified as well as complicated connections between them.

Of the construction systems that we have been thinking about, two –

'card-houses' and 'bundles of sticks' – might have been suited to direct expression of genetic information by 'naked genes'. That is to say, genetic crystals themselves might have come together in such ways to make higher order functional objects.

We decided in Chapter 7 that flexible structures would probably be unsatisfactory for primitive genetic materials – they would not be copied sufficiently faithfully. So construction systems that depend on thin flexible layers would have to be based on other materials that were somehow being produced by the genetic material. One way would be for individual layers sometimes to separate from a genetic crystal to provide what, from then, would be separate phenotype structures. Another possibility, that we considered earlier, would be for layers of some other clay to grow epitaxially on the surfaces of genetic crystals. (Rather like DNA making an RNA strand rather than another DNA one.)

'Printed circuits' and 'integrated circuits'   Pure metal oxides, and silicates such as mica and kaolin, are generally insulators at ordinary temperatures; but many minerals, such as ZnO, TiO_2 and Fe_2O_3, are **semiconductors** when they contain suitable defects. Small amounts of cations in a higher or lower valency state than the norm can be particularly effective in creating either **n-type** or **p-type** semiconductors respectively. In *n*-type materials conduction arises from the electrons introduced by the impurities that are over and above those required for the covalent bonds in the structure. In *p*-type semiconductors there is a deficiency of electrons but these 'positive holes' can also act as charge carriers.

In so far as semiconductivity is a defect-dependent property one might speculate that patterns of electrical connectivity, 'printed circuits', might be particularly readily specifiable by genetic information that was itself in the form of a crystal defect structure. More subtle solid state devices that depend on patterning of *n*- and *p*-type semiconductive zones might also be imagined – some kind of 'integrated circuit'.

More indirect specification of micro-environments   One can see a number of shells of influence exerted by DNA in modern organisms. DNA controls its surroundings via RNA and then protein, and then more remotely through molecules, such as lipids, that proteins help to make. In higher organisms the chains of command are still more stretched – in the specification of tissues and organs for example. We saw this trend towards indirect genetic control as perhaps the most significant long-term trend in evolution (page 285).

And we have been supposing that this sort of thing would have been happening during the early evolution of clay organisms. With respect to organic reaction control, we have been seeing genetic crystals as having possibly exerted primary control – as 'naked genes'. But control via flexible structures would have been a more indirect secondary control – at the same level as the folded RNA molecule. And it seems likely that there would have been still higher order effects on organic reactions – further shells of genetic influence.

For example, any of the pieces of apparatus that we have been discussing would have tended to create new micro-environments which would have influenced subsequent mineral precipitations – whatever effects there might have been on organic reactivity. Perhaps a zeolite tends to form because a piece of (secondary) apparatus makes a rather alkaline pocket with an inner surface feature that nucleates that zeolite. And then perhaps that zeolite is a catalyst, or an adsorbent (or whatever) for organic molecules in the surroundings. This would then be tertiary control – the level at which protein operates in modern organisms.

At a much later stage, once organic molecules were already being synthesised under genetic control, a whole range of new techniques for the further control of organic reactions would become accessible. There would be new ways of joining clay particles together – via specially designed molecules with sections that would adhere to layer surfaces or edges. And new kinds of organophilic environments could be made between and around clay layers. In particular, it would become possible to specify hydrophobic regions through the synthesis of suitable amphiphilic molecules. For example, silicate layers that have been suitably propped open by long-chain alkyl ammonium ions may then take up the corresponding alcohols (figure 6.53). Here one set of molecules – the organic cations – is collaborating with the clay to create a set of micro-environments that has a more or less specific affinity for the other set of molecules – the long-chain alcohols. In so far as the cations will locate near the negative layer charges one can see perhaps how the disposition of such micro-environments might come under genetic control.

Osmotic construction Apparatus specified at primary, secondary and tertiary control levels might often include compartments that developed different solute concentrations and that were separated by semipermeable membranes. This would create various osmotic effects – and the possibility of another kind of construction system: that of the 'silica garden', where a vesicle containing a high salt concentration may grow or embark on a complicated budding. This happens because not only is there an osmotic pressure tending to swell or burst the vesicle, but the material of the semipermeable membrane is continuously being precipitated.

It is, of course, one of the necessary conditions for a putative 'vital mud' that it should be actively precipitating (although not necessarily very quickly) in its environment. In some respects ordinary muds and soils are models for the stuff that we have been trying to imagine. Here also, very often, clays are actively forming and there may be hierarchies of effect. As we discussed at the end of Chapter 6, the minerals that form at a particular place and time may depend, among other things, on the minerals that have already formed. The difference is that among the mineral species of a 'vital mud' there would have to be at least one that could hold and replicate potentially large amounts of information. That, of course, would make all the difference – because, through it, the mud could develop an evolutionary memory, of how to survive and propagate.

In this section I have been trying to show how, given the means to accumulate the know-how, there might have been a great variety of techniques available for handling organic molecules, via precipitating inorganic mineral structures. In the next section we will start to think about the sources of power needed to run this putative apparatus.

Power supplies

Radioactive heating inside the Earth is the main power source for the tectonic engine that reforms rocks at high temperatures and pressures inside the Earth and then pushes them up to the surface where they are then metastable. For an inorganic mineral life no other power supplies would have been needed to begin with. Given the slightly supersaturated solutions derived from the weathering of high-energy rocks, the units for the very first organisms could have put themselves together without any further inputs of energy. It would only have been when organic molecules became involved that energy would have become a problem. Or perhaps we should say that it was only when the mineral organisms started to trap energy from the environment that synthetic organic chemistry became possible. If we have abandoned the idea of a primordial soup as a supplier of useable organic fuels, and if we see the primordial carbon source as having been CO_2, then, presumably, some primitive technique of photosynthesis was at the beginnings of organic biochemistry.

According to the classical way of looking at it, the first such photosyntheses would have been strictly non-biological, contributing to a primordial soup. But there is also a 'neo-classical' view, as proposed by Granick (1957), Sillen (1965) and Hulett (1969): first photosynthesis was still non-biological, but in place of a general primordial soup, one should imagine instead numerous 'microsoups' being generated continuously by the action of sunlight in localised regions. We will come back to Granick's particular suggestion that organic molecules were synthesised around impure magnetite particles acting as light transducers.

There is a great advantage already in seeing organic molecules as having been produced locally like this (where required, as it were). The organic molecules did not then have to be concentrated from very dilute solutions; they did not have to survive for millions of years, and they could have been made in relatively minute amounts. And a rather inert atmosphere would be a distinct advantage – inert atmospheres are often used by organic chemists to minimise unwanted side reactions.

Mineral organisms that occupied areas close to localised sources of organic molecules might then have become adapted to use these molecules in the kinds of ways that we have already discussed. They might even have evolved mechanisms for entraining photoactive particles (magnetite or whatever).

On our story, however, we can go a step further and imagine that the phototransducing minerals were not simply found in the environment but were *made* by early inorganic organisms with the nature of the active minerals, the shapes and sizes of the particles, and especially their defect structures, contrived by natural selection.

As we shall discuss shortly, mineral phototransducers are very conceivable, but to be at all competent they would have to be contrived to a considerable extent, with a number of components connected together appropriately. I will be suggesting that mineral organisms would have been pre-adapted to the discovery of photosynthesis because structures that would have been typical for such organisms – especially mineral membranes and semiconductors – are precisely the components needed. I will be assuming, with Hartman (1975a), that the first organisms to use organic molecules were autotrophs with respect to organic molecules, that is, these organisms synthesised their organic molecules from CO_2, etc. (These organisms were heterotrophs with respect to their inorganic constituents – they depended on supersaturated inorganic solutions.)

Catching the energy of sunlight

Let us first consider briefly four ways in which solar energy might have been converted into chemical energy on the primitive Earth.

1. Photothermal effects The simple heating effect of sunlight, by causing water to evaporate, might encourage condensation reactions in ways such as those investigated by Lahav, White & Chang (1978) discussed in Chapter 1 (page 48). One might imagine such effects being contrived by inorganic organisms: for example, a 'vital mud' that had light and dark coloured patches would develop transverse thermal gradients in the sunlight. Such gradients might be useful, if only to drive water pumps.

2. Photochemical fuels There is no great difficulty in principle in imagining the localised photochemical production of high energy organic materials on the primitive Earth, or in primitive organisms. All that might be needed would be a reaction such as:

$$A + B \underset{\longleftarrow}{\overset{h\nu}{\longrightarrow}} C + D$$

such that $C + D$ was a metastable combination whose energy could be recovered by catalysing the (usually slow) back reaction. The prospect of making such solar fuels is a topic of considerable current interest that has been reviewed by Bolton (1978). The technical problems are, however, very considerable (and, incidentally, no commercially satisfactory reactions have yet been discovered). One of the main difficulties is in preventing the back-reaction from taking place at once. Quite high activation energy barriers would be needed, or alternatively there would have to be suitable secondary reactions to remove one or other of the immediate products from the scene. In either of these cases, though, some of the light energy would be lost as a fee for the security of that energy. As always, for organic molecules, there are also problems associated with side-reactions. Then again it is necessary that the back-reaction should be somehow harnessable to a useful synthetic process.

3. Photovoltaic effects A typical solar cell for converting sunlight into electricity contains a junction between an *n*-type and a *p*-type semiconductor. On making such a junction in the first place, some electrons would spread across the junction from the *n*- to the *p*-region, creating a localised potential difference which would soon stop further flow. Energy

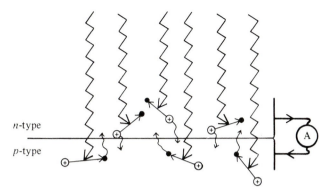

Figure 8.11. Solar cell design. Photons make new mobile charge carriers – electrons and positive holes – near a junction between semiconductors. A localised potential gradient across such a junction tends to separate the charge carriers, opposing immediate recombination and allowing instead recombination via an external circuit (right).

could not be drawn from this potential – the system would now be in equilibrium. On illuminating such a junction, electrons are excited so that new mobile electrons and positive holes are created. The standing electric field serves to separate these immediate photoproducts to some extent – electrons preferentially cross the junction into the *n*-region while positive holes go the other way. Hence energy can be drawn through electrons flowing back in an external circuit (figure 8.11).

4. *Photoelectrochemical devices* We can take a step nearer to the photosynthetic apparatus of modern plants by considering another approach to the trapping of solar energy that has attracted interest recently (e.g. Bard, 1980) – through photoelectrochemical devices that might be said to combine the photochemical with the photovoltaic approach.

If an *n*-type semiconductor is immersed in a solution, some electrons may move from the semiconductor solid into the solution so that at equilibrium there is a potential gradient between the bulk of the solid and the solution. If the semiconductor is illuminated with sufficiently energetic photons, electron–hole pairs will be generated – and they will tend to separate because of the potential gradient, the mobile electrons moving into the bulk of the solid while the positive holes migrate towards the surface. As in the photochemical and electrovoltaic approaches this separation of the products of the primary photo-event is a crucial part of the process of catching the photon energy. But to continue indefinitely

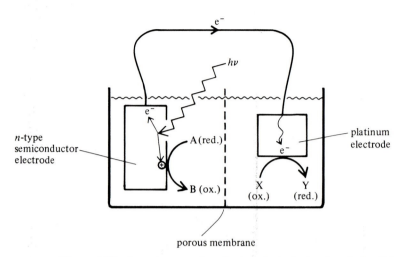

Figure 8.12. An *n*-type semiconductor photoelectrochemical cell in which a redox reaction, $A + X \rightarrow B + Y$, is driven against a thermodynamic gradient. The free electrons and holes produced by photoexcitation tend to separate due to a localised potential gradient, the electrons moving away from the surface of the semiconductor electrode to be transferred to the electrode on the right.

there must be somewhere for the electrons and holes to go. A suitable sink can be provided for the holes if the solution contains some electron-donating (reduced) species which can fill the holes, in becoming oxidised. The electrons, on the other hand, can be taken via, say, a platinum wire to another electrode some distance away (which need not be a semiconductor and need not be illuminated) where a suitable reduction reaction takes place. A formalised photoelectrochemical cell of this sort is illustrated in figure 8.12. An analogous cell can be set up using a *p*-type semiconductor.

It is often necessary in such devices to use an externally applied electrical bias in which case the redox processes are assisted rather than driven by the light. Halmann (1978) reduced CO_2 to formic acid, formaldehyde and methanol using an electrical bias and ultraviolet radiation. In one experiment the bias was provided by a solar cell to make thus a more complicated but wholly photosynthetic device. Inoue, Fujishima, Konishi & Honda (1979) have reported another such system.

These are particularly interesting cases in that they involve not only the trapping of energy but the fixation of carbon from the atmosphere. Nitrogen too can be fixed by inorganic photosynthetic means, as Schrauzer

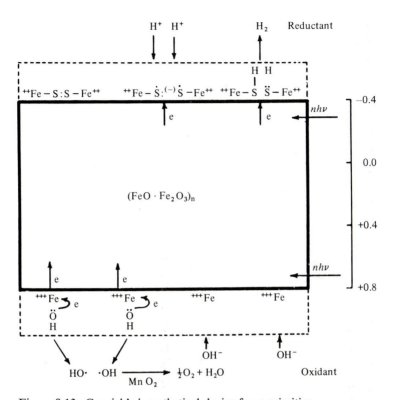

Figure 8.13. Granick's hypothetical device for a primitive photosynthesis consists of a very small impure crystal of the n-type semiconductor magnetite. Reduction and oxidation reactions take place on different surfaces. In this case water is being split in two stages: disulphide impurity on one surface mediates the transfer of photo-excited electrons to protons. On the other surface electrons are taken off hydroxyl groups in a monolayer of ferric hydroxide. The OH radicals are then converted to O_2 with MnO_2. (From Granick, 1957.)

& Guth (1977) have shown: ammonia is produced when damp iron-doped TiO_2 (rutile) is irradiated with ultraviolet light:

$$N_2 + 3H_2O + n \text{ photons} \rightarrow 2NH_3 + 1\tfrac{1}{2}O_2$$

Smaller amounts of hydrazine are also formed under these conditions.

The ideal arrangement It is a serious limitation of many photo-electrochemical systems that they only work, or only work well, with ultraviolet photons. While it is very possible that there was much more

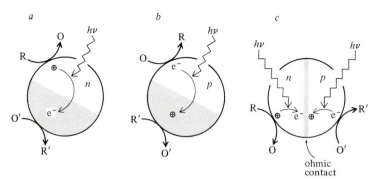

Figure 8.14. Bard's models for semiconductor particles acting as photosynthetic devices. (After Bard, 1979.) (*a*) Mobile electrons in an *n*-type semiconductor move from an illuminated surface into the bulk of the particle. Oxidations may then take place at the illuminated surface while reductions take place elsewhere. (*b*) A similar scheme for a *p*-type semiconductor. (*c*) A two-photon system with *n*- and *p*-type semiconductors in contact.

ultraviolet light on the primitive Earth than now, availability would not be the only problem for a high-energy photosynthesis. Ultraviolet photons are very destructive – it would have been difficult for a well organised organic biochemistry to have evolved in organisms that had to live in the ultra-violet sunlight. One can imagine ways round this problem – say by having only part of the organism exposed to the light – but there is no doubt that modern plants enjoy a double advantage when they use the softer photons of visible light. They have the keys to unlock an abundant source of chemical energy which, without those keys, is comparatively harmless. That seems the ideal arrangement.

The plant's technique is to combine the effects of two or more lower energy photons. Similar synthetic systems are being considered (Balzani *et al.*, 1975; Bolton, 1978; Bard, 1980). One of the ideas is to use an *n*-type and a *p*-type cell back to back – to have two illuminated semiconductor electrodes each giving an uphill push to the electrons. Bard (1979) has discussed a 'dual *n*-type semiconductor model' and compares this to plant photosynthesis. Granick's hypothetical photosynthetic magnetite particle was a similar idea similarly comparable to modern photosynthesis in its two-stage operation (see figure 8.13). Bard also considers some much simpler particulate semiconductor systems that are also possible in prin-ciple. Examples are given in figure 8.14. The great recurring practical difficulty with such schemes is in the prevention of back-reactions – the

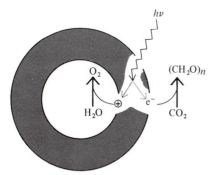

Figure 8.15. A minimal view of the reduction of CO_2 in plant photosynthesis. Photons create electrons and positive holes within the membranes of closed vesicles (thylakoids). The electrons eventually reduce CO_2 to carbohydrate, while the holes oxidise water.

oxidised and reduced species that are being made always tend to re-combine. Let us see how modern plants overcome such problems.

The thylakoid membrane

The specialised organelles of plant photosynthesis, the chloro-plasts, contain numerous membrane-bound vesicles called thylakoids. The thylakoids are the engines of photosynthesis and the thylakoid membranes the main sites of activity. Here photons generate electrons and positive holes. These are separated, the reducing effect of the electrons and the oxidising effect of the holes being directed to opposite sides of the mem-branes. The thylakoid membranes are highly structured and necessarily asymmetric – and they serve as barriers against back-reactions. A first very simplified view of photosynthesis is given in figure 8.15.

A photon is caught in the first place by an array of a few hundred chlorophyll and carotenoid molecules (figure 8.16a and b). These are located in the thylakoid membrane, and act as an antenna. The light energy is rapidly passed between the molecules by successive re-emissions and absorptions until it arrives at a special chlorophyll molecule attached to a protein, where an excited electron and positive hole are created. In O_2-producing photosynthesisers there are two classes of such reaction centres, designated I and II, with somewhat different arrays of light-catching pigments.

The following is an account of what happens next in terms of Mitchell's

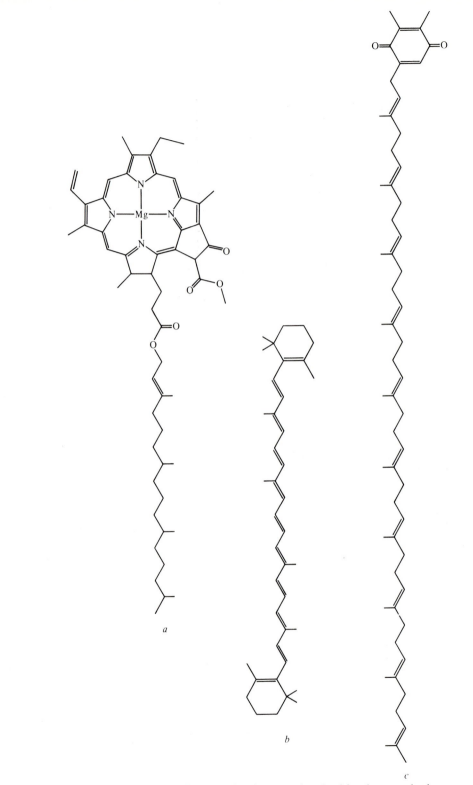

Figure 8.16. Three molecular types involved in photosynthesis.
(a) Chlorophyll a; (b) β-carotene; (c) plastoquinone.

now generally accepted chemiosmotic hypothesis. (See Hinkle & McCarty, 1978; and Miller, 1979, for more detailed accounts.)

Positive charges that are generated at a type-II reaction centre oxidise water and thus appear on protons – inside the thylakoid:

$$4 \oplus + 2H_2O \rightarrow 4H^+ + O_2$$

The high-energy electrons are taken through the thylakoid membrane to its outer edge, and react there with a hydrogen carrier, plastoquinone (figure 8.16c):

$$4e^- + 4H^+ + 2PQ \rightarrow 2PQH_2$$

picking up protons from outside the thylakoid. Plastoquinone then moves through the thylakoid membrane to the inner surface to give up its reducing power now to an electron carrier, the iron-containing protein cytochrome f, depositing protons inside the thylakoid:

$$2PQH_2 + 4 \text{ cytochrome } f \text{ (oxid.)} \rightarrow 4 \text{ cytochrome } f \text{ (red.)} + 2PQ + 4H^+$$

The electrons have by now lost much of their original energy and pass via the copper-containing protein plastocyanin to fill the positive holes of a type-I reaction centre, allowing it to keep up a supply of the somewhat higher energy electrons generated by light energy in this type of reaction centre. These newly excited electrons run through a 'wire' of two further iron-containing proteins in the membrane – iron sulphur protein and ferredoxin – arriving at the outer surface where they reduce the hydrogen carrier FAD (for structure see figure 9.1):

$$4H^+ + 4e^- + 2FAD \rightarrow 2FADH_2$$

taking up protons from the outside to do this, although two protons are put back in the final transfer of reducing power to the hydride carrier NADP (for structure see figure 9.1):

$$2NADP^+ + 2FADH_2 \rightarrow 2NADPH + FAD + 2H^+$$

Hence light energy from several photons is used to make a molecule of a reducing agent. It seems a long haul. Why so many transfers? Partly, no doubt, to help to prevent back-reactions by physically separating oxidised and reduced species. But the chemiosmotic hypothesis gives also a more subtle explanation: the picking up and dropping of protons concomitant with the changing back and forth between electron and hydrogen carriers, and the way in which these carriers are arranged in the membrane, allows protons to be picked up mainly on one side of the thylakoid membrane and deposited on the other. Much of the energy of the excited electrons thus ends up as a proton potential difference across this membrane. ATP is then

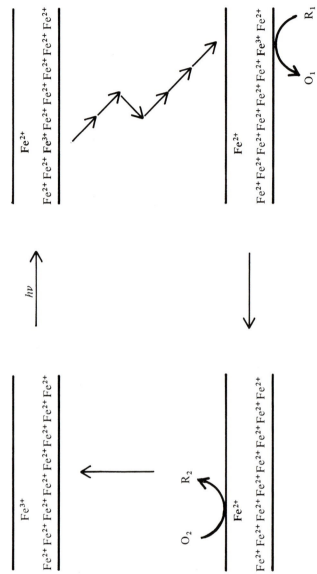

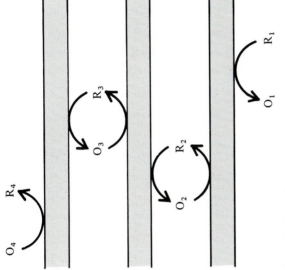

Figure 8.19. (*a*) A hypothetical photoredox machine based on a 1:1 layer silicate which has a few tetrahedral Fe^{3+} ions and many octahedral Fe^{2+} ions. After a light-induced charge transfer, an Fe^{3+} 'hole' migrates among the octahedral sites until it brings about an oxidation. The unstable Fe^{2+} is a reducing agent. (*b*) A more powerful multi-photon device with photomembranes and redox pairs at different redox levels. R_4 is a powerful reducing agent and O_1 a powerful oxidising agent.

In place of the thylakoid membrane

If mineral semiconductors seem likely elements of the very first photosynthetic apparatus, mineral membranes seem at least as likely. I have argued already for an important role for membranous materials for very early organisms; but of all places it is in the design of primitive energy transducers, and most particularly for phototransducers where excited photoproducts must be quickly chemically separated, that membranes – asymmetric membranes – are most called for.

The modern thylakoid membrane is decidedly a post-protein invention. There have to be proteins to structure the membrane, to lay out the charge conduits, to hold the right metal ions in the right places, and to maintain the essential asymmetry. Without proteins how could all this be arranged in an organic membrane? How could genetic information have set it up?

The case for clay membranes for the first photosynthetic apparatus depends partly on the whole idea that, anyway, first genes and first organic-synthetic apparatus would have been best made from clays: but it depends especially on the naturally membranous character of so many clays, whose structures are often necessarily asymmetric. Such mineral membranes could hold transition metal cations to catch light and conduct charge – as well as acting as inert barriers to separate photoproducts.

One might think particularly of clay minerals such as smectite, halloysite, imogolite or allophane, which often form membranes, tubes and vesicles. Iron atoms are common in clays, and these might have served both as light catchers and as charge conduits. Nontronite with Fe^{3+} in it – a yellow smectite clay – develops a dark blue-green colour when the iron is partly reduced chemically. This is attributed to charge-transfer transitions which then become possible between adjacent iron atoms in the lattice (Anderson & Stucki, 1979), and it indicates that, at least in the sunlight, charges could move along rows of iron atoms embedded in the unit layers of smectite, or any suitably 'doped' 2:1 layer clay.

Even 2:1 silicate layers are sometimes asymmetric. There can be a difference in the charges associated with each of the tetrahedral sheets – as in rectorite (figure 8.5). A 1:1 unit layer is asymmetric anyway and I have used such a layer in the speculative photoredox machines illustrated in figure 8.19. It is worth noting here that structural Fe^{3+} in 2:1 layer silicates can be a particularly effective oxidising agent both for benzidine (McBride, 1979) and for adsorbed Fe^{2+} (Gerstl & Banin, 1980).

The stability of the oxidation state of a transition metal atom is easily affected by the field of ligands around it (Nyholm & Tobe, 1963). This is in line with the observations of Yershova *et al.* (1976) on differently oxidisable iron atoms in berthierine (page 215). So we can take it that the redox potential of a given iron atom in a clay layer would be influenced by the pattern of charges and lattice strains generated by the cations in its immediate surroundings. In any case, different transition metals could provide alternative redox levels for the membranes in a device such as that shown in figure 8.19*a*. A more powerful multi-photon redox machine might then be envisaged, as shown in figure 8.19*b*.

Making use of phosphate

It has often been suggested (e.g. by Lipmann, 1965) that simple condensed phosphates such as pyrophosphate, triphosphate or polyphosphate might have been the forerunners of ATP and other energy carriers. Various possible primitive environmental sources of condensed phosphates have been considered (Griffith, Ponnamperuma & Gabel, 1977; Ponnamperuma, 1978), but it must be said that condensed phosphates are unusual minerals on the Earth today and that even monophosphate is generally present in oceans and freshwaters in concentrations of only around 10^{-7} M (Halmann, 1974).

Schwartz (1971) has suggested an ingenious way of overcoming the problem of low environmental phosphate concentrations. The basic problem is the ubiquity of calcium in the environment and the insolubility of the main phosphate source, apatite $(Ca_{10}(PO_4)_6(F,Cl,OH)_2)$. Noting that many biochemical molecules, particularly the di- and tricarboxylic acids of the Krebs cycle, are excellent complexing agents for calcium – and that indeed a widely used analytical method for rock phosphate depends on the solubilisation of apatite with citrate – Schwartz suggests that such organic molecules might have served to increase local phosphate concentrations by removing calcium from solution. Although not now a 'biochemical', oxalate would have been particularly effective because of the great insolubility of calcium oxalate.

Here we are trying to avoid using probiotic sources of organic molecules, but we might surmise that if phosphate was important for early organisms, then so might calcium-complexing agents – and that this provided selection pressures for the first synthesis of Krebs acids.

But was phosphate important for very early organisms? We might note first that nowadays phosphorus is an inconvenient element for organisms – unlike other major biological non-metals it cannot be cycled in volatile form through the atmosphere and it often becomes the limiting nutrient

(Emsley, 1977). Of course we can now see many reasons why our biochemistry is committed to phosphorus; but when did it become so? In view of the central part played by phosphorus in the energy transactions of present-day life, and the initiating role, according to our story, of energy-transducing mechanisms for the organic part of early biochemistry, one might suppose that phosphate, and the ability to make condensed phosphate, came in early.

A second point to note is that although concentrations of phosphate are generally low in bulk environmental waters, phosphate is adsorbed by clays, and may be present in interstitial waters of the sediments of lakes, estuaries and oceans at concentrations that are three orders of magnitude higher (Halmann, 1974).

Were polyphosphates among the first 'organic' biochemicals? A third point is that phosphates, and particularly polyphosphates, might have been useful for an early clay-based life – as a means of preventing sideways growth of genetic crystals, or as metal-complexing agents or as strings to hold clay platelets together. And phosphate itself would have been a much cleaner starting material for making simple water-soluble polymers than abiotic organic molecules, such as sugars, which if they had been there at all would have been present in a tar of some sort. And at least to make phosphate polymers you do not need to inject energy to make the monomers. It is only in condensing the phosphate units together that you begin a kind of 'organic chemistry', and build persistent metastable molecular structures. Perhaps, then, pyrophosphate provided one of the first hinges between the crystal-inorganic and the molecular-organic way of life: perhaps it was among the very first 'organic' biochemicals.

But is it plausible that wholly inorganic organisms could have driven the formation of phosphoanhydride bonds using some available energy source? Could one imagine such machinery being made from, say, clay?

In thinking about this question we may note first that the amount of energy needed to form phosphoanhydride bonds is not exceptional. As discussed by Mahler & Cordes (1971) pyrophosphate and ATP are 'energy-rich' – but then so are many other molecules. For example, at pH 7 and 25 °C the standard free energies of hydrolysis of pyrophosphate and ATP (to ADP) are about 34 and 31 kJ mole^{-1} respectively. This is rather less than for the active ester p-nitrophenyl acetate (55 kJ mole^{-1}), about the same as for acetyl-coenzyme A (32 kJ mole^{-1}), and not so very different from ethyl acetate (20 kJ mole^{-1}).

Nor is the origin of pyrophosphate energy particularly mysterious even if it has turned out to be difficult to pin-point. Among the contributing factors that have been suggested are restricted delocalisation of electrons in condensed phosphates and differences in solvation between simple and condensed phosphates. Such factors could be greatly affected by binding on a mineral surface. For example, binding on the edges of a layer silicate with exposed positive charges (figure 7.9) would tend to localise the electrons on monophosphate as well as pyrophosphate, and the solvation energy difference would also be expected to change. Hence when bound to a suitable mineral surface the condensed phosphate might very well be stabilised.

Such an effect would not in itself be sufficient to provide a means of altering the position of the equilibrium between monophosphates and condensed phosphates in solution: it could not have provided a means of production of pyrophosphate – that would be getting energy for nothing (compare figure 8.21a). But for a surface in the sunlight one could imagine a number of ways in which photon energy could be used to alter the position of equilibrium favourably. For example, photons might in effect knock the pyrophosphate groups off the surface through photo-induced charge transfers in the mineral, temporarily reducing the edge charges.

Energy-coupling devices Given a source of condensed phosphates the energy of their hydrolysis might be coupled to synthetic reactions that would normally be energetically unfavourable. For example Rabinowitz, Flores, Krebsbach & Rogers (1969), starting with a solution of glycine and 0.1 M sodium trimetaphosphate, made diglycine in 31% yield after 359 h at room temperature and pH of between 7 and 8. A mechanism of the kind suggested for this homogeneous condensation is shown in figure 8.20: here the condensed phosphate is acting as a coupling agent. The simpler diphosphate, sodium pyrophosphate, was much less effective (0.7% yield) although Burton & Neuman (1971) have used this reagent successfully in the presence of apatite crystals. They suggest that the reaction takes place in grooves on the crystal surfaces.

Heterogeneous reactions for using the energy of condensed phosphates would seem to offer the best prospects for control, particularly if the solid surface in question is a clay whose detailed structure has been engineered by natural selection. As remarked earlier, coupling agents are not very selective in themselves – the selectivity has to come from somewhere else. One might imagine a number of coupling mechanisms via coadsorption.

The three-stage cycle shown in figure 8.21a would *not* work, because although it is quite feasible that the equilibrium for the reaction should be

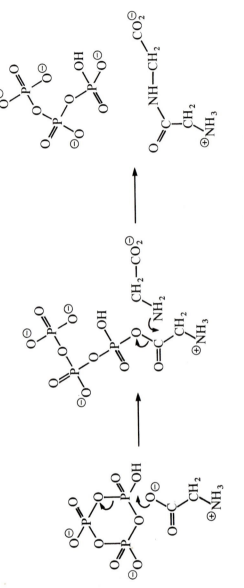

Figure 8.20. Simplified mechanism for a condensed phosphate acting as a coupling agent for amino acids (cf. figure 1.10).

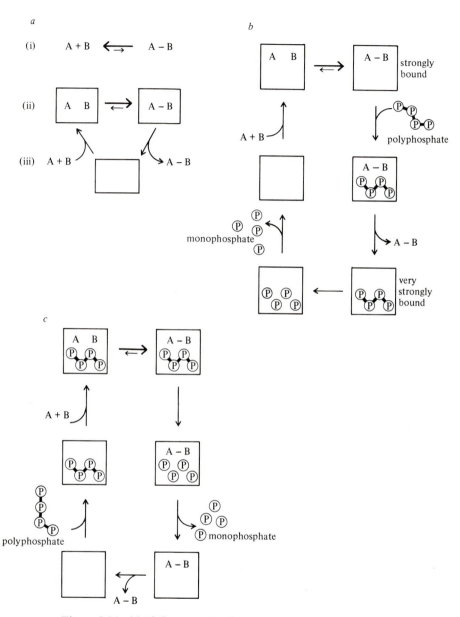

Figure 8.21. (*a*) If (i) represents the position of an equilibrium
between species in solution and (ii) the situation when these species
are adsorbed on a surface, then the cycle shown in (ii) and (iii) could
not operate spontaneously. (*b*) Elaboration of the cycle in (*a*) in
which condensation of A and B is coupled with hydrolysis of
polyphosphate. Such a cycle could in principle operate
spontaneously. (*c*) Similar to (*b*) except that A–B is bound more
strongly, not less strongly, when polyphosphate is bound nearby.

shifted in favour of synthesis when the reacting species are adsorbed, it would be necessary in that case that the product should be more strongly bound than the starting materials. (To keep the cycle going you would have to supply energy to remove A–B from the mineral surface – otherwise you could make a fuel for nothing.) In the scheme shown in figure 8.21*b* the product A–B is, in effect, displaced by polyphosphate which must now be at least as strongly bound. So energy is still needed to get the polyphosphate off – and is provided in this scheme by the hydrolysis of the polyphosphate *in situ* to monomers which are less strongly adsorbed.

Notice that it would not be necessary that in displacing the synthetic product the polyphosphate should compete for the same site; all that is needed is that when polyphosphate binds, the binding of the product is weakened. Such an effect might well operate *through* a clay membrane. Suppose, for example, that the binding of some (anionic) product was at an anomalously positively charged region, resulting from local high-charge octahedral cations. Then adsorption on the opposite side of the clay membrane by a phosphate polyanion might very well serve to displace the synthetic product. In that case, reagent and reactant would never need to mix and 'work-up' would be that much easier.

Figure 8.21*c* is a similar scheme more suited to cationic reaction products – I will leave the reader to see how the energy of polyphosphate hydrolysis is coupled to the synthesis in this case.

Figure 8.22 represents conceivable forms of energy coupling that again might operate through a clay membrane. This scheme depends on the idea that the redox potential of a structural transition metal cation would generally be altered by the proximate binding of (multiply negatively charged) polyphosphate. This might be a direct electrostatic effect, or it might operate through 'allosteric' distortions of the geometrical arrangement of the ligand atoms around the cation. Since R and O would be a metastable pair it might be as well if they were generated on opposite sides of the coupling membrane, with perhaps also the charge on M^+ causing a local 'allosteric' adjustment to the crystal structure that made M^+ less accessible from the upper surface.

The machine in figure 8.22 is a sort of electron pump, and one might try to imagine other kinds of pumps similarly driven by polyphosphate hydrolysis. One might think of an asymmetric clay membrane with pores – clusters of octahedral vacancies – that open when polyphosphate binds . . .

Of course all these designs that we have been considering are highly speculative – the more so as we do not even know in detail how the modern equivalents work. It is too easy, you may say, to make such paper schemes.

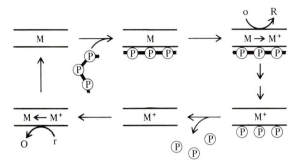

Figure 8.22. M represents a structural transition metal cation in its normal valency state in a crystal – for example an octahedral Fe^{2+} in a trioctahedral layer silicate. M becomes a reducing agent for o when (highly negatively charged) polyphosphate binds. (M 'wants' to become more positively charged now.) Polyphosphate, which is now very strongly bound, must fall apart to come off again (as in figure 8.21*b* and *c*): when it does so, M^+ is left as an oxidising agent for r. Hence the reaction $o + r \rightleftharpoons O + R$ is driven to the right against the thermodynamic gradient at the expense of condensed phosphate bonds.

Yet really that is my point. In clays, only somewhat contrived clays, you have the elements for many kinds of energy-transducing machines that would otherwise be difficult to come by.

Evolving biochemical pathways

Like any material with an engineering purpose, clay minerals would have had their strengths and weaknesses as materials of construction for organic synthesisers. As well as for making energy-transducing machinery, the clay membrane might have been particularly suitable for feedback control devices – depending again on the 'allosteric' properties of silicate structures. Their ability to create interconnecting compartments might have been particularly useful too in making chemical oscillators (for example as in figure 8.23) for time sequencing devices.

Following from earlier discussions in this chapter, the main weak point for clays would have been in molecular recognition. This would have been much less effective before there was protein to provide well engineered sockets. To mitigate this lack of discrimination I would suggest three rules for organisms embarking on organic synthesis.

The first rule would be to use a simple feedstock, not a complicated soup of molecules. The ideal environment would be one that had only a few

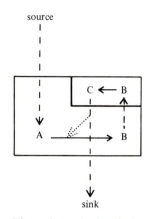

Figure 8.23. A chemical oscillator. Reaction A → B is self-inhibiting, indirectly, via C which is generated in a separate compartment. Because of the delays introduced by diffusion into and out of the subcompartment, the concentrations of A, B, and C do not reach a steady state, but oscillate instead. (Solid arrows, chemical reactions; dashed arrows, diffusion or other form of translocation; dotted arrow, inhibitory control.)

things to offer as potential organic feedstock – perhaps just water, carbon dioxide, nitrogen, phosphate and sulphide. There would be much less recognising to do in an environment like that.

The second rule would be to be tidy, to prevent as far as possible different types of molecule that were being made from getting arbitrarily mixed up with each other. As far as possible everything should be kept in its proper place so that a molecule would be recognised by where it was. (That is, after all, the chemist's day-to-day way of recognising compounds: he knows that he has, say, acetic anhydride, partly because it looks like acetic anhydride and smells like it; but mainly because it came out of the acetic anhydride bottle.)

The third rule would be to be unambitious to begin with – to specialise in the production of just one kind of molecule or of some limited set. After all, in starting the adventure into organic chemistry, inorganic organisms would have had one huge advantage: they did not really need organic molecules at all. The first organisms that made those first simple organic products would have done so under selection pressures that added somewhat to their viability – that increased, perhaps, the range of the environment that could be occupied. But whatever they were for, these organic products must have been 'optional extras' to begin with.

Two modes of metabolic evolution

1. Symbiosis A favourable site for the start of organic biochemistry would have been at some junction between the ancient elements: where there was not only Earth and Water to provide the main materials, but also Air, and the Fire of sunlight, to provide materials and energy for the new evolutionary adventures.

For this part of our story we might imagine the kind of site that Darwin (1871) imagined for the origin of life: 'some warm little pond with all sorts of ammonia and phosphate salts, light, heat, electricity etc.' We might translate Darwin's image to a later stage of evolution and think of Darwin's pond as an ecological system consisting of a community of well evolved clay-based organisms living in shallow waters exposed to the sunlight.

In this pond there are various species that have recently evolved photosynthetic competences – independently and for different reasons. They are strictly speaking plants: they are organisms using sunlight to make metastable molecules.

One species is a *phosphate-condensing plant* of the kind that we discussed earlier. Polyphosphates are a relatively simple class of linear or cyclic polymer that would have profoundly modified the immediate environment of clay particles that were making them (see page 309). Such control could have been all the more effective since polyphosphates are relatively unstable polymers. This is an important general point. For a biological control molecule to be effective it is usually necessary that the molecule is not only being made, but also being destroyed or otherwise disposed of. One might think of messenger RNA molecules, or neurotransmitters, or hormones – it is no good if such molecules simply go on accumulating. In terms of our primitive example, the consistency of a clay–polyphosphate assemblage would be controllable by controlling the relative rates of synthesis and hydrolysis of the polyphosphate molecules.

We might also suppose that there were *carbon dioxide reducing plants* in Darwin's pond – using the 'electricity' that Darwin saw as part of the story. Not lightning, but those photovoltaic and photoelectrochemical effects which must happen when the sun shines on any assemblage of clay particles. Usually the energy from the potentials that are generated will be quickly converted to heat, in a chaos of tiny currents, before there is any lasting chemical effect. But that is because usually the photoelectrochemical cells in an ordinary patch of mud are arbitrarily arranged with electrical effects cancelling out and any redox products quickly recombining. But here we are talking about clays designed by natural selection to make the kinds of machines that were discussed in the last section – with localised but powerful oxidising and reducing sites and the means to prevent back reactions.

One can think of a number of modes of control open to a clay organism that reduced carbon dioxide to simple organic molecules. In so far as there were changes in acidic and basic species present in solution there would be changes in pH – and this might have provided a means of optimising pH for clay synthesis even if the organic materials had no other catalytic use. Or perhaps the carbon dioxide was simply a sink for reducing power. In this way, perhaps, Fe^{2+} in solution could be converted to Fe^{3+} so that a ferric oxide-hydroxide membrane could be made, for example:

$$\tfrac{1}{2}CO_2 + H^+ + e^- \rightarrow \tfrac{1}{2}HCO_2H$$
$$Fe^{2+} \rightarrow Fe^{3+} + e^-$$
$$2H_2O + Fe^{3+} \rightarrow FeO(OH) + 3H^+$$

Nitrogen-reducing plants might have accounted for that ammonia in Darwin's pond – perhaps through the photoreaction found by Schrauzer & Guth (1977), if titanium dioxide had been among the minerals in some patch of vital mud (page 333).

One can see simple ways in which the ability to control the production of ammonia might have been useful: again for pH control or as a sink for reducing power; but also to make a new kind of interlayer cation to replace Na^+, etc., and so adjust the cohesion between clay layers. Here would be a new kind of exchangeable cation that could be picked out of the air.

Then again there might have been various *sulphur plants* driving redox reactions one way or the other between sulphide, disulphide and higher oxidation levels.

Control of the oxidation level of sulphur in solution is part of the means of controlling the crystal forms of iron sulphides (Rickard, 1969 – see page 178). This might have provided sulphur plants with part of their machinery for photosynthesis: Hall, Cammack & Rao (1974) have suggested that mineral iron sulphides were precursors of ferredoxins, comparatively simple proteins whose role in modern photosynthesis has already been referred to. A ferredoxin molecule can be thought of as a minute 'crystal' of a few iron and sulphur atoms (often four of each) that is stabilised by being held in a folded polypeptide frame.

Different photosynthetic organisms living in the same pond could hardly help interacting with each other – if only through kinetic effects on pH and on E_h (oxidation–reduction level). In addition, each would be a source of more or less long-lived metastable molecules. Communities of organisms would tend to persist if the effects of the products of the members were on the whole mutually beneficial.

For example, a carbon dioxide reducing plant that was making a smudge of Krebs-type carboxylic acids might help to dissolve apatite in the environment (page 343) – to the benefit of phosphate plants, perhaps – while the phosphate plants' products would be sources of energy for organic synthesis that could be used in the dark by yet other organisms that were deeper in the mud at the bottom of the pond. Similarly, sulphur plants might inadvertently be exporting redox energy in soluble form or as a metastable colloidal iron sulphide.

In so far as two or more organisms were mutually beneficial there would be advantages if they were held together in some more positive way than simply sharing a small pond. More particularly, for more open waters we might imagine small heterogeneous colonies enclosed by a mutually contrived barrier of some sort. These organisms would now have their own tiny 'pond', much more sensitively contrivable to the mutual advantage of the organisms in it.

For a suitable barrier material we might think of minerals that readily precipitate from solution as a result of pH and/or E_h shifts. Such are the materials of chemical sediments: for example silica and metal carbonates, sulphides, oxides and oxide-hydroxides (Krumbein & Garrels, 1952). We might see such materials being deposited at the interface between a general solution, containing appropriate metal and other ions, and the solution entrained by a floccule of actively photosynthesising organisms which through their activities were shifting their localised pH and/or E_h (figure 8.24a and b).

Such a barrier would tend to remain permeable to small molecules and ions – its deposition would depend directly on some of the species being able to diffuse through. Also the pH and E_h gradients would only be maintained so long as photosynthesis, and clay crystallisation, etc., was continuing within the colony: if the nutrient flow was cut down too much the barrier would start to redissolve – and continue to do so until it was thin enough to let in sufficient nutrients.

From these considerations we might predict that such a barrier would grow to the point at which it just let in the essential nutrients. In that case it might very well not let out larger photosynthetic products such as polyphosphates or the larger organic molecules. A positive osmotic pressure would then develop inside to provide a driving force for growth and reproduction in the manner of a 'silica garden' (page 327) – see figure 8.24.

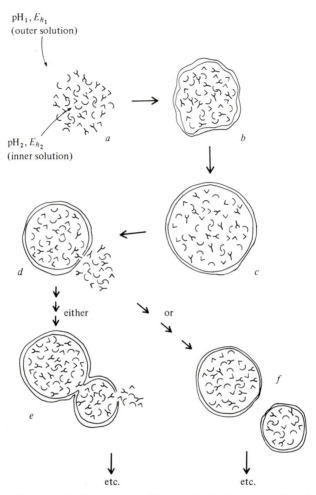

Figure 8.24. (*a*) A group of inorganic plants create a localised pH and E_h different from that in the surroundings. (*b*) This precipitates ions present in the outer solution and makes an interfacial barrier. (*c*) Photoproducts generate an internal osmotic pressure which (*d*) bursts the wall, leading either to (*e*) multicompartments, or to (*f*) 'cell division' (or both).

Not only would the potential for more elaborate organic syntheses be improved now that separately evolved subsystems were in closer collaboration, but this near-cellular form of life would also produce new reasons for making metastable molecules. There would be jobs to be done in controlling the osmotic pressure; in suitably structuring the internal solutions (a gel consistency might be a good idea for the explosive budding stage of figure 8.24*d*); in lining the inside of the barrier to adjust its permeability and pliability; or to nucleate the kinds of crystal growth in the barrier membrane that would optimise such properties.

2. Forward extension of pathways Another form of metabolic elaboration would have been from within, through a kind of exploratory process based on metabolites whose production had already been perfected. This would have been an earlier enactment of a general evolutionary process that is still visible to us. I am thinking of the way in which secondary metabolic routes are built up in modern organisms. Whole classes of secondary plant products are made by joining and modifying very limited sets of standard molecules – the morphine alkaloids are derived from tyrosine for example. For a primitive organism, 'experiments' based on a pre-existing metabolite would usually have to be performed in a separate compartment to prevent the various products and intermediates of the new reactions from becoming embroiled with the pre-existing reactions. So the outcome from a wholly internal origin of a metabolic path might be a set of suborganelles connected together. Such a production line would have to include purification stages, in keeping with the general rule for multi-step organic syntheses that at least some of the intermediates must be purified to stem the slide into tar.

I am suggesting, then, that a metabolic pathway would be built up *forwards* through a series of experimental episodes with each episode being more or less completed before the next was embarked on. The very first episode in a series would start from a simple feedstock from the environment, but each episode in the series would begin with the making of some new crude product through a mutation in the clay-held information. Perhaps a new catalytic site was generated; or a new interconnection made between compartments that brought materials together that had not previously been mixed. The next long stage would be a gradual refinement of the new product through natural selection operating on its means of production: specialist organelles were perfected which contained suitable sub-compartments, catalytic sites, adsorption sites, pumps, etc. suitably connected up. The original smudge of reactions was thus gradually transformed into just one reaction or one short clean sequence of reactions. This general mode of evolution of a biochemical pathway is illustrated in figure 8.25.

In part this progression would have been through improved design of apparatus for controlling the conditions for the reaction – improved pH control, catalyst specificity, and so on – but partly too it would have been

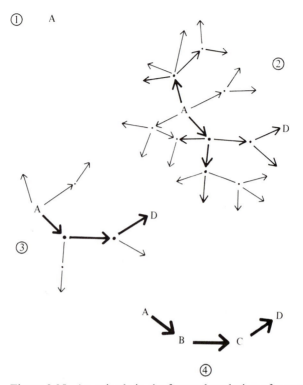

Figure 8.25. An episode in the forward evolution of a metabolic pathway. 1. Some established metabolite, A, is induced to react in a variety of ways – for example because of the introduction of a new catalytic site, or because it becomes mixed with some high-energy reagent. The result is a semi-chaotic mixture of products. 2 and 3. Selection pressures favouring mixtures rich in one of the products (D) improve the conditions for reactions leading to that product. Other reactions tend to be suppressed. This mechanism is only plausible for short sequences, but the intermediates B and C would need to have no function in themselves. Longer (straight or branched) pathways would become possible through further episodes starting with D, B, or C.

through improvements in purification procedures (work-up). Yields would fall short of 100% so that in addition to necessary waste products there would be other unwanted materials to be dealt with. Much of this would be insoluble tar. For organisms that were actively synthesising clays, intractable wastes might be dealt with by being enclosed in clay membranes and eventually dumped. Or, looking again at figure 8.24, suppose that intractable materials tended to stick to the outer wall: they would then be left out of the newly budding 'cells'.

Fixing carbon dioxide

A carbon dioxide molecule is only properly fixed when its carbon atom has been joined to another carbon atom. Just to make formic acid or formaldehyde or methanol is not quite good enough, because the next step of putting two such things together can still be difficult.

Maintaining C_1 receptors Modern organisms go out of their way to avoid joining two C_1 units. In the fixation of carbon dioxide in plant photosynthesis, for example, the C_1 unit, carbon dioxide, joins to a C_5 sugar – and the Calvin cycle can be seen as a device for keeping up the supply of the appropriate receptor molecule (figure 8.18).

The reductive carboxylic acid cycles of bacterial photosynthesis similarly regenerate carbon dioxide receptors which have more than one carbon atom in them (Buchanan, 1972). The short reductive carboxylic acid cycle, a reversal of the Krebs cycle, is shown in figure 8.26.

Interestingly, the mechanism of formaldehyde uptake in the formose reaction involves the same idea: once the formose reaction has started there is a network of reactions (pages 28–31) which continuously regenerate sugars that can fix formaldehyde. The Breslow cycle (figure 1.7) is thus analogous to the carbon dioxide fixing cycles in organisms.

The evolution of biochemical cycles The formose reaction may provide us with a model of more primitive, more chaotic versions of the carbon-fixing cycles of modern biochemistry. The real formose reaction is far, far more complex than Breslow's minimum mechanism. There are dozens, perhaps even hundreds, of formaldehyde-fixing sugars involved. It is formose as a mixture that fixes formaldehyde to produce more of the mixture. One might surmise that the first carbon dioxide fixing systems evolved from some such smudge of molecules to be refined later as catalysts became more specific. Such an evolutionary progression would be a cyclic version of the mechanism shown in figure 8.25: one or two closed cycles selected from what had originally been a dense network (Cairns-Smith & Walker, 1974).

Formaldehyde as an early biochemical? Although we will have doubts about the idea later, let us pursue for the moment the suggestion that the formose reaction might have provided a primitive form of metabolism for clay-based organisms that could photoreduce carbon dioxide down to formaldehyde.

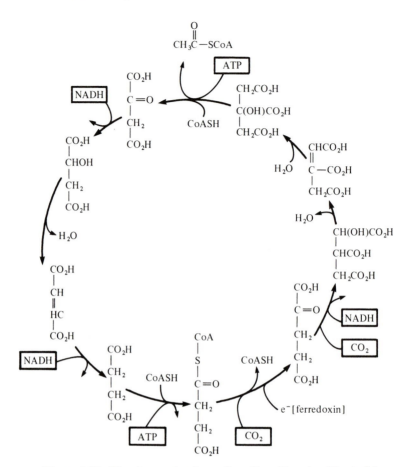

Figure 8.26. The short reductive carboxylic acid cycle of bacterial photosynthesis. This is the Krebs cycle driven in reverse by high-energy photoproducts ATP, **NADH** and reduced ferredoxin.

There are micro-organisms on the Earth today that can handle formaldehyde and some of these may help us to bridge the gap between a putative formaldehyde-based metabolism and the CO_2-based metabolism of modern plants (Quayle & Ferenci, 1978). But formaldehyde is hardly a convenient biochemical now. It is too reactive, especially with amines and phenols: apart from anything else it resinifies protein. But before protein, formaldehyde might have been an easier reaction intermediate. Hence the seemingly curious feature might be explained that a molecule so inimicable to our present biochemistry should be so closely related to sugars – because there was a time when formaldehyde was a key biochemical, the internal precursor of sugars.

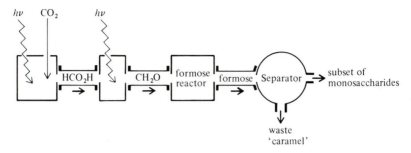

Figure 8.27. Hypothetical organelle for making mixtures of monosaccharides (see text).

One might then imagine the forward evolution of an organelle for making monosaccharides (figure 8.27). First there was a formic acid plant photoreducing carbon dioxide. Some of the product was then further reduced to formaldehyde in an adjoining compartment to give, more or less at once, a sticky caramellising mess – formose. This was useful (as a glue or something) and so formose production became more efficient through the evolution, now, of another attached suborganelle – a new compartment or set of compartments which provided improved conditions for the reaction. In the surroundings of such formose reaction vessels there would be other compartments (typical of any clay assemblage) with flows and counterflows and weak binding sites that had slightly different binding constants for different monosaccharides. As a result there were, from time to time, accidental chromatographic effects with more or less particular sugars accumulating in certain regions. Hence separators began to evolve and with them different kinds of formose mixtures became available suited to somewhat different functions.

In view of the complexity and instability of formose it is doubtful if such a system could ever have generated particular sugars efficiently, if at all. Yet it might have provided a starting point for the evolution of something better. As a variant of the system in figure 8.27, one might imagine that some carbon dioxide was fixed in the formose reactor to give now somewhat more carboxylated mixtures. These would have to be reduced sooner or later to maintain a C_1-fixing capacity – which would require a powerful, presumably photoproduced, reducing agent. But it would be a way of getting rid of formaldehyde – and a step towards our present system (cf. figure 8.18).

Carboxylic acids first? Alternatively, the set of carbon dioxide fixing molecules might have been more oxidised than sugars – a mixture of (mainly) carboxylic acids, a disorganised version of the reductive carboxylic acid cycles of bacterial photosynthesis.

Buchanan (1972) has argued that this bacterial photosynthesis is the more primitive kind, and Hartman (1975a) would see clay organisms as having picked up carbon dioxide with carboxylic acid receptor molecules.

Carboxylations with carbon dioxide are chemically more difficult than hydroxymethylations with formaldehyde. This might be seen as a point against a reversed Krebs cycle for early carbon fixation. Any such cycle, even a semi-chaotic one, would have to be driven. α-Carboxylations are an essential part of carboxylic acid cycles and are particularly difficult. For example, the reaction:

$$
\begin{array}{ll}
\text{CoAS} & \text{CO}_2\text{H} \\
\;\;\mid \quad +\text{CO}_2 & \;\;\mid \quad +\textbf{CoASH} \\
\text{C}{=}\text{O} & \text{C}{=}\text{O} \\
\;\;\mid & \;\;\mid \\
\text{CH}_2 + \text{ferredoxin} \;\rightarrow\; \text{CH}_2 + \text{ferredoxin} \\
\;\;\mid \qquad (\text{red.}) & \;\;\mid \qquad (\text{oxid.}) \\
\text{CH}_2 & \text{CH}_2 \\
\;\;\mid & \;\;\mid \\
\text{CO}_2\text{H} & \text{CO}_2\text{H}
\end{array}
$$

is being driven by the powerful reducing agent, reduced ferredoxin. Also the succinyl group is in a reactive form as a thioester.

But such objections are by no means clear. Referring back to those rules for organisms embarking on organic chemistry, there would be something to be said for reaction types that were not too easy – and could be better controlled for that reason – and for first organic products that belonged to a smallish set of possibilities. And we know that mixtures of carboxylic acids ('fulvic acid') can catalyse clay synthesis (page 251) – so we can see an immediate use. And in any case we are imagining clays that are quite capable of generating powerful reducing sites: α-carboxylations might thus have been quite possible (perhaps via the photogenerated radical anion $\cdot\text{CO}_2^-$ – compare Getoff's reaction sequences on page 32). If it is so, as Hall, Cammack & Rao (1974) suggest, that the ancestors of ferredoxins were iron sulphide minerals, then the critical role of ferredoxin in the reductive carboxylic acid cycle could be taken as further evidence in support.

Yes, I think there is much to be said for the idea that our present Krebs acids are examples of the very earliest carbon compounds in our biochemistry. Such molecules might have been the more convenient as they are mainly achiral. Sugars were perhaps too unstable and various to have been easily dealt with at the start.

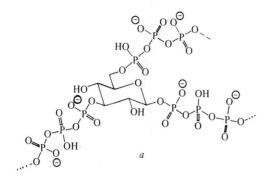

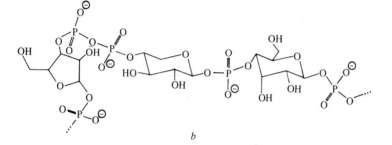

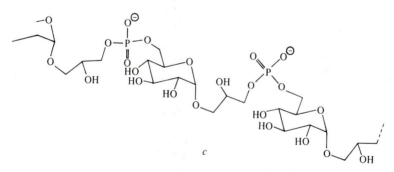

Figure 8.28. (*a*) A possible early use for sugars: to give a few branch points in mainly polyphosphate polymers. (*b*) A somewhat random teichoic acid-like polymer. (*c*) A modern teichoic acid: polyglucosyl glycerol phosphate.

Uses for sugars Eventually, though, some pre-nucleic acid organisms must have made sugars and been able to generate particular enantiomers of particular sugars. Only then would the subsequent consistent synthesis of nucleotides have become feasible.

Whether via formaldehyde or not, the first sugary products would have been semi-chaotic mixtures with functions of the sort that semi-chaotic mixtures can perform, for example as an anchor, diffusion controller, or osmotic pressure controller. Then there would be selection pressures to make different mixtures for different purposes and to be less chaotic, to specialise in mixtures that were richer in more effective components.

This pressure might have been strongest where monosaccharides were being used to make polymers. For example, the properties of polyphosphate that were discussed earlier could be greatly modified if sugars were to be incorporated as connectors. Then branched structures would become possible (figure 8.28*a*) as well as more evenly balanced sugar–phosphate copolymers – teichoic acid-like molecules (figure 8.28*b* and *c*). Such molecules could have provided new ways of controlling viscosity, holding clay particles together and generally structuring the aqueous environment around the clay particles. And they could also have had a more definite structural use. Perhaps, like the modern teichoic acids, such molecules helped to structure the outer cell walls.

Once monosaccharides were being incorporated into polymers – and particularly where these polymers had structural functions that involved polymer chains aligning to any extent with each other – then there would be strong selection pressures on the control of the synthesis of the monomers. It would matter now if all the sugars in a polymer chain were the same or not – and chiral uniformity, or lack of it, would matter too. Polysaccharides would often be particularly sensitive. Much of the point of a polysaccharide is lost unless it is well made. For example, cellulose owes its excellent inertness to having lath-like molecules that stack compactly. But that shape depends on cellulose being made up from glucose molecules of a definite chirality β-linked through 1- and 4-hydroxyl groups. There are also more subtle properties of polysaccharides that depend on the shapes of the molecules. As we will discuss in the next chapter, three-dimensional gel structures can be formed by precise intertwinings and other interactions between some kinds of polysaccharide molecules. As with cellulose, one mistake – one wrong sugar or a wrong connection between two sugars – could be very disruptive.

The entry of nitrogen

Whether you put sugars or Krebs acids at the very start, and wherever you put lipids, there are reasons for thinking that nitrogen came in relatively late. This was the view expressed by Lipmann (1965) who, on the basis of the present relationships between the molecules in our biochemistry, suggested the general sequence: carbon dioxide, formate, carbohydrate, acetate, fats and lipids, amino acids and then nucleotides.

There are two ways of looking at the place of nitrogen in our biochemistry. By paying attention to the now central genetic–catalytic machinery – the nucleic acids, the proteins, the coenzymes – one sees nitrogen as essential. Nitrogen is everywhere at the centre of control. But it is not at the centre of the biochemical pathways. Nitrogen is absent from *this* centre. Leaving aside coenzymes, there is no nitrogen in the glycolytic pathway or the Krebs cycle. And carbohydrates and lipids represent great tracts of the biochemical map that are virtually nitrogen-free. What exceptions there are here – for example glucosamine and ethanolamine derivatives – can easily be seen as later elaborations.

This apparently ambivalent position of nitrogen can be explained in terms of the idea that many of our metabolic pathways were invented, or at least sketched in, under the control of clay genetic–catalytic machinery, in organisms which to begin with need have contained no nitrogenous molecules. We might suppose that nitrogen had yet to be incorporated by the time Krebs acids and carbohydrates were already quite well established.

It is not only formaldehyde and simple amines that might have been difficult to deal with together, but even ordinary sugars and amino acids tend to misreact, especially through the Maillard or 'browning' reactions. It would seem that, generally speaking, special measures have to be taken if both sugars and amino acids are to be handled within the same system. It is perhaps significant that reactions between sugars and amino acids are early signs of deterioration of biological materials – this is one of the first things to go wrong as biological organisation breaks down (Feeney, Blankenhorn & Dixon, 1975).

Abelson (1966) had pointed to the difficulty that these reactions present to the whole idea of a primordial broth. An evolving organism would also be constrained by this mutual reactivity which might indeed be the explanation for the absence of nitrogen from the centre of the biochemical map – it might have been necessary to keep clear of amino acids until a C, H, O and phosphate organic biochemistry was well under control.

A first set of amino acids But eventually, we must suppose, it became possible to incorporate nitrogen – perhaps through a symbiosis with a simple ammonia plant of the kind that we were thinking about earlier. In any case, ammonia is the effective form in which nitrogen is incorporated into biochemical molecules today. In only a few steps, eight of the twenty protein amino acids are derived from central non-nitrogen

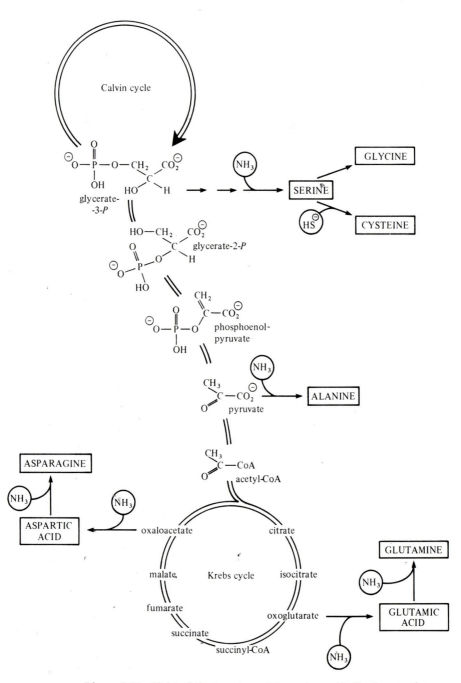

Figure 8.29. Eight of the twenty protein amino acids lie close to the central non-nitrogen pathways.

metabolites (figure 8.29), which suggests that amino acids, or at least some of these amino acids, were among the first organic nitrogen compounds.

The metabolic onion

Following on ideas of Bernal (1960), Lipmann (1965) and Szent-Györgyi (1972), Hartman (1975a) would see our biochemistry in terms of concentric shells that formed on top of each other, each new shell tending to fix the shells underneath ('the onion heuristic'). Looking at figure 8.29 it is easy to see those eight amino acids as part of a new shell that was added after the centre was established.

We can see some uses for amino acids from the point of view of a clay organism. As zwitterions they would be a new class of highly water-soluble molecules that were reasonably stable. They would act as buffers, and as a new range of ligands for metal cations. And their interactions with clay would be strongly conditional on such factors as pH and metal ion concentrations. Clay genes would extend their control of their immediate aqueous environment if they could control the production of molecules like these.

One might predict that an advanced organo–clay biochemistry would have found uses too for some heterocyclic nitrogen compounds, since these too are often both water-soluble and stable, and since they often interact with clays (Chapter 6). Pyridines, pyrimidines and purines, for example, commonly intercalate 2:1 layer silicates serving as alternative exchangeable cations or (un-ionised) in exchange for interlayer water. Hence, complex organophilic micro-environments might be contrived in regions between clay layers.

Purines and pyrimidines Heterocyclic nitrogen compounds are, of course, a key class for modern biochemistry – with purines and pyrimidines at the very centre of the modern control machinery. But to judge from their positions on the metabolic map, heterocyclic nitrogen compounds were comparatively late additions. The purines and pyrimidines are at least the next shell out from those eight amino acids in figure 8.29. Their present biosyntheses presuppose, respectively, glycine and aspartic acid as structural units.

The modern biochemical route to purines is shown in figure 8.30. In line with the forward evolutionary strategy, we would see this pathway as a final refinement of a maze of earlier routes to purines and other molecules like them. To begin with there would have been many paths to many

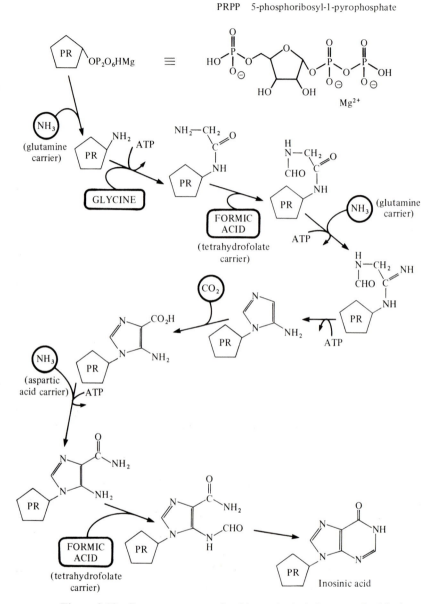

Figure 8.30. Contemporary purine biosynthesis is integrated with the synthesis of nucleotides. The activated RNA nucleotides adenosine triphosphate (ATP) and guanosine triphosphate (GTP) are derived from inosinic acid in four further steps each.

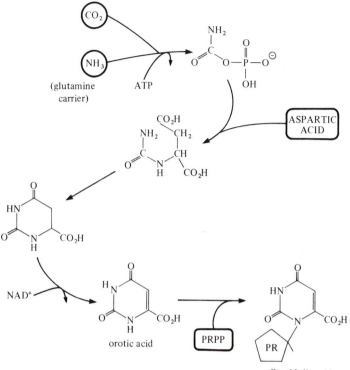

Figure 8.31. Pyrimidine biosynthesis (as far as orotic acid) is separate from nucleotide synthesis. Orotidylic acid is converted into the activated RNA nucleotide uridine triphosphate (UTP) in three further steps. Then one step more is needed to make cytidine triphosphate (CTP).

products. That would have been so not only *within* organisms that could only at first make semi-chaotic products, but it would be so also *between* different organisms, where there would be differences of style. Just because these adventures were peripheral they would be varied – in line with Darwin's generalisation (page 105).

But I will assume that the modern purine pathway still retains some information about the conditions under which its first semi-chaotic versions evolved. It looks as if a rather sophisticated sugar phosphate biochemistry was already established to provide by far the most complicated component, PRPP (figure 8.30, top), which is at the start of the synthesis. One might surmise that the original evolutionary experiments were in making nitrogen derivatives of sugars and sugar phosphates, using materials that were

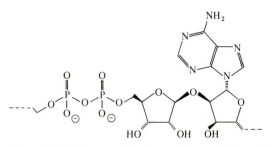

Figure 8.32. An example of a modern nucleic acid-like biopolymer, poly(ADP-ribose).

available at the time. Most of these materials were very simple – only glycine has more than one carbon and/or nitrogen atom in it. Two other 'first-shell' amino acids – glutamine and aspartic acid – are also involved in the modern synthesis as carriers for ammonia. This may also be a reflection of the molecules available early on, although it is fairly clear that the formate carrier, tetrahydrofolic acid, would have been a much later invention. (The structure of this coenzyme is given in figure 9.1.) The ATP reagent would also have to be seen as a later replacement.

The modern biochemical route to pyrimidine nucleotides is shown in figure 8.31. It is more ambiguous as to whether the sugar or heterocyclic unit should have come first since these units only join together after the pyrimidine unit has been made. But again the only component molecules with more than one carbon and/or nitrogen atom in them are PRPP and an amino acid – this time aspartic acid.

Why attach molecules with nitrogen in them to sugars? Surely it was not so that a new kind of genetic material would be found: replicating molecules were still far in the future. The reasons would have been more immediate – to make positively charged polysaccharides; to make polymers that would stick between clay layers; to make ultraviolet filters – you can think of many unsophisticated reasons. Even today there are molecules that are somewhat like nucleic acids but which seem to have much less sophisticated functions, for example teichoic acids (figure 8.28c) or poly (ADP-ribose) (figure 8.32).

Outer shells Several of the more complex amino acids are at much the same distance from the central pathways as the nucleotides. On Nicholson's (1974) metabolic map it takes about ten steps to make phenylalanine or tyrosine, for example, and about fourteen for tryptophan. Histidine seems the most recent of all – in a shell beyond the nucleotides – about 26 steps from the central pathways.

Many of the functional group types for coenzymes are also near the outer edges of the biochemical map. Among these, even the simple pyridine nucleus is apparently difficult to put together. For example, pyridine-2,3-dicarboxylic acid is six steps beyond tryptophan from which it is derived; and there are then four and five more steps to get to NAD and NADP respectively. Many of the coenzymes (figure 9.1) are evidently later than the RNA nucleotides of which they are elaborations.

Two ambiguous classes Porphyrins and other tetrapyrrols are more difficult to place since although they are derived from glycine and succinate – molecules that would have been available early on – tetrapyrrols are not themselves on the routes to other types.

A forward evolution is nevertheless strongly implied from the biosynthetic relationships between different tetrapyrrols, as Granick (1957) pointed out. Chlorophylls, haems, cytochromes and vitamin B_{12} are all derived from a single tetrapyrrol intermediate (Battersby, Fookes, Matcham & McDonald, 1980). The nucleotide portion(s) of vitamin B_{12} coenzymes implies a post-nucleotide origin for this subgroup, but the porphyrins cannot be placed more accurately than 'after glycine and succinate' unless other considerations are introduced.

Lipids are similarly ambiguous being derived from a central molecule, acetate, and yet being a more or less self-contained class. Towards the end of the next chapter I will suggest a late entry for both lipids and porphyrins.

Conclusion

While I think that it is broadly correct that our metabolic pathways were built up from the centre towards the periphery, we should not imagine that all the details of the present pathways reflect detailed sequences of events. No doubt there were some major revisions. For example, the present Calvin cycle involves not only aldol condensations but also ketol transfers – which are much less normal reactions. These depend on the particular action of a coenzyme – thiamine pyrophosphate. Similarly, pyridoxal phosphate mediates reactions that would not otherwise have been expected. There would have been a great deal of updating possible once such 'magicians' had come onto the scene: and then again when enzymes had appeared.

So no doubt an evolved clay-based life would have used significantly different tactics in the design of pathways, in view of the very different means of organic-molecular control available to it. Perhaps cyanide, or its more accessible hydrate formamide, was used in those early days rather

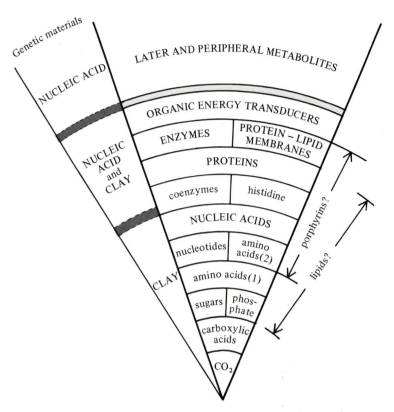

Figure 8.33. From the ways in which they now depend on each other, most of the components of our central biochemistry can be seen as having been added in a succession of layers (central sector). The classes shown here have been greatly simplified and there would have been a considerable spreading of some of them, such as the coenzymes, into layers above and below. Lipids and porphyrins are especially ambiguous, there being a wide range between the time when the components for their synthesis were available and the time when these classes would themselves have been needed for subsequent developments (right). A model like this only makes sense on the supposition that there was a takeover of genetic control. The timing of the genetic takeover, assumed to have been a one-step takeover between clay and nucleic acid, is indicated in the sector on the left.

than the less reactive ammonia and formic acid – as conceivably formaldehyde had an earlier biochemical role as discussed above. In that case our nitrogen pathways must have been quite radically redrawn. The question is very open, but I am inclined to think that the earliest successful syntheses of such molecules as adenine were not so different from present biosyntheses: because it is such a good idea, if you want to keep things under control, *not* to use building blocks that are too reactive – especially ones that react multifariously with themselves.

But even the broadest view of the biochemical sequence of events that can be read into the structure of today's pathways provides an argument for genetic takeover. So much had to evolve before our genetic material could have been made that there must have been other genetic material(s) to have supported that evolution. If the evolution of our biochemical pathways had depended from the start on nucleic acids – and if the onion idea is correct – then nucleic acids should be near the metabolic centre as well as at the control centre of present life of Earth. The present position of nucleic acids in the metabolic scheme can most easily be interpreted as that of an usurper that had come late onto the scene.

Figure 8.33 summarises the broad view of biochemical evolution given over the last few pages. I have added a few extra layers to the onion in anticipation of the next chapter.

9

Revolution

In the last chapter we were trying to see why and how clay-based organisms might have evolved a competence in organic chemistry – as part of a trend towards a more extended control of the environment by (clay-held) information. According to our overall story, the stage was now set for revolution. Control would pass from the old centre of government to the provinces. At least one new genetic material was to appear in the organic parts of at least one species of evolved clay-based organisms; and this new material was to take over entirely the control of the means of its own reproduction.

According to the story so far, clay-based life has evolved a very considerable competence in organic chemistry. It can fix carbon dioxide and nitrogen. It can discriminate between enantiomers and can carry through quite long reaction sequences. It can use the energy of sunlight to make reagents to drive energetically uphill reactions. Clay life has arrived at a level of technological development at which the consistent synthesis of particular molecules of the complexity of nucleotides has become possible. It is expert at small-molecule organic chemistry – and is indeed now exploring the uses of polymers such as polysaccharides. Perhaps even small peptides – dipeptides, tripeptides – are being put together. But such molecules do not yet carry genetic information in the kind of way that our nucleic acids and our proteins may be said to. Nor do they yet directly contrive the synthesis of other molecules. Mainly the functions of these polymers and oligomers are structural – making a suitable gel with suitable conditional properties perhaps: in general helping to provide the milieu for metabolism rather than an intricate molecular control.

That, I think, is how the stage was set for the invention of our biochemical control machinery. This invention was to require the appearance of some polynucleotide-like replicating polymer – and then this polymer was to come to control peptide synthesis. Clearly there are several more 'how' questions ahead. But at least such an eventuality becomes more plausible within an evolved biochemistry of some sort, as we discussed in Chapter 3 and 4.

But *why*, I hear Dr Kritic say. Why should natural selection ever stumble on such an extraordinary thing as a replicating molecule? If you remember the load of design restraints on any such thing (Chapter 5), it hardly looks like the kind of molecule that you come on by accident. Yet how else? Natural selection could not have *aimed* at molecular replication. It may be true that a replicating molecule is much more likely to have appeared in an evolved phenotype than a primordial soup, but is not such an accident still wildly improbable all the same?

If you recall the inventive strategies of evolution that we discussed in Chapter 3 you may see that such a conclusion is much too hasty. How was the lung stumbled on, or the inner ear, or the ability to fly? There was no forethought there either. The answers to such questions lie very largely in the terms 'preadaptation' and 'functional ambivalence' (Chapter 3). An object designed for one purpose – or evolved under one set of selection pressures – often turns out to have other uses. Indeed when we concluded that a replicating polymer would be more likely to appear in an evolved phenotype than in a primordial soup, we were making use of this idea. It is just because evolved apparatus, coupling agents, and so on, can have many uses that this would be so. Organic chemistry is generally easier in a well stocked laboratory even if the particular piece of research now being embarked on was not in mind when the laboratory was set up.

So our question is not a new one in kind. It is a question of whether a more specific preadaptation is plausible: whether some of the more detailed subsidiary functions needed for molecular replication could have been arrived at and perfected under selection pressures for functions that were initially more accessible. I will be suggesting in this chapter that a pre-protein system for locating and connecting together organic molecules provided preadaptations for molecular replication – and that this same expertise in joinery was also to predispose organisms to the discovery of protein synthesis.

The seeds of revolution; nucleic acid

Locating and connecting

Clays could have controlled the locations of organic molecules in ways that we have already discussed – by controlling their synthesis and hence where they were produced, and by providing more or less specific edge and surface sites where organic molecules could congregate. Hence genetic clays could have controlled organic molecules either directly or more indirectly via secondary clays.

The next shell of influence would start to form when, to some extent, organic molecules were communicating with each other – when, for example, molecules A and B, made in different places, could find each other and connect up without the further intervention of clays.

Today, proteins are very much involved in intermolecular communications. An enzyme brings molecules A and B together to react; or the protein subunits of a microtubule or muscle fibril 'self-assemble'. This is because, with protein, molecular surfaces can evolve which will key specifically with other molecular surfaces. A protein can be engineered to choose a molecule from its surroundings by having a socket that fits; or a collection of proteins can put themselves together in a way that is specified by their structures – a jigsaw puzzle shaking itself together by kinetic motion.

But before there was protein – before there was a generalised technique for forming the inverse image of another molecule – how did organic molecules recognise each other? How did they communicate? How could an organic molecule that had been made in one place be located precisely somewhere else?

Let us take that last, more specific question. The location of components is clearly an important part of the building of machines of any sort – and that includes biomolecular machinery. Think of an enzyme, for example, or a photosynthetic membrane. Several or many functional groups, molecules and ions have to be appropriately arranged. Indeed the arrangement, you might say, is everything.

We might get some insight into pre-protein locating devices for organic molecules by thinking about some of the non-protein devices that are in our present biochemistry. Some of these may reflect techniques that were more prominent in earlier times.

Lipophiles The clustering together of amphiphilic molecules is perhaps the most obvious example of where molecules seem to know how to arrange themselves. To some extent at least their structure tells them. If clay organisms had been able to make such molecules, then, by specifying the synthesis of some particular set, they would have been able to control the making of higher order organic structures – micelles, membranes, vesicles perhaps.

Then, given a system that has hydrophilic and lipophilic regions in it, other molecules would tend to partition themselves according to their hydrophilic or lipophilic character. In modern organisms the location of a molecule may be determined largely by its possessing a suitable 'handle'. For example the (lipophilic) phytyl chain in chlorophyll (figure 8.16*a*) is a locating device – it helps to hold the molecule in the thylakoid membrane. A similar device is used in the synthesis of bacterial cell walls. There, large and mainly hydrophilic monomer units have a long polyisoprenoid chain attached to them while they are being elaborated:

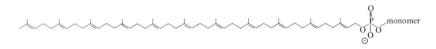

Coenzyme 'handles' There seem to be 'handles' too on most co-enzymes (figure 9.1). Many of these molecules are much bigger than they would seem to have to be to account for their chemical functions. The seemingly non-functional parts can be understood as devices that help the enzyme to hold some key group securely without blocking its functionality. They can be understood as 'handles', that is, or in some cases as 'handles plus tie-lines' (for example in coenzyme A).

That coenzyme 'handles' now interact with proteins does not preclude them from the category of possible pre-protein locating devices. It is very striking how many of our coenzymes have nucleotide or nucleotide-like 'handles'. This led Orgel & Sulston (1971) to suggest that some of our coenzymes were designed originally to link up with polynucleotides. White (1976) has suggested that perhaps a metabolic system composed of nucleic acid enzymes existed before the evolution of ribosomal protein synthesis. Our coenzymes were part of this system: so were precursors of our transfer or tRNAs, which can be seen as large coenzymes participating in the group transfer of amino acids. These are attractive ideas.

Labels There is a certain ambiguity between handles and labels. A fancy handle may help in identification – and you might say that the tRNAs are like this. They turn an amino acid into an object that can be recognised by nucleic acid – the point being that a stretch of polynucleotide is very good at recognising another (complementary) stretch of poly-nucleotide, but it is not very good at recognising anything else. So one might surmise that when coenzymes were associating with nucleic acids rather than with protein enzymes – before there was protein to do most of the recognising – the coenzyme handles had to be more explicit about what

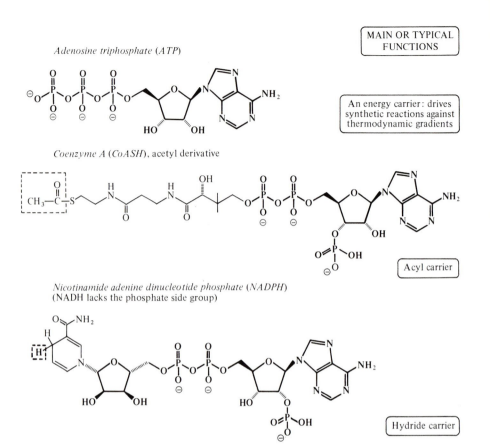

Adenosine triphosphate (ATP)

MAIN OR TYPICAL FUNCTIONS

An energy carrier: drives synthetic reactions against thermodynamic gradients

Coenzyme A (CoASH), acetyl derivative

Acyl carrier

Nicotinamide adenine dinucleotide phosphate (NADPH)
(NADH lacks the phosphate side group)

Hydride carrier

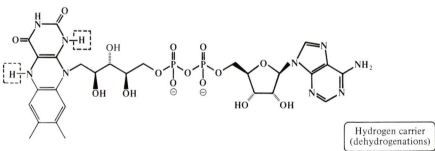

Flavin adenine dinucleotide (FAD)
(In FAM, flavin adenine mononucleotide, there is only a
phosphate group in place of the right hand adenosine diphosphate part)
Hydrogenated form (FADH₂) :

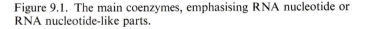

Hydrogen carrier
(dehydrogenations)

Figure 9.1. The main coenzymes, emphasising RNA nucleotide or
RNA nucleotide-like parts.

Uridine diphosphate coenzymes, e.g. UDP-glucose

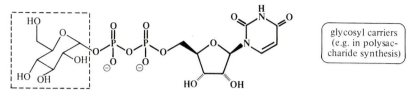

> glycosyl carriers
> (e.g. in polysac-
> charide synthesis)

Cytidine diphosphate–alcohols, e.g. CDP-ethanolamine

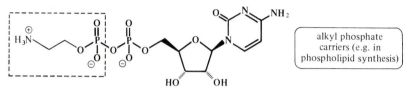

> alkyl phosphate
> carriers (e.g. in
> phospholipid synthesis)

Thiamine pyrophosphate (TPP)

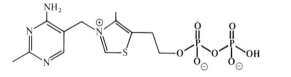

> decarboxylations of
> α-keto acids:
> $-\overset{\displaystyle O}{\underset{\parallel}{C}}-CH_2OH$ group
> carrier in the
> Calvin photosynthetic
> cycle (figure 8.18)

Pyridoxal phosphate

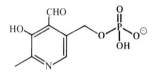

> various trans-
> formations of
> amino acids

Tetrahydrofolic acid (FH$_4$), N^{10} formyl derivative

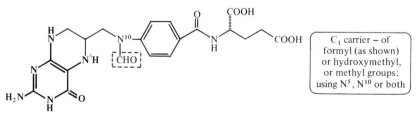

> C$_1$ carrier – of
> formyl (as shown)
> or hydroxymethyl,
> or methyl groups;
> using N^5, N^{10} or both

Figure 9.1 (*contd.*)

Vitamin B$_{12}$ coenzymes, e.g. Co-5′-deoxyadenosylcorrinoid

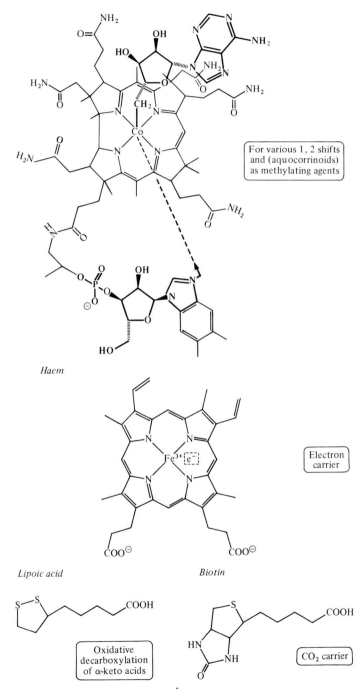

For various 1, 2 shifts and (aquocorrinoids) as methylating agents

Haem

Electron carrier

Lipoic acid   *Biotin*

Oxidative decarboxylation of α-keto acids

CO$_2$ carrier

Figure 9.1 (*contd.*)

they were carrying. We now see, perhaps, only the stub ends of earlier much fancier devices: in place of the adenosine or other nucleotide 'handle' in ATP, coenzyme A, etc., there were longer identifying oligonucleotide 'labels'. In our present tRNAs, then, we are perhaps seeing examples of what all coenzymes used to be like.

Carbohydrate systems Carbohydrates are involved in a number of identifying and locating devices in modern organisms. Plant glycosides, such as saponins, are at least more water-soluble by virtue of their mono-saccharide or oligosaccharide attachments, even if the sugar units have no more precise locating functions.

Oligosaccharides attached to the outer surfaces of cells are important for communications between cells – in recognition and cohesion. The multiplicity of possible monosaccharide units and of the ways in which they can be linked together to make straight or branching chains makes it possible in principle for a great deal of information to be conveyed in the choice of just a few kinds of oligosaccharide – although proteins are also involved in these very sophisticated intercellular interactions.

Interactions between polysaccharide molecules may give us a better indication of pre-protein techniques. Polysaccharide molecules can often connect specifically with each other, if they contain like chains of mono-saccharide units. This idea can be used to explain the properties of several gel-forming polysaccharides (Rees, 1977; Rees & Welsh, 1977; Winterburn, 1974).

ι-Carrageenan is one of the gel-forming polysaccharides in agar, the intercellular matrix material of red seaweeds. The ι-carrageenan molecule contains stretches of alternating D-galactose 4-sulphate and 3,6-anhydro-D-galactose 2-sulphate which are interrupted every so often with a D-galactose 2,6-disulphate, or 6-sulphate, unit in place of the anhydrogalactose (figure 9.2). The regular stretches twist together to form double helices which are well stabilised by regular hydrogen bonding between the chains – through the two hydroxyl groups of the galactose sulphate units. Hence the molecules become tied together into an open three-dimensional net-work (figure 9.3). A great deal of water is entrained in such networks – a typical agar gel is more than 99% water.

The double helical tie-points in agars can be 'melted' and reformed by heating and cooling. With alginates – major components of brown sea-weeds – the cation environment controls the state of gelation. Here there is a rather different way in which the polysaccharide chains lock together. The alginates are copolymers of D-mannuronic acid and L-guluronic acid.

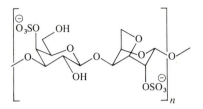

Figure 9.2. A regular sequence such as this in an *ι*-carrageenan molecule will form a double helix with a similar stretch in another molecule.

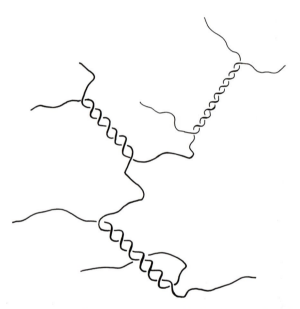

Figure 9.3. *ι*-Carrageenan molecules joining together through stretches of intermolecular double helix.

Stretches of pure L-guluronic acid are self-cohesive in the presence of calcium ions and it is suggested that the metal ions are located as shown in figure 9.4, in 'egg boxes' between the, rather rigid, puckered chains (Grant, Morris, Rees, Smith & Thom, 1973).

The cellulose fibre owes its rigidity and stability to the locking together of the identical flat polyglucose molecules that make it up. Albersheim (1975) has suggested that at a higher level, in the organisation of the fibres in a primary plant cell wall, there are matchings of a more complicated

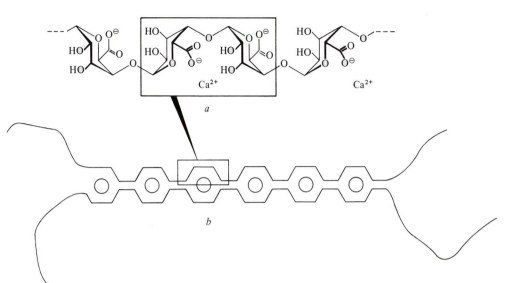

Figure 9.4. (*a*) A stretch of 1,4-linked α-L-guluronate showing a mode of co-ordination with calcium ions (co-ordinating oxygens shaded), as suggested by Rees (1977). (*b*) Longer view of a pair of such chains held together by calcium ions in ten-fold co-ordination. (Redrawn from Rees, 1977.)

kind. A complex branched polysaccharide in the cell wall contains stretches of 1,4-β-linked glucose (as in cellulose itself) and it is thought that these stretches lock onto the cellulose fibres, holding them in place.

The invention of nucleic acid

I am not imagining that the very polysaccharides that we now find in our seaweeds (or anywhere else) were present as components of pre-protein locating mechanisms. What these studies of modern polysaccharides show is the range of control devices that could have become available to primitive organisms that had arrived at the ability to make particular monosaccharides and put them together in particular ways. We might suppose that molecules like polysaccharides – big molecules that were rather regular, water-soluble, and not too flexible – were among the first that could lock with each other at all precisely.

In the last chapter we were imagining an era of early evolution when a number of species of organisms were exploring a variety of water-soluble polymers – for directly connecting clay particles together. Now we might suppose that clay organisms could make polysaccharides with suitably contrived self-cohesive stretches in them. The organisms could now specify a pattern of association that did not directly involve clay particles – they could make, now, a new set of micro-environments among the twisted and twisting molecular threads. Here would be a new kind of matrix with places in it. Other molecules could be located in this matrix if they had oligomers attached to them that were of the right sort.

Before protein appeared, to do away with most of them, there might have been quite a number of kinds of organic locating mechanisms. Poly-nucleotide was perhaps only one of the more sophisticated ideas that were around. No alarm bells rang when heterocyclic side chains first started to be attached to various sugar-phosphate polymers. There was, I daresay, some duly pedestrian reason for it – to make some polymer stretches positively charged, perhaps, to hold them between clay layers, or to provide an ultraviolet filter.

Nitrogen heterocyclic molecules often stack into micelles in aqueous solutions (for example acridine orange, cynanine dyes) and, in any case, rigid and somewhat hydrophobic side groups would tend to cluster to-gether within a gel matrix. Interesting hydrophobic regions might be made like this through inter- and intramolecular associations that might be further stabilised by hydrogen bonds between the heterocyclic side groups.

For a molecule readily to fold on itself, the rather rigid polysaccharide backbone would be less satisfactory than one that included more flexible links, such as open-chain alcohols or phosphate groups. The modern teichoic acids have both these kinds of flexible groups in them (figure 8.28c).

But there is a snag about flexibility here. It may allow self-folding but it will weaken the tendency for stretches of polymer to lock with each other. So an increasing flexibility – arising, say, through an increasing frequency of interruptions by phosphate in polysaccharide chains – might call for side groups that fitted into each other – pairing bases for example. Hence a molecule with a flexible hydrophilic backbone plus more hydrophobic side chain that could pair with each other might be seen as a sophisticated way of getting the best of both worlds – for making a particularly effective structure-forming system in which flexible molecular segments could never-theless be located precisely. In addition, other molecules that had suitable (polynucleotide-like) tails attached to them could also be precisely placed.

These molecules were not yet nucleic acids. Structurally they were getting close, but functionally they would have been more like our proteins. With purely structural roles to begin with, they were to lead to that era domin-ated by the 'nucleic acid enzymes' that White talks about. Not all the side

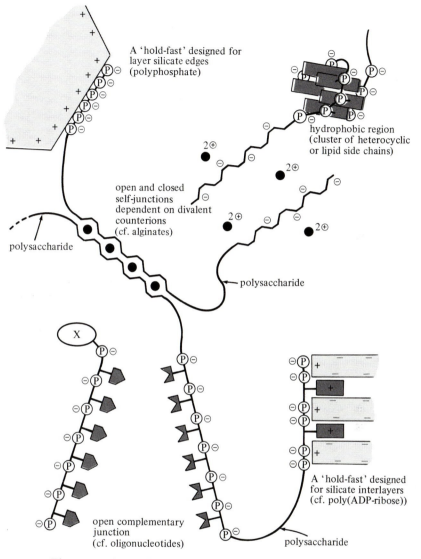

A 'hold-fast' designed for layer silicate edges (polyphosphate)

hydrophobic region (cluster of heterocyclic or lipid side chains)

open and closed self-junctions dependent on divalent counterions (cf. alginates)

polysaccharide

polysaccharide

X

A 'hold-fast' designed for silicate interlayers (cf. poly(ADP-ribose))

open complementary junction (cf. oligonucleotides)

polysaccharide

Figure 9.5. Some elements in an imaginary form of 'molecular connective tissue' for clay-based organisms. There are three main functions for this material: to hold mineral components together, to structure the intercrystalline aqueous environment, and to locate particular smaller organic components in particular places within that structure – for example, lipophiles might concentrate in polymer-micelle regions (top right), while a special functional grouping (X) might be located by virtue of an attached oligomer tail.

chains would have been heterocyclic bases; and, of the bases, not all would have been pairing types. Cohesion between side chains was only part of the structure-forming technique – and no one had yet whispered the word 'replication'. A sketch of the sorts of molecules that we might be thinking about now is given in figure 9.5.

Questions of synthesis One might imagine clay organisms polymerising organic monomers by lining them up in specially engineered grooves and tubes. But even with such apparatus it is questionable whether it would have been possible to specify a long copolymer sequence unit-by-unit. In Chapter 7 we decided that genetic information would have been written on a rather larger scale in clays than it is written in nucleic acids. It would be especially difficult to see how a clay template could be sufficiently finely contrived to specify the positioning of individual amino acids in a polypeptide chain since the information density here is so high. But copolymers mainly made of blocks (with an occasional special group perhaps) – molecules like carrageenans or alginates – might have been much more accessible.

With such molecules the genetic information is expressed firstly in the kinds of blocks and special groups used. For primitive organisms this would have been determined by the nature of the clay synthetic apparatus. Secondly, the genetic information is expressed in the arrangement of the blocks and special groups.

Referring to the hypothetical molecules in figure 9.5, we might imagine a clay organism that has one organelle specialising in the production of (something like) hexa(U), another specialising in the production of teichoic acid-like material, and so on. These organelles are connected up with a final one that has in it a template for aligning the pieces appropriately before the final assembly. That template would contain part of the information needed to make the final products. But because most of the units being handled were fairly large, this information could still be quite coarse grained – it might be, for example, an arrangement of grooves defined by twin composition planes with occasional special sites at edges or intersections.

Given the biochemical machinery for making stretches of polymer that stick to each other, the discovery of an organic replicating molecule would have come much closer. Oligomers such as hexa(U) would, on their own, tend to stick to complementary regions of a pre-existing polymer. This – part of their *raison d'être* as structure-formers – would also provide an alternative mode of alignment for the final assembly of special connector molecules that were made up entirely from such oligomer units. A molecule like this we might now call a 'protopolynucleotide'. (It would be like nucleic acid but with, perhaps, different pairing bases and/or a somewhat different backbone.)

The 'classical' hypothesis for the replication of nucleic acid involved the idea of single nucleotides locating on a preformed strand and then polymerising (Watson & Crick, 1953). It is doubtful if such a system based on single nucleotides would work. Single nucleotides do not stick well to polynucleotide strands. But long enough stretches do and indeed the kind of mechanism that we have been imagining has been demonstrated to the extent that hexa(U) segments have been joined together on a poly(A) template without enzymic assistance (Naylor & Gilham, 1966).

So that is one way, perhaps, in which organic replication might have eased itself into an evolved clay-based life. One aspect of one kind of organic polymer was now directly transmissible between the organic polymers themselves: a copolymer block structure was a new kind of genetic information that was independent of clay. The clay templates previously needed to specify such block structures could now be dropped. This was only a small devolution, no doubt, since most of the genetic information needed to make these copolymers would have been involved in the specification of the necessarily complex apparatus needed for the synthesis of the monomers and for making the blocks. The assembly of these blocks was a comparatively trivial affair. But it was a start.

We discussed in Chapter 4 another possible easy route to organic genes – through non-genetic replication – where it was important only that two sequences were more or less matching, but where it did not matter much what the sequences were. We were thinking of structure-forming systems, but one might also imagine labelling systems that were like this. If you want to write an oligomer label that will seek out a particular stretch of polymer then it might be an idea to build the oligomer on the polymer so that it has a complementary sequence. The label will then remember where to come back to.

A third easy route that might be imagined would be through organic genes that, to start with, held no sequence information. For example, suppose that ABAB . . . is a self-cohesive polymer, then it might be best synthesised on a pre-existing strand – so that the *length* of the daughter strand is defined. Length is the only genetic information being transmitted in this case, since the polymer is always ABAB. To begin with, perhaps, only this sequence is replicable although evolving replication apparatus subsequently allows other sequences, eventually any sequence, to be replicable.

When for any reason there was a selective advantage in being able to copy copolymer blocks, or features of any sort, there would have appeared selection pressures to have improved the efficiency of that replication. New kinds of catalytic grooves and tubes might be expected to appear now and be modified by selection. Such a piece of apparatus would no longer be a template transferring information to a phenotype polymer. It would be a jig, designed mainly to hold: to locate daughter units, and to help to join them up – indifferent to the features (e.g. sequences) being copied. In short, *any* use for organic replication would be a reason for the evolution of some sort of 'replication apparatus'.

To begin with, and from the clay's point of view, replication was simply an alternative way of finishing off the synthesis of a certain rather limited class of phenotype polymer. The technique would have been limited too in that it would first have appeared in only one species (with perhaps rather odd requirements). These polymers would in any case have been part of the 'secondary', or outer, metabolism – useful no doubt, but specialist structures adapting a particular organism to a particular life style. In this respect they would have been like our modern carrageenan, or lignin, or melanin.

Indeed up to now all the organic chemistry of the evolving clay organisms was 'secondary metabolism' in the sense that it was only very indirectly concerned with the central task of transmitting genetic information to the next generation. This has been an important part of our story. So placed, the organic part of primitive metabolism could be variable under evolutionary pressures. A great variety of products and possibilities could be explored – as they are explored still in the vast numbers of secondary metabolic products of present-day life.

These new organic genes would have been peripheral to begin with (in spite of a slight smudging now between primary and secondary metabolic categories). They were recent elaborations – optional extras if you like. Only some organisms had them at all. In line with more general evolutionary considerations (pages 93–104) you might have expected that soon there would have been a considerable structural variability between the organic replicating polymers in different organisms that were occupying different niches. There would be room, still, for evolutionary experimentation with these new materials – and a chance to discover a new molecular biology.

In fact the main work had already been done. It might have seemed a small step to have introduced those first crudely replicating polymers, but it was fatal to the whole clay system. In any competition between organic and inorganic control systems, the organic ones would be almost bound to win in the end – because metastable structures can be engineered more finely with the use of organic molecules. Anything clay could do organic molecules would eventually come to do better.

Improvements There was a long road ahead, though, even if it was downhill overall. First of all the new genetic information itself would have to come to be more finely written and more faithfully copied – say through copolymer blocks becoming shorter, and single units being included more often, till eventually information was replicated unit-by-unit. This would be possible in principle now that the templates were the protopolynucleotides themselves rather than crystal surfaces.

To begin with we might imagine 'replication apparatus' made from clays. Positively charged grooves on the edge surfaces of layer silicates, or grooves along sepiolite crystals, might have been particularly well suited (look at figures 5.7, 6.29, 6.48 and 7.9). There would also by now be some quite sophisticated organic structures to be brought in – if they happened to be of marginal assistance – and then improved under natural selection. Later, protopolynucleotides might collaborate, or even take over as the main material for the 'replication apparatus'.

Of the various control polymers now around these protopolynucleotides would be star performers, their flexibility and high information density allowing them to fold into functioning objects with particularly well specified tertiary structures. But these new and versatile machines were to set the stage for another more dramatic breakthrough. This would have happened through the normal tendency for evolving genetic information to extend its control. Rather as clay genes had extended their control by assisting the growth of secondary clays, so these new organic genes were to come to cause the formation of another kind of organic polymer – only this time the secondary control material was to be very precisely specified – more intricately indeed than the genetic material itself. With this, the invention of protein, control was to shift decisively in favour of the protopolynucleotide genes. Protein was the way that nucleic acid won.

The agents of revolution; protein

Amazing machinery

Our understanding of protein synthesis in modern organisms has developed rather similarly to that of DNA replication (described in Chapter 5). In each case formalised schemes were drawn up early on in which monomer units were seen to line up against a premade polymer

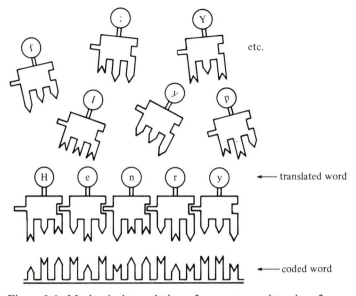

Figure 9.6. Mechanical translation of a message written in a four-symbol code can be imagined that depends on adaptors that fit sets of symbols – here triplets – in a message. Each adaptor that fits a particular triplet has a characteristic symbol of a much longer alphabet attached. The set of adaptors thus constitutes the 'code book' for the translation. There would be 64 kinds of adaptors possible in this case, corresponding to 64 possible kinds of symbol for the translated message. (This might be, for example, upper and lower case letters, plus eight punctuation marks, spaces, etc.) Such a system would correspond quite closely to the way in which (four-symbol) genetic messages are used to determine the sequences of the twenty different kinds of amino acids in proteins. In protein synthesis, however, no more than two adaptors are put in sequence at the same time (see figure 9.7). With only twenty amino acids, together with 'stop' and 'start' punctuation marks, the coding capacity appears to be underused by the protein-synthetic machinery.

acting as template. For protein synthesis, schemes of this sort were necessarily complicated by the lack of correlation between the number of kinds of monomers in the supposed template (DNA with four kinds of nucleotide units), and in the polymer that was supposedly being produced (protein, with twenty kinds of amino acid units). It was difficult in any case to see why amino acids, with their diverse side chains, should tend to line up against a polynucleotide template.

Crick's adaptor hypothesis, and the idea of a triplet code embodied in a set of adaptors, was a neat solution. With such an idea a line-up mechan-

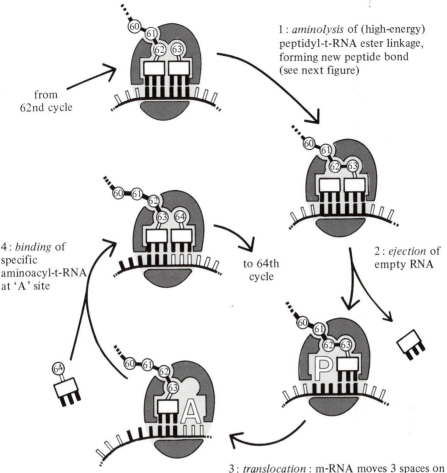

1: *aminolysis* of (high-energy) peptidyl-t-RNA ester linkage, forming new peptide bond (see next figure)

from 62nd cycle

4: *binding* of specific aminoacyl-t-RNA at 'A' site

to 64th cycle

2: *ejection* of empty RNA

3: *translocation*: m-RNA moves 3 spaces on and peptidyl t-RNA locates in the 'P' site

Figure 9.7. A formalised and highly simplified view of events in the elongation of a peptide chain by a ribosome. Whereas the amino acid units being put together contain 7–24 atoms, the stretch of message tape needed to specify one amino acid has 80–90 atoms in it. The whole ribosome (stippled) is a machine built from some 170000 atoms. In addition several soluble proteins are needed for this machine to work.

ism can be formulated for translating messages written with a four-symbol alphabet into messages written in an alphabet of up to 64 symbols (figure 9.6). The postulated adaptors were soon identified with a set of relatively small RNA molecules of about 80 nucleotide units each – the tRNAs.

For protein synthesis, as for DNA replication, a line-up mechanism, although correct at a certain formal level, is not literally correct. Here again the true mechanism, although still not completely understood, is a sequential unit-by-unit one: it is more like the operation of a typewriter than that of a printer's block. And it is not exactly DNA, but offprints of one of its strands that act as templates – the messenger or mRNAs.

The third kind of RNA – ribosomal or rRNA – is the major constituent of the ribosome. This is the device that brings the other two kinds of RNA together. Again ideas about it have changed somewhat. At one time the ribosome was thought of as a static sort of jig. Now it is seen rather as a GTP-powered coupling machine of great sophistication.

All the ribosomes in all the organisms on Earth seem to work in much the same way. Ribosomes are large and complex structures. The *E. coli* ribosome, for example, contains three single-stranded RNA molecules with about 3000 nucleotide units in the biggest of them, and about 1600 and 120 in the other two. There are also 55 protein molecules – all of them different from each other and mostly very different. There are about 8000 amino acid units altogether in these structural proteins.

In operation, GTP and several non-structural proteins are needed usually – for initiation, for chain elongation, and for termination. Each of these stages has special requirements. A formal picture of the main synthetic operation – the cycle of chain elongation – is given in figure 9.7.

Looking at Merrifield's machine for peptide synthesis (figure 1.11), or indeed reading the experimental section of any paper describing a laboratory synthesis of a long peptide, we should not really be surprised that the protein-synthetic equipment in organisms is so elaborate. It is (quite simply) *difficult* to put amino acids together into long defined sequences.

Preadaptations? How did such expertise evolve? There was no forward planning. The ribosome was not designed first and then evolved towards. It was arrived at, as usual, through successive small modifications or new combinations of pre-existing machines – but machines that had not necessarily been concerned, to begin with, with anything like protein synthesis.

Already, according to our story, some key pieces for a molecular translating machine have been invented. These organisms are attaching labels to molecules as part of a generalised technique for putting things in their proper place. Particular enantiomers of particular amino acids are among the molecules that are being made consistently – and they are being located in this way. So something very like the tRNAs is already there. For a long time there have been selection pressures to perfect the means of attaching amino acids (among other things) to their particular RNA-like oligomers.

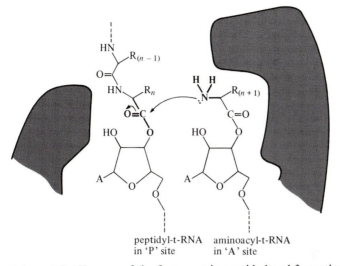

peptidyl-t-RNA   aminoacyl-t-RNA
in 'P' site   in 'A' site

Figure 9.8. Close-up of the first event in peptide bond formation (in stage 1 of figure 9.7).

One might suppose that sometimes the purpose of locating a particular molecule somewhere was to bring it next to another similarly located molecule with which it was to react – that this was part of the technique of the 'nucleic acid enzyme', and a means whereby dipeptides, among other things, could be conveniently made. Then, somehow, this technique was extended to allow the synthesis of polypeptides of indefinite length.

However this happened, the site where the amide links were forming had to be appropriately made. It had to be accessible to the reactants – the amino acids with their identifying labels – and yet kept dry to avoid the ever present threat of hydrolysis in place of amide-bond formation. And the site should have been such as to favour the un-ionised form of the amino group that was reacting (see figure 9.8).

Following our earlier discussions, clays might have been particularly useful in helping to provide suitable reaction sites. That is to say, clays might have been components of protoribosomes.

Speculations about protoribosomes would be less wild, perhaps, if we knew more about the details of how the modern ribosome works. But we can see already one most characteristic feature of this huge 'enzyme', which can be used to illustrate yet again how the pre-existence of a well evolved organism can soften the edges of the most difficult looking puzzles.

Translocation, the shifting of the message tape in relation to the site in the ribosome where the peptide bonds are formed, is especially what allows this 'enzyme' to judder on indefinitely – so as to make a polymer instead of releasing its product after each cycle of operations as most enzymes do (see figure 9.7). There is a mechanism here equivalent to the gate in a cine camera; or the mechanism that shifts the carriage of a type-writer between letters. It seems very contrived and mechanical. And yet, can we not imagine preadaptations for that sort of thing?

Here is one possibility. Although the subsidiary energy carrier, GTP, is generally used to help to drive the cycle of chain elongation in peptide synthesis, there appear to be some cases where GTP is not absolutely required (Spirin, 1978). The energy derived from the aminolysis (figure 9.8) is seemingly sufficient to drive the rest of the cycle, including transloca-tion. So perhaps translocation came first – the aminolysis (or hydrolysis) of active esters of tRNA-like molecules being used simply to *move* RNA molecules.

We will touch again on the intriguing question of the evolution of the ribosome. But for the rest of this section we will be concerned mainly with two related questions that are perhaps even more critical (and which we skated over rather). These are questions about how amino acids were attached to the correct adaptors in the early days, and how the genetic code evolved.

Matching adaptors to amino acids

One of the tangles in the problem of the origin of our protein-synthetic machinery is that, now, to make a protein there must be many proteins already made (cf. figure 3.2). Yet so much of the present day protein-synthetic machinery is RNA, in one form or another, that it is tempting to take a simple line and suppose that there were earlier versions of our protein-synthetic machinery that did not use proteins – that were made entirely of RNA-like molecules or of RNA-like molecules in con-junction with other non-protein materials (Rich, 1965). That way we can dispense with some of the twists in the tangle. We do not then have to suppose that there was ever any other way of making protein.

Of all the present-day involvements of protein in protein synthesis the most critical is in that large set of large and highly sophisticated enzymes that select amino acids, activate them, and then join them to correct tRNA adaptors – the aminoacyl-tRNA ligases. This is precisely the point in pro-tein synthesis today at which the amino acids are distinguished from each other. It is the only point: once a tRNA molecule is loaded up there are no further checks. If it has not been properly loaded then a wrong amino acid will duly be incorporated in some growing polypeptide chain.

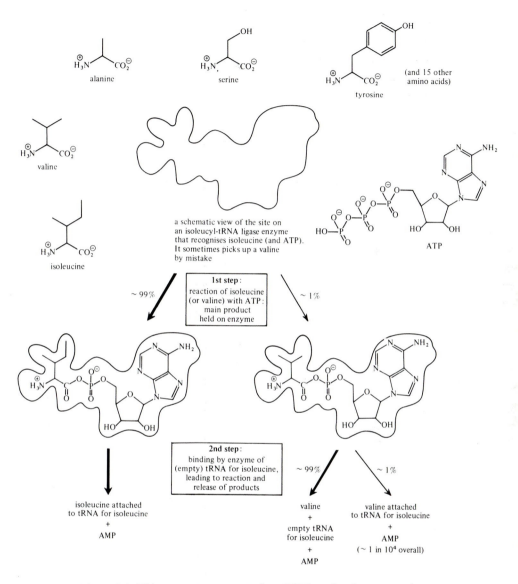

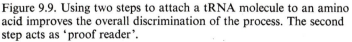

Figure 9.9. Using two steps to attach a tRNA molecule to an amino acid improves the overall discrimination of the process. The second step acts as 'proof reader'.

One might try to imagine aminoacyl-tRNA ligase enzymes made out of RNA. There should be no great problem about recognising the RNA adaptors in that case. The difficulty would be to imagine an RNA device that could also distinguish one amino acid side chain from another. It might be that a suitably folded RNA-like molecule could have made a sufficiently discriminating socket (Smithies, in Crick, 1968). In the folding of a long copolymer, information may become, in effect, concentrated in a small active region. But the overall information density in RNA-like molecules is much less than for proteins. And even protein seems only just good enough in some cases. There is a rather elaborate two-step mechanism for joining amino acids to their adaptors which appears to be needed to achieve sufficient discrimination. The isoleucyl-tRNA ligase, for example, makes a mistake every 100 or so times and picks up a valine instead. The second step acts as 'proof reader', almost always hydrolysing a wrongly activated amino acid rather than completing the operation by forming an ester bond with the isoleucine-coding tRNA (figure 9.9).

Perhaps pre-protein 'ligase machines' were similar in this respect to the modern enzymes in using series of events of relatively low discrimination to achieve a sufficiently high discrimination overall. This would be conceivable for evolved organisms of the kinds that we have been thinking about that can set up production lines, pump solutions through tubes, and so on. But one can also think of other ways in which a pre-protein biochemistry might have contrived to attach the correct labels to particular molecules such as amino acids. Here are four possibilities (that would not have been mutually exclusive).

1. Evolved specificity This would have been a forerunner of the way in which today's proteins acquired their discriminatory powers – through successive fine adjustments to a socket of some sort. Genetically determined RNA tertiary structures might have been able to evolve like this to some extent, as remarked on above. In Chapter 7 we imagined evolving crystal defect structures as having possibly been able to engineer binding sites to a limited extent.

2. Fortuitous specificity It is quite possible that a surface site on a clay mineral, or a crevice in a folded RNA molecule, might just happen to fit some molecule in the environment – and that this accident was then made use of, whether or not the site was subject to adjustment subsequently through natural selection. It was thought at one time that there might be such a fortuitous relationship between triplets of bases and corresponding amino acids, but that does not seem to be the case. (It is perhaps too much to suppose that twenty different amino acids should happen to fit exactly with a rather limited set of RNA triplet sequences.) But, nevertheless, one or two fortuitous matchings of some sort may have played a part in the early evolution of the genetic code.

3. Identification by place It was an idea introduced earlier that clay organisms would have 'kept tabs' on molecules by keeping them localised. If, let us say, there is an amino acid in this vesicle then it must be glycine because this is the glycine output vesicle. So if glycine is to be attached to a piece of RNA with a particular word written in it then the glycine vesicle should be connected to the vesicle that is making this label. Hence the connection is made, in effect, by the apparatus which need not be capable of individual molecular discrimination (any more than the organic chemist's apparatus). Quite coarse-grained genetic information might thus be able to make the required connections. It might even be possible for the same label (adaptor) to be used over and over. A label could find its way home, perhaps, back to the glycine vesicle, because in or near that vesicle there was a polymer with the appropriate complementary sequence (the original templates for the original labels perhaps).

4. Choice of amino acids Another way of minimising problems of recognition is by limiting the number of kinds of objects to be recognised – and by making them very different from each other. Of course if you have both isoleucine and valine in your system you may have difficulties telling them apart; but the first systems for making long polypeptides need not have used either such a large or such an *ambiguous* set of amino acids as we now have. Here is another enormous advantage in having well evolved organisms as the source of amino acids rather than a primordial soup. A soup would contain a very ambiguous set of molecules (Chapter 1).

Genetic codes

The present genetic code, that relates particular mRNA triplets to particular amino acids, is shown in figure 9.10. The question of the origin of this code had been well discussed already by the late nineteen sixties (for example by Woese, 1967; Crick, 1968; Orgel, 1968). Several of the ideas put forward then will be used in the following discussion. The main difference will be in the setting that is assumed – not a primordial soup.

FIRST POSITION	SECOND POSITION				THIRD POSITION
	U	C	A	G	
U	phenylalanine	serine	tyrosine	cysteine	U
	phenylalanine	serine	tyrosine	cysteine	C
	leucine	serine	Ⓣ	Ⓣ	A
	leucine	serine	Ⓣ	tryptophan	G
C	leucine	proline	histidine	arginine	U
	leucine	proline	histidine	arginine	C
	leucine	proline	glutamine	arginine	A
	leucine	proline	glutamine	arginine	G
A	isoleucine	threonine	asparagine	serine	U
	isoleucine	threonine	asparagine	serine	C
	isoleucine	threonine	lysine	arginine	A
	methionine Ⓘ	threonine	lysine	arginine	G
G	valine	alanine	aspartic acid	glycine	U
	valine	alanine	aspartic acid	glycine	C
	valine	alanine	glutamic acid	glycine	A
	valine	alanine	glutamic acid	glycine	G

a

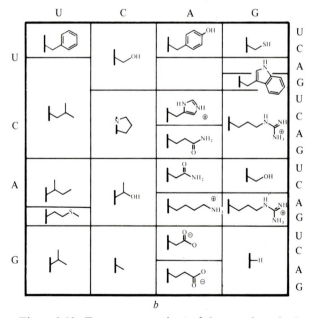

b

Figure 9.10. Two representations of the genetic code. In (*a*) the letters I and T stand for initiating and terminating signals. (*b*) Shows the structures of the side chains of the amino acids, emphasising a non-random distribution of types.

Which amino acids first? It is usual to suppose that the earliest amino acids that were coded for would have been 'primordial': that is to say, they would have been from among that set that we discussed in Chapter 1 that are common products of various non-biological syntheses (glycine and alanine most prominently, but aspartic acid, glutamic acid, serine, valine and leucine might be included). The variety of ways in which these molecules form suggests that they are rather generally chemically accessible. Certainly this would be a point in their favour for inclusion at any time. Why not make use of easily made molecules?

To begin with, in the invention of a system for peptide synthesis, there might have been a good reason. Because what is easily made may not be easily recognisable – which, in this yet protein-less biochemistry, might have been much more to the point.

Ease of synthesis and recognisability are two factors: a third would be usefulness. I do not mean usefulness for the individual amino acids – they must already be useful. But there must be a use too for simple polymers that incorporate the amino acids in question.

Proproteins We might ask what the simplest useful **proproteins** might have been like – defining a poly-α-amino acid as (at least) a pro-protein if it is both sufficiently accurately specified and long enough to fold, or otherwise 'self-assemble', into some distinct higher order structure.

Following Orgel (1972) we could imagine some very simple first pro-protein structures based on just two kinds of amino acid, one hydrophobic and one hydrophilic. Orgel suggested that, with such an alternation, co-herent β-structures would tend to form, that is, sheets made from aligned polyamino acid chains in which (here) one surface would be covered with hydrophilic and the other with hydrophobic groups (figure 9.11a). These β-structures might then be expected to assemble further into water-dispersible bilayers (figure 9.11b). Furthermore, there might be a connec-tion between this and a previously noted feature of our genetic code, namely that there is a distinct tendency for more hydrophobic amino acids to have a pyrimidine (especially U) in the middle position of their code words, while hydrophilic amino acids usually have a purine (especially A) (Woese, 1967; Dickerson, 1971). This point can be seen clearly in figure 9.10b. The connection that Orgel suggests here arises from studies of the interactions between mononucleotides and polynucleotides. Without any 'replication apparatus' it would be particularly difficult to replicate poly-nucleotides in which purines were adjacent. (Because single pyrimidine nucleotides would not stick.) Hence primitive nucleic acids might have been alternating pyrimidine–purine copolymers in which case (given a

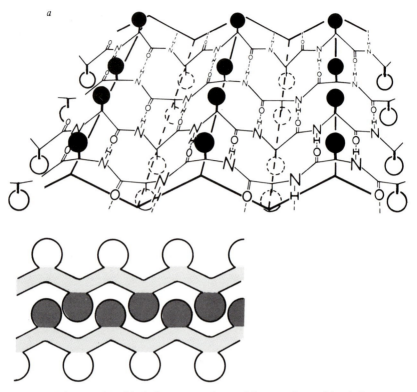

Figure 9.11. (*a*) A β-structure formed from polypeptide chains arranged parallel to each other and connected by hydrogen bonds. Here the chains are running in opposite directions ('antiparallel'). Alternate side groups point above and below the (pleated) sheet. (*b*) If the black circles and open circles in (*a*) represent hydrophobic and hydrophilic side groups respectively, then pairs of these β-pleated sheets may form bilayers that are dispersible in water, as indicated in this simplified end view.

similar code to the present one) the early polypeptides would have been predominantly alternating hydrophobic–hydrophilic copolymers.

The tendency for such polypeptides to form β-structures has been confirmed by a number of studies with synthetic copolymers. For example:

poly(valine–lysine)

poly(leucine–lysine)

and   poly(leucine–glutamic acid)

These all form optically clear solutions containing stable, antiparallel β-pleated sheets – most likely of the bilayer type (Brack & Orgel, 1975; Brack, 1977; Brahms, Brahms, Spach & Brack, 1977). Moderate salt concentrations are generally favourable to making β-structures, by providing counterions to off-set the mutual repulsion between the charged groups.

Brack reports an interesting effect with montmorillonite and poly-(valine–lysine). The clay becomes hydrophobic, releasing about 90 % of its interlayer sodium ions – the polymer intercalating instead, seemingly in the β-form.

As expected from the classical β-structure, these polymers are generally sensitive to substitutions that interrupt the alternation of side chains. And they are sensitive too to chiral purity. Even 5 % of D-isomers in an L-chain noticeably reduces the tendency to form β-structures. Again this is to be expected from the normal β-structure where the characteristic up-and-down orientation of the side groups (figure 9.11) depends on the polypeptide being chirally homogeneous.

To our question 'which amino acids came first?' let us take as a provisional and partial answer that there were two – one hydrophobic and one hydrophilic. Let us name them phobine and philine.

What were the first code words? Let us accept also what we might call Orgel's First Rule – that the first polynucleotide sequences coding for proprotein were of the general form:

$$\ldots x\text{–}Pu\text{–}x\text{–}x\text{–}Py\text{–}x\text{–}x\text{–}Pu\text{–}x\text{–}x\text{–}Py\text{–}x\text{–}\ldots \tag{1}$$

where x is any nucleotide and Pu and Py are respectively A *or* G and U *or* C.

It would simplify our starting point, and make it more realistic perhaps, if we follow the idea (e.g. Crick, 1968) that the very first prototypes of this new way of specifying non-genetic polymers was based on polynucleotides with two rather than four kinds of bases. It is not that the organisms were incapable of producing many different kinds of nucleotides by this stage; but for any novel machine that depends to any extent on a fortuitous coming together of components, the fewer such components the better. Since U-centred and A-centred code words show the hydrophobic–hydrophilic dichotomy most clearly (figure 9.10*b*) we can, for the sake of discussion, choose those two. A typical sequence then becomes:

$$\ldots x\text{–}A\text{–}x\text{–}x\text{–}U\text{–}x\text{–}x\text{–}A\text{–}x\text{–}x\text{–}U\text{–}x\text{–}\ldots \tag{2}$$

where x is A *or* U. There are sixteen possible sequences like this, although if we accept for the present Orgel's Second Rule – the preference for Pu–Py alternations – then there would only be one possibility:

$$\ldots A\text{–}U\text{–}A\text{–}U\text{–}\ldots \tag{3}$$

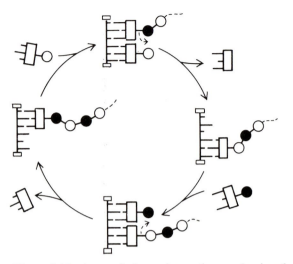

Figure 9.12. A speculative reciprocating mechanism for a (protoribosomal) peptide synthesis avoiding the need for translocation. A very short fixed 'message tape' generates an alternating copolymer of indefinite length.

This is not a very interesting sequence admittedly, and with no potential variability (except in length) it is hardly a 'message' at all. But being generally self-cohesive it could have been useful in itself; and as a first 'message tape', for the synthesis of a first proprotein, it would have had the advantage at least of being unambiguous in two important senses: the same pair of code words alternate wherever you start reading – and whether you read a + or − strand. Furthermore, molecules incorporating such strands would tend to fold on themselves. This might have provided, perhaps, a pair of suitable adaptors for attachment to phobine and philine. With such adaptors and the ... AUAUA ... tape, a sequence of operations equivalent to those of the modern ribosome (figure 9.7) would now produce poly(phobine–philine). But the sequence of operations might have been different to begin with. Orgel suggested that peptidyl-tRNAs might sometimes have detached themselves from the message tape and reattached elsewhere on the tape. In that case the tape would still have gone on making the same polypeptide – and quite a short message could have been enough to make a long chain. Alternatively, the rather difficult looking step of translocation might have been avoided to begin with, as illustrated in figure 9.12.

Neither of these devices would be very good for making proproteins of controlled length, so they might soon be superseded; but the evolution of such simpler modes could nevertheless have helped to perfect some of the basic machinery (tRNAs, etc.) that would be required for those later systems that could make alternating polypeptides of defined lengths by running along given stretches of polynucleotide.

Why a triplet code? Our putative table of code word assignments now reads as follows:

		U	A	
		U	A	
U		–	philine	U
		–	–	A
		–	–	U
A		phobine	–	A

The dashes stand for 'not applicable' or rather 'not replicable'. We are assuming a triplet code from the outset – for simplicity of discussion more than anything else, because at the very beginning, for certain very simple 'message tapes', one might imagine an easy evolution from one word length to another. For example, our ... AUAUAUAU ... can be read not only as an alternating sequence of three-letter words but of words with one letter or five letters – or indeed any odd number. But it is probably fair to say that a triplet code was settled on fairly early. Formally it could have been a singlet code if there were only two amino acids and two nucleotides; but, mechanically, at least triplets might have been needed – simply to get the adaptors to stick to the message tapes. One might guess that a triplet coding system was selected because, with the protoribosomal apparatus available early on (when the decision was made as it were) a triplet of bases was the shortest sequence that would locate properly.

Code word multiplicity There would be much to be said for an immutable alternating 'message tape' (of some sort) if all that was required was an alternating polyamino acid. Then no code words other than, say, AUA and UAU would ever appear, to confuse things. Such a set-up could still evolve in so far as the choice of which amino acids were alternating was made by apparatus that was specified by existing genetic systems – the apparatus for making the amino acids and for attaching them to adaptors. More significantly, very much of the expertise now embodied in our protein-synthetic machinery could have begun to appear already at this stage under selection pressures provided by the usefulness of an accurately alternating polyamino acid. For example, the triplet code idea and a translocating machine of some sort might have been arrived at by organisms that had a use for an accurately alternating polyamino acid of defined length.

But of course we know that the future lay with more interesting sequences. Several things would be needed for such an adventure. The 'replicase apparatus' would have to be competent enough by now to print accurately more complex sequences. This would provide extra coding capacity – and a number of new problems. Every so often mutations would produce non-code words that would block proprotein synthesis because there were no adaptors to fit these words. There would be strong selection pressures at once to make adaptors that would fit any of the words that were now bound to turn up – and to attach each to *some* amino acid to prevent blocking or premature termination of proprotein synthesis (cf. Sonneborn, 1965). The shortest step would be to assign existing amino acids – in our model phobine and philine – to the new adaptors.

The middle letter It is clear that in principle (and however long the code word) only one definitive letter in a code word – say the nth letter – would be needed to translate between a two-nucleotide message and a two-amino acid proprotein. There might be a convention, for example, that if the nth letter is A then it is a philine code word (whatever the other letters are). Similarly, the nth letter is a U in any phobine code word. And we might add that it seems to be the case that the middle letter of some set of early (triplet) code words was chosen to be definitive – to account for the hydrophobic–hydrophilic dichotomy in our genetic code referred to earlier (figure 9.10). Perhaps the middle letter was chosen because it was energetically the most critical for locating purposes.

The table of code word assignments would now have read:

	U	A	
U	phobine	philine	U
	phobine	philine	A
A	phobine	philine	U
	phobine	philine	A

The big advance was that now phobine and philine did not always have to be strictly alternating. The polynucleotide sequences were now real messages able to define a more subtle and varied range of phobine–philine proproteins.

Amino acid multiplicity Now suppose that a mutation occurs (in the clay genetic information) that causes an organism that has been faithfully producing pure L-philine to make a product that contains L-pseudophiline as well. The adaptor ligase apparatus sees little difference between the two, so this new amino acid becomes more or less randomly substituted for philine in the proprotein being made. For some purposes this might not make much difference – particularly if philine itself is still in large excess. Indeed for certain structural purposes it might be a positive advantage for a proprotein bilayer to be softened up a bit by an occasional anomalous amino acid.

Nevertheless it would always be better to have such things under control: to be able to insert a pseudophiline at genetically determined positions. Such an outcome could be approached gradually through the evolution of an increasing discrimination in the adaptor ligase apparatus. We might suppose, for example, that adaptors that had A-starting, U-centred anticode words would be more easily attached to philine, while adaptors with U-starting, U-centred anticode words would be more easily attached to pseudophiline. That might be due to direct recognition of the newly definitive base in the adaptor. But not necessarily. There could be other more prominent differences between the two kinds of adaptors (remember that we are talking about organisms that could evolve such differences).

The difficult part for the adaptor ligase apparatus would have been in discriminating between the amino acids. This would have been a point in favour of philine and pseudophiline being rather distinctly different if they were to be able to acquire their own code words. On the other hand, as Crick (1968) points out, a similar amino acid would have a better chance of taking over a code word because such an amino acid would be less functionally disturbing to the proteins into which it would be incorporated. The balance between these two contrary factors will be discussed shortly.

Identifying strands, starts, stops and phases Now that the ... AUAUAUAU ... tapes were being superseded there would be several new problems of identification. The complementary strands of replicating polynucleotides would not now generally be the same and nor would the messages read the same wherever the reading started. How would the protoribosomes know a message tape when they saw one? How would they know roughly where to begin reading it – and when to stop? How would they know which bases represented the beginnings of words so that the messages could be read in phase?

These are serious problems that are solved nowadays very largely by means of protein machinery. One might suspect that before there was protein the message tapes had to be a good bit more explicit in themselves – that they contained very clear internal markers.

One way to be clear would be to use another pair of bases exclusively as markers, say G and C. For example, suppose that in the coding strand the third position of the code words was always G. The coding strand would be immediately identifiable and so would the positions of the code words on it. Starts and stops might be signalled by some break in this patterning. It is a well noted feature of our present genetic code that the third position of the code words has a much smaller coding role than the other two positions. Very often it does not matter at all what the third letter is and, even when it does, either of the purines or either of the pyrimidines will usually do (see figure 9.10). It makes some sense to suppose that this now half-used position used to have a vital significance – although not in a coding role.

Extending the vocabulary By now our imaginary evolving code table reads as follows:

	U	A	
U	phobine	philine	G
A	phobine	pseudophiline	

The maximum capacity here would be for four amino acids only, with the third position as a 'semicolon'. But it might not be long before G's and C's started to leak into coding positions on the message tapes; at first this could increase the number of redundant code words for the existing set of amino acids, and then it could let in new amino acids – through some mechanism of the kind already described.

Making the rather arbitrary assumption that G leaked first (and that a code word once assigned never changed) we might imagine the following code table at some stage:

	U	A	G	
U	Leu	term.	Try	
A	Met (init. ?)	Lys	Arg	G
G	Val	Glu	Gly	

Such a varied set would fit with the idea that the constraint of recognisability was more important in deciding which new amino acids were to come in than that of not upsetting too much the existing proteins (page

403). (In view of all the assumptions that have been made we should not take this table too seriously. Its main virtue is in its variety. There *is* a great variety among our protein amino acids as well as similarities between many of them: there must have been some way for that variety to have become established.)

After that, on our scheme, C was introduced into the coding strands. (This was now possible because there were now other means of distinguishing between coding and non-coding strands than by their having or not having G's or C's in them.) To begin with, we might guess, C's were mistaken for U's, so that the new C-centred code words were assigned at first to incumbent hydrophobic amino acids. They were later to be differentiated, but by now it was much more important not to upset the existing proteins too much. So this time the differentiation was more marginal with the new amino acids more like those already there. Nevertheless it took place, and under conditions that were still, it seems, quite fluid. Seemingly it was almost completed as far as the 'semicolon' system would allow. At least, to judge from present assignments, there was eventually little redundancy in the table:

	U	C	A	G	
U	Leu	Ser	term.	Try	
C	Leu	Pro	Gln	Arg	G
A	Met/init.	Thr	Lys	Arg	
G	Val	Ala	Glu	Gly	

Finally the possibilities were multiplied by four as the code word marker became unnecessary. The third letter started to acquire a coding significance. In two cases A and G came to be distinguished in this position, and in seven cases a pyrimidine could be distinguished from a purine (cf. figure 9.10). But the evolution of our genetic code never went further than that. For some reason it got stuck.

Under what circumstances can a genetic code evolve?

Other accounts of the origin of our code have been given (for example, Hartman, 1975b; Eigen *et al.*, 1981). A frequent theme is that to begin with message tapes had a low information capacity (but were relatively easily replicated and read), and that they then transformed into tapes with much higher information capacity (but which needed specifically evolved machinery to deal with them).

The crucial point, though, is still more general – and rather generally agreed. Surely the code *evolved*. It was built up in stages of some sort. *First* there must have been organisms, *then* the evolution of the code.

It has been suggested indeed that our code is designed to minimise the effects of mutations (Dickerson, 1971) and/or errors in protein synthesis (Woese, 1967). In that case, if our genetic code is in any sense optimised, there must have been innumerable trials and experiments with different arrangements. But in any case, in violent contrast to the present staid state of affairs, there must have been a time when code word assignments were more or less fluid. How could that have been?

Crick (1968) insists that code word alterations would only have been possible within organisms that used 'a small number of proteins and especially proteins which were somewhat crudely constructed'. Once there were more than a few proteins present it would become increasingly difficult for a code word to change its meaning – too many proteins being arbitrarily affected. Hence sooner or later the evolution of the code would get stuck.

If we accept this view then the code's evolution must have been virtually complete within a line of organisms for which protein was not nearly so important as now. We have to suppose that even quite late on, by which time the machinery must have been already quite complex, all this machinery was devoted to the production of a relatively small number of proteins. Perhaps some of these became involved in the protein-synthetic machinery during the later phases. But such help could not have been very great: as soon as the protein-synthetic machinery came to depend on more than a very few proteins, the evolution of the code would have stopped. Is it too much to suppose the protein-synthetic machinery was evolved under selection pressures that favoured just a few, perhaps crudely constructed, products?

Not necessarily: a material does not have to have multiple central uses to be vital nevertheless to a particular organism. The colour of a flower's petals, or the stretchiness of a spider's silk, or the olfactory attractions of a female moth might in each case depend on just one kind of molecule that has nevertheless been elaborated under strong selection pressures. To get such things just right may be critical to the particular success of an organism and worth a lot of trouble in synthesis.

So we should not be supposing that protein was a sort of universal invention of all life on Earth, even if it did become universal. It seems more sensible to suppose that it arose in some subclass of organisms, perhaps quite a small subclass with special requirements and for which the elaboration of a few key proproteins was well worth the trouble.

On the line that we were pursuing before, even the earliest proprotein need not have been crudely made in the sense that sequence was unimportant; but it must have been true, if the code was to remain fluid, that the nature of the amino acids present in the proprotein did not affect function too critically. β-Structure-forming materials seem to be like this. Poly(leucine–lysine) for example has rather similar physical properties to poly(leucine–glutamic acid) (Brack, 1977); and Brack & Orgel (1975) comment on two natural silks – one with an alanine–glycine alternation and one with alanine and glutamine in roughly equal proportions (presumably alternating). Often, it seems, it is the alternation that matters more than the detailed character of the amino acids that are alternating. With materials like that there could be selection pressures to define (simple) sequences very carefully while allowing, sometimes, quite major amino acid changes. The unthinkable for a modern organism – a changeover (in all the proteins) from lysine to glutamic acid – would be quite conceivable for structural materials of this sort, especially if only one or a few proproteins were present in the organism.

It was under such circumstances, perhaps, that the early 'big decisions' were made about the general types of amino acids to be used – hydrophobic, positively charged, negatively charged, large polar, and so on. The recognition problem might have encouraged at this stage a 'contrasty' kind of code book such as that on page 405 (compare discussion on page 397). After that the changes and new introductions were less radical.

There would always be a tendency for lines of organisms to evolve multiple uses for their proteins, and we might be tempted to think that the first organisms to do so were on the way to greater things. But quite the reverse – they would have put themselves out of the race. The code would have stopped evolving in those lines, as it stopped evolving eventually in every line. The future lay with the late developers, with lines that had kept going for a long time with proteins of a limited type, so that a complex code had had time to appear.

The new technology

By now we are imagining that there are many codes that have evolved in different lines of descent from those organisms that first made simple proproteins. In each line the code is fixed, which keeps each line separate. With hindsight we know that only one line is going to win – they all lie before the last common ancestor of all life on Earth (see figure 3.9 on page 103).

The winner would have been one that had gone quite far before its code became frozen in. More precisely it had evolved a rather sophisticated code book that would allow, later, the synthesis of such a variety of proteins that all clay machinery could be dispensed with.

That was the way the revolution was to end, but the final dynamic phase could only have started after the code, our code, was virtually fixed in our ancestral line. It could only have been then that protein began seriously to displace clay – as an alternative material for the membranes, catalysts and so on, on which organic synthesis depended. (This follows from the argument given earlier: when protein started to diversify very much the code would have frozen.)

Why? How? The excellence of protein is hardly in doubt – we are all living proof of it. So perhaps it is hardly necessary to argue further for the superiority of protein over clay synthetic machinery. In a nutshell, the new technology would allow a more intricate control.

The 'why' questions raise no great difficulties then. But what of the 'how' questions? How would protein get a foothold? However sophisticated its potential, the first fortuitous unevolved protein version of any piece of machinery would be a poor competitor for the already well evolved clay equivalent – or so it would seem.

The genetic view Let us recast our thinking for a moment in terms of genes. We can think of any of our clay–organic organisms, as we can think of any other organism, as a set of genes, a community living in a symbiosis: each gene generates effects that improve the survival chances of the community and hence of the gene itself. In the organisms that we are thinking about, most of the genes are still clay, specifying clay machinery, and through that the synthesis of organic molecules. Some of the genes are nucleic acid. These depend largely on the clay community to maintain monomer supplies, etc.; but they also help the community as a whole – now very largely through this new stuff, protein.

We know what has to happen next. With time, over many generations, the nucleic acid genes in the community must tend to become more numerous while the clay genes become less numerous. It is necessary that, on average, mutations in the nucleic acid genes are more likely to catch on.

In part there is nothing exceptional about such an idea. It is the way evolution works still – genes come and genes go, and some classes of genes are more actively evolving than others. The difference is that for these organisms there are very different kinds of genes in the community, genes that are made differently and work differently – and there is an overall trend in favour of the more recently invented kind (the kind that can specify proteins).

Some threads from Chapter 3 There are two ways in which a sub-system, and the genes specifying it, can be radically changed through evolution. The first is the most generally recognized. It depends on a succession of structural modifications. This was how, for example, the fins of fish transformed into the limbs of land animals. Underlying such transformations there must be corresponding successions of genetic change. Clearly such a mechanism would be no good for changing a clay gene into a nucleic acid one. But, as discussed in Chapter 3, there is also an evolutionary mechanism that depends only on functional connections. Such a mode of evolution is not so much a transformation as a replacement or takeover. The gills of fish were replaced by the lungs of land animals. True, there were long successions of small modifications in this – part of the oesophagus transformed into lungs – but, nevertheless, the element of takeover broke the overall connection: gills and lungs needed to have no structural relationships. Clearly *this* is the sort of way that nucleic acid must have usurped clay.

The key ideas in understanding such radical evolutionary processes are preadaptation, functional overlap and redundancy – as discussed at length in Chapter 3.

First consider preadaptation: it would not be at all odd if, say, a β-structure bilayer that had evolved under selection pressures as a barrier should happen to find other uses – say to orientate acidic and basic groups so as to give some kind of catalytic activity. That sort of thing – preadaptation – is part of normal procedure in evolution. Again and again a feature – an organ, a tissue, a molecule – that has evolved under one set of selection pressures has turned out to have other uses.

Then again functional overlap is commonplace. There can *easily* be a place for an incompetent beginner beside the evolved expert. This may be because the beginner is not actually doing quite the same thing. Part of its function may be similar – there is an overlap perhaps – but there is also an extension of function. As for example when a fish gulps air. This is a very unsubtle way of picking up oxygen when compared with the beautifully designed gas-exchange systems of the gills. But there are times when air gulping is what is wanted. Crude techniques can have their uses – under special circumstances or as a back-up – and hence a crude technique can get a foothold. Once there, if its evolutionary potential is greater in the circumstances, then the crude technique may transform (this time through a succession of small structural modifications) into a highly sophisticated one that makes the original technique redundant.

Updating the control of metabolism

A step-wise mechanism Suppose that an organism contains organelles that are specified by clay genes and which bring about the reaction sequence:

$$A \rightarrow B \rightarrow C \rightarrow D$$

There are many of these organelles in the organism – perhaps 100 or so. D is an amino acid that has to be well made – one enantiomer and free of impurities – because it is being used for protein synthesis. But D has other uses as well, and for some of these a racemic, indeed semi-chaotic product that is merely rich in D is still effective. This gives a chance for an alternative design for a percentage of the organelles in which one of the steps – say B $\rightarrow$ C – is catalysed by a Mark-I protein catalyst specified by nucleic acid. The new catalyst is a fairly complex β-structure that had evolved for another use but which happened to have some B $\rightarrow$ C catalytic activity. Although clumsy, it is perhaps faster or can work at a pH at which the clay machinery tends to dissolve. Hence a new subclass of D-making organelles appear and they are located close to where their product is to be used. The original wholly clay-specified organelles are still the source of D for protein synthesis.

But protein now has a foothold in the organism as a B $\rightarrow$ C catalyst. This catalyst is now improved under natural selection so that C is not only made faster, and under a wider range of conditions by the protein, but it is made more cleanly than ever before. The protein catalyst had been able to evolve a much higher specificity than the clay catalytic surfaces, and most of the clay separator equipment could now be dispensed with. Like a single silicon chip replacing a maze of wires, capacitors, valves, etc., this now perfected piece of protein technology could replace much cumbersome machinery. The engineering, although as complex, was on a smaller scale.

The modified A $\rightarrow$ D organelles could now displace the original ones even as the source of D for protein synthesis. Eventually, through other similar series of events, the whole train of reactions A $\rightarrow$ B $\rightarrow$ C $\rightarrow$ D is taken over and the A $\rightarrow$ D organelle is simplified to hardly more than a bag of three enzymes with primary structures specified wholly by nucleic acid.

CO_2, N_2, etc.

complex products

Figure 9.13. See text.

A stage-wise mechanism Alternatively, we might imagine evolutionary excursions that started from a quite new set of organelles that were specified by nucleic acid from the start. The metabolite (or environmental molecule) A is being transformed within this new class of substructure to make a variety of (at first) semi-chaotic products. Some of these are useful – and a variety of pathways are built up, in different lines of organisms, according to the forward evolutionary strategy given earlier (figure 8.25).

D is among the molecules that are sometimes being hit on. Since D is already useful in a number of ways, it is particularly likely that this will make a useful contribution as some kind of back-up or alternative mode of synthesis. Selection pressures then further improve the efficiency of the A → D reaction train in this new class of organelles – which eventually outstrip the old ones.

This stage-wise story is similar to the step-wise one except that here there would be no need for the whole route A → B → C → D to have been copied. The updated control technique might use a different pathway.

Starting points, goals and stepping stones Figure 9.13 (heavier arrows) might represent synthetic activities of an organism using clay-specified machinery to transform a set of environmental starting materials (left-hand side) to a set of end products (right-hand side). In addition there are numerous 'stepping stones' – intermediates, such as glycine and aspartic acid, that are needed for nucleic acid replication and protein synthesis. With the code now fixed such stepping stones would represent fixed points for the future evolution of any line of organisms that was to retain the nucleic acid–protein machinery. These points, as well as many of the end points, would have to remain even if routes between these points were altered (lighter arrows in figure 9.13).

In spite of these restrictions on what could happen there would have been a considerable freedom as to the order of events. Broadly speaking, it would be an advantage for any of the steps or stages to be updated in any order. Whereas the original pathways under their original control had had to be built up forwards, this takeover of control could have been piecemeal.

The place of lipids

Not only enzymes but some kind of membrane-forming material would have been needed before clays could have been dropped entirely. We discussed earlier the suggestion that protein was one of the original organic membrane-forming materials (as β-structures). Even today protein forms an essential part of the most important kinds of biological membranes (as reviewed by Wickner, 1980). But lipids are also essential components of these modern membranes, and one might guess that the protein–lipid membrane had evolved in its essentials before clay could be dropped.

Again, although the details are difficult to make out, there is no great paradox here. One kind of barrier can stand in for another even if made of a quite different material. One could imagine gradual transformations. A purely inorganic barrier (for example as in figure 8.24) is first lined with polypeptides, etc., and then the inorganic parts become less important and then, in some cases, disappear. Particularly for more subtle functions – for selective barriers, or for well insulated energy-transducing membranes – one might suppose that clay–organic structures would often be better than pure clay structures. Then it would be the old story: because the organic components were subject to more precise specification than the inorganic ones, there would always be a tendency for organic structures to be preferred. And one might guess that it was very largely the invention of lipids that made a final step possible – to purely organic structures.

Which lipids first? Of the two main classes of lipids there are some indications that the polyisoprenoids (for example chains like those in figure 8.16), although more complex, are of more ancient origin than the polyketides (for example the straight chains of lipids such as that in figure 8.2).

One of these indications is in the important and somewhat diverse roles of polyisoprenoids in energy transducers: carotenoids are among the photon-catching pigments; plastoquinone is part of the electron-transporting system of photosynthesis, as are the similar ubiquinones in mitochondria; and then the 'handle' on chlorophyll is a polyisoprenoid structure (see figure 8.16). The 'handle' used in bacterial cell wall biosynthesis (page 375) is also polyisoprenoid, and so are the hydrophobic side chains on some tRNAs. One might suppose that these various structures were invented in the first place to locate in lipid membranes at a time when all lipids were polyisoprenoids. In support of this it has been found that membranes of bacteria that are regarded for other reasons as belonging to ancient classes are often rich in polyisoprenoids (Oró, Sherwood, Eichberg & Epps, 1978; Woese, Magrum & Fox, 1978).

Perhaps polyisoprenoids were, at the time, easier to make. Certainly there is an enormous variety of isoprene-based molecules among modern secondary plant metabolites (Nicholas, 1973), indicating a rather general chemical accessibility. On the other hand, the modern biochemical machinery for polyketide synthesis is particularly sophisticated, involving a multi-enzyme complex and a special carrier protein. It looks as if those seemingly simple straight polymethylene chains are not very easy to put together.

One might suppose, then, that the first organic energy-transducing membranes were built largely from proteins and polyisoprenoid lipids – with perhaps also porphyrins, by now, to help to locate metal ions within the hydrophobic interiors of these membranes.

Again we do not have to assume a sudden switch from one subsystem to another. An organic patch or plug in an otherwise inorganic membrane might contribute in a minor way to the pumping of protons (or whatever) and then gradually increase its role and its extensiveness. Or separately evolved and purely organic organelles might independently contribute to the production of some high-energy molecule, to become later the sole source. After all if the function of organelle A is, say, to use photons to make ATP from ADP, then it can in principle be displaced by organelle B that does the same thing. The inner workings of A and B are not the main point – and might be altogether different. In so far as a subsystem is a module having simple connections with the rest of the system, its replacement can be easily imagined (cf. page 92) – even its gradual replacement, in the kinds of ways that we have discussed.

In figure 8.33 (on page 370) I implied that it was the discovery of organic energy-transducing membranes that allowed, at last, the clay machinery to be left behind. We might now put 'lipids' (meaning isoprenoid lipids) quite late in the sequence, porphyrins later still, and polyketides at some stage after the revolution was all over.

The conservative consequences of revolution

Sooner or later the circle was complete: organisms appeared that used only the nucleic acid–protein–lipid control machinery to make all the components needed for that machinery. Eventually free living organisms appeared that had no clay in them, and eventually too these included autotrophs. With that the connection was made between very simple environmental atom sources, CO_2, N_2, etc., and a minimum but inevitably complex set of components – built from what we now call 'the molecules of life'.

There was something quite new about this way of living. Up to then at least some of the monomers for the central control machinery – the units for clays – had been provided by the environment. Now much more complex organic monomers had to be made and activated. But no longer was life tied to weathering solutions, and to the time scales of clay synthesis. These less fettered organisms could now evolve faster and further to fill a whole new range of niches. They were still somewhat tied to the ground, as life still is, in the need for phosphates and metal cations; but mostly the atoms for life were now taken out of the air.

As with so many revolutions there were conservative consequences. Now there was no simple way of living for these organisms and no simplifying evolutionary paths. They had lost the scaffolding on which they had been built. Now at their centre they had a minimum fixed complex machine that was to be inherited by that organism that was to become the last common ancestor of all life on Earth. Biochemistry was still to evolve, but only outward, now, on the basis of what had been fixed when the clay disappeared.

Coda

How the problem will be resolved

The method of overlap A number of general considerations taken together can often lead to surprisingly particular conclusions. For example in the game of '20 Questions', where the aim is to discover some particular object that is being thought of, it is actually a mistake to start with particular questions. Rather you should start by finding out whether the unknown object belongs to certain large sets. (Through questions such as: Is it in this room? Can you eat it? Is it green? And so on.) By overlapping such sets you can quickly narrow the field.

This way of playing '20 Questions' represents a very general strategy of enquiry. Western science started with the asking of general questions – it started in philosophy. That can still be the right starting territory for new or immature fields, although now there is a vast stock of general ideas available – laws, principles, rules, recurring observations – that have been generated by past scientific enquiry.

For most sciences the game is well advanced, and for most scientists the crucial questions are of a particular sort to be answered by cleverly designed experiments. In the end it must be possible to ask the particular questions; but to insist too soon that only experiment will do as the means of resolving such a remote and complex subject as the origin of life is, to my mind, to misplay the game of '20 Questions' – and quite possibly to miss the chance of solving the problem at all.

This book has been mainly an attempt to look for the place where the answer to the problem of the origin of life may lie – by asking a large number of general questions. There were questions of general biology: what kind of system is it (in the abstract) that can evolve under natural selection? How can evolution give rise to highly co-operative organisation? And so on. At a somewhat more specific level there were chemical questions – about the most geochemically accessible means of holding and printing large amounts of information, or of making membranes, or of trapping the energy of sunlight in chemical form, or of controlling organic reaction sequences without enzymes. Particular experimental results were often referred to in these discussions, and innumerable observations and experiments lie behind general ideas on, say, the nature of crystal defects, or the characteristics of chemical bonds, or of later evolutionary processes. But our principal preoccupation has been with general considerations, and a region of overlap defined by them.

One of the main conclusions of this book is that the relevant region of overlap is, surprisingly, centred on colloidal inorganic crystalline minerals rather than on 'the molecules of life'. I tried to define this region more tightly by thinking about the sorts of minerals that might have been most suited to genetic and to other functions. As the relevant areas of science advance (as clay structures are known in greater detail, for example) it should be possible to see the critical region more clearly – and more clearly still with the sharper tools of specially contrived experiments.

Alternatively, this region of overlap may be fined away to nothing – if, say, an experiment or a theoretical consideration makes it clear that a crystal gene could not work because no defect structure would be stable enough and at the same time be replicable through crystal growth. These speculations are disprovable in principle.

But perhaps a more likely form of failure for any hypothesis about a remote subject is not through direct refutation, but because the region of overlap fails to sharpen up: the hypothesis remains too vague. Remoteness as such – that is, inaccessibility to the senses – should not prevent such a sharpening with time if the hypothesis is indeed on the right lines. Consider, for example, the question of the existence of atoms. This was a very remote idea for the Greeks, but it was to become a virtual certainty by the end of the nineteenth century – before anyone had in any sense seen an atom. Even hypotheses about remote entities can become very distinct if they continue to work, and especially if they fit in with new and unexpected effects. No, the remoteness of the origin of life on Earth should not prevent an hypothesis about it becoming clearer – although progress may be slow and require the consideration of very many diverse data and ideas.

Experimental accessibility In any case it is probably only the historical question that is essentially remote – that is to say the question of how in fact our ultimate ancestor arose. This is a fascinating piece of genealogy, but it is perhaps not the most important question. The important question is rather more general; it is to delineate the set of possible ultimate ancestors. What kinds of systems, among those that could have been generated by geochemical processes on the primitive Earth, could have formed the basis for an indefinite evolution through natural selection? Whatever those systems are like, when you have refined the possibilities sufficiently to be able to make a guess at how they might be made, then your guess should be accessible to experiment. Let me expand a little on this.

We cannot be sure, of course, because God or spacemen or gigantic luck may have put us here. But if you really believe that life arose spontaneously on this planet as a natural outcome of physicochemical processes, then you should expect to be able to re-enact in the laboratory something like the very earliest stages of its evolution. You should expect to be able to make microsystems of some sort that can evolve under natural selection in the fully Darwinian sense.

There are still some provisos. Maybe the experiment would take a million years because some necessary crystal growth process is very slow. Perhaps Nature's original experiments took that long. Perhaps the first organisms had generation times measured in millennia. But Nature had to succeed with materials and conditions that happened to be on the Earth, and commonplace enough to be consistently maintained over very long periods. In the laboratory we can contrive conditions: we can control solution concentrations, adjust temperatures, choose exotic materials for models, and so on. In the laboratory we may lack time, but we can set up and maintain situations whose *a priori* probability in an unevolved world would have been vanishingly small. As discussed in Chapter 5, you do not have to do very much to arrive at the point at which the contrivance of your experimental set-up is far beyond what Nature might have been expected to arrive at (never mind maintain) even given two billion years. Well, Nature *did* contrive the first evolving systems and took less than two billion years to do it.

So I think we should be able to make a primary organism. The difficulties here are not at all like the difficulties of synthesising, say, *E. coli* from scratch. The difficulties in making an organism built along modern lines are technical and enormous – but for *E. coli* we would just about know what to do. By contrast, the technical problems of making something like our ultimate ancestor cannot be very great if Nature overcame them without natural selection. In this case the main problems are theoretical – of knowing *what* to do. When we know that, the experiments may not be so very difficult.

Whenever and however such experiments succeed they will then presumably prove successful under a range of conditions and with a range of somewhat different materials. (To suppose otherwise would be to substitute the postulate of a fluke event by that of a fluke Universe in which there was only one rather precise way of bringing off an origin of life.) So I think that when we are near to the answer we will find many similar answers, and so we will not be able to say exactly how life must have started on Earth. The genealogical problem will not be solved precisely. But the main problem may be effectively resolved all the same – when we have a clear view of a set of possible solutions and experimental demonstrations of some members of that set.

How close are we to this adventure? Weiss (1981) is already well embarked: he reports experiments that clearly demonstrate hereditary effects during the growth of smectite layers through the mechanism of 'intercalating synthesis'. Here new layers crystallise *between* layers of seed crystals, and in doing so they pick up information in the form of (at least) Al for Si substitution densities. By re-seeding new solutions, the process can be repeated for more than 30 generations before the characteristic charge density of the original seed is 'forgotten'. Catalytic and adsorptive properties can concomitantly be transmitted for a similar number of generations.

If it proves possible to transmit clay-held information accurately enough over many generations – with very occasional mistakes – then an experiment along the lines of figure 7.11 (referred to on page 287) becomes imaginable, with selection operating on rheological properties. As discussed in Chapter 7, samples of clay made up of virtually identical particles (because derived from a single seed) might have some very special properties – 'magical' properties almost – if they had been produced through successive selections of replicating, mutating crystallites.

Less direct experiments would still be needed, though, to help to delineate the set of possible primary genetic materials from among the microcrystalline minerals likely to have been forming on the primitive Earth. For reasons discussed in Chapter 7, attention might be concentrated on well crystallised materials – clays such as kaolinite perhaps – to look for elements of order and disorder. Any kind of domaining within layers or ordering between them is of interest, as would be regularities of overall morphology – the appearance again and again, in a sample, of some rather odd complex shape.

The relevant minerals are clays in the most general sense (page 165), including, for example, microcrystalline metal oxides, oxide-hydroxides and sulphides, as well as feldspars, zeolites, or silica where any of these have formed from aqueous solution.

Continuous crystallisers Studies could be done on clays made in the laboratory in continuous crystallisers. These are closely analogous to chemostats (see figure 5.1) and are used for producing crystal suspensions under controlled conditions. In a continuous crystalliser, as in a chemostat, constant conditions can be maintained indefinitely through nutrient (supersaturated) solutions flowing in and product – crystals in suspension – flowing out. Such a device might be used fo discover genetic crystals.

Consider first an ordinary crystallisation process for which the difficult step is nucleation (seeding). At first nothing would happen in a continuous crystalliser, the supersaturated solution would simply flow in and flow out. If one seed was added this might still not be enough if all that happened was that that seed grew bigger. Eventually it would be carried out through the waste pipe. To keep going it would be necessary that a seed crystal not only grew but broke up, and furthermore that the rate of production of new seeds in this way was sufficient to maintain the population. It would be necessary that an average crystallite, during its average residence in the crystallising vessel, had grown and broken up into at least two new seeds.

These properties – seed dependence and break-up – would be necessary but insufficient for replicating genetic crystals. We can take one step further by considering an experiment in which there were two different kinds of seeds that were individually replicable – let us suppose that they were two different polymorphic forms of the material that was crystallising out. As in a chemostat containing mutant micro-organisms with slightly different reproduction rates, if one of the polymorphs grew and broke up even marginally faster it would sooner or later take over *completely*. (After every generation time the slower one would constitute a smaller proportion of the population, and that trend would continue however small the amount of the slower reproducer, which would thus, sooner or later, be wiped out.) Under such circumstances kinetic control is everything: it is not necessarily the most stable type that wins in the end, but the one that reproduces most quickly. Such a set-up would be an improved model for a primitive biological system, but still too limited.

Now suppose that there was some immense number, n, of possible kinds of seeds that were individually replicable. In this case the individuality of the seeds could hardly be due to polymorphism, to their each having a distinct crystal structure. Here the individuality would have to come from some characteristic defect structure, some kind of patterning superimposed on the crystal structure. And of course this patterning would have to be replicated as the crystals grew, and be maintained as the crystals broke up (as discussed in Chapter 7). But the same general arguments would apply: if for any reason one of a set of patterns present initially could reproduce even marginally faster than the others, it would eventually displace the others entirely. Occasional mutations would then allow the most successful modification of the most successful modification of the most . . . to catch on. That way, seemingly highly improbable systems could be arrived at. We would be watching a Darwinian evolution. As discussed in Chapter 3, it need not take very long for reproducing, mutating systems to arrive at exceedingly improbable, exceedingly minute, subsets through successive selections (pages 87–8).

Looking for clay-based life on the Earth today Nature may be doing experiments like this for us all the time. Look again at those scanning electron micrographs in figures 6.34 and 6.54 which show clays crystallising from solutions percolating through pores in hard rocks. Near the surface of the Earth there are innumerable regions into which barely supersaturated solutions are flowing; within which solutions are being 'stirred' (by turbulence, back-flow, cyclic flows); within which clays are crystallising; and from which clays in suspension are emerging. Nature sets up continuous crystallisers.

An alternative approach, then, would be to look for crystal genes in Nature. In Chapter 7 we discussed in general terms how this might be done by looking for appropriate morphologies: kaolinite vermiforms were conceivable examples. Do such vermiforms breed? Are there conditions where the growth of such crystals is wholly through elongation in the stacking direction plus break-up?

We might look for signs of a limited kind of evolution having taken place. Looking at kaolinite vermiforms, for example (figure 6.29), we might ask if there were selective factors at work that favoured one kind of thickness and cross-section over another, so that one type became predominant. If so (and if the breeding condition was also satisfied) then such vermiforms would be low-level organisms, able to pass on to offspring a rather particular way of surviving and propagating.

There ought to be such 'origins of life' going on now if conditions on the Earth now are broadly similar to conditions on the primitive Earth. You would not necessarily expect starter organisms to be particularly common, in view of the highly stabilised conditions that would be needed to begin with; and you would not expect the higher, photosynthesising forms to be able to compete with modern micro-organisms. But there ought to be at least the beginnings of evolutionary processes to be found somewhere.

Indeed one would expect that such processes would have gone far enough, in some cases, for minerals to be crystallising in certain environments *by virtue* of their being evolved. That is to say because they have some very particular kind of defect structure that was a product of natural selection, they were able to grow at a significant rate in the environment in question.

One might look for this sort of thing in minerals that crystallise readily in Nature, have rather complex morphologies, and yet seem to have been difficult to make in the laboratory. (That would be because in the laboratory conditions had not been set up to allow the gradual evolution of efficiently breeding seeds.) Kaolinite might fit here; it is one of the very commonest clays in Nature, where it comes in a great variety of forms (Keller, 1978); and yet kaolinite is particularly difficult to synthesise (page 253).

More advanced organisms – the kinds of organo–clay systems that we were discussing in Chapter 8 – would have to be sought as fossils. Objects that look like fossils of modern-type cells, or of modern-type colonies of micro-organisms, have been found in some of the most ancient rocks on Earth. The older ones (from about 2500–3500 million years ago) are less sure, and the microstructures have rather abnormal size distributions (Schopf, 1975, 1976). Conceivably these are cells of an altogether earlier form of life.

Any odd-looking structures in very ancient rocks might be interesting, particularly if similar structures appear in a number of different localities within rocks of similar age. They might be fossils of precellular colonies – fossilised vital mud.

A singular case Sherlock Holmes preferred what he called singular cases – seemingly baffling ones. 'As a rule,' he said (in *The Red-Headed League*) 'the more bizarre a thing is the less mysterious it proves to be. It is your commonplace featureless crimes that are really puzzling . . .'

The singular aspect of 'The Case of the Origin of Life on Earth' is in the nature of the first genetic material(s). (How on Earth could information printing machinery have generated spontaneously?) But the very difficulty of this puzzle is to be welcomed, as Holmes would have welcomed it: there cannot be so many solutions. If we can find *any* solution it is likely to be near the mark. Hence the concentration in this book on the nature of the primary genetic materials and, in this short chapter, on experimental possibilities based on ideas in Chapters 5 and 7. There is a promise here of a direct demonstration of what ought to be demonstrable: a synthetic system that is able to evolve under natural selection. (It would not take many generations to demonstrate this ability – compare again page 87).

In direct contrast to the views expressed by Dr Kritic in Chapter 4, it seems to me that the idea of genetic takeover, by widening the initial possibilities, allows us to consider systems that really might be starter organisms. They might seem biochemically unfamiliar, but their mode of evolution would be Darwinian – not some 'chemical evolution' interposed between conditions that one might set up and evolving organisms that subsequently appear.

Other experimental approaches The classical experimental approach to the origin of life (through abiotic organic syntheses) is still very relevant. In relation to the later invention of our kind of biochemistry it is important to understand the chemical paths of least resistance. What kinds of organic reactions take place easily at around ordinary temperatures in aqueous solutions?

Reactions in the presence of clays of all sorts are interesting – although not so much general catalytic effects of clays as effects that depend on particular defect structures. (Compare enzymes: a reaction that was catalysed by just *any* polypeptide sequence could have no critical place in our biochemistry.) It is interesting, for example, that only *some* structural Fe^{3+} in a hectorite was effective in oxidising benzidine and that these processes were very sensitive to pretreatment of the clay (McBride, 1979).

Part III of this book suggests many research possibilities: some extending work that has already been done, others more particularly aimed at testing the plausibility of the story being told in Part III. Let us consider very briefly some other questions that we might want answers to.

In Chapter 7 we took it that one of the first ways in which clay genes would have extended the control of their environment would have been through controlling to some extent the conditions for the synthesis of secondary (non-genetic) clays. This gives us another reason for being interested in the processes of clay crystallisation, especially in the ways in which, in mixed assemblages, one clay can affect the growth of another (through epitaxy, stabilising ion concentrations, binding inhibitors, and so on).

We would like to know more about the effects of crystal size, shape, and defect structure on colloidal properties of clays; about how silicate layers 'self-assemble' into mixed layer or card-house structures; how such layers fold on themselves; how well they stick to each other, and how much that depends on internal cation arrangements, etc. Adsorption of polyphosphate at layer silicate edges, and of organic molecules at edges and interlayers are other topics of interest. How selective are adsorption sites on layer silicates, zeolites – clays of all sorts? Are they ever chirally discriminating? To what extent are they defect dependent?

And we would like to know more about organic catalysts for clay synthesis. How does fulvic acid work? Do any modern plants engineer the kinds of clay around them by producing substances that catalyse clay synthesis? Could one design an organic catalyst for clay synthesis?

How do organic molecules affect the microporosity of clay assemblages and how does microporosity affect distributions of organic molecules? Are there chromatographic effects when solutions flow through clay assemblages? To what extent can (unevolved, uncloned) clays exert a selective influence on the course of organic chemical reactions so as to limit the variety of products? (Mordenite can have such an effect on the formose reaction under somewhat artificial conditions; Weiss *et al.*, 1979.) What are the roles of defect structures and of compartmentalisation? Can one set up oscillations in open clay–organic reactors?

Some of the elements for evolved energy transducers could be looked for in ordinary unevolved clays. There are the proton potentials that can be generated by temperature gradients (Freund & Wengeler, 1978). There is an active interest in clays as photocatalysts and energy stores (Coyne *et al.*, 1981). There is the whole field of semiconductor minerals which include many in the broad category of clays. There are the charge-transfer spectra of transition metals in clays and the effects on redox potential of the crystal environment of such ions. Then there are synergistic binding effects that might be looked for and which would be relevant to some of the formalised energy transducers discussed in Chapter 8. How does the binding of polyphosphate affect the binding of other things?

Already we might begin to think of fabricating energy transducers by orienting asymmetric clay membranes. Can one produce an asymmetric effect – any kind of potential – by illuminating an asymmetric clay membrane?

Related to devices for catching energy there are mineral semiconductor devices for fixing carbon dioxide and nitrogen. Studies on these lines are already well under way.

Further advances in our knowledge of primary and secondary metabolic pathways will help to establish (or tend to refute) the idea that our primary metabolism, like our secondary metabolism, was built from the centre out.

Evolutionary studies at all levels should give us further insights as to general evolutionary mechanisms, for example the roles of symbioses and of takeovers.

Bahadur and his school (see, for example, Bahadur & Ranganayaki, 1970; Smith, Folsome & Bahadur, 1981) have been studying microstructures – complex budding vesicles that Bahadur calls 'jeewanu' – which are generated when certain solutions are precipitated in sunlight (for example solutions containing molybdate, ammonia, phosphate and formaldehyde). Synthesis of a number of our biochemicals has been claimed. It will be interesting to know how these complex structures form. Is this a case of osmotic construction (page 327)? To what extent is their compartmentalised structure relevant to photosynthetic activity? Perhaps the earliest photosynthesising organisms were a subclass of jeewanu, a kind whose structure was guided to some extent by inorganic crystal genes.

By the start of Chapter 9 we were beginning to think about the kinds of polymers that would have been available for structural purposes once the consistent synthesis of particular chirally homogeneous monomers could be presupposed. Insights into possible first uses for homopolymers and regular copolymers can come, as we saw, from studies of polysaccharides, sugar-phosphate polymers, polynucleotides and 'proproteins' – both natural and synthetic. Could we design molecules for holding together clay particles (as in figure 9.5)? Or could we even invent a 'nucleic acid enzyme' (with oligonucleotide coenzymes to go with it)?

In considering first forms of nucleic acid replication and the evolution of our protein-synthetic machinery, we are coming to the final stages of the evolution of our kind of life. But advances in understanding here could still be relevant to the question of genetic takeover. Such advances should tend to favour or refute the contention that the invention of our nucleic acid–protein control machinery could *only* have been made by organisms, highly evolved organisms, that had had to manage without any such machinery to begin with.

A useful life? To come back to the very beginning, there are practical reasons for being interested in biogenesis if indeed first forms of life were mineral and could be recreated. Here would be a new technique of fabrication at the otherwise difficult, colloidal, size level – through artificial selection of replicating defect structures. For example, exceedingly finely engineered semiconductor devices would become available, made from the cheapest possible materials. It is such devices that are perhaps most needed to solve the problem of converting cheaply and efficiently solar energy into other forms.

And there is a curious advantage in this form of engineering. If you use artificial selection to perfect your product you do not have to know in detail how it works (any more than a rose breeder has to know in detail how plants work). The replicating pattern with its special properties can be arrived at with an eye only to its performance. And once you have arrived at this valuable structure you can make as many copies of it as you please.

The existence of life on Earth is the best clue to the existence of this undiscovered field of technology.

How did life originate on Earth? That question will be resolved, if never precisely answered, when we have found out how to make primary organisms of various sorts. The problem of the origin of life on Earth will then have lost its mystery.

References

Abelson, P. H. (1956). Amino acids formed in 'primitive atmospheres'. *Science,* **124**, 935.

Abelson, P. H. (1957). Some aspects of paleobiochemistry. *Annals of the New York Academy of Sciences,* **69**, 276-85.

Abelson, P. H. (1966). Chemical events on the primitive Earth. *Proceedings of the National Academy of Sciences USA,* **55**, 1365-72.

Adams, J. M. (1978). Unifying features relating to the 3D structures of some intercalates of kaolinite. *Clays and Clay Minerals,* **26**, 291-5.

Albersheim, P. (1975). The walls of growing plant cells. *Scientific American,* **232**, no. 4, 80-95.

Alberts, B. & Sternglanz, R. (1977). Recent excitement in the DNA replication problem. *Nature,* **269**, 655-61.

Alcover, J. F., Gatineau, L. & Mering, J. (1973). Exchangeable cation distribution in nickel- and magnesium-vermiculites. *Clays and Clay Minerals,* **21**, 131-6.

Aleksandrova, V. A., Drits, V. A. & Sokolova, G. V. (1972). Structural features of dioctahedral one-packet chlorite. *Kristallographie,* **17**, 525-32

Alexander, G. B., Heston, W. M. & Iler, R. K. (1954). The solubility of amorphous silica in water. *Journal of Physical Chemistry,* **58**, 453-5.

Allaart, J. H. (1976). The pre-3760 m.y. old supracrustal rocks of the Isua area, Central West Greenland, and associated occurrence of quartz-banded ironstone. In *The Early History of the Earth,* ed. B. F. Windley, pp. 177-89. London: John Wiley & Sons.

Allen, F. J. (1899). What is life? *Proceedings of the Birmingham Natural History and Philosophical Society,* **11**, 44-67.

Allen, W. V. & Ponnamperuma, C. (1967). A possible prebiotic synthesis of monocarboxylic acids. *Currents in Modern Biology,* **1**, 24-28.

Anderson, C. S. & Bailey, S. W. (1981). A new cation ordering pattern in amesite-$2H_2$. *American Mineralogist,* **66**, 185-95.

Anderson, W. L. & Stucki, J. W. (1979). Effect of structural Fe^{2+} on visible absorption spectra of nontronite suspensions. In *International Clay Conference 1978,* ed. M. M. Mortland & V. C. Farmer, pp. 75-83. Amsterdam: Elsevier.

Atkinson, R. J., Posner, A. M. & Quirk, J. P. (1977). Crystal nucleation and growth in hydrolysing iron (III) chloride solutions. *Clays and Clay Minerals,* **25**, 49–56.

Aumento, F. (1970). Serpentine Mineralogy of Ultrabasic Intrusions in Canada and on the Mid-Atlantic Ridge. *Geological Survey Canada, paper 69–53.*

Bada, J. L. & Miller, S. L. (1968). Ammonium ion concentration in the primitive ocean. *Science,* **159**, 423–5.

Baes, C. F. & Mesmer, R. E. (1976). *The Hydrolysis of Cations.* New York: Wiley.

Bahadur, K. & Ranganayaki, S. (1970). The photochemical formation of self-sustaining coacervates. *Journal of the British Interplanetary Society,* **23**, 813–29.

Bailey, S. W. (1963). Polymorphism of kaolin minerals. *American Mineralogist,* **48**, 1196–209.

Bailey, S. W. (1966). The status of clay mineral structures. *Clays and Clay Minerals,* **14**, 1–23.

Bailey, S. W. (1975). Cation ordering and pseudosymmetry in layer silicates. *American Mineralogist,* **60**, 175–87.

Baird, T., Cairns-Smith, A. G. & MacKenzie, D. W. (1973). An electron microscope study of magnesium smectite synthesis. *Clay Minerals,* **10**, 17–26.

Baird, T., Cairns-Smith, A. G., Mackenzie, D. W. & Snell, D. S. (1971). *Clay Minerals,* **9**, 250–2.

Baly, E. C. C., Heilbron, I. M. & Barker, W. F. (1921). Photocatalysis. Part I. The synthesis of formaldehyde and carbohydrates from carbon dioxide and water. *Journal of the Chemical Society,* **119**, 1025–35.

Baly, E. C. C., Heilbron, I. M. & Hudson, D. P. (1922). Photocatalysis. Part II. The photosynthesis of nitrogen compounds from nitrates and carbon dioxide. *Journal of the Chemical Society,* **121**, 1078–88.

Balzani, V., Moggi, L., Manfrin, M. F., Bolletta, F. & Gleria, M. (1975). Solar energy conversion by water photodissociation. *Science,* **189**, 852–6.

Bard, A. J. (1979). Photoelectrochemistry and heterogeneous photocatalysis at semiconductors. *Journal of Photochemistry,* **10**, 59–75.

Bard, A. J. (1980). Photoelectrochemistry. *Science,* **207**, 139–44.

Barker, W. C., McLaughlin, P. J. & Dayhoff, M. O. (1972). Evolution of a complex system: The immunoglobulins. In *Atlas of Protein Sequences and Structure,* vol. 5, ed. M. O. Dayhoff, pp. 31–9. Washington, D.C.: National Biomedical Research Foundation.

Bar-Nun, A., Bar-Nun, N., Bauer, S. H. & Sagan, C. (1970). Shock synthesis of amino acids in simulated primitive environments. *Science,* **168**, 470–3.

Bar-Nun, A., Bar-Nun, N., Bauer, S. H. & Sagan, C. (1971). Shock synthesis of amino acids in simulated primitive environments. In *Molecular Evolution I: Chemical Evolution and Origin of Life,* ed. R. Buvet & C. Ponnamperuma, pp. 114–22. Amsterdam: North-Holland.

Bar-Nun, A. & Hartman, H. (1978). Synthesis of organic compounds from carbon monoxide and water by U.V. photolysis. *Origins of Life,* **9**, 93–101.

Barrer, R. M. (1978). *Zeolites and Clay Minerals as Sorbents and Molecular Sieves.* London: Academic.

Barrer, R. M. & Kerr, I. S. (1959). Intracrystalline channels in levynite and some related zeolites. *Transactions of the Faraday Society,* **55**, 1915–23.

Barrer, R. M. & Sieber, W. (1977). Hydrothermal chemistry of silicates, part 21: Zeolites from reaction of lithium and caesium ions with tetramethylammonium aluminosilicate solutions. *Journal of the Chemical Society Dalton*, pp. 1020–6.

Barshad, I. (1948). Vermiculite and its relation to biotite as revealed by base exchange reactions, X-ray analysis, differential thermal curves and water content. *American Mineralogist*, **33**, 655–78.

Battersby, A. R., Fookes, C. J. R., Matcham, G. W. J. & McDonald, E. (1980). Biosynthesis of the pigments of life: formation of the macrocycle. *Nature*, **285**, 17–21.

Beadle, G. W. (1949). Genes and biological enigmas. *Science in Progress*, **6**, 184–249.

Becquerel, P. (1924). La vie terrestre provient-elle d'un autre monde? *Bulletin de la Société astronomique de France*, **38**, 393–417.

Bennett, C. D. (1974). Similarity in the sequence of *Escherichia coli* dihydrofolate reductase with other nucleotide-requiring enzymes. *Nature*, **248**, 67–8.

Bernal, J. D. (1951). *The Physical Basis of Life*. London: Routledge & Kegan Paul.

Bernal, J. D. (1960). The problem of stages in biopoesis. In *Aspects of the Origin of Life*, ed. M. Florkin, pp. 30–45. Oxford: Pergamon Press. Reprinted from Clark, F. & Synge, R. L. M. (eds) (1959). *The Origin of Life on the Earth*, pp. 38–53. London: Pergamon.

Bernal, J. D. (1967). *The Origin of Life*. London: Weidenfeld & Nicolson.

Bernal, J. D., Dasgupta, D. R. & Mackay, A. L. (1959). The oxides and hydroxides of iron and their structural inter-relationships. *Clay Minerals Bulletin*, **4**, 15.

Blout, E. R. & Idelson, M. (1956). The kinetics of strong-base initiated polymerisations of amino acid-*N*-carboxyanhydrides. *Journal of the American Chemical Society*, **78**, 3857–8.

Blum, H. F. (1951). *Time's Arrow and Evolution*. Princeton: Princeton University Press.

Bodansky, Y. S., Klausner, M. & Ondetti, A. (1976). *Peptide Synthesis*. New York: Wiley.

Bolton, J. R. (1978). Solar fuels. *Science*, **202**, 705–11.

Bondy, S. C. & Harrington, M. E. (1979). L-amino acids and D-glucose bind stereospecifically to a colloidal clay. *Science*, **203**, 1243–4.

Bonner, W. A. (1972). Origins of molecular chirality. In *Exobiology*, Chapter 6, ed. C. Ponnamperuma, pp. 170–234. Amsterdam: North-Holland.

Bonner, W. A. & Kavasmaneck, P. R. (1976). Asymmetric adsorption of DL-alanine hydrochloride by quartz. *Journal of Organic Chemistry*, **41**, 2225–6.

Bonner, W. A. & Kavasmaneck, P. R. (1977). Adsorption of amino acid derivatives by *d*- and *l*-quartz. *Journal of the American Chemical Society*, **99**, 44–50.

Bonner, W. A., Kavasmaneck, P. R., Martin, F. S. & Flores, J. J. (1974). Asymmetric adsorption of alanine by quartz. *Science*, **186**, 143–4.

Bonner, W. A., Kavasmaneck, P. R., Martin, F. S. & Flores, J. J. (1975). Asymmetric adsorption by quartz: a model for the prebiotic origin of optical activity. *Origins of Life*, **6**, 367–76.

Bonner, W. A., Van Dort, M. A. & Yearian, M. R. (1975). Asymmetric degradation of DL-leucine with longitudinally polarised electrons. *Nature*, **258**, 419–21.

Brack, A. (1976). Polymérisation en phase aqueuse d'acides amines sur des argiles. *Clay Minerals*, **11**, 117–20.

Brack, A. (1977). β-Structures of alternating polypeptides and their possible role in chemical evolution. *Biosystems*, **9**, 99–103.

Brack, A. & Orgel, L. E. (1975). β-Structures of alternating polypeptides and their possible prebiotic significance. *Nature*, **256**, 383–7.

Bradley, W. F. & Serratosa, J. M. (1960). A discussion of the water content of vermiculite. *Clays and Clay Minerals*, **7**, 260–70.

Brady, G. W., Kurkjian, C. R., Lyden, E. F. X., Robin, M. B., Saltman, P., Spiro, T. & Terzis, A. (1968). The structure of an iron core analog of ferritin. *Biochemistry*, **7**, 2185–92.

Bragg, W. L. (1937). *Atomic structure of minerals*. Ithaca: Cornell University Press.

Bragg, W. L. & Brown, G. B. (1926). Die Struktur des Olivins. *Zeitschrift für Kristallographie*, **63**, 538–56.

Brahms, S., Brahms, J., Spach, G. & Brack, A. (1977). Identification of β-turns and unordered conformations in the polypeptide chains by vacuum ultraviolet circular dichroism. *Proceedings of the National Academy of Sciences USA*, **74**, 3208–12.

Brauner, K. & Preisinger, A. (1956). Struktur und Entstehung des Sepioliths. *Tschermaks Mineralogische und Petrographische Mitteilungen*, **6**, 120–40.

Breslow, R. (1959). On the mechanism of the formose reaction. *Tetrahedron Letters*, **no. 21**, 22–6.

Brindley, G. W. (1951). *X-ray Identification and Crystal Structures of Clay Minerals*. London: The Mineralogical Society.

Brindley, G. W. & Maksimovic, Z. (1974). The nature and nomenclature of hydrous nickel-containing silicates. *Clay Minerals*, **10**, 271–7.

Brown, B. E. & Bailey, S. W. (1962). Chlorite polytypism: I. Regular and semi-random one-layer structures. *American Mineralogist*, **47**, 819–50.

Brown, B. E. & Bailey, S. W. (1963). Chlorite polytypism: II. Crystal structure of a one-layer Cr-Chlorite. *American Mineralogist*, **48**, 42–61.

Buchanan, B. B. (1972). Ferredoxin-linked carboxylation reactions. In *The Enzymes*, vol. 6, ed. P. D. Boyer, pp. 193–216. New York: Academic.

Buerger, M. J. (1945). The genesis of twin crystals. *American Mineralogist*, **30**, 469–82.

Bull, A. T. (1974). Microbial growth. In *Companion to Biochemistry*, ed. A. T. Bull, J. R. Lagnado, J. O. Thomas & K. F. Tipton, pp. 415–42. London: Longmans.

Bungenberg de Jong, H. G. (1932). Die Koazervation und ihre Bedeutung für die Biologie. *Protoplasma*, **15**, 110–73.

Bursill, L. A., Thomas, J. M. & Rao, K. J. (1981). Stability of zeolites under electron irradiation and imaging of heavy cations in silicates. *Nature*, **289**, 157–8.

Burton, F. G. & Neuman, W. F. (1971). On the possible role of crystals in the origins of life. V. The polymerisation of glycine. *Currents in Modern Biology*, **4**, 47–54.

Butlero[v], A. (1861). Bildung einer zuckerartigen Substanz durch Synthese. *Annalen der Chemie und Pharmacie*, **120**, 295–8.

Buvet, R. & Le Port, L. (1973). Non-enzymic origin of metabolism. *Space Life Sciences*, **4**, 434–47.

Bystrom, A. & Bystrom, A. M. (1950). The crystal structure of hollandite, the related manganese oxide minerals, and α-MnO_2. *Acta Crystallographica*, **3**, 146–54.

Cady, S. S. & Pinnavaia, T. J. (1978). Porphyrin intercalation in mica-type silicates. *Inorganic Chemistry*, **17**, 1501–7.

Caillère, S. & Hénin, S. (1948). Sur la preparation et quelques caractères d'une série d'aluminates hydratés. *Comptes Rendus de L'Académie des Sciences Paris*, **226**, 580–2.

Caillère, S. & Hénin, S. (1949). Experimental formation of chlorites from montmorillonite. *Mineral Magazine*, **28**, 612–20.

Cairns-Smith, A. G. (1966). The origin of life and the nature of the primitive gene. *Journal of Theoretical Biology*, **10**, 53–88.

Cairns-Smith, A. G. (1968). An approach to a blueprint for a primitive organism. In *Towards a Theoretical Biology, I. Prolegomena*, ed. C. H. Waddington, pp. 57–66. Edinburgh: Edinburgh University Press.

Cairns-Smith, A. G. (1971). *The Life Puzzle: On crystals and organisms and on the possibility of a crystal as an ancestor*. Edinburgh: Oliver & Boyd.

Cairns-Smith, A. G. (1974). The methods of science and the origins of life. In *The Origin of Life and Evolutionary Biochemistry*, ed. K. Dose, S. W. Fox, G. A. Deborin & T. E. Pavlovskaya, pp. 53–8. New York: Plenum.

Cairns-Smith, A. G. (1975a). A case for an alien ancestry. *Proceedings of the Royal Society B*, **189**, 249–74.

Cairns-Smith, A. G. (1975b). Ambiguity in the interpretation of abiotic syntheses. *Origins of Life*, **6**, 265–7.

Cairns-Smith, A. G. (1977). Takeover mechanisms and early biochemical evolution. *BioSystems*, **9**, 105–9. Also in *Origin of Life*, ed. H. Noda (1978), pp. 399–404. Tokyo: Center for Academic Publications, Japan.

Cairns-Smith, A. G. (1979). Organisms of the first kind. *Chemistry in Britain*, **15**, 576–9.

Cairns-Smith, A. G. (1981). Beginnings of organic evolution. In *Study of Time*, IV, ed. J. T. Fraser, pp. 15–33. New York: Springer-Verlag.

Cairns-Smith, A. G. & Davis, C. J. (1977). The design of novel replicating polymers. In *The Encyclopaedia of Ignorance: Life Sciences and Earth Sciences*, ed. R. Duncan & M. Weston-Smith, pp. 391–404. Oxford: Pergamon.

Cairns-Smith, A. G., Ingram, P. & Walker, G. L. (1972). Formose production by minerals, possible relevance to the origin of life. *Journal of Theoretical Biology*, **35**, 601–4.

Cairns-Smith, A. G. & Walker, G. L. (1974). Primitive metabolism. *Biosystems*, **5**, 173–86.

Calvin, M. (1969). *Chemical Evolution: molecular evolution towards the origin of living systems on the earth and elsewhere*. Oxford: Clarendon.

Calvin, M. (1974). Solar energy by photosynthesis. *Science*, **184**, 375–81.

Campbell, J. H., Lengyel, J. A. & Langridge, J. (1973). Evolution of a second gene for β-galactosidase in *Escherichia coli*. *Proceedings of the National Academy of Sciences USA*, **70**, 1841–5.

Chaussidon, J. (1973). Le spectre infrarouge des biotites: Vibrations d'élongation basse frequence des OH du reseau. In *Proceedings of the International Clay Conference 1972*, ed. J. M. Serratosa, pp. 99–106. Madrid: Division de Ciencias, CSIC.

Chittenden, G. J. F. & Schwartz, A. W. (1976). Possible pathway for prebiotic uracil synthesis by photodehydrogenation. *Nature*, **263**, 350–1.

Choughuley, A. S. U., Subbaraman, A. S., Kazi, Z. A. & Chada, M. S. (1977). A possible prebiotic synthesis of thymine: uracil–formaldehyde–formic acid reaction. *Biosystems*, **9**, 73–80.

Chukhrov, F. V., Gorshkov, A. I., Rudnitskaya, E. S., Beresovskaya, V. V. & Sivtsov, A. V. (1980). Manganese minerals in clays: a review. *Clays and Clay Minerals*, **28**, 346–54.

Clark, S. P. Jr, Turekian, K. K. & Grossman, L. (1972). Model for the early history of the Earth. In *The Nature of the Solid Earth*, ed. E. C. Robertson, pp. 3–18. New York: McGraw-Hill.

Clemency, C. V. (1976). Simultaneous weathering of a granitic gneiss and an intrusive amphibolite dike near São Paulo, Brazil, and the origin of clay minerals. In *Proceedings of the International Clay Conference 1975*, ed. S. W. Bailey, pp. 15–25. Wilmette, Illinois: Applied Publishing Ltd.

Coleridge, S. T. (*ca* 1820). *Theory of Life*, p. 42. Quoted in Snyder, A. D. (1929). *Coleridge on Logic and Learning*. (1816 and 1823 are alternative suggestions for the publication date of *Theory of Life* – see Snyder, A. D. (1928). Coleridgeana. *Review of English Studies*, **4**, 432–4.)

Corbridge, D. E. C. (1978). *Phosphorus*. Amsterdam: Elsevier.

Cotton, F. A. & Wilkinson, G. (1972). *Advanced Inorganic Chemistry*, 3rd edn. New York: Interscience.

Coyne, L. M., Lawless, J., Lahav, N., Sutton, S. & Sweeney, M. (1981). Clays as prebiotic photocatalysts. *Origin of Life*, ed. Y. Wolman, pp. 115–24. Dordrecht: D. Reidel.

Cozzarelli, N. R. (1980). DNA gyrase and the supercoiling of DNA. *Science*, **207**, 953–60.

Cradwick, P. D. G., Farmer, V. C., Russell, J. D., Masson, C. R., Wada, K. & Yoshinaga, N. (1972). Imogolite, a hydrated aluminium silicate of tubular structure. *Nature Physical Science*, **240**, 187–9.

Cram, D. J. & Cram, J. M. (1978). Design of complexes between synthetic hosts and organic guests. *Accounts of Chemical Research*, **11**, 8–14.

Crick, F. H. C. (1968). The origin of the genetic code. *Journal of Molecular Biology*, **38**, 367–79.

Crowther, J. P. & Westman, A. E. R. (1954). The hydrolysis of the condensed phosphates: I. Sodium pyrophosphate and sodium triphosphate. *Canadian Journal of Chemistry*, **32**, 42–8.

Darwin, C. (1871). In *Life and Letters of Charles Darwin*, vol. 3, ed. F. Darwin, p. 18. London: John Murray.

Daubrée, A. (1879). *Études Synthétiques de Géologie Expérimentale*. Paris. (Quoted in Deer, Howie & Zussman (1962), vol. 4, p. 399.)

Dauvillier, A. (1938). *Astronomie*, **52**, 529.

Dauvillier, A. (1965). *The Photochemical Origin of Life*. New York: Academic.

Dawkins, R. (1976). *The Selfish Gene*. Oxford: Oxford University Press.

De Kimpe, C. R. (1976). Formation of phylosilicates and zeolites from pure silica–alumina gels. *Clays and Clay Minerals*, **24**, 200–7.

Deer, W. A., Howie, R. A. & Zussman, J. (1962). *Rock forming Minerals*, vols I–V. London: Longmans.

Degani, C. & Halmann, M. (1967). Chemical evolution of carbohydrate metabolism. *Nature*, **216**, 1207.

Dent, L. S. & Smith, J. V. (1958). Crystal structure of chabazite, a molecular sieve. *Nature*, **181**, 1794–6.

Dickerson, R. E. (1971). Sequence and structure homologies in bacterial and mammalian-type cytochromes. *Journal of Molecular Biology*, **57**, 1–15.

Dickerson, R. E. (1978). Chemical evolution and the origin of life. *Scientific American*, **239**, no. 3, 62–78.

Dixon, J. B. & McKee, T. R. (1974). Internal and external morphology of tubular and spheroidal halloysite particles. *Clays and Clay Minerals*, **22**, 127–37.

Dobzhansky, T. (1973). Nothing in biology makes sense except in the light of evolution. *American Biology Teacher*, **35**, 125–9.

Dobzhansky, T., Ayala, F. J., Stebbins, G. L. & Valentine, J. W. (1977). *Evolution*. San Francisco: Freeman & Co.

Dose, K. (1971). Catalysis. In *Theory and Experiment in Exobiology*, vol. 1, ed. A. W. Schwartz, pp. 41–71. Groningen: Wolters-Noordhoff.

Dose, K. (1974). Peptides and amino acids in the primordial hydrosphere. In *The Origin of Life and Evolutionary Biochemistry*, ed. K. Dose, S. W. Fox, G. A. Deborin & T. E. Pavlovskaya, pp. 69–77. New York: Plenum.

Dose, K. (1975). Peptides and amino acids in the primordial hydrosphere. *BioSystems*, **6**, 224–8.

Doty, P. & Lundberg, R. D. (1956). Configurational and stereochemical effects in the amine initiated polymerisation of *N*-carboxyanhydrides. *Journal of the American Chemical Society*, **78**, 4810–12.

Dousma, J. & de Bruyn, P. L. (1976). Hydrolysis-precipitation studies of iron solutions: I. Model for hydrolysis and precipitation from Fe(III) nitrate solutions. *Journal of Colloid and Interface Science*, **56**, 527–39.

Dousma, J. & de Bruyn, P. L. (1978). Hydrolysis-precipitation studies of iron solutions: II. Ageing studies and the model for precipitation from Fe(III) nitrate solutions. *Journal of Colloid and Interface Science*, **64**, 154–70.

Dowty, E. (1980). Crystal-chemical factors affecting the mobility of ions in minerals. *American Mineralogist*, **65**, 174–82.

Drake, J. W. (1969). Comparative rates of spontaneous mutation. *Nature*, **221**, 1132.

Dustin, P. (1980). Microtubules. *Scientific American*, **243**, no. 2, 58–68.

Eberl, D. & Hower, J. (1975). Kaolinite synthesis: the role of the Si/Al ratio and the (Alkali/H+) ratio in hydrothermal systems. *Clays and Clay Minerals*, **23**, 301–9.

Eigen, M., Gardiner, W., Schuster, P. & Winkler-Oswatitsch, R. (1981). The origin of genetic information. *Scientific American*, **244**, no. 4, 78–94.

Emsley, J. (1977). Phosphate cycles. *Chemistry in Britain*, **13**, 459–63.

Ewing, F. J. (1935a). The crystal structure of lepidocrocite. *Journal of Chemical Physics*, **3**, 420–4.

Ewing, F. J. (1935b). The crystal structure of diaspore. *Journal of Chemical Physics*, **3**, 203–7.

Farmer, V. C. & Fraser, A. R. (1979). Synthetic imogolite, a tubular hydroxy-aluminium silicate. In *International Clay Conference 1978*, ed. M. M. Mortland & V. C. Farmer, pp. 547–53. Amsterdam: Elsevier.

Farmer, V. C., Fraser, A. R. & Tait, J. M. (1977). Synthesis of Imogolite. A tubular aluminium silicate polymer. *Journal of the Chemical Society, Chemical Communications*, 462–3.

Farmer, V. C., Russell, J. D., McHardy, W. J., Newman, A. C. D., Ahlrichs J. L. & Rimsaite, J. Y. H. (1971). Evidence for loss of protons and octahedral iron from oxidised biotites and vermiculites. *Mineral Magazine*, **38**, 121–37.

Feeny, R. E., Blankenhorn, G. & Dixon, H. B. F. (1975). Carbonyl–amine reactions in protein chemistry. *Advances in Protein Chemistry*, **29**, 135–203.

Feitknecht, W. & Schindler, P. (1963). Solubility constants of metal oxides, metal hydroxides and hydroxide salts in aqueous solution. *Pure and Applied Chemistry*, **6**, 125–99.

Ferris, J. P. & Chen, C. T. (1975a). Chemical evolution. XXVI. Photochemistry of methane, nitrogen and water mixtures as a model for the atmosphere of the primitive Earth. *Journal of the American Chemical Society*, **97**, 2962–7.

Ferris, J. P. & Chen, C. T. (1975b). Photosynthesis of organic compounds in the atmosphere of Jupiter. *Nature*, **258**, 587–8.

Ferris, J. P. & Edelson, E. H. (1978). Chemical evolution. 31. Mechanism of the condensation of cyanide to HCN oligomers. *Journal of Organic Chemistry*, **43**, 3989–95.

Ferris, J. P., Donner, D. B. & Lobo, A. P. (1973). Possible role of hydrogen cyanide in chemical evolution: The oligomerisation and condensation of hydrogen cyanide. *Journal of Molecular Biology*, **74**, 511–18.

Ferris, J. P., Joshi, P. C., Edelson, E. H. & Lawless, J. G. (1978). HCN: a plausible source of purines, pyrimidines and amino acids on the primitive Earth. *Journal of Molecular Evolution*, **11**, 293–311.

Ferris, J. P. & Nicodem, D. E. (1974). Ammonia: Did it have a role in chemical evolution? In *The Origin of Life and Evolutionary Biochemistry*, ed. K. Dose, S. W. Fox, G. A. Deborin & T. E. Pavlovskaya, pp. 107–17. New York: Plenum.

Ferris, J. P., Sanchez, R. A. & Orgel, L. E. (1968). Synthesis of pyrimidines from cyanoacetylene and cyanate. *Journal of Molecular Biology*, **33**, 693–704.

Ferris, J. P., Wos, J. D., Nooner, D. W. & Oró, J. (1974). Chemical evolution. XXI. The amino acids released on hydrolysis of HCN oligomers. *Journal of Molecular Evolution*, **3**, 225–31.

Ferris, J. P., Zamek, O. S., Altbuch, A. M. & Freiman, H. (1974). Chemical evolution: XVIII. Synthesis of pyrimidines from guanidine and cyanoacetalde-hyde. *Journal of Molecular Evolution*, **3**, 301–9.

Fischer, W. R. & Schwertmann, U. (1975). The formation of haematite from amorphous iron III hydroxide. *Clays and Clay Minerals*, **23**, 33–7.

Fitzpatrick, R. W., le Roux, J. & Schwertmann, U. (1978). Amorphous and crystalline titanium and iron–titanium oxides in synthetic preparations, at near ambient conditions, and in soil clays. *Clays and Clay Minerals*, **26**, 189–201.

Fleischer, R. L., Price, P. B. & Walker, R. M. (1975). *Nuclear Tracks in Solids.* Berkeley: University of California Press.

Force, E. R. (1980). The provenance of rutile. *Journal of Sedimentary Petrology*, **50**, 485–8.

Foster, M. D. (1963). Interpretation of the composition of vermiculites and hydrobiotites. *Clays and Clay Minerals*, **10**, 70–89.

Fournier, R. O. & Rowe, J. J. (1962). The solubility of crystobalite along the three phase curve, gas plus liquid plus crystobalite. *American Mineralogist*, **47**, 897–902.

Fox, S. W. (1956). Evolution of protein molecules and thermal synthesis of biochemical substances. *American Scientist*, **44**, 347–59.

Fox, S. W. (1969). Self-ordered polymers and propagative cell-like systems. *Naturwissenschaften*, **56**, 1–9.

Fox, S. W. & Dose, K. (1972). *Molecular Evolution and the Origin of Life.* San Francisco: Freeman.

Fox, S. W. & Harada, K. (1960). The thermal copolymerisation of amino acids common to protein. *Journal of the American Chemical Society*, **82**, 3745–51.

Fox, S. W. & Windsor, C. R. (1970). Synthesis of amino acids by the heating of formaldehyde and ammonia. *Science*, **170**, 984–5.

Francombe, M. H. & Rooksby, H. P. (1959). Structure transformations effected by dehydration of diaspore, goethite and delta ferric oxide. *Clay Minerals Bulletin*, **4**, 1.

Frank, F. C. (1953). On spontaneous asymmetric synthesis. *Biochimica et Biophysica Acta*, **11**, 459–63.

Fraser, R. D. B. (1969). Keratins. *Scientific American*, **221**, no. 2, 86–96.

Freund, F. & Wenegler, H. (1978). Proton conductivity in hydroxides as related to transport phenomena in soils. *Book of Summaries for the 6th International Clay Conference (Oxford)*, p. 610. London: The Mineralogical Society.

Frey, E. & Lagaly, G. (1979). Selective coagulation and mixed layer formation from sodium smectite solutions. In *International Clay Conference 1978*, ed. M. M. Mortland & V. C. Farmer, pp. 131–40. Amsterdam: Elsevier.

Friebele, E., Shimoyama, A., Hare, P. E. & Ponnamperuma, C. (1981). Adsorption of amino acid enantiomers by Na-montmorillonite. *Origins of Life*, **11**, 173–84.

Gabel, N. W. & Ponnamperuma, C. (1967). Model for origin of monosaccharides. *Nature*, **216**, 453–5.

Galbraith, S. T., Baird, T. & Fryer, J. R. (1979). Structural changes in β-FeOOH caused by radiation damage. *Acta Crystallographica*, **A35**, 197–200.

Gardner, M. (1976). Mathematical games: On the fabric of inductive logic, and some probability paradoxes. *Scientific American*, **234**, no. 3, 119.

Garrison, W. M., Morrison, D. C., Hamilton, J. G., Benson, A. A. & Calvin, M. (1951). Reduction of carbon dioxide in aqueous solutions by ionising radiation. *Science*, **114**, 416–18.

Gatineau, L. (1964). Structure réelle de la muscovite. Répartition des substitutions isomorphe. *Bulletin de la Société française de Minéralogie et Crystallographie*, **87**, 321–55.

Gerstl, Z. & Banin, A. (1980). Fe^{2+}–Fe^{3+} Transformations in clay and resin ion-exchange systems. *Clays and Clay Minerals*, **28**, 335–45.

Getoff, N. (1962). Reduktion der Kohlensaure in wässeriger Lösung unter Einwirkung von UV-Licht. *Zeitschrift für Natûrforschung*, **17B**, 87–90.

Getoff, N. (1963a). Einfluss der Ferroionen bei der Bildung von Oxalsaure aus wässeriger Kohlensaure durch UV-Bestrahlung. *Zeitschrift für Naturforschung*, **18B**, 169–70.

Getoff, N. (1963b). Role of iron II ions in carbonic acid reduction in aqueous solutions by means of ultraviolet light. *Oesterricher Chemiker-Zeitung*, **64**, 70–2.

Giese, R. F. (1978). The electrostatic interlayer forces of layer structure minerals. *Clays and Clay Minerals*, **26**, 51–7.

Giese, R. F. & Datta, P. (1973). Hydroxyl orientation in kaolinite, dickite, and nacrite. *American Mineralogist*, **58**, 471–9.

Gildawie, A. M. (1977). *Electron Microscope Studies of Colloidal Metal Oxides.* PhD Thesis, University of Glasgow.

Gimblett, F. G. R. (1963). *Inorganic Polymer Chemistry.* London: Butterworths.

Goodwin, A. M. (1976). Giant impacting and the development of continental crust. In *The Early History of the Earth*, ed. B. F. Windley, pp. 77–95. London: John Wiley & Sons.

Gottardi, G. & Meier, W. M. (1963). The crystal structure of dachiardite. *Zeitschrift für Kristallographie*, **119**, 53–64.

Granick, S. (1957). Speculations on the origins and evolution of photosynthesis. *Annals of the New York Academy of Sciences*, **69**, 292–308.

Granick, S. (1965). Evolution of heme and chlorophyll. In *Evolving Genes and Proteins*, ed. V. Bryson & H. J. Vogel, pp. 67–88. New York: Academic.

Granquist, W. T. & Pollack, S. S. (1959). A study of the synthesis of hectorite. *Clays and Clay Minerals*, **8**, 150–69.

Grant, G. T., Morris, E. R., Rees, D. A., Smith, P. J. C. & Thom, D. (1973). Biological interactions between polysaccharides and divalent cations: the egg box model. *FEBS Letters*, **32**, 195–8.

Greenland, D. J. (1956). The adsorption of sugars by montmorillonite, I and II. *Journal of Soil Science*, **7**, 319–28 and 329–34.

Griffith, E. J., Ponnamperuma, C. & Gabel, N. W. (1977). Phosphorus, a key to life on the primitive Earth. *Origins of Life*, **8**, 71–85.

Grim, R. E. & Güven, N. (1978). *Bentonites: Geology, Mineralology, Properties and Use.* Amsterdam: Elsevier.

Guggenheim, S. & Bailey, S. W. (1975). Refinement of the margarite structure in subgroup symmetry. *American Mineralogist*, **60**, 1023–9.

Güven, N. (1971). The crystal structure of $2M_1$ phengite and $2M_1$ muscovite. *Zeitschrift für Kristallographie*, **134**, 196–212.

Güven, N. & Burnham, C. W. (1967). The crystal structure of 3T muscovite. *Zeitschrift für Kristallographie*, **125**, 163–83.

Güven, N., Pease, R. W. & Murr, L. E. (1977). Fine structure in the selected area diffraction patterns of beidellite and its dark field images. *Clay Minerals*, **12**, 67–74.

Haldane, J. B. S. (1929). The origin of life. *Rationalist Annual* **3**. (Reprinted in Bernal, 1967, pp. 242–9.)

Hall, D. O., Cammack, R. & Rao, K. K. (1974). The iron–sulphur proteins: evolution of a ubiquitous protein from model systems to higher organisms. *Origins of Life*, **5**, 363–86.

Hall, S. H. & Bailey, S. W. (1979). Cation ordering pattern in amesite. *Clays and Clay Minerals*, **27**, 241–7.

Halmann, M. (1974). Evolution and ecology of phosphorus metabolism. In *The Origin of Life and Evolutionary Biochemistry*, ed. K. Dose, S. W. Fox, G. A. Deborin & T. E. Pavlovskaya, pp. 169–82. New York: Plenum.

Halmann, M. (1978). Photoelectrochemical reduction of aqueous carbon dioxide on p-type gallium phosphide in liquid junction solar cells. *Nature*, **275**, 115–16.

Harada, K. & Fox, S. W. (1964). Thermal synthesis of natural amino acids from a postulated primitive terrestrial atmosphere. *Nature*, **201**, 335–6.

Harada, K. & Fox, S. W. (1965). The thermal synthesis of amino acids from a hypothetically primitive terrestrial atmosphere. In *The Origins of prebiological systems*, ed. S. W. Fox, pp. 187–201. New York: Academic.

Harder, H. (1977). Clay mineral formation under lateritic weathering conditions. *Clay Minerals*, **12**, 281–8.

Harder, H. (1978). Synthesis of iron layer silicate minerals under natural conditions. *Clays and Clay Minerals*, **26**, 65–72.

Hartman, H. (1975a). Speculations on the origin and evolution of metabolism. *Journal of Molecular Evolution*, **4**, 359–70.

Hartman, H. (1975b). Speculations on the evolution of the genetic code. *Origins of Life*, **6**, 423–7, and (1978), **9**, 133–6.

Hartmann, W. K. (1975). The smaller bodies of the solar system. *Scientific American*, **233**, no. 3, 142–59.

Hay, R. L. (1966). *Zeolites and Zeolitic Reactions in Sedimentary Rocks*. Geological Society of America Special Paper no. 85, New York: Geological Society of America.

Hayatsu, R., Studier, M. H. & Anders, E. (1971). Origin of organic matter in early solar system. IV. Amino acids: confirmation of catalytic synthesis by mass spectrometry. *Geochimica et Cosmochimica Acta*, **35**, 939–51.

Hazen, R. M. & Wones, D. R. (1972). The effect of cation substitutions on the physical properties of trioctahedral micas. *American Mineralogist*, **57**, 103–29.

Henmi, T. & Wada, K. (1976). Morphology and composition of allophane. *American Mineralogist*, **61**, 379–90.

Hey, M. H. (1932). Studies on zeolites, II: Thomsonite and gonnardite, and III: Natrolite and metanatrolite. *Mineral Magazine*, **23**, 51–125 and 243–89.

Hinkle, P. C. & McCarty, R. E. (1978). How cells make ATP. *Scientific American*, **238**, no. 3, 104–23.

Hinshelwood, C. N. (1951). *The Structure of Physical Chemistry*, p. 452. Oxford: Clarendon.

Hocart, M. R. (1962). *Genèse et Synthèse des Argiles*. Paris: CNRS.

Hoffmann, R. W. & Brindley, G. W. (1960). Adsorption of non-ionic aliphatic molecules from aqueous solutions on montmorillonite. Clay-organic studies II. *Geochimica et Cosmochimica Acta*, **20**, 15–29.

Holland, H. D. (1962). Model for the evolution of the Earth's atmosphere. In *Petrologic Studies; a volume in honor of A. F. Buddington*, ed. A. E. J. Engel, H. L. James & B. F. Leonard, pp. 447–77. New York: Geological Society of America.

Hopfield, J. J. (1974). Kinetic proofreading: A new mechanism for reducing errors in biosynthetic processes requiring high specificity. *Proceedings of the National Academy of Sciences USA*, **71**, 4135–9.

Horiuchi, S. (1978). Visualising atoms in inorganic compounds by 1 MV HRTEM. *Chemica Scripta*, **14**, 75–81.

Horowitz, N. H. (1945). On the evolution of biochemical synthesis. *Proceedings of the National Academy of Sciences USA*, **31**, 153–7.

Horowitz, N. H. (1959). On defining life. In *The Origin of Life on the Earth*, ed. F. Clark & R. L. M. Synge, pp. 106–7. London: Pergamon.

Hsu, P. H. (1972). Nucleation, polymerisation and precipitation of FeOOH. *Journal of Soil Science*, **23**, 409–19.

Hubbard, J. S., Hardy, J. P. & Horowitz, N. H. (1971). Photocatalytic production of organic compounds from CO and H_2O in a simulated Martian atmosphere. *Proceedings of the National Academy of Sciences USA*, **68**, 574–8.

Hubbard, J. S., Hardy, J. P., Voecks, G. E. & Golub, E. E. (1973). Photocatalytic synthesis of organic compounds from CO and water: Involvement of surfaces in the formation and stabilisation of products. *Journal of Molecular Evolution*, **2**, 149–66.

Hubbard, J. S., Voecks, G. E., Hobby, G. L., Ferris, J. P., Williams, E. A. & Nicodem, D. E. (1975). Ultraviolet gas phase and photocatalytic synthesis from CO and NH_3. *Journal of Molecular Evolution*, **5**, 223–41.

Hulett, H. R. (1969). Limitations on prebiological synthesis. *Journal of Theoretical Biology*, **24**, 56–72.

Hull, D. E. (1960). Thermodynamics and kinetics of spontaneous generation. *Nature*, **186**, 693–4.

Hutchison, J. L., Jefferson, D. A. & Thomas, J. M. (1977). The ultrastructure of minerals as revealed by high resolution electron microscopy. In *Surface and Defect Properties of Solids, Vol. 6, Specialist Periodical Report*, ed. M. W. Roberts & J. H. Thomas, pp. 320–58. London: The Chemical Society.

Huxley, J. (1974). *Evolution: The Modern Synthesis*. London: Allen & Unwin.

Hynes, M. J. (1975). Amide utilisation in *Aspergillus nidulans*: Evidence for a third amidase enzyme. *Journal of General Microbiology*, **91**, 99–109.

Inoue, T., Fujishima, A., Konishi, S. & Honda, K. (1979). Photoelectrocatalytic reduction of carbon dioxide in aqueous suspensions of semiconductor powders. *Nature*, **277**, 637–8.

Israelachvili, J. (1978). The packing of lipids and proteins in membranes. In *Light Transducing Membranes*, ed. D. W. Deamer, pp. 91–107. New York: Academic.

Jasmund, K., Sylla, H. M. & Freund, F. (1976). Solid solution in synthetic serpentine phases. In *Proceedings of the International Clay Conference 1975*, ed. S. W. Bailey, pp. 267–74. Wilmette, Illinois: Applied Publishing Ltd.

Jensen, R. A. (1976). Enzyme recruitment in evolution of new function. *Annual Review of Microbiology*, **30**, 409–25.

Johnson, J. S. & Kraus, K. A. (1959). Hydrolytic behaviour of metal ions. X. Ultracentrifugation of lead (II) and tin (IV) in basic solution. *Journal of the American Chemical Society*, **81**, 1569–72.

Johnson, L. J. (1970). Clay minerals in Pennsylvania soils in relation to lithology of the parent rock and other factors – I. *Clays and Clay Minerals*, **18**, 247–60.

Jonas, E. C. (1976). Crystal chemistry of diagenesis in 2:1 clay minerals. In *Proceedings of the International Clay Conference 1975*, ed. S. W. Bailey, pp. 3–13. Wilmette, Illinois: Applied Publishing Ltd.

Kalt, A., Perati, B. & Wey, R. (1979). Intercalation compounds of $KHSi_2O_5$ and $H_2Si_2O_5$ with alkylammonium ions and alkylamines. In *International Clay Conference 1978*, ed. M. M. Mortland & V. C. Farmer, pp. 509–16. Amsterdam: Elsevier.

Katchalski, E. (1951). Poly-α-amino acids. *Advances in Protein Chemistry*, **6**, 123–5.

Kay, W. W. (1971). Two aspartate transport systems in *Escherichia coli*. *Journal of Biological Chemistry*, **246**, 7373–82.

Keller, W. D. (1976). Scan electron micrographs of kaolins collected from diverse environments of origin – I. *Clays and Clay Minerals*, **24**, 107–13.

Keller, W. D. (1977). Scan electron micrographs of kaolins collected from diverse environments of origin – IV. *Clays and Clay Minerals*, **25**, 311–45.

Keller, W. D. (1978). Classification of kaolins exemplified by their textures in scan electron micrographs. *Clays and Clay Minerals*, **26**, 1–20.

Keller, W. D. & Hanson, R. F. (1975). Dissimilar fabrics by scan electron microscopy of sedimentary versus hydrothermal kaolins in Mexico. *Clays and Clay Minerals*, **23**, 201–4.

Kenyon, D. H. & Nissenbaum, A. (1976). Melanoidin and aldocyanoin microspheres: Implications for chemical evolution and early precambrian micropaleontology. *Journal of Molecular Evolution*, **7**, 245–51.

Kenyon, D. H. & Steinman, G. (1969). *Biochemical Predestination*. New York: McGraw-Hill.

Kerr, R. A. (1980). Origin of life: New ingredients suggested. *Science*, **210**, 42–3.

Kirkman, J. H. (1977). Possible structure of halloysite disks and cylinders observed in some New England rhyolitic tephras. *Clay Minerals*, **12**, 199–216.

Kirkman, J. H. (1981). Morphology and structure of halloysite in New Zealand tephras. *Clays and Clay Minerals*, **29**, 1–9.

Kirsten, T. (1978). Time and the solar system. In *The Origin of the Solar System*. ed. S. F. Dermott, pp. 267–346. Chichester: John Wiley & Sons.

Klein, H. P. (1978). The Viking biological investigations: Review and status. *Origins of Life*, **9**, 157–60.

Koshland, D. E., Némethy, G. & Filmer, D. (1966). Comparison of experimental binding data and theoretical models in proteins containing subunits. *Biochemistry*, **5**, 365–85.

Krasnovsky, A. A. (1974). Chemical evolution of photosynthesis. In *The Origin of Life and Evolutionary Biochemistry*, ed. K. Dose, S. W. Fox, G. A. Deborin & T. E. Pavlovskaya, pp. 233–44. New York: Plenum.

Krumbein, W. C. & Garrels, R. M. (1952). Origin and classification of chemical sediments in terms of pH and oxidation–reduction potentials. *Journal of Geology*, **60**, 1–33.

Kuhn, H. (1976). Model consideration for the origin of life: Environmental structure as stimulus for the evolution of chemical systems. *Naturwissenschaften*, **63**, 68–80.

Kuhn, T. S. (1970). *The Structure of Scientific Revolutions*, 2nd edn. Chicago: University of Chicago Press.

Kunze, G. (1956). Die gewellte Struktur des Antigorits, I. *Zeitschrift für Kristallographie*, **108**, 82–107.

Lagaly, G. (1979). The 'layer charge' of regular interstratified 2:1 clay minerals. *Clays and Clay Minerals*, **27**, 1–10.

Lagaly, G. (1981). Inorganic layer compounds. *Naturwissenschaften*, **68**, 82–8.

Lahav, N. & Chang, S. (1976). The possible role of solid surface area in condensation reactions during chemical evolution: Re-evaluation. *Journal of Molecular Evolution*, **8**, 357–80.

Lahav, N., White, D. & Chang, S. (1978). Peptide formation in the prebiotic era: Thermal condensation of glycine in fluctuating clay environments. *Science*, **201**, 67–9.

La Iglesia, A. & Galen, E. (1975). Halloysite-kaolinite transformation at room temperature. *Clays and Clay Minerals*, **23**, 109–13.

La Iglesia, A. & Martin-Vivaldi, J. L. (1973). A contribution to the synthesis of kaolinite. In *Proceedings of the International Clay Conference 1972*, ed. J. M. Serratosa, pp. 173–85. Madrid: Division de Ciencias, CSIC.

La Iglesia, A. & Martin-Vivaldi, J. L. (1975). Synthesis of kaolinite by homogeneous precipitation at room temperature. *Clay Minerals*, **10**, 399–405.

La Iglesia, A., Martin-Vivaldi, J. L. & Lopez Aguayo, F. (1976). Kaolinite crystallisation at room temperature by homogeneous precipitation III: Hydrolysis of feldspars. *Clays and Clay Minerals*, **24**, 36–42.

La Iglesia, A. & Van Oosterwyck-Gastuche, M. C. (1978). Kaolinite synthesis I: Crystallisation conditions at low temperatures and calculation of thermodynamic equilibria. Application to laboratory and field observations. *Clays and Clay Minerals*, **26**, 397–408.

Lailach, G. E. & Brindley, G. W. (1969). Specific co-absorption of purines and pyrimidines by montmorillonite (clay–organic studies XV). *Clays and Clay Minerals*, **17**, 95–100.

Lailach, G. E., Thompson, T. D. & Brindley, G. W. (1968a). Absorption of pyrimidines, purines and nucleosides by Li-, Na-, Mg-, and Ca-montmorillonite (clay–organic studies XII). *Clays and Clay Minerals*, **16**, 285–93.

Lailach, G. E., Thompson, T. D. & Brindley, G. W. (1968b). Absorptions of pyrimidines, purines and nucleosides by Co-, Ni, Cu-, and Fe(III)-montmorillonite (clay–organic studies XIII). *Clays and Clay Minerals*, **16**, 295–301.

Lambert, R. St J. (1959). The mineralogy and metamorphism of the Moine schists of the Morar and Knoydart districts of Inverness-shire. *Transactions of the Royal Society of Edinburgh*, **63**, 553–88.

Laswick, J. A. & Plane, R. A. (1959). Hydrolytic polymerisation in boiled chromic solutions. *Journal of the American Chemical Society*, **81**, 3564–7.

Lawless, J. G. (1980). Organic compounds in meteorites. In *Life Sciences and Space Research*, vol. 18, ed. R. Holmquist, pp. 19–27. Oxford: Pergamon.

Lawless, J. G. & Levi, N. (1979). The role of metal ions in chemical evolution: Polymerisation of alanine and glycine in a cation-exchanged clay environment. *Journal of Molecular Evolution*, **13**, 281–6.

Lawless, J. G. & Peterson, E. (1975). Amino acids in carbonaceous chondrites. *Origins of Life*, **6**, 3–8.

Lawless, J. G. & Yuen, G. U. (1979). Quantification of monocarboxylic acids in the Murchison carbonaceous meteorite. *Nature*, **282**, 396–8.

Lawless, J. G., Zeitman, B., Pereira, W. E., Summons, R. E. & Duffield, A. M. (1974). Dicarboxylic acids in the Murchison meteorite. *Nature*, **251**, 40–2.

Leach, W. W., Nooner, D. W. & Oró, J. (1978). Abiotic synthesis of fatty acids. In *Origin of Life*, ed. H. Noda, pp. 113–22. Tokyo: Center for Academic Publications Japan.

Lemmon, R. M. (1970). Chemical evolution. *Chemical Reviews*, **70**, 95–109.

Lewontin, R. C. (1978). Adaptation. *Scientific American*, **239**, no. 3, 156–69.

Liecester, H. M. (1956). *The Historical Background of Chemistry*. New York: John Wiley & Sons.

Lin, E. C. C. (1970). The genetics of bacterial transport systems. *Annual Review of Genetics*, **4**, 225–62.

Linares, J. & Huertas, F. (1971). Kaolinite: synthesis at room temperature. *Science*, **171**, 896–7.

Lipman, C. B. (1924). The origin of life. *Scientific Monthly*, **19**, 357–67.

Lipmann, F. (1965). Projecting backward from the present stage of evolution of biosynthesis. In *The Origins of Prebiological Systems*, ed. S. W. Fox, pp. 259–80. New York: Academic.

Lister, J. S. & Bailey, S. W. (1967). Chlorite polytypism: IV. Regular two-layer structures. *American Mineralogist*, **52**, 1614–31.

Lodish, H. F. & Rothman, J. E. (1979). The assembly of cell membranes. *Scientific American*, **240**, no. 1, 48–63.

Loeb, W. (1913). Über das Verhalten des Formamids unter der Wirkung der stillen Engladung. *Berichte der chemischen Gesellschaft*, **46**, 684–97.

Lohrmann, R. (1976). Formation of nucleoside 5′-polyphosphates under potentially prebiological conditions. *Journal of Molecular Evolution*, **8**, 197–210.

Lohrmann, R., Bridson, P. K. & Orgel, L. E. (1980). Efficient metal ion catalysed template-directed oligonucleotide synthesis. *Science*, **208**, 1464–5.

Lohrmann, R. & Orgel, L. E. (1971). Urea–inorganic phosphate mixtures as prebiotic phosphorylating agents. *Science*, **171**, 490–4.

Lohrmann, R. & Orgel, L. E. (1978). Template-directed polynucleotide condensation as a model for RNA replication. In *Origin of Life*, ed. H. Noda, pp. 235–44. Tokyo: Center for Academic Publications Japan.

Lowe, C. U., Rees, M. W. & Markham, R. (1963). Synthesis of complex organic compounds from simple precursors: Formation of amino acid polymers, fatty acids and purines from ammonium cyanide. *Nature*, **199**, 219–22.

Mackay, A. L. (1960). β-Ferric oxyhydroxide. *Mineral Magazine*, **32**, 545–57.

Mackay, A. L. (1964). β-Ferric oxyhydroxide – akaganeite. *Mineral Magazine*, **33**, 270–80.

Mackenzie, D. W. (1971). *Studies on the Synthesis of Layer Silicates*. Ph.D Thesis, University of Glasgow.

Mackinney, G. (1932). Photosynthesis *in vitro*. *Journal of the American Chemical Society*, **54**, 1688–9.

Mahler, H. R. & Cordes, E. H. (1971). *Biological Chemistry*, 2nd edn. New York: Harper & Row.

Mansfield, C. F. & Bailey, S. W. (1972). Twin and pseudotwin intergrowths in kaolinite. *American Mineralogist*, **57**, 411–25.

Margulis, L. (1970). *Origin of Eukaryotic Cells*. New Haven: Yale University Press.

Mason, B. (1966). *Principles of Geochemistry*, 3rd edn., p. 167. New York: Wiley.

Mathews, C. N. & Moser, R. E. (1967). Peptide synthesis from hydrogen cyanide and water. *Nature*, **215**, 1230–4.

Mathieson, A. McL. & Walker, G. F. (1954). Crystal structure of magnesium-vermiculite. *American Mineralogist*, **39**, 231–55.

Matijevic, E. (1976). Preparation and characteristics of monodispersed metal hydrous oxide sols. *Progress in Colloid and Polymer Science*, **61**, 24–35.

Matijevic, E. & Scheiner, P. (1978). Ferric hydrous oxide sols: III. Preparation of uniform particles by hydrolysis of Fe(III) chloride, nitrate, and perchlorate. *Journal of Colloid and Interface Science*, **63**, 509–24.

Maynard Smith, J. (1969). The status of neo-Darwinism. In *Towards a Theoretical Biology, 2. Sketches*, ed. C. H. Waddington, pp. 82–9. Edinburgh: Edinburgh University Press.

Maynard Smith, J. (1970). Time in the evolutionary process. *Studium generale*, **23**, 266–72.

Maynard Smith, J. (1975). *The Theory of Evolution*, 3rd edn. Harmondsworth: Penguin.

McBride, M. B. (1979). Reactivity of adsorbed and structural iron in hectorite as indicated by oxidation of benzidine. *Clays and Clay Minerals*, **27**, 224–30.

McMurchy, R. C. (1934). The crystal structure of the chlorite minerals. *Zeitschrift für Kristallographie*, **88**, 420–32.

Megaw, H. D., Kempster, C. J. E. & Radoslovish, E. W. (1962). The structure of anorthite, $CaAl_2Si_2O_8$, II: Description and discussion. *Acta Crystallographica*, **15**, 1017–35.

Meier, W. M. (1960). The crystal structure of natrolite. *Zeitschrift für Kristallographie*, **113**, 430–44.

Merrifield, R. B., Stewart, J. M. & Jernberg, N. (1966). Instrument for automated synthesis of peptides. *Analytical Chemistry*, **38**, 1905–14.

Meunier, A. & Velde, B. (1979). Biotite weathering in granites of western France. In *International Clay Conference 1978*, ed. M. M. Mortland & V. C. Farmer, pp. 405–13. Amsterdam: Elsevier.

Meyers, P. A. & Quinn, J. G. (1971). Fatty acid–clay mineral associations in artificial and natural sea water solutions. *Geochimica et Cosmochimica Acta*, **35**, 628–32.

Miller, K. R. (1979). The photosynthetic membrane. *Scientific American*, **241**, no. 4, 100–11.

Miller, S. L. (1953). A production of amino acids under possible primitive Earth conditions. *Science*, **117**, 528–9.

Miller, S. L. (1955). Production of some organic compounds under possible primitive Earth conditions. *Journal of the American Chemical Society*, **77**, 2351–61.

Miller, S. L. (1957). The mechanism of synthesis of amino acids by electric discharges. *Biochimica et Biophysica Acta*, **23**, 480–9.

Miller, S. L. & Orgel, L. E. (1974). *The Origins of Life on the Earth*. New Jersey: Prentice-Hall.

Miller, S. L. & Urey, H. C. (1959). Organic compound synthesis on the primitive Earth. *Science*, **130**, 245–51.

Millot, G. (1973). Data and tendencies of recent years in the field 'Genesis and synthesis of clay minerals'. In *Proceedings of the International Clay Conference 1972*, ed. J. M. Serratosa, pp. 151–7. Madrid: Division de Ciencias, CSIC.

Mills, D. R., Kramer, F. R. & Spiegelman, S. (1973). Complete nucleotide sequence of a replicating RNA molecule. *Science*, **180**, 916–27.

Misawa, T., Hashimoto, K. & Shimodaira, S. (1974). The mechanism of formation of iron oxide and oxyhydroxides in aqueous solutions at room temperature. *Corrosion Science*, **14**, 131–49.

Mizuno, T. & Weiss, A. H. (1974). Synthesis and utilisation of formose sugars. *Advances in Carbohydrate Chemistry*, **29**, 173–227.

Monod, J., Wyman, J. & Changeaux, J. P. (1965). On the nature of allosteric transitions: a plausible model. *Journal of Molecular Biology*, **12**, 88–118.

Moorbath, S. (1977). The oldest rocks and the growth of continents. *Scientific American*, **236**, no. 3, 92–104.

Moorbath, S., O'Nions, R. K. & Pankhurst, R. J. (1973). Early archean age for the Isua iron formation, West Greenland. *Nature*, **245**, 138–9.

Moore, B. (1913). *The Origin and Nature of Life*. The Home University Library of Knowledge, No. 63. London: Henry Holt.

Moore, B. & Webster, T. A. (1913). Synthesis by sunlight in relationship to the origin of life. Synthesis of formaldehyde from carbon dioxide and water by inorganic colloids acting as transformers of light energy. *Proceedings of the Royal Society B*, **87**, 163–76.

Morey, G. W., Fournier, R. O. & Rowe, J. J. (1962). The solubility of quartz in water in the temperature interval from 25° to 300 °C. *Geochimica et Cosmochimica Acta*, **26**, 1029–43.

Morgan, T. H. (1926). *The Theory of the Gene*. New Haven, Connecticut: Yale University Press.

Morimoto, S., Kawashiro, K. & Yoshida, H. (1978). The asymmetric adsorption of α-aminopropionitrile on D-quartz. In *Origin of Life*, ed. H. Noda, pp. 349–53. Tokyo: Center for Academic Publications Japan.

Morita, K., Ochiai, M. & Marumoto, R. (1968). A convenient one-step synthesis of adenine. *Chemistry and Industry*, p. 1117.

Muller, G. (1961). Die rezenten Sedimente im Golf von Neapel: 2. *Beitrage Mineralogie und Petrographie*, **8**, 1–20.

Muller, H. J. (1929). The gene as the basis of life. In *Proceedings of the Inter-*

National Congress of Plant Sciences 1926, ed. B. M. Dugger, pp. 897–921. Menasha, Wisconsin: George Banta.

Mumpton, F. A. & Ormsby, W. C. (1976). Morphology of zeolites in sedimentary rocks by scanning electron microscopy. *Clays and Clay Minerals*, **24**, 1–23.

Murphy, P. J., Posner, A. M. & Quirk, J. P. (1976a). Characterization of partially neutralised ferric nitrate solutions. *Journal of Colloid and Interface Science*, **56**, 270–83.

Murphy, P. J., Posner, A. M. & Quirk, J. P. (1976b). Characterization of partially neutralised ferric chloride solutions. *Journal of Colloid and Interface Science*, **56**, 284–97.

Murphy, P. J., Posner, A. M. & Quirk, J. P. (1976c). Characterization of partially neutralised ferric perchlorate solutions. *Journal of Colloid and Interface Science*, **56**, 298–311.

Nagasawa, K. & Miyazaki, S. (1976). Mineralogical properties of halloysite as related to its genesis. In *Proceedings of the International Clay Conference 1975*, ed. S. W. Bailey, pp. 257–65. Wilmette Illinois: Applied Publishing Ltd.

Nakai, M. & Yoshinaga, N. (1978). Poster display: occurrence of a fibrous iron mineral in some soils from Japan and Scotland. *Book of Summaries for the International Clay Conference 1978*, p. 653. London: The Mineralogical Society.

Nakazawa, H. & Morimoto, N. & Watanabe, E. (1975). Direct observations of metal vacancies by high-resolution electron microscopy. Part I: 4C type pyrrhotite (Fe_7S_8). *American Mineralogist*, **60**, 359–66.

Nalovic, Lj., Pedro, G. & Janot, C. (1976). Demonstration by Mössbauer spectroscopy of the role played by transitional trace elements in the crystallogenesis of iron hydroxides (III). In *Proceedings of the International Clay Conference 1975*, ed. S. W. Bailey, pp. 601–10. Wilmette Illinois: Applied Publishing Ltd.

Naylor, R. & Gilham, P. T. (1966). Studies on some interactions and reactions of oligonucleotides in aqueous solution. *Biochemistry*, **5**, 2722–8.

Negrón-Mendoza, A. & Ponnamperuma, C. (1978). Interconversion of biologically important carboxylic acids by radiation. In *Origin of Life*, ed. H. Noda, pp. 101–4. Tokyo: Center for Academic Publications Japan.

Newnham, R. E. (1961). A refinement of dickite structure and some remarks on polymorphism of kaolin minerals. *Mineral Magazine*, **32**, 683–704.

Nicholas, H. J. (1973). Terpenes. In *Phytochemistry, vol. II*, ed. L. P. Miller, pp. 254–309. New York: Van Nostrand Reinhold.

Nicholson, D. E. (1974). *Metabolic Pathways*. Colnbrook, England: Koch-Light Laboratories Ltd.

Ninio, J. (1975). Kinetic amplification of enzyme discrimination. *Biochimie*, **57**, 587–95.

Nissenbaum, A. (1976). Scavenging of soluble organic matter from the prebiotic oceans. *Origins of Life*, **7**, 413–16.

Norrish, K. (1973). Factors in the weathering of mica to vermiculite. In *Proceedings of the International Clay Conference 1972*, ed. J. M. Serratosa, pp. 417–32. Madrid: Division de Ciencias, CSIC.

Nyholm, R. S. & Tobe, M. L. (1963). The stabilisation of oxidation states of the transition metals. *Advances in Inorganic Chemistry and Radiochemistry*, **5**, 1–40.

Odom, D. G., Rao, M., Lawless, J. G. & Oró, J. (1979). Association of nucleotides with homoionic clays. *Journal of Molecular Evolution*, **12**, 365–7.

Oparin, A. I. (1924). *Proiskhozhdenie Zhizny*. Moscow: Izd. Moskovskiĭ Rabochiĭ. (Reprinted in translation in Bernal, 1967, pp. 199–234.)

Oparin, A. I. (1938). *Origin of Life*. Reprinted 1953. New York: Dover.

Oparin, A. I. (1957). *The Origin of Life on the Earth*. Edinburgh: Oliver & Boyd,

Orgel, L. E. (1968). Evolution of the genetic apparatus. *Journal of Molecular Biology*, **38**, 381–93.

Orgel, L. E. (1972). A possible step in the origin of the genetic code. *Israel Journal of Chemistry*, **10**, 287–92.

Orgel, L. E. (1979). Selection *in vitro*. *Proceedings of the Royal Society B*, **205**. 435–42.

Orgel, L. E. & Lohrmann, R. (1974). Prebiotic chemistry and nucleic acid replication. *Accounts of Chemical Research*, **7**, 368–77.

Orgel, L. E. & Sulston, J. E. (1971). Polynucleotide replication and the origin of life. In *Prebiotic and Biochemical Evolution*, ed. A. P. Kimball & J. Oro, pp. 89–94. Amsterdam: North Holland.

Oró, J. (1960). Synthesis of adenine from ammonium cyanide. *Biochemical and Biophysical Research Communications*, **2**, 407–12.

Oró, J. (1965). Stages and mechanisms of prebiological organic synthesis. In *The Origins of Prebiological Systems*, ed. S. W. Fox, pp. 137–71. New York: Academic.

Oró, J. & Kamat, S. S. (1961). Amino acid synthesis from hydrogen cyanide under possible primitive Earth conditions. *Nature*, **190**, 442–3.

Oró, J. & Kimball, A. P. (1961). Synthesis of purines under possible primitive Earth conditions. 1. Adenine from hydrogen cyanide. *Archives of Biochemistry and Biophysics*, **94**, 217–27.

Oró, J., Sherwood, E., Eichberg, J. & Epps, D. (1978). Formation of phospholipids under primitive Earth conditions and the role of membranes in prebiological evolution. In *Light Transducing Membranes*, ed. D. W. Deamer, pp. 1–21. New York: Academic.

Österberg, R., Orgel, L. E. & Lohrmann, R. (1973). Further studies of urea-catalysed phosphorylation reactions. *Journal of Molecular Evolution*, **2**, 231–4.

Paecht-Horowitz, M. (1974). Micelles and solid surfaces as amino acid polymerisation propagators. In *The Origin of Life and Evolutionary Biochemistry*, ed. K. Dose, S. W. Fox, G. A. Deborin & T. E. Pavlovskaya, pp. 373–85. New York: Plenum.

Paecht-Horowitz, M., Berger, J. & Katchalsky, A. (1970). Prebiotic synthesis of polypeptides by heterogeneous polycondensation of amino acid adenylates. *Nature*, **228**, 636–9.

Palm, C. & Calvin, M. (1962). Primordial organic chemistry. 1. Compounds resulting from irradiation of $^{14}CH_4$. *Journal of the American Chemical Society*, **84**, 2115–21.

Parfitt, R. L. & Henmi, T. (1980). Structure of some allophanes from New Zealand. *Clays and Clay Minerals*, **28**, 285–94.

Parham, W. E. (1969). Formation of halloysite from feldspar: low temperature artificial weathering versus natural weathering. *Clays and Clay Minerals*, **17**, 13–22.

Pasteur, L. (1848). Sur les relations qui peuvent exister entre la forme cristalline, la composition chimique et le sens de la polarisation rotatoire. *Annales de Chimie Physique*, **24**, 442–59.

Penrose, L. S. (1959a). Automatic mechanical self-reproduction. *New Biology*, **28**, 92–117.

Penrose, L. S. (1959b). Self-reproducing machines. *Scientific American*, **200**, no. 6, 105–14.

Perutz, M. F. (1978). Haemoglobin structure and respiratory transport. *Scientific American*, **239**, no. 6, 68–86.

Pfeil, E. & Ruckert, H. (1961). Über die Formaldehydkondensation. *Annalen Chemie*, **641**, 121–31.

Phillips, D. C. (1966). The three-dimensional structure of an enzyme molecule. *Scientific American*, **215**, no. 5, 78–90.

Pincock, R. E., Bradshaw, R. P. & Perkins, R. R. (1974). Spontaneous and induced generation of optical activity in racemic 1,1″-binaphthyl. *Journal of Molecular Evolution*, **4**, 67–75.

Pirie, N. W. (1951). In *The Physical Basis of Life*, by J. D. Bernal, p. 71. London: Routledge & Kegan Paul.

Pirie, N. W. (1957). Some assumptions underlying discussions on the origins of life. *Annals of the New York Academy of Sciences*, **69**, 369–76.

Pirie, N. W. (1959). Chemical diversity and the origins of life. In *The Origin of Life on the Earth*, ed. F. Clark & R. L. M. Synge, pp. 76–83. London: Pergamon.

Plançon, A. & Tchoubar, C. (1977). Determination of structural defects in phyllosilicates by X-ray powder diffraction – II. Nature and proportion of defects in natural kaolinites. *Clays and Clay Minerals*, **25**, 436–50.

Ponnamperuma, C. (1965). Abiological synthesis of some nucleic acid constituents. In *The Origins of Prebiological Systems*, ed. S. W. Fox, pp. 221–42. New York: Academic.

Ponnamperuma, C. (1978). Prebiotic molecular evolution. In *Origin of Life*, ed. H. Noda, pp. 67–81. Tokyo: Center for Academic Publications Japan.

Ponnamperuma, C. & Peterson, E. (1965). Peptide synthesis from amino acids in aqueous solution. *Science*, **147**, 1572–4.

Ponnamperuma, C., Shimoyama, A., Yamada, M., Hobo, T. & Pal, R. (1977). Possible surface reactions on Mars: implications for the Viking biology results. *Science*, **197**, 455–7.

Quayle, J. R. & Ferenci, T. (1978). Evolutionary aspects of autotrophy. *Microbiological Reviews*, **42**, 251–73.

Rabinowitz, J., Flores, J., Krebsbach, R. & Rogers, G. (1969). Peptide formation in the presence of linear or cyclic polyphosphates. *Nature*, **224**, 795–6.

Radoslovich, E. W. (1960). The structure of muscovite, $K\,Al_2(Si_3Al)O_{10}(OH)_2$. *Acta Crystallographica*, **13**, 919–32.

Radoslovich, E. W. (1963). The cell dimensions and symmetry of layer-lattice silicates. IV. Interatomic forces. *American Mineralogist*, **48**, 76–99.

Rausch, W. V. & Bale, H. D. (1964). Small angle X-ray scattering from hydrolysed aluminium nitrate solutions. *Journal of Chemical Physics*, **40**, 3391–4.

Rausell-Colom, J. A., Sanz, J., Fernandez, M. & Serratosa, J. M. (1979). Distribution of octahedral ions in phlogopites and biotites. In *International Clay Conference 1978*, ed. M. M. Mortland & V. C. Farmer, pp. 27–36. Amsterdam: Elsevier.

Rautureau, M. & Tchoubar, C. (1976). Structural analysis of sepiolite by selected area electron diffraction – relations with physico-chemical properties. *Clays and Clay Minerals*, **24**, 43–9.

Rees, D. A. (1977). *Polysaccharide Shapes*. London: Chapman Hall.

Rees, D. A. & Welsh, E. J. (1977). Secondary and tertiary structure of polysaccharides in solutions and gels. *Angewandte Chemie, International Edition*, **16**, 214–24.

Reiche, H. & Bard, A. J. (1979). Heterogeneous photosynthetic production of amino acids from methane–ammonia–water at Pt/TiO_2. Implications in chemical evolution. *Journal of the American Chemical Society*, **101**, 3127–8.

Reid, C. & Orgel, L. E. (1967). Synthesis of sugars in potentially prebiotic conditions. *Nature*, **216**, 455.

Rein, R., Nir, S. & Stamadiadou, M. N. (1971). Photochemical survival principle in molecular evolution. *Journal of Theoretical Biology*, **33**, 309–18.

Rensch, B. (1959). *Evolution above the Species Level*. London: Methuen.

Reynolds, R. C. (1965). An X-ray study of an ethylene glycol–montmorillonite complex. *American Mineralogist*, **50**, 990–1001.

Rich, A. (1965). Evolutionary problems in the synthesis of proteins. In *Evolving Genes and Proteins*, ed. V. Bryson & H. J. Vogel, pp. 453–68. New York: Academic.

Richmond, M. H. & Smith, D. C. (1979). The cell as habitat: A discussion organised by M. H. Richmond & D. C. Smith, F.R.S. *Proceedings of the Royal Society B*, **204**, 113–286.

Rickard, D. T. (1969). The chemistry of iron sulphide formation at low temperatures. In *Stockholm Contributions in Geology*, vol. 20, ed. I. Hessland, pp. 67–95. Stockholm: The University of Stockholm.

Rimsaite, J. (1973). Genesis of chlorite, vermiculite, serpentine, talc and secondary oxides in ultrabasic rocks. In *Proceedings of the International Clay Conference 1972*, ed. J. M. Serratosa, pp. 291–302. Madrid: Division de Ciencias, CSIC.

Ring, D., Wolman, Y., Friedman, N. & Miller, S. L. (1972). Prebiotic synthesis of hydrophobic and protein amino acids. *Proceedings of the National Academy of Sciences USA*, **69**, 765–8.

Rohlfing, D. L. (1976). Thermal polyamino acids: synthesis at less than 100 °C. *Science*, **193**, 68–70.

Rosen, B. P. (1978). *Bacterial Transport*. New York: Dekker.

Ross, M., Takeda, H. & Wones, D. R. (1966). Mica polytypes: Systematic description and identification. *Science*, **151**, 191–3.

Rousseaux, J. M., Gomez Laverde, C., Nathan, Y. & Rouxhet, P. G. (1973). Correlation between the hydroxyl stretching bands and the chemical composition of trioctahedral micas. In *Proceedings of the International Clay Conference 1972*, ed. J. M. Serratosa, pp. 89–98. Madrid: Division de Ciencias, CSIC.

Rouxhet, P. G., Samudacheata, N., Jacobs, H. & Anton, O. (1977). Attribution of the OH stretching bands of kaolinite. *Clay Minerals*, **12**, 171–9.

Roy, D. M. & Roy, R. (1954). An experimental study of the formation and properties of synthetic serpentines and related layer silicate minerals. *American Mineralogist*, **39**, 957–75.

Rubey, W. W. (1951). Geologic history of seawater: An attempt to state the problem. *Bulletin of the Geological Society of America*, **62**, 1111–47.

Ruckert, H., Pfeil, E. & Scharf, G. (1965). Der sterische Verlauf der Zuckerbildung. *Chemische Berichte*, **98**, 2558–65.

Sagan, C. & Khare, B. N. (1971). Long wavelength ultraviolet photoproduction of amino acids on the primitive Earth. *Science*, **173**, 417–20.

Sanchez, R. A., Ferris, J. P. & Orgel, L. E. (1966a). Conditions for purine synthesis: Did prebiotic synthesis occur at low temperatures? *Science*, **153**, 72–3.

Sanchez, R. A., Ferris, J. P. & Orgel, L. E. (1966b). Cyanoacetylene in prebiotic synthesis. *Science*, **154**, 784–5.

Sanchez, R. A., Ferris, J. P. & Orgel, L. E. (1967). Synthesis of purine precursors and amino acids from aqueous hydrogen cyanide. *Journal of Molecular Biology*, **30**, 223–53.

Satir, P. (1974). How cilia move. *Scientific American*, **231**, no. 4, 44–52.

Saunders, P. T. & Ho, M. W. (1976). On the increase in complexity in evolution. *Journal of Theoretical Biology*, **63**, 375–84.

Schidlowski, M. (1978). Evolution of the Earth's atmosphere: current state and exploratory concepts. In *Origin of Life*, ed. H. Noda, pp. 3–20. Tokyo: Center for Academic Publications Japan.

Schmalhausen, I. I. (1968). *The Origin of Terrestrial Vertebrates*. New York: Academic.

Schopf, J. W. (1975). Precambrian paleobiology: Problems and perspectives. In *Annual Review of Earth and Planet Sciences*, vol. 3, ed. F. A. Donath, F. G. Stehli & G. W. Wetherill, pp. 213–49. Palo Alto, California: Annual Reviews Inc.

Schopf, J. W. (1976). Are the oldest 'fossils' fossils? *Origins of Life*, **7**, 19–36.

Schrauzer, G. N. & Guth, T. D. (1977). Photolysis of water and photoreduction of nitrogen on titanium dioxide. *Journal of the American Chemical Society*, **99**, 7189–93.

Schwartz, A. W. (1971). Phosphate solubilisation and activation on the primitive Earth. In *Chemical Evolution and the Origin of Life*, ed. R. Buvet & C. Ponnamperuma, pp. 207–15. Amsterdam: North-Holland.

Schwartz, A. W. (1981). Chemical evolution – the genesis of the first organic compounds. In *Marine Organic Chemistry*, ed. E. K. Duursma & R. Dawson, pp. 7–30. Amsterdam: Elsevier.

Schwartz, A. W. & Chittenden, G. J. F. (1977). Synthesis of uracil and thymine under simulated prebiotic conditions. *Biosystems*, **9**, 87–92.

Schwertmann, U. (1979). Non-crystalline and accessory minerals. In *International Clay Conference 1978*, ed. M. M. Mortland & V. C. Farmer, pp. 491–9. Amsterdam: Elsevier.

Schwertmann, U. & Thalmann, H. (1976). The influence of [Fe(II)], [Si] and pH on the formation of lepidocrocite and ferrihydrite during oxidation of aqueous FeCl₂ solutions. *Clay Minerals*, **11**, 189–200.

Serratosa, J. M. (1979). Surface properties of fibrous clay minerals (paygorskite and sepiolite). In *International Clay Conference 1978*, ed. M. M. Mortland & V. C. Farmer, pp. 99–109. Amsterdam: Elsevier.

Sheppard, R. A. & Gude, A. J. (1973). *Zeolites and Associated Authigenic Silicate Minerals in Tuffaceous Rocks of the Big Sandy Formation, Mohave County, Arizona.* U.S. Geological Survey Professional Paper, No. 830.

Sherrington, C. S. (1940). *Man on his Nature*, p. 88. Cambridge: Cambridge University Press.

Sherwood, E., Nooner, D. W., Eichberg, J., Epps, D. E. & Oró, J. (1978). Prebiotic condensation reactions using cyanamide. In *Origin of Life*, ed. H. Noda, pp. 105–11. Tokyo: Center for Academic Publications Japan.

Shigemasa, Y., Sakazawa, C., Nakashima, R. & Matsuura, T. (1978). Selectivities in the formose reaction. In *Origin of Life*, ed. H. Noda, pp. 211–16. Tokyo: Center for Academic Publications Japan.

Shirozu, H. & Bailey, S. W. (1966). Crystal structure of a two-layer Mg-vermiculite. *American Mineralogist*, **51**, 1124–43.

Sieskind, O. & Ourisson, G. (1971). Interactions argile–matière organique: formation de complexes entre la montmorillonite et les acides stéarique et béhénique. *Comptes Rendus de L'Académie des Sciences Paris*, C272, 1885–8.

Siffert, B. (1979). Genesis and synthesis of clays and clay minerals: recent developments and future prospects. In *International Clay Conference 1978*, ed. M. M. Mortland & V. C. Farmer, pp. 337–47. Amsterdam: Elsevier.

Siffert, B. & Wey, R. (1973). Contribution à la connaissance de la synthèse des kaolins. In *Proceedings of the International Clay Conference 1972*, ed. J. M. Serratosa, pp. 159–72. Madrid: Division de Ciencias, CSIC.

Sillen, L. G. (1965). Oxidation state of the Earth's ocean and atmosphere. I. A model calculation on earlier states. The myth of the 'probiotic soup'. *Arkiv för Kemi*, **24**, 431–56.

Simon, H. A. (1962). The architecture of complexity. *Proceedings of the American Philosophical Society*, **106**, 467–82.

Sipling, P. J. & Yund, R. A. (1974). Kinetics of Al/Si disordering in alkali feldspars. In *Geological Transport and Kinetics*, ed. A. W. Hoffmann, B. J. Giletti, H. S. Yoder & R. A. Yund, pp. 185–94. Washington: Carnegie Institute Publication 634.

Smith, A., Folsome, C. & Bahadur, K. (1981). Carbon dioxide reduction and nitrogenase activity in organo–molybdenum microstructures, *Experientia*, **37**, 357–9.

Smith, J. V. & Yoder, H. S. (1956). Experimental and theoretical studies of mica polymorphs. *Mineral Magazine*, **31**, 209–35.

Smith, R. W. & Hem, J. D. (1972). Effect of ageing on aluminium hydroxide complexes in dilute aqueous solutions. U.S. Geological Survey Water Supply Paper 1827-D.

Solomon, D. H., Loft, B. C. & Swift, J. D. (1968a). Reactions catalysed by minerals. IV. The mechanism of the benzidine blue reaction on silicate minerals. *Clay Minerals*, **7**, 389–98.

Solomon, D. H., Loft, B. C. & Swift, J. D. (1968b). Reactions catalysed by minerals. V. The reaction of leuco dyes and unsaturated organic compounds with clay minerals. *Clay Minerals*, **7**, 399–408.

Sonneborn, T. M. (1965). Degeneracy of the genetic code: extent, nature and genetic implications. In *Evolving Genes and Proteins*, ed. V. Bryson & H. J. Vogel, pp. 377–97. New York: Academic.

Spiegelman, S. (1970). Extracellular evolution of replicating molecules. In *The Neurosciences: Second Study Program*, ed. F. O. Schmitt, pp. 927–45. New York: Rockefeller University Press.

Spiegelman, S., Mills, D. R. & Kramer, F. R. (1975). The extracellular evolution of structure in replicating RNA molecules. In *Stability and Origin of Biological Information*, ed. I. R. Miller, pp. 123–72. New York: John Wiley & Sons.

Spirin, A. S. (1978). Energetics of the ribisome. *Progress in Nucleic Acid Research and Molecular Biology*, **21**, 39–62.

Spiro, T. G., Allerton, S. E., Renner, J., Terzis, A., Bils, R. & Saltman, P. (1966). The hydrolytic polymerisation of iron (III). *Journal of the American Chemical Society*, **88**, 2721–6.

Steinfink, H. (1958a). The crystal structure of chlorite: I. A monoclinic polymorph. *Acta Crystallographica*, **11**, 191–5.

Steinfink, H. (1958b). The crystal structure of chlorite: II. A triclinic polymorph. *Acta Crystallographica*, **11**, 195–8.

Stillinger, F. H. & Wasserman, Z. (1978). Molecular recognition and self-organisation in fluorinated hydrocarbons. *Journal of Physical Chemistry*, **82**, 929–40.

Stoeckenius, W. (1976). The purple membrane of salt-loving bacteria. *Scientific American*, **234**, no. 6, 38–46.

Stroud, R. M. (1974). A family of protein-cutting proteins. *Scientific American*, **231**, no. 1, 74–88.

Suquet, H., Iiyama, J. T., Kodama, H. & Pezerat, H. (1977). Synthesis and swelling properties of saponite with increasing layer charge. *Clays and Clay Minerals*, **25**, 231–42.

Szent-Györgyi, A. (1972). *The Living State*. New York: Academic.

Takeda, H. & Ross, M. (1975). Mica polytypism: Dissimilarities in the crystal structure of coexisting 1M and $2M_1$ biotite. *American Mineralogist*, **60**, 1030–40.

Taylor, P. (1980). The stereochemistry of iron sulphides – a structural rationale for crystallisation of some metastable phases from aqueous solution. *American Mineralogist*, **65**, 1026–30.

Taylor, W. H. (1933). The structure of sanidine and other felspars. *Zeitschrift für Kristallographie*, **85**, 425–42.

Taylor, W. H., Darbyshire, J. A. & Strunz, H. (1934). An X-ray investigation of the felspars. *Zeitschrift für Kristallographie*, **87**, 464–98.

Tazaki, K. (1979a). Scanning electron microscope study of imogolite formation from plagioclase. *Clays and Clay Minerals*, **27**, 209–12.

Tazaki, K. (1979b). Micromorphology of halloysite produced by weathering of plagioclase in volcanic ash. In *International Clay Conference 1978*, ed. M. M. Mortland & V. C. Farmer, pp. 415–22. Amsterdam: Elsevier.

Theng, B. K. G. (1974). *The Chemistry of Clay-Organic Reactions*. Bristol: Hilger.

Thiemann, W. (1974). Disproportionation of enantiomers by precipitation. *Journal of Molecular Evolution*, **4**, 85–97.

Thiemann, W. & Darge, W. (1974). Experimental attempts for the study of the origin of optical activity on Earth. *Origins of Life*, **5**, 263–83.

Thilo, E. (1962). Condensed phosphates and arsenates. *Advances in Inorganic Chemistry and Radiochemistry*, **4**, 1–75.

Thilo, E. & Wieker, W. (1957). Die anionenhydrolyse kondensierter phosphate in verdünnter wässriger lösung. *Zeitschrift für anorganische und allgemeine Chemie*, **291**, 164–85.

Thompson, T. D. & Brindley, G. W. (1969). Absorption of pyrimidines, purines and nucleosides by Na-, Mg- and Cu(II)-illite (clay–organic studies XVI). *American Mineralogist*, **54**, 858–68.

Topal, M. D. & Fresco, J. R. (1976). Complementary base pairing and the origin of substitution mutations. *Nature*, **263**, 285–9.

Toupance, G., Raulin, F. & Buvet, R. (1971). Primary transformation processes under the influence of energy for models of primordial atmospheres. In *Chemical Evolution and the Origin of Life*, ed. R. Buvet & C. Ponnamperuma, pp. 83–95. Amsterdam: North-Holland.

Troland, L. T. (1914). The chemical origin and regulation of life. *The Monist*, **24**, 92–133.

Tseng, S-S. & Chang, S. (1975). Photochemical synthesis of simple organic free radicals on simulated planetary surfaces – an ESR study. *Origins of Life*, **6**, 61–73.

Turner, R. C. & Ross, G. J. (1970). Conditions in solution during the formation of gibbsite in dilute Al salt solutions. 4. Effect of Cl concentration and temperature and a proposed mechanism for gibbsite formation. *Canadian Journal of Chemistry*, **48**, 723–9.

Urey, H. C. (1952). *The Planets*. New Haven, Connecticut: Yale University Press.

Van Olphen, H. (1966). Maya blue: A clay-organic pigment? *Science*, **154**, 645–6.

Van Olphen, H. (1977). *An Introduction to Clay Colloid Chemistry*, 2nd edn. New York: John Wiley & Sons.

Van Oosterwyck-Gastuche, M. C. & La Iglesia, A. (1978). Kaolinite synthesis. II: A review and discussion of the factors influencing the rate process. *Clays and Clay Minerals*, **26**, 409–17.

Van Wazer, J. R. (1958). *Phosphorus and its Compounds*, vol. I. New York: Interscience.

Vedder, W. (1964). Correlations between infrared spectrum and chemical composition of mica. *American Mineralogist*, **49**, 736–68.

Veitch, L. G. & Radoslovich, E. W. (1963). The cell dimensions and symmetry of layer-lattice silicates: III. Octahedral ordering. *American Mineralogist*, **48**, 62–75.

Velde, B. (1965). Experimental determination of muscovite polymorph stabilities. *American Mineralogist*, **50**, 436–49.

Velde, B. (1977). *Clays and Clay Minerals in Natural and Synthetic Systems.* Amsterdam: Elsevier.

Verwey, E. J. W. & Heilmann, E. L. (1947). Physical properties and cation arrangement of oxides with spinel structures. I. Cation arrangement in spinels. *Journal of Chemical Physics*, **15**, 174–80.

Wada, K. (1979). Structural formulas for allophanes. In *International Clay Conference 1978*, ed. M. M. Mortland & V. C. Farmer, pp. 537–45. Amsterdam: Elsevier.

Wada, S. I. & Wada, K. (1977). Density and structure of allophane. *Clay Minerals*, **12**, 289–98.

Waddington, C. H. (1957). *The Strategy of the Genes: A Discussion of Some Aspects of Theoretical Biology.* London: Allen & Unwin.

Waddington, C. H. (1968). The basic ideas of biology. In *Towards a Theoretical Biology, 1, Prolegomena*, ed. C. H. Waddington, p. 3. Edinburgh: Edinburgh University Press.

Wagener, K. (1974). Amplification processes for small primary differences in the properties of enantiomers. *Journal of Molecular Evolution*, **4**, 77–84.

Wakamatsu, H., Yamada, Y., Saito, T., Kumashiro, I. & Takenishi, T. (1966). Synthesis of adenine by oligomerisation of hydrogen cyanide. *Journal of Organic Chemistry*, **31**, 2035–6.

Wald, G. (1957). The origin of optical activity. *Annals of the New York Academy of Sciences*, **69**, 352–68.

Walker, G. L. (1971). *On the Formation of Sugars from Formaldehyde.* Ph.D. Thesis, University of Glasgow.

Walker, J. C. G. (1976a). Implications for atmospheric evolution of the inhomogeneous accretion model of the origin of the Earth. In *The Early History of the Earth*, ed. B. F. Windley, pp. 537–46. London: John Wiley & Sons.

Walker, J. C. G. (1976b). *Evolution of the Atmosphere.* New York: Hafner.

Watson, D. M. S. (1946). The evolution of the proboscidea. *Biological Reviews*, **21**, 15–29.

Watson, J. D. (1976). *Molecular Biology of the Gene*, 3rd edn. Menlo Park, California: Benjamin.

Watson, J. D. & Crick, F. H. C. (1953). Genetical implications of the structure of deoxyribonucleic acid. *Nature*, **171**, 964–7.

Weir, A. H., Nixon, H. L. & Woods, R. D. (1962). Measurement of thickness of dispersed clay flakes with the electron microscope. *Clays and Clay Minerals*, **9**, 419–23.

Weiss, A. (1963). Mica-type layer silicates with alkylammonium ions. *Clays and Clay Minerals*, **10**, 191–224.

Weiss, A. (1981). Replication and evolution in inorganic systems. *Angewandte Chemie, International Edition*, **20**, 850–60.

Weiss, A. & Roloff, G. (1966). *Proceedings of the International Clay Conference (Jerusalem)*, vol. 1, 263–75.

Weiss, A. H., Trigerman, S., Dunnells, G., Likholobov, V. A. & Biron, E. (1979). Ethylene glycol from formaldehyde. *Industrial and Engineering Process Design and Development*, **18**, 522–7.

Wells, A. F. (1975). *Structural Inorganic Chemistry.* Oxford: Clarendon.

Wells, N., Childs, C. W. & Downes, C. J. (1977). Silica Springs, Tongariro

National Park, New Zealand – analyses of the spring water and characterisation of the alumino-silicate deposit. *Geochimica et Cosmochimica Acta*, **41**, 1497–1506.

Wey, R. & Siffert, B. (1962). Réactions de la silice monomoléculaire en solution avec les ions Al^{3+} et Mg^{2+}. In *Genèse et Synthèse des Argiles*, ed. M. R. Hocart, pp. 11–23. Paris: CNRS.

White, H. B. (1976). Coenzymes as fossils of an earlier metabolic state. *Journal of Molecular Evolution*, **7**, 101–4.

Wickner, W. (1980). Assembly of proteins into membranes. *Science*, **210**, 861–8.

Wilkins, R. W. T. (1967). The hydroxyl-stretching region of the biotite mica spectrum. *Mineral Magazine*, **36**, 325–33.

Williams, D. G. & Garey, C. L. (1974). Crystal imperfections with regard to direction in kaolinite minerals. *Clays and Clay Minerals*, **22**, 117–25.

Wilson, D. B. (1978). Cellular transport mechanisms. *Annual Review of Biochemistry*, **47**, 933–65.

Wilson, M. D. & Pittman, E. D. (1977). Authigenic clays in sandstones: Recognition and influence on reservoir properties and paleoenvironmental analysis. *Journal of Sedimentary Petrology*, **47**, 3–31.

Winterburn, P. J. (1974). Polysaccharide structure and function. In *Companion to Biochemistry*, ed. A. T. Bull, J. R. Lagnado, J. O. Thomas & K. F. Tipton, pp. 307–41. London: Longman.

Woese, C. R. (1967). *The Genetic Code*. New York: Harper & Row.

Woese, C. R., Magrum, L. J. & Fox, G. E. (1978). Archaebacteria. *Journal of Molecular Evolution*, **11**, 245–52.

Wolfe, R. W. & Giese, R. F. (1973). A quantitative study of one-layer polytypism in kaolin minerals. In *Proceedings of the International Clay Conference 1972*, ed. J. M. Serratosa, pp. 27–33. Madrid: Division de Ciencias, CSIC.

Wolman, Y., Haverland, W. J. & Miller, S. L. (1972). Nonprotein amino acids from spark discharges and their comparison with the Murchison meteorite amino acids. *Proceedings of the National Academy of Sciences USA*, **69**, 809–11.

Wolman, Y., Miller, S. L., Ibanez, J. & Oró, J. (1971). Formaldehyde and amonnia as precursors to prebiotic amino acids. *Science*, **174**, 1039.

Wright, S. (1932). The roles of mutation, inbreeding, crossbreeding and selection in evolution. *Proceedings of the 6th International Congress of Genetics*, **1**, 356–66.

Yamagata, Y. (1966). A hypothesis for the asymmetric appearance of biomolecules on Earth. *Journal of Theoretical Biology*, **11**, 495–8.

Yershova, Z. P., Nikitina, A. P., Perfil'ev, Yu. D. & Babeshkin, A. M. (1976). Study of chamosites by gamma-resonance (Mössbauer) spectroscopy. In *Proceedings of the International Clay Conference 1975*, ed. S. W. Bailey, pp. 211–19. Wilmette, Illinois: Applied Publishing Ltd.

Yoder, H. S. & Eugster, H. P. (1955). Synthetic and natural muscovites. *Geochimica et Cosmochimica Acta*, **8**, 225–80.

Youatt, J. B. & Brown, R. D. (1981). Origins of chirality in Nature: A reassessment of the postulated role of bentonite. *Science*, **212**, 1145–6.

Zeeman, E. C. (1976). Catastrophe theory. *Scientific American*, **234**, no. 4, 65–83.

Zuckerkandl, E. (1975). The appearance of new structures and functions in proteins during evolution. *Journal of Molecular Evolution*, **7**, 1–57.

Index

After a page entry, *d* indicates that a definition is to be found on that page, *e* indicates an electron micrograph and *s* a structure.

Abelson, P. H., 16, 17, 22, 46, 363
'abiotic' origin of chirality, 37, 39, 43
abiotic syntheses of 'biochemicals',
    9–33, 134, 422; reinterpretation of,
    vii, 64–9
accuracy: of crystallisation processes,
    155–6; of DNA replication, 147,
    156–6; evolution of, in replication,
    137
acetic acid, 33
acetyl-coenzyme A, 313, 358, 360, 364,
    376*s*
achiral machines, 35
acrolein, 22*s*
'active hydrogens' in carbonyl
    compounds, 28*d*
active transport, 60, 339; in primitive
    organisms, 312–13
Adams, J. M., 241
adaptation, 2, 88
adaptive landscapes, *see* functional
    landscapes
adaptive peaks, 88*d*
adaptors: matching to amino acids,
    392–5; in protein synthesis, 388
add-and-subtract evolution, 97
adenine, 10*s*, 23–4, 66
adenosine triphosphate, 376*s*, *see* also
    ATP
agar, 379–80
akaganéite (*β*-FeO(OH)), 169, 170*e*, 175,
    176*s*
alanine, 13, 15, 18, 20, 22, 24, 34*s*, 66,
    393, 396–7, 407
β-alanine, 15, 18, 20, 24, 27*s*
alanylalanine half life, 48
Albersheim, P., 380
Alberts, B., 147
albite, 188

alchemists and gold synthesis, 63–4
Alcover, J. F., 233
aldocyanoin, 55–6
aldol condensations, 28–9, 369
aldopentose, 29–30*s*
aldotetrose, 30*s*
Aleksandrova, V. A., 232
Alexander, G. B., 182
alginates, 379–81*s*
alkyl ammonium ions, 242–3, 327
Allaart, J. H., 17
Allen, F. J., 11, 13
Allen, W. V., 33
alloisoleucine, 18
allophane, 236–40, 238*e*, 257, 342
'allosteric effects' in layer silicates,
    223–5, 229, 348–9
allostery, 116, 224–5*d*, 316
allothreonine, 18
Altbuch, A. M., 26
aluminium clays, synthesis of, 251–3,
    256, 418
aluminium hydroxide, 168, 171–4, 250,
    see also cliachite and gibbsite
aluminium oxyanion, 179
ambiguous amino acid sets, 395
amesite, 199, 212, 213*s*; interlayer
    ordering in, 212–14, 276; twinning
    in, 213–14, 279
amino acids, 10*s*, 396*s*; abiotic
    production of, 3, 13–24; A-class
    biochemicals, 66–7; ambiguity of,
    395; C-class biochemicals, 67;
    central set of, 363–5; clay
    interactions with, 47–8; earliest set,
    397; hydrophilic and hydrophobic,
    397–9; later set, 368; ligands for
    metal ions, 310; metabolites of
    primary organisms, 123; in

455

amino acids (*cont.*)
    meteorites, 19–20; not among first
    organic biochemicals, 363–5, 370;
    polymerisation of, on clays, 47–51;
    reactions with sugars, 363; uses of,
    for clay organisms, 365
amino acyl adenylates, 51*s*, 393
amino acyl tRNA ligase enzymes,
    392–4; preprotein equivalents,
    394–5
α-amino-n-butyric acid, 15, 18, 20
β-amino-n-butyric acid, 20
γ-aminobutyric acid, 18, 20
4-aminoimidazole-5-carboxamide, 24
α-aminoisobutyric acid, 15, 18, 20, 24
β-aminoisobutyric acid, 20
ammonia: in biochemical pathways,
    363–7; in Darwin's pond, 351–2;
    limits to in primitive atmosphere,
    17; mineral photoproduction, 333,
    352
amphibole, 186, 187*s*
analcite (analcime), 189, 190*s*, 194*e*;
    authigenic synthesis, 193, 197
anatase, 177
Anders, E., 17
Anderson, C. S., 213–14, 279
Anderson, W. L., 342
3,6-anhydro-D-galactose 2-sulphate, 379
anion polymerisation, 178–84
anorthite, 188
antenna molecules in photosynthesis, 335
antigorite, 199, 209, 212*s*
Anton, O., 203
apatite, 110, 243, 345
apparatus: control of reactions with,
    310–11; and evolution of
    biochemical pathways, 355–6, 359,
    'self-assembly' of clays into,
    319–29
aquo ions, 166
arabinose, 29–30, 32
arch model for evolution, 95–9, 106,
    116, 124
arginine, 22, 67, 396*s*
Aristotle, 117
artificial selection, 424–4
asbestos: blue, 187, white, 209, 212
asparagine, 22, 364, 396*s*
aspartic acid, 15, 18, 20, 22, 24, 38, 48,
    364, 366–8, 396*s*, 397
assemblages: of clays, 257–8, of minerals,
    165
asymmetric carbon atom, 28*d*, 29
Atkinson, R. J., 170
atmosphere, primitive: advantages of an
    inert, 329, composition of, 11–17

ATP, 310, 313, 344, 358, 366–7, 376*s*,
    393, 413, forerunners of, 343–8,
    production and use of in photo-
    synthesis, 337–9
Aumento, F., 209
authigenic minerals, 193*d*, in sandstones,
    246–9*e*, 250–1, in tuffs, 193–7
'autoacquisative' catalysts, 73
autocatalysis: in ester hydrolysis, 71, in
    formose reaction, 30–1, limitations
    for hereditary mechanisms, 72
automatic organic synthesis, 312–17
automatic peptide synthesis, 53–4
autotrophs, 329, 414
Ayala, F. J., 88

Babeshkin, A. M., 215
back reactions, photochemical, 330–8
Bada, J. L., 17
Baes, C. F., 168
Bahadur, K., 424
Bailey, S. W., 204, 212–14, 218, 226,
    227, 229, 230–3, 247, 279–82, 292
Baird, T., 254–5, 270
Bale, H. D., 168
Baly, E. C. C., 11, 13
Balzani, V., 334
band silicates, 186–7
Banin, A., 342
Bard, A. J., 17, 331, 334, 338
Barker, W. C., 107
Barker, W. F., 11
Bar-Nun, A., 13, 17
Bar-Nun, N., 17
Barrer, R. M., 191–2
Barshad, I., 233
base pairing, 146–8, 273, 382
basins in functional landscapes, 85, 88,
    109, 111
Battersby, A. R. 369
Bauer, S. H., 17
bauxite, 172
Baveno twin law, 270–1
Beadle, G. W., 11
Becquerel, P., 11
beginnings as special, 110, 263
beidellite, 199, 235, 324*e*; dislocations in,
    269*e*, 322*e*
Bennett, C. D., 108
bentonite, 38, 49–50, 235
benzidine, 422
γ-benzylglutamic acid anhydride, 41
Beresovskaya, V. V., 177
Bernal, J. D., 4, 9, 37–8, 47, 62, 175,
    365
Berger, J., 51
berthierine, 199, 214

Big Sandy Formation, zeolites etc. in, 193
bilayers: lipid, 302–3; polypeptide, 397–9, 403
bimodal organisms, 113, genetically, 131
binding sites in primitive organisms, 292, 310, 314
biochemical control machinery, evolution of: modern, 372–414, primitive, 261–359
biochemical economy, 66*d*, 66–9
biochemical pathways, early evolution of, 349–71; to amino acids, 363–5, 368; ammonia and, 352, 363–8, apparatus and, 353–6, 359; backward mechanism for, 59, 128; carbon dioxide fixation and, 357–60; carboxylic acids and, 359–61; central non-nitrogen, 357–62; to coenzymes, 369; of cycles, 357; forward mechanism for, 355–6; Horowitz mechanism for, 59, 128; of least resistance, 66–8, 263; to lipids, 369, 412–13; nitrogen in, 362–5; onion model and, 365–71, phosphate and, 343–5, 351; piecemeal updating and, 410–12; to porphyrins, 369, 413; to purines and pyrimidines, 365–8; secondary metabolism and, 355; sugars and 357–9; 361–2; sulphides and, 352; symbiosis and, 351–5; updating mechanisms in, 410–12
biochemical stone age, 166
biochemicals: first kinds of, 164–258; modern, binding of by clays, 242–3; modern, three classes of, 66–7
biochemistry, interdependence of central, 94; as different to begin with, 164
biopolymers, trouble with abiotic synthesis, 42, 45–59
'biotic' origin of chirality in organisms, 37*d*, 39, 43
biotin, 378*s*
biotite, 199, 227–9, 233, 244
'black box' first genes, 138–9
Blankenhorn, G., 363
block copolymers, 384–5
Blout, E. R. 41
Blum, H. F., 11
Bodansky, Y. S., 51
boehmite, 172, 173*s*, 174*e*
Bolton, J. R., 330, 334
Boltzmann equation, 84
bond types, in genetic crystals, 155, 160–1, 283–4; in inorganic and

organic materials, 160–3, 264, 302–4
Bondy, S. C., 38
Bonner, W. A., 37–8, 44
booster rocket analogy for primary life, 126, 133
borate, 179
Brack, A., 51, 398–9, 407
Bradley, W. F., 233
Bradshaw, R. P., 39
Brady, G. W., 170
Bragg, W. L., 185, 189–90
Brahms, J., 398
Brahms, S., 398
branched-chain sugars, 31
Brauner, K., 236
Bremsstrahlung photons, 38–9
Breslow cycle, 30–1, 357
Breslow, R., 31
Bridson, P. K., 58
Briggs, M. H., 44
Brindley, G. W., 198, 212, 241–2
Brown, B. E., 230–1
Brown, G. B., 185
Brown, R. D., 38
'Browning reaction', 363
brucite, 171, 172*s*, 199, 227, 230–1
Buchanan, B. B., 357, 360
Buerger, M. J., 277
Bull, A. T., 138
'bundle of sticks' morphology, 195, 319
Bungenberg de Jong, H. G., 12
Burnham, C. W., 225
Bursill, L. A., 267
Burton, F. G., 345
Busenberg, E., 174
Butlero[v], A., 27
Buvet, R., 14, 21
Bystrom, A., 176
Bystrom, A. M., 176

Cady, S. S., 243
Caillère, S., 233, 251
Cairns-Smith, A. G., vii, 27, 31, 141, 150, 152, 254–5, 357
calcium hydroxide, 31, 171
Calvin cycle, 339, 357, 364, 369
Calvin, M., 9, 12, 17, 338
Cammack, R., 178, 352, 360
Campbell, J. H., 111
Cannizzaro reaction, 31, 33
'caramel' as formose product, 30–1
carbodiimides, 51–3*s*, 54
carbon and life, 163, 300–71
carbon dioxide: first carbon source for organisms, vii, 11–13, 328–9, 350–2; fixation of, 335–9, 357–62,

carbon dioxide (*cont.*)
372; photoreduction of, 32, 332, in primitive atmosphere, 6, 11–17, in purine synthesis, 366; in pyrimidine synthesis, 367

carbon monoxide, 12–13, 22, 26, 32–3

carbonate in most ancient rocks, 17

carbonyl shifts in formose reaction, 30

α-carboxylations, 360

carboxylic acids: abiotic synthesis of, 15, 32–3; catalysts for clay synthesis, 251, 360; first organic biochemicals?, 359–60, 370; fixation of carbon dioxide with, 357–60; in Murchison meteorite, 33; phosphate mobiliser, 343

'card house' structures in clays, 286–7, 319

β-carotene, 336*s*

carotenoids, 335–6, 412

ι-carrageenan, 379–80*s*, 386

catalysts: clays as, 49–51, 318, 422; defect structures as, 292, 318, 422; organic, for clay synthesis, 251, 310, 360, 423; in primitive organisms. 4, 70–3, 313–14; for replicating organic polymers, 146–50, 387

cation order and disorder: in layer silicates, 203–5, 212–14, 217–33, 275–7, 279–83, 418; stability in crystals, 247–9, 270–2

cation polymerisation, 166–78

cation substitutions in minerals, 171, 270; amphiboles, 186; chlorite, 230–2; framework silicates, 188; micas, 215–16, 225–9, 247–8; olivine, 184–5; serpentines, 212–15; smectites, 234–5, 270, 418; vermiculite, 232–3, 247–8

CDP-ethanolamine, 377*s*

cells: as needed for DNA life, 82–3, 298; precursors for, 353–5

cellulose, 362, 380–1

central biochemistry: evolution of, 93–118, 300–414; as 'high technology', 114–16, 121; interdependences in, 93–4; 'layers' in, 365, 370–1; present invariance of, 3–6, 103–4, 414

chabazite, 191*s*, 193

Chada, M. S., 27

chain silicates, 185–6

'7Å chamosite' *see* berthierine

chance: and engineering, 144–6; role in evolution, 87

Chang, S., 32, 47–9, 330

Changeaux, J. P., 224

channels in crystals, 236–7, 31

charges: in polymers, 141, 308–9, 348, 382–3, 399, 407; in silicate layers, 212, 215–16, 230–5, 276, 286–7, 308–9, 319–20, 322–3, 325–6, 340–3, 348–9

Chaussidon, J., 228

check-list genome, 74

chemical accessibility, 62–9, 116, 397, 413

'chemical evolution' and difficulties, 9*d*, 9–77, 131–3

chemical oscillator, 349–50, 423

chemical sediments, 353

chemiosmotic hypothesis, 335–9

chemostats, 137–40, 419

Chen, C. T., 22

chicken-and-egg questions, 93–5

Childs, C. W., 240

chiral discrimination: absence of in polymerisations, 41–5; origin of in organisms, 36–45, 305–8; presence of in crystallisations, 37–43

chirality, 28, 34*d*; in kaolinite, 204, 306; in machines, 34–6, 40–1; need for control of in biopolymers, 41–4, 122, 305–6, 362; in organisms, 34–45; of quartz, 37, 184, 306; of screw dislocations, 306

Chittenden, G. J. F., 27

chlorite, 199, 229–32, 230*s*, 247–8*e*, 250

chlorophyll, 66–7, 335, 336*s*, 412

chloroplasts, 335

Choughuley, A. S. U., 27

chromic oxide-hydroxide, 172

chromium chlorite, 231, 233

chrysotile, 199, 209, 212, 214

Chukhrov, F. V., 177

chymotrypsin, 112

cilia, 108–9

citrate, 343, 358, 364

Clark, S. P., 16

classes of biochemicals, 66–7

classification and evolution, 99–100, 106

clay, 165*d*

clays: alive, how they might become so, 293–4, 419–20; assemblages of, 257–8; chirality of, 38–9, 204, 306; classification of layer silicate, 199; colloid chemistry of, 286–9; 293–4, 308–10; conditional properties of, 292–4; crystal structures of, 196–241; defect structures in, 269–70*e*, 275–83, 322*e*, 318–28; evolutionary possibilities with, 287–99, 417–21; experimental synthesis of, 168–71, 169*e*, 170*e*, 207, 212, 214–15, 240, 251–7, 253–5*e*, 418–19, 422–5;

clays (*cont.*)
> flexibility of layers in, 235*e*, 246–9*e*, 253–5*e*, 321*e*, 324*e*; genesis in Nature of, 172–8, 174*e*, 192–3, 194–5*e*, 197, 200–1*e*, 205–9, 208*e*, 210–11*e*, 233, 240, 243–51, 246–9*e*; hereditary effects in, 273–85, 300, 418; organic catalysts for synthesis of, 251, 310, 360, 423; organic interactions with, 4, 38–9, 47–51, 241–3, 308–10, 318, 383, 422–3; phosphate and polyphosphate binding to, 286–7, 309, 344–9; unsymmetrical layers in, 196–215, 227, 282–3, 320, 342; *see also* under specific names

clays as materials for first organisms, 4–5, 165; assemblages of, 165, 294–9; as catalysts, 313–14, 318, 422; charge patterns on, 319–26; conditional properties and, 294; disappearance of, 370, 387, 408; discovery of, on Earth, 420–1; for energy transducing membranes, 340–9; for excretory devices, 356; experimental approaches to, 418–23; as genes, 264–301, 370, 418–21; inertness and, 318, 342; organic polymers and, 372, 379–87; for organic-synthetic apparatus, 318–28, 349; for phenotypes, 294–8; for photosynthetic apparatus, 340–3; as protoribosomes, 391; for 'replication apparatus', 387; 'self-assembly' and, 319–25, 422–3; for separators, 356, 359, 410

cleavage of genetic crystals, 157–8, 284–5, 419
Clemency, C. V., 174, 258
cliachite, 249*e*, 250
clinoptilolite, 190, 194*e*
clock as co-operative machine, 91
coacervates, 12, 70–3
code word assignments, evolution of, 401–5
coenzymes, 10, 376–8*s*; as before protein, 370, 375; as C-class biochemicals, 66–7; functions of, 313, 375–9; 'handles' and 'labels' on, 375–9; as metabolic late entries, 369–70; as reagents, 313; *see also* under specific names
coenzyme A, 313, 358, 360, 364, 376*s*
Coleridge, S. T., 78, 91, 95
collector devices in primitive organisms, 314

colloidal minerals, 178; *see also* clay and clays
common ancestors, 36, 75, 100–4, 262
communications between molecules, 374
compartments: in authigenic clays, 246–50; for chemical oscillators, 349–50; for control of reactions, 311, 423; and evolution of biochemical pathways, 354–5, 359
complexity, 89*d*, 90–1, 98
composition plane, 266, 267*d*, 270–1, 280
concentrating mechanisms in primitive organisms, 313
concentrating ('focussing') information, 291–2, 394
concentrations of molecules in a 'primordial soup', 25, 31–2, 46–7, 62
condensation reactions (for biopolymers) in aqueous solution, 48–53
conditional properties, 294, 372
confirmation in science, 62–5
connecting molecules, 373, 383
conservatism in evolution, 105–6, 414
continuous crystallisers, 419–20
co-operation between parts, 91–9; arch model for systems with, 95–7, 116; in central biochemistry, 94, 116, 262; as characteristic of life, 2–3, 79, 93, 117; hierarchy in, 92; in machines, 2, 91–3
copolymer crystal genes, 151–3
copolymer block structures, 385
Corbridge, D. E. C., 309
Cordes, E. H., 67, 311–12
corundum, 176
Cotton, F. A., 166
coupling agents, 51–3, 345–6
Coyne, L. M., 423
Cozzarelli, N. R., 147
Cradwick, P. D. G., 236, 239
Cram, D. J., 60
Cram, J. M., 60
'crazy-paving' domains, 280–3
Crick, F. H. C., 13, 42, 147, 159, 385, 388, 394–5, 399, 403, 406
cristobalite, 187, 249*e*, 250
crocidolite, 187
cronstedtite, 199, 214
Crowther, J. P., 180
crystal: 'breeding', 284; cleavage, 157–8, 284–5, 419; defects, 237*e*, 265–85, 268–70*e*, 290–1, 304–6, 318–35, 322*e*, 418–423; genes, 151–63, 265–99, 418–20, *see also* genetic crystals; growth mechanisms, 155–9, 222–3, 256–7; habit, 284, 319, *see also*

crystal (*cont.*)
  morphologies of crystals; structure,
    265*d*; structure of minerals, 171–241,
    see also individual mineral names
crystallisation: of clays, 168–71, 192–7,
    205–9, 210–11*e*, 214–15, 240,
    243–58, 246–9*e*, 253–5*e*, 279–82;
    and error correction, 45, 155–9;
    mechanisms of, 155–9, 222–3,
    256–7; replication through, 45,
    151–60, 418–20
crystals as information stores, 265–72
Cu²⁺, polycondensation of glycine on
    clays with, 49–50
cyanamide, 51–3*s*
cyanate, 26
cyanide: abiotic syntheses and, 21–7,
    52–3, 55; as early biochemical?,
    369–71
cyanoacetaldehyde, 21*s*, 26
cyanoacetylene, 21*s*, 26
cyanogen, 26
cycles, biochemical, evolution of, 357
cysteine, 364, 396*s*
cytidine triphosphate, biosynthesis, 367
cytochrome *c*, 117
cytochrome *f*, 337–8
cytosine, 10*s*, 26*s*

Darbyshire, J. A., 270
Darge, W., 39, 44
Darwin, C., 1, 99, 103, 105–6, 125, 264,
    351–2, 367
Darwinian evolution, 12, 287, 300, 417,
    420, 422, *see also under* evolution
Darwin's pond revisited, 351–5
Dasgupta, D. R., 175
Datta, P., 203
Daubrée, A., 193
Dauvillier, A., 11–12
Davis, C. J., 141, 150, 152
Dawkins, R., 83, 263
Dayhoff, M. O., 107
de Bruyn, P. L., 170
De Kimpe, C. R., 192
Deer, W. A., 173, 175, 187, 190, 227–8
defect structure, 267*d*; as analogue of
    molecule structure, 304–5; chirality
    and, 204, 306; types, 266–70
defects in crystals: of beidellite, 269*e*,
    322*e*; disordered polytypes, 204, 217,
    274–5; domains, 160, 204, 226, 233,
    266–7, 272, 277, 278–83, 418; edge
    dislocations, 266; of feldspar, 267,
    270–1; of β-FeO(OH), 269*e*; as
    information stores, 265–72;
    intergrowths, 266–7*d*, 277–83; in

kaolinite, 204, 280–3; line, 267;
    ordering of, 266, 275–82, 418;
    point, 267; in pyrrhotite, 270*e*;
    replication of, 274–85, 305, 419–20;
    screw dislocations, 222, 266–7, 275,
    306; of sepiolite, 237*e*; stability of,
    271–2; substitutions, 266–7*d*; of
    tungsten-niobium oxide, 267, 268*e*;
    twinning and pseudotwinning,
    213–14, 266–7*d*, 270*e*, 271, 277–83;
    ultrastructures, 267*d*; vacancies,
    266–7*d*; of zeolites, 267
deflocculating (peptising) agents, 286–7,
    308–10
Degani, D., 68
Dent, L. S., 191
deoxyribose, 10*s*
Descartes, 91
design approaches, evolution of new,
    91–118, 262
design constraints on replicating
    polymers, 68, 140–6
design specifications for primary
    organisms, 136
diaspore, 172, 174, 175*s*
α-γ-diaminobutyric acid, 18
diaminofumaronitrile, 23*s*
diaminomaleonitrile, 23*s*
α-β-diaminopropionic acid, 18
diaminosuccinic acid, 24
diastereomers, 28*s*, 35, 306
dicarboxylic acids in Murchison
    meteorite, 33
Dickerson, R. E., 9, 397, 406
dickite, 199, 200*e*, 202–5, 203*s*
dicyanamide, 53*s*
dicyandiamide, 53*s*
dicyclohexylcarbodiimide, 52*s*
dihydrouracil, 27*s*
dihydroxyacetone, 30*s*
4,5-dihydroxypyrimidine, 24
dihydroxyuracil, 27
dilute solutions and clay synthesis, 251–7
dimethyl sulphoxide, 241
dioctahedral layer silicates, 196–9*d*, *s*;
    micas, 215–17
diopside, 185–6*s*
dislocations in crystals, 222, 266, 267*d*,
    275, 306
disordering of Al/Si in feldspar,
    activation energies, 272
disproportionation in formose reaction,
    31
Dixon, H. B. F., 363
Dixon, J. B., 205–6
DNA, 10*s*, 81–5, 117, 125, 132, 164, 272,
    276, 291, 297, *see also* nucleic

DNA (*cont.*)
    acid(s); replication of, 1, 42, 44, 94, 74, 146–9, 155–60, 273, 283–4; as starting model for primary genetic material, 136, 140
Dobzhansky, T., 1, 88
domains in crystals, 160, 204, 226, 233, 266–7*d*, 272, 278–83, 418
Donner, D. B., 23
Dose, K., 47–8
Doty, P., 41
double helix in polysaccharides, 379–80
doubts, 61–77; about Miller's experiment, 64–9; about 'molecules of life', 3–4, 75–7, 104, 116–17; about 'probiotic evolution', 69–75
Dousma, J., 170
Downes, C. J., 240
Dowty, E., 272
Drake, J. W., 147
Drits, V. A., 232
Dustin, P., 108

Earth: conditions on primitive, 3–6, 9–17; continuous crystallisation of clays on, 251, 420; cycles maintained by, 138–40; radioactive heating in, 328; tectonic cycle of, 251, 328
easily made molecules, 66–9
Eberl, D., 209
Edelson, E. H., 23–4
edge dislocations, 266
edge-face associations, 286–7, 319
edges of clays, binding polyanions at, 286–7, 309
efficiency as product of evolution, 85–9
efficiency range, 114*d*, 115–16
E_h control by mineral organisms, 352–4
Eichberg, J., 413
Eigen, M., 82, 405
electrons as reducing agents, 332–8
elephant's trunk, evolution of, 97
Emsley, J., 344
enantiomeric crystals, 37, 44, 184, 204, 306
enantiomeric defect structures, 306
enantiomeric machine components, 34–6, 40–1
enantiomeric molecules, 28–9, 34–45
enantiomers, 28*d*, 34–45; difficulties in resolving, 37–44, 56, 63; *see also* chirality
energy: for abiotic syntheses, 14, 17–19, 21–2, 25, 32–3, 48–53; coupling devices, 56–9, 345–9, 364 and defect structures, 270–2, 282–3

devices for catching solar, 330–3, *see also* photosynthesis problems for genetic crystals, 158–9; source for mineral organisms, 328
energy transducers, 312*d*, 313, 328–49; from clays, 329, 340–3, 423; membranes and, 335–43; in photosynthesis, 312, 335–9; polyisoprenoids in, 412–13
environment for life, 80, 137–40
environmental evolution, 69
environmental programming, 150–1, 315
enzymes: apparatus in place of, 310–28; late evolution of, 370, 408–12; nucleic acid replication as needing, 146–9; organic chemistry without, 310–17; tRNA ligase, 392–4; specificity of, 60, 311
epitaxy, 245, 295, 326, 422
Epps, D., 413
erionite, 191*s*, 193, 195*e*
error correction: in clay synthesis, 256; in crystal growth processes, 42, 155–9, 302; in DNA replication, 42, 147–8, 156–7
erythrose, 32
essential amino acids, 67, 114
ethanolamine, 10*s*, 363
ethylene glycol in clay interlayers, 240–1
N-ethyl-β-alanine, 18
ethylene glycol, 241
N-ethylglycine, 18, 20
Eugster, H. P., 218
evolution (biological), 1*d*, 2–6; absence of foresight in, 3–4, 97, 114, 262–3, 308, 310, 373, 390; accumulation of 'know how' during, 328; add-and-subtract processes and, 97; arch model for, 95–9, 106, 116, 124; beginnings of, 116; of biochemical cycles, 357; of biochemical pathways, *see* biochemical pathways, early evolution of; of central biochemistry, 357–71; and classification, 99–100, 106; of code word assignments, 401–5; and conservatism, 105–6, 414; of elephant's trunk, 97; of enzymes, 107–8; fixing central sybsystems in, 99; of functional interdependence, 93–9; of genetic codes, 395–408; of immunoglobulins, 107; and indirectness of genetic control, 285, 387; invariance of subsystems and, 104; invention during, 105–18, 262, 362–3; of keratin, 106; locus of active, 105; of lungs, 5, 110,

evolution (*cont.*)
    112–13, 373, 409; of microtubules,
    108–9; of multifunctionality, 99;
    products and prerequisites, 2, 79,
    98; of proteins, 110; of protein-
    synthetic machinery, 387–407, 424
    radical mechanisms, 104–18, 409;
    and 'recruitment' of enzymes, 108;
    of the ribosome, 390–2; of RNA
    molecules, 81–2; rope analogy for,
    5, 263; of secondary genetic
    materials, 121–4, 373–87; of serine
    proteases, 108; of subsystems, 105,
    116; takeovers in, 5, 112–18, 409;
    technological analogies, 110; trees
    and, 99–104; of triplet code, 401;
    two kinds of, 69; what does it do?,
    83–93; 'zero-technology' organisms
    at start of, 121; *see also* natural
    selection
'evolution' of mineral assemblages,
    358
evolutionary landscapes, *see* functional
    landscapes
evolvable systems *see* organism
evolved systems *see* life
excretion, 314, 356
expanding clays, 234*d*
experimental possibilities, 416–25
extinctions, 103
Ewing, F.J., 173, 175

FAD, 337, 376*s*
Farmer, V. C., 228, 240
fatty acids, 10*s*, 33, 243
faujasite, 191
fayalite, 184
Fe^{3+} in smectite clays, 342, 422
feedback control, 315–17
feedstocks for primary organisms, 314,
    349–50
Feeny, R. E., 363
Feitknecht, W., 177
feldspar, 187–8, 188*s*; authigenic
    synthesis of, 188, 193, 250;
    stability of Al/Si arrangements in,
    272; twinning of, 267, 270–1;
    weathering of, 174, 244
feldspathoids, 188–90, 189–90*s*
α-FeO(OH) (goethite), 168–9, 169*e*, 175*s*
β-FeO(OH) (akaganéite), 169, 170*e*, 175,
    176*s*, 269*e*
γ-FeO(OH) (lepidocrocite), 169, 173–4*s*
δ-FeO(OH), 175
Fernandez, M., 228
Ferenci, T., 358
ferredoxin, 337–8, 352, 358, 360

ferric oxide-hydroxide, 168–71, 169–70*e*,
    174–5, 352, *see also* FeO(OH)
Ferris, J. P., 17, 21–7
FH_4 377*s*, *see* tetrahydrofolic acid
Filmer, D., 224
fine and coarse adjustments in protein
    evolution, 86
fire analogy for life, 263–4
Fischer, W. R., 171
Fischer–Tropsch reactions, 33, 68
fishmonger's slab, idol of the, 101
fitness, 88–9*d*, 109
Fitzpatrick, R. W., 177
fixation of carbon dioxide, 335–9,
    357–62, 372
fixation of nitrogen, 332–3, 352, 362–5,
    372
fixing of subsystems through evolution,
    98–9, 103–6
flavin adenine dinucleotide, 337–8, 376*s*
Fleischer, R. L., 271
flexibility: of clay layers, 246–9*e*,
    253–5*e*, 321*e*, 324*e*; of organic
    polymers, 143–4, 382, 387
flocculating agents for clay suspensions,
    286, 309
Flores, J., 345
Flores, J. J., 37
'focussing (concentrating) information,
    291–2, 394
Folsome, C., 424
Fookes, C. J. R., 369
Force, E. R., 177
forethought (foresight, plan), absence of
    in evolution, 3–4, 97, 114, 262–3,
    308, 373, 390
form and substance, 117
formaldehyde: in abiotic syntheses,
    21–33; aldocyanoin from, 55; as an
    early biochemical?, 357–9, 370–1;
    from photochemical reduction of
    CO_2, 332; sugars from, 27–32, 358
formamide, 23, 241–2, 369–71
formic acid (formate), 15, 25, 32–3, 66,
    332, 366, 368, 370
formose reaction, 27–32, 357–9, 423
forsterite, 184
Foster, M. D., 232
Fournier, R. O., 184
Fox, G. E., 413
Fox, S. W., 17, 19, 48
framework silicates, 187–97, *see also*
    'feldspar' and 'zeolites'
Francombe, M. H., 175
Frank, F. C., 44
Fraser, A. R., 240
Fraser, R. D. B., 106

Freiman, H., 26
Fresco, J. R., 157
Freund, F., 212, 214
Frey, E., 319–20
Friebele, E., 38
Friedman, N., 17–18
fructose, 32
Fryer, J. R., 270
Fujishima, A., 332
fulvic acid as catalyst for clay synthesis, 251, 310, 360, 423
fumarate, 364
functional ambivalence, 107*d*, 108–12, 373
functional interdependence of subsystems, 2, 91–9, 116, 130; evolution of, 93–9, 124, 132
functional landscapes, 85–91; basins in, 85, 88, 109, 111; catchment areas in, 110–11, 289; 'earth movements' in, 88; interpenetrating, 108–11; multiple minima in, 88
functional overlap, 111*d*, 112–13, 116, 409
furanose ring forms, 29*d*, 57

Gabel, N. W., 27, 343
galactose, 32
D-galactose 2,6-disulphate, 379
D-galactose 4-sulphate, 379
D-galactose 6-sulphate, 379
Galbraith, S. T., 270
Galen, E., 209
Gardiner, W., 82
Gardner, M., 65
Garey, C. L., 280
garnet, 185
Garrels, R. M., 353
Garrison, W. M., 12, 32
Gatineau, L., 226, 233, 276
genealogy, 416–18
genes, 2*d*, 6, 70, 78, 136, 286–7, 408–9, *see also* naked genes
genesis of clay minerals in Nature, *see under* clays and authigenic minerals
genetic code, 396*s*; circumstances for evolution of, 129–30, 405–7; evolution of, 307, 395–408; as an evolved entity, 406; non-randomness in, 396–8; similarity of, in all organisms now, 36, 307
genetic control: direct and indirect, 285–9, 300–1, 326–7; primitive and advanced forms of, 285–6
genetic crystals, 151–63, 300–1, 326; block models for, 154–6, 160; bonding in, 160–1, 283–4; defect

structures in, 159–60, 265–94, 305–6, *see also* under dislocations, domains, twinning, substitutions; discovery of, 416–21; epitaxial growth on, 295; evolution of, 285–99; forming with non-genetic ones, 294–9; indirect control by, 295–8; 'intercalating synthesis' of, 418; interlayer ordering and, 275-7; kaolinite vermiforms as?, 279–85; morphology of, 278–9; polytypism and, 274–5; replication cycles for, 155, 157–8, 273–85, 419–20; type-1 and type-2, 154–8, 155*d*, 274–83; what millions might do, 286–91, 298
genetic definitions of life, 78–9, 124, *see also* genetic view of life
genetic information, 1*d*; copolymer block structure as, 385; defect structure as, 159–60, 264–72, 305–6; direct effects of, 81–2, 286–94; in DNA (or RNA), 1–2, 80–2, 84–5, 117, 276, 283–4, 372, 387–90; grain size of, 160, 163, 384, 395; indirect effects of, 285; meaning of, 81–2, 125, 285; as polymer length, 385; as software, 80; transmission of, 1–2, 57, 70–5
genetic instructions, 79–80, *see also* genetic information
genetic materials, 2*d*; essential requirements for, 137; evolution of (continuous), 77, 117, 125; evolution of (through takeovers), *see* genetic takeover; nucleic acids as, 1–2, 75, 83, 129–32, 136, 146–50, 164, 370; as prerequisites for evolution, 73–4, 121; primary, 6, 130, 136–40, 162, 164, 264, 300–1, 416–21; secondary, origin of, 120–5, 129–32, 372–87; *see also* genetic crystals
genetic takeover, vii, 6, 118, 120*d*, 121–35, 263; background to idea of, 1–118; change of view allowed by, 120–258; earliest evolution as seen using, 261–414; by nucleic acid, 121–5, 372–414; objections to, 125–35; plot of story of origin of life according to, 301; structure of metabolic pathways as evidence for, 370–1
genetic view of life, 11–12, 62, 70, 74, 408
genetically bimodal organisms, 123
genotype, 79*d*, 80–1
Gerstl, Z., 342
Getoff, N., 12, 32, 360

gibbsite, 171–4, 173*s*, 196–7, 202, 215, 231–2, 236–7, 239, 251
Giese, R. F., 203–5, 319
Gildawie, A. M., 170, 174
Gilham, P. T., 385
Gimblett, F. G. R., 181
'glassware' in organisms, 11, 311–12
glauconite, 234
glucose, 10*s*, 32, 38, 55, 68, 107, 362, 381
glutamic acid, 15, 18, 20, 22, 41, 48, 364, 396*s*, 397, 407
glutamine, 364, 366–8, 396*s*, 407
glyceraldehyde, 28*s*, 29–30, 32, 34
glyceraldehyde-3-phosphate, 339*s*
glycerate-1,3-diphosphate, 339*s*
glycerate-2-phosphate, 364*s*
glycerate-3-phosphate, 339*s*, 364
glycerol, 10*s*, 33
glycine, 13, 15, 18, 20, 21*s*, 22, 24, 33–4, 46, 48–50, 66, 345, 364–5, 366, 396-7, 407
glycolaldehyde, 22, 28–32
glycolic acid, 15, 33
glycolytic pathway, 68, 363–4
gmelinite, 191
God, assumption of non-involvement, 3, 417
goethite, 168, 169*e*, 175*s*
Golub, E. E., 13
Gomez Laverde, C., 228
Goodwin, A. M., 17
Gorshkov, A. I., 177
Gottardi, G., 192
'grain size' of organisms, 160, 163, 291–2, 384
Granick, S., 178, 329, 333–4, 338, 369
granite, 244
Granquist, W. T., 255
Grant, G. T., 380
greenalite, 199, 214
Greenland, D. J., 243
greigite, 178
Griffith, E. J., 343
Grim, R. E., 235, 324
grooves in crystals: of apatite, 345; of kaolinite, 200–1*e*, 280; in mica layers, 220–2; for organic-synthetic apparatus, 319; polymerisation in, 345, 384; for 'replication apparatus', 149–53, 387; in sepiolite, 237*e*; twinning as cause of, 278–80
Grossman, L., 16
groutite, 175
GTP (guanosine triphosphate): biosynthesis, 366; in protein synthesis, 390
guanidine, 24, 26

guanidinoacetic acid, 24
guanine, 10*s*, 24–6
guanosine-5′-phosphorimidazolide, 58
Gude, A. J., 193
Guggenheim, S., 227
L-guluronic acid, 379–80
Guth, T. D., 333, 352
Güven, N., 220–1, 235, 270, 322, 324
gyrase, 147

haem, 378*s*
haematite, 170-1, 176
Haldane, J. B. S., 11–12, 78, 91, 93
Hall, D. O., 178, 352, 360
Hall, S. H., 212–13, 279
halloysite, 199; experimental synthesis of, 207–9; genesis of in Nature, 205–9, 210–11*e*, 257; morphologies of, 205, 206–8*e*; in primitive photosynthesis?, 342
Halmann, M., 68, 332, 343–4
Halobacteria, 338
'handles' on molecules, 375, 412–13
Hanson, R. F., 200, 202, 209
Harada, K., 17
Harder, H., 214, 251–2, 256
Hardy, J. P., 13
Hare, P. E., 38
harmotome, 191, 193
Harrington, M. E., 38
Hartman, H., 13, 329, 360, 365, 405
Hartmann, W. K., 17
Hashimoto, K., 175
Haverland, W. J., 17–18
Hay, R. L., 193
Hayatsu, R., 17
Hazen, R. M., 229
hectorite, 199; catalysis by, 422; synthetic, 253–5*e*
Heilbron, I. M., 11
Heilmann, E. L., 177
α-helix, 41
Hem, J. D., 174
Hénin, S., 233, 251
Henmi, T., 237, 239–40
hereditary machinery: and chiral discrimination, 39–45; model evolution for, 70–5; origin of as key question, 4, 69–70, 264; *see also* genetic code, genetic crystals, genetic information *and* genetic materials
heredity, 1*d*; and long-term evolution, 69–75, 78–9
Heston, W. M., 182
heterocyclic nitrogen compounds: abiotic synthesis of, 23–7;

heterocyclic nitrogen compounds (*cont.*)
biosynthesis of, 94, 365–9; in
coenzymes, 376–8; interactions with
clays, 51, 241–3; uses of, for clay
organisms, 310, 365, 367–8, 382–7
heterotrophs, 59*d*, 329
heulandite, 190
hexa(U), oligomerisation of, 384–5
Hey, M. H., 190
hierarchical systems, 92
'high technology' systems, 111, 114–16
Hinkle, P. C., 337, 339
Hinshelwood, C. N., 136, 159, 274
histidine, 22, 67, 368, 370, 396*s*
Ho, M. W., 90
Hocart, M. R., 251
Hoffmann, R. W., 241
Holland, H. D., 16
Holmes, S., 421
Honda, K., 332
Hopfield, J. J., 157
Horiuchi, S., 270
hornblendes, 187
Horowitz, N. H., 11, 13, 59, 72, 78–9,
128
Hower, J., 209
Howie, R. A., 173, 175, 187, 190, 227–8
Hsu, P. H., 171
Hubbard, J. S., 13, 26, 32
Hudson, D. P., 11
Huertas, F., 251
Hulett, H. R., 46, 329
Hull, D. E., 46
Hutchison, J. L., 267
Huxley, J., 90
hydrargillite, *see* gibbsite
hydration of cations, 166
hydrazine, 241, 333
hydrogen bonds: in layer minerals,
172–5, 202, 204, 230–1, 241; in
organic polymers, 160, 283, 379,
382, 398
hydrophobic and hydrophilic amino
acids, 397–9, 402
hydrophobic (lipophilic) regions, 374–5,
382
hydrothermal synthesis, 171*d*, 192, 202,
206, 212
hydroxide crystal structures, 171–5*s*
hydroxo complexes, 166
hydroxy acids, abiotic synthesis of, 15,
21, 33
α-hydroxy-γ-aminobutyric acid, 18
α-hydroxybutyric acid, 15
hydroxyl (hydroxo) groups in minerals,
167–8, 170–5, 182, 197–8, 202–3,
221, 227–8

2-C-(hydroxymethyl)glyceraldehyde,
31*s*, 32
hydroxyuracil, 24, 27
Hynes, M. J., 112
hypoxanthine, 24–5

Ibanez, J., 19
Idelson, M., 41
Iiyama, J. T., 235
Iler, R. K., 182
illite, 199, 234, 244–5, 246*e*, 250
imidazole, 58
imidazole amides, 51
iminoaceticpropionic acid, 15
iminodiacetic acid, 15
immunoglobulins, evolution of, 107
imogolite, 236–9, 238*e*, 342; genesis of, 240
inert materials, usefulness of, 312, 318, 329
information: concentration ('focussing')
of, 291–2, 394; density, 152, 163,
384, 394; transfer in silicates,
221–6; transmission of, indefinitely,
1; *see also* genetic information
information capacity: of clays, 276–7,
290–2; of early polynucleotides,
405; of a primary genetic material,
125; of proteins, 291–2
information storage: by crystals, 265–72;
metastability and, 270–2, 283; by
organic polymers, 265
infrared spectroscopy of micas, 227–8
Ingram, P., 27
inorganic v. organic materials for
organisms, 4, 162–6, 302–8; for
first organic-synthetic machinery,
302–8, 317–18, 349; for first
photosynthesis, 342–3; for primary
genes, 160–3, 264
inosinic acid, 366*s*
Inoue, T., 332
'integrated circuits' in clays, 326
'intercalating synthesis' of smectites, 418
interdependences in central biochemistry,
94, 116
intergrowth, crystal, 266–71, 267*d*,
277–83
interlayer forces: in amesite, 212; in
beidellite, 322–3; charges and,
215–16, 230–5, 286, 319–26; in
chlorite, 230–2; in hydroxides and
oxide-hydroxides, 171–5; in kaolinite
minerals, 202–5; in micas, 215–29,
319; in rectorite, 320–2; and
replication, 282; and 'self-
assembly', 319–20, 324–5; in
smectites, 234–5; in vermiculite,
232–3

interlayer ordering: in amesite, 212–14, 279; in chlorite, 230–2; in kaolinite and dickite, 203–5, 279–81; in type-2 genetic crystals, 275–7, 418; in vermiculite, 233
internucleotide bond formation, 58, 148, 385
interstratified clays, 234
intralayer ordering: in amesite, 212–13; in chlorite, 231–2; in genetic crystals, 277; in kaolinite, 203–4, 280–1; in micas, 216, 220, 226–9; in vermiculite, 233; *see also* domains in crystals
invention in evolution, 105–16, 262, 372–3
iron oxides, 175–6, 326, 333
iron oxide-hydroxides, 168–71, 169–70*e*, 174–5, 269*e*, 352
iron sulphides, 178, 270*e*, 333, 352, 360
iron sulphur protein, 337
iron-bearing layer silicates, 214–15, 235, 252, 422; for photoredox machine?, 340–3, 422–3
isocitrate, 364
isoleucine, 18, 67, 393*s*, 394, 396
N-isopropylglycine, 18
isoserine, 18
isovaline, 18, 20
Israelachvili, J., 302

Jacobs, H., 203
Janot, C., 171
Jasmund, K., 212, 214
jeewanu, 424
Jefferson, D. A., 267
Jensen, R. A., 108, 112
Jernberg, N., 53–4
jig-saw puzzle machines, 41
Johnson, J. S., 178
Johnson, L. J., 257
Jonas, E. C., 245
Joshi, P. C. 23–4

Kalt, A., 184
Kamat, S. S., 22
kaolin, 202, 242; *see also* kaolinite
kaolinite, 196–8*s*, 199, 200-1*e*, 247*e*, 249*e*; 'books', 200–1*e*, 202; chirality of, 204, 306; disordered, 204–5, 281; domains in, 280–3; experimental synthesis of, 209, 251–3, 421; genesis of, in Nature, 197, 202, 207, 209, 244–50, 257–8, 421; interlayer forces in, 202–5; interlayer ordering in, 203–5, 279–81; intralayer ordering in,

203–4, 280–1; organic molecules in, 241–2; as primary genetic material?, 279, 300, 420–1; pseudotwinning in, 280–2; twinning in, 281–2; vacancy ordering in, 203–5, 280–1; vermiforms, 200–1*e*, 202, 279–85, 420
Katchalski, E., 48
Katchalsky, A., 51
Kavasmaneck, P. R., 37
Kawashiro, K., 37
Kay, W. W., 112
Kazi, Z. A., 27
Keller, W. D., 200, 202, 209–10, 280, 421
Kempster, C. J. E., 223
Kenyon, D. H., 9, 17, 44, 55
keratin, 106, 113
Kerr, I. S., 191
Kerr, R. A., 15
ketopentose, 30*s*
ketotetrose, 30*s*
Khare, B. N., 17
Kimball, A. P., 23
Kirkman, J. H., 205, 207–8
Kirsten, T., 17
Klausner, M., 51
Klein, H. P., 47
Kodama, H., 235
Konishi, S., 332
Koshland, D. E., 224
Kramer, F. R., 81
Krasnovsky, A. A., 338
Kraus, K. A., 178
Krebs acids: as early biochemicals?, 360, 370; as ligands, 310; as phosphate mobilisers, 343, 353
Krebs cycle, 357–8, 360, 363–4
Krebsbach, R., 345
Dr Kritic's objections, 125–35, 161–3, 373, 422
Krumbein, W. C., 353
Kuhn, H., 69, 150, 315
Kuhn, T. S., 61
Kunze, G., 212

La Iglesia, A., 209, 251–3
'labels' on molecules, 375–9, 385
lactic acid, 15, 33
ladder copolymers, 140–9
Lady G's diamond, 64
Lagaly, G., 243, 319–20
Lahav, N., 47–9, 330
Lailach, G. E., 242
Lambert, R. StJ., 227
landscape of evolution, *see* functional landscapes
Langridge, J., 111

last common ancestor, 36, 102–3, 262, 407

Laswick, J. A., 168

Lawless, J. G., 19, 20, 23–4, 33, 49–50, 243

layer hydroxides and hydroxides, 171–5, 172–3s, 174e

layer silicates, 196–258; *see also* under names given on *p.* 199

Le Port, L., 21

Le Roux, J., 177

Leach, W. W., 33

Lemmon, R. M., 9

Lengyel, J. A., 111

lepidocrocite, 169, 171, 173–4s

leucine, 18, 22, 38, 67, 396s, 397

leucite, 188–9

Levi, N., 49–50

levyne, 191

Lewontin, R. C., 88

Liecester, H. M., 63

life, 2d, 78–9d; fire analogy for, 263–4; machines and, 2, 5, 34, 76, 78–9, 85–7, 91–3, 107–11, 114–16, 261–2; as product of evolution, 2–3, 6, 79, *see also* under evolution; synthesis of, 416–25; when minerals would be a form of, 293–4, 298–300

'ligase machines', pre-protein forms of, 394–5, 403

Lin, E. C. C., 60, 112

Linares, J., 251

lipids: as C-class biochemicals, 66–7; as difficult to synthesise, 33, 132, 263, 317, 413; late entry for?, 369, 412–13; membrane-forming, 10s, 303s, 374; polyisoprenoid, 375, 412; polyketide, 412–13; -protein membranes, 338, 342, 412

Lipman, C. B., 11

Lipmann, F., 12, 343, 362–3, 365

lipoic acid, 378s

lipophilic groups as locating devices, 374–5, 382–3

Lister, J. S., 231

lizardite, 199, 209

Lobo, A. P., 23

locating systems, molecular, 373–84

Lodish, H. F., 312

Loeb, W., 17

Loft, B. C., 318

Lohrmann, R., 42, 57–8, 149

Lopez Aguayo, F., 251

Lowe, C. U., 23

Lundberg, R. D., 41

lung, evolution of, 5, 110, 112–13, 373, 409

lysine, 22, 48, 67, 396s

lysozyme catalysis, 153

lyxose, 29, 32

McBride, M. B., 342, 422

McCarty, R. B., 337, 339

McDonald, E., 369

machines, evolved organisms as, 2, 5, 34, 76, 78–9, 85–7, 91–3, 107–11, 114–16, 261–2

Mackay, A. L., 170, 175

McKee, T. R., 205–6

Mackenzie, D. W., 253–5

mackinawite, 178

Mackinney, G., 13

McLaughlin, P. J., 107

McMurchy, R. C., 230

Maddrell's salt, 179s

magnetite, 176–7s; and primitive photosynthesis, 329, 333–5

Magrum, L. J., 413

Mahler, H. R., 67, 311–12, 344

Maillard ('browning') reaction, 363

Maksimovic, Z., 212

malate, 364

Manebach twin law, 270–1

manganese oxides, 176–7; α-MnO_2, 176s; in primitive photosynthesis?, 333

mannose, 32

D-mannuronic acid, 379

Mansfield, C. F., 280–2, 292

marcasite, 178

margarite, 199, 227

Margulis, L., 100

marine sediments, 244–5, 299, 344

Markham, R., 23

Mars, destruction of molecules on, 47

Martin, F. S., 37

Martin-Vivaldi, J. L., 251

Marumoto, R., 23

Mason, B., 183

Matcham, G. W. J., 369

Mathews, C. N., 23

Mathieson, A. McL., 233

Matijevic, E., 170

Matsuura, T., 31

Maynard Smith, J., 11, 78–9, 90, 97, 110

mechanical models for silicates, 223–5

Megaw, H. D., 223

Meier, W. M., 190, 192

melanoidin, 55

membranes, 4; 'allosteric' effects in clay?, 348; asymmetry in photosynthetic, 335, 342, 423; clay, 235e, 246–9e, 250, 253–5e, 318–26,

membranes (*cont.*)
321*e*, 324*e*, 412, 423; clay-organic, 412; in energy coupling schemes, 345–9; lipid, 33, 302–4, 412; for organic-synthetic apparatus, 312; photosynthetic (modern), 335–9; photosynthetic (primitive), 329, 340–3, 423–4; primitive barrier, 327–8, 353–5, 412; proteins in, 342, 412
Mering, J., 233
Merrifield, R. B., 53–5, 59, 315, 390
Mesmer, R. E., 168
metabolic 'onion' model, 365–71
metal oxides, 175–7
metal sulphides, 178
metastability and information storage, 270–2
methane, 11, 13–18, 22, 25
methanol, 332
methionine, 67, 396*s*
method of overlap, 415–16
N-methyl urea, 15
N-methylalanine, 15, 18, 20
N-methyl-*β*-alanine, 18
N-methylglycine, 20
Meunier, A., 248
Meyers, P. A., 243
micas, 199, 216*s*; 'allosteric' effects in, 225, 229; dioctahedral, 215–21, 217, 225–7, see also muscovite and margarite; in granite, 244; infrared spectroscopy, 227–8; ordering of cations in, 225–7, 229; polytypism of, 217–23; trioctahedral, 227–9, see also biotite and phlogopite
micelles, 33, 374
microcline, 188
microfossils, 421
micromorphology of minerals, see under morphologies of crystals
microtubules, 108–9
Miller, K. R., 337
Miller, S. L., 9, 13–22, 26, 32, 44, 48, 56, 61, 64, 75, 132
Miller's experiment, significance of, 64–9, 75
Millot, G., 245, 257
Mills, D. R., 81
mineral organic v. inorganic materials, 161–2, 304
minerals, 164–256; as formose catalysts, 27; as materials for first organisms, 4–6, 161–6, 258, 261–414; see also clays
'mirror life', why none, 36
Misawa, T., 175
Mitchell, P., 335

mitochondria, 339
mixed layer clays, 234, 246*e*, 319–20
Miyazaki, S., 205, 209
Mizuno, T., 29, 31–2
modules, 92, 413
molecular cost effectiveness, 67
molecular life v. crystal life, 305
molecular recognition (discrimination): by adaptor ligase apparatus, 394–5, 403–5, 407; of amino acids, 308, 392–5, 397, 403–5, 407; chiral, 36–44, 305–8; with labels, 375–9, 383, 385, 395; and location, 373–87; by place, 350, 395; by polynucleotides, 375–9, 394; by primitive heterotrophs, 59–60; by proteins, 43–4; and reading message tapes, 400, 403–5; by tRNA ligases, 392–4; by sockets, 43–4, 59–60, 305, 308, 311, 374
'molecules of life', 9, 10*s*, 61, 104, 116, 132, 262, 414, 416
molybdate, 178, 243, 424
Monod, J., 224
montmorillonite, 199, 234–5, 235*e*, 245; in abiotic pyrimidine synthesis, 27; in bentonite, 235; cation substitutions in, 234–5, 270; organic molecules in, 51, 240–3, 399; polymerisation on, 51 (on bentonite, 48–50)
montrosite, 175
Moorbath, S., 17
Moore, B., 11, 13
mordenite, 191, 192*s*, 194–5*e*; and formose reaction, 423; synthesis in Nature, 193–5
Morey, G. W., 184
Morgan, T. H., 11
Morimoto, N., 270
Morimoto, S., 37
Morita, K., 23
morphologies of crystals: 'bags', 253*e*; 'barnacles', 249*e*; 'books', 247*e*, 200–1*e*; 'bundles', 195*e*; 'cabbage-heads', 248*e*; 'card houses', 247*e*; 'crazy-paving', 268–9*e*; crumpled sheets, 174*e*; 'doughnuts', 253; folded ribbons, 235*e*, 321*e*, 324*e*; 'holey', 237*e*; hollow spheres, 207*e*; membranes, 235*e*, 246–9*e*, 253–5*e*, 321*e*, 324*e*; 'picture framing', 237*e*; plates, 194*e*; pods, 238*e*; 'rat's nest', 195*e*; 'rosettes', 249*e*; 'seaweed', 249*e*; spiral discs, 208*e*; threads, 194–5*e*; tubes, 206*e*, 238*e*; vermiforms, 200–1*e*; webs, 247*e*; 'wet grass', 211*e*

morphology: control by genetic crystals, 295–8: of genetic crystals, 157, 278–9
Morris, E. R., 380
Moser, R. E., 23
mud: virtues of, 162, 258; vital, 298, 328, 330, 352
Muller, G., 197
Muller, H. J., 11, 70, 81
multifunctionality, 107–11; of bone, 108; of glucose, 107; of objects, 109–10; outcome of evolution, 99; of proteins, 107–8, 131
Mumpton, F. A., 194
Murchison meteorite, 19–20, 33
Murphy, P. J., 168–9
Murr, L. E., 270, 322
muscovite, 199, 215–16, 220*s*; corrugated layers in, 220-1; in granite, 244; and illite clays, 234, 244–5; K+ ions in, 220–4; ordering in, 219, 225–6, 276; polytypism of, 218–23
mutations, 2*d*, 108, 159, 403, 408

NAD(H), 313, 358, 367–9, 376*s*
NADP(H), 337–9, 369, 376*s*
Nagasawa, K., 205, 209
Nakai, M., 238
Nakashima, R., 31
Nakazawa, H., 270
'naked genes', 81*d*; of clay, 287–94, 326; as first organisms, 120–1, 138–9; of nucleic acid, 81–3, 127, 130; as simplest organisms, 81–3, 138–9
Nalovic, Lj., 171
Nathan, Y., 228
natrolite, 189–90, 190*s*
natural selection, 1–2*d*; conserver, 88, 107; engineer, 3, 55, 78, 130; and heredity, 1, 69–73; improver, 88, 107; inventor, 107–118; optimiser, 85–9; *see also* evolution, biological
Naylor, R., 385
Negrón-Mendoza, A., 33
Némethy, G., 224
nepheline, 189
Neuman, W. F., 345
Newnham, R. E., 203–4
Ni^{2+}, polycondensation of glycine on clays with, 52–3
Nicholas, H. J., 413
Nicholson, D. E., 368
Nicodem, D. E., 17
nicotinamide adenine dinucleotide (NAD(H)), 313, 358, 367–9, 376*s*
nicotinamide adenine dinucleotide phosphate (NADP(H)), 337–9, 369, 376*s*

Nikitina, A. P., 215
Ninio, J., 157
niobate layers, 243
Nir, S., 47
Nissenbaum, A., 47, 55
nitrogen (N_2): fixation of, 332–3, 352, 362–5, 372; in primitive atmosphere, 6, 16–19, 350
nitrogen-containing molecules, 13, 25, 362–3
p-nitrophenyl esters, 51
Nixon, H. L., 321
non-genetic clays, 294–7, 328, 422
non-genetic replication, 123, 385
non-layer silicates, 184–97
nontronite, 199, 235, 342
Nooner, D. W., 23, 33
Norleucine, 18
Norrish, K., 225
norvaline, 18, 20
nucleation in crystal growth, 155, 170–1, 158–9, 253, 307, 419
nucleic acid(s), 10*s*; bases, 10*s*, 23–7; dependence on (now), 3, 94; as 'enzymes', 375, 382, 391, 424; evolution (invention) of, 6, 122–4, 127–8, 381–7; implausibility of prevital, 56–60, 62; for minimal organisms?, 13, 45, 81; non-genetic uses for, 123, 375, 379, 385, 389–92, 394–5; precursors of, 368, 373–86; replication, 1, 42, 44, 57–8, 60, 67–8, 74, 146–9, 155–60, 273, 283–4; as secondary genetic material, 121; takeover by, 122–5, 370, 381–414; as usurpers, 371
nucleoside 5'-phosphorimidazolides, 58
nucleosides, 56–8
nucleotides: automatic synthesis of, 316; biosynthesis of, 366–7; as C-class biochemicals, 66; clay binding of, 242–3; coenzymes and, 369, 375–9; as difficult to make, 56–9, 67–80, 263, 316–17; late entry of, 361, 370
Nyholm, R. S., 342

objections: to genetic takeover, 125–35, 161–3, 373, 422; to inorganic organisms, 161–3
Occam, 76, 116, 263
Ochiai, M., 23
octahedral sites in crystals, 172–7, 184–5, 197, 203–4, 212–36, 280–2, 340–1, 348
Odom, D. G., 243
olation, 167*d*, 170
oligoanions, 178

oligomerisation: of cyanide, 23–7; of formaldehyde, 27–32; of glycine, 48–50; of hydrated cations, 167–8
olivine, 184–5, 185*s*
Ondetti, A., 51
onion model for metabolism, 365–71
O'Nions, R. K., 17
opal, 193
Oparin, A. I., 11–14, 62
Oparin-Haldane hypothesis, 11; *see* 'primordial soups'
open systems: and clay synthesis, 257, 420; and genetic crystals, 159; and life, 137–40; on the primitive Earth, 137–40, 159, 298–9; Sherrington's stream and, 80
O_2-producing photosynthesis, 335
optimising procedures, 85–9
'optional extras' in organisms, 99, 106, 307, 350, 386
ordered cations in layer silicates, *see* interlayer ordering, intralayer ordering *and* cation order and disorder
orderliness, 83–5; and entropy, 84; maximum (biological) of *E. coli*, 84–5; minimum (physicochemical) of *E. coli*, 84; as trivial product of evolution, 85
organic cations and clays, 241–3
organic molecules: as catalysts for clay synthesis, 251, 310, 360, 423; clays, interactions with, 4, 38, 47–51, 241–3; first uses for, 308–10, 351–5; v. inorganic crystals, vii, 4–5, 160–3, 302–5; as ligands, 310; mineral, 162; as 'optional extras', 307, 350; semi-chaotic mixtures of, 302, 360, 362, 367
organic polymers in primary organisms, 122–4, 308, 344, 361–2, 368, 372–3, 379–87, 390–407
organic replicating polymers: designs for, 140–53; discovery of, by primitive organisms, 384–7; preadaptations for, 384–7
organic synthesis: difficulties of, 27, 31–3, 45–60; rules for, 349–50; *see also* abiotic syntheses of 'biochemicals'
organic-synthetic apparatus: components for, 312–17; minerals for, 317–28; need for, 310–11; organic molecules and, 317–18
organisation, four aspects of, 83; complexity, 89–91; co-operation between parts, 91–3; efficiency, 85–9; orderliness, 83–5

organism, 1*d*; genetic definition, 79; nature of, 79–83; *see also* evolution (biological), organisms (evolved) and and primary organisms
organisms, evolved: previous kind needed, 124, 263–4, 405–7; teleonomic character of, 79; *see also* life
Orgel, L. E., 9, 15, 18, 21, 27, 32, 42, 44, 48, 57–8, 82, 149, 375, 395–400, 407
origami, 325*d*; micro-, 319–26
origin of life, 2–6; change of view on, 119–258; new experimental approaches to, 415–25; new story of, 259–414; problem of, 7–118
Origin of Species, 103, 105
original biochemical similarity, 9, 75–7, 261
Ormsby, W. C., 194
Oró, J., 19–23, 33, 243, 413
orotic acid, 24, 367*s*
orotidylic acid, 367*s*
orthosilicates, 184
osmotic construction, 327–8, 353–5, 424
Österberg, R., 57
'Ostwalt ripening', 249
Ourisson, G., 243
overlap, method of, 415–16
oxalate (oxalic acid): as catalyst for haematite synthesis, 171; as catalyst for kaolinite synthesis, 251; as phosphate mobiliser, 343; as ultraviolet photoproduct, 12, 32
oxaloacetate, 364
oxoglutarate, 364
oxolation, 167*d*, 170
oxyanions, polymerisation of, 178–84

Paecht-Horowitz, M., 51
Palm, C., 17
palmitate, 10
Pankhurst, R. J., 17
paramolybdate anion, 178*s*
para-montrosite, 175
Parfitt, R. L., 239–40
Parham, W. E., 207, 295
parsimony of Nature, 134–5
Pasteur, L., 37
Pease, R. W., 270, 322
Pedro, G., 171
Penrose, L. S., 151
peptide synthesis, 50–3, 387–407
peptides, 51, 59, 372
peptising (deflocculating) agents, 286–7, 308–10
Perati, B., 184
Perfil'ev, Yu D., 215

peripheral metabolites, 367, 370, 386; *see also* secondary metabolism
Perkins, R. R., 39
Perutz, M. F., 224
Peterson, E., 19–20, 52
Pezerat, H., 235
Pfeil, E., 31
pH control by primitive organisms, 352–4
phenotype, 79–80*d*; early kinds of, 294–9; functions of, 79–83; role in genetic takeover, 120
phenylalanine, 22, 67, 396*s*
philine, 399*d*, 400–4
Phillips, D. C., 153
phillipsite, 191, 193
phlogiston theory, 63
phlogopite, 199, 227, 233
phobine, 399*d*, 400–4
phosphate, 10*s*; concentrations of in Nature, 343–4; coupling agents from, 52–3, 345–6; in early organisms, 308–10, 343–51, 361–2, 370; mobilising agents for, 343–4, 353; in polymers, 382–4; *see also* polyphosphate and polyphosphate hydrolysis
phosphatidylethanolamine, 303*s*
phosphoenolpyruvate, 364*s*
photocatalysts, clays as, 423
photochemistry: abiotic production of organic molecules, 11–13, 17, 22, 26–7, 32–3; destruction of organic molecules, 17, 46–7, 58; fixation of carbon, 332, 351–2, 357–60; fixation of nitrogen, 332–3, 352; fuels, 330
photoelectrochemical devices, 331–5
photosynthesis: bacterial, 357–8; magnetite and primitive, 329; non-biological, 328–9; plant, 335–9; as source of first organic biochemicals, vii, 11–12, 328–335, 340–5, 351–4, 357–60
photosynthetic membrane: modern, 335–9; primitive, 329, 340–3, 423–4
photosynthetic pigments, 335–6
photothermal effects, 330
phototransducers, 328–45, 351–5
photovoltaic effects, 330–1, 351
phytyl chain, 375
Pincock, R. E., 39
Pinnavaia, T. J., 243
pipecolic acid, 20
Pirie, N. W., 75, 110
Pittman, E. D., 246–50
plagioclase, 188
Plançon, A., 204
Plane, R. A., 168

plants, 351*d*; first kinds of?, 351–5
plastocyanin, 337–8
plastoquinone, 336*s*, 337, 412
Pollack, S. S., 255
poly(A), 57, 385
poly(ADP-ribose), 368*s*
polyalcohols in formose, 28, 31
polycarboxylic acids, 33
poly(G), 57
polyisoprenoid chains, 336*s*, 375*s*; as earlier lipid type?, 412–13
polyketide chains, 303*s*; as post-revolutionaries?, 413
poly(leucine-glutamic acid), 398, 407
poly(leucine-lysine), 398, 407
polymerisation (inorganic), 161; of anions, 178–84; of cations, 166–78; and clay synthesis, 256–7
polymerisation (organic), 160–1; abiotic, trouble with, 45–59; in primitive organisms, 362, 384–7
polynucleotides: as 'handles', 375–9; implausibility of prevital, 56–60, 62; as labels, 390; as locating devices, 382–4; *see also* nucleic acids
polypeptides: alternating, 397–9; from γ-benzyl glutamic acid anhydride, 41; *see also* protein, proprotein
polyphosphate: clay binding of, 286–7, 309, 423; as first 'organic' biochemical?, 344–5; in Maddrell's salt, 179*s*; in nucleic acid fore-runners?, 361, 368, 383; peptising effect of, 309; stability of, 179–81
polyphosphate hydrolysis, early uses for?: as ATP forerunner, 343; in coupling agents, 345–8; for making oxidising and reducing agents, 348–9; for pumps, 348; as rheology controller, 351
polysaccharides, 10; double helices in, 379–80; gel-forming, 379–81; implausibility of prevital, 55; as locator/connectors, 379–83; structural uses for, 362, 382–3
polyethene as information store, 265
polytypism: and 'allosteric' effects, 224–5, 276; in chlorites, 230–1; crystal growth mechanism for, 222–3; and genetic crystals, 274–5; in kaolinite clays, 203–5, 225; in micas, 217–19
poly(valine-lysine), 398–9
pond, Darwin's revisited, 351–5
Ponnamperuma, C., 26–7, 33, 38, 47, 52–3, 57, 343
pores: in allophane, 239; in bundles of crystals, 237, 319; in imogolite, 239;

pores (*cont.*)
octahedral vacancy clusters as?,
348; in sepiolite, 236–7*e*; in
zeolites, 189–92
porphyrins, evolution of, 369, **413**
positive holes, 326, 331–8
Posner, A. M., 168–70
preadaptation(s), 107*d*, 116, 409; of
environment, 124, 151; for primary
and secondary genetic materials,
124–5; for protein synthesis, 123–4,
373, 390–2; for replicating organic
polymers, 122–3, 149, 373, 384–7
Preisinger, A., 236
Price, P. B., 271
primary genes, 120, 129
primary genetic materials, 121*d*, 136–63;
clays as, 4, 264–99; crystals as,
151–60; discovery of, 418–20; as
inorganic, 160–3; as key to origin of
life, 130, 136; as minerals, 4–6, 124;
'preadaptations' for, 124, 151; size
of writing in, 160, 163, 291–2; *see
also* genetic crystals
primary organisms, 121*d*, 122; as made
of clay, 4–6, 164–5, 258, 287–98;
'naked genes' as, 81–3, 120–1,
138–9, 287–94; places for, 298–9;
search for fossils of, 421; search, for
on Earth, 420–1; size of, 298; as
source of first organic biochemicals,
vii, 329; synthesis of, 416–18; as
'vital mud', 298, 328, 330, 352
primitive and advanced technologies,
5–6, 110–11, 114–16
primitive Earth, conditions on, 3–6,
11–17
'primordial soups', 11–60, 69–70; doing
without, 68, 328–9; 'microsoups',
47, 329; scepticism about, vii, 3,
46–7, 52–60, 65, 67–8, 137–8, 363
'printed circuits' on clays, 326
probiotic evolution, 69
'probiotic pathway', 68
production lines in primitive organisms,
313–15, 355, 359
proline, 18, 20, 396*s*
'proof reading': in amino acid
activation, 393–4; in crystal growth
processes, 155–8; in DNA
replication, 147–8, 156–7
propionic acid, 15, 33
proproteins, 397*d*, 398–407
N-propylglycine, 18
protein(s), 10*s*, 98*d*; in first organisms?,
13, 45, 116; evolution of, 86–7,
107–8; invention of, 128, 387–408;

in membranes, 59–60, 342, 412; and
molecular recognition, 305–6, 374;
multifunctionality of, 106–8, 131;
as 'optional extra', 106; organisms
without, 4, 6, 98, 406; revolution,
98, 124, 387–414; tertiary structure
specification in, 305–6
proteinoid, 48
protein-synthetic machinery: evolution
of, 97–8, 390–407, 424; in modern
organisms, 387–96; preadaptations
for, 373, 390–2; as prerequisite now,
59, 93–4; RNA and, 389–95
proton potentials, 312, 337–9, 423
proton pumps, 339, 413
protopolynucleotides, 385–7
protoribosomes, 391, 400
PRPP, 366*s*, 367–8
pseudophiline, 403
pseudorutile, 177
pseudotwinning in kaolinite, 280–2
pumps in primitive organisms, 312–13,
330, 348, 413
purification, abiotic, trouble with, 45–6
purines: abiotic synthesis of, 23–6;
biosynthesis of, 365–8
puzzles and anomalies, 61
PVC as information store, 265
pyranose ring forms, 29
pyridoxal phosphate, 369, 377*s*
pyridine (pyridinium), 241, 369
pyridine-2,3-dicarboxylic acid, 369
pyrimidines: abiotic syntheses of, 23–4,
26–7; biosynthesis of, 367–8
pyrite, 178
pyrophyllite, 198*s*, 199, 215
pyroxene, 185, 186*s*
pyrrhotite, 178; microtwin boundaries in
270*e*
pyruvate, 364

quartz, 181–4, 183*s*, 187, 193, 197, 244,
250; chirality of crystals of, 37, 306
Quayle, J. H., 358
Quinn, J. G., 243
Quirk, J. P., 168–70

Rabinowitz, J., 345
racemic mixtures, 29*d*, 37–45
Radoslovich, E. W., 223–4, 226, 229
Ranganayaki, S., 424
Rao, K. J., 267
Rao, K. K., 178, 352, 360
Rao, M., 243
Raulin, F., 14
Rausch, W. V., 168
Raussel-Colom, J. A., 228

Rautureau, M., 237
reagents in primitive organisms, 313
recognition of molecules
'recruitment' of enzymes, 108
recrystallisation of clays in Nature, 244–51
rectorite, 320–2, 321e
redox potentials of cations in clays, 342–3, 348
redox processes: model for polyphosphate coupled, 348–9; models for light-driven, 331–5, 340–3, 351–3; in plant photosynthesis, 335–8
reducing atmosphere, was there ever one?, 11–17
reduction of CO_2: in bacterial photosynthesis, 357–8; in hypothetical primitive organisms, 351–2, 357, 359; in plant photosynthesis, 335–9; by ultraviolet light, 32, 332
redundancy in evolution, 99, 112–15, 409
redundancy of information, 151–2, 160, 163
Rees, D. A., 379–81
Rees, M. W., 23
Reiche, H., 17
Reid, C., 27
Rein, R., 47
Rensch, B., 110, 112
'replicase catalysts', 123
replicating molecules: are they needed?, 70–4; design principles for, 140–51; difficulties in making, 144–6, 317; preadaptations for, 122–3, 149, 373–87; *see also* nucleic acids
replicating structures, expectation of in organisms, 73–5
replication, 72d; through crystallisation, 151–61, 273–85, 305, 418–20; of DNA, 146–9; phenotypic, 123, 385; of RNA, 148; in smectite clays, experimental demonstration of, 418
replication apparatus, 147d, 148–9; primitive versions of, 149–50, 386–7, 402
replication cycles, 150–1, 157–8, 419–20
residence time for molecules in oceans, 47
resolution of enantiomers by crystallisation, 37
reversible bonding: and crystal growth processes, 42, 153–7, 304; in genetic crystals, 155, 157–61; and 'self-assembly', 304
revolution, the biochemical, 372–414; conservative consequences of, 414; lipids as finishing, 413; nucleic acids

as seeds of, 373–87; proteins as agents of, 387–414
Reynolds, R. C., 240
ribose, 10s, 29, 30, 32, 56
ribosomes, 55, 389–91; preadaptations for, 390–2; primitive versions of, 391, 400
ribulose-1,5-diphosphate, 339s
ribulose-5-phosphate, 339s
Rich, A., 392
Richmond, M. H., 100
Rickard, D. T., 178, 352
riebeckite, 186
Rimsaite, J., 209
Ring, D., 17–18
ring structures of sugars, 29, 57
RNA, 10s; evolution of, 81; as genetic material, 148; roles in protein synthesis, 389–94; *see also* nucleic acids
RNA-like molecules: and coenzymes, 375–9; in primitive organisms, 375–9, 381–7, 392–5, 399–405
Rogers, G., 345
Rohlfing, D. L., 48
Roloff, G., 243
Rooksby, H. P., 175
rope analogy for evolution, 5, 263
Rosen, B. P., 60
Ross, G. J., 168
Ross, M., 218, 225
rotation of tetrahedra in layer silicates, 202–3, 219, 229
Rothman, J. E., 312
Rousseaux, J. M., 228
Rouxhet, P. G., 203, 228
Rowe, J. J., 184
Roy, D. M., 212
Roy, R., 212
Rubey, W. W., 16
Ruckert, H., 31
Rudnitskaya, E. S., 177
rutile, 177, 333

Sagan, C., 17, 19
Sakazawa, C., 31
Samudacheata, N., 203
Sanchez, R. A., 21, 23, 25–6
sandstone, clay synthesis in, 246–51, 288
sanidine, 188s
Sanz, J., 228
saponite, 199, 235
sarcosine, 15, 18
sauconite, 199
Saunders, P. T., 90
'scaffolding' for building arches, 97–8, 116, 124, 414

scale of primary organisms, 163, 298
Scharf, G., 31
Scheiner, P., 170
Schidlowski, M., 17
Schindler, P., 177
Schmalhausen, I. I., 112
Schopf, J. W., 421
Schrauzer, G. N., 332, 352
Schuster, P., 82
Schwartz, A. W., 17, 27, 343
Schwertmann, U., 71
scissors, chirality of, 34
screw dislocations, 266*d*; chirality of,
    306; crystal growth through, 222–3;
    in genetic crystals?, 275; replication
    of, 275, 306
secondary genes, 120, 129
secondary genetic materials: nucleic
    acids as, 126; preadaptations for,
    122–5, 149, 373–87; primary life as
    search device for, 131; as
    structurally unlike primary
    materials, 4, 118, 124, 130–1,
    160–3, 302
secondary metabolism, 113, 355, 386, 423
seeding in crystal growth, 245, 418;
    v. polymer growth, 42–5; *see also*
    nucleation in crystal growth
'self-assembly', 302–5; with clays,
    319–26, 422; with lipids, 320–3; with
    lipids, 320–3; with proproteins, 397;
    with proteins, 374; requirements
    for, 304
self-replicating organic polymers, 140*d*;
    DNA as not one, 140, 146–9;
    search for (in vain), 140–6
'self-starting' organisms, *see* primary
    organisms
semi-chaotic organic mixtures, 302,
    317–18, 356, 360
semiconductors, 326*d*; in light
    transducers, 330–4; minerals as,
    326, 333, 338; for primitive
    photosynthesis?, 332–4, 338–43, 345
separators in primitive organisms, 314,
    359, 410
sepiolite, 236*s*, 237*e*, 387
sequencing in organic synthesis, 46, 55,
    313, 315, 355, 359
serial arguments, 162
serine, 18, 22, 364, 393*s*, 396–7
serpentines, 199, 209, 212–15
Serratosa, J. M., 228, 233, 236
Sheppard, R. A., 193
Sherrington, C. S., 80
Sherwood, E., 52, 413
Shigemasa, Y., 31

Shimodaira, S., 175
Shimoyama, A., 38
Shirozu, H., 233, 247
shock waves, 17
short cut thinking, 243–4
Sieber, W., 192
Sieskind, O., 243
Siffert, B., 182, 251, 256
silica, 181–4; in clay synthesis, 251–6;
    crystalline forms of, 184; *see also*
    cristobalite and quartz; 'garden',
    327–8, 353–4; gels, 181–2;
    inhibitor for lepidocrocite, 171;
    monomeric, 251
    Silica Springs, 240
silicate minerals, 184–258
silicate polymers, 181–4
silicic acid, 4, 181–4; polymer, 184
Sillen, L. G., 46, 329
siloxane network, 219, 223
Simon, H. A., 92
Sipling, P. J., 272
Sivtsov, A. V., 177
size of primary organisms, 298
smectite(s), 199, 234–5; authigenic,
    246–7*e*, 250; experimental synthesis
    of, 254–5*e*, 418; genesis in Nature,
    250, 257–8; -illite, 246*e*; illites from,
    245; 'intercalating synthesis' of,
    418; mixed layer, 319–20; organic
    interactions with, 241–3; as
    photosynthetic membrane?, 342;
    replication in, 418; 'self-assembly'
    of, 319–20
Smith, A., 424
Smith, D. C., 100
Smith, J. V., 191, 217–18, 222
Smith, P. J. C., 380
Smith, R. W., 174
Smithies, O., 394
smythite, 178
snarks, problems in purchasing, 145
Snell, D. S., 254
sockets, molecular, 43–4, 59–60, 305,
    308, 311, 374
sodalite, 189*s*
sodium ammonium tartrate, 37
'software', persistance of, 80, 89; *see
    also* form and substance
Sokolova, G. V., 232
solar cells, 330–1
solar fuels, 330
solid state: biochemistry, vii; *see also*
    crystal
Solomon, D. H., 318
somatoids, 170*e*
Sonneborn, T. M., 402

sorbose, 32
space, protein, 85–7
Spacemen, assumption of non-
    involvement, 3, 417
Spach, G., 398
sparking experiments, 13–18
specificity, problems about, 52, 73, 314;
    see also molecular recognition
Spiegelman, S., 81, 128
spinel, 176, 177*s*
Spirin, A. S., 392
Spiro, T. G., 169
stability: of defect structures, 270–2; of
    substance v. form, 117
Stamadiadou, M. N., 47
standardisation of parts, 93
stannic oxide, 178
starter organisms, *see* primary
    organisms
Stebbins, G. L., 88
Steinfink, H., 231
Steinman, G., 9, 17, 44
stereoisomers, 28*d*
Sternglanz, R., 147
Stewart, J. M., 53–4
Stillinger, F. H., 150
Stoeckenius, W., 338
stone age biochemistry, 166
stone arch model, 95–9
stresses and strains in silicates, 223–4
Stroud, R. M., 108
structural polymers: alternating poly-
    peptides as, 397–9, 407; chiral
    purity and, 122, 362, 399; nucleic
    acid precursors as, 122–3, 381–4;
    polysaccharides as, 122, 362, 372,
    379–83; proteins and peptides as,
    342, 372, 397–9, 407, 409–10, 412
β-structures, 397–9, 398*s*, 407, 409–10,
    412
Strunz, H., 270
Stucki, J. W., 342
Studier, M. H., 17
Subbaraman, A. S., 27
substance and form, 117
substitutional crystals, 159
substitutions in crystals, 171, 266, 267*d*
subsystems, evolution of, 5, 91–9, 105,
    111–12
succinate (succinic acid), 15, 33, 364
succinyl-CoA, 360, 364
sugars: as B-class biochemicals, 66;
    early uses for, 361–2; from
    formaldehyde, 27–32, 357–9;
    instability of, 31, 360; reactions
    with amino acids, 363; variety of,
    31, 360

sulphides, 178; roles for, in primitive
    organisms?, 178, 333, 350, 352, 360
sulphur plants, 352
Sulston, J. E., 375
supersaturation levels: for clay
    synthesis, 251–8, 295; and crystal
    growth, 155; for genetic crystals,
    159, 273; for primary organisms,
    295, 328, 420; stabilisation of, 257,
    295; of weathering solutions, 257,
    273, 328
Suquet, H., 235
surface nucleation, 155–8
'survival machines', 263, 298
Swift, J. D., 318
symbiosis: and evolution of biochemical
    pathways, 351–5; and genetic
    takeover, 129, 408
synthesis (formation, genesis) of
    minerals in Nature: allophane,
    238*e*, 240; analcite, 193, 194*e*, 197;
    anatase, 177; berthierine, 214;
    boehmite, 172–3; brucite, 177;
    chabazite, 193; chlorite, 229,
    247–8*e*, 250; cliachite, 249*e*, 250;
    clinoptilolite, 193, 194*e*;
    cristobalite, 249*e*, 250; cronstedtite,
    214; diaspore, 172–3; dickite, 200*e*;
    diopside, 186; erionite, 193, 194*e*;
    feldspars, 188, 193, 250; gibbsite,
    172, 174; goethite, 175; greenalite,
    214; greigite, 178; halloysite,
    205–11, 204–8*e*, 210–11*e*, 257;
    harmotome, 193; illite, 244–5, 246*e*,
    250; imogolite, 238*e*, 240;
    kaolinite, 200–1*e*, 202, 244–5, 247*e*,
    249*e*, 257–8; lepidocrocite, 174;
    mackinawite, 178; manganese
    oxides etc., 177; marcasite, 178;
    montmorillonite, 245; mordenite,
    193, 194–5*e*; muscovite, 215; opal,
    193; phillipsite, 193; pseudorutile,
    177; pyrite, 178; pyroxenes, 186;
    pyrrhotite, 178; quartz, 193, 197, 250;
    rutile, 177; sepiolite, 236; serpen-
    tines, 209; smectite, 246–7*e*, 250,
    257–8; smythite, 178; vermiculite,
    233, 245, 247–8; zeolites, 193–7
synthetic minerals (from aqueous
    solutions): akaganéite (β-FeO(OH)),
    169, 170*e*; chrysotiles, 212, 214;
    clays, 251–7; goethite, 168, 169*e*;
    greenalite, 214; greigite, 178;
    haematite, 170–1; halloysite, 207,
    209; hectorite, 253–5*e*, 255;
    imogolite, 240; iron-bearing
    serpentines, 214–15; kaolinite, 209,

synthetic materials (*cont.*)
251–3; lepidocrocite, 169–71, mackinawite, 178; marcasite, 178; pyrite, 178; pyrrhotite, 178; serpentines, 212, 214–15; smectite, 254–5*e*, 418; smythite, 178; sols of, 168, 170; zeolites, 192–3
system invariance, 262
Sylla, H. M., 212, 214
Szent-Györgi, A., 365

tagatose, 32
Tait, J. M., 240
Takeda, H., 218, 225
takeovers in evolution, 104, 112–14, 117, 263, 409, *see also* genetic takeover
talc, 199, 227
tar as an organic product, 13, 20, 27, 302, 304, 355–6
Taylor, P., 178
Taylor, W. H., 188, 267, 270
Tazaki K., 240, 295
Tchoubar, C., 204, 237
technological analogies for evolution, 2, 5, 110, 114–16, 129, 262
tectonic cycle, 328
teichoic acid, 361*s*, 362, 382
teleonomic definitions of life, 78–9, 124
tetrahedral sites in crystals, 172–7, 184–5, 212–16, 220, 226–7, 230–5, 247–8, 276, 322–3, 240–1
tetrahydrofolic acid, 366, 368, 377*s*
tetrapyrrols, 369
Thalmann, H., 171
Theng, B. K. G., 242, 318
thermodynamics, 83, 244, 253, 257
thiamine pyrophosphate, 369, 377*s*
Thiemann, W., 39, 44
Thilo, E., 179–80
Thom, D., 380
Thomas, J. M., 267
Thompson, T. D., 242
threonine, 18, 67, 396*s*
thylakoid membrane, 335–8, 342
thymine, 10*s*, 27
titanate layers, 243
titanium dioxide, 17, 177, 333
Tobe, M. L., 342
Topal, M. D., 157
Toupance, G., 14
translation, protein synthesis as, 388–90
translocation in protein synthesis, 389; preadaptations for?, 391–2; way of avoiding?, 400
transport proteins, 59–60, 112, 311, 339
trees, evolutionary, 99–104, 131

tremolite, 186
tricarballylic acid, 33
triglycerides on clays, 243
trimetaphosphate coupling agent, 345–6
trioctahedral layer silicates, 197, 199; micas, 227–9
triplet code evolution, 401
Troland, L. T., 11
trypsin, 112
tryptophan, 22, 368–9, 396*s*
Tseng, S-S, 32
tubular crystals: chrysotile, 209, 212; halloysite, 205, 206*e*; imogolite, 236–9, 238*e*, 239*s*
tuff, 193*d*, 299
Turekian, K. K., 16
Turner, R. C., 168
'twin laws', 267, 270–1
twinned crystals, 213–14, 266, 267*d*, 270–1, 277–83, 384
type-1 and type-2 genetic crystals, 154–8, 155*d*, 274–85
typewriters, problems of manufacture, 35, 40–1
tyrosine, 22, 393*s*, 396

UDP-glucose, 377*s*
ultimate ancestors, 101, 103, 417
ultrastructure(s), 267*d*
ultraviolet light: destructive effects of, 47, 58, 333–4; synthesis of molecules using, 11–13, 17, 27, 332–4
unambiguous polynucleotides, 400
unity of biochemistry as artefact, 104, 262
unthought-of explanations, 64–5
updating control of metabolism, 410–12
uracil, 10*s*, 26–7; half life of, 26
urea, 15, 27, 241
Urey, H. C., 11, 13–14
UTP (uridine triphosphate) biosynthesis, 367

vacancies in crystals, 266, 267*d*
vacant octahedral sites: in chlorites, 231; in dioctahedral micas, 216–20; in gibbsite, 172–3; in kaolinite clays, 202–5, 280–1; in trioctahedral micas, 227–9
Valentine, J. W., 88
valine, 18, 20, 67, 393*s*, 396–7
Van Dort, M. A., 38
Van Olphen, H., 286–7, 309, 318
Van Oosterwyck-Gastuche, M. C., 252–3
Van Wazer, J. R., 179–81

variable characteristics in evolution, 105–6, 386
Vedder, W., 228
Veitch, L. G., 229
Velde, B., 218, 248, 250
vermiculite, 199, 232–3, 245, 247–8
Verwey, E. J. W., 177
'vital mud', 298, 328, 330
vitamins: B_{12}, 378s; as C-class biochemicals, 67; and redundancy, 114
Voecks, G. E., 13

Wada, K., 237, 239
Wada, S. I., 239
Waddington, C. H., 78–9, 89
Wagener, K., 39
Wakamatsu, H., 23
Wald, G., 37–44
Walker, G. F., 233
Walker, G. L., 27, 31, 357
Walker, J. C. G., 16
Walker, R. M., 271
Wasserman, Z., 150
Watanabe, E., 270
water: in allophane, 239; in crystal growth, 161, 283; cycle, 80; interlayer (clays), 232–5, 241; as medium for mineral syntheses, *see under* synthesis and synthetic minerals; pumps, 312–13, 330, 348, 413; in sepiolite, 236s; in zeolites, 189, 190s; *see also* weathering of rocks
Watson, D. M. S., 97
Watson, J. D., 13, 42, 108, 159, 385
weathering of rocks, 4, 6, 165–6, 172–7, 207, 240, 244–51, 257, 273, 299, 328
Webster, T. A., 11
Weir, A. H., 321
Weiss, A., 242–3, 418
Weiss, A. H., 29–32, 423
Wells, A. F., 183
Wells, N., 240
Welsh, E. J., 379
Wengeler, H., 423
Westman, A. E. R., 180
Wey, R., 182, 184, 251, 256
White, D., 48–9, 330
White, H. B., 375, 382
Wickner, W., 412

Wieker, W., 180
Wilkins, R. W. T., 228
Wilkinson, G., 166
Williams, D. G., 280
Wilson, D. B., 60
Wilson, M. D., 246–50
Windsor, C. R., 19
Winkler-Oswatitsch, R., 82
Winterburn, P. J., 379
Woese, C. R., 395, 397, 406
Wolfe, R. W., 204–5
Wolman, Y., 17–19
Wones, D. R., 218, 229
Woods, R. D., 321
work-up procedures, 46, 52, 61, 314
Wos, J. P., 23
Wright, S., 88
Wyman, J., 224

xanthine, 24–5
xylose, 29, 32

Yamagata, Y., 39
Yearian, M. R., 38
Yershova, Z. P., 215, 343
yields in abiotic syntheses, 15, 18, 22, 24–5, 27, 31–3, 49–50
Yoder, H. S., 217–18, 222
Yoshida, H., 37
Yoshinaga, N., 238
Youatt, J. B., 38
Yuen, G. U., 33
Yund, R. A., 272

Zamek, O. S., 26
Zeeman, E. C., 63
zeolites, 189–97, 190–2s; in early organisms?, 327, *see also* clays as minerals for first organisms; morphologies of, 194–5e; ultrastructures in, 267
'zero-technology' systems, 114–16; first organisms as, 115–16, 121
zircon, 185
Zn^{2+}, catalytic effects, 73; in forming 3′–5′ internucleotide bonds, 58; in forming oligomers of glycine, 49–50
Zuckerkandl, E., 98, 107, 110
Zussman, J., 173, 175, 187, 190, 227–8
zwitterions, stability of, 50, 66

The Socialist Third World

The Socialist Third World

Urban Development and
Territorial Planning

Edited by

Dean Forbes and Nigel Thrift

Basil Blackwell

British Library Cataloguing in Publication Data
The Socialist Third World : urban
  development and territorial planning.
  1. Socialism—Developing countries
  2. Developing countries—Economic
  conditions
  I. Forbes, Dean II. Thrift, Nigel
  330.9172′4   HC59.7

  ISBN 0-631-13442-5
  ISBN 0-631-15616-X Pbk

Library of Congress Cataloging in Publication Data
The Socialist Third World.

  Bibliography: p.
  Includes index.
  1. Developing countries—Economic policy.
  2. Developing countries—Economic conditions.
  3. Regional planning—Developing countries.
  4. Socialism—Developing countries.
  I. Forbes, D. K. (Dean K.) II. Thrift, N. J.
  HC59.7.S5857   1987   338.9′009172′4   87-5165
  ISBN 0-631-13442-5
  ISBN 0-631-15616-X (pbk.)

Typeset in 10 on 11½ pt Plantin
by Photographics, Honiton, Devon
Printed in Great Britain by Billing and Sons Ltd., Worcester

Contents

Preface   ix

1   Introduction
Dean Forbes and Nigel Thrift   1

Socialism   1
The heterogeneity of socialist developing countries   4
Territorial organization in socialist developing countries   5
The chapters   19

2   The Urban Question in Socialist Developing Countries
Enzo Mingione   27

Introduction: industrialization strategies and the urban question   27
The phases of industrial development, the world system and the urban question   30
Socialist industrialization, underdevelopment and the urban question   35
A preliminary confrontation with the Chinese case: from a descriptive model to some interpretative assumptions   44

3   Chinese Socialism and Uneven Development
Chung-Tong Wu   53

Introduction   53
A brief chronological review of Chinese development policies   55
Impacts of policies   61
Uneven regional development   66
Backward areas   85
Recent regional development policies   86
Conclusions   91

4 Territorial Organization, Regional Development
and the City in Vietnam
Dean Forbes and Nigel Thrift 98

Introduction 98
Historical patterns of territorial organization 98
The spatial organization of the South, 1954–75 101
Economy and spatial structure in the North, 1954–76 104
Economic and social planning, 1976–85 108
The regional question in Vietnam 113
Population distribution and the New Economic Zones 114
The district and the small town 119
The large cities: Hanoi and Ho Chi Minh City 121
Conclusions 126

5 Regional Disparities and Regional Development in Algeria
M'Hamed Nacer and Keith Sutton 129

Introduction: Algerian socialism and regional development 129
Regional disparities and regional development in Algeria 132
The national development plans: regional and social issues 146
The contribution of the *Plans Communaux de Developpement*
(PCD) 155
Recent industrial development – deconcentration and
decentralization 158
Conclusions 166

6 Sand Reform or Socialist Agriculture? Rural Development in PDR
Yemen 1967–1982
Jim Lewis 169

Introduction 169
The nature of socialist rural development 169
PDR Yemen: background information 171
Rural development policies 175
Evaluation of the policies 189
Socialist transformation? 192

7 Urban and Regional Development in Zimbabwe
David Drakakis-Smith 194

Introduction 194
The evolution of socio-spatial inequality 197
Post-independence changes in the Zimbabwean space
economy 204

Present and future socialist policies towards spatial and
urban development   208
Conclusion   212

8 Guyana: Co-operative Socialism, Planning and Reality
   Lesley M. Potter   214

   Is Guyana socialist?   214
   Socialism and territorial planning   219
   The territorial structure of Guyana in 1970   220
   Planning in the Co-operative Republic, 1970–7   223
   Assessment of the Second Development Plan, 1973–7   228
   Central planning, 1978–81   235
   The new regional system – local democratic organs   242
   Economic realities, 1981–4   246
   Conclusions – co-operative socialism, planning and reality   247

9 Spatial Equality and Socialist Transformation in Cuba
   Paul Susman   250

   Introduction   250
   Overcoming inequalities   255
   Increased democratization post-1970: production equality   273
   Conclusions   278

10 Socialism, Democracy and the Territorial Imperative:
   A Comparison of the Cuban and Nicaraguan Experiences
   David Slater   282

   The diverse faces of socialism   282
   On the historical specificity of revolutionary change –
   or why Nicaragua is not a second Cuba   289
   From the territory of revolution to the revolution
   of territory   292
   Concluding implications for revolutionary change in the
   Third World   299

Bibliography   303

Notes on Contributors   323

Index   326

Preface

By any standards, the socialist developing countries are an important group of countries in the world today. Yet, it sometimes seems, social and economic researchers have been inclined to put them to one side, or tended to look at them as unique (China), as extensions of Eastern Europe (Cuba), or of interest solely for the role they have played in international politics (Vietnam). In this book we try to demonstrate that the socialist developing countries are an interesting group of countries in their own right. We believe they have a number of characteristics in common, and that these characteristics can help us to understand better the process of urbanization and territorial development in developing countries. The task of studying these countries' urban and regional development has, however, not been an easy one. We had originally hoped to include chapters on Angola and Mozambique, but were unable to do so, and the extreme political instability of some of these countries – an example being the *coup* in PDR Yemen – has already changed the situation from that described in some of the chapters in this volume.

We wish to acknowledge assistance in the preparation of this volume. Barbara Banks compiled the bibliography and edited the text with her usual thoroughness and attention to detail. Carol McKenzie typed several chapters on the word processor and interpreted the editors' scrawl with great patience. Ian Heyward drew the maps and diagrams for chapters 1, 3 and 4. David Slater's chapter first appeared in *Antipode*, *18*(2), pp. 155–85, and is used with permission.

1
Introduction

Dean Forbes and Nigel Thrift

Socialism

Two very different development strategies can be identified in the post-war experience of the developing countries. On the one hand, the 'newly industrializing countries' represent the export-led strategy so strongly advocated by bodies such as the World Bank and the International Monetary Fund. These so-called 'miracle economies' have been the subject of much attention in the development studies literature and are frequently put forward as a model for all developing countries. On the other hand, there are the 'socialist developing countries'. Often shunned by the major international institutions, many of these countries have in the past been relatively neglected in the literature.

In recent years, however, socialist developing countries have increasingly come to be seen as important in their own right (e.g. Mittelman, 1981; Ottaway and Ottaway, 1981; Chase-Dunn, 1982; Wilber and Jameson, 1982; Wiles, 1982; White, Murray and White, 1983; White, 1984; Giddens, 1985). Why should this be so? There are at least two answers. First, there is the rapid growth of socialist developing countries in the post-war period: Mongolia (1924), North Korea (1945) and China (1949) were the first, followed by North Vietnam (1954), Guinea (1958) and Cuba (1959). Another seven came into existence in the 1960s, and a further seventeen in the 1970s (see figure 1.1). Only two socialist developing countries – Surinam and Burkina Faso – have so far come into being in the 1980s. Throughout the post-war period socialist regimes in developing countries like Chile and Grenada have also flickered out of existence, emphasizing the precarious position in which these regimes often find themselves. The second answer is that the development strategies of socialist developing countries provide an important counterpoint to the orthodoxy of much of the development literature.

Before we can come to grips with the importance of development strategies in socialist developing countries, it is necessary to try to define what is to be understood by the term 'socialist'. White (1983, p. 1) has

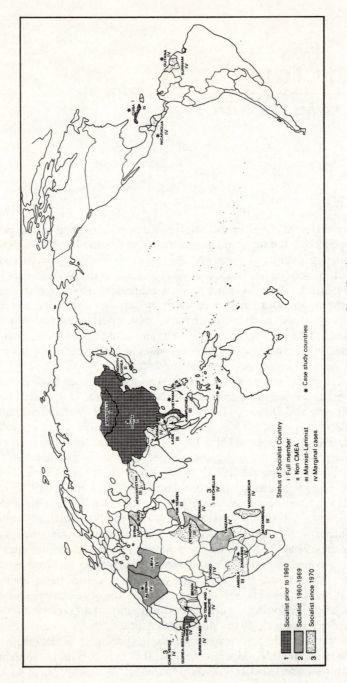

Figure 1.1 Socialist developing countries in 1985 according to the Wiles Classification

argued that there are two structural features of a socialist system. The first is that the regime has:

broken ... the autonomous power of private capital over politics, production and distribution, abrogated the dominance of the law of value in its capitalist form, and embarked upon a development path which does not rely on the dynamic of private ownership and entrepreneurship.

A second requirement is that some fundamental transformations have been brought about to society and economy:

most notably, the nationalisation of industry, socialisation of agriculture, abolition or limitation of markets, and the establishment of a comprehensive planning structure and a politico-ideological system bent on the transition to an ultimate communist society.

However an attempt such as this to determine the 'essence' of socialism is often of only limited use in deciding which developing countries are socialist and which are not. The rhetoric of socialist regimes in developing countries can often conceal an economic and political situation not so very different from that to be found in developing countries which do not lay claim to instituting socialism. For example, a number of developing countries have a high degree of State ownership of industry and aspire to a comprehensive planning structure but would clearly not claim to be socialist!

Another, more pragmatic way of defining socialism is through a series of indicators of a socialist State including effective one-party rule, socialist goals in the constitution, a high and increasing degree of State ownership of industry and agriculture, the beginnings of a centralized command economy, and the direction of external relations (Wiles, 1982). A categorization of developing countries according to these criteria is shown in figure 1.1. Group I countries are full member States of the Council for Mutual Economic Assistance (CMEA) or Comecon. Group II countries are all well-established communist or socialist States, but outside the CMEA and strong Soviet influence. Group III is composed of countries with self-proclaimed hard-line socialist governments, generally closely aligned to the USSR or its allies. Finally Group IV contains the marginal cases, many of which are one-party socialist States. This category includes countries which are unstable and liable to significant shifts in ideology. The country case studies in this volume are purposely drawn from all four categories – Vietnam and Cuba from I, China from II, PDR Yemen from III, and Algeria, Guyana and Nicaragua from IV. Zimbabwe has a problematical status, but seems to be on the verge of breaking into Group IV.

The heterogeneity of socialist developing countries

Clearly then, socialist developing countries are, at least according to the Wiles indicators, many in number. But, at the same time, they are far from homogeneous. There are at least three sources of heterogeneity. First of all, the initial conditions under which socialism is instituted will vary widely. These conditions will include: (1) the particular history and geography of the country concerned (and especially the colonial heritage and natural resource endowment) leading to a particular level and pattern of wealth or poverty; (2) the prevailing class structure (which may be strongly peasant-oriented and may include a considerable urban middle class); (3) the structure of civil society (especially tribal and regional divisions); (4) the state of economic relations (particularly, the degree of exposure to the capitalist world economy); and (5) prevailing international relations (including, not least, the geopolitical sphere in which the country finds itself and the degree of militarization needed to maintain territorial integrity in what will often be a situation of war). The second source of heterogeneity will be in the brand of socialism adopted. A strongly centralist Leninist ideology will have quite different effects on the process of development than a more decentralist one. A third source of heterogeneity will be the point in socialist development which a country has reached. It is not uncommon for a post-revolutionary period to be characterized by turmoil and rapid transformation, for the State bureaucracy to grow and consolidate power, and then at a later stage to implement limited reforms and a degree of 'market' socialism. Yet despite the difficulties of sustaining any generalizations about the overall experience of the socialist developing countries in the face of heterogeneity, they all have three characteristics in common (as in the earlier quotations from White, 1983).

First, a socialist regime comes to power intent on doing away with all or most capitalist structures, particularly the control over resources by privately owned capital, the central role of the market-place in allocating resources, and the class structure which is part and parcel of this arrangement. This is very clearly an extremely complex process. The capitalist class structure does not dissolve overnight. It is replaced by another class structure based on the existence of a State class which has its own requisites. And, throughout the process, conflicts of interest, sectoral struggles and conflicts over ideology persist.

Second, in socialist developing countries the State has the potential power to make a substantial contribution to the transformation of society, although, of course not even revolutionary governments can start with a *tabula rasa*. In doing away with capitalist structures of ownership, socialist States are presented with the opportunity to demonstrate the contribution which alternative forms of organization of production and socialization can bring to society. But these achievements can be problematic:

> the principal contradiction of socialist societies . . . is between the planned organisation of production, mediated through the state, and the mass participation of the population in decisions and policies that affect the course of their lives. (Giddens, 1981, p. 248)

Third, many socialist developing countries have opted for the direct path to socialism without first achieving capitalist development. In other words, their task is to combine the destruction of capitalist structures and the transformation of society *and* simultaneously to achieve economic growth through industrialization. As White (1983, p. 7) has noted, the revolution may have been fought primarily for freedom, but freedom soon becomes subservient to economic necessity.

Territorial organization in socialist developing countries

Territorial organization is an integral part of the planned socio-economic development of all developing countries, whether capitalist or socialist and is bound to have impacts on the character of cities and regions. However, socialist developing country experience in the urban and regional arenas has not, with some notable exceptions, been widely reported in the literature (see Lo and Salih, 1978; Roberts, 1978; Honjo, 1981; Renaud, 1981; Stöhr and Taylor, 1981; Gilbert and Gugler, 1982). There are, of course, a number of detailed studies of individual country experiences, especially of Cuba (e.g. Susman, 1974; Barkin, 1980; Gugler, 1980; Slater, 1982) and China (Ma and Hanten, 1981; Diamond, Hottes and Wu, 1984; Kirkby, 1985), as well as some of the longer established socialist countries like Vietnam (Nguyen Duc Nhuan, 1978; Thrift and Forbes, 1986), Algeria (Sutton, 1981), Tanzania (Lundqvist, 1981; O'Connor, 1984) and Guyana (Potter, 1984). But the overwhelming weight of knowledge on urbanization and territorial planning in developing countries still stems from the experience of capitalist market-based societies. The purpose of this book is to begin to correct this imbalance by focusing on some of the key urban and regional issues confronting socialist developing countries.

As noted above, the socialist developing countries are heterogeneous and their attempts at territorial organization of cities and regions can be expected to vary according to the three sources of heterogeneity noted above. First of all, the initial conditions of urban and regional development vary widely. These conditions will include: (1) the particular urban history and geography of the country concerned (with special emphasis being laid in many countries on the colonial heritage), leading to a particular level of resources with which to influence the rate and pattern of urbanization; (2) the prevailing class structure (which may include a comparatively large urban middle class or a fairly small urban middle class, with different urban implications following); (3) the structure of civil society and especially tribal and regional divisions (which may cause considerable problems of internal pacification to which urban strategies may be a partial solution); (4) the state of economic relations (and particularly the degree of exposure to the capitalist world economy which, if high, will exact considerable pressures for urban growth); and (5) prevailing international relations and especially warfare (which may mean that population is directed to sensitive border areas, for example).

The second source of heterogeneity will be the brand of socialism adapted. A strongly centralist Leninist ideology will sacrifice participation for control with consequent urban effects such as the desire to use cities as explicit mechanisms of surveillance (Giddens, 1985) to keep the population in check. A third source of heterogeneity will be the point in socialist development which a country has reached. There is some evidence that in the earlier phases of a socialist developing country's life drastic urban strategies, such as deurbanization, are more likely to be instituted.

Nevertheless, despite all these sources of heterogeneity, many socialist developing countries do have common urban characteristics, of which the most remarked upon is their *slower rate of urban growth*, compared to many 'capitalist' developing countries (see figure 1.2). Whilst a slower rate of urban growth is not typical of all socialist developing countries, it is typical of enough – and especially those which have been in existence for a comparatively long period of time – to make this a significant and important phenomenon. A second phenomenon which is important is the number of socialist developing countries within which the degree of urban primacy seems to be being reduced (see figures 1.3 and 1.4). Again this phenomenon of *polarization reversal* seems to be more typical of those socialist developing countries which have been established for a comparatively long period of time.

In the socialist developing countries the State is the main sphere of determination. It initiates changes to the social and economic structure in a rather more independent way than is possible in capitalist societies. Economic constraints are still important, particularly in those societies which embrace 'market socialism' either voluntarily or by necessity, but it is the impact of the state to which we need to look first. Explanations of polarization reversal in the socialist developing countries can usefully be divided into two types. One argues that it is an intended outcome which draws its strength from ideological convictions. The other sees urban outcomes as unintended consequences of the pursuit by the State

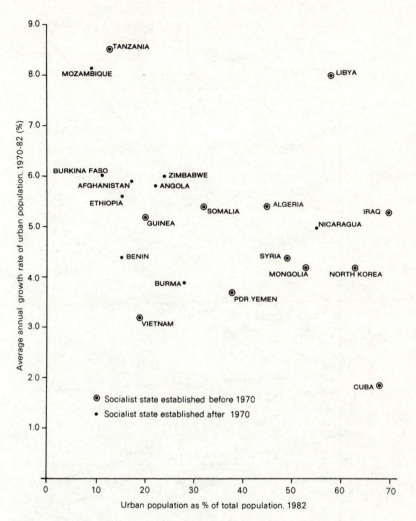

Figure 1.2 Urban growth in socialist developing countries, 1970–82

of more fundamental interests. Common to both types of explanation is the key role played by the State *vis-à-vis* both the economy and civil society.

The ideological explanation rests on the assumption that socialism should generate within a society a radical alternative social organization and with it a radically different socio-territorial system. Mingione (1981, pp. 167–8) is worth quoting at length on this point:

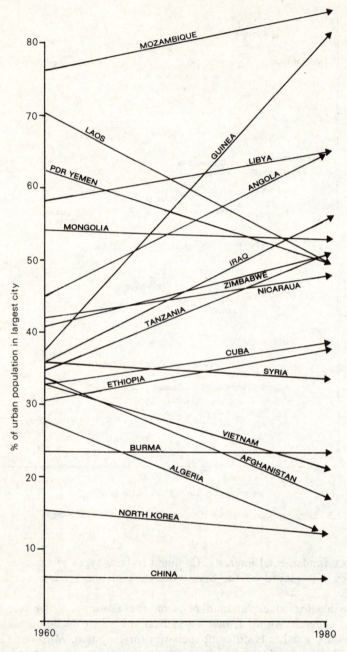

Figure 1.3 Proportions of urban populations in largest cities, 1960–80

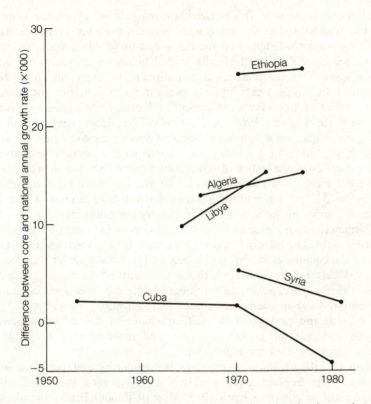

Figure 1.4 The difference between core-region growth and national annual growth rates in selected socialist developing countries

The socialist territorial order will not only be against the forms of concentration and division of labour developed by capitalism, but will also oppose any form of regression to the old, agricultural, rural and pre-capitalist community. It is a totally new order, neither urban nor rural, based on undivided, polyvalent productive unities ... it is possible and necessary for societies in transition to begin to establish counter-trends (at least experimentally). As soon as possible, new polyvalent communities must be created with a high degree of socialist decentralization and development of uninhabited regions ... industrialization of the countryside should be pursued by means of new small-scale technology, and diffusion of equal levels of services and knowledge.

As Mingione (1981, p. 164) and others recognize, however, the foundations for this view were only very superficially articulated in the writings of Marx and Engels. The latter, in both *The Housing Question* and *Anti-Duhring*, argued briefly for getting rid of towns and spreading population and industry uniformly. In *The German Ideology* however, while recognizing the need to abolish the antagonism between town and country, Marx and Engels acknowledged that the rift would inevitably deepen before it was closed. Mao Zedong, the leader of the Chinese revolution, is probably better known for his ideological commitment to the rural areas. Not only was the Chinese revolution a peasant revolution, but the Maoist development strategy favoured agricultural co-operativization and rural industry (see Schram, 1963, pp. 233–54). The Maoist strategy had an important impact on a number of socialist developing countries, particularly those which, like China, had experienced a peasant revolution. There have been a number of other contemporary writers as well, such as Frantz Fanon, who have espoused an anti-city strategy, and Nelson (1979, pp. 342–60) has pointed out the relative lack of interest of Marxist political parties in the urban poor in developing countries.

Characteristically, large cities in socialist developing countries are often accused of being 'consumer cities', especially if they were an important part of the colonial economy, and retain a large tertiary sector workforce. The task before policy-makers is therefore to convert them into 'producer cities'. This requires achieving a number of different objectives, including reducing the size of the largest cities (deurbanization), containing large-city growth and promoting the self-sufficiency of the urban economy, and encouraging the growth of small- and medium-sized towns and, where necessary, creating new towns.

Without a doubt, the most dramatic contemporary example of an anti-urban ideology in practice occurred in Kampuchea soon after the Khmer Rouge came to power in April 1975. Most of Phnom Penh's population of 2.5 million was driven out of the city and towards the rural areas where they were expected to work on rehabilitating rural infrastructure, building new irrigation systems and housing, and rejuvenating the rural economy. The city's population dropped to some 50,000 in the process, and many of the former residents died from hunger or brutal treatment (Ponchaud, 1978). The new government in Vietnam in 1975 took a similar position with regard to the 'consumer' role of the large city, particularly Saigon (Ho Chi Minh City), but the deurbanization process was far less dramatic. However, the urban population of the country as a whole did decrease from 21.5 per cent of the total population in 1975, to a little over 19 per cent in 1980, before steadily climbing back to the 1975 level in 1984 (Khong Dien, 1983).

Shifts of huge numbers of people to and from cities have also been a hallmark of China's development. Mao Zedong first articulated the need for urban youth to acquire a taste of rural hardship in the mid–1950s, and by the end of 1957 two million school-leavers had gone to live in villages. However, during the Great Leap Forward (1958–9) some 20 million rural dwellers moved into the cities, but the industrialization programme proved an economic disaster. It prompted another phase of the youth rustication programme, which was announced in 1963, and resulted in millions of youth moving back to the rural areas. The thrust of the programme intensified during the Cultural Revolution, when more than 17 million 'heeded the call'. Shanghai excelled itself – between 1968 and 1976 some 1.29 million youth left the city, equivalent to just under a quarter of the population of the city in 1976 (Kirkby, 1985, pp. 37–49).

Another characteristic of socialist developing countries in the attempt to convert large 'consumer' cities into 'producer' cities have been policies to limit large-city growth. These include direct measures aimed at population redistribution such as state investments in State-owned industry, infrastructure and investment subsidies for industry, and land colonization and resettlement schemes. Coupled with these are various measures to monitor and regulate the movement of individuals; included are residence registration, discriminatory access to jobs, education and services, the rustication programmes alluded to earlier, as well as various restrictions on occupations in particular areas (Fuchs and Demko, 1981). Socialist developing countries, because of the strength of the State *vis-à-vis* the economy, are able to regulate both the labour market and most of the investment in infrastructure and production more efficiently than in other developing countries.

It is not uncommon to link the objective to make the city self-sufficient in foodstuffs with the strategy of urban containment. This objective is usually pursued through the expansion of the urban boundary to incorporate a productive rural hinterland, and in some cases a greening of the suburbs through the development of household fruit and vegetable gardens. A plan to 'urbanize the country and ruralize the city' through the development of peri-urban agriculture and the allocation of urban labour to this region was introduced in Havana in the late 1960s. By the early 1970s, however, attention had once more turned to the more conventional industrial development within the city, although by then the city's hinterland had become a 'collective orchard' (Stretton, 1978, pp. 125–8).

It would be wrong to conclude that socialist developing countries have only been concerned with containing the large cities and developing the rural areas. They have also paid attention to developing the small- and medium-sized cities and, where necessary, to creating new urban areas. In this context, the Soviet and East European 'socialist cities' concept has proved an important model. The debate about the significance of the urban in Marx and Engels was resolved by Stalin, who argued that the creation of new cities was a prerequisite for cultural and economic development. Cities are required because they assist the formation of economies of agglomeration and permit the creation of economies of scale, which are essential to the development of the forces (or means) of production. According to this argument, the State has greater power in socialist societies due to the elimination of potential competitors such as the church and unions. Moreover, the greater degree of public ownership of the means of production in socialist economies, and the centralized nature of accumulation ensure that the State is able to develop an effective spatial policy (See Gruchman, 1982). The result however, is often a process of 'underurbanization', at least compared to cities in developing countries, because of the tight controls on investment and residential location.

The other main type of explanation for polarization reversal rests on the assumption that socialist countries do not have strong attitudes to urbanization *per se*. Rather, it is put forward that deurbanization is the unintended (though not necessarily undesirable) outcome of processes and policies of a higher priority. These processes can be of two types: primarily political, or primarily economic. The most sophisticated presentation of the political argument is in the model advanced by Murray and Szelenyi (1984) (see figure 1.5). Their key proposition is that the main characteristic of the city is a large middle class or petty bourgeoisie. In socialist states, the emergent State-class of bureaucrats and revolutionaries opposes the urban bourgeoisie because of the latter's control over the economy. The ensuing class struggle has unintended anti-urban consequences but is not primarily an anti-urban strategy. Furthermore, in

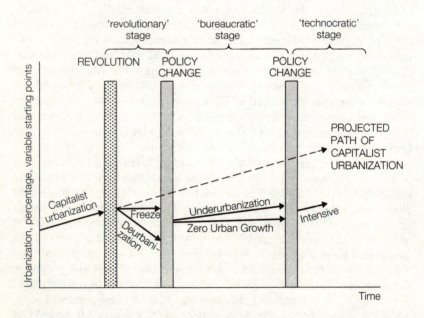

Figure 1.5 The Murray–Szelenyi model of urbanization in socialist countries

trying to break the economic power of the urban middle class and bring economic activity into public ownership, the economic diversity of cities is reduced, which is in turn another constraint to city growth.

In the 'revolutionary stage' the socialist country undergoes a freeze on urban growth, or, in some situations, deurbanization. As the country evolves it moves into a 'bureaucratic stage' in which long-term socialist goals become more important. Conscious ideological choices based on either the Chinese or Soviet models figure more importantly in the bureaucratic stage.

Other unintended consequences also shape the structure of population distribution. Using a case study of Vietnam, Thrift and Forbes (1986) argue that the Murray–Szelenyi model over-emphasizes the importance of the State, and neglects critical processes in the realm of politics and civil society. Recent urban development in socialist Vietnam cannot be adequately explained without taking into account such factors as warfare and more lasting external security threats. The latter, particularly, helps explain at least part of Vietnam's recent land resettlement programme. In addition, ethnic rivalry and conflict have played a part in Vietnam, particularly where the Chinese–Vietnamese are concerned. Although a socialist society, social conflict cannot be explained by control of the economy alone. Tensions over aspects such as ethnicity – inherited from a past society in which civil society was the central arena of class struggle – continue to have an impact on social process.

The influence of the struggle for territorial integrity on urbanization in other socialist developing countries must also be emphasized (Giddens, 1985; Shaw, 1984; Thrift and Forbes, 1985, 1986). Regional separatist movements (themselves often socialist in character) threaten a number of socialist developing countries from within, and population movements can often be explicitly directed towards combating these separatist movements. For example, in Ethiopia current population movements from Eritrea are not simply the result of drought and the relocation of population to more fertile areas – they are also a useful way of denuding a dissident province of population.

More importantly, many socialist developing countries are threatened from without. The example of South Africa's continual harassment of Mozambique and Angola is a case in point. Many socialist developing countries, such as Vietnam, have instigated population movements to threatened borders with consequent effects on the pattern of urbanization. But the effects of warfare, or the threat of warfare, can be more subtle than this: for example, resources for urbanization, such as concrete, can be diverted to war purposes with consequent problems for urban housing programmes. Then again, standing armed forces can mean the diversion of manpower resources desperately needed to build up urban economies (table 1.1). Of course, large armed forces can also be a solution to a

Table 1.1   Some 'formal' military participation ratios (ratio = population total/'formal' armed forces total)

Country	Ratio
Iraq	23.2
DR Korea	24.9
Syria	28.6
Libya	47.8
Vietnam	47.9
Mongolia	49.3
Nicaragua	51.7
Cuba	65.1
Albania	69.3
Laos	74.5
PDR Yemen	80.0
Ethiopia	130.7
Algeria	166.9
Angola	181.4
Congo	195.4
Burma	214.9
China	259.7
Afghanistan	304.3
Tanzania	506.0
Mozambique	766.7

Source: Institute for Strategic Studies, 1984.

problem, since many of those in armies might otherwise join the ranks of the urban unemployed. The list of effects of armed struggle is long and diverse and constitutes a major research frontier.

Yet while political struggles and conflicts are undoubtedly important in shaping the spatial structure of socialist developing countries, we cannot ignore the fact of strong centralized control of the economy and the obvious impact that this has on the space economy. In the case of China, industrial development has been the main priority of the socialist government, Thus it has been argued that the pursuit of industry-centred policies has been largely responsible for the structure of the space economy, and not abstract concepts of anti-urbanism. The Communist Party has not been opposed to cities, as such – it has been concerned to convert 'consumer' cities into 'producer' cities, but it has also been concerned not to expend money on non-productive activities in cities. Included within this category is social overhead capital – housing, roads, public transport facilities, drains, electricity supply and shops. Thus, restrictions on city growth are a way of minimizing investments in social overhead capital. Optimal industrial growth is ensured because city-growth restrictions avoid depriving the rural areas of labour and prevent the build-up of an impoverished urban mass capable of posing a political threat (Kirkby, 1985, pp. 13–18).

This strategy, however, is currently leading to the escalation of tensions in China. Since the economic reforms of 1978 and 1984, China has opened up to foreign investment but retained, largely intact, restrictions on labour migration and an intention to contain large-city growth. However, these two policies are somewhat at odds: In the case of the large industrial city of Tianjin, one of the four key 'open' coastal cities, planners are endeavouring to develop a large port and industrial estate in the coastal satellite at Tanggu, whilst limiting the growth of Tianjin and moving population from the city to Tanggu. If, as might be expected, it proves difficult and expensive to attract a skilled workforce to Tanggu, which policy goal will have the top priority, industrial development or city-size distribution? Given China's current commitment to economic development one might reasonably expect labour-market deregulation and, hence, restrictions on urban growth, to occur as a means of facilitating the 'open door' policy (Forbes and Wilmoth, 1986).

While the development of heavy industry has been the top economic priority of most socialist developing countries, where other priorities have been determined these also have had an impact on the organization of space. Cuba in 1963 was suffering the effects of the American economic blockade, while inside the country large numbers of rural residents were shifting to Havana. The government decided on a radical reversal of policy – all urban and industrial development was stopped or suspended and policy re-oriented to agriculture, in particular large, mechanical State farms and the development of the sugar industry for exports. After 1963 there was no further developmental work or investment in Havana and the city began a long period of physical decay. The Havana plan was released in 1965, and was meant to resolve the problems of the city (such as the inequalities between city centre and suburbs) and to arrest city growth. However, due to the government's emphasis on agricultural growth there were insufficient resources to finance the city's development and besides it was realized that expenditure on ameliorating intra-urban problems would only exacerbate inequalities between city and country (Stretton, 1978, pp. 120–31).

The country case studies that follow have been chosen to show both the diversity of socialist developing country experience of urbanization and regional development and, also, the common features in such a way as to be comprehensive. There is, however, one topic that is missing from this volume which is of sufficient importance to have attention drawn to it here. In many countries around the world socialist–nationalist independence movements either currently or in the past have controlled large tracts of territory which they have organized in some detail. The

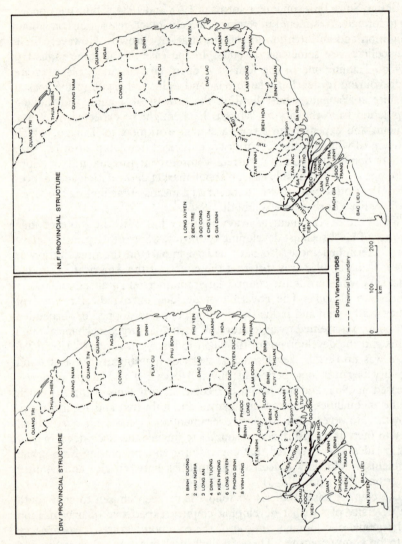

Figure 1.6 Territorial organization of South Vietnam by the DRV and the NLF, 1968

National Liberation Front (NLF) operating in the south of Vietnam during the main war years from 1965 to 1975 is an historical example. The Eritrean People's Liberation Front (EPLF), operating in the Eritrean province of Ethiopia, is a contemporary one. These movements have organized their space in quite significant ways with subsequent impacts – strongly mediated by the conflict – on urban and rural life. The NLF had a complex provincial structure (mirroring, to some extent, the provincial structure of the Republic of Vietnam) and an active policy of urban and rural reorganization (Fall, 1967; Pike, 1966) (figure 1.6). The EPLF has instituted land reform in its liberated areas, as well as various industrial and socialization projects which together make for a kind of urban policy (Firebrace, 1984) (figure 1.7). The territorial impacts of movements like these remain chronically under-researched.

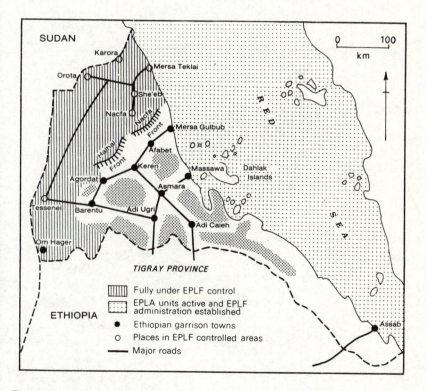

Figure 1.7   Eritrea, 1984

Table 1.2 Case study countries: selected statistics

Country	Population mid-1982 (millions)	Average annual population growth, 1970–82 (%)	GNP per capita 1982 ($US)	Annual average growth 1960–82 (%)	Manufacturing share of GDP 1982 (%)	Industrial labour force 1980 (%)	Urban population as proportion of total 1982 (%)	Average annual growth rate 1970–92 (%)	Socialist category (after Wiles)
Algeria	19.9	3.1	2,350	3.2	10	25	45	5.4	4
China	1,008.2	1.4	310	5.0	2	19	21	—	2
Cuba	9.8	1.1	—	—	43b	31	68	2.1	1
Guyana	0.9a	2.0a	—	2.9b	19b	—	22c	—	4
Nicaragua	2.9	3.9	920	0.2	26	14	55	5.0	4
PDR Yemen	2.0	2.2	470	6.4	—	15	38	3.7	3
Vietnam	57.0	2.8	—	—	—	10	19	3.2	1
Zimbabwe	7.5	3.2	850	1.5	25	15	24	6.0	4

a United Nations, 1985b.
b United Nations, 1985a.
c United Nations, 1980.
Source: World Bank, 1984.

The chapters

This volume consists of ten chapters, and the case study countries include Algeria, China, Cuba, Guyana, Nicaragua, PDR Yemen, Vietnam and Zimbabwe (see figure 1.1). Some brief introductory statistics for these countries are displayed in table 1.2. Case study nations vary in size from the largest country in the world (China), with a population of over a billion, to a small country like Guyana, with a population under a million. Equally, these countries span the range of income levels used by the World Bank in 1984. Two of them, China and Vietnam, are 'low-income economies', another five, PDR Yemen, Zimbabwe, Cuba, Guyana and Nicaragua are 'middle-income economies', and another falls in the 'upper middle-income' bracket, Algeria. The share of manufacturing in Gross Domestic Product is highest in Cuba (43 per cent) and lowest in China (2 per cent), and the industrial labour force proportional to the total workforce is largest in Cuba (31 per cent) and lowest in Vietnam (10 per cent). Cuba is also the most urbanized of these countries, with over two-thirds of the population living in urban areas, followed by Nicaragua (55 per cent). On the available evidence, Vietnam is the least urbanized, with only 19 per cent of the population living in urban areas. In contrast, between 1970 and 1982 Vietnam had the highest annual average urban rate of growth (6 per cent). The following paragraphs contain a short précis of the content of the chapters.

The relationship between industrialization and urbanization is considered by Mingione in chapter 2. In particular the author asks how has industrial development affected the structure and function of cities, and what sort of impact has urban development had on industrial structure? There is, as yet, no general theory of socialism and so there are limitations to a theoretical discourse on this topic, but there is some limited evidence from European history which can be used to enlighten the discussion of industry and the city. Mingione recognizes that other processes, such as cultural factors and the pre-industrial inheritance, need to be taken into account in explaining urban development, but largely confines himself to industrialization in this chapter.

Mingione identifies a key contradiction in the industrialization process: industry's need for cheap labour on the one hand, and its need to cultivate a domestic market in order to avoid underconsumption on the other. The disequilibrium that industrial development brings about can cause a deterioration of urban infrastructure and an urban fiscal crisis, as the author demonstrates by looking at phases of industrial development in the industrialized countries. In developing countries rapid urbanization puts a great strain on city budgets and ultimately leads to increased overseas borrowing by the State in order to finance urban development.

Moreover, even in modern industrializing cities in developing countries the informal sector and squatter settlements continue to grow and function as a means of minimizing the costs of reproduction of labour.

Defining socialist countries as those characterized by strong central political control of relations with the world system and centralized control of production and consumption, Mingione argues that agricultural rationalization has been slowed down in socialist developing countries in order to place a brake upon rapid city growth. Within the city, the emphasis on heavy industrial production and the relatively scant attention paid to the manufacture of consumer durables has resulted in a scarcity of consumer goods. He also argues that State control of land and housing, the priority the State gives to satisfying basic needs, and the relationship between urban and rural classes has been important in socialist countries.

Chapter 3, by Wu, examines the experience of China. Spatial development, the author argues, is best approached by looking at three key policy areas: industrialization, rural development, and population growth and distribution policy. The chapter therefore opens with a brief summary of the six Chinese five-year plans since 1953. It highlights the main aspects of each, from the Maoist strategy of industrialization, through the agricultural failures of the 'Great Leap Forward' and the 'Cultural Revolution', to the implementation of the responsibility system in agriculture and the period of 'readjustment and reform' since 1979. Wu highlights the fact that – Maoist rural strategy rhetoric notwithstanding – the majority of State investment was channelled into heavy industry and not agriculture.

The consequence of this is that the municipalities of Shanghai, Beijing and Tianjin, and the province of Liaoning still dominate China's regional distribution of per capita industrial output, but there is some evidence that this domination is declining as a result of growth of industrial production in inland provinces. The same four municipalities/provinces dominate total production by brigades, and it is clearly the coastal provinces where distributed collective income is highest. Wu shows that the poor counties (*xian*) are concentrated disproportionately in the inland regions of China, whereas around one-half of the rich brigades in the country are located in the suburban rural areas of the large municipalities, and some three-quarters overall are located in the coastal region. He concludes that significant regional inequalities persist in China, the chief disparities being between coastal and inland regions on the one hand and between the urban/suburban and rural regions on the other. Moreover, there are a number of regions officially declared 'backward areas', typically peopled by non-Han Chinese, which it is realized will increasingly require special attention.

In the last few years the Chinese government has implemented a series of policies directed at the spatial structure of China. These include a new thrust based on developing regions according to their comparative advantage, a focus on special economic regions (such as the Shanxi coal region and the industrial agglomeration in the Shanghai–Changjian Delta), a core-cities programme designed to improve the linkages between urban and rural areas, and the development of special economic zones in fourteen coastal cities. Yet Wu concludes that these regional strategies are principally growth-oriented and are likely to worsen regional equity, because they focus disproportionately on the growth prospects of the coastal provinces and the urban areas in particular.

Chapter 4, by Forbes and Thrift, documents the evolution of spatial structure in Vietnam, highlighting the contribution of regional policy to social and economic organization. The first part of the chapter looks at pertinent themes from the pre-socialist period. Shifts in power between the mandarin-staffed bureaucracy and the traditionally autonomous commune (or village) are shown to be an enduring theme in Vietnamese history. Population distribution policies have also been an important tool of spatial integration and a means of establishing territorial economic and cultural control over the country. During the period from 1954 to 1975 spatial policies were an important means by which the government of South Vietnam sought to maintain political control of the rural areas. Although ostensibly a mechanism for improving rates of economic development, the various programmes intended to create agglomeration centres, agrovilles and strategic hamlets were hastily conceived and implemented, and given little real chance of success.

The governments of socialist Vietnam have given first priority to national goals and objectives. These include the development of an industrial base, the co-operativization of agriculture and national security. Since 1978 the so-called New Economic Policies have been introduced in order to stimulate the private sector by providing new incentives for individual enterprise. Spatial problems including the imbalance between north and south, regional uneven development, the problems of the large cities, and the gaps between centralized planning and local areas were also of some significance to Vietnamese social and economic development. Three major responses to Vietnam's spatial planning requirements are discussed in the chapter: population redistribution policies and the development of New Economic Zones; proposals to develop the district as the key agro-industrial economic unit; and policies to convert the larger cities, particularly Ho Chi Minh City, from 'consumers' to 'producers'.

M'Hamed Nacer and Keith Sutton are the authors of chapter 5 on the development of regional policy in Algeria. They point out that Algerian socialism has sometimes been equated with strong centralized planning, but that from the mid-1970s there has been some relaxation of this thrust. The State has become concerned about under-utilization of productive capacity, and is trying to divide up some of the larger State-owned firms in order to improve efficiency. But it is definitely not heading down the capitalist path. In the first half of the chapter Nacer and Sutton briefly summarize some of the Special Programmes which the Algerian government adopted in the early 1970s, and assess the impact of these programmes through an analysis of patterns of regional convergence and divergence between 1966 and 1977. They conclude that the marked regional disparities had reduced during this period, due to some development in the lagging and especially in the intermediate regions. However, they note that regional disparities were still serious and that the trend in industrial development remains towards increasing regional divergence.

Since the mid-1970s the Algerian government has initiated a number of programmes designed further to reduce spatial disparities. These include the *Plans Communaux de Developpement*, which involves all 704 communes in planning, with a particular focus on hydraulics and social infrastructure. The Five-Year National Development Plan launched in 1980 also provided evidence of government commitment to regional development, particularly through the *Options Hauts-Plateaux* targeted on central plateau *wilayate*, and some increase in the power of the city governments over urban planning. In contrast to the social planning goals of the commune planners, the national plan has concentrated on industry, economic infrastructure, housing and education.

Nacer and Sutton stress that the spatial development of Algeria should be examined in the context of the government's commitment to industrialization, which has seen the industrial workforce grow from 160,000 in the late 1960s to 439,000 in 1981. Currently industry is overwhelmingly concentrated along the coastal belt and the company headquarters in Algiers. Government plans through this decade are to disperse industry to the interior, and especially into small towns, by means of a variety of mechanisms such as through the spatial restructuring of the large state companies (e.g. Sonatrach) and the development of small industry.

Chapter 6, by Lewis, examines and evaluates rural development issues and policies in the People's Democratic Republic of Yemen. According to the 1978 Constitution, PDR Yemen aims to develop the national economy using the principles of 'scientific socialism'; this has involved a heavy emphasis on nationalization of the economy. Agriculture and fishing are the main sectors of the economy, accounting for 43 per cent of the workforce in 1978, though agricultural production is increasing very slowly and food imports are still significant. PDR Yemen has very little industry. An Agrarian Reform programme was initiated in 1970, and a series of three- and later five-year plans beginning in 1971 have been central to rural policy. There has been a strong emphasis on the development of irrigation, some larger integrated rural development projects, and priority given to the development of communications and social services, especially education and health. However, price controls on food have been maintained and therefore have provided no incentive for increased agricultural production.

Evaluation of these programmes, Lewis argues, is not easy, but can be tackled in terms of the government's three stated goals: improved worker productivity and incomes; an increased contribution by agriculture to the economy; and a more egalitarian access to resourcs. While rural incomes are increasing relative to urban incomes, Lewis cautions that this is at least in part due to remittances from labour migration to the Gulf, rather than local rural development *per se*. Similarly, the government has had only limited success in promoting agriculture, for its share of GDP actually decreased between 1969 and 1979. Finally, while there is evidence to suggest a convergence in the distribution of land due to the Agrarian Reform programme, and more even access to education and health services, early successes in reducing inequalities of income distribution are being replaced by a divergence in income distribution. Lewis concludes that PDR Yemen has had limited success, but that recent debate has begun to question the Soviet model of State-farm development adopted in PDR Yemen, and raised the possibilities of a greater role for the market and local participation in the rural development process.

Chapter 7, on Zimbabwe, is authored by Drakakis-Smith. In the first part of the chapter the historical evolution of the Southern Rhodesia/ Zimbabwean space economy is documented. Of particular importance was the process of settlement by the exogenous settler capitalist classes under the aegis of the British South Africa Company. The Land Apportionment Act of 1930, which followed soon after annexation by Britain, divided land between the races out of all proportion to the population profile of the colony, thereby creating serious problems for the future. It was not very long before increased population pressure in the tribal trust lands led to landlessness and a drift of dispossessed people to the white-controlled towns and cities. This coincided with the growth of manufacturing industries in the urban areas and the creation of a black urban proletariat.

With independence the rural sector has continued to stagnate, resulting in an increased shift of population to the larger cities, particularly Harare and Bulawayo. Overall in-migration has, however, been partly off-set by the shift of white Zimbabweans out of the city and out of the country, frequently to South Africa. Concurrent with the change in the ethnic structure of the urban areas has been a growth of urban subsistence and petty commodity production.

Mugabe's government, according to Drakakis-Smith, is ideologically oriented to rural Maoism and the African socialism of Zimbabwe's neighbours, such as Mozambique and Tanzania. However, it has been constrained in its ability to alter the social relations of production in Zimbabwe by the continued presence of an economically powerful white bourgeoisie and the promises of compensation made as part of the Lancaster House agreements. Despite this, some progress has been made in transferring rural land from white to black control, although the longer term goal of establishing collective, co-operative and State farming enterprises has had to be postponed. Urbanization strategies have been confined to a concern about containing the growth of Harare and Bulawayo and some incipient discussion of the viability of growth centres. Drakakis-Smith believes the Zimbabwean government's tendency to ignore the system of urban settlements is indicative of its failure to grasp the relationship between capitalism and space.

The experience of Guyana since the declaration of the Co-operative Republic in 1970 is the focus of Chapter 8 by Potter. While Guyana has modelled its development planning on Tanzania and later Cuba, and in 1983 suspended its negotiations with the International Monetary Fund (IMF), it remains on the periphery of the socialist world. Potter's chapter documents the process of decolonization in Guyana after 1970 and notes the optimism with which socialist goals, such as the development of co-operatives, were pursued. However, by the late 1970s the performance of the economy of Guyana had deteriorated considerably and has yet to recover. This put a serious brake on reform and socialist goals have receded into the background.

Potter argues that control over population distribution is important in socialist developing countries, and discusses the series of attempts in Guyana to manage and alter the structure of the space economy. Population is concentrated along the coastal strip and in a few interior enclaves, but the first attempt in the early 1970s to draw up a regional plan for the country with ambitious population resettlement goals failed when the plan was not implemented. The Second Development Plan, scheduled to run from 1973–7, also failed to have much influence on population distribution. The co-operatives programme was intended as the mechanism for settling the poorly populated interior, but it also was not successful. Ultimately, the success of these programmes has been impossible to judge accurately because of restrictions on access to information from the 1980 population census.

A State Planning Commission was formed in Guyana in 1978, coinciding with increasing economic problems for the country as a whole. A new regional structure for the country has been set into place, based on ten local government areas, designed to assume real power and responsibility for regional economic planning. Recent evidence suggests, however, that Guyana is placing a renewed importance on IMF austerity measures for the economy, which may mean a trade-off between economic growth and social and regional equity. In these circumstances, regionalist policies are likely to be given a lower priority than Guyanese planners might have hoped.

Chapter 9, by Susman, looks at spatial policy in Cuba since 1959. The author argues that Cuba's socialist goals include an emphasis on equality in production, consumption and political participation. The chapter assesses Cuba's success in achieving these goals by looking at government programmes to reduce the primary role of the capital city, Havana, to diminish the differences between the urban and rural areas, and to develop the provinces in an equitable way that is consistent with each province's size and resource base. It is argued in the chapter that there is evidence that Cuba is achieving its equity goals in consumption and, to a lesser extent, in production. Political participation is also being broadened, but Susman questions the significance of this to fundamental decisions about national priorities.

The goal of Cuban spatial restructuring is argued to be the creation of a 'landscape of socialist integration', characterized by an equality of access to production, consumption and politics across both the population and space. In particular, the inequalities between urban and rural areas would need to be reduced in order for such a goal to be achieved. The author warns that spatial restructuring is, of course, not the only, or even the main, goal of the Cuban revolution, but one of many features of society and economy transformed since 1959. Nor is spatial reorganization without inherent contradictions. Susman points out that centralized planning is a prerequisite for the creation of a landscape of socialist integration, but it does partially contradict the goal of decentralization and increased local public participation in planning and policy-making.

The concluding chapter 10, by Slater is a critical examination of socialism, democracy and territoriality, drawing on the comparative experiences of Nicaragua and Cuba. In the opening part of the chapter a distinction is drawn between four different approaches to socialism: First is the style of socialism which the author calls social democracy, with its emphasis on parliament and the welfare role of the State, and its tendency to downplay or ignore fundamental contradictions between capital and labour. The second is a form of radical nationalism, exemplified by Tanzania, anti-imperialist and oriented to social reform, but without a structural break from internationalized capital. Third is what Slater labels the Leninist model, in which the state and party embody the interest of the proletariat, controlling the organization of production and strongly authoritarian in practice. Fourth is an approach, based on the work of Gramsci, that emphasises the centrality of popular hegemony and the democratization of 'social existence' as crucial to the meaning of socialism. In a post-revolutionary context, as exists in most socialist developing countries, this requires a guarantee of pluralism. Although they share similarities, Slater argues that Cuba and Nicaragua are different types of socialist societies on these measures, Cuba being of the Leninist model and Nicaragua an example of the pluralist popular hegemony strategy.

Slater compares and contrasts Nicaragua and Cuba in terms of the issue of decentralization, a comparison which allows him to combine the discussion of democracy and territoriality. He argues that in Cuba power is highly centralized, with representation at the local level being for the purposes of efficiency and not for the expression of political alternatives. Although both Cuba and Nicaragua share a fear of the United States and have organized the administration of their territory with this threat firmly in mind, Slater argues that the decentralization process in Nicaragua has been much different to that in Cuba. The example of moves towards the autonomy of the Atlantic-coast region is given to illustrate the seriousness with which the Sandinista government views the need for popular hegemony and decentralized administration.

Acknowledgement

We would like to thank David Marr for loan of the maps which formed the basis of figure 1.6.

2
The Urban Question in Socialist Developing Countries

Enzo Mingione

Introduction: industrialization strategies and the urban question

Socialist experiences cannot be evaluated through a general theory of socialism, mainly because such a theory is either impossible or not yet possible at the world scale at the present stage of development. A scientific theory of socialism cannot be built on a set of political ideals or on vague, abstract and occasional remarks by Marxist theorists. Nor can the specific experience of one, or even a few countries, be assumed to provide a general theoretical model. Starting from these assumptions means it is necessary to confront the different socialist experiences with a pragmatic descriptive approach. This chapter examines the complex interrelation between industrial development patterns and the different processes which characterize the urban question. If we do not discuss the theoretical assumptions of this interrelation, we are forced to rely on a descriptive approach, which means there is no possibility of generalization and little significance to comparative analysis. Moreover, descriptive 'ideal type' approaches often tend to incorporate hidden causation assumptions which are neither explicated nor discussed. This is the case with the very interesting recent hypothesis of Szelenyi (Murray and Szelenyi, 1984; Konrad and Szelenyi, 1977), in which the basic assumed interrelation between industrial and urban processes is never fully discussed. This omission raises serious problems, as I will try to explain in the following pages.

The use of terms like underurbanization or overurbanization is based on the assumption that there is a 'normal' urbanization process typical of industrial development in other experiences. Consequently, these terms appear to have a hidden comparative meaning. They raise questions, even now insufficiently discussed, such as how much industrial development, at different times and in different situations, explains urbanization, and how does the evolution of the urban question feed back into the different patterns and conditions of industrial development?

There are three points to be clarified, at least at a preliminary stage, on the duo-directional relation between industrial development and urban processes: (1) industrial development does not completely explain the urban social question at any time and place, as the latter is also influenced by pre-industrial events and specific socio-cultural factors; (2) industrial development processes are differentiated in time and space so that the impact of industrialization on the urban question is bound to be different at different times and in different regions; and (3) the internal balance between the urbanization and industrialization processes is often only apparent and may well be equilibrated by external factors. These three points, together, need to be carefully explored and clarified otherwise our theorization/interpretative/comparative capacity is substantially weakened.

There is ample historical evidence that cities existed before the industrial age to various extents and in various forms in different pre-industrial social systems. However, even the accurate urban history available for Europe has seldom been utilized to show how much industrial take off relied on and transformed already existing urban systems, and how much it dismantled them in favour of new industrial cities.[1] This argument is more complicated if we use it at the world scale where we find a wide range of pre-industrial urban structures and cultures (from the Japanese system of cities to various tribal nomadic situations) and of times and conditions of industrial take-off and development (or underdevelopment). The simplified assumption that the European mercantile urban structure has been easily incorporated and instrumentalized to the advantage of industrial development does not work even in the few countries where it can be easily traced. While the original English experience of industrialization was mainly directed towards new industrial cities, the French experience incorporated the mercantile pre-industrial urban structure (mainly in Paris and Lyons), while the Italian experience reacted in a selective way towards its rich pre-industrial inheritance (Milan and Genoa were incorporated but not Florence and Venice, and even less the Southern mercantile cities).

This assumption becomes totally meaningless if we take into consideration other areas of the world which have not been involved in the mercantile transition and which had different histories through the industrial age. For instance, latecoming industrial countries, like Japan, or differently colonized countries (most of Africa and Asia), or countries semi-controlled by colonial powers, like China and those in Latin America. We may also assume that the pre-industrial inheritance has not only had an impact on the transition to industrialization but has also contributed to shaping the attitude of the population towards the urban style of life.[2]

Although I maintain that industrial development can explain a large part of the urbanization/urban question in our age, the importance of pre-industrial inheritances and of cultural factors needs to be recognized and taken into consideration. For instance, a combination of fast, labour intensive industrial growth and slow urbanization may mean either that the new industrial workers commute from non-urban areas where they continue to live (i.e. Szelenyi's underurbanization) or that a large urban pre-industrial population is converted into industrial jobs from other forms of employment. This point is relevant when we consider the cases of socialist developing countries where the urban structures and cultures inherited from the previous periods are largely different.

Before going any further, I would like to highlight some technical warnings on the use of urbanization, urban growth and urban structure data in comparative analysis. The data are collected in different ways in different countries so that it becomes difficult to compare them. Most scholars who have made comparative data surveys have had to spend a lot of time trying to adapt the series to each other. But this is only an initial difficulty. The statistical delimitation of what is urban and what is not remains arbitrary and problematic, even within a country, and more so at the international level.

Let us take a clear example from the Italian case (Mingione, 1981). The official population size of the two largest cities, Milan and Rome, does not match the reality. Milan has narrow administrative boundaries and has expanded outside them since the beginning of the century. It is now a large metropolitan city, surrounded by an urban continuum incorporating more than 100 administrative communes. Rome, on the contrary, has relatively large (by Italian standards) administrative boundaries so that it includes even now some non-urbanized or agricultural lands. Italian urbanization trends have always been underestimated, as new citizens settled by thousands in small formally independent communes which in reality were incorporated in the Milan or Turin urban continuum. The results are even more distorted when we come to urban growth trends for such cities as, for instance, the Milan metropolitan area, for there the suburban communes have been increasing at a much higher rate than the central city for a very long time.

When we come to the international situation the difficulties are even more apparent. For instance, it is practically impossible to compare the Chinese data on urbanization and urban growth with those of other countries and even within China. In fact, many large cities in socialist China have been surrounded by agricultural green belts where tens of thousands are active in agricultural communes. This has not happened everywhere in China; for instance, it is not the case in many middle-scale cities – nor is it so in many other socialist developing countries. In most cases the incorporation has meant an enlargement of the administrative boundaries of the city and a sudden increase in the population, i.e. a fictitious urban growth and urbanization process.

I have not raised this problem in order to advocate the complete rejection of the use of urbanization data but to stress their ambiguous relevance and to underline the importance of qualifying them with other sources of information. This problem will become even more difficult in the future when suburbanization reaches further stages of development, and the computer – as well as fast transport and communication technologies – will generate possibilities less bound to specific locations. These events may ultimately revolutionize concepts of what is urban and what is not.

The phases of industrial development, the world system and the urban question

The duo-directional relations between industrial development and urbanization (defined as the increase in the percentage of population living in cities, particularly large cities, and the problems this causes with regard to housing, commuting and transport, land costs and speculation, standard of living, etc.) have been interpreted in various ways. It has been difficult to distinguish the activation of urbanization by new industrial requirements, from the feedback of urbanization in terms of new possibilities for, or constraints on, industrial development. The general idea is that industrial development requires high rates of urbanization for three fundamental reasons: First there is the necessity to concentrate a large number of workers in the expanding, large industrial factories production system. Second, there is the need for a concentrated market oriented towards the consumption of industrial goods. This often requires the parallel progressive dismantling of self-consumption economies which are typical of rural and transplanted semi-rural traditional milieux and cannot easily be sustained in large cities. Third, there are the other conditions-of-scale economies on which the majority of industrial location theories have been based. These three main factors may be considered in the duo-directional sense: industry pushes forward urbanization, and urbanization has a positive feedback on further industrial development.

However, as we have anticipated, the possibility of an internal equilibrium which promotes a balanced accelerating cycle is only apparent for two opposing reasons. First, if the dismantling of self-consumption economies produces enough market expansion, the parallel increases in the cost of reproduction of the labour force (which is no longer reproduced in surviving cheap milieux) can create serious difficulties for further industrial development. Second, if the cost of the labour force is kept under control by depressing the living conditions of the workers (urban poverty, high degree of self-consumption also in the cities, minimization of the urbanization process, etc.) strong underconsumption tendencies put the industrialization process in difficulties.

The scale-economy argument is particularly controversial as a point of balance between urbanization and industrialization. The advantages of urban growth for the industrial units are, in the long run, negatively compensated by high social costs of urbanization and concentration, which have to be paid for by the workers, the state and the industrial units themselves in different terms. Thus, at a certain stage of growth, the scale-economy advantages of a further increase in the size of the city become much less, through diseconomies of congestion, than the costs borne by the community.

These kinds of disequilibria have been historically compensated by exogenous (not immediately related to urbanization and industrial development) factors. That is to say, the industrial/urban development process has always produced high internal incongruities. This has been manifest in either a deficit in the market – in urban terms insufficient infrastructure or poor housing – and/or a deficit in the profitability/competitiveness of the socio-economic system such as indebtedness, urban fiscal crisis, competition and high costs of survival, etc. Therefore, we aim to demonstrate how, at different times of industrial development, the internal incongruities have been compensated provisionally by external factors: there is no balanced combination of industrial development and urbanization. Urban social problems – even if we succeed in isolating those which are connected immediately with industrial development – vary significantly in different times and places.

The *first phase* applies strictly to early English industrialization. Other countries industrialized after this phase was already mature (Mingione, 1981). The available documentation – mainly the Blue Books, i.e. the parliamentary inquiries on lifestyles of the urban poor in nineteenth-century Britain – shows fast urbanization in extremely poor conditions: overcrowded housing in slums, absence of any hygienic services, high mortality, etc. We may even assume that for some decades the urban standard of life remained worse than the rural one. Industrial development was characterized by the indiscriminate exploitation of all kinds of labour for very long working hours at a very low undifferentiated rate of productivity. The industrialization process was slow – because of the low level of productivity the increase in production was not becoming accelerated – producing an internal market deficit. This was temporarily overcome by the international demand for English industrial goods by foreign middle classes, and the demand for English technology and tools by foreign industrialists. The urbanization costs were partially contained by the use of pre-existing structures – for instance, if we take housing, the old poor areas, the servants floors and stables in the palaces of decayed aristocrats became overcrowded with newly urbanized workers.

The feedback from this circuit is only provisionally positive: the increases of the costs of labour and of urbanization are minimized at the expense of the intensive expansion of the aggregate demand up to when the competition increases and pushes for accelerated productivity gains.

In the *second phase* aggregate demand has become insufficient and the equilibrium is broken. After a critical period, characterized in every industrialized country by frequent overproduction crises and by relevant social, political and technological changes – like the legal abolition of child employment, the birth of modern trade unions and socialist parties, a decrease of female employment in manufacturing industries, a strong decrease in costs and times of water transport, an enormous expansion of the railway system, etc. – a new provisional balance becomes established. This transition has had enormous consequences, which are too often forgotten in the prevailing evolutionary interpretation of historical changes, for the urban style of life in the industrialized countries. But the new internal equilibrium between industrial development and urbanization is only apparent because it is largely subvented by external resources.

It is assumed that the new balance achieved by western industrialization represents an optimal or normal or ideal degree of urbanization able to sustain a continuous industrial growth. The increase in the numbers of industrial workers, selectively homogenized (Edwards, Gordon and Reich, 1982), produced a consistent rate of urbanization (also because a large part of the female and child workforce was no longer working) and the increases of aggregate demand towards mass-consumption standards appeared sufficient to absorb even more consistent increases in productivity. From another point of view, the increases of labour productivity appeared sufficient to grant profitability, even in view of substantial increases in the cost of labour which are less and less likely to be reproduced in a cheaper self-consumption rural environment.

In reality the industrialization process is accompanied by an enormous expansion of international operations – it is the age of early imperialism – to the disadvantage of Third World food and raw materials production. Thus, on one side, the difference between the more consistent increase of productivity and less consistent cost of labour is absorbed by the horizontal expansion of the aggregate demand for industrial goods in the colonies and in Latin America. On the other side, the deficit produced by increasing urbanization costs is largely paid for by the high profits of the imperialist operations. The world crisis of 1929 put an end to this phase and opened a relatively long transition stage. Eventually, the uncontrolled increase of industrial production was no longer compensated for by the increase of aggregate demand.

The world industrial system had to be largely destroyed by the crisis, and by the war, before it could find another kind of equilibrium capable of expansion, which has led to a different interrelation between industrial development and urbanization. The specific characteristics of the *third phase* are more difficult to synthesize as they are articulated in many different horizontal (international and socio-geographical) and vertical (labour productivity, tertiarization, technological gap and stratification, etc.) situations. However, I will first raise some points concerning industrialized free-market countries and developing countries, and following this I will translate these observations into a preliminary interpretation of the situation of socialist countries.

In the *industrialized areas*, further industrial expansion requires both a segmentation of the working processes and labour markets, and a progressive increase in the size of the modern service sector. Moreover, the segmented reorganization of working processes in manufacturing industries is oriented towards consistent increases in the productivity of the primary dependent segment, which means also a progressive tendency towards labour savings and the decrease of employment in manufacturing industries. The tendencies towards overproduction are massively counterbalanced by market expansions in the internal (complete disappearance of differences between the styles of life and consumption in the city and in the countryside) and Third World agricultural areas and by the progressive increases in state social expenditure and economic intervention. The levels of aggregate demand are increasingly financed by the uncontrolled rise of the state deficit and international debt of most Third World countries.

On the other hand, the progressive increase of labour costs in industrialized milieux (in the suburbanized, congested and expensive city) is partially kept under control because of the pronounced segmentation of labour processes and markets which means potential savings in restructuring operations (industrial decentralization and relocation, shifts from more expensive to less expensive working processes without parallel losses in productivity, subcontracting, etc.). Urbanization proceeds by extension. The city becomes a very differentiated region without much continuity. The extremely high costs of this kind of urban development (restructuring, rapid obsolescence of buildings and infrastructures, congestion, long and complicated commuting systems, etc.) are among the means used to prevent overproduction, at least provisionally, so that they are welcomed until the fiscal crisis of cities (in parallel to that of the state) intervenes to break the dream of unlimited expansion. *Ex post* it has been noticed how this kind of balance has been at the same time spectacular and fragile. But it remains difficult to say where the current transition is leading.[3]

In the *developing countries* the world-system operations of this phase have meant massive dismantling of subsistence, rural or tribal economies and a strong acceleration of the urbanization process. Urbanization tendencies in the Third World are not exclusively activated by migration from the countryside. However, the penetration of capital in agriculture and the dismantling of subsistence agriculture, integration into the world food market, and so on, create high rates of rural overpopulation and help force villagers to migrate to the cities. No matter how poor and difficult appear the working and living conditions of Third World masses in large cities in the slums, *barrios*, *favelas*, and squatter areas, they remain comparatively better than those of the countryside.[4] This is contrary to what happened in the industrialization and urbanization process in western countries. The point is that a large part of the international debt of these countries is created by the need to finance urbanization and urban standards of life. The deficit is largely absorbed by the quantitative multiplication of the problem, and by the uneven internal distribution as the middle and upper income groups absorb the largest share. It does not produce a visible amelioration of the condition of survival of the urban masses, a fact which has often misled commentators on the phenomenon. Even the relatively poorest urbanization processes are substantially financed by the foreign debt of Third World countries (even more so in countries which are involved in industrial relocation processes, such as South Korea, Taiwan, Brazil, Mexico, etc.) (Gilbert and Gugler, 1982). From another point of view this factor has been essential to absorb the surplus industrial production of industrialized countries and prevent a massive, generalized overproduction crisis.

If this is true then the urbanization process in the Third World is a vicious circle, reproducing very poor conditions of survival, insufficient employment at very low incomes and uncertain working conditions (although much more expensive than in the countryside which is reflected in relatively higher standards of living). Yet it is an essential moment of the equilibrium of the industrialized powers which matured in this phase. The so-called overurbanization of the Third World countries is the negative aspect of the growing difficulties in maintaining a balance in mature industrialization; that is, keeping control over increasing urbanization costs, controversial employment levels and the insufficient capacity of the world market to absorb the growing surplus produced by a decreasing number of workers. These difficulties are shifted towards the Third World in terms of urban under- and unemployment, and the incapacity to provide and pay for the costs of urbanization.

It is interesting for our purposes to add here a few observations on the character of underdevelopment, urbanization and the urban question. Greater attention needs to be devoted to the roles of the building sector and to the so-called 'informal sector' (see Mingione, 1984).

A large part of the growing demand for manufactured goods – either expensive durables for middle classes, or cheap industrial subsistence goods for the worse-off, or production and service technology – is inevitably addressed in a free world market to the mature producers of the industrialized areas. Urban growth is activated and further activates the expansion of the building and informal sectors, mainly in local service production such as transport, trade, restaurants, crafts, etc. The building operations take place at very low productivity and wage levels, which means large numbers of workers are employed irregularly at uncertain and low wages. Urbanization promotes an uncontrolled increase of these workers, and eventually the settlement of their families. They are neither able to pay the full costs of their urban habitat (relatively high costs of regular housing, shopping, transport, etc.), nor are they able to acquire more stable and higher income jobs. This means that there is an increasing necessity for the expansion of an informal sector to provide cheap goods and services for the increasing population without access to regular market consumption.

The enormous pressure of this kind of urban surplus population freezes working and living conditions, but at the same time increases the demand for industrial goods produced in the more advanced areas (which is reflected in the increase of the foreign debt). Even when industrial relocation and decentralization is consistent, low wages, the persistence of surplus population, and the necessity to buy and sell in the international market to supplement an internal production which remains incomplete and/or export oriented and/or technologically dependent, prevents a sufficient tendency to accumulation.

It is interesting to note how the planning of new towns, even those monumental capitals intended as mirrors of a new modern era, like Brasilia and Chandigarh, do not escape the problems of the underdevelopment urbanization cycle.[5] Thus, self-built slums and barracks areas and the informal trade and craft sector characterize these cities as well (see, for instance, Sarin, 1982). The building conditions necessarily activate the situation in combination with unequal market exchange, unless the building workers are paid relatively high wages and then shifted into administration or expelled from the city as soon as the building process is complete.

Socialist industrialization, underdevelopment and the urban question

After the Second World War a number of Third World countries decided to follow a socialist path of development. Following our consideration of the industrial/urban question in advanced and developing free-market countries, we shall now discuss what we assume to be socialist regimes, in their multiplicity of pragmatic forms, and whether socialist regimes have been able to modify the constraints of underdevelopment and of poor and unequal urbanization paid for by an increasing foreign debt.

The decision to adopt a pragmatic evaluation of socialism, instead of an idealistic one, differs from the approach I have followed in previous works on the subject (Mingione, 1977a; b; 1980, 1981). The reason is that the analysis of different socialist experiences has revealed such a great distance between the reality and hypothetical abstract ideals of socialism, that the latter approach is of limited use in understanding the character and specificities of any socialist experience. On the other hand, any uncritical acceptance of the self-definition of a regime as socialist could be arbitrary and create serious problems.

Is it possible to spell out minimum pragmatic orientations which characterize more or less every current regime to be defined as socialist? I would suggest two criteria:[6] first, a relatively high degree of central political control of the exposure of the local economy to world-market relations; second, a centralized system of redistribution of resources – to both consumption and production ends – along lines which are not only different (for this is a characteristic of any modern mixed-welfare economy) but largely divergent from the market distribution system. I am aware that these criteria are not fully convincing and remain problematic along lines we will shortly discuss, but they seem the only ones to be easily traced in different forms and degrees in every country which is commonly considered a socialist one.

The political control of the exposure of the local economy to international market relations is a problematic factor. It is subject to a set of independent constraints, such as the size of the country, the availability or otherwise of a wide range of raw materials and survival resources, the phase of development of the local economy and the attitude of the world market itself towards the socialist country. Historical experience shows that the control of international exposure may be very high if the country is either large, with a great variety of raw material reserves, or very small, without raw materials but relatively self-sufficient in food. Again, the capacity to control international market exposure is higher at a relatively undeveloped stage and it becomes more and more difficult to achieve when the country reaches a stage of varied industrial production. Finally, the capacity to control exposure may be externally supplemented by the hostility of the world market towards the country. Or, on the contrary, it may be weakened by world market strategies of penetration at whatever conditions are posed by the regime (we can take the example of China during the period 1949–72 which was marked by hostility, and China afterwards was characterized by penetration; or the current cases of Cuba, which faces hostility or Yugoslavia or Poland which have penetration).

Middle-size countries at an intermediate stage of industrial development, favourably considered by the international market, have a degree of exposure to the world market which is not very different from countries which are not considered socialist. This may be the present case in

Yugoslavia when it is compared with Italy, Spain or Greece. While the resulting *de facto* situation may not be very different in terms of foreign indebtment and balance of trade, the *de jure* institutional difference plays a relevant role in international capital operations. In the case of socialist regimes, every single operation abroad (import–export finance and investments), is authorized and inspected by the central government, which is nearly always the national contractor, while in the case of free-market countries these operations are only bound to respect general rules and laws. Thus, international capital operations with socialist countries are bound to be rather complicated. In particular, central socialist governments will trust some foreign contractors more than others, for reasons other than the purely economic, and prevent operations which they consider to be damaging in the long run for national interests, even when immediately advantageous and feasible.[7]

But the more a socialist country needs foreign resources for its own development and survival, the less it is able to keep its operation abroad under control because in international trade it becomes subject to international market rules in terms of prices and profitability. Complete isolation from the international market, or at least from its main capitalist stream, is not a political option open to every socialist country; indeed, is an exceptional event. This factor should be borne in mind when considering the internal development/industrialization strategies of socialist countries. In fact, different strategies, independently from other consequences and ratios, may produce increasing or decreasing interdependency with regard to foreign resources or markets.

As we have mentioned, the second characteristic of a pragmatic socialist country is the redistribution of internal resources along non-market lines.[8] This question becomes problematic if we consider it a long-run process. Relatively limited or insufficient resources characterize socialist developing countries, so how does a centrally planned system expand production and increase the availability of resources in comparison with a market system? How is the socialist redistribution different from the market one in promoting economic growth and industrial development? As we do not want to enter into an extremely complicated debate on socialist accumulation we will limit ourselves to a few observations which appear immediately relevant to the analysis of the urbanization question in socialist countries.

In practically every case, socialism emerged in countries still predominantly characterized by a large agricultural sector and by relatively undeveloped industry. One of the most important socialist goals has thus been to increase the surplus produced in agriculture and use it to promote industrial expansion. The fact that this shift did not pass through the hands of landlords or agricultural capitalists makes a great difference in the balance of power and directions of development, but not in some of their important economic and social consequences. Increases in agricultural production through productivity gains have meant a decrease in agricultural employment and a shift of workers to industrial production in those sectors which became favoured by the state planning officers in their redistribution of the available resources. The original shift has happened in most cases in peculiar conditions: (1) the initial substitution of agricultural commodities by goods and raw materials imported at decreasing prices – as happened in most other late-industrializing countries (Mingione, 1981) – was not possible because of difficult international relations; (2) for the same reason imports of foreign investment and technology were rather limited (in many cases to those available from the Soviet bloc); (3) the transition took place when world average industrial production was already mature in many sectors, even more so in the Third World countries which have joined the socialist club in the last thirty years; and (4) socialist countries had different pre-socialist industrialization bases but none was built upon the *tabula rasa* of early industrial development in Great Britain.

If we sum up these observations we can see that different socialist regimes have had to face and manage an extremely problematic situation: agricultural accumulation was at the same time too fast and too slow to achieve the goal of a rapid, balanced industrial development. Agricultural accumulation proved too fast in creating a relative surplus population, which could not be immediately absorbed by industrial development at a relatively high productivity ratio, and too slow to keep pace with a much faster industrialization process, where even relatively backward technologies allowed consistent productivity gains.

The centrally planned economies have been able to keep this increasing incoherence under control from the redistributive point of view – avoiding increasing differentials in productivity and consequently increasing distance in income, profitability and convenience – but they could not avoid paying the relevant social costs attached to the redistributive solutions they adopted. In many cases, agricultural rationalization was slowed down through protection of farmers' income and food prices to prevent the formation of too large a surplus population. This means that few socialist countries are food exporters, that many of them have food shortage problems from time to time, and that the *per capita* average productivity of socialist agriculture is well below the capitalist advanced one, although above that of subsistence agriculture in Third World free-market countries.

The plan for industrial accumulation has usually given privilege to some sectors such as heavy industries, machinery and energy production, primary subsistence goods, tools and means of production and, more and more in the last decades, military production, to the detriment of others, particularly individual mass consumption durable goods.[9] This strategy initially worked as an accelerator of industrial development, but it later created substantial difficulties. In fact, the policy to undermine the development of mass-consumption goods production and the slow growth of productivity in these sectors was reflected by limited growth of the internal market, accompanied, in many cases, by an increasing waste of consumption/accumulation potentials.

Consumers have the money to buy but there are not enough goods of the required quality which means that, as prices are fixed and imports are under control, a substantial part of the consumption capacity is captured by black/secondary illegal markets which are not under the direct control of the centrally planned economy, or by non-accumulative or low accumulation sectors like internal tourism, restaurants, artificially expensive foods, etc. Thus, socialist industrialization matches the overproduction tendencies of free-market industrialization, but becomes unable to keep pace with productivity in every sector of an industrial mix of growing complexity and faces increasing dangers of underproduction.[10] Paradoxically, the model, as sketched here, appears oriented to keeping a much stricter control on the cost of labour than any free-market industrialization process. This fact is confirmed by the evidence that a growing number of corporations find it convenient to decentralize or subcontract their activities in socialist countries, even when they have to go through very complicated bureaucratic procedures. Furthermore, we will find some confirmation of this idea in the general and specific relations between socialist industrialization and urbanization.

There are two preliminary points we wish to raise in general on the pragmatic relation between socialist industrialization and the dynamic of the urban question: the first concerns the assumed absence of land-rent and housing speculation; the second concerns the political attitude within the specific socialist revolutionary process towards agricultural classes in respect of non-agricultural ones (industrial workers, intellectual urban *elites*, etc.).

Under a centrally planned system we can assume that the production/distribution/consumption of the 'built environment' – residential stocks, administrative or collective consumption building of various kinds, transport facilities, etc. – is not disturbed by land, property or building speculation. In the free-market system it has generally been noticed that this speculation plays an important role in artificially increasing costs and prices and in generating the 'housing question'. This difference appears important. But the problem is whether the absence of speculation alone is able to solve the 'housing question' in the long run. In other words, can we assume that the absence of rent speculation automatically permits production of a sufficient stock of built environment resources at prices which are easily accessible to every stratum of citizens, independent of their productivity/income? In addition, we must check whether in centrally planned systems there is a degree of persistence in the priority allocation of scarce land resources (e.g. central areas of cities, areas which are better off from the existing transportation or infrastructure network point of view, etc.) to certain ends independently of the profitability of the venture.

The absence of land speculation certainly makes a great difference. Prices of housing do not sky-rocket artificially to unbearable levels, confining the poor to very difficult housing and living conditions; urban colonization paid by collective investments is not to the uncontrollable private advantage of a few rich land speculators and the real costs of urbanization are kept under control. But these differences do not mean that the question is completely cancelled out. In any case, urbanization costs absorb an important share of resources. Central planning agencies redistribute the advantage equally among different social groups (which is extremely difficult if the situation continues to be one of scarce resources) and some less productive urban groups receive more than they produce while others may receive less. In addition, we should note that the high priority that centrally planned economies have usually given to a model of fast industrialization has penalized some urbanization investments in the allocation of land resources. In fact, if the strategy is to increase at all costs the production of heavy machinery, means of production or military industries, housing and some urban infrastructures are bound to be penalized.[11]

Within this framework we have also given a partial answer to the second part of the question. If existing land resources are scarce and different, in terms of advantages, it is clear that socialist planners will allocate them according to a system of priority which may not appear to be totally different from the market priority of profitability. The reason for this is that both systems are, at the present stage, oriented to industrial accumulation, i.e. to the increase of productivity in the industrial system to make it capable of producing an increasing quantity/quality of resources. The fact that they have different internal priorities and a different way of working does not mean that the outcome for the city is any different. For instance, the more heavily infrastructure-endowed areas of the inner city in both capitalist and socialist countries, host sophisticated tertiary and bureaucratic activities, because in one system it is more profitable and in the other it is a more productive/accumulative destination.

But in socialist regimes there is another absolute priority which often works in the completely opposite direction in respect of socialist industrialization goals. It is the necessity to achieve and maintain a minimum level of survival for every inhabitant. This is an absolute priority because otherwise the specific popular consensus on which a socialist regime is based fades away. In a pragmatic way the existence of this kind of priority, and its different combinations with the socialist industrialization priorities, contributes much to the understanding of different urban and urbanization policies in different times and situations. It is more likely that the survival-level priority plays a role largely unrelated to industrialization goals at the initial stages and in more undeveloped countries.[12] The redistribution of existing resources and short-term investments is oriented to the rehabilitation of survival capacities of millions of peasants and urban dwellers, who were previously under the minimum survival levels. Once the rehabilitation has reached an acceptable level, there is a consistent shift towards socialist industrialization goals, although attention is still devoted to the necessity of maintaining and developing the minimum survival level.

The second question – that concerning agricultural classes – has been largely underestimated. One reason for this is the assumption that socialist revolutions are moved by the same precodified 'Marxist' goals and are achieved by an urban proletarian movement. It is assumed that peasant movements are indifferent to socialist development. From a pragmatic point of view it is true that the large majority of the socialist revolutions were led by an 'urban' intellectual elite and achieved through the essential intervention of urban workers and students (Gugler, 1982), even in the underdeveloped countries where the majority of the population was living and operating in the countryside (not only in the early case of the Soviet Union but also in the more recent cases of Cuba, Nicaragua, and Mozambique). The only clearly opposite example is China. But there are also intermediate cases, such as Vietnam and Tanzania.

The point is that it is wrong to assume that there will never be an important peasant role in the revolutionary process. The priorities of a newly constituted socialist regime may change a great deal if the revolution was also based on a peasant movement which became essential for the success of the revolution itself. The minimum survival-level priority may be more urgently pursued in the countryside than in urban 'non-revolutionary' contexts. The resources used for basic socialist industrialization may be allocated in a different way so that the agricultural milieu benefits directly from investment in means of production, new infrastructure and services, and basic consumption goods. A comparison of the Chinese experience with others, even during the initial stage when China was strictly allied with the USSR and in theory adopted the Soviet model, shows that this difference played an important role.

Now we are prepared to consider very briefly the urban/urbanization question in general. Here we can assume as correct at least part of the Murray and Szelenyi (1984) periodization of socialist development. I would distinguish the following three periods: first, the stage of consolidation of the revolution and pursuit of immediate goals; second, the stage of achievement of socialist industrialization; and third the stage of more sophisticated industrial development in a wide range of sectors. We will not consider the third stage, as few countries have entered it and we do not have sufficient documentation to do so. In the initial stage various alterations of functioning of socialist industrialization trends – as they occur in different typologies during the second phase – may be recorded. Initial conditions of importance can be summarized as follows:

1. the industrial and urban structure inherited from the presocialist situation, inclusive of the level of industrialization and the role of the national economy in the international world system within the system-comprehensive stage of development (in this sense a certain situation in industrial development and a certain urban structure will make a difference in 1917, in 1949 or in the 1970s, for example);
2. the role played by the policy priority of guaranteeing the urgent achievement of a minimum level of survival conditions in both urban and rural milieux;
3. the balance of revolutionary bases in urban and rural classes;
4. the possible existence or absence of foreign inputs and/or of a strong world-market conditioning the newly constituted socialist regime.

The mix of these disturbances has produced on the urban/urbanization question different situations which range from forms of radical de-urbanization (in the extreme case of Kampuchea) to the massive influx of refugees into cities during the civil war period in the Soviet Union.

I believe that the interpretation of this *first stage* should be based mainly on the priority of achieving the minimum level of survival possible in the existing national and international economic conditions. There is not only a policy of redistribution in favour of rehabilitation of urban survival levels. If the existing urban resources are insufficient to achieve this, it becomes necessary to shift population to the countryside where it is likely to be easier to achieve minimum survival levels and to contain the rural–urban migration process by stopping migration to large cities, where life is more expensive. This explains most of the differences at the initial stage and why Murray and Szelenyi (1984) find a large number of typologies in the first phase of socialist development.

The character of the revolution may also be important. The extreme cases are an urban insurrection on the one hand and a national civil or independence war fought mainly in the countryside and with a decisive input of the peasants on the other. The former case will give priority to the rehabilitation of survival in the large cities and, by doing so, will not be able to prevent a certain degree of migration to the cities themselves. The latter will be generally hostile to large cities and give a certain priority to the rehabilitation of survival in the countryside and will consequently minimize, for quite some time, any urbanization tendencies.

The *second phase* is interesting because it is closely associated with socialist industrialization processes. However, it is difficult to comprehend for the very reason that we do not completely understand what the general mechanisms of socialist accumulation are. We have already mentioned some ideas that can be deduced from the various long-term socialist experiences. In synthesis they are:

1. the redistribution of resources in favour of forced industrial accumulation in some sectors (heavy industry, means of production, energy production, military and defence) against some light mass-consumption sectors;
2. the possibility that there is indifference to the levels of productivity – which can remain widely differentiated – in favour of absolute increases in production and the attainment of full employment;
3. the possibility that the previous indifference to 'market' redistributive criteria is weakened by exposure to the world-market system.

There may also be present two tendencies which appear to be quite different from those taking place in free-market countries: a tendency towards underproduction (the mass consumption sectors expand slowly and a growing part of the wages cannot be spent within the socialist consumption sector and is therefore spent outside it on petty commodities or in the black market); and a tendency to maintain at a relatively low level the reproduction costs of the labour force. These tendencies are immediately reflected in the development of the urban question in socialist countries, but of course not everywhere in the same form. It is common for socialist developing countries to provide the minimal standards of survival in cities assured by state welfare provisions (low-cost housing – even if overcrowded and very basic – transport, education and health facilities for every citizen), but not to spend heavily in the cities.

As the centrally planned systems have to maintain the minimum survival standards and update them to keep in line with development trends they also have to face the danger of erosion of accumulation potentials by the urbanization deficits. This is usually achieved by a controversial (because it produces social costs) combination of savings options. Those options that have been more clearly identified include: discouraging the urbanization of new industrial workers who are forced to maintain a non-urban residence and commute to work; decentralization of new industrial initiatives in small- or medium-scale towns where the urbanization costs remain lower; decentralization of a wide range of industrial activities in the countryside; and the use of some hidden low productivity/low-cost resources in the large cities to produce some urban services (such as the assistance, on a voluntary basis, of the old, young and female population). As Szelenyi (1983) made extremely clear the redistribution criteria of urban and, particularly, housing resources in favour of attracting into expanding cities certain strata of bureaucrats, managers and industrial technicians and professionals may create unplanned serious social unbalances and discriminations. Depending on the balance and degree of these savings operations, these policies produce situations which have been described as underurbanization (Konrad and Szelenyi, 1977), contained urbanization or zero-urban growth. But also in socialist countries the industrialization and urbanization processes are not free from problematic social costs and contradictions.

Here we have to face the opposite question to the one concerning the free-market countries: to what extent is the incoherence between socialist urbanization and industrialization internally produced, and how much is it produced by the degree of exposure to world-market relations? Unfortunately, we cannot yet answer this question because we do not fully understand in theory the mechanisms of socialist accumulation. From observation we can argue that world market exposure is responsible for a part of the incoherence, but there may also be internal reasons which are difficult to detect in a complex situation. The externally induced part of the contradiction is due to the necessity to develop a productivity-oriented industrialization in some modern sectors. This means more concentration and more urbanization, an acceleration process which is out of socialist control, both as regards urban costs (and the reproduction costs of a growing part of the labour force) and as regards the consumption structure.

The mix of incoherences of different paths of socialist industrialization and different world-system situations makes comparative analysis and generalization extremely difficult. However, it is useful to check our hypothesis through a preliminary schematic interpretation of the Chinese case. The specifics of the Chinese experience are both interesting and unique, and perhaps easier to explain than are other developing countries' experiences which are insufficiently explored or which have been established for only a comparatively short time.

A preliminary confrontation with the Chinese case: from a descriptive model to some interpretative assumptions

It may be useful to begin by stressing the methodological limits of an analysis of a case such as China. We have three different series of inputs to the study of the Chinese experience (approximately the same as in other cases where we cannot rely on accurate extensive local research on the subject). First there is macro-institutional data originating either from the Chinese government or from foreign and international corrections of Chinese data. For instance, there are at least three population series for the period 1950–70. This was when western demographers did not trust the Chinese information, and it was estimated that the 'real' situation varied by between ten and twenty per cent from Chinese assessments. This caused great changes to be made to both the *per capita* economic and demographic data. The second source is the direct but occasional scientific observations of foreign visitors and the few local studies. The third input is the Chinese accounts in the politicized style of *Peking Information* or *China Quarterly*. As I have stressed in previous works on the subject (Mingione, 1977a,b; 1981), to obtain an approximately acceptable result we shall combine carefully all these three inputs and place a guarded reliance on the Chinese reports. If we do not adopt a very careful critical approach it is easy to misinterpret macro-data if they are not accompanied by a deep knowledge of the social reality, particularly in the Chinese case.[13]

The Chinese post-revolutionary experience can be subdivided into three periods: immediate post-revolutionary reconstruction (1949 to mid-fifties); the phase of internationally isolated industrial development (mid-fifties to the mid-seventies); and the phase of industrial development which includes greater exposure to the world system (mid-seventies–mid-eighties) (see also Basso Farina, 1980).

We can understand the urban question in the *immediate post-revolutionary* period as a combination of problems posed by the pre-revolutionary character of Chinese urban and social structure, the stage of world industrial development, and the specific character of the Chinese revolution which involved the peasants to a much greater extent than other socialist revolutions. China not only had a low level of urbanization (ninety per cent of the population was living in the countryside) but was also very unevenly urbanized with the urban population concentrated in a double system of cities. The city system consisted of the administrative capitals and centres inherited by the empire, and the commercial coastal or river cities where foreign powers had established their bases (mainly Shanghai, Guangzhou, Tianjin). Both networks of cities were considered largely unproductive by the revolutionary leadership and both contained a large population surviving in slums or in very poor areas under minimum survival standards (this was more pronounced in the commercial cities than in the administrative/traditional ones). The initial policy was to devote an increasing quantity of resources to create industrial activities in the existing cities and to expand those already located there. The emphasis was on labour-intensive processes in order to absorb as much of the workforce as possible into productive activities. At the same time there was an effort to rehabilitate the housing, urban infrastructure and facilities, but with great attention to saving as much as possible and to using hidden and semi-hidden resources and the voluntary activities of the population itself.

The available accounts of housing rehabilitation of the slums of Shanghai resemble the strategies which have become fashionable and practised recently in Third World countries with World Bank support. The state provided materials and restructuring projects and the population itself did the job in a relatively short time. Housing remained overcrowded and basic, by western standards, but the condition of survival quickly reached a level which was well above the previous level and that prevailing in most Third World cities.

Although we do not have exact figures, some of the urban population was forced to move back to the countryside. The migration was more consistent in the very large cities of Shanghai and Guangzhou and less consistent in the administrative centres, in cities surrounded by already overcrowded or poor agricultural settlements and in medium-size cities. This movement was mainly directed towards the nearby countryside, when possible, and oriented towards the establishment of an intensive vegetable growing agricultural green belt around major cities. In a way the programme (which was also pursued in the second phase) resembles the spontaneous mix of urban and agricultural survival economies which can be found in the informal sector of most Third World cities. But in the Chinese case it was highly organized and more divided, so that suburban subsistence agriculture tended to become very productive and specialized in the long run. By the early 1970s the suburban communes surrounding Shanghai, Guangzhou and Beijing were among the richest and most productive agricultural ventures in China.

Although during this phase China was a close ally of the Soviet Union and seemed intent on following a model of forced industrialization through the building up of heavy industries, the dominating rural character of the revolution was reflected in Chinese plans for the increase of agricultural output to be utilized to ameliorate the conditions of survival of the great majority of peasants. This was also reflected in attempts to halt almost completely migration from the countryside (which has not been the case in other countries where the dominating urban character of the revolution was reflected in increasing differentials in the standards of life in favour of the cities and finally in strong urbanization pressures).

The two decades (approximately) of the second phase of *internationally isolated industrial development* can hardly be described as a homogeneous period. They are characterized by various experiments in different directions, including the Great Leap forward, the following critical period (due mainly to the combination of natural negative events – poor harvests and serious floods – with the negative effects of the too radical and unbalanced policies of the Great Leap forward and of the sudden total international isolation of China), the consolidation of the communes' system, the Cultural Revolution, and the transition period before the death of Mao. The discontinuities of these sub-periods are also reflected in the urban question and policies. This was particularly important in the first decade, when the failure of the agricultural development policies of the Great Leap forward generated an uncontrolled wave of rural–urban migration which, in turn, resulted in policies aimed at sending large numbers of migrants back to the villages. In general, these twenty years have been described as a period of zero-urban growth, but the reality is in many cases different, not only because of the time discontinuities – which we cannot describe here in detail – but also because of the results. I will try to explain how the urban/urbanization question changed during this long period of socialist industrialization in isolation from the world market.

The emphasis of industrial development was differently articulated in three different kinds of location: the large cities; the medium and small peripheral (in relation to the coastal urban structure) cities and regions; and the countryside. The large cities and the north-eastern region had already become at the beginning of the period the favoured location of heavy industries. There they could find an already established working class and a more favourable economy without attracting large waves of new migrants because the area was already 'overurbanized'. Here the Chinese leadership devoted a great deal of attention to discouraging the multiplier effects of the industrial concentration and keeping control of the development scale.

In fact these are the only zero-growth (and, in Shanghai's case, eventually de-urbanizing) cities.[14] Once they had reached the stage where they could absorb the presocialist population, new industrial investments were shifted elsewhere, even if this penalized the general rate of economic growth. In these areas great attention was given to the exploitation of hidden resources even when they were at a lower rate of productivity, like housewives' workshops, mixed agricultural–industrial settlements, small-scale, light-labour intensive ventures etc. Economic activities not strictly profitable from a capitalist point of view, but contributing to absolute increases of output and maintaining full employment, became established. In this way cities achieved a non-specialized industrial mix at a wide range of productivity levels. Moreover, the agricultural green belt was strictly enforced to make the cities largely self-sufficient in food. Thus, these cities assumed a physiognomy different from the suburbanized western-style metropolitan areas and also from socialist cities in the eastern European countries.

Workers in Beijing or Shanghai did not get better housing, transport or health facilities than in the smaller cities, and a great deal of saving and shifting resources from these areas to others was continuously practised. Improvements in living conditions have been slow but more equally distributed. This is reflected in different 'generations' of housing programmes in different locations: space and facilities *per capita* improved slowly and in a relatively undifferentiated way. Workers living in a 1970 block are not substantially better off than their colleagues living in one built in 1960. Overcrowding was absorbed to a certain extent but housing remained basic: a room for a couple, one for the children, a bathroom for every two families, a large kitchen to every three or four families, various community facilities in every block and other leisure/cultural/ health facilities for every number of blocks. This standard was set independent of the size of the city and the income or profession of the citizens.

The most important share of the industrialization resources was shifted to small- and medium-sized peripheral cities, and to the peripheral countryside, at the expense of the growth rate, but avoiding risks of overproduction. Some medium- and small-scale cities in the peripheral areas grew at a tremendous rate during the two decades[15] and absorbed a very large part of the surplus agricultural population created by rationalization of the agricultural labour processes.[16] Another large part was absorbed by the industrial activities of the communes. This took place in every industrial sector independent of the productivity ratio. Thus, small-scale steel, chemical, mechanical factories and workshops, some of them not registered in the national accounts because they were oriented only to local needs (like producing fertilizers with local raw materials for the exclusive use of the commune or the repair of agricultural tools and machines), became established even in the most remote mountainous and poor regions.

Visitors to China in the early 1970s were surprised to notice the degree to which the consumption system was equally distributed: large shops in Beijing often had the same goods in stock as shops in remote villages. Consumption in the remote areas was largely subsidized by the state at the expense of the rate of development. This reduced the attraction of the large cities and helped to avoid the long queues in some central areas which have been noticed, for instance, in cities like Moscow, Warsaw and Havana. However, these diffusion policies did not completely prevent a certain degree of concentration and attractiveness of cities. Higher education, advanced research and tertiary and political activities could not be diffused, so that the government periodically had to face the resistance of young people not wanting to go back to the countryside or to remote regions. Some were deported back by force, some remained and some returned illegally, so that the rate of urbanization was not totally under control and, in any case, it produced a certain degree of conflict and dissatisfaction.

The present phase of *greater exposure to the world system* is not well known enough to produce mature reflection, but it stimulates some questions. The first is whether the opening and exposure (although strictly controlled in the sense that even now China has avoided acquiring a huge foreign debt) to the world market has been an open political option or, on the contrary, the natural consequence of the expiry of the previous decades' development character. In other words, is it possible that the Chinese closed industrialization process had reached a deadlock, so that to gain further achievements it became necessary to become more exposed to the world system? Probably the answer to this question is, at least in part, positive. The tendencies towards underproduction within a closed system had become so great that the Chinese leadership nurtured strong convictions about the necessity to become exposed, to achieve a higher productivity and technology ratio and to increase the production capacities of mass consumption goods so as to promote further development. But in history it is quite impossible to say whether alternative options can be practised and where they lead if they are not effectively practised.

The other question is where the new policies will lead in terms of urban/urbanization perspectives? Although there is very little evidence[17] at the present stage, we may assume that renewed urbanization and concentration tendencies are taking place. An increased orientation towards the productivity ratio and individual differentiated mass consumption produces unavoidably selective effects on territory. The areas where central planning agencies, technological research, higher education and advanced industries are located become naturally more attractive: an increasing number of young people from all over the country, but particularly from the remote peripheral areas, will desperately want to move to the 'centres' of progress and modernity. On the other hand, the partially exposed economy will become more unevenly distributed and different ranges of productivity will necessarily be reflected in income levels (given the necessity to stimulate higher productivity by monetary incentives) and the rationalization of some low-productivity/hidden resources ventures, which become inconvenient in terms of production costs and prices.

Eventually, however, the long-term results of this new stage will largely depend on the redistributive policies of the central planning agencies. In fact, we may assume that much of what has been achieved by Chinese development in the previous decades will not be completely dismantled, although it may contradict the pure productivity ratio logic. For instance, it appears unlikely that the agricultural green belt will be invaded by more productive industrial enterprises. Priority attention to agriculture and food production and a relatively. high degree of petty commodity industrial diffusion in the countryside will also probably survive, and will help to moderate the radical effects of a more productivity-oriented system. Industrial decentralization towards medium cities in peripheral regions is also bound to continue for security and productivity reasons.

The combination of opposite tendencies will probably produce an articulation of the urban/urbanization question that we are unable to forecast. Also, if our assumptions on the socialist industrialization process are correct, China has entered a stage when a great deal of 'stop and go' will take place. The general tendency towards underproduction and relatively slow but balanced and controlled economic growth, will from time to time again become dominant as overproduction tendencies produced by the world-system exposure and high productivity ratio increase. The Chinese socio-economic system is likely to become more and more a mixed economy, where the dominant underproduction effects of state redistributive policies are paralleled by controlled exposure to accelerating/selective effects operating in the world system.

But such a system as this, operating at this stage of world development, is totally unexplored from the theoretical point of view. We will have to build a new interpretative scale to understand the system and its consequences such as the urban/urbanization question.

Notes

1. Even if we limit ourselves to European history, we can trace at least four different types of cities in the pre-industrial age: (a) capital/administrative cities; (b) commercial and artisan centres; (c) commercial cities, and (d) large agricultural towns. In addition, many settlements had a complex, mixed character. These variegated pre-industrial urban systems became differently involved in the industrialization process. See Mumford (1966), Mingione (1981) and Dobb (1951, 1954).
2. For instance we can wonder how much Anglo-Saxon anti-urbanism has pre-industrial roots. See Glass (1955). The fact that anti-urban feelings – including the preference for one-family houses and the dislike of blocks of flats – are more diffused in the Anglo-Saxon culture than in the Italian, French or Japanese ones may very well have pre-industrial roots.
3. We have a few controversial elements on which we can base our hypothesis on the perspectives of the present transition in capitalist and socialist countries. Moreover at the present stage of knowledge, we are tempted to overemphasize the role of technological and scientific innovation, as we know better their absolute potentials, and forget the social forces which use technology in different ways, i.e. the balance of power among different social groups, the labour and productive processes, the new and old social movements, etc.
4. Amongst various other studies, the well-documented study by Gilbert and Gugler (1982) proves this point. See mainly chapter 3 on rural–urban migration in Third World countries. Unfortunately, too often we evaluate the living conditions in Third World cities by industrialized country standards so that we arrive at misleading conclusions.
5. The experience of new-towns in developing countries has now been sufficiently explored to give us a clear idea of how they develop into a complex mix of planned/formal and unplanned/informal realities. See, as examples, research on Brasilia (Epstein, 1973), or Chandigarh (Sarin, 1982).
6. I have not mentioned the political system criterion to identify socialist countries because I consider it very ambiguous and dangerous. The experience of some Arab countries – Algeria, Syria, Iraq, Libya and even Egypt – proves

the point. Their political systems are or have been rather similar to those of socialist countries but it is difficult to classify them on this basis.

7. It is possible that in socialist countries the contradictory relations between Capital (with its increasingly international interests) and Nation State (as acknowledged by Singer, 1984) persist but with a reversed balance and largely different consequences.

8. Socialist industrial accumulation has been studied by various authors. In my observations I have been inspired by the following works: Lange (1966), Dobb (1969), Kalecki (1969) and Bettelheim (1975; 1976). This literature, although brilliant and accurate in the description of industrialization processes, shows how far we are from a general theory of socialism.

9. Socialist economic planners give priority to heavy energy and means of production industrial sectors to speed up the industrialization process and, at the same time, avoid underconsumption tendencies. In the long run, this orientation has shown dangerous limits because technological progress has been confined to these sectors and not sufficiently diffused to consumption goods production. From this point of view, socialist planners have consciously constrained the expansion of industrial production of mass consumption goods in the western style, but have unconsciously failed to promote the expansion of alternative collective consumption industries with advanced technologies. The degree of distortion and incoherence produced by the importance and growth of the military advanced sector, mainly in the Soviet Union, has not yet been sufficiently explored.

10. The use of the term 'underproduction' has no deep theoretical emphasis like its opposite – overproduction – in the Marxist theory of capitalist development. This is because we do not yet have a mature theory of socialist development.

11. In reality the housing question continues to be dramatic in many socialist countries (see Szelenyi 1983). This is not surprising if we take into account the low priority given to housing investment and the parallel incapacity to promote the fast development of building technologies.

12. Although I do not know the situation well, I would speculate that the cases of Angola and Mozambique are persistently dramatic because of the difficulties of achieving minimum survival standards. Thus the great majority of existing resources are frozen by this priority and cannot be used for socialist industrialization. In these cases the second phase is delayed until the rehabilitation of the urban and rural conditions of survival has taken place.

13. We have already mentioned our doubts on the urbanization and urban growth series. Per capita production data are ambiguous, not only because the population base is not certain, but also because they do not include a large productive sector for immediate local self-consumption, such as commune workshops. The same problems occur with employment data, where the part-time activities of agricultural workers in tertiary or manufacturing industries are not accounted for, nor is the voluntary agricultural work of students, intellectuals and others. For instance, growth series do not mean much if the city is expanded to include suburbs or an agricultural belt; per capita steel-production series do not mean much if we know that they do not include the expanding section of petty local steel production by 'communes" small factories; employment series by sector do not mean much if we assume that

an increasing number of agricultural workers are employed part-time in 'communes" industrial activities and so on.

14. The population of the ten largest Chinese cities increased by approximately 57 per cent from 1953 to 1970 (from less than twenty million to more than thirty, see Pannell, 1981). The rate of growth does not vary much (between a minimum of 45 per cent and a maximum of 65 per cent) with the exception of the two largest cities. Shanghai has a minimal growth (12.8 per cent), well below the natural increase of the population, while Beijing registers a growth rate (80.6 per cent) well above the average accompanied by a substantial expansion of its administrative territory (more than doubled).

15. Among medium-sized cities, the rate of growth in the two decades has varied from case to case but has been substantial. We may quote four cases of peripheral cities: Xian (+103 per cent); Luda (+115 per cent); Lanzhou (+265 per cent); Baotou (+516 per cent).

16. Part of the surplus agricultural population has been absorbed by agriculture itself as the Chinese have expanded cultivated land by several million acres during the 1950s and 1960s. This is also a sign that they have not been interested in productivity rates but only in output in absolute terms. In the same period the rest of the world registered a consistent opposite movement towards abandoning the least productive lands.

17. The first superficial impressions of a recent (summer 1985) visit to China confirm approximately what I wrote in the conclusions of this chapter. Although I had neither time to study the research material collected during the visit nor have the space here to develop my feelings as a tourist, it is worthwhile to mention that the accelerated social and physical change taking place in large cities – and particularly in Beijing – is impressive. It is likely that huge social problems are emerging which will enter the agenda of the Chinese government very soon. These include the combination of an already enormous and still growing bicycle traffic with fast increasing truck and car traffic, the acceleration of inner cities' renewal and tertiarization, etc.

3
Chinese Socialism and Uneven Development

Chung-Tong Wu

Introduction

Uneven development, increasing regional inequality and rapid urbanization are considered hallmarks of Third World development under capitalism, but it is unclear whether the same phenomena follow from socialist
development strategies. Studies of eastern European socialist countries
(French and Hamilton, 1979) have tended to emphasize the similarities
of their urbanization pattern and urban development with those found
in western countries. Studies of Cuba by Slater (1982) suggest that while
the Cuban policies are intended to foster spatially more even development,
a host of conditions have prevented this from being achieved to any
significant degree.

China is often considered a nation which has been pursuing policies
specifically aimed at minimizing regional inequalities and controlling
urbanization. Certainly Mao (1969), in his noted speech on the 'Ten
Great Relations', made specific statements about the development of
inland areas of China and about the relative emphasis on the development
of heavy industries, light industries and agriculture. Mao directed that
over ninety per cent of the new industrial investments should be in the
inland areas. In the same discussion, Mao argued for more investments
in the agricultural and light industries but not to the point of taking way
the priority given to the heavy industries. On the subject of urbanization,
Mao had earlier noted that the long-term process of economic development of China would involve peasants moving into the industrial sector
and into the cities and towns (Mao, 1969).

Such were the rhetoric and policies under Mao. In recent years, some
observers of Chinese development have raised questions about the practice, intentions and effectiveness of the spatial dispersion policies under
Mao (Paine, 1978; Wu and Ip, 1980). On the other hand, there are others
who suggest that China has been able to achieve many of her goals of
controlling urbanization and more equal regional development (Mingione,
1977b; Szelenyi, 1981). Others further suggest that as far as the rural

sector is concerned, there have been great strides in the directions that Mao indicated (Aziz, 1978; Maxwell, 1975).

The development of the Chinese spatial economy and concomitant urbanization pattern could be examined in the light of the sets of policies dealing with three areas: those concerned with industrialization; with rural development; and with population growth and distribution. These three sets of policies have profound impacts on where development investments are located, in what sectors and what impacts might be experienced by the individual Chinese.

Two aspects of the policies concerned with industrial development are particularly germane to the present discussion: the sectoral priorities and the spatial/locational priorities. Since the founding of the People's Republic, the industrial sector of China has commanded the lion's share of total State investments and the decisions with respect to how much should be invested in each sector and where to locate these investments have certainly made significant impacts on the spatial economy of China.

Along with industrial policies, policies concerning agriculture and the leadership's expectations of agriculture's contribution to the national economy vitally affect peasants' welfare. With about eighty per cent of the total population of China residing in the rural areas, what affects their welfare and income clearly has decisive impacts on the country as a whole. More directly affecting individual incomes, were the policies concerning the distribution of collective income to individual commune members. Finally, policies which inhibit or encourage the diversification of the rural economy affect the opportunities available to the commune members to expand their income and, of equal importance, to have access to non-agriculture employment.

With a population close to 1,032 million, China is the most populous nation in the world, but this population is unevenly distributed with most of the population in and near the coastal provinces.[1] Policies of population control, of spatial distribution of population and the growth of large cities, and of population growth in the minorities area, affect *per capita* income and have impacts on economic growth and impose changes on the individual.

This chapter first presents a review of the changes in these three major policy areas during the last three and a half decades according to each of the five-year plan periods. The next section discusses the impacts of these policies on overall economic growth, regional development, urbanization and individual income and welfare. This will be followed by a review of the situation in a group of less-developed areas. The chapter concludes with a prognostication on the likely impacts of the post-1979 policies.

A brief chronological review of Chinese development policies

Since the establishment of the People's Republic, six five-year plans have been implemented. A review of the policy changes in the three major areas can be conveniently organized by the plan periods, partly because of their chronological order but also because each period tends to reflect the changing policy emphasis and directions. Extensive discussions of these economic policies are available elsewhere,[2] so the following discussion merely highlights the key points which are related to the theme of this chapter.

The First Five-Year Plan (1953–7)

The First Plan[3] dealt mostly with rebuilding the national economy and providing directions for the emerging socialist economy. The industrial policies promoted the heavy industrial sector which was considered the key to national economic growth and the basis for the development of other sectors. During this period, industries commanded over 46 per cent of total State investments, of which 87 per cent went to the heavy industrial sector (Sun, 1980) (see table 3.1). Iron and steel industries and the energy industries took up most of the investments. Energy projects had to be developed where the resources could be found, and much of the investment during the First Plan period was in the northeast. However, following Mao's directives, there was an explicit policy of decentralizing industries to the inland areas of the country and to some of the minorities areas where the resources were available.

Mao, in his 1956 speech on the 'Ten Great Relationships' (Mao 1969) identified the relationship between the development of the coastal regions and the development of the inland regions as a key concern. Though he advocated decentralization, he was careful not to be construed as advocating the neglect or abandonment of the industrial capacity in the coastal areas. Industrial decentralization to the inland regions remained a stated goal of the Chinese policies until the 1970s.

Providing food to the population was another of the chief concerns during the First Plan period and considerable attention was given to raising productivity and expanding agriculture though a relatively small portion of the total capital investments were devoted to the agricultural sector (see table 3.1). As far as the individuals in the rural areas were concerned, the most significant steps were those taken to progressively organize production under various types of agricultural co-operatives. Agricultural co-operatives retained some features of individual ownership and at first income was divided according to the original land ownership.

Table 3.1 Percentage share of state investments by sector, 1953 to 1972

	Percentage of total state capital investments			
	Agriculture	Light manufacturing	Heavy industry	Metallurgical industries[a]
1st FYP period				
(1953–7)	7.8	5.9	46.5	10.5
1958	10.5	7.3	57.0	17.7
1959	10.5	5.2	56.7	16.1
1960	13.0	4.0	53.3	16.2
2nd FYP period				
(1958–62)	12.3	5.2	56.1	14.8
1963–6	18.8	3.9	49.8	9.3
3rd FYP period				
(1963–7)	11.8	4.0	57.4	11.8
4th FYP period				
(1968–72)	11.3	5.4	54.8	10.6
1952–79	12.0	5.0	50.0	11.0

[a] Metallurgical industries are included in the 'Heavy industries' sector. The data in this column give details on the metallurgical industries alone.
Between 1952 and 1979, the accumulated State investments in capital construction totalled Y630,000 million.
Source: Liang 1981.

Though the attention was on increasing agricultural production, handicraft industries were also encouraged (Sun, 1980, p. 180). Rural markets were allowed and some forms of private ownership were still possible.

At first Mao regarded China's future growth and strength in terms of the size of its population and little control was placed on population growth until 1956. Due to the development of industries and the lack of control over population movements, large numbers of peasants began moving into towns and cities (Wu, 1981, p. 97) leading to a significant increase of the number of towns and cities to a total of 157.[4]

The first birth-control campaign began in 1956 and mass transmigration of population from the coastal areas to the inland provinces was also initiated. One example of this was the effort in 1956 to relocate over 69,000 residents of coastal cities and provinces to the inland province of Qinghai (Feng, 1983, p. 53). Transmigration from the more populous coastal areas to inland areas was to continue in other forms during subsequent years.

Second Five-Year Plan Period (1958–62)

The Second Plan began with a programme of 'Three Years of Great Leap'. One of the goals was the doubling of the output of the iron and steel industry (Xu et al., 1981). A consequence of this policy was the sharp disequilibriums in State investments devoted to different sectors of the economy while the spatial focus continued to be one of decentralization.

The rural sector was expected to release labour to assist in boosting iron and steel output and peasants were encouraged to develop small refineries. The withdrawal of labour from the agricultural sector meant that in the short run agricultural productivity suffered (Xu et al., 1981, p. 9).

At the same time, the collectivization of the means of production and organizing the peasants in large communes proceeded apace in the rural areas. Agriculture was, however, also expected greatly to expand production during the 'Great Leap Forward' and to concentrate on the production of food grain. Organizational blunders and disastrous weather prevented the 'Great Leap Forward' from succeeding. The agricultural failure led to a review of the commune system and reorganization which was initiated in 1962 at the end of the Second Plan period.[5] Organizational reforms and policies which recognized the importance of diversification of agriculture were instituted.

The Second Plan also heralded programmes which sought to reclaim wasteland for cultivation. Massive programmes to transfer educated youth from coastal provinces to the interior to participate in the programme were implemented. Between the years 1958 to 1960, as part of this campaign, the province of Henan sent over 53,000 educated youths to Qinghai (Feng, 1983).

However, the agriculture failures experienced at the beginning of this plan period led to large numbers of peasants fleeing to nearby cities looking for employment. Appeals were made to the peasants to move back to their villages and when persuasion failed, some of them were removed back to the villages anyway. With that experience in mind, the government instituted severe restrictions on change of residence, on movement between cities and especially on moves from the rural areas to the urban areas (Ma, 1983, p. 127).

Third Five-Year Plan Period (1963–7)

The first three years of this period, 1963 to 1965, was considered a 'period of adjustment' during which the heavy industrial sector was deemphasized and some attempts were made to achieve a more balanced investment policy. This more balanced approach followed Mao's call for regarding 'agriculture as the base, and industries as the key' (Xu et al., 1981, p. 10). With the restructuring of the communes into smaller units and individuals organized in brigades and work teams for production, the agricultural sector slowly returned towards full production and more diversification.

The government continued its programmes of sending back to the home villages those peasants who had migrated into the cities during the aftermath of the 'Great Leap Forward'. At the same time, selective migration programmes sending educated youth into the more remote regions continued. For example, again, close to 10,000 individuals were moved to the Qinghai province. The policy of moving the population of the cities and coastal areas to the more remote and sparsely populated regions of the country was extended during the late 1960s (Bernstein, 1977). In spite of population decentralization, by 1964 there were officially 169 cities – twelve more than in 1952 (Wu, 1981).

Fourth Five-Year Plan Period (1968–72)

The Fourth Plan began with the onset of the 'Great Cultural Revolution' which again placed emphasis on the heavy industries, particularly the iron and steel industry which was expected to double its output to between 35,000 and 40,000 tonnes. During this period, the percentage of capital investments devoted to the heavy industrial sector reached over fifty-two per cent (see table 3.1).

The rural population was affected by three key changes in agricultural policies: an emphasis on the cultivation of food grain; more centralized planning of agriculture and control over the distribution of collective income. The food-first policy was an attempt to achieve self-sufficiency in food, but it was often implemented with little regard for local conditions. More rigid centralized planning of agriculture meant that decisions with respect to production were taken out of the hand of the peasants. Private plots were largely eliminated and the rural markets were also closed.[6] Furthermore, restrictions as to what percentage of the rural collective income might be distributed to the commune members – seen as a means of achieving more equitable income distribution in the rural areas – were also imposed. An associated policy which affected individual income was the change in the accounting unit from the production teams to the commune level, following the lead of the model commune, Dazhai (Maxwell, 1975).

Self-reliance was considered a paramount goal, and the regime emphasized the need for small-scale industries in the rural areas to supply the necessary fertilizers and small machineries which support agriculture.

As far as population policies were concerned, the 1962 campaign on birth control was abandoned during the 'Great Cultural Revolution'. This was not restored until 1971 when the new birth control programme had three announced goals: 'late marriage, longer spacing between births and fewer children' (Chen and Kols, 1982). However, the programme to send youths to the countryside accelerated (Bernstein, 1977). Some 30 million youths from the coastal and urban areas were sent to the inland and remote regions (Wu, 1981).

Fifth Five-Year Plan Period (1976–80)

The disruptions of the 'Great Cultural Revolution' and the power struggle which ultimately led to the demise of the 'Gang of Four' meant that national economic planning was in a state of confusion. It was not until 1978 that the next national economic plan was unveiled. This was a Ten-Year Economic Development Plan announced by the new chairman, Hua Guofeng. It identified the industrial sector as the priority and set goals of expansion which demanded significant increases in the capital construction expenditures of the nation. Within a year, this was abandoned as overly ambitious and scaled down considerably.

The disruptive effects which evolved from the extremely centralized agricultural policies of the 'Great Cultural Revolution' were slowly diminishing by 1977 (Wu and Ip, 1980). The brigades and work teams were again given the autonomy with respect to production and distribution of income.

By 1979, recognizing the disruptions of the 'Great Cultural Revolution' on the countryside, the government announced new policies with respect to the agricultural sector (Zhonggong Zhongyang, 1979). These included higher prices for agricultural production, incentives and subsidies for rural industries and the encouragement of rural diversification (Wu 1980; Wu and Ip, 1982). Private plots were restored, free markets were again permitted, and the peasants were encouraged to engage in sidelines.

Due to the massive transfer of population from the coastal areas to the rural, inland and remote regions and the effective controls over change in residence, a gradual lowering of the percentage of population in the coastal region began to be detected by the mid-1970s (Wu, 1981, p. 98). However, the growth of population in cities continued and, by 1976, there were officially 186 cities.

The birth-control programmes instituted in 1971 were seen to be inadequate as national planners realized that, in order to achieve significant national economic growth, zero population growth policy was required. Population control was incorporated into the constitution in 1978 as a mark of the serious attention that the government gives to birth control. A renewed effort of birth control with a one-child per family goal was launched in 1979 (Chen and Kols, 1982). As in other campaigns and birth-control policies in the past, the minorities and nationalities were exempted.

The Sixth Five-Year Plan (1981–85)

The Ten-Year National Economic Development Plan (1976–85) was officially abandoned and a period of 'readjustment and reforms' was announced. This began with reviews of capital spending and scaling down of many ambitious capital investment projects which required large amounts of foreign aid or foreign currency. Much more importantly, sectoral priorities were placed in the order of agriculture, light industries and heavy industries. At the same time, recognizing the problems of some of the inland regions, the policies did not abandon the priority of investing in the inland regions, but sought to encourage the co-operation between coastal cities and coastal regions with the inland regions (Ma, 1982; Yu, 1982). The Plan made special reference to economic regions which are seen to be different from the economic co-operative regions of the past (Zhonghua Renmen Gongwoguo, 1983, p. 107). Two economic regions, one in Shanxi and one centring around Shanghai and the Yangtze delta are identified as key regions where specific planning programmes are to be formulated.[7]

In agriculture, the policies of 1979 are maintained but the rapid introduction of the 'responsibility system' must be considered a fundamental change, the significance of which may be equal to the restructuring of the commune system in 1961 (Ip and Wu, 1983). The various forms of the 'responsibility system' generally involve assigning work or a piece of land to households or individuals according to what might be called a 'contract system', in which the individuals or households are held to produce a set quota amount for a set price, but what they produce over and above the quota is to be disposed of by them in whatever way they see fit.[8] The 'responsibility system' is being applied to agriculture production, to the small-scale industries in the rural areas as well as to the urban industrial sector.

Rural diversification based on the principle of 'according to what suits the land' and the development of forestry, fisheries and husbandry are high on the national agenda. Small-scale industrialization in the communes and rural small towns are considered important not only to diversify the rural economy, but also to provide non-agricultural employment for the rural population.

The population campaign of 'one child per family' continues but since the 1982 Population Census, and the discovery of the tremendous growth of population in the minorities areas, it has been decided that the birth-control programmes would apply to all in the nation beginning in 1985.

The above brief summary of the major policies in industrial development, agriculture and population is intended to present the key policies implemented within each of these main areas. The next section outlines the impacts of these policies in terms of national economic change, regional development, urbanization and individual income.

Impacts of policies

Overall economic growth

Large segments of the Chinese population are still engaged in agriculture, or reside in the rural areas. Consequently, the relative growth of the agricultural sectors of the economy and the location of growth have important impacts on the development of the spatial economy and the resultant regional differences.

Over the years from 1952 to 1979, the percentage share of total State capital investment devoted to heavy industries was about 50 per cent, light industries about 5 per cent and agriculture 12 per cent (Liang, 1981). Agriculture received little relative to other sectors and in *per capita* terms the amount is certainly very small (tables 3.2a and 3.2b). Since the mid-1970s and in the first half of the 1980s, the percentage share of total State capital investments assigned to the heavy industries has declined to roughly 40 per cent. Agriculture's share has also declined to just less than 6 per cent, though the share enjoyed by the light industries has made a steady but small gain.

Another characteristic of the last three decades of Chinese economic planning is the high rate of accumulation demanded of the economy for reinvestment in production. Too high a rate of accumulation has made it difficult for industries to invest in development markets or to try out new products and has generally led to overcapacity in sectors which do not keep pace with the demands of the overall economy (Chen, 1981; Liang and Tian, 1983). In agriculture, it has meant that some of the accumulated surplus was sent out of the rural areas to further industrial development. In some ways then, part of the surplus created by agriculture have been used to support the development of industries, especially heavy industries (table 3.3).

While it is significant that much of the investment was in the heavy industrial sector, particularly the metallurgical industries, the agriculture sector benefited indirectly. Part of the investment in the heavy industrial sector had been in fertilizers, pesticides and agricultural machineries – all of which are closely related to the development of agriculture. The share of agriculture-related industrial investment though by no means insignificant, was, however, relatively small (see table 3.4).

Table 3.2a Sectoral share in State capital investment (percentage of total)

Sector	1952–7	1958	1959	1960	1961	1962	1963	1964	1970	1975–7	1978–9
Agriculture	7.8	10.5	10.5	13.0	14.4	21.3	24.6	20.3	1.59	11.5	12.1
Light industry	5.9	7.3	5.2	4.0	na	na	na	na	na	na	na
Heavy industry	46.5	57.0	56.7	55.3	na	55.0	49.2	49.0	na	na	na

Sources: Chen, 1981, p. 259 ff. World Bank, 1981a, Annex C, p. 58. Yang and Li, 1980, p. 24

Table 3.2b Sectoral share in State capital investments (percentage of total of all sectors)

Sector	1952–7	1957–8	1958–62	1963–5	1966–70	1971–5	1976–80	1981–5
Agriculture	7.8	7.1	11.3	17.7	10.7	9.8	10.5	5.81
Light industry	5.9	6.4	6.4	3.9	4.4	5.8	6.7	7.32
Heavy industry	46.5	36.1	54.0	45.9	51.1	49.6	45.9	39.94

Source: Statistical Yearbook of China 1985, p. 424

Table 3.3 Rates of accumulation

	1953–7	1958–62	1963–7	1970	1971	1972	1978	1982
Rate of accumulation	24.2	30.8[a]	22.7[b]	32.9	34.1	31.6	36.6	29[c]
Accumulation as % of national income	33.6	na	na	na	na	na	32.3[d]	na
National income generated by each Y100 of accumulation	Y33.6	na	na	na	na	na	Y32.3(1978) Y36.6(1979)	na

[a] 1969 rate was 43.8%.
[b] Refers to 1963–5.
[c] Zhao, 1983. The rate recommended for the Sixth Plan is 25%.
[d] In 1979, this was reported to be 36.6%.

Sources: Xu, 1981, p. 205ff; Chen, 1981, p. 259ff; Yang and Li, 1980, p. 24; Zhang, S. 1981, p. 43

Table 3.4   National investments in agriculture-related heavy industries

	Total investment in heavy industries (Y100 million)	Amount invested in agriculture-related heavy industries (Y100 million)	Percentage of total
1953–7	255.99	7.58	2.9
1958–62	666.27	38.27	5.7
1963–5	201.26	19.68	9.7
1963–7	524.77	47.73	8.0
1968–72	920.58	93.38	10.1
1976–8	657.83	73.16	11.1
1952–1978	3241.63	279.93	8.6

Source: Yang and Li, 1980, p. 31

The overall implication conveyed by these data is that agriculture had been relatively neglected during the first thirty years of the People's Republic and that much of the available resources went into the heavy industrial sector and not where the great majority of the population reside and are employed. This generalization is supported by the available national economic development indicators (table 3.5).

Recent data provided by various official publications and Chinese scholars confirm that the Chinese economy has been growing steadily but due to the selective policies of the government, this growth had been lopsided in favour of the heavy industrial sector. The significance of this bias on the development of the spatial economy will be discussed in the next section on regional development but the effects of these sectoral investment policies can be gauged by the available information on percentage share of total output value by sector (see table 3.6).

Another important set of data refers to the fiscal stability of the government during the major periods under review. In an extensive analysis of the revenues and expenditures of the nation during its first three decades, Chen (1981) characterized the state of the national economy at the end of 1976 as 'near a state of collapse'.

Although Chen does not provide detailed and comprehensive data for the past three decades, table 3.7, which includes data available more recently, provides some idea of the fiscal deficits and surpluses of the government during much of the past thirty years. While the sums (approximately $US 8,030 million in 1979, the year with the largest deficit) involved in the deficit years are small compared to the fiscal deficits of many countries, it is important to note what the analysts

Table 3.5 Economic indicators – national income and production (current values)

| | Five-Year Plan Periods | | | | |
	1st 1953–7	2nd 1958–62	3rd 1963–7	4th 1968–72	5th 1976–80
GNP ($US million)	na	87,143	137,571	146,642	215,000
GNP per capita ($US)[a]	na	132	167	159	224
GNP average annual increase (%)[b] (1952–78 = 6.5%)	8.9	5.4	6.8	5.4	6.4[d]
Annual average % increase of total value of industrial output[c]	18	na	17.9	na	na
Annual average % increase of total value of agriculture[c]	4.5	na	11.1	na	na
Annual average % increase of productivity in State-owned industrial enterprises[c]	8.7	na	23.1	na	na
National average annual wage (Yuan)[b]	446 (1952)	na	na	na	644 (1978)
% change in per capita consumption[b] (adjusted for inflation)	5.7	3.3	2.1	1.7	2.1

Sources:
[a] Zhang, 1982 – data on GNP refers to the years 1960, 1970, 1975 and 1978 respectively;
[b] Zhou, 1982; [c] Liang, 1981; [d] According to Xu (1982), the 1976 national income was 2.7% less than 1975; and the 1967 gross value of industrial output was lower by 3.8% than the year before, where as the 1968 gross value of industrial output was 5% lower than that of 1967. He also reported that during the 4th Plan period, the total gross value of agriculture and industrial output was lowered by 7.8% compared to the 3rd Plan period

suggested were the causes of the deficits. The 'Great Leap Forward' and the 'Great Cultural Revolution' were considered to be the main causes of the fall in revenues during much of the 1960s and early 70s. The identified deficits for the year 1979 were said to be due to the rise in prices paid to the peasants for agricultural production and the overly ambitious capital-development projects initiated in 1978. Thus the subsidies and active transfer of income to the agricultural population are seen by some to be detrimental to the fiscal health of the nation.[9] If this view prevails and indeed events have proved that they have, the type of development policies to be followed by the Chinese government would be quite different from the ones which have been implemented up to the mid-1980s and the regional impacts may become even more inequitable.

Table 3.6   Percentage share of total output value by sector

	Agriculture	Light industry	Heavy industry
1949	70.0[a]	22.1	7.9
1957	43.3	31.2	25.5
1962	38.8	28.9	32.3
1965	37.3	32.3	30.4
1970	33.7	30.6	35.7
1975	30.1	30.8	39.1
1978[c]	27.8	31.1	41.1
1980	30.8	32.6[b]	36.6
1982[c]	33.6	33.4	33.0
1983	33.9	32.1	34.0
1984	34.8	30.9	34.3

[a] For values of agriculture output, see Sun (1980), pp. 208–9. Sun provided a detailed analysis of the various growth rates and changes during the First Five-Year Plan period. [b] Xu (1982) reported the share by light industries in the total value of production as being 43% in 1979 and 46.7% in 1980. If Xu is correct, then the changes in percentage share of output by 1980 and 1982 are remarkable. [c] Zhao, 1983.

Sources: Liang and Tian, 1983, p. 103; *Statistical Yearbook of China*, 1985, p. 29.

Uneven regional development

While the Chinese economy has not been expanding at the spectacular rate of some Asian nations, its national growth rate of about six per cent per annum has been adequate to provide the population with basic needs. Much improvement could be made of course, for China is still a relatively poor nation. The sectoral biases embodied in the development policies of the last three decades have resulted in selective regional impacts. The Maoist rhetoric of 'even' regional development notwithstanding, it is clear the present leadership recognizes the issue of less-developed regions (Anon., 1983). The extent of uneven development in China could be assessed by examining the regional distribution of industrial output, by examining agricultural income (particularly the extent to which non-agriculture activities have contributed to the overall income of the commune), and by examining the terms of trade between rural and industrial sectors.

Terms of trade

From the perspective of those peasants who have the opportunities to earn higher income, times have never been better. Between 1952 and 1977, even before the more recent rises in the price of agricultural

Table 3.7   State revenues and expenditures, 1952–84 (Y100 million)

	Revenues	Expenditures	Balance
1952	183.7	176.0	+ 7.7
1953	22.5% over 1952	52% over 1952	
1956	5.7% over 1955	58% over 1955	
1957	310.2	304.2	+ 6.0
1958	na	2 × that of 1958	
1961–2	continue to lower	na	
1963–5	avg. annual increase of 14.7%	na	
1965	473.3	466.3	+ 7.0
1974–6	3 years of deficit	na	
1978	1,121.1	1,111.0	+ 10.0
1979	1,103.3	1,273.9	− 170.6
1980	1,085.2	1,212.7	− 127.5
1981	1,089.5	1,115.0	− 25.5
1982	1,124.0	1,153.3	− 29.34
1983	1,249.0	1,292.5	− 43.5
1984	1,465.0	1,515.0	− 50.0

Source: Chen, 1981, *Almanac of China's Economy*, 1983, p. III–33; *Statistical Yearbook of China*, 1985, p. 523

production, the prices paid by the government to the peasants for the purchase of quota production increased 68.8 per cent, while during the same period, the retail price of industrial goods dropped 0.5 per cent. It is reported that the value of the industrial goods which could be purchased by the peasants with the same amount of agricultural products increased 70 per cent (Jiang, 1981).[10]

Jiang illustrated his findings with a highly selective list, but the general changes appears to have been for the terms of trade to improve in favour of agriculture. Table 3.8, which reports a similar pattern also, provides a more up-to-date picture of the changes in the terms of trade for major commodities. What remains to be seen is whether the general improvements are experienced relatively equally by all the rural areas.

Spatial distribution of industrial and agricultural output

Since the largest share of the State investments went to the heavy industrial sector, data on industrial output throughout the years can be used

Table 3.8   Parity price ratio of selected major commodities 1952, 1982, 1984

Commodity (100 kg)	Manufactured processed goods	1952	1982	1984
Wheat	Table salt (kg)	70	116	116
	Refined sugar (kg)	11	21	21
	White cloth (metres)	19	35	29
	Kerosene (kg)	15	46	48
Rice	Table salt (kg)	43	96	84
	Refined sugar (kg)	8	17	15
	White cloth (metres)	13	29	21
	Kerosene (kg)	9	38	35
Maize	Table salt (kg)	38	80	76
	Refined sugar (kg)	6	15	14
	White cloth (metres)	10	25	19
	Kerosene (kg)	9	32	31
Soya bean	Table salt (kg)	63	258	221
	Refined sugar (kg)	9	43	40
	White cloth (metres)	13	75	55
	Kerosene (kg)	13	100	91
Peanut	Table salt (kg)	82	249	248
	Refined sugar (kg)	16	44	45
	White cloth (metres)	19	72	61
	Light diesel oil (kg)	16	96	102

Source: Statistical Yearbook of China, 1983, p. 467; Statistical Yearbook of China, 1985, pp. 540–2

to assess the relative share of each administrative unit in the overall growth of the economy. Given the nation's announced desire to achieve a more even distribution of industrial growth, the extent to which this has been achieved is therefore an important indicator of regional equity. Table 3.9 details the *per capita* total agricultural and industrial output and the *per capita* industrial output by province.

The dominance of Shanghai, Tianjin, Beijing, and Liaoning are unchallenged throughout the past three decades. Their dominance is particularly complete when considering the *per capita* industrial output values alone. When agriculture is included, those provinces with total agricultural and industrial output value above the national average also include Jiangsu, Jilin, Zhejiang, Heilongjiang, Shandong and Hubei. At the bottom end of the distribution, the relative positions of the provinces of Anhui, Guizhou, Yunnan, Xizang, Guangxi and others remain virtually unchanged throughout the three decades.

While the dominance of the four major industrial centres which existed at the beginning of the People's Republic continues, this does not mean that the regional imbalances in industrial output have remained unchanged.

Table 3.10 shows clearly that on a *per capita* basis, the share of the top four province/municipalities of the national gross value of industrial output (GVIO) has declined. While the absolute size of GVIO has expanded greatly, in relative terms, the four key areas' share of the total has also declined significantly since the mid-1970s, indicating an overall regional shift.

The relative distribution of industrial output amongst the provinces can be assessed by examining the declines in the value of the coefficient of variation of the provincial GVIO from 1.16 in 1952, 0.8143 in 1981 to 0.771 in 1984. This steady decline indicates that the average growth of industrial output had grown faster than the standard deviation, indicating a dispersal of industrial growth.[11] However, the extent of dispersion does not threaten the dominance of the coastal centres and never approached the goal of 'even' distribution of industrial development amongst the regions of the nation.

For an overall assessment of the relative industrial growth of different regions of China, the regional share of total national GVIO can be examined. For the purposes of this exercise, the nation is divided into three zones – the coastal zone, which includes all the provinces and nationally administered municipalities on the east coast; the intermediate zone, consisting of the set of provinces immediately west to the coastal zone; and the inland zone, consisting of the provinces conventionally referred to as the north-west and south-west regions of China (see figure 3.1).

The data in table 3.11 show the coastal region's share of the national total GVIO has dropped almost ten per cent between 1952 and 1984. The larger gain has been made by the intermediate region with the inland region experiencing a shift of about four per cent. The gradual contraction of the coastal region's share is obvious by the mid-1970s, but it is in the last decade that the changes have been more pronounced. The coastal region's share of the national total dropped from 68 per cent in 1952 to 65 per cent in 1974 and to 58.3 per cent by 1984. While the relative shares are changing, it is important to emphasize that, in both absolute and relative terms, the dominance of the major coastal industrial centres is still very much ahead of their nearest rivals as the *per capita* GVIO data in tables 3.9 and 3.12 show.

Another indicator of the unevenness of development can be gleaned from examining the data on agricultural output and particularly the percentage of agricultural income which is contributed by non-agricultural activities.[12] During the 'Great Cultural Revolution' years, agriculture largely concentrated on the production of grain – in other words, rural diversification was neglected and even, in some locations, seen as anti-revolutionary. This approach has been rejected by the present leadership.

Table 3.9 Gross value of industrial and agricultural output by province and administrative area, 1984 (1980 constant prices)

Province/administrative area	1984 gross value of industries and agriculture (RMB 100 million)	Per capita GVI and A (Yuan)	Gross value of industries (RMB 100 million)	Per capita GVI (Yuan)	Per capita GVI index (national average = 100)
China	10,406.82	1,005.73	7,029.85	679.38	100.00
Beijing	315.68	3,333.47	281.72	2,974.87	437.88
Tianjin	281.39	3,521.78	251.49	3,147.56	463.30
Shanghai	792.74	6,578.76	744.37	6,177.34	909.26
Hebei	476.90	869.15	290.25	730.01	107.45
Shanxi	257.78	991.46	171.04	657.85	96.83
Nei Monggol	139.45	702.52	82.13	413.75	60.90
Liaoning	705.56	1,930.40	577.66	1,580.47	232.63
Jilin	276.91	1,212.39	187.47	820.80	120.82
Heilongjiang	437.73	1,328.47	316.19	959.61	141.25
Jiangsu	1,003.84	1,626.71	680.13	1,102.14	162.28
Zhejiang	518.02	1,297.32	333.85	836.09	123.07

Anhui	339.61	665.51	185.04	362.61	53.37
Fujian	197.21	736.68	115.76	432.42	63.65
Jiangxi	221.51	647.50	121.19	354.25	52.14
Shandong	772.82	1,011.94	456.24	597.41	87.93
Henan	491.53	642.86	267.81	350.26	51.56
Hubei	536.30	1,099.88	359.33	736.94	108.47
Hunan	400.02	719.33	229.42	412.55	60.72
Guangdong	535.53	868.52	366.90	595.04	87.59
Guangxi	189.18	497.06	104.19	273.75	40.29
Sichuan	671.10	663.67	391.00	386.67	56.92
Guizhou	134.93	460.20	73.35	250.17	36.82
Yunnan	185.48	551.70	104.97	312.22	46.00
Xizang (Tibet)	8.07	409.64	1.34	68.02	10.01
Shaanxi	220.23	742.52	144.57	487.42	71.74
Gansu	137.01	679.61	97.99	486.06	71.54
Qinghai	26.20	651.74	16.22	403.48	59.39
Ningxia	28.71	707.14	18.36	452.22	66.56
Xinjiang	105.38	784.08	59.87	445.46	65.57

Source: Statistical Yearbook of China, 1985

Table 3.10 Percentage share of total gross value of industrial output for four ranking province/municipalities

Province/ municipality	1952	1957	1965	1974	1982	1984
Shanghai	19.3	16.3	17.1	15.4	11.42	10.59
Liaoning	14.1	14.4	13.5	12.0	8.54	8.22
Tianjin	5.5	5.6	4.9	5.1	3.80	3.58
Beijing	2.1	3.0	4.0	5.4	4.10	4.01
Total	41.0	39.3	39.5	37.9	27.86	26.39

Sources: Wu, 1979; and Statistical Yearbook of China, 1983, 1985

Present policies stress the potential increases to rural income from diversification – cash crops, fisheries, forestry, husbandry, sidelines and small-scale industries. Rural enterprises in general and rural industries in particular, are expected to generate employment for the rural labour force and to provide a potential market for rural products. The value of the output of commune and brigade industrial enterprises represents an important indicator of agricultural diversification – of special importance because rural industries are regarded by the present leadership as one of the chief means of raising rural income. Historical data on commune and brigade industrial enterprises are difficult to obtain, and table 3.13 presents a comparison of available data on the value of industrial output of commune and brigade industrial enterprises for 1979 and 1983.[13]

The data in table 3.13 indicate three important features of the process of rural industrialization now underway in China. First, there has been significant growth in the absolute values of the gross value of industrial output of commune and brigade enterprises, signalling the rapidly rising importance of this sector.

Second, the dominance of the provinces which are also the major industrial centres of the nation is also very strong. On a *per capita* basis, Shanghai, Beijing and Tianjin again lead the nation. However, the province of Jiangsu (the province neighbouring Shanghai), in fact outpaced Liaoning and Zhejiang for third place.[14] The same provinces which have low industrial output are again near the bottom of the list.

Third, on a regional basis, a comparison of the 1979 and 1983 data point to the coastal region's increasing dominance of the share of value of commune and brigade industrial enterprises output (see table 3.13). Furthermore, the gap is widening. This trend is contrary to the overall share of the coastal region in the nation's total gross value of industrial

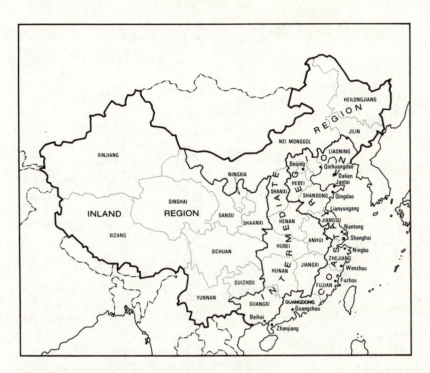

Figure 3.1   China: Provinces, regions and 14 coastal cities with Special Economic Development Zones

output. The available data again reinforce the fact that in rural industrialization, as in the heavy and light industrial sectors, it is those localities which are in the coastal provinces and those proximate to large industrial centres which have developed the most rapidly. The increasing share of the coastal region, in absolute value, was achieved even though there have been significant gains by several provinces in the intermediate region.

If the provinces were to be ranked using data on *per capita* basis, the coastal region contains the top seven ranks in 1979 and 1984. Only one province, Fujian, in the coastal zone, had a *per capita* value of commune brigade industrial output significantly lower than the national average in both years. The available data show that the gaps between the three major regions are widening. In 1979, on the basis of average *per capita* value of commune and brigade industrial output, the coastal region was 2.62 times that of the intermediate region and 5.15 times that of the inland region. By 1984, the same ratios have become 3.16 and 6.71 respectively.

Table 3.11   Percentage share of national GVIO by regions[a] (percentage of total)

Region	1952	1982	1984
Coastal			
Liaoning	14.0	8.54	8.22
Beijing	2.0	4.10	4.01
Tianjin	6.0	3.80	3.58
Hebei	4.0	4.21	4.13
Shandong	6.0	6.59	6.49
Jiangsu	8.0	9.02	9.67
Shanghai	19.0	11.42	10.59
Zhejiang	3.0	4.14	4.75
Fujian	1.0	1.57	1.65
Guangdong	5.0	4.88	5.22
Subtotal	68.0	58.27	58.31
Intermediate			
Jilin	3.24	2.58	2.67
Heilongjiang	5.55	4.81	4.50
Nei Monggol	0.56	1.23	1.17
Shanxi	1.89	2.39	2.43
Henan	2.59	3.92	3.81
Anhui	1.85	2.61	2.63
Hubei	2.81	4.89	5.11
Hunan	2.26	3.44	3.26
Jiangxi	1.69	1.74	1.72
Guangxi	1.01	1.58	1.48
Subtotal	23.45	29.19	28.78
Inland			
Sichuan	4.85	5.40	5.56
Guizhou	0.79	0.94	1.04
Yunnan	0.98	1.43	1.49
Xizang	0.00	0.02	0.02
Shaanxi	1.12	2.03	2.06
Gansu	0.68	1.45	1.39
Qinghai	0.01	0.25	0.23
Ningxia	0.03	0.25	0.26
Xinjiang	0.51	0.83	0.85
Subtotal	8.97	12.60	12.90

[a] In the 7th Five-Year Plan, the nation is divided into three economic regions: east, central and west. With one exception, these correspond to the coastal, intermediate and inland divisions used in this chapter. The province of Guangxi is included in the 'east' region in the 7th Plan.

Sources: Field, Lardy and Emerson, 1975; and Statistical Yearbook of China, 1983, 1985.

Table 3.12 Provincial and administrative area by *per capita* GVIO for selected years

Province	Per capita *GVIO* as percentage of the first rank					
	1952	1957	1965	1974	1982	1984
Beijing	24.53	31.69	33.06	44.20	46.15	48.16
Tianjin	65.37	73.50	48.96	48.96	50.83	50.95
Hebei	3.37	3.63	7.62	9.16	8.08	8.56
Shanxi	4.26	6.24	5.39	5.44	9.83	10.65
Nei Monggol	2.30	4.15	21.54	14.42	6.62	6.70
Liaoning	22.00	25.32	28.91	22.76	24.84	25.58
Jilin	9.18	10.39	12.53	8.11	11.88	13.29
Heilongjiang	15.09	14.83	16.72	10.28	15.30	15.53
Shanghai	100.00	100.00	100.00	100.00	100.00	100.00
Jiangsu	6.00	5.52	7.54	8.43	15.49	17.84
Zhejiang	4.57	5.10	6.15	4.32	11.05	13.53
Anhui	1.96	2.43	3.42	2.83	5.45	5.87
Fujian	3.00	4.55	4.89	3.73	6.31	3.67
Jiangxi	3.26	3.58	5.00	3.64	5.45	5.73
Shandong	4.07	4.08	4.97	5.79	9.20	9.67
Henan	1.03	1.94	3.32	3.16	5.47	5.67
Hubei	3.27	4.94	5.49	3.74	10.62	11.93
Hunan	2.19	2.73	3.52	3.34	6.62	6.68
Guangdong	4.52	6.02	8.15	6.44	8.55	9.63
Guangxi	1.85	2.15	2.80	3.90	4.51	4.43
Sichuan	2.38	3.59	4.55	3.13	5.63	6.30
Guizhou	1.70	2.18	6.59	2.66	3.42	4.05
Yunnan	2.42	3.13	2.88	2.47	4.56	5.05
Xijang (Tibet)	na	0.43	0.68	1.06	1.20	1.10
Shaanxi	2.28	3.78	5.82	4.91	7.31	7.89
Gansu	1.12	3.28	7.67	9.79	7.69	7.87
Qinghai	2.04	3.45	5.69	6.33	6.75	6.53
Ningxia	na	0.77	1.57	2.43	6.80	7.32
Xinjiang	3.30	4.99	4.28	3.64	6.60	7.21

Sources: Wu, 1979; and *Statistical Yearbook of China*, 1983, 1985

Some of the factors which contribute to flourishing rural industries include the availability of markets, transportation, raw materials and technology or technical assistance. The province of Jiangsu, neighbouring Shanghai, for example, is one of the areas of the country best served by all modes of transportation, has a tradition of handicraft industries, has very good backward and forward links with the largest industries, is near a major market, and is close to the most technologically advanced centres in the nation. In 1979, of the seventy-three *xians* which reported income from commune and brigade enterprises totalling Y100 million or more,

Table 3.13   Value of commune and brigade industrial output 1979 and 1983 (current value)

Area	1979 Total value of B/C Industrial Output (Y million)	1979 Per capita value of B/C industrial output (Yuan)	1983 Total value of B/C Industrial Output (Y million)	1983 Per capita value of B/C industrial output (Yuan)
National	42,455	52.58	68,675.41	82.21
Coastal				
Beijing	912	243.46	2,082.50	542.32
Tianjin	960	266.59	1,462.22	394.77
Shanghai	2,075	484.47	4,914.89	1,129.08
Liaoning	2,358	105.79	3,393.96	153.28
Hebei	3,296	73.23	3,211.70	68.07
Shandong	4,554	69.66	6,445.16	96.15
Jiangsu	7,487	148.79	14,701.99	285.30
Zhejiang	2,752	83.03	7,248.56	212.87
Fujian	823	38.67	1,315.75	58.25
Guangdong	2,461	51.79	4,512.68	89.63
Subtotal[a]	27,678 (65.2)	93.33	49,289.41 (71.8)	160.67
Intermediate				
Jilin	569	38.51	771.48	52.22
Heilongjiang	970	52.32	1,038.30	57.44
Shanxi	1,471	72.38	1,793.13	86.18
Nei Monggol	301	22.94	335.57	24.58
Anhui	860	20.21	1,240.33	27.97
Jiangxi	869	31.09	1,157.94	41.86

(continued)

thirty-one of these were in Jiangsu. The highest placed *xian*, Wuxi with Y510 million and Jiangming with Y450 million, were both located in Jiangsu province (*Year Book of Chinese Agriculture 1980*). In the same year, it was noted that villages (xiang) with more than Y100 million income from rural industries numbered twenty-one, the highest placed being in Guangdong and the next two ranks located in Jiangsu province.

Experience in China shows commune and brigade enterprises can contribute significantly to higher rural income. Between 1975 and 1980, the total industrial output of the nation rose by 9.2 per cent, but the total value of output of the nation's commune and brigade enterprises rose by twenty-four per cent. The commune and brigade enterprises obviously had to improve on a much smaller base. However, these are the enterprises which are located in the rural areas and therefore could

Table 3.13 (continued)

Area	1979 Total value of B/C Industrial Output (Y million)		1979 Per capita value of B/C industrial output (Yuan)	1983 Total value of B/C Industrial Output (Y million)	1983 Per capita value of B/C industrial output (Yuan)
Henan	2,460		37.60	2,819.10	41.34
Hubei	1,323		33.80	2,486.32	62.66
Hunan	1,934		41.90	2,340.79	48.63
Guangxi	580		18.83	662.38	20.13
Subtotal[a]	11,337	(26.71)	35.56	14,645.34 (21.3)	50.81
Inland					
Sichuan	1,882		21.83	2.699.55	30.42
Guizhou	244		10.15	234.70	9.19
Yunnan	284		10.27	450.20	15.38
Xizang (Tibet)	na		na	na	na
Shaanxi	547		22.84	828.11	33.84
Gansu	261		16.17	265.60	15.78
Qinghai	59		22.06	56.74	19.93
Ningxia	54		64.06	46.20	15.09
Xinjiang	109		15.60	159.56	21.92
Subtotal[a]	3,440 (8.09)		18.10	4,740.66 (6.9)	23.94

[a] Figures in brackets are % of total.
Source: Yearbook of Chinese Agriculture, 1980, 1984

potentially benefit the peasants more directly. By 1980, the income from commune and brigade enterprises constituted thirty-four per cent of the total agriculture income and the number of individuals working in these enterprises constituted 9.4 per cent of the total labour force in the rural areas (Ma, 1983). Since 1980, the pace of development of rural (brigade and work-team) industries continued to flourish and by 1983 comprised 11.8 per cent of the total gross value of agriculture output (*Year Book of Chinese Agriculture 1984*, p. 79). However, a more inclusive definition of rural industries which takes into account township (commune) and village (brigade) industrial enterprises would raise this to 15.3 per cent.

The policy of encouraging the diversification of agriculture has brought about significant changes in the structure of the rural economy and raises the incomes of those peasants who reside in an area where such rural enterprises flourish. Since the conditions conducive to the establishment and development of these enterprises are not the same everywhere, the overall effects have tended to increase regional income differences.

A related policy instituted in the 1977–9 period was an increase in the prices paid to the rural population for grain and for the production of sidelines. Although this was reported to be one of the chief causes of fiscal deficits in 1979 and the years since, did the policy itself achieve the goal of lifting the overall income of the rural population? The conditions for developing rural sidelines, like those conducive to rural enterprises, are unevenly distributed and so the impacts of the policy could be expected to be highly skewed. Indeed, Jiang (1981) reported that the 1979 price rises had benefited only some of the peasants. In the areas rich in sideline production, the price rises have been a bonanza. Shanghai suburban rural population benefited about five times more than the national average.[15] The rural population in poorer areas not only did not benefit, they suffered due to the inflation in prices for these goods (Jiang, 1981, pp. 656–67).

Though tremendous strides have been made, 'poor' teams still constitute a significant portion of the total. In 1965, 42 per cent of the teams were considered poor, while in 1978, about 16.8 per cent are considered poor (Ge, 1981). According to the central government, in 1978, about 25 per cent of the work teams had *per capita* annual collective income of less than Y50 (Zhonggong Zhongyang, 1979). The government admitted that some of the work teams had collective accumulation of less than Y10,000 (approximately $US5,000) and some had difficulties maintaining production from year to year. These poor brigades/teams are, however, spatially concentrated, with approximately one-third of them in the hilly regions of China and amongst the minorities regions.

While in 1983, Premier Zhao stated that all of the 240 'poor' *xian* have solved the basic problems of feeding their population, the full dimensions of 'poor' brigades and 'poor' *xian* warrant investigation. As used by Premier Zhao and in official publications, the term 'poor' is defined by two indicators: The first relates to *per capita* distributed collective income and the second is related to the *per capita* grain ration (unprocessed grain).[16] 'Poor' is defined nationally as those brigades which have *per capita* distributed collective income of less than Y50 a year, or *per capita* grain ration of less than 200 kilograms per year. Tables 3.14, 3.15, 3.16 and 3.17 provide some data of recent changes in *per capita* income and grain rations.

Table 3.15 shows that the administrative areas well-endowed in terms of industrial output and commune and brigade industries are generally the same as those which have high *per capita* distributed collective income. The sparsely populated provinces of Xizang and Xinjiang are the exceptions though, as they ranked fourth and tenth nationally in terms of *per*

Table 3.14 Average annual *per capita* distributed collective income for commune population (current values)

	Per capita distributed collective income (Yuan)	Of which cash income (Yuan)
1957	40.5	14.2
1962	46.1	—
1965	52.3	14.5
1970	59.5	—
1975	63.2	12.4
1976	62.8	12.0
1977	65.0	12.8
1978	73.8	18.97
1979	83.4	23.23
1980[a]	85.9	na
1981[b]	101.32	na

[a] World Bank, 1981, Annex A, p. 107
[b] *Yearbook of Chinese Agriculture*, 1982, p. 91.
Source: Nongye Bu, 1982, tables 6–12, p. 202

capita distributed collective income but ranked near the bottom by industrial output.

Poverty is a relative concept and the definition of 'poor' adopted by the Chinese government refers to the basic minimum which a person must have in order to survive. These could be compared with other policies and with definitions used by the provinces – an example from Shandong will suffice to illustrate the differences.

In the 1979 statement on agricultural policy, the government made the commitment not to impose quota purchases on those brigades which had less than 150 kilograms grain rations (Zhonggong Zhongyang, 1979). This can be taken as tacit recognition that 150 kilograms *per capita* grain ration per year was very low indeed and is required for basic survival. This might be considered an indicator of *abject* poverty. By contrast, Shandong province defined a 'poor' brigade in 1980 as one with Y100 or less average *per capita* distributed income. The basis for this poverty line was the Provincial Agricultural Bureau's calculation of an average grain ration of 20 kilograms to 25 kilograms per month (or 240 kilograms to 300 kilograms per year) being required for an able-bodied, adult labourer working in agriculture (Wu and Ip, 1982). The young and the old would need less. If one were to accept this as a standard of adequate grain ration, then the national definition of 'poor' – less than 200 kilograms of grain ration per person – the *abject* poverty level of 150 kilograms *per capita* is really very low indeed.

Table 3.15   1979 and 1981 distributed collective income *per capita* by province

	Per capita *distributed collective income (Yuan)*	
	1979	*1981*
Nation	83.4	101.32
Beijing	150.8	203.34
Tianjin	145.1	153.66
Hebei	83.2	86.15
Shanxi	85.8	79.57
Nei Monggol	76.8	93.18
Liaoning	115.0	125.29
Jilin	116.1	144.05
Heilongjiang	110.2	120.81
Shanghai	214.0	177.59
Jiangsu	99.2	120.77
Zhejiang	104.4	111.57
Anhui	70.3	122.64
Fujian	67.9	91.84
Jiangxi	89.3	96.62
Shandong	79.2	126.29
Henan	63.4	100.82
Hubei	106.2	102.45
Hunan	92.3	107.36
Guangdong	74.7	130.01
Guangxi	88.4	82.94
Sichuan	69.2	75.38
Guizhou	46.4	65.71
Yunnan	64.6	74.53
Xizang (Tibet)	127.5	157.35
Shaanxi	79.6	62.42
Gansu	56.6	56.79
Qinghai	97.6	83.21
Ningxia	68.7	81.81
Xinjiang	102.4	113.99

Sources: World Bank, 1981a, Annex A, p. 150; *Yearbook of Chinese Agriculture*, 1982, p. 91

As to the definition of poverty by the annual average *per capita* distributed collective income, the sum of Y50 is again very low. The grain ration figures used by the Shandong Province, in its definition of a 'poor' brigade, meant a cost to the able-bodied individual of about Y58 to Y72 a year for his/her grain ration. Officials of the Prefecture of Yantai in Shandong thus calculated that Y100 per year of distributed collective income would just allow enough extra to cover the other daily essentials,

Table 3.16   National average annual *per capita* grain ration for selected years

	Average per capita *grain ration (kg)*
1957	203
1975	207
1977	208
1978	231.5
1979	244

Source: Nongye Bu, 1982, p. 204

Table 3.17   *Xian* by categories of average annual *per capita* grain ration

	1978	*1979*
Below 200 kg	695 (30.3)[a]	600 (25.6)
200–250 kg	886 (38.6)	823 (35.1)
Above 250 kg	716 (31.2)	925 (39.4)

[a] Figures in brackets represent percentage of total number of *xian*.
Source: Nongye Bu, 1982, p. 205.

such as clothing, required by each individual. Compared to the definition of poverty used by Yantai Prefecture, the national average of Y50 therefore would barely be enough for each individual to purchase the minimum grain rations without any cash distribution. By the standard of Shandong province, nineteen provinces in 1979 had average *per capita* distributed collective income of less than Y100, and can be considered 'poor'.

However, even if one were to accept the Y50 poverty line, 221 *xian*, or 9.8 per cent of the total, were reported to be below this line for three consecutive years (1977–9) – representing a total of 87.87 million people or 11.3 per cent of the total population in what might be called *abject* poverty (Nongye Bu, 1982). The number of individuals involved being close to 90 per cent of the population of the province of Sichuan, the most populous province in China. By the same standard, 27.5 per cent of the production teams, some 1,417,350 teams, were in abject poverty in 1979 (see table 3.18). According to more recent estimates made by the World Bank, using expenditures on wheat and rice, the incidence of poverty in rural areas in 1982 would average about 13 per cent of the rural population (World Bank, 1985).[17]

Where are the poor located? Available data do not allow a detailed or specific answer. Provincial data are an average of many smaller units and

Table 3.18 Distribution of *xian* and brigades by categories of average annual *per capita* distributed collective income

	1977	1978	1979
No. of *xian*			
Below Y40	182 (8.51)[a]	97 (4.54)	81 (3.79)
Y41–Y50	333 (15.58)	284 (13.28)	202 (9.45)
Y51–Y99	—	1,210 (56.59)	1,011 (47.29)
Y100–Y150	1,623 (75.91)	423 (19.78)	621 (29.05)
More than Y150	—	124 (5.08)	223 (10.43)[b]
No. of brigades (in 000s)			
Below Y50	1,800 (39.0)	1,390 (29.5)	1,370 (27.2)
Above Y300	na	na	1,622 (0.23)
Percentage of production teams[c]			
Below Y40	na	na	16.1
Y41–50	na	na	11.4
Y51–80	na	na	31.7
Y81–100	na	na	15.6
More than Y100	na	na	25.2

[a] Numbers in brackets refer to percentage of the total number of *xian*, which was 2,138 in 1979.
[b] In 1979, two *xian* were reported to have average annual *per capita* distributed collective income of over Y300, the highest being in Jiangsu province.
[c] World Bank, 1981a.
Source: Calculated from data in Nongye Bu, 1982, p. 203

table 3.18 presents data by *xian* (county), brigade and work team. The available information is not disaggregated by province and can provide but a glimpse into the large variations within each administrative grouping. However, table 3.19 provides data on twenty-five provinces which had reported *xian* with Y50 or less distributed collective income during 1977 to 1980. While the number of *xian* reporting such low income in 1980 have been reduced nearly by half that of 1977, the data must be examined with caution. Some provinces clearly have very large percentages of the poor. For example, the World Bank (1985) estimated that forty-one per cent of the rural population in Gansu province were poor.

Using Y50 to define the poverty line over a period of years, as the national government has done, does not provide a satisfactory comparison. Although detailed inflation rates are not available for the specific years in the table, one could use as a reference the rates available for 1975 and

Table 3.19 *Xian* with Y50 or less distributed collective income in 1977–80 (number of *xian*)

Province	1977	1978	1979	1980
National	515	376	283	278
As percentage of all *xian*	22.5	16.8	12.4	11.5
Coastal				
Shandong	63	46	26	8
Hebei	51	17	13	21
Fujian	23	22	12	13
Jiangsu	18	6	2	2
Zhejiang	15	5	3	2
Guangdong	7	11	7	2
Liaoning	5	2	0	2
Intermediate				
Henan	49	45	31	14
Shanxi	22	24	10	15
Anhui	20	15	11	10
Guangxi	8	8	6	6
Heilongjiang	3	0	1	1
Nei Monggol	3	12	11	23
Jiangxi	2	5	1	6
Hunan	2	0	0	0
Hubei	2	1	0	1
Inland				
Guizhou	52	58	53	44
Yunan	45	28	32	32
Sichuan	39	7	3	3
Gansu	35	27	32	37
Shaanxi	30	24	13	23
Xinjiang	13	9	9	7
Ningxia	6	4	5	5
Qinghai	2	0	2	0
Xizang	0	0	0	1

Source: *Zhongguo Baike Nianjian*, 1982, p. 363

1981 which indicate a fourteen per cent rise in consumer prices (*Statistical Yearbook of China*, 1983, p. 572). If this is taken as a guide, then the Y50 poverty line in 1977 should be close to Y54.5 by 1980. Consequently, the number of *xian* reported in the table must represent a slight under-reporting.

If the provinces reported were to be grouped according to the three regions adopted earlier, then in 1980 the coastal areas would have 50, the intermediate 76 and the inland region 152 of the poor *xian*. Part of the reason why the coastal region have so few may be due to the new boundaries for the province of Nei Monggol, which took in parts of Jilin and Liaoning provinces, both of which are in the coastal region, although under the new boundaries, only Liaoning province is reported to have counties with less than Y50 income.

One could also examine the location of the more prosperous rural units to gain additional insight into the pattern of distribution of rural income in China. Of the brigades reported to have an average *per capita* income of Y300 and above (table 3.18), locational information is available for 1,162 or 72 per cent of them (table 3.20).

In terms of overall distribution, in 1979 57.7 per cent of the high-income brigades were located in the suburban rural areas of large munici-palities, 26.1 per cent in agricultural areas, 8.5 per cent in fishing areas, and only 1.9 per cent in forestry areas (Nongye Bu, 1982). A similar pattern is discernible for 1980. In other words, close to three-quarters of the high-income brigades are located in the coastal region and most of these are those near large cities and towns.

In summary, regional inequality in China exists not only between coastal and inland regions but also between suburban and remote rural regions. In spite of the policies of decentralizing development to the inland regions and policies aimed at improving the development of rural areas, regional inequalities are still marked. For a peasant living in a rural suburban *xian* near or within the boundaries of a large metropolitan area in the coastal region of China, improvements of income and welfare have

Table 3.20 Distribution of brigades with 1979 and 1980 annual average *per capita* distributed collective income of Y300 and above

Province or municipality	Number of brigades (1979)	Percentage of total No. of brigades in province	
		(1979)	(1980)
Shanghai	485	16.60	19.9
Guangdong	164	0.63	2.8
Heilongjiang	132	0.97	2.2
Zhejiang	129	0.31	1.2
Liaoning	117	0.76	2.4
Jiangsu	42	0.12	na
Beijing	22	0.55	5.2

Source: Nongye Bu, 1982; *Yearbook of Chinese Agriculture* 1980, p. 156; and Zhongguo Baike Nianjian, 1982, p. 363

been far greater than for those who live in an entirely rural area, located in the inland hilly regions or in remote regions far from any major cities or towns. Since most of those who reside in the coastal region already have relatively high incomes, the differential impacts of the more recent policies would widen the gap further.

Backward areas

The previous section presented data to detail a highly uneven pattern of regional development, which it is argued is due partly to past sectoral investment policies and partly to unanticipated impacts of more recent rural policies. That poverty is so widespread in parts of the rural areas is of deep concern to the Chinese government (Xu, 1982, p. 81). In the government's 1979 review of rural policies, it was made plain that the leadership deemed it unacceptable that, after thirty years of struggle, the standard of living for so much of the population is still so low. Those in the hill regions of China are often poor – a fact which is especially burdensome to the leadership since many of the first supporters of the revolution were the population in the hilly regions of central–west China. In the 1980 budget, the government established a special fund to assist the less-developed areas. However, the funds, amounting to Y500 million, were too limited to tackle even the basic resource and environmental problems which hamper the development of agriculture in part of the backward regions.

Nine areas are generally recognized as 'backward, these include the provinces of Xizang, Nei Monggol, Ningxia, Xinjiang, Guangxi, Qinghai, Gansu, Yunnan and Guizhou. Five of these are autonomous *zhou* and the other four provinces include a significant number of minorities and nationalities.[18] If one were to examine indicators such as gross value of agriculture and industrial output (see table 3.9) and the number of poor *xian* in each province (see table 3.19), it would be obvious that at least the hilly areas of the provinces of Anhui, Fujian, Jiangxi, Henan, Human, Sichuan and Shaanxi must also be included in the same group (see table 3.21).

Table 3.22 presents more detailed data on a number of provinces and several minorities areas within them (Zhang, T., 1981). The data, for two different years, provide some idea of the population pressures and economic changes in the rural sector of these less-developed regions.

Tables 3.22 and 3.23 confirm the low level of rural development in the minorities' and nationalities' areas. However, the less-developed areas include more than the areas occupied by the minorities and nationalities. This fact can be gleaned from additional, though still fragmented information, for individual less-developed areas – mostly the hilly regions of China.

Table 3.21   Selected indicators of the less-developed regions for 1981

Indicators	As percentage of national total
Value of industrial output	10.0
Population	15.7
Area	62.8
Pasture land	90.0
Timber reserves	50.0
Hydro-electricity resources	50.0

Source: Anon., 1983

Since the last section already dealt with the issue of poor brigades and poor *xian*, only two examples will be given here.[19] The Linyi prefecture of Shandong province, in the south-western part of the province, ranked last in Shandong province in terms of average annual *per capita* distributed collective income, but 3,577 of its brigades had less than Y40 distributed income, and another 399 brigades had no cash distribution from the collective income at all in 1981. A total of 559,600 households were permanently in debt (Li et al., 1982). The same report cited the Hezhang *xian* of Geizhou province in which 83 per cent of its brigades had less than Y50 distributed income and 57.2 per cent of the households in the *xian* relied permanently on government assistance for food and necessities (Li et al., 1982).

While the Chinese leadership is anxious to publicize the economic gains of the backward areas (particularly the areas inhabited by the nationalities/minorities), the past mistakes, the lack of transportation, the lack of education opportunities and other physical as well as social facilities, are readily acknowledged. The gap between the level of economic development of the Han people and the nationalities/minorities is acknowledged to be vast (Hua, 1981). It appears that backward areas are beginning to be recognized as such and not merely subsumed under the problems of uneven development between the coastal and inland regions.

Recent regional development policies

Uneven development and rapid urbanization are issues high on the agenda of the national government. The 7th Five-Year Plan (7FYP) (1986–90) contains specific references to these issues (Zhonggong Zhongyang, 1986). Within the past several years, a number of policies concerned with urban and regional development have been promulgated. The relevant new

Table 3.22 Population density and *per capita* grain output for selected minorities' areas

Area	Year	Population density (person per km²)	Per capita cultivated land (ha)	Per capita grain output (kg)
Guangxi (province)	1950	79	0.09	230.5
	1979	151	0.05	338
Guangxi (Longsheng Gezu Zizhixian)	1949	37.1	0.09	na
	1979	61.7	0.05	269
Guangxi Sanjiang Dongzu Zizhixian	1949	68.3	0.06	191.5
	1979	111.7	0.04	250
Yunnan Dehong Daizu Jingpozu Zizhizhou	1952	30	0.13	378.6
	1979	63.6	0.07	376 (1977)
Nei Monggol	1949	5	0.71	349.5
	1978	15	0.29	272
Qinghai	1949	2	0.31	230.5 (1952)
	1979	5.2	0.15	220.5
Xinjing	1952	1.0	0.31	135
	1978	1.6	0.21	287
Ningxia	1949	10.6	0.49 (1957)	436 (1958)
	1978	55	0.24	164.5
Heilongjiang	1979	—	4.1	na
Shandong	1979	—	1.5	na
Hubei	1979	—	1.3	na
Guizhou	1979	—	1.2	na
Shaanxi	1979	—	2.1	na
China	1979	105 (1982)	1.5	342[a]

[a] This figure refers to average *per capita* based on total population. If this were to be calculated based on rural population alone, then the *per capita* grain output is 408 kg.

Sources: Zhang, T., 1981, p. 116. Figures for China and other provinces from the *Yearbook of Chinese Agriculture* 1980

policies can be classified into two groups: those dealing with the general issue of comprehensive balance in national economic planning, and those more specific policies of establishing economic regions, central cities, coastal open cities and 'economic and technical development zones'.

Comprehensive balance in national development

One of the key criticisms of the development policies during the 'ten years of disturbance' – that is, the 'Great Cultural Revolution' and its

Table 3.23   Minorities/nationalities areas compared to rest of China, 1982

Indicator	Minorities/nationalities' areas[a] [b]	Nation[c]	
Per capita GVIA[d]	Y470.50	Y805.93	58.4
Per capita GVIO[e]	Y190.64	Y564.09	33.8
Per capita irrigated land	0.057ha	0.044ha	129.5
Per capita grain output	274.8kg	350.5kg	78.4

[a] Minorities/nationalities' area figure is given as a percentage of the national figure.
[b] Refers to the sub-provincial areas of the provinces of Hebei, Guangdong, Guangxi, Heilongjiang, Nei Monggol, Hunan, Hubei, Guizhou, Yunnan, Sichuan, Gansu, Xinjiang, Ningxia, Liaoning, Qinghai, Xizang.
[c] Refers to data on the rest of China excluding the areas mentioned in [b].
[d] Gross value of industrial and agricultural output.
[e] Gross value of industrial output.
Source: Zhongguo Baike Nianjian, 1982.

aftermath – was the one-sided approach it took to the role of the heavy industrial sector, the rigid controls which were imposed on accumulation, the policies regarding the role of agriculture and the general disregard for fiscal balance. Consequently, a key idea behind much of the present national policy is the concept of 'comprehensive balance'. In the usage by various Chinese scholars and policy-makers, this concept takes on a number of meanings, but the general idea is that national development must be achieved within a context of overall balance of sectoral development, of resource inputs, of fiscal balance and of regional balance (Liu, 1981; Liu and Wang, 1980; Tian and Li, 1981). Within this context, the national government is expected to encourage specializations where comparative advantages are available. Planning is expected to be decentralized to the regional level. Regions are defined not merely by existing administrative divisions, but on the basis of either available resources or on the basis of highly integrated economic systems.[20] Within each region, the resource inputs should be as self-reliant as possible, and all sectors should receive proper attention, exploiting comparative advantages where appropriate (Li, 1981; Lu, 1982).

Economic Regions

The regional development policies of late are very much related to the concept of 'comprehensive balance'. The Sixth Five Year Plan (1980–5) specifically identified two regions, the Shanxi coal region and the Shanghai–Changjiang Delta region, as the key regions where studies will be made to implement this concept (Zhonghua Renmen Gongwoguo, 1983). Since that announcement, several other economic regions including the Beijing–Tianjin–Tangshan Region, the Zhujiang Delta (Guangzhou) region, the Minnan Delta (Xiamen) region, the Liaodong Peninsular (Dalien) region and the Jiaodong Peninsula (Qingdao and Yantai) region have been announced but few details are available.[21]

The boundaries of the two more publicized regions have been defined in terms of the resource availability and joint exploitation programmes in Shanxi, but in the Shanghai–Changjiang Delta region, the basis of inclusion is one of existing and potential economic relations.

As the two regions are conceived to be different functional types, the planning for the two regions is expected to be quite different. The Shanxi region is more concerned with the efficient exploitation of the coal resources, and the development of new industries which could best make use of the newly available energy. The Shanxi coal region is therefore expected to be a major new energy–industrial complex. Part of the development will be for coal which could be exported and therefore the establishment of a new capacity in the transport system to facilitate that role will be an important aspect of the regional development programme.

The Shanghai–Changjiang Delta region is already a well-established industrial area, with Shanghai being the premier industrial centre of the nation. The regional development programme is therefore one of making use of the industrial technology and expertise of Shanghai to further the development of technologically advanced industries and to forge further links with the other industrial activities of the rest of the region. The Delta area is already served by possibly the best transportation network in the nation, there are a number of smaller but significant industrial centres in the region, and a great deal of traditional expertise (Japan–China Economic Association, 1982). That the region includes part of the most advanced agricultural regions of China is expected to lead to significant improvements in the income of the rural as well as the urban population.

The 7FYP further refines the concept of economic regions by suggesting a three-tier system. The first tier would include the Shanghai, Dongbei (Liaodong), Shaanxi, Beijing–Tianjin–Tangshan and south-west economic regions. These are considered of major national importance (*Wenhuibao*, 15 April, 1986). The second tier consists of economic regions with provincial capitals, major coastal ports and major transportation nodes as centres. The third tier consists of province-administered municipalities and their economic regions. It is, however, unclear what are the expected linkages between these economic regions and how these are expected to enmesh with the rural areas.

Core cities

A third policy which has both economic and overt spatial consequences is the 'core city' (*zhongxin chengshi*) policy which embodies an expression of economic policy as well as administrative reforms (Ma, 1982). This policy makes the implicit assumption that economic development of the rural areas must be linked with those more diverse urban economies in close proximity to the rural areas (Sun, 1983). For most rural areas, this could be a county town or a major district town which has a higher level of industrialization.

The administrative reforms which accompanied this policy have largely abolished the prefecture which used to be the administrative unit immediately subordinate to the provincial government, and administratively placed the counties (*xian*) under the 'leadership' of the core cities. These administrative reforms are largely in place in most of the country but the exact implications for the rural areas are still unclear. Selected evidence from parts of Guangdong, Jiangsu and Shandong provinces suggests the following preliminary conclusions.[22] It would seem that instead of having to deal with a prefecture administration, the rural areas now deal with a 'core-city' administration. For the designated core cities however, it would mean that they now deal directly with the provincial government and not through the intermediary of the prefecture. Part of the rationale for this reform has to do with the implementation of local elections. As the prefectural officials were appointed by the provincial government, it would have been anomalous for the elected county and city government officials to have to answer to appointed officials.

The core cities are expected to administer the rural counties and the smaller towns within those counties. The core cities' administration is expected to provide the leadership to promote the economic development of the entire area under their administration and not simply to exploit the resources of the rural areas (Tao, 1983; Zhonggong Weihai Shiwei, 1983). No special funds are earmarked by either the central or the provincial governments to implement this general policy, but some assistance may be available for specific projects which have a multi-county function, e.g. large reservoirs and power plants. Since the provincial government often relies on the *xian* and core-city governments for its revenue, the sums available for local development will be limited.[23]

Economic and technical development zones

A fourth policy announced in April 1984 is the decision to allow fourteen coastal cities to designate an area within the city as an 'economic and technical development zone' (ETDZ) in order to attract foreign investments (*Wenhui Bao*, 7 April, 1984).[24] The cities are in turn given autonomy in the use of contract systems for hiring workers, in the adoption of a variable wage system and in negotiations with foreign investors. This move is aimed at expanding the areas of the country which could directly attract foreign investors and thus hasten the 'modernization' of the Chinese economy.[25] However, by early 1986, the leadership realized that the capital investments required for the simultaneous development of all the ETDZ in the fourteen cities would be enormous and have since scaled this down to four coastal cities focusing on Dalien, Tianjin, Shanghai and Guangzhou.

As these policies have been implemented only between 1983 and 1986, it would be presumptuous to attempt an evaluation. The direction of these policies is towards reliance on existing industrially advanced centres to generate the impulses for future growth in the nation and particularly for the rural areas. Given what is known of the existing spatial pattern of development in China, a general discussion of the implications of this approach could be attempted.

Conclusions

Thirty-six years have passed since the founding of the PRC. China's development policies and experiences have been largely independent from most other less-industrialized nations. More recently, the Chinese leadership has embarked on an approach which promotes trade with the rest of the world, encourages foreign investments and seeks to lift the income levels of the population, particularly those in the rural areas.

At the end of the 1970s, it was observed that the policies of the past left a legacy consisting of two sets of inequalities. First, the sectoral investment policies which favoured the heavy industrial sector were partly responsible for an impoverished agriculture sector and, as a consequence, widened the income gap between rural and urban regions.

Second, in spite of the policies aimed at a more 'even' distribution of industries, the coastal industrial centres continued to dominate. Though a gradual but small erosion of this domination could be detected, it was clear that the regional differences between coastal, intermediate and inland regions were still marked.

The policies adopted since 1978, particularly those related to core cities, the establishment of economic and technical development zones, even though they have been cast within the framework of 'comprehensive balance', can be said to reinforce the coastal and inland inequalities. The overall impact seems likely to be one of further polarization of development in the existing centres. The absence of additional specific policies and programmes to assist the backward areas is likely to reinforce the patterns of uneven development rather than to build on the modest beginnings of a more even spatial pattern of development.

Continuing polarization

The sections above on industrial development in the provinces and in the rural areas have already stressed that the pattern of development in the past three decades has indicated a modest decentralization from the coastal regions but, overall, the coastal region and the existing centres are still the dominant industrial centres. The regional schemes, the core-cities policy and the latest designation of economic and technical development zones in fourteen coastal cities are not designed to change this pattern of development, rather they would seem to reinforce the coastal region's dominance. The areas which will benefit most from these policies are largely the coastal regions and the areas which are close to the industrial centres – areas which are already enjoying relatively high income and well developed rural industries and agriculture.

Widening intra-regional differences

While the new rural policies have contributed to a small overall reduction of income differences between urban and rural regions, the intra-regional differences within the rural areas are becoming more obvious. New policies which encourage sidelines and rural industrialization highlighted the differences between suburban rural areas and the more remote rural areas. The observable trends of rural industrialization again indicate a widening gap between the coastal region and the intermediate and inland regions. This is in sharp contrast to the pattern of regional shares indicated by the data on total industrial output, which is indicating a pattern of gradual closing of the gaps.

Rural–urban links

The core-cities policy is supposed to develop and reinforce rural–urban links. Whether these links will develop will depend on a variety of conditions – among these are: the natural resource base, the stage of development of the local economies and the existing transportation system. In most of the core-city regions, the development of new rural–urban links will depend largely on the state of the economic development of the core cities themselves. The levels of economic development in many of the smaller core cities are relatively low. Whether the future development of the core cities and their rural counties will contribute to a significant change in the overall spatial pattern of development is difficult to predict. Since specific policies or programmes aimed at ensuring a diffusion of economic growth are not evident, then past experience in other Third World nations would lead one to suspect that the backwash effects may be so strong that many of the benefits would flow to the core cities rather than to the rural counties.

Backward regions

Although the present leadership is frank in admitting that there are backward areas and that many of them are areas where minorities and nationalities reside, there is little in the new development policies which constitute a specific concerted programme to redress the situation. Though the 7FYP pays specific attention to what it calls the 'old, minorities and border' regions, no specific policies beyond general statements on the central government's capital assistance and reduction of taxation are available (*Wenhuibao*, 15 April, 1986). While the advanced centres, such as Shanghai and Wuxi, have been asked to provide technical, manpower and financial assistance to the backward areas (Sun, 1984, p. 348), this assistance can only be described as too little and too dispersed to have significant impacts.

The regional schemes, if successful, could further widen the gap between the already high-income industrialized regions and the backward areas. The approach of the present leadership is perhaps one of expecting the overall economic growth to filter through to the rest of the country, including the backward regions. Past experience in other Third World nations would caution against optimism that their expectation will come to fruition.

Chinese socialist development policies during the past three and a half decades can be summarized in three phases: the pursuit of collectivization; of equity; and of higher growth (Ip and Wu, 1983). The over-emphasis of collectivization was criticized for being administratively inefficient, organizationally unwieldy and unrealistic in its goals. The single-minded pursuit of equity was criticized for sapping individual incentives, for its unreasonable interference with production and for being economically inefficient. The present phase promotes the theme of a socialist society which need not be poor, one which must incorporate the positive values of individual incentives, one in which economic efficiency is paramount and one in which self-reliant need not mean autarchy. Throughout these phases, three constants can be discerned: that of collective ownership of the means of production; that of extensive state direction; and that of relatively strict controls over population movements – features which are rare in non-socialist countires.[26]

The present leadership is embarking on a course of action few other socialist countries have taken. They suggest a socialist society can make use of features of capitalism (profits, economic efficiency ... etc.) and still remain largely socialist. This chapter has presented data on the present pattern of development and suggests that many of the recent policies, like similar policies adopted by other Third World nations, seem to be heading towards the reinforcement of uneven development. This is not to suggest that the present pragmatic leadership cannot recognize the potential problems in time. Indeed, it has displayed a high degree of flexibility in dealing with identified problems. However, at this stage, the recently announced policies have incorporated few features to ensure the avoidance or minimization of the potential that some of the undesirable features of Third World development might be replicated in China.

Acknowledgement

The assistance of Albert Fan, Cho Ming Lee, Chi Kin Leung, K. C. Lo and Mi Yin Wu in literature search and update of data files is gratefully acknowledged. Part of the information presented in this paper was gathered during field-work, supported by a grant from the Australian Research Grant Scheme.

Notes

Currency conversion: $US1.00 = 2.15 Yuan (1984); 2.70 Yuan (1986)

1. According to the July 1982 Census, the total population of continental China (29 provinces and municipalities) was 1,003,937,078.
2. See for example, US Congress, 95th Session, *Chinese Economy Post-Mao*, US Government Printing Office, Washington DC, 1978.
3. To simplify referencing, each of the five-year plans are referred to as the First Plan, Second Plan ... etc.
4. China's cities are officially designated; that is, Chinese cities and towns are not recognized as such purely by size alone, official designation is required. By the end of 1984, there were 295 *shi* of which 50 are over 1 million in population. 239 *shi* have a total population of 145 million, of which 97 million are non-agriculture population (*Almanac of China's Economy*, 1983, p. III–5).

   In China, population is generally divided into two categories, agriculture and non-agriculture. These are designations on the basis of whether the individuals receive their grain ration from the communes. Those who receive their grain ration from the communes are categorized as 'agriculture population'. Consequently, it is possible to have 'agriculture population' within the boundaries of the large municipalities.
5. The reorganization of the commune was chiefly to reduce its size and to use

natural villages as the basis for delimiting its boundaries. Communes were further subdivided into brigades and work-teams.

6. Private plots refer to the small strips of land assigned to each household for cultivating its vegetables and fruits chiefly for its own consumption. Where there is a surplus, peasants often take the excess to the rural markets, often called 'free markets', for sale.

7. See later discussion on other 'economic regions'.

8. For a more detailed discussion of the 'responsibility system', see Ip and Wu (1983).

9. The deficits in 1983 and 1984 were also largely attributed to bumper crops.

10. Jiang illustrated his findings with data on a sample of regions south of the Chanjiang. The table shows the amount of three commodities exchanged for each 50kg of unhusked rice

Commodity	Beginning of the PRC	1977	Percentage change
Ammonia sulphate	21.5 kg	42 kg	193.0
Kerosene	8.2 kg	13.8 kg	59.6
White cotton cloth	10.485 m	13.25 m	126.5

11. Calculation of standard deviation and coefficient of variation was made with 1952 and 1981 GVIO current values converted into 1970 prices (1952 = 1.29683 and 1981 = 0.99349). The conversion of the 1981 GVIO to 1970 prices was by the use of the 1979 sectoral price indices compiled by the World Bank (1980, table A.5) as an index for 1981 is unavailable. The effect of this is probably a slight exaggeration of the actual value of 1981 GVIO.

12. 'Agriculture income' refers to the total income of the communes/brigades. It would therefore include income from non-agriculture sources.

13. Available statistical compendiums published by China often include non-comparable data from year to year. Data on industrial output of commune and brigade industrial enterprises was published in 1979 and 1983, as data published other years in between refer to all commune and brigade enterprises, that is including services, commerce and other activities. Commune and brigade enterprises are now called township and village enterprises following the dismantling of the commune system.

14. The province of Jiangsu is one of the most agriculturally developed as well as one of the more industrialized provinces of China. On the basis of per capita gross value of industrial and agricultural output, it ranked fifth in 1984, but in terms of the absolute value of gross value of industries and agriculture, it ranked first.

15. In 1979, of the 73 xian in the whole of China (or 3.4 per cent of all xian) which reported income over Y100 million from commune and brigade enterprises, all ten of the suburban xian of Shanghai were included.

16. Collective income refers to the income generated through collective farming, enterprises and other activities. Prior to the dismantling of the commune system in 1984 and prior to the introduction of the 'responsibility system' in 1980, much of a commune member's income (in cash and in kind) was derived from distribution of the collective income. With the reforms in the

economic system and the dissolution of the commune system, the notion of 'distributed collective income' becomes somewhat obsolete and future comparisons will have to use another index. Much of the following section therefore cannot include data beyond 1982.

17. For details of the calculations used by the World Bank, see tables I.4 to I.9 appendix J, annex 5: 'China Economic Structure in International Perspective' in World Bank, 1985.

18. In the nine provinces considered 'backward', the percentage of their population being minorities and nationalities are as follows: Nei Monggol 15.5%; Guangxi 37.8%; Guizhou 18.2%; Yunnan 25.2%; Xizang 95.3%; Gansu 6.2%; Qinghai 26.3%; Ningxia 31.6%; Xinjiang 59.2%.

19. A prefecture is a sub-provincial administrative unit which consists of a number of *xian* (counties) and *shi* (towns, municipalities). In the 1983–4 administrative reforms which changed communes into 'administrative villages' or townships, brigades to 'villages' and put *xian* under the administration of certain *shi* (core cities), prefectures are largely (but not all) abolished.

20. 'Agriculture regionalization' is being implemented in the country and is supposed to be completed by 1985. It emphasizes the study of natural resources as a basis for agriculture planning.

22. The nine economic designs and limited information about them are as follows. Changjiang Delta: Administrative areas included are Shanghai and parts of Jiangsu and Zhejiang provinces involvng 10 *shi* and 57 *xian*. The 1980 population of this area was 50.8 million. The announced specialization of this region is industrial development with emphasis on new technology and foreign trade.
    Shanxi: Parts of Shanxi province with a population of 25.46 million in 1980. The announced specialization is that of an energy base with energy intensive type of industries.
    Zhongqing: One of the major municipalities of Sichuan province.
    Beijing/Tianjin/Tangshan: Includes the municipalities of Beijing and Tianjin and parts of the Hebei province, with a total of 36 *xian*. The 1980 population of this region was 20 million. It is already one of the largest industrial complexes of China and will continue to be one with more attention on environmental management and development of transportation networks.
    Pearl River Delta: Parts of Guangdong province including the municipality of Guangzhou (one of the 14 coastal cities to develop a 'special economic development zone') and involves 24 *xian*. The 1980 population was 17 million. It is expected to develop its agriculture and industries and foreign trade.
    Northeast: Involving the provinces of Jilin, Liaoning and Heilongjiang.
    Min River Delta: Parts of Fujian province.
    Poihai: Parts of Jiangxi province involving 25 *xian* and a 1980 population of 11.72 million.
    Poyang Hu: Parts of the province of Jiangxi around the Poyang Lake.
    Source: Almanac of China's Economy, 1983, p. VI–130.

22. Fieldwork in Wuxian in Jiangsu, Xinhui and Panyu counties in Guangdong and Huangxian in Shandong in 1983–4. The administrative changes also meant the designation of new *shi* such as Zhongshan *shi* in Guangdong. An example of a core city might be Jiangmen *shi* in Guangdong which now leads

five *xian*, some of which were previously under the Fushan prefecture. In turn, Fushan *shi* becomes another core city.

23. Core cities lack of financial resources to reinvest in their own development and the lack of resources to stimulate the development of the counties are topics which exercise the mayors of the large and small cities. Even relatively well-developed cities such as Hangzhou are beginning to find the burden difficult to bear (*Wenhuibao*, 29 July, 1984). The lack of suitably trained individuals to provide the technical backup for such economic development task is another obstacle.

24. These are to be distinguished from the 'special economic zones' of Shenzhen, Zhuhai, Shantou and Xiamen.

    The 14 cities include: Dalien (Liaoning); Qinghuangdao (Hebei); Tianjin; Yantai, Qingdao (Shandong); Lianyungang, Nantong (Jiangsu); Shanghai; Ningbao, Wenzhou (Zhejiang); Fuzhou (Fujian); Guangzhou, Zhangjiang (Guangdong); and Beihai (Guangxi).

25. A recent concession is that foreign investors or joint ventures in the 'special economic zones' involved with energy development, transportation and port facilities will enjoy a concession tax rate of 15 per cent with the period of concession negotiable (*Wenhui Bao*, 18 August 1984).

26. The previously strict control of migration from the rural areas to urban areas was relaxed in 1984. Where rural population have arranged employment or where they can demonstrate ability and sufficient capital to start an enterprise, they can move to the townships.

4
Territorial Organization, Regional Development and the City in Vietnam

Dean Forbes and Nigel Thrift

Introduction

The post-Second World War history of Vietnam, at least in outline, is better known in the west than the recent history of most socialist developing countries because of the widespread media interest in the Vietnam War. In contrast, we know or understand surprisingly little about the nitty-gritty of pre-socialist or socialist development in that country, and even less about regional problems and regional development.

The purpose of this chapter, therefore, is to examine the relationship between spatial structure and development in Vietnam since 1954, highlighting those planning initiatives targeted at the organization of space. The chapter opens with sections summarizing patterns of territorial organization in Vietnamese history, and in the non-socialist south of the country between 1954 and 1975. In the following two sections socialist economic planning is examined in the north between 1954 and 1976, and in the whole of the country since 1976. The remainder of the chapter canvasses the way in which regional issues have been tackled in socialist Vietnam, concentrating in particular on population distribution and the New Economic Zones, the role of the district and district towns in the interface of urban and rural, and the problems and prospects of the large- and medium-sized cities, notably Hanoi and Ho Chi Minh City.

Historical patterns of territorial organization

The eighth century BC is a watershed in Vietnamese history. It saw the dissolution of the primitive communes and the emergence of a hereditary monarchy, accompanied by the formation of a State, a bureaucracy and a class-divided feudal structure (Hodgkin, 1981, pp. 11–12). The authority of the king was relatively weak, though, and public finances were raised through a system of tributes (Nghiem Dang, 1966, p. 37). With the Chinese invasion in 207 BC the administrative structure was gradually assimilated with the Han dynasty model current in China. By the sixth century AD the Vietnamese structure bore many of the characteristics of the Han: power was centralized in the hands of the Chinese and the functions of authority were exercised by mandarins chosen in examinations based on the Confucian classics. With the ending of Chinese occupation in the year AD 939, the establishment of the Ly dynasty saw a further strengthening of central authority with a concomitant State-building (for instance the formation of a national army) and the development of a stronger feudal system of territorial organization (Nguyen Khac Vien, 1980, p. 243; Hodgkin, 1981, p. 35). The country was divided into twenty-four provinces, at first controlled and administered by princes from the ruling dynasty or by military leaders, and later to be governed by civil mandarins. The provinces were then divided into districts (*phu*), these were divided into sub-districts (*huyen*) and the sub-districts divided further into villages (*xa*). Each level was administered by civil mandarins (Nghiem Dang, 1966, p. 22).

Vietnam was briefly invaded by Mongols (in the thirteenth century) and Ming Chinese (in the fifteenth century), and in the aftermath the formation of the Le dynasty (1428–1788) represented the climax of the development of the centralizing medieval state (Hodgkin, 1981, p. 64). Two major neo-Confucianist changes to the structure of administration occurred: first, the landlord class (the *chua* or hereditary princes descended from the Trinh family in the north and the Nguyen in the south) became the ruling class (Nguyen Khac Vien, 1980, p. 243); second, a form of communal decentralization occurred as the village mandarins (*xa quan*) were replaced by village authorities (*xa truong*) chosen by the villagers themselves (Nghiem Dang, 1966, pp. 22–3).

The Tay Son movement, in the late eighteenth century, was the culmination of several decades of revolutionary activity. It achieved once again the reunification of the north and the south of the country under the Nguyen dynasty at the beginning of the nineteenth century. Under the Nguyen, Vietnam was divided into three main regions, of which the central region was placed under the immediate authority of the court, and the northern and southern regions were the responsibility of imperial delegates. As before, the regions were divided into provinces, districts, sub-districts and villages. The villages retained local autonomy, and all levels of administration were entrusted to mandarins, who at first were chosen from among the collaborators with the new regime but who were later recruited from those chosen from examinations based on the Confucian classics (Nghiem Dang, 1966, p. 24). During this period there was a bringing together of the ruling class and the bureaucracy, but being 'essentially a counter-revolutionary regime', the army tended to occupy itself with suppressing internal rebellion rather than external aggression (Hodgkin, 1981, pp. 100–10).

The administrative structure of Vietnam impressed many members of the colonial ruling class in the early days of French rule, with the result that they urged a policy of association which would preserve the mandarinate and the communal organization of administration. The French seemed to underestimate the resistance to their rule, though, and many mandarins, for instance, refused to co-operate. However, with the establishment of the Indochinese Union in 1887, a co-ordinated system of administration began to emerge. As the requirements of the colonial power towards the end of the nineteenth century became more and more apparent, the administrative structure changed to suit the new demands. Power over financial matters (customs' duties, monopolies, etc.) and administration (e.g. the delivery of services) was increasingly concentrated in the hands of the Governor-General, although some did try to involve Vietnamese as consultants to government, particularly at the regional level. On balance, then, French colonial policy, geared towards the development of an export economy based on *colon* settlement and corvée labour, was of necessity highly centralized and repressive: in opposition, though, there emerged Vietnamese nationalist and communist movements which eventually overturned the colonial power and established an independent, united Vietnam.

Running through Vietnamese history are three themes pertinent to the pattern of territorial organization. First, Vietnam has a long history of bureaucratic administration which is unified, well-organized and deeply rooted in a Confucian philosophy of administration. This 'great tradition', as it is sometimes referred to, was highly centralized,

but this centralization did not result in too complicated an organization of the administrative machinery since the need for administrative measures in such an agrarian society were not too great. In the colonial period, the centralization necessary to the unity of the colonial policy brought about the creation of agencies only at the central level, and did not require the addition of services of execution. (Nghiem Dang, 1966, p. 3)

Yet being based on a Chinese model of public administration also created problems, for Vietnam differed from China in substantial ways: it was smaller and less urbanized, and ethnic minorities were proportionately larger, so in fact, particularly when it came to provincial and district administration, there was a 'gulf between institutional aspirations on the one hand and social realities in the other' (Woodside, 1971, p. 112).

Second, decentralized power has historically rested in the commune. The headman was responsible for public administration, both in the implementation of general policy as well as in the management of local affairs. This principle of communal autonomy meant the decisions were made at the commune level by the headman, who was highly respected by the people. The implementation of these decisions was then the responsibility of the villagers (Nghiem Dang, 1966, p. 3). The commune has been a fundamental unit in Vietnamese society 'which has preserved its essential character from the Bronze Age to the present day, like a fortress behind its bamboo fences, the common ricelands stimulating respect for the common life, fostering ideas of solidarity and mutual assistance' (Hodgkin, 1981, p. 5).

Third, government-sponsored population resettlement programmes have often occurred in Vietnamese history and have been effective means of deconcentrating the Vietnamese population and spreading Vietnamese culture and control. The push southward into the conquered territories of the Champa empire and the setting up of military colonies (*don dien*) in the fifteenth century, and the seventeenth-century southern expansion of Vietnamese under the protection of the Nguyen dynasty were early examples of significance. More recently during the last century Nguyen dynasty rulers reintroduced military settlements as a means of soaking up floating populations, as a means of cultivating more land and as a political strategy designed to produce regional cores of loyal subjects. Resettlement strategies were also used by French colonial administrators as a means of breaking political movements such as the Can Vuong (Hodgkin, 1981, pp. 112 and 161).

The spatial organization of the South, 1954–75

Following the Second World War, Vietnam was administered by a series of short-term governments. In the independent Vietnamese State established with Bao Dai as head in March 1949, power was officially divided among three regional governors, but the co-existence of widespread communist control in the North and a continuing French presence shows that Bao Dai's power was more nominal than real. Following the partition of Vietnam at the Geneva Accords, the Constitution of the First Republic was promulgated on 26 October 1956. It specified a new government structure, but with the events of 1 November 1963 and the formation of the Second Republic, a new constitution was prepared.

In the years after the 1954 partition of Vietnam, the government of South Vietnam, at first under Ngo Dinh Diem and later under Nguyen Van Thieu, tried to wrest control of the countryside of South Vietnam from communist cadres by means of a series of 'community development programmes' which focused on territorial reorganization and economic development. These programmes included the land development programme of 1956, agglomeration centres (1959), agrovilles (1959–61), the strategic hamlet plan (later renamed 'new life' hamlets) (1961), programmes of village self-development (VSD) and the formation of a provincial development fund (PDF) in the late 1960s. Late in the war the government of South Vietnam formed a Central Urban Development Committee, charged with responsibility for forming a national urban policy, but like many of the arguments and plans for decentralized economic development formulated in the late 1960s and early 1970s (e.g. the Thuc–Lilienthal study (Lilienthal, 1969), the joint Brookings Institute–State Department study (Haviland et al., 1968), Rondinelli's (1973) study), it had very little impact on policy in the last few years before reunification. Despite the ultimate lack of success of these plans a number of them are interesting for two reasons. First, they were based on principles of territorial organization aimed at generating rural-based 'development from below' through principles of self-help. Second, the purpose of territorial reorganization and economic development at the village level was transparently obvious: it was an attempt to regain social, political and eventually military control of a population that could not be won over by the force of arms alone.

The plan to develop *agrovilles* – or rural cities – was announced in July 1959 as part of a large security plan for the whole country. Focused on the southern delta, it was planned to build eighty central agrovilles between 1960 and 1963 (plus satellites), catering for roughly 500,000 people. The policy had four elements:

(1) Regroupment of the population into agrovilles, linked by a new strategic route system ... (2) Development of competent cadres for village councils and administrative posts ... (3) Improvement of village self-finance resources, especially the development of public lands ... (4) The formation of a vigorous youth movement. (All in italics in text.) (Zasloff, c.1963, pp. 1–10)

By 1961 the programme had disappeared, only twenty-three agrovilles having been formed with a total of less than 40,000 inhabitants. Communist party cadres had mounted an intensive propaganda campaign against agrovilles, calling them 'disguised military bases' (Race, 1972, p. 173). More importantly, the peasantry had two main objections: first, they had been uprooted and moved into small towns, usually far from their fields; second, self-help was achieved by forced labour, which the peasantry resented, and though they could escape it by paying a levy of 200 piasters, this was regarded as an unjust tax (Nghiem Dang, 1966, pp. 157–9; Race, 1972, p. 69).

The *strategic hamlet programme*, initiated in 1961, shifted the focus of territorial organization from towns to hamlets. As their name implies, they were an attempt to settle the population into defensible hamlets (likewise the communists had organized 'combat villages' – Race, 1972, p. 193), internally composed of five-family groups. Originally set up ostensibly for mutual assistance, in practice they came to be part of the security, propaganda, and information organization of the village (Nghiem Dang, 1966, p. 159). The strategic hamlets were a focus for the development of self-reliance and a basic unit of territorial administration, handled by an elected hamlet chief. Later to be known as 'new life hamlets', the traditional autonomous village was the key concept underpinning the policy. Yet despite the linking of the 'new life hamlet' to traditional village autonomy, and despite the fact that in 1963 seventy per cent of the population lived in strategic hamlets, they had faded out by the mid-1960s. There are many reasons: like the agrovilles they caused resentment by relocating peasants away from their land and they were similarly dependent on forced labour; and they were poorly supported by the central government, receiving little or no financial help (inhabitants had 'few defensive weapons or obstacles, and with the occupants themselves often forced to buy barbed wire and pickets', Race, 1972, p. 133).

Community development programmes underwent another change after the Tet offensive in 1968 had revealed their ineffectiveness, with the formation of the *Village Self-Development Programme* (VSD). Once again the thrust of the programme was a move away from strong central government management towards decentralization and increased village participation in development. The programme was based on five main principles: political development, local economic development through the promotion of self-sufficiency, government aid and support for self-sufficiency, popular participation in decision-making and the strengthening of local government (Rondinelli, 1971, pp. 163–5). A Provincial Development Fund was set up as well and charged with co-ordinating and consolidating developments in sectors such as education, fisheries, animal husbandry, health and public works (Rondinelli, 1971, p. 169). Yet like their predecessors, these Community Development programmes ended up achieving very little. Beset with technical delays, distrust of the central government and corruption, the VSD, on the one hand, 'made substantial progress toward expanding popular participation in local decision-making, crystallizing the interests of local residents, inculcating principles of self-help and strengthening local leadership and organisation'. On the other hand it achieved nothing in the cities and ultimately failed as a tool of 'pacification'. All that could be said was that community development 'was successful in achieving a limited number of goals in selected areas of the country' (Rondinelli, 1971, pp. 171–4). This, however, should not be underestimated.

In general decentralization policies (as for instance manifest in the community development programmes), were an attempt to counteract the growing alienation between town and country in Vietnamese society which French colonial rule had brought about. The development by the French of a town-based indigenous administrative elite, together with the partial phasing out of the exams' system for mandarins, alienated the peasantry. With the beginnings of the second Indo-China war the gap between city and village widened: the cities of South Vietnam became military and economic enclaves supported by infusions of United States' money, commodities and military technology 'isolating the urban population – and the leaders and cadres recruited from them – and alienating them further and further from the peasantry' (Ton That Thien, 1967, p. 463). The focus of the attempt to counteract the alienation was the idea of community development based on 'personalist' development and a focus on small groups. Citizens were to assume responsibility for community projects such as the construction of dykes and the building of drains, schools and kindergartens.

Clearly the promotion of regional territorial units echoes through the community development plans put forward in the 1950s and 1960s. South Vietnam was a social laboratory for trying out a whole series of plans for territorial reorganization, ostensibly for economic and community development, but more centrally for social control. One of the major stimuli to territorial reorganization and decentralization is a political counter to communist insurgency (the Philippines and Malayan strategic hamlet programmes could also be included). As a result, the experience of South Vietnam by no means provides an adequate measure of the value of territorial reform in relation to economic development. Despite the rhetoric, real decentralization of power did not occur. The failure of the community development programme in Vietnam could not simply be attributed to poor project design and implementation: there was a war going on. The skill, organization and determination of the Vietnamese communists, not to mention their system of economic planning and territorial organization, were an important part of their appeal to the South Vietnamese and were crucial in undermining any strategies the South Vietnamese government may have developed. It is to the question of territorial organization in a State collectivist society such as existed in North Vietnam after 1954 and Vietnam as a whole after 1976 that we now turn.

Economy and spatial structure in the North, 1954–76

The Democratic Republic of Vietnam (DRV) – colloquially known as North Vietnam – came into existence in 1954. In fact, members of the DRV had had effective control of the countryside since the 1940s *despite its occupation by a large western army* – through the establishment of small but efficient administrative units that duplicated the existing Franco-Vietnamese administration' (Fall, 1967, p. 133). This was the so-called 'parallel hierarchy'. By 1955 an interim administrative structure had been created and in 1958 a State Commission of Planning was formed. In 1960 a Constitution was promulgated that tended to reinforce a centralized hierarchy of control, with limited regional self-government (Fall, 1967, p. 151). Planning within this structure was also highly centralized. The State Planning Commission prepared the development plan, the Council of Ministers then went on to approve it, targets and strategies were devised by the Vietnam Workers Party (as the Vietnam Communist Party (VCP) was then called), and Ministries were responsible for implementation. Planning and implementation was top-down, production quotas being determined by the government with bonuses constituting the major incentives for performance (Cao, 1978, p. 265).

By and large up until 1960 the main focus of economic planning was reconstruction, particularly of industry. However, land reform had commenced in the early 1950s (1953–6) and from mid-1958 through to the mid-1960s the transfer of land ownership from individuals to groups represented the first major attempt at any form of decentralized management of the economy. The strategy was the formation of co-operatives and it occurred in stages, based on the Chinese model. At first, peasants who pooled their land for collective cultivation were paid a total dividend based on the size of their land contribution, other capital inputs and their labour input. Over time, the proportion of the dividend to land and capital reduced, and the return to labour increased (White, 1982, p. 13).

The co-operativization programme took a long time to complete. Nine years after the commencement of the programme in 1958, eighty-five per cent of peasant families belonged to 'high-level co-operatives' (where all the means of production were cooperatively owned) and ten per cent to 'low-level co-operatives' (where land and tools were collectively used but remained in private ownership, the owners receiving rent for their use). By comparison, it took the Soviet Union seven years to reach an equivalent stage and China three (Gordon, 1981, pp. 20–1). The co-operatives represented a significant attempt to bring about public participation in economic planning, for they would be run by their members, who would devise production plans based on government guidelines (Elliot, 1975, p. 167). Co-operatives were also a measure for promoting political integration. The personnel resources of the Workers' Party were insufficient to oversee the daily workings of the co-operatives, which were thus entrusted with responsibility for ensuring public participation in day-to-day decisions affecting individual welfare (Elliot, 1975, p. 185).

The co-operativization programme was also expected to contribute substantially towards economic development. The Central Committee of the Vietnam Workers Party advocated three revolutions: the revolution in the relations of production; the technical revolution; and the cultural and educational revolution. With the establishment of socialist relations of production within the co-operative, improvements in agricultural productivity were expected to follow. The evidence about their success is ambiguous. In the first place the targets set for agricultural production in the early 1960s were impossibly high (e.g. initial targets were for an annual growth of 10 per cent, when the 4.4 per cent achieved in previous years was regarded as highly satisfactory). Yet it seems clear that agricultural production in the early 1960s was static and even declining (Gordon, 1981, p. 32).

Overall there were a number of reasons for this. Not least was the hierarchy of priorities which was formalized in the First Five-Year Plan (1960–5). The plan took an orthodox socialist line, giving priority to heavy industry, supported by State investment. (Incidentally, the State budget was drawn from foreign aid and industrial profits – agriculture neither contributed nor benefited in any significant way). Paradoxically, heavy industry was inevitably a centralizing factor. Modern manufacturing was located around Hanoi and into pre-existing cities like Haiphong, though there was an attempt to create new industrial centres, such as the textiles and light industry focus at Nam Dinh (Buchanan, 1967, pp. 152–3). This process, combined with the importance of the raw materials sector (coal, iron ore, etc.) and shortages of electricity reinforced the centralizing tendency of heavy industry (Fall, 1967, pp. 173–4).

There were of course many other reasons for the decline of agricultural productivity following co-operativization other than the emphasis given to industry. For instance, investment by co-operatives was poor, fragmented farm plots led to the under-utilization of irrigation, fertilizer supplies were inadequate and economic management on the whole was poor. However, production by the co-operatives started to increase along with the establishment of high-level co-operatives after 1963 and overall there occurred a slow improvement in production *per capita* (Gordon, 1981, p. 36; Elliot, 1975, pp. 183–6). But with the beginnings of American bombing of the north in 1965, the requirements of the co-operatives' movement underwent a change. The cooperatives' network was to play a crucial part in the war effort, for it provided a form of territorial organization that maximized the possibilities of achieving local-level economic self-sufficiency and assisted decentralized political decision-making. Both were critical ingredients in the ultimate success of the 'people's war' (White, 1982, p. 16).

Population resettlement schemes were also an important part of the territorial restructuring of the north. Between 1961 and 1966 a mass movement of people was planned from the Red River delta to the upland New Economic Zones in Bac Kan, Son La, Lai Chau and Nghia Lo (see figure 4.1), as well as some further movement from the highlands down into the 'intermediate upland areas'. The aim was an ambitious one: to move numbers equivalent to the entire estimated natural increase of the population (Jones and Fraser, 1982, p. 7). Movement increased in 1965 and 1966, partly because of the American bombing – a half a million people were evacuated from the urban centres and over the decade, the

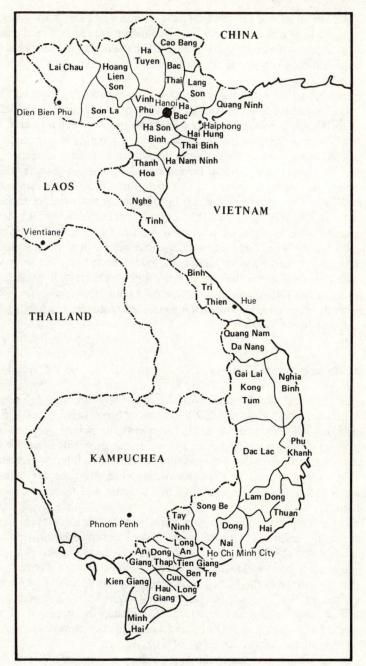

Figure 4.1 The provinces of Vietnam

total number of people resettled rose to one million. The primary motiv-
ation for these planned movements of population was to reduce the high
population pressures in the north, especially in the delta, and to spread the
population more evenly over the country and thus bring into cultivation
previously unworked areas.

During the 1970s, attitudes towards the co-operatives deteriorated. The
dangers of isolating collectives from the control of the proletarian State,
the possibilities of rivalry inherent in competing collectives, and the
perils of autarky signalled the decline of the self-reliant co-operatives. Of
particular concern was the co-operatives' contribution to economic
growth, for crop multiplication slowed down, and although crop yields
increased, the increase in population per unit of cultivated land was
faster. The net outcome was that land productivity improved, but labour
productivity did not. A move was undertaken in the early 1970s to zone
the country into agricultural regions as a basis for developing regional
comparative advantage. It met with resistance from villagers reluctant to
abandon self-reliant cropping and depend on the State's pricing, transport
and distribution system. However, events quickly overtook the economic
problems of the DRV, and with the vanquishing of the opposition in the
south in 1975, a new and much larger set of problems confronted the
government in the north.

Economic and social planning, 1976–85

Reunification and the formation of the Socialist Republic of Vietnam
(SRV) took place formally in 1976. It was a more rapid process than
either side had expected as southern opposition barely survived the
withdrawal of American troops in 1973. Yet the problems faced by the
SRV seemed overwhelming – they included the need to integrate economies separate for decades, rebuilding to overcome war-time destruction,
little and declining international economic support, and fears for national
security.

The planning structure and procedures used in the SRV were based
on the model used in the north. The State Planning Commission was
responsible for the preparation of the national development plans. At
reunification it was headed by Le Thanh Nghi, a Deputy Premier and
Politburo member of the Vietnam Communist Party (VCP). The Central
Committee of the VCP set the targets and strategies for the plans, the
State Planning Commission prepared the plans, the approval of the
Council of Ministers was required and the Ministries were then responsible for implementation (cf. Cao, 1978).

Five-year plans have been the key guiding documents. The Second
Plan ran from 1976 to 1980, the Third 1981–5, and the Fourth was

released in 1986. Congresses of the VCP are timed to coincide with the release of the Five-Year Plans, so that there is an opportunity for discussion of the key themes of the plans. Annual State Plans are prepared, incorporating revised targets and announced by the Chairman of the State Planning Commission. Regional plans are also prepared, as well as provincial plans and district master plans.

Post-reunification development strategy in Vietnam can be conveniently divided into two phases: The first, associated with the Second Five-Year Plan, plotted a conventional centralized bureaucratic socialist strategy of economic management. The second phase, following the Sixth Plenum of the VCP Central Committee in September 1979, saw the adoption of the so-called 'New Economic Policies' allowing for new production incentives and decentralized management of parts of the economy.

It is not surprising that the programme for integrating the economy of the SRV, which had been prepared in a rush, was ambiguous in intent. The programme was presented in the Second Five-Year Plan (1976–80) and received the approval of the Fourth National Congress of the VCP at its sessions in December 1976 (see Central Committee, 1977). Party Secretary-General, Le Duan, announced the target was to transform Vietnam from a backward agrarian economy into a modern socialist economy. He stressed the importance of self-reliance adding that 'each socialist country has to rely mainly on its own efforts' (*Far Eastern Economic Review Yearbook*, 1978, p. 329). In the south the goal was to transform privately owned agriculture and industry through the development of mutual aid teams of peasants and joint State–private enterprises.

Agriculture and industry were the two key sectors. White (1982, pp. 20 and 24) affirms that the basic strategy was orthodox and pro-industrialization, and that 'to Vietnamese leaders it is axiomatic that "building socialism" means first and foremost industrialization; this has never been questioned as a primary aim'. Yet it is also clear that there was a great deal of concern about agriculture and light industry and a fear that they had been neglected in the past. Both the Economist Intelligence Unit (1981) and the *Far Eastern Economic Review Yearbook* (1978) argue that agriculture was the main focus, with light industry the second priority.

Initially the government made clear that it would not nationalize private property if people were willing to co-operate with the new administration by maintaining production. However, at the same time there occurred increasing government intervention, for example, through attempts to control the black marketeers, hoarders and speculators. This sort of control was unsuccessful, though, and so the VCP responded with its own version of a 'great leap forward'.

In August 1977 it was announced that southern agriculture would be collectivized beginning with the planting of the following year's crop, and that capitalist trade would be abolished throughout the country by the end of 1978 (Nyland, 1981, p. 442). It was the government's intention to exploit the traditional 'rice bowl' in the Mekong Delta, hoping that the establishment of co-operatives would improve the procurement of rice. Indeed, by May 1980 some 1,747 co-operatives, 16,801 agricultural production collectives and 303 collectives responsible for agricultural machinery had been established. In some areas such as Central Vietnam, the destruction caused by the war assisted co-operativization, for the restoration of agriculture on a family basis would have been very difficult. However, overall by mid-1980 only fifty per cent of families using thirty-six per cent of the available area had been induced to join co-operatives (Nyland, 1981, p. 444).

It is generally accepted by the SRV and its critics that the main goals of economic development set in the Second Five-Year Plan have not been achieved (see *Vietnam Courier*, 1982a; Economist Intelligence Unit, 1981; Quinn-Judge, 1985a). The country's population growth rate to 1979 was around 2.5 per cent per annum (Fraser, 1985). However, the grain harvest scarcely increased between 1976 when it was 13.5 million tonnes and 1979 when it reached 13.8 million tonnes. In the intervening years it actually declined to 12.5 million tonnes in 1977 and 12.2 million tonnes in 1978 (Nguyen Xuan Lai, 1983, p. 54; Quinn-Judge, 1985a, p. 31). The value of industrial and agricultural production between 1975 and 1980, at 1970 prices, is shown in table 4.1. According to these figures, production has fluctuated in both sectors with an average per annum increase of 2.9 per cent in agriculture and 3.2 per cent in industry. Finally, trade data reflect the structural problems in the Vietnamese economy during this period. In 1977 exports were valued at 29.6 per cent of imports, but by 1980 this had slipped back to 23.7 per cent (*Indochina Chronology*, 1984, p. 28).

In all, few targets in industry, agriculture or mining were met during the Second Plan. One of the new regime's harshest critics charged the VCP with a loss of moral authority: the 600,000 refugees that have fled Vietnam, the 30,000 (or so) still in 're-education' camps, and the constant

Table 4.1   Industrial and agricultural production in Vietnam, 1975–80

	1975	1976	1977	1978	1979	1980
Value of industrial production (million dong, 1970 prices)	7,288	8,209	9,029	9,520	—	8,549
Change (1975 = 100)	100	113	124	131	—	117
Value of agricultural production (million dong, 1970 prices)	6,430	7,088	6,710	6,743	—	7,400
Change (1975 = 100)	100	110	104	105	—	115

Source: Adapted from Nguyen Xuan Lai, 1983

need for 'cultural purity' campaigns have combined with economic failure to create a crisis of confidence in the regime (Pike, 1982, p. 72). Another observer argued that restiveness in the north was a very clear signal to the government of the failure of economic planning. It was noted that there had been food riots in Haiphong and the Nghe Tinh during the autumn of 1980, and in Hanoi citizens had openly expressed discontent. While such things were expected in the south, protest in the north was not (Chanda, 1981a, p. 28). Moreover, a key intellectual and supporter of the VCP strategy, Nguyen Khac Vien, is reported to have strongly criticized the policies of the regime, implying that Truong Chinh, the Chairman of the State Council and a noted 'ideologue' (see Chanda, 1981b, pp. 14–16) was responsible (Quinn-Judge, 1982, pp. 14–16).

The government, in response, cited four reasons for its economic difficulties. First of all there was the change in China's attitude to Vietnam. It had resulted in the withdrawal of all assistance from Vietnam after the latter's incursion into Kampuchea. Second, Vietnam had been subject to a long sequence of natural calamities. These included drought (1977), floods (1978 and 1979) and a serious typhoon in 1980 (cf. Spragens, 1980). Third, it was admitted that the difficulties of increasing production had been underestimated, particularly during a period of external threat. And fourth, State investment had been inadequate (see *Vietnam Courier*, 1982a, pp. 6–8).

It was clear, though, that a significant group within the VCP believed that Vietnam's economic problems stemmed from a more fundamental problem of strategy. Discussions about a shift in strategy within the Central Committee first surfaced in a significant way in late 1979 with the publication of the outcome of the Sixth Plenum, but it was not until the following year (1980) that the formal directives for implementing the reforms were put into concrete form (Nguyen Xuan Lai, 1983, pp. 37–8). The reforms were accepted at the Fifth Congress of the VCP (March 1982) and were incorporated into the Third Five-Year Plan, which was due to run from 1981–5, but was not released until 1982.

A cornerstone of the New Economic Policies – termed 'market socialism' – was the introduction of production incentives. These took the form of a product-based contract system in agriculture, which allowed peasants to sell their surplus, and a piece-rate wages' system in industry. The best-known measure of the programme was Resolution 6: 'Resolution 6 of the Party Central Committee – affirmed the legality and legitimacy of private individual enterprise in the period of transition to socialism' (Nguyen Khac Vien, 1982, p. 21). It proclaimed that privately organized small-scale production has an important role to play in the transition to socialism, particularly in the production of consumer goods. (The residents of Ho Chi Minh City were among the first to respond to liberalization with a proliferation of tea-houses, restaurants, etc.). State-run enterprises were permitted to produce goods not in the plan by making use of surplus raw materials, and these goods could be sold on the open market. Members of co-operatives could likewise grow additional crops, on unused land, for sale in the local market. Along with production incentives, new systems of economic management were introduced. Managers of State enterprises were given increased power and responsibility for day-to-day decisions, and a new system of profit–loss accounting, designed to improve economic efficiency, was introduced (see Spragens, 1980; Chanda, 1981c; Davies, 1981; Nguyen Lam, 1982; White, 1982).

Apart from these management initiatives, the main sectoral thrust of the Third Five-Year Plan was oriented towards agriculture and light industry, especially the production of consumer goods (Le Duan in BBC, 1 April 82, FE6993/C/13–C/26). A new set of production targets were outlined in detail (see Nguyen Xuan Lai, 1983, pp. 39–49). Consistent with the new thrust, smaller, practical projects were to be given greater emphasis than the larger and more grandiose projects of the previous unsuccessful plan (cf. Morrow, 1982, p. 49). A primary aim was:

to coordinate the development of industry and agriculture. We're not talking simply about the coordination of a few big, heavy industries. If you go down to the local level, to a district, to a cooperative, right there the people will have industrial units ...

In the North every district has a machine shop. This machine shop will make small machines for agriculture, hand tools, and carts; it will make all the basic necessities for agriculture in that district. (Le Hong Tam, 1980, p. 24)

On both economic and political grounds it is too early to make any final judgements about the long-term contribution of the reforms to Vietnamese development. Population growth rates have been slowed to around 2.4 per cent, but there is still a long way to go before they are brought down to the target of 1.7 per cent per annum (Quinn-Judge, 1984a, pp. 46–7; Nguyen Cong Thang and Nguyen Thi Xiem, 1983). Grain production, admittedly, has increased spectacularly. The grain harvest in 1984 was estimated at 17.1 million tonnes, which represents an annual average increase of 4.6 per cent per year since 1980, compared to 1.5 per cent per year in the four years prior. Trade remains a serious problem. Imports in 1983 were still only valued at 29.7 per cent of exports, and the USSR continued to dominate trade links, accounting for 53.9 per cent of Vietnam's imports and 47.1 per cent of exports (see the data in *Indochina Chronology*, 1984, p. 28).

Of at least equal, perhaps even greater, importance, is the fact that not all of Vietnam's leaders accept the value of the reform policies. The man most closely associated with these policies is Vo Van Kiet, Chairman of the State Planning Commission since 1982, and a Deputy Premier and

Politburo member (Quinn-Judge, 1983a, pp. 32–3). His key supporter is Party Secretary-General, Le Duan. Believed to be among the opponents of the reforms at the most senior levels of government in Vietnam are Truong Chinh, Chairman of the State Council, and influential Politburo Members Le Duc Tho and To Huu (cf. *Far Eastern Economic Review Yearbook*, 1984). While some noted commentators of Vietnamese politics believe that the liberal orientation of economic policies will remain (Chanda, 1984, p. 32), others have noted some 'backtracking' (Quinn-Judge, 1983b, p. 23).

The eventual outcome is impossible to anticipate accurately. Since the Ninth Plenum of the VCP in December 1980, the pragmatists in the Party appear to have held sway. However, there are many such as Le Duc Tho who believe the answer to Vietnam's economic problems lies with increased control of the economy by the State, not less (White, 1982, p. 23). The free market now accounts for about seventy per cent of goods in circulation. In contrast, a communiqué issued at the end of the Sixth Plenum of the VCP Central Committee (July 1984) 'comes down on the side of the reforms', stressing 'increased local initiative, a streamlined and reduced central bureaucracy and a gradual transition to socialism' (Quinn-Judge, 1984b, p. 15). This position was reaffirmed at the Eighth Plenum (June, 1985). As the daily *Nhan Dan* announced: 'in order to control and transform the free market, we must know how to use it and not subjectively reject it' (Quinn-Judge, 1985b, p. 37). With the Sixth Party Congress held in early 1986, this is at least an indication that the reforms will continue into the next Five-Year Plan.

The regional question in Vietnam

Spatial problems figured large in the SRV after reunification. Perhaps the major problem was the differences between the north of the country and the south. The former had been effectively 'socialist' for three decades, while the economy of the latter had come to depend increasingly on the support of American aid and spending by the American armed forces (Wiegersma, 1983).

Second, there were particular regional problems on which to concentrate. Among these were the need to restore the Mekong Delta to its pre-eminent role in rice and food production, the security problems of the peripheral border areas and the need to rehabilitate the large area of agricultural land devastated during the War and bring new areas into production.

Third, the very special problems of the large cities in the south, particularly Ho Chi Minh City but also Da Nang and the other swollen secondary cities, needed to be addressed. The population size of these cities had to be restored to appropriate levels, and the urban economy and infrastructure needed to be developed to shift their role from one of consumption to one of production.

Fourth, a balance needed to be struck between centralized planning and local planning. This was a reflection of the shortages of human resources in Hanoi and the magnitude of the planning problem. It probably also owes something to a belief in the importance of decentralized economic management of the economy, though this sentiment only became clear after 1979.

A number of measures initiated by the Government of Vietnam have addressed these problems. The following parts of the chapter focus on three inter-related sets of initiatives intended to address the problems within the Vietnamese space economy. These include the attempt to redistribute population away from the large cities (particularly in the south) and overcrowded regions such as the Red River Delta and towards New Economic Zones in sparsely settled regions or in the 'suburbs' surrounding the large cities; the choice of the district as the basic territorial unit for agro-industrial production and planning, and the key role to be played by the district town as the interface between agriculture and industry; and the development of urban policies designed to tackle the problems of the large cities and improve the contribution of the system of large and medium cities in Vietnam to socialist goals and objectives.

Population distribution and the New Economic Zones

Population distribution in Vietnam had long been a concern of the government, as we have shown earlier in this chapter, and so the redistribution programmes implemented after 1976 must be considered in the light of this historical interest. A number of aspects of population distribution were of particular concern to the socialist government (see Che Viet Tan in BBC, 21 Jan 1982, FE/6933/G/1–C/3). There was the problem of an 'irrational' distribution of population *vis-à-vis* cultivable land. For example, population densities in the delta areas, particularly of the Red River, were very high, whereas there were significant mountain and plateaux regions only sparsely settled. Authorities were also concerned about the distribution of the nomadic population and the continued practice of shifting cultivation techniques in agriculture. Finally, they were concerned with the concentration of population in the 'hypertrophied' cities of the south (figure 4.2). Although there was significant movement out of these cities in the first few months of peace, it was decided a planned movement of people out of the cities was necessary to reduce the population and convert the urban economies from a 'consumer' to a 'producer' orientation (cf. Dao Van Tap, 1980).

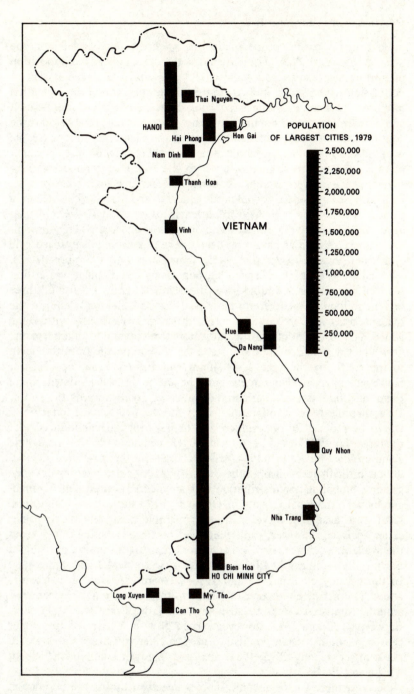

Figure 4.2 The population of large Vietnamese cities, 1979

Four main streams of population resettlement were targeted in the Second Five-Year Plan. These included the reduction in the population of northern provinces and the Red River Delta, the redistribution of population in the south from the cities to the rural areas, the formation of more New Economic Zones (particularly in the north and south-west), and a redeployment of population within districts and provinces and the consolidation of small settlements. During the Plan, 1.47 million people were moved (although the target was four million), while forty per cent of movements were from the north to the south. A number of New Economic Zones were established along the Kampuchean border, primarily composed of soldier–civilian settlements, and designed to act as a buffer. Another major focus of resettlement was the Central Highlands, again with security reasons in mind for this is an area where the non-Vietnamese minorities have long been active in resisting centralized rule (Jones and Fraser, 1982, pp. 119–25; Tran Dang Van, 1978, pp. 10–12).

During the Third Five-Year Plan population resettlement was at first planned to continue at a rapid pace. Between 1981 and 1985 around three or four million people were to be moved into the Mekong Delta, to the plateau areas and into the eastern rubber growing areas, with some additional movement of shifting cultivators from upland to valley regions. Overall, it was planned to move some ten million people from the north to the south by the year 2000 (Jones and Fraser, 1982, pp. 125–7). However, by mid-Plan targets have begun to be scaled down. State Planning Commission Chairman Vo Van Kiet noted towards the end of 1982 that the internal migration target for the five-year period 1983–7 would be reduced to one million and then a year later announced that the target for 1984 would be to shift 120,000 people to the New Economic Zones (*Indochina Report*, 1984, p. 15).

It is difficult to determine the overall impact of the resettlement programme on population distribution with any degree of precision. Vietnam conducted a national Census in October 1979, and although the main results are available, there is little comparable accurate data for time-series analysis. However, some tentative conclusions can be drawn from the available evidence. First it is apparent that urbanization rates slowed down quite significantly after reunification (see table 4.2 and figure 4.3). In the period between 1975 and 1980 Vietnam experienced deurbanization. The urban population of Vietnam grew at only 0.1 per cent per annum, compared to an overall population growth rate of 2.5 per cent. Between 1980 and 1984, however, urban growth accelerated to 5.7 per cent per annum, while population growth rates remained stable at 2.6 per cent (cf. Khong Dien, 1983, pp. 26–7; Nguyen Duc Nhuan, 1984b; Thrift and Forbes, 1986).

Second, there has been some success in redistributing the population from the north to the south. In the years between 1975 and 1979 the

Table 4.2   Urbanization and population in Vietnam, 1975–84

	Total population ('000s)	Urban areas ('000s)	Urban (%)
1975	47,638	10,242	21.5
1976	49,160	10,127	20.6
1977	50,413	10,108	20.1
1978	51,421	10,130	19.7
1979	52,462	10,094	19.2
1980	53,772	10,301	19.1
1982	55,540	11,441	20.6
1984	59,980	12,836	21.4

Sources: Khong Dien, 1983, p. 27; Fraser, 1983, p. 8

population of the north grew at 1.6 per cent per annum compared to 3.6 per cent per annum in the south. The net result is that in this period the south increased its share from 48.5 per cent to 50.4 per cent of the total population (cf. Monnier, 1981).

Third, emigration has played an important role in restricting overall population growth in Vietnam. It has been estimated that up to 300,000 Vietnamese of ethnic Chinese extraction from the northern parts of the country have been resettled in China, and between 750,000 and 1,000,000 people have emigrated elsewhere from Vietnam (Fraser, 1984, p. 6). Estimates of the number of Vietnamese settlers in Kampuchea alone vary between 56,000 and 640,000 (*Indochina Report*, 1984, p. 11).

One of the key problems of the resettlement programme has been the conditions in the New Economic Zones (NEZs). It has been estimated that during the Second Plan 1,472,000 people were shifted, 625,000 in the north and 847,000 in the south, though data on the precise location of settlements is scarce (cf. Hill, 1983). There are many problems in the zones which still need to be overcome. These difficulties include shortages of food and equipment for new settlers, unstable agricultural output, low crop yields and inappropriate crop choice, insufficient management support for farmers, poor co-ordination between resource development and environmental protection, and various social problems in new communities (Dao Van Tap, 1980, p. 510; Stern, 1981, p. 367). One manifestation of the difficult conditions in the zones during 1980 was large numbers clandestinely leaving the NEZs to return to the cities (*Far Eastern Economic Review Yearbook*, 1981, p. 263).

In the Third Five-Year Plan there has been some attempt to improve conditions in the NEZs: the rights and entitlements of movers have been clarified and improved, training and management has received some

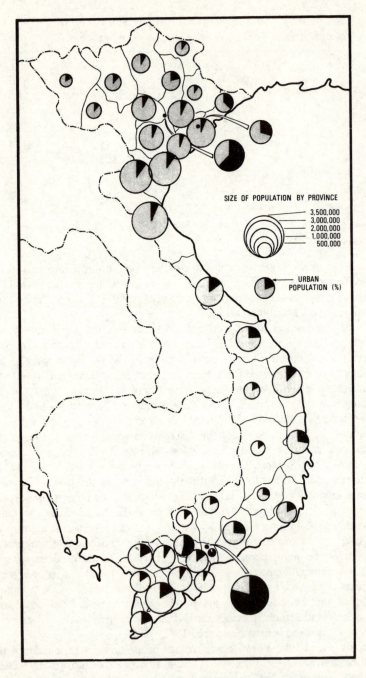

Figure 4.3 Urban population of Vietnam by province, 1979

attention, and the new incentives scheme may make the Zones more attractive. Furthermore, the Minister for Agriculture in the SRV, Nguyen Ngoc Triu, has recently expressed the view that settlements in the NEZs have improved in the last few years. The resettlement areas are mainly used for animal husbandry and cash crop production, organized around State farms. Production is improving because the NEZs are located in fertile regions, settlement not yet having been forced into the marginal lands (personal communication, Canberra, 16 May 1985).

The district and the small town

Government in the north of Vietnam after 1954 was divided between State and Party. The former was the principal authority in Hanoi, while local administration was in the hands of the Party and the co-operatives (Elliott, 1979, p. 350; see also Ginsburgs, 1963). In 1972, however, the district was resurrected as an experimental administrative unit, though without budget or planning responsibilities. Then following the Fourth VCP Congress in 1976 the Central Committee announced that the district would take on a larger role and become the key local economic and political unit. The resolution stated the need:

To firmly build the district into a real agro-industrial economic unit and an area for reorganization of production, organization and redivision of labour, combining industry with agriculture, combining the national economy with collective economy, and workers with peasants. (Central Committee, 1977, p. 234)

The village would no longer be the key planning unit, and direct administrative control of the population was to be gradually withdrawn from the province. Provincial officials were instead to assume leadership responsibilities, provide technical assistance and supervise the implementation of jobs assigned to the district directly by central government (Elliott, 1979, p. 355; Dong Thao, 1983, pp. 6–7; Che Viet Tan in BBC, 21 Jan 1982, FE/6933/C/3–C/6).

Some progress, albeit uneven, was made on building up the role of the district as the key territorial unit for the management of economy and society between 1976 and 1982. However, achievements fell well short of creating the full planning and budgetary echelon intended for the district level. Around two-thirds of Vietnam's over 400 districts developed comprehensive planning programmes, particularly regarding water conservation. Most districts were zoned for crops, and outline plans for industry, handicraft, transport, communications and construction were commenced. Districts were authorized to draw up a comprehensive socio-economic plan, and though it was noted that the standard of district cadres improved, additional assistance was often required from organizations outside the district (Nguyen Tien Loc, 1983).

The Fifth Party Congress in 1982 reaffirmed the importance of the district in Vietnam. It was noted that planning programmes in 210 districts had been revised, and that the devolution of responsibility for farming, small industry and handicraft co-operatives had been completed. In some districts, a number of State-run service industries (e.g. tractor stations) had also been brought under local control, while other districts had set up food funds and transferred health care, education and cultural facilities such as cinemas, into district control. District people's committees had also been reorganized, the number of internal departments increasing from sixteen to twenty-one, at the same time as staff cuts reduced the employment establishment of each from around 160 to 120 (To Huu, 1984, pp. 16–18).

The district town was intended to play a crucial role within the district and the transition to socialism within Vietnam. According to planners, it was the means of eliminating the differences between town and countryside, a method of promoting controlled urbanization, and a basis for facilitating the industrialization of agriculture. It was intended that district towns grow to a population of around 10,000 (or between five and ten per cent of the population of the district), and provide an urban style of life, with a full range of public amenities and services such as electricity, water, drainage and communications (Pham Tri Minh, 1978, pp. 9–13).

The formation of an urban system involves firstly 'transforming and rebuilding existing towns; [and] second in implanting evenly distributed small-sized new towns in the various regions of our national territory' (Dao Van Tap, 1980, p. 511). While in the past it is conceded that each district had a principal town, it is argued that its population was small and its primary function generally administration. Thus the town's role in relationship to economy and society was very limited:

The time has come to enlarge and redevelop these towns, to turn them into urban areas of several thousand inhabitants with all the characteristics of economic, political and cultural centres. Several industrial plants will be built, as well as trade and shopping facilities and communication links within the district and outside it. There will be administrative offices, economic enterprises and facilities catering for the material and cultural welfare of the population: a school, hospital, library, club, theatre, cinema, restaurant, scientific and technical research station directly serving the local population, etc. (Dao Van Tap, 1980, p. 513)

The economic specialization in town and district would be determined by the natural environment and the comparative advantage which it determines the region should have. For instance, districts in mountain regions were intended to be larger economic and administrative units than on the coast due to the less-dense settlement pattern, and have a product specialization built around the highland environment (for regional specialization in Vietnam, see Vo Van Kiet, 1983).

District towns have five main functions: First, they are to be political and planning centres and the base for specialized government departments. Second, the district towns are to become cultural centres, being the locus for servicing the office and factory population in the town as well as the population in surrounding villages. Third, district towns are to concentrate the research and technical development capacity of the district, particularly the research servicing agriculture and husbandry. Green belts are also planned around the district town, enabling the urbanization of the countryside. Fourth, the towns are to be a means of linking agriculture and industry, providing markets for the sale of agricultural produce and, through the development of small industries, creating a demand for surplus agricultural labour and the raw material output of the rural sector. Fifth and finally, small towns are expected to develop as significant commercial centres. Thus, there is a need to develop department stores to promote consumption, warehouses for agricultural inputs and outputs, transport terminals and commercial institutions to encourage commerce and trade (Pham Tri Minh, 1978, pp. 9–13).

At a conference in Hanoi in October 1984, the Head of the District Building Commission and Politburo member, To Huu, outlined some of the problems with the district programme in Vietnam. These included an excessive reliance on rice monoculture in district agriculture, a failure to develop linkages between agriculture and industry, and inadequate attention being paid to the circulation and distribution of goods. In addition, a criticism of the excessive 'bureaucratic subsidy-based management' systems used in the district neglected the growing concern in Vietnam that local spatial units should become financially self-reliant (To Huu, 1984, p. 16). On a more fundamental note, there would also appear to be a shortage of appropriately trained technical staff, particularly economists and engineers, at the district level, and a need to up-grade the qualifications of district heads. Manpower shortages and budgetary constraints are also responsible for the difficulties with infrastructure development which many districts face. Finally, some doubt the districts plan can ever be very successful, because 'it would necessitate surrender of power by well-established provincial and cooperative (village) political and economic institutions' (White, 1982, p. 17).

The large cities: Hanoi and Ho Chi Minh City

Urban centres, especially the large cities, play a very important role and must promote their ever increasing impact on the economy, especially in industry, small industries and handicrafts. Central and local authorities are duty-bound to help build Hanoi into a political, economic, scientific, technological and cultural centre worthy of representing the whole country. Ho Chi Minh City with its varied possibilities must become an important centre for economic, cultural and foreign trade actvities. (Pham Van Dong, quoted in BBC, 30 March 1982, FE/ 6991/c/5)

Warfare in Vietnam in the 1960s and 1970s had a damaging effect on both Hanoi and Ho Chi Minh City. Hanoi had been extensively bombed in 1972 and 1973, and substantial numbers of the population evacuated. Following the end of the bombing there was a rush of people back to the city. However, the bombing had taken its toll, and around 17,000 housing units had been destroyed, creating an acute housing shortage. Moreover, the dispersed city administration could not keep up with the influx of people. Apart from the housing shortages there was corruption and the misuse of State property and unauthorized and illegal markets began to appear. Moreover, the large decentralized industries were disrupted by the evacuation of the city and the efficiency of their operations declined. More self-sufficient industries, such as handicrafts, were not nearly so vulnerable to social disruption and continued to perform adequately but still it was several years before the city had returned to normal (Nguyen Vinh Vien, 1974; Turley, 1975).

In contrast to the bombing and evacuation of Hanoi, Ho Chi Minh City (or Saigon, as it was then called) had experienced rapid population growth and the development of a large tertiary sector workforce. It suffered many of the problems of similar primary Asian cities including a poor standard of housing and services, air pollution, extensive under-employment and a neo-colonial economy, highly dependent upon American aid and spending. These problems were compounded in the last months of the war in early 1975, as soldiers left the army and returned to the city and set about looking for work (Lang and Kolb, 1980, p. 14).

It is clear that following reunification, the problems of the southern cities had priority over those in the north, though there were few resources available for urban development overall. The most pressing issue was the sheer size of the southern cities, which led to the formation of a deurbanization strategy as part of a more wide-reaching population redistribution programme (discussed earlier in this chapter). While not successful in bringing the population down to the 2.5 million level hoped for, the size of Ho Chi Minh City has been contained, reaching 3.29 million in 1985, well below its estimated population in 1975. Hanoi has grown to 2.67 million (see tables 4.3 and 4.4).

Concurrent with the reduction of the population of Ho Chi Minh City, the Vietnamese have been concerned with converting the balance of the city's economy from consumption to production. This is a common element to the development of a 'socialist city'. Hanoi had achieved the

Table 4.3 Population of Hanoi and Ho Chi Minh City, 1985

	Hanoi	Ho Chi Minh City
Urban population	826,900	2,338,134
Percentage	31	71
Suburban population	1,947,500	955,012
Percentage	69	29
Total	2,674,400	3,293,146

Source: Indochina Chronology, 1985, p. 3

Table 4.4 Hanoi population distribution, 1982

	Population	Area (km²)	Density persons per km²	Administrative Structure Wards	Communes	Towns
Inner districts						
Hoan Kiem	154,000	4.5	34,222.2	18	—	—
Dong Da	223,000	14.0	15,928.6	25	—	—
Ba Dinh	160,770	10.5	15,304.8	15	—	—
Hai Ba Trung	238,000	11.0	21,636.4	23	—	—
Suburban districts						
Gia Lam	224,700	175.7	1,278.9	—	31	4
Dong Anh	169,100	184.2	918.0	—	23	1
Me Linh	177,100	254.9	694.8	—	22	2
Soc Son	143,200	313.3	457.1	—	23	—
Tu Liem	179,500	109.7	1,636.3	—	25	3
Thanh Tri	160,300	100.2	1,599.8	—	26	1
Hoai Duc	151,400	122.1	1,240.0	—	27	—
Dan Phuong	87,900	76.8	1,144.5	—	15	1
Thach That	87,500	93.4	936.8	—	19	—
Phuc Tho	98,900	102.5	964.9	—	22	—
Ba Vi	207,300	543.3	381.6	—	32	—
Towns						
Son Tay	33,900	14.6	2,321.9	3	9	—
Ha Dong	64,200	14.7	4,367.4	—	2	—
Total	2,560,700	2,245.6	1,140.3	84	276	12

Source: Adapated from Vietnam Courier, 1982b, p. 18

transformation, or so it is argued, by 1961. Hanoi in 1954 had a population of 270,000, but contained only nine industrial establishments, most providing repairs and services. The industrial workforce in 1955 amounted to 1,369, with another 5,000 employed in handicraft establishments. Planning in the period 1954–60 concentrated on rehabilitating the economy, introducing land reform and bringing privately owned production into ownership by the State or the collective. The result was that the value of output by large and small industry and the handicraft sector increased several times between 1955 and 1960, thus earning Hanoi the unofficial title of a 'producer' city (Nhan Dang, 1984).

Ho Chi Minh City has proved a more difficult city to convert to socialism. At the beginning of 1975, 360,000 of the city's 550,000 households depended on income earned through trade and services (trade accounted for 55 per cent of the total), and this proportion increased in the early months after reunification. The years between 1975 and 1978 saw a struggle between the extant domestic bourgeoisie and the State for control over the urban economy. A series of laws were promulgated to try to bring trade and services under greater government control. These included measures to confiscate certain goods, a surtax on excess profits and measures to give government enterprises a monopoly in the distribution of staples and primary goods (The Nguyen, 1978, pp. 11–12).

However, the crackdown on the private sector proved insufficiently effective, and with the implementation of the New Economic Policies between 1979 and 1981, a new tack was adopted in Ho Chi Minh City. In fact, residents of the city were very quick to react to Resolution 6 and it has been often remarked that the resurgence of small privately owned enterprises has been more noticeable in Ho Chi Minh City than practically anywhere else in Vietnam. The immediate impact was described in the following way:

Shops have mushroomed, some opened by government services to manufacture bicycle tyres, medicines, etc. Everywhere people try to turn out some kind of product with makeshift means. Export companies have been set up ... Overseas Vietnamese send capital, equipment, ideas. The movement has spread throughout the country but possibilities are naturally the greatest in Ho Chi Minh City where most of the unused capital and stocks are concentrated. (Nguyen Khac Vien, 1982, p. 22)

The enduring transformation of Ho Chi Minh City from a 'consumer' economy to a 'producer' economy is one of the major problems facing both city and national authorities. It was admitted in 1978 that the State controlled only thirty per cent of the volume of goods traded, and it is unlikely to have increased very rapidly in the years since (The Nguyen, 1978, p. 18).

A plan had been compiled in 1978 for the commercial sector in the city. It envisaged a three-tier structure. Ho Chi Minh City was to have three major trading centres located in Saigon, Cholon and Gia Dinh, sharing sixty large department stores retailing high-order consumer durables and processed foods. At the district level, another 120 stores and 1,641 shops were planned. These would be located along major thoroughfares and concentrate on small household requirements such as utensils, appliances and textiles. The third tier was focused on wards, each with a population of about 10,000, in which it was planned to build in total 657 stores and 3,234 shops. The ward's commercial centre would concentrate on lower-order needs such as tobacco, vegetables and other perishable foodstuffs. At the same time, the city's markets were to be rationalized, the numbers being reduced from 125 to 50 (The Nguyen, 1978).

A series of policy initiatives are designed to back-up the 'socialization' of the urban economy. The first of these is directed at 'life's dust' – the social problem groups in the city (Nguyen Khac Vien, 1978, pp. 18–20). At reunification it was estimated Ho Chi Minh City contained 100,000 prostitutes, a similar number of drug addicts, and at least 30,000 who suffered from venereal disease. Aside from this, the city provided a home for many disabled war veterans, orphans and numerous others who had suffered during the war. Although priority was given to the relocation of population, serious efforts were made to rehabilitate drug addicts and prostitutes, re-establish the network of educational institutions and improve the health system.

A second initiative included a plan to develop a cluster of significant satellite cities around Ho Chi Minh City. The most important would be Bien Hoa, which has an industrial base, and the nearby coastal city of Vung Tau, which specializes in oil and gas. This cluster of large cities, with each characterized by its own economic specialization, is intended to assist in reducing the population growth of Ho Chi Minh City, which it is planned will reach no more than five million by the year 2000 (Viviani et al., 1981, p. 27). In other words, Ho Chi Minh City is to be restricted to an annual growth of around 2.8 per cent between 1985 and 2000.

A third innovation is the development of a green belt around the city, designed to be used for agricultural production. These green 'suburbs' of Ho Chi Minh City contain 151,000 hectares of cultivable land and are planned to be used by 1.2 million people. The intention is that they will provide the opportunity for market gardening, industrial crop production and stockbreeding. Although it is planned that suburban production will enable Ho Chi Minh City to become self-sufficient in certain types of foodstuffs, planned production from the zones is also for export (Le Dan, 1983, pp. 4–6; see also Lang and Kolb, 1980; Thien Anh, 1983).

Although the problems of Ho Chi Minh City appear more serious than Hanoi, and consequently it has been the focus of more attention, it is clear that conditions in Hanoi in some sectors may be poorer than in the south. This is certainly the case in terms of housing provision. There is a severe shortage of housing in Hanoi, the average allocation being only 1.5 square metres per person. Despite the significant groups of homeless in Ho Chi Minh City, in 1976 there was some fourteen square metres of housing space per person. Hanoi has experienced no new inner-city housing since 1955, although there have been low-rise apartments built around the outskirts of the city (Nguyen Vinh Vien, 1974, p.5). However, it has been estimated that around one-third of the Hanoi workforce still live in rural dwellings, with commuting distances in the order of 20–40 kilometres (Nguyen Duc Nhuan, 1984a, p. 83). Serious attention has been given to improving the quality of housing supply, particularly the standards of design and construction (*Vietnam Courier*, 1982c, p. 24).

Conclusions

This chapter has attempted to provide a synoptic overview of the spatial aspects of social and economic planning in a socialist developing country. More specifically, it has demonstrated and documented aspects of the evolution of socialist strategies of territorial management and their relationship to regional planning and urban development in Vietnam. The topic is a large one in chronological depth and thematic breadth. Therefore the treatment accorded it here can only be considered a summary of some of the more important trends. Understanding of Vietnam is hampered by the same constraints that hamper research in many other countries, both socialist and capitalist. The information available to the researcher is restricted and access to the country is difficult. Conclusions are, therefore, based on data sources emanating from government agencies and distributed through government-owned and operated print and broadcast media, such as the broadcasts monitored by the BBC or the monthly journal *Vietnam Courier*.

With these limitations in mind, it is possible to draw out three themes which have been developed in the chapter: First, centralization and decentralization are recurrent themes in the history of territorial organization in Vietnam. Even while Vietnam was subject to strong centralized rule, such as under the Nguyen dynasty at the beginning of the nineteenth century and had a powerful bureaucracy controlled by mandarins, the village retained autonomy over local affairs. More recently, during the 1950s and 1960s, a specific series of spatial programmes were introduced in the south in order to achieve a political solution to the guerilla warfare of the time. A further series of spatial policies were implemented by the socialist governments of the north and later of the reunified Vietnam. On the one hand, some element of these programmes were similar. For instance, the size of the spatial units – districts or communes – varied little, there was always an emphasis on local autonomous decision-making and each strategy was intended to achieve economic and political goals simultaneously, the latter being the security of the State. On the other hand, the economic, social and political context in which these territorial strategies were proposed and implemented were entirely different. The key conclusion is that decentralized territorial planning and management is an enduring theme in Vietnamese development.

Second, it has been shown that socio-economic planning in the SRV has shifted direction since 1979. Up until the Sixth Plenum of the Central Committee of the VCP, in September 1979, the Vietnamese adopted a conventional socialist path. The greatest emphasis was placed on the transformation of the relations of production and distribution. Attempts were made to co-operativize agriculture in the south, nationalize the large firms and take control of the urban economy of Ho Chi Minh City. A major crackdown on the private sector occurred as late as 1978. But debates within the ranks of Vietnam's rulers were first made public at the time of the Sixth Plenum, and the 'New Economic Policies' were eventually incorporated in the Third Five-Year Plan. 'Market socialism', as it was termed, promoted production incentives, affirmed the legality of private enterprise, and gave a new impetus to decentralized autonomous management of enterprises and regions. The pragmatic emphasis of economic and social planning, closely identified with State Planning Commission Chairman, Vo Van Kiet, continues to hold sway in Vietnam, despite some strong opposition at the top.

The third theme developed in the chapter is a demonstration of the significance of urban and regional planning to socialist strategies of socio-economic development in Vietnam. It is quite clear that spatial planning is not the most important priority of the Vietnamese government. National security, economic development (i.e. industrialization and agricultural development) and the fostering of socialism all take precedence as key national goals. However, the spatial distribution of population and the organization of space for planning have been given considerable attention, the former through the population resettlement programme and the latter through the promotion of the district as the primary agro-industrial unit. On the evidence available it would appear that the resettlement programme, despite its setbacks, has been relatively more successful than the district development programme, which despite being in existence since 1976, is still in its early stages. Urban development, however, appears to have been a lower priority. The dominance of Ho Chi Minh City within the urban system has drawn attention to it, and important initiatives include the containment of urban growth and the attempt to develop a green ring around the city. However, the most important formative policies as far as the cities in the south are concerned have been those associated with the struggle by the State to control the economy, or in other words, to convert the cities from consumers to producers.

To summarize, spatial strategies and policies have been an important feature of socio-economic development planning in socialist Vietnam. Moreover they do not seem to have been affected very much by the shifts in ideological hegemony manifest in changes of direction at senior levels of government in Vietnam.

Acknowledgement

Earlier drafts of this chapter were presented at seminars in the Department of Human Geography, Australian National University, and the Department of Geography, Flinders University of South Australia. Suzie Jeffcoat drew the maps.

5
Regional Disparities and Regional Development in Algeria

M'Hamed Nacer and Keith Sutton

Introduction: Algerian socialism and regional development

Algeria's claim to be a socialist State is much debated by the purists just as it is repeatedly promoted by the Algerian government (Chaliand and Minces, 1972). While self-proclamations of socialist policies and values should not be taken at their face-value, there is no denying the rejection since independence in 1962 of the liberal capitalist approach to economic development by successive Algerian governments. Foreign mining, business and financial interests were nationalized and turned into State companies. The 1971 take-over of French oil interests marked the culmination of this exercise and meant that by 1972 State companies controlled ninety per cent of the industrial sector and employed seventy per cent of the industrial workforce (Economist Intelligence Unit, 1977). Colonial agricultural land was also socialized into a self-management or *autogestion* sector of agriculture, within the framework of participatory workers' control, but all too often bureaucratic dominance and a rigid centralism stifled any spontaneous peasant mobilization. A subsequent agrarian reform of Algerian private agriculture during the 1970s again started in a welter of socialist participatory rhetoric though the eventual results and interim proceedings left many with misgivings (Abdi, 1975; Sutton, 1982). In particular Algerian 'socialism' has been equated with the heavy centralization of powers and planning, and central National Plans. Four National Plans (1967–9, 1970–3, 1974–7, 1980–4) have set out priorities with the aim of achieving a more socialist society and of distributing the benefits of development more equally. However, this has also resulted in side effects embodied in the growth of a cumbersome bureaucracy which stifles initiative, inhibits the best use of resources, and so, in the long run, goes against reducing disparities and, indeed, runs counter to some of the fundamental ideas of socialism as advocated by the Algerian leaders and media outlets.

In an appraisal of Algerian socialism Nellis stresses that the policies of the 1965–71 period in particular were thoroughly state capitalist, with a single-minded pursuit of centralism and industrialization. However, by the mid-1970s a re-emphasis on popular mobilization and participation prompted Nellis to re-evaluate earlier strategies (Nellis, 1977, p. 533). He hypothesized that an initial reversion to centralization by the post–1965 Boumedienne Government was a calculated risk to create the strong economic and political base on which later to build the more decentralized and participatory economy and society which had always been desired; a case of *reculer pour mieux sauter*. This hypothesis will receive some support from the evidence in this chapter of reduced regional disparities following active policies of regional development, and of increased decentralization of activity and decision-making.

In many world and especially Third World assemblies Algeria's socialist stance is generally accepted. Its support for anti-colonialist movements, its membership of Third World or socialist 'blocs' and its activities in the Organization of Petroleum Exporting Countries (OPEC) and the United Nations Committee on Trade and Development all give credence to its socialist vocation, even if they contrast somewhat uncomfortably with its recent reliance on the United States market for an increasing proportion of its essential exports of oil and gas, now exceeding ninety-five per cent of its total exports by value.

While the 'industrializing industries' approach to development and the State companies' dominance of those industries have transformed the Algerian economy there has emerged increasing evidence of difficulties and shortfalls in economic performance. Whereas industrial investment has superficially been impressive, the production reality has been well behind targets. Between 1967 and 1978 300 billion dinars[1] were invested in the Algerian economy but additional production was only 46.4 billion dinars. Especially poor capital–output ratios were found in industry where 88 billion dinars were invested for 7.8 billion dinars additional production. So in manufacturing the capital–output ratio attained the high level of 11.3, roughly seven times higher than in normal industrial countries (Elsenhans, 1982, p. 62; MPAT, 1980a). The under-utilization of industrial capacity has seriously reduced the developmental effectiveness of much industrial investment. Sid Ahmed evaluates the 1980 levels of utilization of capacity to be 70.1 per cent in tractor production, 11.2 per cent in the sugar beet industry, 72.7 per cent in cast iron production, 59.5 per cent in the steel industry, and 50 per cent in cement production (Sid Ahmed, 1982, p. 55). Another source even suggests that 'most factories were operating at around 30 per cent capacity in 1978 and are probably still doing so' (*Arabia*, 1983, p. 54). The same anonymous source suggests that by 1980 the cement industry, a key indicator, was producing a mere 1.5 million tonnes a year against an annual capacity of 8 million tonnes. Official Algerian statistics put 1980 cement production at 4.16 million tonnes (SEP, 1983, p. 15).

Certainly Algeria has suffered from excessive production costs as scale economies are not achieved and inflationary over-staffing and other pressures play havoc with earlier estimates. The result is a range of import substitution industries producing more expensive products than the same goods imported.

In terms of Algeria's avowed socialist option these industrial development problems appear to have prompted the Chadli Bendjedid Government to implement certain modifications since 1979–80. An effort is being made to divide up the big state companies, like Sonatrach, to improve efficiency. Perhaps more significantly, private enterprise is rehabilitated with encouragement being given to the private sector, particularly in certain branches of industry like textiles, plastics and television sets, which are complementary to the State sector. Restrictions on buying land and building private houses have been lifted (*Arabia*, 1983). Arguably this change of emphasis stems from the realization that the 315,000 or so private businesses, which have survived earlier restrictions, now provide about one-third of all jobs in Algeria and have considerable potential to provide many more (Ghiles, 1983). Consequently, the 1980–4 Plan opens up to private investment the final transformation of intermediate products, the mechanical, electrical and metal manufacturing industry and the production of spare parts even for the public sector. This should add to the private sector's established importance in textiles, food processing, shoe and leather goods, and the construction industry which in 1978 accounted for 29 per cent of production and 45 per cent of employment. However, it should be emphasized that 'Algeria is not on the way to capitalism' (Elsenhans, 1982, p. 69). There is no tendency apparent for private enterprise to become powerful enough to restrict the government's investment policies which still favour the state sector. Indeed, the newly structured, smaller, decentralized units of the state companies may well compete with private enterprise.

Perhaps in identifying the socialist nature of Algeria's development strategy less emphasis nowadays ought to be given to the nationalization of large foreign concerns and the promotion of State companies, and more account taken of government attempts to reduce regional socio-economic disparities through regional development policies and of the increased willingness to decentralize both economic activity and decision-making about that activity. Arguably, a concern that all regions and classes of Algerian society should participate in the organization and results of economic development is a more socialist approach than the earlier centre-down State centralism which was too rigid, bureaucratic and non-participatory. The contributions of regional development and decentralization to transforming the economic development strategy of a socialist Algeria will be the main themes pursued in this chapter.

Regional disparities and regional development in Algeria

The two post-independence population censuses of 1966 and 1977 allow the objective assessment of regional disparities in Algeria and permit the effectiveness of early regional development policies to be examined in terms of regional convergence or divergence (Nacer, 1979; Williamson, 1965). When the 1966 data are recalculated for the post-1974 administrative divisions (see figure 5.1): it was found by Nacer (1979) that the top five *wilayate* of Algiers, Oran, Annaba, Constantine and Blida contained 24 per cent of the population but concentrated 46 per cent of industrial jobs; 55 per cent of vehicles in circulation; 55 per cent of the urban population; 61 per cent of the value of deposits in savings accounts, and 65 per cent of telephone subscribers.

These grave regional imbalances were engendered by the environmental and resource base of Algeria, and were exacerbated by spatial organization during the French colonial period which favoured the coastal regions for European investment and settlement. For example, the dualism of modern and traditional agriculture displayed the same spatial disparities with its contrast between the modern export-oriented sector near the coastal urban centres and the neglected, largely traditional interior. The initial years of independent government worsened this regional imbalance. Within Algeria's increasingly planned economy, the heavy centralization of decision-making, the sectoral and macro-approach adopted, all furthered regional disparities.

In the late 1960s and early 1970s some attempts at regional development were made by means of Special Programmes, each for a specific lagging region, (figure 5.2) and through the Economic Regions postulated in the Second Plan, 1974–7 (Sutton, 1981). The latter approach suggested six Economic Zones (figure 5.3) cutting across established administrative divisions, each zone having certain planning objectives aimed to lessen

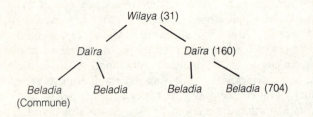

Elected Bodies:

Assemblée populaire
de la *wilaya*

Wilaya (31)

Daïra   Daïra (160)   —

Beladia   *Beladia*   *Beladia*   *Beladia* (704)   Assemblée populaire
(Commune)   communale

Figure 5.1 The administrative structure of Algeria
(Source: SEP, *Annuaire Statistique de l'Algérie* 1979, 1980, p.3.)

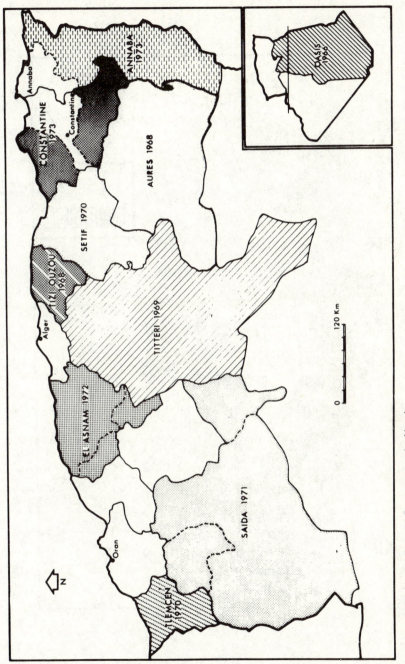

Figure 5.2 Regional Special Programmes in Algeria

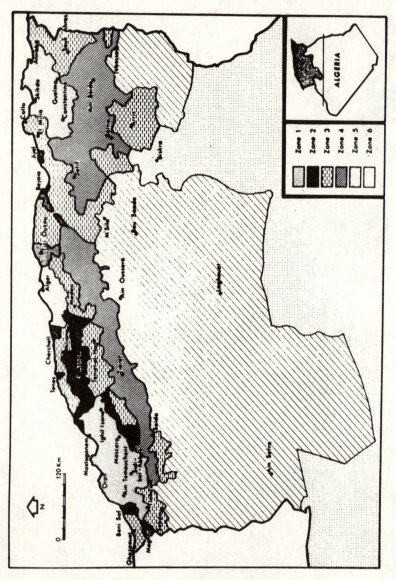

Figure 5.3 The Second Plan's Economic Zones of Algeria

the growing concentration of population and activity in the coastal poles. However, these imaginative proposals were not fully accepted by the main powerful economic ministers and so did not consequently become operative. More success was met by the ten Special Programmes launched intermittently between 1966 and 1973 (see table 5.1). Initially these regional development programmes applied to complete *wilayate*. In 1971–2, the programmes for Saïda and El Asnam incorporated lagging parts of adjacent *wilayate*, whereas the 1973 programmes for Constantine and Annaba focused only on those communes with particular economic problems so excluding the main urban poles (see figure 5.2). As table 5.1 indicates, the allocation to the ten programmes totalled 11,304 million dinars. Another estimate suggests that between 15,000 million and 20,000 million 'current' dinars had been allocated to the programmes up to 1978 (Prenant, 1978). Set against the concurrent National Plans these regional programmes were quite substantial. The first four programmes amounted to 46.36 per cent of the 1967–9 'Pre-Plan' but expenditure declined to 21.16 per cent of the First Plan, 1970–3. The sectoral allocation of these regional investments illustrates their social objectives (table 5.2). Investments in housing, health and other social services amounted to over half of the total expenditure. Agriculture was particularly assisted, with more than a quarter of the funds.

This contrasts with the pattern of investment in the parallel National Plans which financed some large industrial investments in the same regions, particularly in Tlemcen, Sétif and El Asnam. From these investment priorities and the changed spatial emphasis within *wilayate* it can

Table 5.1   The Special Programmes

Special Programme	Year	Total allocation (million dinars)
Oasis	1966	1,000.0
Aurès	1967	1,100.0
Greater Kabylia	1968	893.0
Titteri	1969	1,236.6
Tlemecen	1970	1,252.3
Sétif	1970	1,248.3
Saïda	1971	1,015.4
El Asnam	1972	1,558.3
Constantine	1973	1,000.0
Annaba	1973	1,000.0
Total		11,303.9

Table 5.2   Investments by sectors in six Special Programmes (1968–72)[a]

	Investments	
Sectors	million dinars	percentage of total
Directly productive investments	2241.3	39.44
Industry	765.0	10.62
Agriculture	1878.3	26.07
Tourism	198.0	2.75
Indirectly productive investments	3401.0	47.53
Infrastructure and transport	1867.6	25.93
Education and training	1368.0	18.99
Administrative infrastructure	165.4	2.30
Social services investments	810.3	11.25
Health	331.0	4.59
Housing	286.0	3.97
Other social services	193.3	2.68
Other non-classified items	151.4	2.10
Total	7204.0	100.0

[a] The table covers the six Special Programmes for Greater Kabylia, Titteri, Tlemcen, Sétif, Saïda and El Asnam.
Source: Nacer, 1979, p. 35

be concluded that the aims of the Special Programmes were more the provision of some transfer of resources and social benefits rather than the development of backward areas. This task was still the responsibility of the national plans but these were not yet ready for that purpose.

By the mid-1970s the Special Programmes' approach was abandoned for Communal Development Plans, the aims and objectives of which will be discussed later. Although the Special Programmes displayed failings in that too much of their funds went to the main towns and the projects lacked co-ordination between each other or with the National Plan's investments locally, they did have some positive results. They demonstrated the need to involve the participation of local authorities in more democratized and decentralized decision-making. Also, the Special Programmes introduced the spatial dimension into development planning and emphasized the need for area-specific policies to overcome regional disparities which threatened otherwise to worsen.

By the time of the 1977 census the replacement Communal Development Plans, examined more fully later in this chapter, had only made a limited impact. Their total allocation of 6,515 million dinars represented only 6 per cent of the investments of the 1974–7 Second Plan and delays in realizing these regional investments meant that by the end of 1977 only 31 per cent of them had been effectively used. Consequently, any evaluation of the changes in regional disparities between 1966 and 1977, is largely a judgement of the impact of the regional Special Programmes together with any regional effects of the National Plans.

To examine regional convergence–divergence a range of socio-economic and demographic variables have been extracted from the 1966 and 1977 censuses and from ministerial publications. The variables have been standardized into z-scores and examined by means of coefficients of spatial variation (CSV)[2]. Most attention is focused on the scores of the 10 top and 10 bottom *wilayate* of the rank order. The 1966 data had to be reorganized into the 31 post-1974 *wilayate* by the use of commune or *daïra* data (Nacer, 1979). Table 5.3 displays the 1966 situation for all 20 variables. Generally very high CSVs are noted; the highest (variable 11) is for savings account deposits *per capita*, a surrogate for income; the lowest is variable 16, an inverse rate of overcrowding, which implies that the problem of overcrowding is prevalent all over Algeria. The conclusion is that levels of regional disparities were very high in 1966 and this applied to almost all of the indicators. A similar pattern emerges for 1977 (see table 5.3). Although most indicators show some progress in the provision of economic and social opportunities, the levels of regional disparity as expressed by the CSV values remain very high. Furthermore, it is largely the same indicators which show the highest regional imbalances. Nevertheless, some decrease in levels of regional disparities can be detected, between 1966 and 1977. The greatest decreases in CSV in absolute terms were recorded by variable 11 (deposits in savings accounts *per capita*), variable 9 (telephone subscribers) and variable 4 (building employment). Of the three indicators which increased in CSV terms, two relate to the industrial sector and the third to population density and migration. So, despite continuing high regional inequalities, this analysis suggests that some regional convergence occurred during the decade of industry-led economic development.

To examine more closely the relative fortunes of individual *wilaya*, two methods of classification were used, both producing comparable results. First, a crude ranking system and then a z-score approach to ranking were tried (table 5.4). Both methods suffer from having to give equal weight to patently unequal indicators and both exercises required the omission of variable 20 (population density) because of the bias against the sparsely populated Saharan regions. This part of the analysis focuses on the leading ten *wilayate* and the lagging ten *wilayate*. The most prosperous *wilayate* in 1966 (by the z-scores method) were Algiers, Blida, Oran, Tlemcen, Sidi-bel-Abbès, Annaba, Constantine, Ouargla, Laghouat, and Béchar (see figure 5.4). Hydrocarbons and small population sizes helped the last three Saharan *wilayate* in their rankings. The lagging

Table 5.3 Regional differentials, 1966 and 1977

Variable	1966			1977			1966–77 CSV ratio[a]
	Mean	Standard deviation	Coefficient of spatial variation	Mean	Standard deviation	Coefficient of spatial variation	
1 Industry as percentage of total regional employment	8.1	4.2	52.0	15.4	8.8	57.3	1.10
2 Non-agricultural employment as percentage of total regional employment	42.5	18.3	42.9	67.3	15.6	23.2	0.54
3 Industrial jobs per 10,000	119.8	78.4	65.4	215.7	164.1	76.1	1.16
4 Building employment per 10,000	67.2	42.0	62.6	206.9	62.2	29.9	0.48
5 Tertiary employment per 10,000	431.6	241.2	55.9	476.3	158.1	33.2	0.59
6 Employment as percentage of rest of working population	66.7	10.7	16.0	80.1	7.3	9.1	0.57
7 Total regional employment per 10,000	1460.3	351.2	24.1	1223.7	237.4	17.9	0.74
8 No. of vehicles per 10,000	132.5	108.0	81.5	210.4	127.3	60.5	0.74
9 No. of telephone subscribers per 10,000	43.6	50.0	114.8	75.4	57.3	75.9	0.66

						CSV ratio[a]
10 No. of savings accounts per 10,000	18.2	14.1	77.4	284.4	207.5	72.9	0.94
11 Deposits in savings accounts *per capita*	2.6	3.0	119.1	52.9	42.2	79.8	0.67
12 Hospital beds per 10,000	29.1	19.7	67.7	24.4	11.7	48.0	0.71
13 Percentage of average school enrolment	44.7	15.9	35.6	67.6	12.4	18.4	0.51
14 Percentage of female school enrolment	33.3	16.9	50.8	54.5	16.3	30.6	0.60
15 Percentage of literacy rate	23.2	9.1	39.1	39.5	11.1	28.1	0.72
16 Percentage of inverse rate of overcrowding	85.1	2.9	3.4	87.2	1.8	2.0	0.61
17 Percentage of dwellings with running water	32.5	20.3	62.5	42.3	20.6	48.8	0.78
18 Percentage of dwellings with electricity and gas	19.0	14.7	75.5	44.0	21.0	47.7	0.63
19 Percentage of urbanization rate	25.9	20.5	79.2	34.9	17.9	51.3	0.65
20 Density rate	102.8	262.8	255.7	163.4	448.8	274.7	1.07

[a] CSV ratio = $\dfrac{1977 \text{ CSV}}{1966 \text{ CSV}}$

Source: Nacer, 1979, pp. 50–3

Table 5.4 Rankings of the *Wilayate*: leading and lagging regions

| 'Crude' ranking method | | Z-scores method | |
| 1966 | 1977 | 1966 | 1977 |
Wilaya code		Wilaya code	
16	16	16	16
31	23	31	31
23	31	23	23
09	30	30	30
25	08	09	08
03	09	25	09
08	25	03	25
22	13	08	13
13	03	22	22
30	22	13	03
06	20	11	20
29	21	29	11
15	29	15	15
21	11	20	29
27	15	06	21
07	19	07	10
20	10	27	19
11	06	21	06
14	07	05	07
05	27	14	24
19	24	24	27
10	05	19	05
02	02	04	02
24	04	17	04
04	26	10	18
17	28	02	26
26	14	12	28
18	17	26	17
28	12	28	14
12	01	18	12
01	18	01	01

Wilaya code (see figure 5.3):

01 Adrar	11 Tamanrasset	21 Skikda
02 El Asnam	12 Tébessa	22 Sidi Bel Abbès
03 Laghouat	13 Tlemcen	23 Annaba
04 Oum el Bouaghi	14 Tiaret	24 Guelma
05 Batna	15 Tizi-Ouzou	25 Constantine
06 Bejaïa	16 Alger	26 Médéa
07 Biskra	17 Djelfa	27 Mostaganem
08 Bechar	18 Jijel	28 M'Sila
09 Blida	19 Sétif	29 Mascara
10 Bouïra	20 Saïda	30 Ouargla
		31 Oran

Source: Nacer, 1979, p. 57

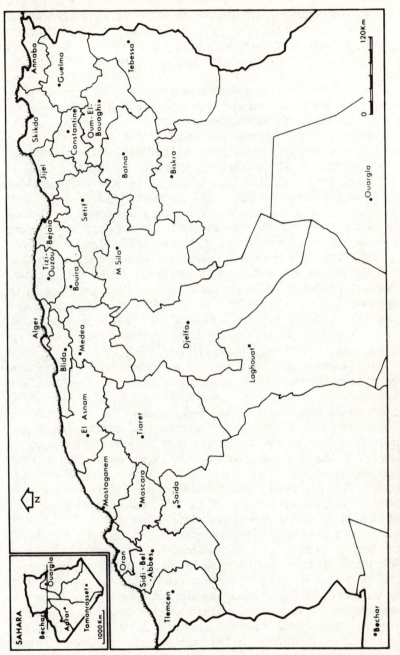

Figure 5.4 The post-1974 *Wilayate* of Algeria

wilayate in 1966 were mainly in the interior in Central and Eastern Algeria. They formed a contiguous zone stretching eastwards from El Asnam plus Adrar in the Sahara which ranked last for most indicators. In 1977 the bottom ten were largely the same *wilayate* with Sétif and Bouïra having improved their rankings, to be replaced by Batna and Tiaret.

These backward areas improved for most indicators much more rapidly than Algeria as a whole, and also more rapidly than the top ten *wilayate*. The two exceptions are quite interesting as these are indicators where the CSV increased 1966–77, namely variable 1 (industrial employment as a per cent of total regional employment) and variable 3 (industrial jobs per 10,000 population). This prompts the reflection that industrial development had played a relatively minor role in the Special Programmes whereas it had dominated the centralized national plans.

One drawback of the analysis so far is that no account is taken of population change between 1966 and 1977. The apparent improvement of an indicator could merely result from a decline in the *wilaya*'s population size, or even from a slower than average growth, and not therefore from any real improvement in the sector measured by the indicator. This is particularly likely as the less endowed *wilayate* recorded lower population growth, between 1966 to 1977, than the national average. Hence an alternative assessment examines the rate of growth of the different *wilayate* in each of the socio-economic sectors without including the population component (Nacer, 1979, pp. 67–79).

These alternative indicators measure the share of each *wilaya* as a percentage of the total for the whole country (table 5.5). When the effect of population size is removed it was found that the CSVs were much greater for the same resources than those found in table 5.3. Two of the *wilayate*, Sétif and El Asnam, which earlier ranked amongst the ten least endowed *wilayate* in 1966, are now amongst the ten leading *wilayate* as their better share of certain resources is no longer cancelled out by their large population size. The reverse applies to Béchar, Laghouat and Ouargla with their small populations. When the changes between 1966 and 1977 in CSV are examined (table 5.5) again regional imbalances appear to be decreasing as measured in this alternative way. This pattern of decrease applies to all sectors except share 04 (regional employment) and share 07 (number of savings accounts). Similar variables had not increased in the earlier analysis because some *wilayate* were also strong areas of population growth which cancelled out share growth. The regional industrial employment share did not increase as it did in the earlier analysis. This is connected with the fact that there has been a small but perceptible diffusion of the spatial distribution of industry, but this limited deconcentration has been hidden by the fact that the largest increases were in *wilayate* with relatively small populations e.g. Ouargla,

Table 5.5 Alternative measures of regional differentials by shares, 1966 and 1977

Variable[a]	CSV 1966	CSV 1977	% change in CSV 1966–1977
Economic resources			
Share 01: Regional industrial employment	139.1	132.7	−4.67
Share 02: Regional tertiary employment	126.3	110.9	−12.20
Share 03: Regional non-agricultural employment	120.9	102.3	−15.54
Share 04: Regional overall employment	62.7	75.7	+20.61
Share 05: Regional number of vehicles	168.7	152.7	−9.48
Share 06: Regional number of telephone subscribers	232.5	198.3	−14.73
Share 07: Regional number of savings accounts	152.1	159.7	+5.00
Share 08: Regional value of savings accounts	230.5	168.1	−27.06
Social resources			
Share 09: Regional number of hospital beds	115.7	91.5	−20.94
Share 10: Regional number of children (6–14) at school	88.1	82.2	−6.65
Share 11: Regional number of girls (6–14) at school	188.1	94.5	−49.78
Share 12: Regional number of literate people[b]	97.3	98.0	+ 0.7
Share 13: Regional number of dwellings	65.8	60.1	−8.62
Share 14: Regional number of dwellings with water facilities	142.3	110.4	−22.46
Share 15: Regional number of dwellings with electricity and gas	164.3	108.2	−34.11
Settlements			
Share 16: Regional urban population	158.5	140.2	−11.53
Regional Population	55.4	62.5	+12.98

[a] As percentage of the national total.
[b] Over ten years.
Source: Nacer, 1979, p. 72

and their extremely high score weighted heavily the evolution of the CSV. It is worth noting that the CSV for population distribution has increased 13 per cent, reflecting migration to the higher population concentrations, in contrast with the decrease in the CSV for most other 'shares'. Also the greatest convergence, between 1966 and 1977, is displayed by share 11 (the number of girls at school), reflecting a genuine improvement in the access to female education. Similar high convergence is found for share 15, the product of programmes of electrification and energy diffusion.

Table 5.6 contrasts the fortunes, from 1966 to 1977, of the top, intermediate, and bottom groups of *wilayate*. For 13 of the 16 variables

Table 5.6 Change in share of resources for the leading, lagging and intermediate groups of *Wilayate*, 1966–1977 as a percentage

Variables[a]	Top 10 wilayate	Intermediate 11 wilayate	Lagging 10 wilayate
Share 01	+2.4	+1.3	−4.0
Share 02	−5.5	+2.6	+3.0
Share 03	−4.8	+3.0	+1.7
Share 04	+3.9	−1.7	−2.2
Share 05	−3.3	+3.2	+0.1
Share 06	−5.6	+4.2	+1.5
Share 07	+4.3	−2.4	−1.9
Share 08	−3.3	+2.4	+0.9
Share 09	−4.8	+3.8	+1.1
Share 10	−2.9	+1.8	+1.1
Share 11	−4.2	+1.7	+2.6
Share 12	−0.5	+1.9	−1.3
Share 13	−0.5	+2.0	−1.3
Share 14	−5.3	+4.2	+1.1
Share 15	−10.0	+5.9	+4.1
Share 16	−4.1	+2.4	+1.7
Population as percentage of total	+3.1	−0.7	−2.4

[a] See table 5.5 for description of shares.
Source: Nacer, 1979, p. 77

the positions of the top ten *wilayate* have clearly been eroded. The bottom ten *wilayate* have slightly improved their position in 11 out of the 16, while the intermediate *wilayate* have displayed more definite improvement in their share of resources in 14 indicators. Moreover, these latter improvements were almost always in diametric opposition to the fortunes of the top ten *wilayate*. This is re-expressed in table 5.7 in terms of the allocation of surplus growth whereby deviations are calculated from the theoretical values which would have been recorded in 1977 if all the *wilayate* had achieved the same rate of change for each sector as the national average. Whereas the top ten *wilayate* have shown negative scores for most sectors, both the bottom ten and especially the intermediate *wilayate* realized better scores than would have resulted from a regionally even rate of change. This suggests that a degree of regional convergence has taken place to the advantage of the lagging and, particularly, the intermediate regions. The trend is only a relative one and it is far from realistic to regard it as resulting in an end of the domination of the prosperous *wilayate*, just a first erosion of their position. Furthermore, within the top ten *wilayate*, there has been a marked decline of the

Table 5.7   Allocation of surplus growth as a percentage[a]

Variables[b]	Top 10 wilayate	Intermediate 11 wilayate	Lagging 10 wilayate
Share 01	+3.9	+5.8	−24.1
Share 02	−9.4	+10.0	+19.4
Share 03	−7.3	+11.6	+32.2
Share 04	+9.5	−5.2	−8.3
Share 05	−5.0	+14.9	+0.9
Share 06	−7.6	+25.3	+15.9
Share 07	+7.1	−9.1	−14.6
Share 08	−9.5	+10.0	+0.9
Share 09	−8.3	+14.9	+6.5
Share 10	−8.7	+5.9	+5.5
Share 11	−7.8	+5.7	+15.2
Share 12	−1.0	+6.4	−6.0
Share 13	−1.4	+5.8	−4.8
Share 14	−8.3	+18.0	+8.0
Share 15	−14.7	+28.6	+36.3
Share 16	−6.2	+12.0	+11.2
Population share	+8.8	−2.0	−8.0

[a] The figures give the difference between the 1977 level and the level which would have been achieved by each group of *wilayate* had they achieved the same rate of growth for each sector as the national growth. The deviations are expressed as a percentage of the 1977 actual scores.
[b] See table 5.5 for description of shares.

primacy of Algiers to the benefit of other leading *wilayate*, including Oran in the west and Annaba in the east.

A note of caution should be injected as regional disparities in Algeria in 1977 were still very high despite this evidence of convergence. Also, there is the possibility that rural–urban differences within *wilayate* could have worsened, as there is evidence that much of the regional development investment was focused on the regional capitals. Furthermore, part of the apparent convergence stems from the way that migration to the main cities has worsened their *per capita* indicators as used in this comparative analysis. A final qualification is prompted by the fact that for industrial employment indices divergence prevailed. As industrialization has remained central to the Algerian national development strategy this could give cause for concern though it can be argued that Myrdal-type 'spread' involves infrastructural and service elements rather than industrial production (Weinand, 1972).

The major regional development effort during the period 1966–77 was the series of Special Programmes discussed earlier. However, the whole problem of lessening regional disparities received due recognition in the Second Plan (1974–7) and the Third Plan (1980–4).

The national development plans: regional and social issues

The first 'decade' (1967–79)

The development strategy pursued by Algeria from 1967 to 1978, was characterized by heavy investments in selected sectors of the economy and, as a consequence of the search for scale economies, in a limited number of localities. The basic aim was to build up rapidly a solid basis for industrial development. However, if left unchecked, such a strategy would increase social and spatial disparities as only a few places would share in the economic growth and these would tend to be in the established better-off regions of northern Algeria, and in the oil-producing but sparsely inhabited regions of the south.

Of the 300 billion dinars of State investments, 1967–78 data are available by *wilaya* for some 252.5 billion dinars. The remaining investments were not assigned to geographical localities (MPAT, 1979, p. 30). Thus the three *wilayate* of Algiers, Oran and Annaba together accounted for 73.2 billion dinars or 29 per cent of the total investments, despite containing only 19 per cent of the 1977 population (see figure 5.5). If the four next-ranking *wilayate*, Laghouat, Constantine, Ouargla and Tlemcen are added, these top seven *wilayate* accounted for 51 per cent of the plans' investments, yet contained only 29 per cent of the population. Furthermore, while industrial investments accounted for 136.7 billion dinars or 54 per cent of the total invested, these seven *wilayate* concentrated 80.6 billion dinars or 59 per cent of the industrial investments. Of these seven *wilayate*, five represent the most urbanized and developed regions of Northern Algeria; the other two are Saharan oil- and gas-producing regions.

The government was well aware of the negative effect on regional disparities of such a concentrated growth strategy. This was expressed in the official communiqué issued after the adoption of the Second Four-Year Plan, 1974–7, which stated that 'it was aware, in this period of early development and building of the economy, of the necessity to prevent the aggravation and the intensification of the regional disparities, and of the rural exodus towards the urban areas, which are not yet ready to receive these migratory flows' (RADP, 1974, p. 16).

Within the constraints of the national development strategy, the government thought that there was scope for diffusing development through enabling local authorities to initiate and implement local projects to develop their resources and meet some basic needs of their inhabitants. These *Plans Communaux de Developpment* (PCD) could also integrate local

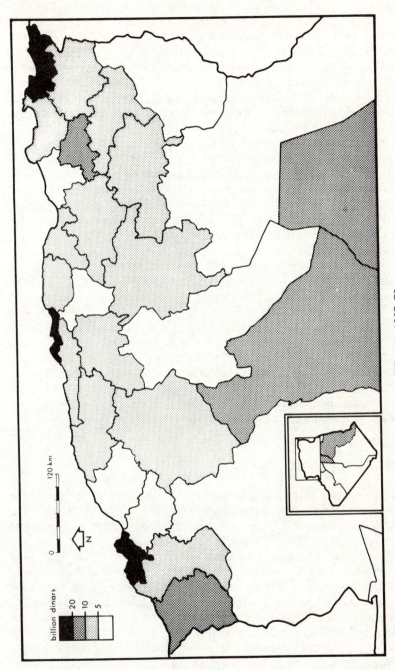

Figure 5.5 Distribution of Algerian State investments by *Wilayate*, 1967–78

authorities into the development planning system and so enlarge what had hitherto been the preserve of central government. Through the PCD system each of the 704 communes could help elaborate and diffuse development. Financial constraints included the modest total size of the PCD programme and its central funding. However, it was hoped to improve what could be done locally. Further, this PCD programme acknowledged the need for a spatial element in Algeria's planning system hitherto mainly sectoral in its approach. In this way the PCD continue the positive effect of the earlier Special Programmes.

Parallel concern was shown for the main urban agglomerations which were experiencing considerable growth of population and economic activity without a corresponding expansion of their basic infrastructures. This resulted in the establishment of *Plans de Modernisation Urbaine* (PMU) or special allocations to 39 selected towns to improve their facilities and their habitability. The PMUs particularly promoted schemes for drinking water, sewerage, urban renewal and redevelopment, and munici- pal buildings.

Another contribution in the 1970s to promoting regional economies as well as popular participation was the reorganization of Algeria's adminis- trative divisions in 1974. The number of *wilayate* was increased from 15 to 31, which immediately brought administrative and related economic activity to several towns and their hinterlands previously neglected or subordinated in regional service hierarchies (see figure 5.4). *Wilaya*-level authorities were sometimes elected, sometimes appointed, and had the dual role of promoting local development and of ensuring the execution of central directives within their areas. Arguably local needs and griev- ances would now receive more consideration at central government level, which still controlled the allocation of public funds.

The 1980–4 Five-Year Plan

The emergence of the new government of President Chadli following the death of President Boumedienne provided the opportunity to assess the social and economic impact of the development strategy and to define new priorities in view of new social needs and earlier shortcomings. To this end the extraordinary congress of the ruling Front de Libération Nationale (FLN) Party in June 1980 recorded some social and economic imbalances which, if left unchecked, would threaten the much-needed social stability required for the continued building of an independent economy. It was argued that the economy was plagued by high social and economic costs of development actions, by the under-utilization of available productive capacity, by weak economic integration, by the growing bureaucracy, and by general limited efficiency. All these shortcomings were attributed to 'the severe negligence in the planning and implementation process' of Algeria's development strategy (FLN, 1980, p. 14).

In terms of the spatial distribution of economic activity, the FLN Congress recorded the lack of an integrated spatial development policy. This had resulted in the location of a majority of development projects in those northern regions with a ready access to markets and infrastructure, all of which hampered the balanced diffusion of development to all regions. Consequently, the Congress called for a more balanced distribution of activities and for the full integration of spatial and regional policies. These objectives would be attained through, first, a real democracy and a full decentralization of managing activities and, second, the strengthening of the planning system through the establishment of new types of plans for areas, at the level of *wilayate* and communes and for functional units, that is, enterprise plans. The plans at the *wilaya* and commune levels are regarded as the important framework within which could be expressed social aspirations and local initiatives developed in accordance with the broad objectives of the national plan. Greater responsibilities would thus be given to local authorities in the conceiving and implementing of development actions.

The 1980–4 development plan was to concretize these resolutions. It planned to invest 400.6 billion dinars, of which nearly half, 196.9 billion, are for rescheduled projects from the previous plan. The structure of investments partly reflected the criticisms made of the previous strategy. More money was made available to meet the growing social needs in housing and social facilities, as well as to improve the basic infrastructures for water, roads and other communications. However, the share of productive investment remained high with 154.5 billion dinars going to enlarge the industrial sector. Table 5.8 demonstrates this shift in development policy through a comparison of the 1980–4 Plan's investment structure with that of earlier Plans.

Spatial and regional components of development policy occupy a full chapter of the *Rapport Général Plan 1980–84*. Major long-term goals for the year 2000 are: the elimination of regional disparities; the putting into production and the preservation of national resources; and a harmonious settlement and occupation of the national territory. In assessing past efforts the 1980–4 Plan made it clear that the spatial pattern of the economy had not changed much since independence. Population was still concentrated in the north, rural exodus was continuing and urban areas were experiencing unplanned growth and severe congestion. This is seen as a consequence of rigidities from the inherited spatial pattern exacerbated by post-independence policies which lacked a global integrated approach to developing the assisted regions and which suffered from the absence of a precise analysis of regional possibilities and from the limited

Table 5.8   The structure of investment in the National Development Plans

| | Plans 1967–78 | | | | 1980–4 Plan | |
| | Planned | | Realized | | Planned | |
Sectors	billion dinars	%[a]	billion dinars	%	billion dinars	%
Agriculture	23.4	5.2	12.82	5.8	24.1	6.0
Industry	257.0	56.7	137.54	62.7	154.5	38.6
Tourism	4.3	0.9	2.53	1.2		
Productive investments	284.7	62.8	152.89	69.7	178.6	44.6
Transport	14.1	3.1	7.66	3.5	13.0	3.2
Hydraulics	18.2	4.0	6.47	2.9	23.0	5.7
Other economic infrastructure	26.5	5.8	11.99	5.5	57.9	14.6
Indirect productive investments	58.8	12.9	26.12	11.9	93.9	23.4
Housing	50.8	11.2	15.26	7.0	60.0	14.9
Education and Training	35.8	7.9	12.79	5.8	42.2	10.5
Public Utilities and other social services	23.3	5.1	12.29	5.6	25.9	6.5
Social investments	109.9	24.2	40.34	18.4	128.1	31.9
Total	453.4	100	219.35	100	400.6	100

[a] Percentage of total resources for designated Plan.
Sources: MPAT, 1980a and 1980b

participation of the local population and authorities in defining and realizing programmes (MPAT, 1980b, p. 235).

The Plan went on to draw a picture of the 'unacceptable scenario' reflecting extrapolated current trends. This suggested that by the year 2000, 70 per cent of the population would be living in just one per cent of the territory. By 1990 the five *wilayate* of Algiers, Oran, Blida, Constantine, and Annaba would concentrate 55–60 per cent of Algeria's urban population which itself would form 55 per cent of the total population. Already in 1977 these five *wilayate* concentrated 25 per cent of Algeria's population and 50 per cent of its urban population. Income differentials by 1990 would range from one to four and rural exodus would be running unchecked at a rate of 170,000 persons a year.

To counter these trends the Plan set out three aims for 1990: a better distribution of settlements; a reduction of income differentials; and a deployment of activities over all regions.

Towards these ends, the major action promoted during the 1980s is the *Options Hauts-Plateaux*, which covers the broad band of interior regions stretching east–west across eight *wilayate* and covering 222,900 square kilometres, or 10 per cent of the country's area. With 4,114,000 inhabitants, 24 per cent of the total population, the Hauts-Plateaux contain only 17 per cent of Algeria's urban population and 22 per cent of its workforce. Between 1966–77 this region had experienced relative depopulation as its share of the national population dropped from 24.6 per cent to 23.6 per cent. The Options Hauts-Plateaux programme aims for an additional 1.5 to 2.0 million people in the region by 1990–5, and for an increased share of the national workforce to 27 per cent. Urban growth in the region would hence be around 20 per cent per year compared with 5.7 per cent nationally.

Although further planning studies are to be undertaken, the present recommendations are for the provision of integrated activities (agriculture–industry–training–services) around medium-sized urban centres. Already work on up-grading the communications infrastructure has commenced with the first stages of a new railway line crossing the Hauts-Plateaux from Tébessa to Saïda through Batna, M'Sila, Aïn Oussera, and Tiaret, and linked to the present rail network at Sidi Bel Abbès (figure 5.6). This new line is expected to open up the interior to the diffusion of development and so back up the establishment of new activities. Criticism centres on the lack of a comprehensive policy for the region despite the launching of important investments such as the railway line, major industrial plant at Aïn Oussera, electricity power stations etc. Each ministerial department is acting on its own, in its particular section, with limited co-ordination in the absence of an overall regional policy. Other shortcomings relate to the policy content. How are nearly two million people to be transferred and located in a region long-considered to be poor and experiencing a difficult climate? Also, there is no comprehensive inventory of the region's resources and potentiality. The objectives for the region are open to question as they relate more to the difficulties being experienced in coastal Algeria than to possibilities in the Hauts-Plateaux themselves. While commendably positive the Options Hauts-Plateaux needs to be backed by objective assessments of just what is possible and desirable. Given some flexibility, this recent regional optior at least represents a willingness to break away from the current imbalanced spatial pattern of development.

This is also expressed by other decisions taken by the Plan on settlement policy whereby population growth is to be checked so that: in the large agglomerations, increase is to be limited to natural growth; in rural areas, population should be stabilized at levels appropriate to their natural resources; productive means are reoriented and transferred towards the disfavoured areas.

As far as urban growth is concerned, the limited impact of earlier urban planning policies[3] has led the government to contemplate new approaches. The most prominent new measure, already underway, has

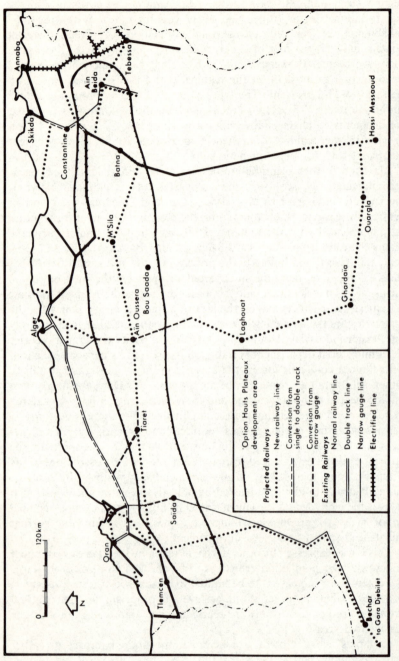

Figure 5.6 Communications – the railway network and the Options Hauts-Plateaux regional development area of Algeria

been the transfer of urban-planning responsibilities from local commune authorities to *ad hoc* committees chaired by the *Wali* or governor, of each *wilaya*. These changes, which have affected the towns of Oran, Constantine, Annaba, Blida, Sidi Bel Abbès, and Skikda, have been justified by the limited capacity of commune authorities to undertake effective policies because of shortages in expertise, poor powers of enforcement and, in particular, the characteristics of sectoral planning which involve the strong dominance by central authorities or their representatives over local decision-making. In the case of Algiers, responsibility for urban planning was removed from the commune authorities as early as 1968. Since then the *Wilaya* of Algiers has exercised planning responsibility for Algiers and is monitoring work on the new *Plan d'Urbanisme Directeur* for Algiers which will replace the 1975 Development Plan the revision of which has been requested by the government (Nacer, 1983).

This growing interest in planning the main urban centres reflects their recent problems. The six cities affected by the new measures total 2,941,000 population, or 43 per cent of Algeria's urban population. Their aggregate increase, between 1966 and 1977, of 1,013,000 people, represented 32 per cent of the national increase in the urban population. Although this rate of increase, 3.97 per cent per year, is lower than the national urban growth rate, it is still of great importance because of its concentration on the six largest cities and because it has not been matched by a corresponding improvement of their facilities in terms of housing, social and other services. Congestion and wasteful unplanned extensions of the built-up area have become the general rule. Highly centralized urban planning procedures have been unable to control development, largely because the inherited legislative framework did not correspond to the situation. Thus, while the system of urban planning was based on the control of development, it did not work where public development became the most important agent transforming the environment. The balance of powers shifted in favour of the public developers, mainly the industrial State enterprises, and against the planners. Also, the urban planners tended to promote a 'specific spatial policy', which seems to counteract the development strategy, and so they became marginalized because, in a socialist country struggling for development, spatial policy could not stand by itself but only as an integrated part of the whole development strategy.

Deconcentration or decentralization?

An important issue taken up by the 1980 FLN Party Congress is that of decentralization. Indeed, some see the recovery of the economy as only possible through a full decentralization of the working of the economy. This implies the dismantling of the rigid sectoral and centre-down approach which has so far characterized Algeria's development strategy. Although the issue of decentralization was ever-present in political speeches, it has recently acquired greater weight as a result of economic difficulties and shortcomings.

Algerian academics draw a pertinent distinction between the two processes of decentralization and deconcentration, both of which are in opposition to centralization. Deconcentration involves the delegation of power by central authorities to subordinate local authorities or to the regional branches of central agencies, who then act on their behalf. Partly this avoids over-burdening central authorities with detailed local matters, but the relationship between higher and lower level authorities remains one of subordination with a hierarchical division of powers. So deconcentrated authorities simply execute centrally originated decisions and operate with delegated powers, which can be removed at will.

In the case of decentralization power is not delegated but devolved to lower level elected authorities which have full powers of their own. Thus decentralization involves the advance of democracy and the participation of people in managing their own affairs. Local initiative and entrepreneurial spirit, both prerequisites for regional development, are made more feasible through decentralization than through deconcentration, which is only a device for putting central directives into operation at local levels. In assessing regional development more weight will be given to actions that further decentralization than those relating to deconcentration, although the development of the latter may also promote the former. Often in Algeria the two processes work together. Local commune and *wilaya* bodies act as deconcentrated authorities for some issues and as decentralized authorities for others. Thus the deconcentration approach applied to the Special Programme of the 1960s and early 1970s, while the more recent PCD actions reflect a decentralized approach. Currently the PCD approach is to allocate a global sum to the *wilaya* authority which then subdivides and reallocates it to the communes. Arguably this implies concentration at *wilaya* level, instead of the earlier and still partly current practice of allocation from central to local commune authorities.

Decentralization procedures would be heightened by the setting up of area plans, but progress so far is slow. Up to now *wilaya* authorities are not responsible for the major programmes located in their territory, although they will have a greater say in the future. According to the MPAT, *wilaya* authorities manage about 35 per cent of the planned investment programmes, under the responsibilities of local authorities (i.e. *wilaya* and commune) (MPAT, 1980a, p. 34).

Decentralization can help regional development only where initiative and entrepreneurship are not hampered by bureaucratic rigidities and the lack of proper resources. Bureaucracy takes the form of a heavy administrative tutelage affecting all fields of intervention by local authorities. This demands that most of the decisions taken by local authorities cannot take effect legally without the full approval of a higher level authority. A procedure to ensure compliance with the prescriptions of the law thus became a bureaucratic means of control over lower level authorities. A better approach would be to use the judiciary system with central authorities taking local authorities to court if they consider that they are misusing their powers. The second factor hampering decentralization is a lack of resources. Most projects undertaken by the communes are centrally financed and further difficulties arise when local authorities have to take charge of facilities originally built with central funds but which now have to be run with local funding. Reforms of local finances, announced since 1971, have produced little progress, so limited independent resources takes away a lot of meaning from decentralization. This problem has been acknowledged by central government with efforts to increase the share of local authorities in the allocation of taxes and duties collected nationally, while insisting on continued central control over the allocation process. This is justified on the grounds that local authorities' revenues vary considerably. A better system is required both to avoid leaving decentralized authorities without proper resources and to ensure greater equality between them.

The reform of the Commune Code and the *Wilaya* Code in 1981 aimed to improve decentralization and foster the participation of local authorities through a clearer definition of their responsibilities. These reforms may be unsatisfactorily implemented unless these questions of local finance can be settled.

The contribution of the *Plans Communaux de Developpement* (PCD)

Between 1975 and 1982 local projects totalling 40.5 billion dinars have been planned within the PCD programme. Of this total, some 20.4 billion dinars have been effectively spent, representing an average annual expenditure of 2.55 billion dinars. Progress was relatively slow up to 1979 and it is only recently that expenditure rates have been increasing. Table 5.9 shows expenditure of 3.75 billion in 1981, and 6.2 billion in 1982. However, in relative terms, the share of the PCD's planned spending remained remarkably steady at around 13 per cent between 1980 and 1983.

The PCDs are undertaken by the 704 communes of Algeria, with additional programmes for the 200 most impoverished communes (Special PCD) and the 39 most important towns (PMU). The PCDs were set up to implement decentralization through three main objectives: (1) to integrate local authorities into the development process; (2) to provide a planning framework within which operations related to the agrarian

Table 5.9   PCDs' expenditure, 1980–3

Provisional expenditure (in billion dinars)	1980	1981	1982	1983
Planned PCD projects (including PMU)	3.2	4.28	5.8	6.05
National capital expenditure (budget)	23.12	31.59	42.6	48.25
PCD programme as % of national budget	13.8	13.6	13.6	12.5
Actual PCD expenditure	3.02	3.75	6.2	—
Rate of realization of PCD expenditure (percentage)	94.4	87.6	106.9	—

reform could be implemented; and (3) to redistribute income to the local areas concerned. Because of practical difficulties only the second and third of these objectives were pursued. So, the main PCD activities aimed to diffuse development by allowing local authorities to elaborate development projects that could use local resources, help to meet identified basic needs such as water supplies, sewerage and local roads, and build social facilities. In time more emphasis was given to meeting local needs than developing local rseources as the changing structure of investment in table 5.10 suggests. While in the initial programmes hydraulics (domestic water supplies, sewerage, etc.) and agriculture (fruit-tree planting, soil improvement) were the prominent sectors of intervention, from 1978 onwards the trend is different. Agricultural investment is largely abandoned to the benefit of other activities like economic infrastructure (the building of local roads, postal agencies, transport,

Table 5.10   Structure of planned investments in PCDs

Sectors	Initial structure up until 1977 million dinars	%	Actual structure 1982 million dinars	%
Agriculture	1151.0	18	1802.9	4
Hydraulics	2468.3	38	17279.2	43
Economic infrastructure	1296.0	20	8154.0	20
Social infrastructure	998.0	15	10570.9	26
Administrative equipment	181.6	3	821.5	2
'Equipment and realization capacities'	422.1	6	1859.4	6
Total	6517.1	100	40487.9	100

Source: Ministry of the Interior, 1983

distribution) and social infrastructure (mainly health centres and rural housing). The share of hydraulics remains steady.

The role of the PCDs in Algeria's development strategy can be more clearly assessed through a comparison with the structure of the National Plan. In the National Plan the main areas of investment are industry, economic infrastructure, housing and education which collectively account for 79 per cent of the total planned expenditure. By 1982 in the PCDs, hydraulics and social infrastructure form the main sectors. This reflects the social transfer aspect of the PCDs, a feature displayed also by the earlier Special Programmes. Funds are allocated where basic needs can easily be identified and met by local authorities. Productive investment is practically non-existent. This is explained by the fact that industry is covered by another type of local programme, the *Petites et Moyennes Industries* programme (PMI) which will be discussed later. For the local commune authorities it may prove more rewarding (not least electorally), and easier to construct a small local road to bring in water supplies than to seek to improve soil conditions or intensify the use of selected seeds by local agriculturalists.

It appears that this transfer policy complements the macro-scale national planning and the spatial allocation of the PCDs confirms this. The five most developed *wilayate* together accounted for only 18.4 per cent of the PCD funds allocated between 1975 and 1982, (table 5.11). That represented just 773 dinars *per capita* compared with 1170 dinars *per capita* nationally. The Hauts Plateaux regions received nearly twice as much PCD funding in both absolute and *per capita* terms. In a limited

Table 5.11   The spatial allocation of PCD funds and National Plan funds

Areas	5 most developed wilayate	High Plateaux	'South'	Rest of Algeria	Total
PCD – planned funding (million dinars)	6176	9172	5677	19517	40542
PCD – actual spending 1975–82 (million dinars)	3750	5476	2537	8621	20384
PCD – *per capita* spending (dinars)	773	1331	1828	1219	1170
National Plans, 1967–78 (million dinars)	96700	42260	39750	73790	252500
National Plans, 1967–78, *per capita* (dinars)	19946	10272	28633	10434	14493

Sources: Ministry of Interior, 1983; MPAT, 1980a

way the PCDs are thus compensating for the way in which successive National Plans have favoured the developed regions of Algeria.

Recent industrial development – deconcentration and decentralization

The attempts at regional development represented by the Special Programmes and the PCDs have operated in the context of an expanding national economy and within the overall strategy of an industry-oriented path to economic development. The dominance of industrial investment within each successive national plan has certainly been translated into a strong growth of employment in Algeria's industry and construction sectors (see table 5.12). Methods of calculation and estimation vary from one source to another, but the strong increase from over 160,000 industrial employees in the late 1960s to 439,000 in 1981 attests to the country's industrialization. An even stronger growth was registered by the construction sector, from 70,000–80,000 in the late 1960s to over 500,000 by 1981. A similar assessment has recently been made by Thiery who considers that the 100,000–117,000 industrial employment in 1967 had expanded to 450,000–488,000 by 1980 and forecasts 600,000 for 1982 (Thiery, 1980, 1981). In view of the scale of the industrial investments involved and of the demand for jobs from Algeria's unemployed and growing cohorts of school-leavers, this steadily increasing industrial employment has been criticized as too small to cope with the country's problems (Schnetzler, 1981). A different criticism has focused on the poor productivity record of many industrial sectors so that industrial value added by employee has declined from 36,800 dinars in 1967 to 31,000 in 1978, if hydrocarbons are excluded from the calculations. In discussing this problem Thiery (1980, pp. 182–3) suggests that the *gestion*

Table 5.12   Growth of industrial and construction employment in Algeria, 1966–81

| | Employment in 000s | |
	Industry	Construction
1966[a]	164	73
1969[b]	161	82
1977[a]	411	356
1979–80[c]	385	427
1981[d]	439	502

Sources: [a] MPAT/DSCN, 1979, p. 14; [b] SEP, 'Annuaire Statistique de l'Algérie 1980', 1982, p. 93; [c] SEP, 1981, p. 1; SEP, 1982, p. 93; [d] SEP, 1983, p. 6

socialiste des entreprises approach to decentralizing management decisions may have had this retrogressive impact.

A more pertinent question, for this study, is whether this decade (or more) of industrial expansion had lessened or increased the regional disequilibria present during the late colonial and early independence periods. In studying industrialization and urbanization in Algeria, Brûlé and Mutin consider that the disequilibrium tendencies of the pre-independence years have been perpetuated with industrialization strengthening the population migration patterns centred on the coastal urban poles (Brûlé and Mutin, 1982; p. 62). They ask whether the new interior industrial centres can establish their own zones of population movement, which would include coastal areas oriented towards the interior, and in this way reverse earlier trends. However, they doubt this eventuality in view of the continued growth of coastal industrial employment at the same time as the spread of the presence of industrialization to many interior towns.

In trying to assess the regional spread and balance of industrial employment, certain data problems are met. Whereas the 1966 and 1977 censuses afford employment statistics they do not accord with data from other official sources for the same years. Furthermore, any census suffers from errors arising from the self-declaratory nature of the household census return and from its place of residence rather than place of work basis. In the early 1970s the annual *Enquête Emploi et Salaires*, carried out by the *Secrétariat d'Etat au Plan* provided useful employment data at the level of *wilayate*. More spatial detail down to the basic level of communes could be painstakingly obtained from the *Direction des Statistique*'s *Fichier des Grandes Unités* even if some discrepancies existed between this source and the *Enquête Emploi et Salaires* (Sutton, 1976). Unfortunately by the late 1970s the annual *Enquête* is based on a sample of 2,500 establishments extracted from the *Fichier des Établissements* in such a way that no regional figures are forthcoming. Still worse, the *Fichier*, which in 1973 was available as computer print-out organized by communes (*Code géographique*), was available in 1983 only for private-sector enterprises. State and local enterprises are in a spatially random order, organized by the listing number acquired by each establishment as it was added to the *Fichier*.[4]

As a result of these data problems, some recent assessments of the regional impact of industrialization are compelled to use other, less precise statistics. Thiery amalgamates several regional programmes running between 1967 and 1980, in suggesting that 750 regional and local *entreprises publiques* had been set up. The period 1973–8 was particularly fruitful for such new creations. Regionally, the developed coastal *wilayate* of Algiers, Oran, Annaba, and Skikda only contained 99 or 15 per cent of these local public enterprises, which usually had been created by the

Assemblées Populaires Communales (Thiery, 1981, p. 51). Another study of the regional results of employment change, in the period 1966–77, used the results of the two population censuses, but reorganized them spatially into the six economic zones outlined in the 1974–7 Plan (see figure 5.3). Both in absolute and percentage growth terms, during 1966–77 the coastal and urban Zone 1 tended to dominate though encouragingly strong growth was also shown by the interior Zones 4 and 6 for industrial employment and by all the other zones, except Zone 6, for construction employment (see table 5.13). It should be noted, however that the dominant Zone 1 does extend well inland in the west around Tlemcen and Sidi bel Abbés and in the east around Constantine and Guelma. Also it should be appreciated that most of the industrial projects of the 1974–7 Plan started production after 1977 and so any contribution towards regional convergence of industrial employment would only be made after that date.

If table 5.13 does not suggest regional convergence in industrial employment, and so confirms the industrial employment counter-trend observed earlier in the wider analysis of regional trends in Algeria's socio-economic development, there is some evidence to suggest that progress towards convergence is imminent and is planned for by the *Ministère des Industries Légères* in its forecasts for the 1976–82 period. Quoting this same source, Thiery suggests that the favoured coastal *wilayate* which concentrated 56 per cent of State industrial employment in 1970 should only contain 29 per cent of it by 1982. This projection is made by the Ministry of Light Industry in a 1977 publication (Ministère des Industries Légères, 1977) which attests to its strong commitment to the deconcentration of industrial

Table 5.13   Employment growth by Economic Zones, 1966–77

Zone	Industry employment (in 000s)			Construction employment (in 000s)		
	1966	1977	Percentage of growth	1966	1977	Percentage of growth
1	94	248	163	22	123	459
2	13	27	107	7	38	443
3	24	51	112	16	82	413
4	13	32	146	11	52	373
5	8	16	100	6	26	333
6	12	35	191	11	35	218
Total	164	411	151[a]	73	356	338[a]

[a] As an average of all zones.
Source: MPAT/DSCN, 'Bilan régional 1967–1978. Instrument d'analyse spatiale et identification des inégalites régionales' 1979 ('Document interne' at DSCN, Algiers), p.14

employment. Employment data for 1976 is given, together with a projected 1982 figure, both with considerable detail about location and type of industrial unit. Obviously some projects have yet to have their precise location decided, though often the *wilaya*, if not the commune, has already been indicated. The spread of this industrialization programme is indicated by figure 5.7 and by the fact that more than 200 projects were under construction in 1977 with a further 700 industrial projects at various stages of planning, and that 150 *daïrate*, or more than 250 communes, are affected. Only ten *daïrate* will not so far receive state industrial units. Regional balance is a major declared aim but not to the neglect of technical and other locational constraints specific to the industrial product. It is anticipated that employment in the state industrial sector will rise from 254,970 at the end of 1976 to 683,270 in 1982. Many of the projects listed were underway in 1977 or were about to have contracts signed. However, the official reports did anticipate certain delays which would prevent the achievement of the 1982 employment target, at least until 1985.

It should be stressed that these projections apply to State industrial employment only and the growing private sector will augment the number of units and jobs, though probably with a greater concentration on the major cities. Also the list of state projects in the Ministry's projection is not exhaustive. It is anticipated that further projects linked to the growth of domestic demand for food products and household electrical goods, etc. will be added to the programme later. As figure 5.7 suggests, the minimum regional objective is to provide a thousand industrial jobs in each *daïra*. In poor, agricultural *daïra*, the Ministry will try to over-achieve and in those with a rich agriculture it will be content with 500–1000 such jobs. Thus the fact of industrial employment will penetrate at least half the communes of Algeria, and not just the administrative centres of *daïrate* and *wilayate*. In broader regional terms, the programme will concentrate jobs in the interior plains and high plateaux (Zone 4 on figure 5.3) where it is intended that the excess population from the over-populated mountain zones should be housed in medium-sized and small towns, rather than going to swell the major agglomerations. Light industries in particular are to be spread to the lagging and even to the mountainous parts of the country, again with the regional development objectives of reversing long-established trends of population migration and of introducing as wide a range of social strata as possible to industrial activity.

These regional development aims of the Ministry of Light Industry would appear to parallel and to reinforce the already established PCD programme discussed earlier. Unfortunately, another parallel may well be their limited achievements, especially in the early years, as the realities of investments and job creation do not always fit the well-intentioned

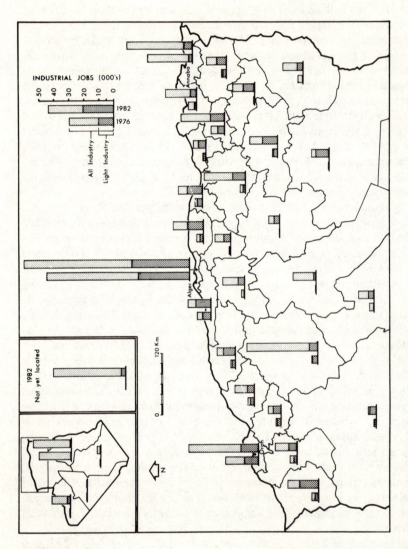

Figure 5.7 Industrial employment in Algeria, 1976 and 1982

forecasts of regional spread. A sample survey of the realizations between 1975 and 1977 of the PCD programme carried out by INEAP revealed certain shortcomings (INEAP, 1979). The sample covered 1169 PCD projects in 133 communes. These projects were much wider than just industrialization, as indicated earlier. Superficially the geographical distribution seems not to favour the coastal Algiers–Mitidja–Oran regions. However, this suggestion of 'spread' is less convincing when the PCD projects are reconsidered in financial terms. The largest and best-funded projects are still in the richer regions, whereas in the poorer lagging regions small, low-investment projects prevail. At the time of the 1977 survey only 6.6 per cent of the total financing of the 1169 PCD projects had been spent and that on only 186 of the projects. As table 5.11 suggests this regional imbalance had been in part rectified by 1982 in that *per capita* PCD expenditure was now decidedly higher in the lagging regions.

Obviously many more PCD projects have been initiated since 1977 but the apparent delays in realizing investment and the concern over the more limited degree of spread than anticipated can both be re-expressed with respect to the larger and more specifically industrial programme being implemented by the Ministry of Light Industries. Figure 5.7 illustrates the projected greater growth planned for many interior *wilayate* like Batna, Sétif, Tiaret, and Oum-el-Bouaghi. Tiaret in particular is planned to leap dramatically into second place in a league of *wilayate* by industrial employment, ahead of Oran and Annaba. However, the 1982 employment data is divided into two categories; that forecast for industrial units 'en cours de réalisation'; and that for units which are merely projected. Nationally, only 82,095 of the new jobs created in the period 1976–82, are in the former category, while 346,205 jobs, the vast bulk, are in the more speculative 'projected' category. In the case of Tiaret only 3,080 new jobs are in projects underway in 1977, while 41,140 are projected. Of the latter 25,000 jobs are in one vast industrial project, a steel-making complex for Western Algeria, the precise location of which was recorded as 'not determined' in 1977, and which was still not commenced in early 1983.

The *wilaya* of Tiaret should however benefit as the location of the long-awaited car manufacturing plant according to an announcement in July 1983 by the Ministry of Heavy Industries (*El Moudjahid*, July 1983). Production of 100,000 vehicles a year is planned to commence in 1987. So, while the data in figure 5.7 reflects the aspiration to regionalize and spread State industrial employment, the results by the early 1980s have fallen short of what were perhaps optimistic projections. Further, the 1982 projections included 49,330 industrial jobs for which no geographical location had been decided by 1977, and these included another massive

25,000 employee engineering complex for the Société Nationale des Constructions Métalliques (SN. METAL). Nevertheless, the very detailed nature of the Ministry of Light Industry's publication, giving information (for each projected industrial plant), of its anticipated number of employees and its *daïra* of location does suggest that deconcentration and regional development considerations have received great priority in this high-spending ministry.

Within the national policy of industrialization, special attention has recently been given to a *Petites et Moyennes Industries* programme (PMI), which focuses on small industries. Initially called the *Programme Développement des Industries Locales*, this initiative was first orientated largely towards craft industries (*artisanat*). From 1974 onwards the PMI programme was reactivated with a larger budget and an expanded agenda of 389 projects. A new orientation emerged as 57 per cent of the budget was now to go to construction materials' industries. However, by the end of 1979, 70 per cent of its credits still remained unallocated, a failure which will have greatly limited the programme's impact. From 1980 to early 1983, 180 projects have been endorsed to a value of 1400 million dinars but, again, 90 per cent of the allocation remains unspent.[5] Taken overall the PMI programme has involved 715 projects in the period 1967–82, to which credits of 3304 million dinars have been granted and for which further credits of 1650 millon dinars remained unclaimed.

The PMI programme can be seen as assisting the spread of industrial development after the initial concentrated phase of heavy industrial projects. Thus, the PMI programme concerns the accession of local *collectivités* (authorities) to industrialization as well as the introduction of rural and semi-rural zones into the national development strategy. This regional bias towards rural and under-industrialized zones meets local social and economic needs and so fits in with broader regional development strategy. However, reports at the first *Séminaire National sur la petite et moyenne industrie*, held near Algiers in April 1983, suggest that these regional development aspirations remain something of a pious hope. While reaffirming the PMI programme, the Minister of Planning and Regional Development frankly recognized the limited results achieved so far (*Revue de Presse*, 1983).[6]

A further element in the regional development impact of deconcentration is the extent to which it is accompanied by decentralization of management functions in what is a State-dominated industrial sector. The centrally planned nature of Algeria's industrializing economy had resulted in the location of the headquarters of most State companies near their related government ministries in the capital, Algiers. A similar overwhelming preference for a location in Algiers was displayed by private capital. The Algiers' *wilaya* accounted for 373 out of the 619 headquarters recorded in the 1973 *Fichier des établissements*. Only Oran figured significantly as a decision-making centre away from the capital (Sutton, 1976, p. 92). Much of this concentration would have resulted from the substantial role of State companies. Criticism of this centralization emerged during the 1976 Charter debate when the relative lack of success of *la gestion socialiste* suggested that a reorganization of the state companies was required. The accession to power of the new Chadli administration in 1979 provided an opportunity for a reassessment and consequently, in 1980, a national committee was set up to restructure State companies. Amongst its recommendations was the decentralization of decision-making. State companies had to give up activities not directly concerned with their own specialization. Also, significantly, they had to regionalize their units of production and to allow those regional units greater autonomy (Elsenhans, 1982, pp. 64–5). By March 1982 the planned future distribution of the head offices of State companies was as shown in table 5.14. This table also reflects the proposed subdivision of the often huge State companies into several smaller enterprises.

With the notable exception of the giant state oil and gas company, the Société Nationale pour la recherche, la production, le transport, la transformation et la commercialisation des hydrocarbures (Sonatrach), tangible progress towards decentralization is rather lacking. Sonatrach's restructuring involves the establishing of thirteen companies. Three of these were set up in 1980, and the decree allowing for a further six companies was published in 1981 (see table 5.15). Each of the new separate companies will have a distinct sphere of activity. For example, ERDP will be responsible for oil and condensate processing and for the export and distribution of refined products throughout Algeria, while the Sétif-based ENPC will handle the promotion, production, and marketing of plastic and rubber products. The significant feature of this restructuring is that several of the new companies are headquartered outside Algiers, apparently as a deliberate attempt at decentralization and regional development. In localities where such headquarters are established new jobs

Table 5.14   Planned distribution of HQs of State companies, March 1982

Area	Number	Percentage of total
Five most developed wilayate	109	57
(of which Algiers)	(48)	(26)
Hauts Plateaux wilayate	19	10
Southern wilayate	15	8
Rest of Algeria	47	25
Total	190	100

Table 5.15 The proposed restructuring of Sonatrach

Companies	Location of HQ
New companies established in 1980	
Entreprise Nationale de Raffinage et de Produits Pétroliers (ERDP)	Algiers
Entreprise Nationale de Grands Travaux Pétroliers (EGTP)	?
Entreprise Nationale de Plastiques et Caoutchoucs (ENPC)	Sétif
New companies decreed in 1981	
Entreprise Nationale de Forage (ENAFOR)	Hassi Messaoud
Entreprise Nationale des Travaux aux Puits (ENTP)	Hassi Messaoud
Entreprise Nationale de Géophysique (ENAGEO)	Hassi Messaoud
Entreprise Nationale de Génie Civile et de Bâtiment (GCB)	Boudouaou (*wilaya* of Algiers)
Entreprise Nationale de Services aux Puits (ENSP)	Hassi Messaoud
Entreprise Nationale de Canalisations (ENAS)	Ouargla

Source: Economist Intelligence Unit, 1982, p. 58

should be created and regional urban expansion should ensue (Economist Intelligence Unit, 1982, p. 58). It remains to be seen how effective Sonatrach's attempt at decentralization proves and whether this example is followed by the other State companies either through a process of restructuring and subdivision or through the devolution of regional decision-making powers.

Conclusions

In this study the aspirations for regional development have been repeatedly expressed and even partly codified in successive government policy statements embodied in National Plans, the 1976 National Charter, and various programmes for regional development including especially the PCD programme. Some significant evidence of regional convergence has emerged from a detailed analysis of regional trends in the period 1976–77. These confirm the assessment made by Thiery – that gains in purchasing power seem to have been made by those social groups most disadvantaged in 1962. He considers that regional disparities, in terms of employment and income, have been at last noticeably reduced (Thiery, 1981, p. 61). The detailed analysis of inter-censal trends during this period confirms this but also indicates a contrary divergence in terms of industrial employment. This might have been expected in view of the early strategy of heavy industrialization embarked on by Algeria.

An evaluation of late 1970s' industrial and other investment together with projected patterns of industrial employment proposed by the Ministry of Light Industries suggest that there is now a strong likelihood that, by the mid-1980s, or at least the late 1980s, industrial employment will conform to the regional convergence trends aspired to by government policy and achieved for many other socio-economic indicators. The strong regional planning elements in the PCD programme together with the patterns on figure 5.7 which remain the objectives of the planners at the Ministry of Light Industries, give cause for optimism that regional development will henceforth mean more than the spread of the fact of industrialization. It will imply a shift in the balance of industrial employment away from its present coastal concentrations.

Such optimism must be tempered by caution in so far as several major regional industrial projects have yet to be set underway, and indeed their precise locations have yet to be chosen. Also, the recent encouragement of private industry may work against regional policy as private entrepreneurs opt for risk-minimizing and familiar locations in and around the major cities, especially Algiers and Oran. Finally, can regional development ever be more than a well-intentioned cosmetic exercise unless it is accompanied by the valid and larger scale devolution of decision-making to those same lagging regions? The degree to which decentralization is accompanying and taking over from deconcentration is as yet very limited. Some evidence is available in the administrative and agrarian reform sectors (Sutton, 1981), and certain State companies like Sonatrach are taking the first steps towards regionalizing their subdivisions. However, a stronger element of decentralization of both productive activities and decision-making is required to redress the regional imbalances long prevalent in Algeria, the result of both its colonial experience and its centrally planned socialism.

If socialism is to be defined in terms of State capitalism and a centrally planned economy then Algeria's development approach could well be classified as socialist. On many world classifications these criteria are sufficient to so designate Algeria. The survival of a private sector of the economy, especially in agriculture, is by itself inadequate to counter this allocation of Algeria to a group of command-economy, socialist countries. However, the early 1980s' government encouragement of this private sector along with parallel support for private house building, or *autoconstruction*, may well raises doubts about Algeria's socialist intent. Further doubts could be expressed by those seeking a more decentralized and participatory form of socialism, where self-management really means the involvement of workers in decision-making about their factory's or farm's activities and where regional development involves the genuine and committed devolution of decision-making to meaningful regional bodies with valid democratic authority and with effective means for putting locally decided policies into operation. The degree to which Algeria's regional development programme during the 1980s demonstrates such decentralization rather than mere deconcentration will be a major indicator of the government's commitment to a socialist path to development.

Notes

1. The value of the Algerian dinar has obviously fluctuated. It was as follows: 1967 – 1 dinar = $US0.20 or £0.08; 1973 (October) – 1 dinar = $US0.24 or £0.10; 1978 (October), 1 dinar = $US0.27 or £0.13. *Source: International Financial Statistics.*
2. The Coefficient of Spatial Variation (CSV) is the quotient of the standard deviation by the mean multiplied by a constant of 100:

$$CSV = \frac{(x - \bar{x})}{S} \times 100$$

3. Two urban planning instruments operate independently: First, the PUD (*Plan d'Urbanisme Directeur*) is required for each agglomeration and is a physical guide to development and control; Second, the PMU (*Plan de Modernisation Urbaine*) provides the financial framework for action in some selected towns.
4. Verbal communication from DGS (Direction Générale des Statistiques), Algiers to M. Nacer, March 1983.
5. 'Petite et Moyenne Industrie, la solution d'avenir', *Révolution Africaine*, no. 1000, 1983; reprinted in *Revue de Presse* (Algiers), no. 274, May 1983.
6. One of the authors, M. Nacer, was present at this seminar in April 1983.

6
Sand Reform or Socialist Agriculture? Rural Development in PDR Yemen, 1967–1982

Jim Lewis

Introduction

This chapter has two objectives. The first and dominant one is to summarize the rural development policies of the People's Democratic Republic of Yemen and to evaluate their impact;[1] the second is to raise the more general issue of the appropriate criteria to use in assessing the ways in which socialist countries tackle the problem of developing rural regions.[2] The structure of the remainder of this presentation reflects this double purpose with the first and last sections forming an 'outer shell', representing an attempt at addressing the question of the nature of socialist rural strategies and evaluating the extent to which PDR Yemen's strategies have laid the foundations for socialist agriculture. The middle section of the chapter constitutes the bulk of the paper and presents a summary of the major features of PDR Yemen, a review of the rural development policies (paying particular attention to agricultural transformation), and an evaluation of their success in their own terms.

The nature of socialist rural development

At first sight it may seem ridiculous to suggest that socialist countries can, or should, have different strategies and goals in rural development. The World Bank's view of rural development as a 'major part of a development strategy' the objectives of which 'include sustained increases in *per capita* output and incomes, expansion of productive employment and greater equity in the distribution of the benefits of growth' (World Bank, 1975, p. 16) seems broad enough to encompass any country's experience. It follows that it is legitimate to conduct exercises such as that of Lele (1975) using standard criteria such as attention to low-income groups or effectiveness of administration to compare rural development programmes in different African countries. Development is development is development.

However, this argument loses much of its force if it is conceded that 'the nature and content of any rural development program or project will reflect the political, social and economic circumstances of the particular country or region' (World Bank, 1975, p. 4) for the self-definition of countries as 'socialist' implies different political and social – if not also economic – circumstances from that of the non-socialist world. Thus, the observation that 'with well designed programs, offering proper incentives to small farmers, development can be much more rapid than is sometimes believed, and the impact on levels of living following the expansion of cash incomes from a subsistence baseline can be dramatic' (World Bank, 1974 pp. 8–9) has little relevance to a policy-maker guided by a political commitment to reduce the economic importance of small farmers, who measures impact in terms of reduced hours of heavy labour rather than increased cash incomes.

Yet there is such diversity within the socialist world that deriving a single set of rural development goals and policies for this group of countries also appears impossible. Instead, it makes more sense to subdivide rural development strategies into two basic sorts in line with the basic political divergence in the socialist world – those close to the model of the Soviet Union and those close to the model of China. Although, as Domes (1980) notes, policy priorities and institutions in the rural areas have changed enormously over the past sixty or so years and both have common concerns such as relative equality in land holdings and the provision of social services in rural areas, it is possible to detect consistently different emphases in the experience of the two countries (for a fuller summary of China's strategy see Aziz, 1978). The main emphases have been related to the organization of collective farming, the diversification of the rural economy and the nature of political management. Whereas the USSR has sought to bring about the productivity increases associated with the classic industrial division of labour by introducing state farms as the dominant form of organization, the Chinese have placed more emphasis on co-operation in production at much smaller scales (the production team or brigade) and placed less stress on the wage relationship. In terms of diversification, the USSR has sought not only to establish a division of labour in terms of regional specialization in types of crop but also to direct industrial and service activities to urban locations, while the Chinese have stressed regional self-reliance and encouraged rural industrialization. Finally, the USSR has increasingly separated the control of agriculture through the central plan from local political activity in contrast to the Chinese who used to maintain a close relationship between political and economic decisions through the commune structure.

In practice, elements of both these models are present in most socialist countries – and indeed in the USSR and China themselves – but these two models provide us with a starting point in trying to define evaluative criteria. It may well be that neither corresponds with your own idea of socialism or that both are of little relevance to certain sorts of countries but they do condense the experience of two of the world's largest countries and provide the basis of the thinking of most governments in the socialist world. In a later section, they will provide the framework for assessing the extent to which PDR Yemen has moved towards a socialist transformation of its countryside.

PDR Yemen: background information

PDR Yemen is located in south-west Arabia, bordered by the Yemen Arab Republic, Saudi Arabia and Oman. This gives it considerable strategic importance due to the naval facilities of Aden harbour (and Socotra Island) and control over southern approaches to the Red Sea. Its size (including Perim and Socotra islands) is 338,100 km^2 (of which only 0.2 per cent is currently used for agriculture) and the only census, in 1973, gave a population of 1.5 million – now estimated at 2.0 million. The poverty of the bulk of the population is indicated in table 6.1. The major settlements are identified in figure 6.1 (which also shows the location of the six governorates that are the most important sub-national units) and their relative size shown in figure 6.2.

The physical environment can be divided into three zones: (1) the coastal plain, 2–25 kilometres wide, annual rainfall 50 millimetres; (2) Intra-montane plains at 300–1000 metres, rainfall 200 millimetres; and (3) Highland plateau at 1500–2500 metres, rainfall 400 millimetres but falling away towards Rub al Khali. In all of these areas a shortage of permanent water limits agriculture to Wadis where spate-irrigation can be practised or wells sunk into underground aquifers. No major terrestrial natural resources are yet available for exploitation (though oil finds earlier in 1982 may change this) but there are rich fishing waters.

Historically, the area was divided into numerous small states which were periodically grouped into empires on the basis of the spice trade. A major change came with the British occupation of Aden in 1839, which – after years as part of Bombay Presidency – became a colony in 1935. The remaining states were grouped into Eastern and Western Protectorates, most of which joined Aden in the Federation of South Arabia in January 1963. On 14 October 1963 the National Liberation Front launched its anti-British offensive in the Radfan and an urban guerrilla war started in August 1964 (for details see Halliday, 1974). The British withdrew on 29 November 1967 and the NLF took power. The loss of the British base and funds coincided with the closure of the Suez Canal,

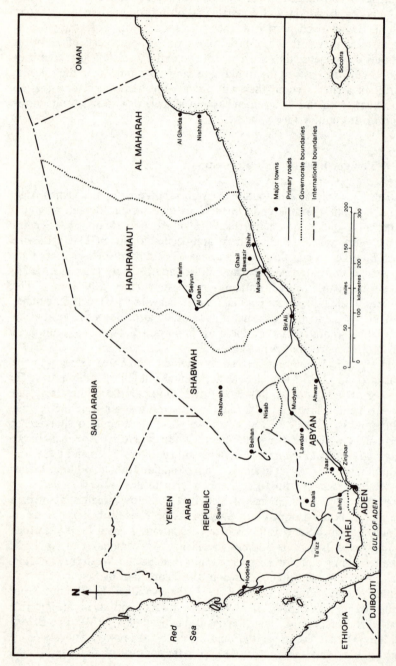

Figure 6.1 Governorates and main towns of PDR Yemen

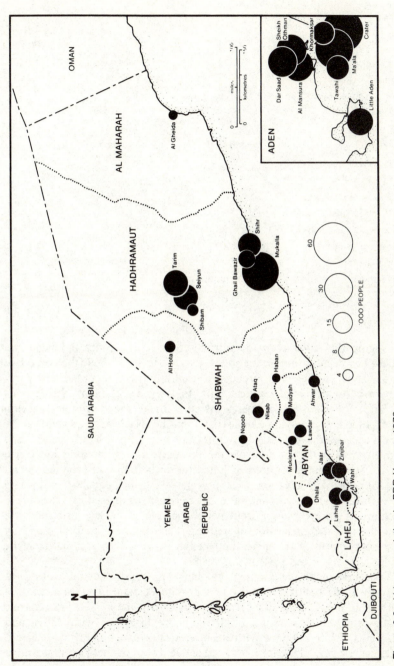

Figure 6.2 Urban population in PDR Yemen, 1978

Table 6.1   Demographic and 'basic need' indices

Item	Index
Population	1.9 million (mid-1979)
Area	338,100 km²
Population density	5.6 km²
Growth rate	2.6% (1973–8)
Cultivable land	231,000 ha
Population density on cultivable land	8.2 per ha
Life expectancy at birth	45 years
Infant mortality	114 per 1,000
Persons per doctor	7,760 (1977)
Access to piped water	27% (1978)
Access to electricity	23% (1978)
Urban population	37%
Adult literacy rates	11% female/66% male
Primary school enrolment (% of age group)	51% female/92% male

Source: World Bank, 1981

plunging the Aden economy into a serious crisis. Since independence there has been an important shift leftwards in June 1969 (the corrective movement) and the NLF broadened itself into the Yemeni Socialist Party in 1967 (for details of political shifts to 1978, see Halliday, 1979).

The 1978 Constitution states that the aim of the state is to 'develop the national economy on the basis of scientific socialism in order to satisfy the needs of the people'. Its efforts are based on a highly centralized government with a major role for the party in determining long-term goals. Its original intentions included the socialization of production and distribution, the creation of a productive sector, equality in income distribution and provision of basic needs. The measures adopted to achieve this have ranged from nationalization of banks, insurance and trading and nationalization of housing to changes in family law and an agrarian reform. Their general effect can be judged by summarizing the achievements of the past fifteen years.

The first major achievement has been the survival of the country at all, given the political hostility to a pro-Soviet State in the area and the economic crisis of the first few years. Ties with the USSR, East Europe and Cuba remain strong, as do the more local ties with Libya and Ethiopia, while relations with neighbours fluctuate – unity with North Yemen remains a long-term goal but is not always pursued – Saudi Arabia is cautious and Oman increasingly hostile after signs of acceptance in the late 1970s. Most aid has come from State-socialist countries but an increasing proportion is from Kuwait, the Arab Fund, the United Arab Emirates (UAE) and international agencies, such as the International Development Association (IDA). In terms of internal politics, there have been elections to People's Local Councils (1976–7 and 1981) and the People's Supreme Council (1977) and not all those elected are party members.

Socially, the major changes have been in the availability of free education (the number of pupils has increased by 11 per cent per annum, 1968–78) and free health care (hospital beds have been doubled in the 1970s, while the number of doctors trebled), the reduction of income inequalities (the ratio of maximum to minimum salaries is down from 11:1 to 6:1) and the improved status of women (especially in urban areas).

Despite the enormous constraints of the poor resource base and collapsing colonial economy, there has been quite rapid growth – partly due to the co-ordination provided by the series of investment plans (Three-Year Plan, 1971–4; First Five-Year Plan, 1974–8; Second Five-Year Plan, 1979–83, since amended and running 1981–5) and the existence of a large public sector. Other contributing factors are the increasing flow of external resources in the form of aid (190 million dinars 1969–79), emigrant remittances (108 million dinars in 1979 alone) and Aden's bunkering trade (some 6 million dinars in 1979). Capital investment has grown from 1 million dinars to 60 million dinars annually between 1970 and 1978 and GDP from 61 million dinars to 189 million dinars over the same period. The structure of the GDP – summarized in table 6.2 – has moved somewhat towards the productive sector but mainly because of the growth of construction. However, this has entailed a worsening balance of payments (in which commodity exports cover only 5 per cent of the import bill) and a growing public debt, as well as the unusual problem of a rural-labour shortage due to emigration to the Gulf (Birks and Sinclair, 1980, estimated that 80,000 (16 per cent of the total labour force) Yemenis are employed there).

Rural development policies

To understand the emphasis of PDR Yemen's strategies on agricultural production it is necessary to appreciate the rural origins of the Yemen Socialist Party leaders and the important place of agriculture in the national economy, both in terms of its overall size and its (potential) contribution to other sectors. Considering size first, both relative employment and production have been declining. Despite the importance of services in the colonial economy, agriculture and fishing employed

Table 6.2   Economic structure[a]

	Structure of GDP (%)			Balance of payments (in millions of dinars)	
	1973	1979	Growth rate (% p.a.)		1979
Agriculture	14.6	7.6	1.8	Exports	15.3
Fisheries	6.6	5.2	8.0	Imports	−141.5
Industry	12.4	13.3	8.9	Workers remittances	108.3
Construction	5.1	11.4	27.6	Other invisible	6.0
Transport	7.5	13.7	12.0	Current account	−20.4
Net indirect tax	6.0	8.7	23.0	Official transfers	8.0
				Official capital	18.3
Finance	4.5	3.1	12.0	Net disburse-	
Personal services	3.1	0.7	−9.2	ment of loans	17.7
				Total	
Government services	22.1	20.6	6.8		23.6

GNP p.c.: $US480 (1979).
[a] Annual average growth rate, 1960–79, 11.8%.
Sources: World Bank, 1981; Sayigh, 1982

138,000 in 1969 (52 per cent of the total) and this figure was estimated at 202,000 in 1967 (43 per cent) (World Bank, 1979; Central Statistical Organization, 1980). Some 25,000 of these are Bedouin but the bulk are settled farmers. The contribution of agriculture to GDP has fallen from 16 per cent in 1969 to 8 per cent in 1979 and production in 1980 was worth around 31 million dinars, most of which was crops.

This agricultural sector has been expected to provide an increasing proportion of the food for the local population, raw materials for industrialization and export earnings. Most of colonial Aden's food was imported and this used to comprise one-third of the cost of non-oil imports. Despite the difficult physical conditions and decades of neglect of local agriculture, the government sought to alter this as part of its self-reliance programme. Although food production has increased, the degree of self-sufficiency remains low – except for fruit and vegetables and milk (see table 6.3) – and FAO estimates that daily food availability *per capita* has been decreasing (figure 6.3). The proportion varies considerably from year to year but food represented 36 per cent of imports in 1979, a relatively typical year.

As regards industrial crops, expansion has been limited by the fact that PDR Yemen has a tiny industrial sector (excluding the ageing oil refinery)

Table 6.3 Agricultural self-sufficiency, 1979

Production	Local production ('000s tons)	Imports ('000s tons)	Self-sufficiency (%)
Wheat and other cereals	26	100	21
Rice	0	57	0
Ghee and oils	6	21	22
Sugar	0	36	0
Milk	48	12	80
Tea	0	4	0
Pulses	0	7	0
Industrial tomatoes	8	5	61
Sesame	1	7	18

but local materials are now used for dairy and leather products, and supply the textile factory at Al Mansura, the cigarette factory nearby and the tomato paste factory in Wadi Tuban. All of these establishments are within 35 kilometres of Aden Harbour and the only important processing plant away from this area is the date-packing factory in Wadi Hadhramaut as there has been little attention to rural industries so far.

In 1967 the only significant agricultural export was long-staple cotton from the Wadi Tuban and Wadi Abyan areas. The value of this has fluctuated recently due to changes in water availability but is generally increasing, reaching 3 million dinars in 1980. Tobacco is the next most important export at 0.6 million dinars. However the value of these exports has been consistently lower than that of imports. In 1967, food imports were mainly of wheat, rice and sugar at about 1.7 million dinars each. Average imports for 1979–80 of these cost 8.8 million, 6.4 million and 6.7 million dinars respectively, while ghee, milk and sesame seeds all cost over 3 million dinars.

It was in order to transform the agricultural sector and change social relationships in the countryside that the government started its programme of Agrarian Reform. Its legal basis was the land reform law of 1968 but this was never properly implemented and so was replaced in December 1970. This law limited land ownership by each family to 20 feddans of irrigated land and 40 feddans of unirrigated land and also confiscated land from sultans, sheiks and absentee landlords. It was implemented partly through NLF-inspired peasant uprisings. Expropriated land was turned into State farms or redistributed to 30,000 families who joined the co-operatives that were established. There are still some

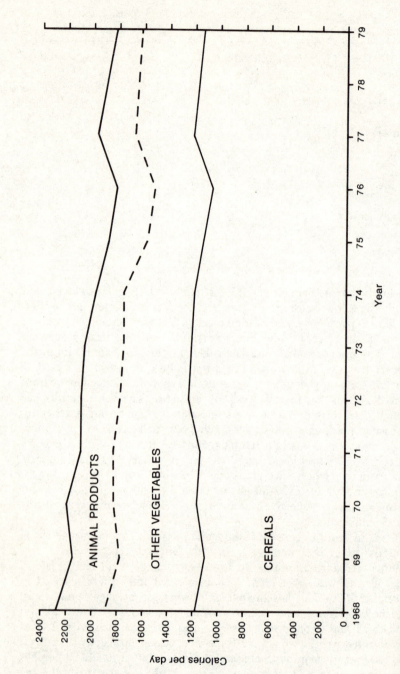

Figure 6.3 Daily food availability in PDR Yemen, 1968–79

2,000 small, private farmers in remote areas and the livestock sector is private. Subsequent legal measures have restricted the marketing of crop products to the Public Corporation for Marketing Fruit and Vegetables but these have recently been relaxed slightly to allow surplus products to be disposed of privately.

There are 35 State farms at present (plus 7 animal farms) with a permanent labour force of 3,000. They own 36,300 feddans and currently cultivate 17,500, usually using an underground water supply released by the drilling of the irrigation department or Yemeni–Soviet project. Amongst crops there is a stress on fruit and vegetables – especially near Aden – and fodder (see table 6.4) while animal production is important more for breeding than for sale.

Table 6.4   Production by type of farm, 1980

Product		State farm	Co-operative	Individual	Total
Crops (1000 tonnes)					
Wheat		0.7	6.0	1.6	8.4
Sorghum and millet		0.9	8.5	2.7	12.1
Maize		0.5	3.1	1.7	5.3
Barley		—	0.1	—	0.1
Fodder		35.2	57.8	27.9	120.9
Sesame		0.3	1.5	0.8	2.6
Cotton		0.1	6.9	—	7.0
Tobacco		—	0.7	0.1	0.8
Coffee		—	0.8	0.1	0.9
Potatoes		0.6	4.9	1.2	6.6
Tomatoes		7.2	9.8	1.8	18.8
Red Onions		0.5	3.7	1.1	5.3
Other vegetables		2.5	8.7	1.7	12.9
Curcubits		4.8	5.7	0.5	11.0
Dates		0.1	6.6	4.6	11.3
Bananas		8.0	3.3	0.8	12.1
Other Fruit		0.6	1.5	0.2	2.3
Total Crops		62.0	129.6	46.8	238.4
Livestock (tonnes)					
Cattle	Meat	160	1440		1600
	Milk	1560	7700		9260
Camel	Meat	—	1600		1600
	Milk	—	3600		3600
Sheep	Meat	4	2900		2902
	Milk	—	6400		6400
Goats	Meat	—	3346		3348
	Milk	—	12200		12200
Poultry	Meat	250	250		500

Table 6.5 Agricultural investment by Plan period, 1971/72–85 ('000s dinars)

	1971–2 Planned	1973–4 Actual	Planned	1974/5–1978/9 Amended	Actual	1982–5 Planned
Irrigation (including studies)	4.9	4.5	16,756.8	32,493.8	26,776.9	44,886
Animal Production	0.5	0.5	1,673.8	6,198.8	5,834.1	4,600
Research and Training	1.0	0.5	1,108.6	8,441.0	6,198.7	—
State farms	0.4	0.4	—	—	—	—
Plant production	—	—	1,257.8	5,423.0	4,572.3	12,500
Total for agriculture	6.7	5.8	20,797.0	52,556.4	43,382	61,935
Total for Plan	32.4	25.1	73,358.8	276,247.0	194,812	425,600

There are 59 co-operatives at present, with 40,000 members cultivating 100,000 feddans. They can be of two major sorts: *production co-operative*, run as a collective farm with members receiving a wage (only one exists, 26th September Farm in Wadi Tuban, but it is seen as the goal for others); *service co-operative*, further subdivided into those 37 where plots are owned and farmed individually and the 21 (mainly in the Hadhramaut) where members are grouped into a number of *feraq* (work groups). A small administration deals with agricultural supplies, services from the Machine Renting Station, seasonal credit and the marketing of produce. Co-operatives' land is 72 per cent spate-irrigated and on average each member cultivates 3 feddans of his 11 feddans. Their production is dominated by cotton, fodder, wheat, dates, tomatoes, sorghum and livestock (see table 6.4).

Within this institutional framework the government's main means of influence is exercised through its control over agricultural investment and over prices, subsidies and taxes. Each of these will be summarized in turn.

The major means of planning the economy is the medium-term investment plan, drawn up by the Ministry of Planning in consultation with the Central Committee of the Yemen Socialist Party, the Politbureau, sectoral Ministries, planning units in the governorates and people's councils. These plans have become increasingly comprehensive over time and have had different sector emphases – Three-Year Plan dominated by transport infrastructure (40 per cent of total); First Five-Year Plan, agriculture and fisheries (37 per cent) and Amended Second Five-Year Plan, industrial infrastructure (33 per cent). Table 6.5 gives the distribution of investment (planned and actual) between activities and the emphasis on irrigation is readily apparent (64 per cent, 62 per cent and 72 per cent of investment in successive plans). This is the basis of the 'horizontal' expansion strategy, attempting to increase the area cultivated from that shown on figure 6.4, and, particularly, to utilize some of the potential agricultural land of the interior wadis as shown on figure 6.5. Irrigation works also contribute to the 'vertical' expansion strategy, intensifying production by improving or replacing spate-irrigation systems. Other important sorts of investment include the creation of seventeen Machinery Renting Stations, which provide equipment for State farms and co-operatives at subsidized rates, and the development of research facilities. Linking into these investments, and increasingly influencing the location, are the 'big projects' that may involve more than just agriculture. The major ones are: (1) The Yemeni–Soviet project which is widespread but for State farms only; it deals with irrigation and land levelling but little information about its cost or effectiveness is available; (2) Wadi Tuban (see figure 6.6), which started in 1979 with assistance

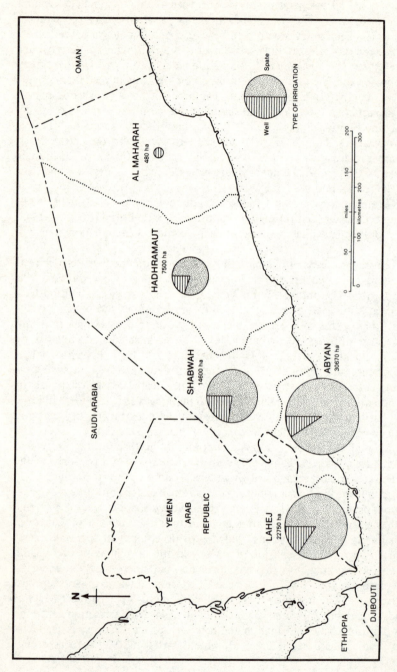

Figure 6.4 Cultivated area of PDR Yemen, 1977

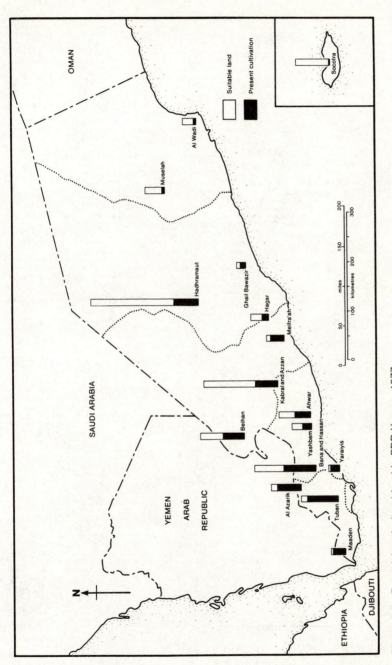

Figure 6.5 Potential agricultural land in PDR Yemen, 1977

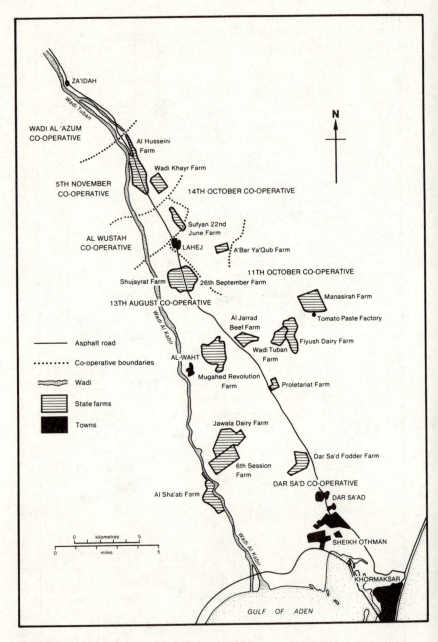

Figure 6.6 Agriculture in Wadi Tuban

from Arab Fund and IDA; this is an integrated project combining irrigation, agricultural extension, village water supplies and feeder roads at a total cost of 4.2 million dinars; (3) Wadi Abyan (see figure 6.7) started in 1973 with aid from Kuwait Fund and was almost entirely irrigation works. Its cost so far has been 6 million dinars but serious damage to the main dam and 16 kilometres of channel in September 1981 and a devastating flood in April 1982 have virtually eradicated the project; and (4) Wadi Hardhramaut (see figure 6.8), started in 1977 with an IDA loan for agricultural mechanization and extension village water supplies, feeder roads and a date-packing plant. There has been an investment of 3.3 milion dinars so far and it was expected that a second phase, with similar objectives, would start in 1982.

Given the broad framework of production expansion set by these investment policies, the government also seeks to influence the details of agricultural output by its control over the economics of production. The main mechanisms for this have been producer price controls and input subsidies. Price control involves the Ministry of Agriculture and Agrarian Reform annually setting prices for almost all crops at a level in between the maximum price that could be paid by an urban consumer on the (official) minimum wage and the average farm-production cost. State farms and co-operatives then contract to supply the relevant Public Corporation with amounts of produce (at the set prices) determined by their managements on the basis of the comparative returns of different crops. For most of the 1970s, prices were virtually constant at levels now regarded as too low to provide incentives for increased production and there have been sharp increases in the producer prices for crops such as wheat and cotton over the past two years in an attempt to stop the widespread move into unregulated crops – especially alfalfa and *qat* (the local narcotic).

For this system to work without rapid urban price rises it has been necessary to subsidize farm input costs more heavily each year. The subsidy on direct inputs – calculated in relation to the import costs of the inputs – is highest on the machinery and petrol available from the Machine Renting Stations, medium on fertilizers and lowest on seeds. As the import and internal transport costs of all of these have risen, so the Ministry has had to devote a growing proportion of its annual budget to subsidies but increasing labour costs continue to push average farm costs upwards. Agricultural credit has also been subsidized in comparison with prevailing interest rates through the operation of the Agricultural Development Fund (which is currently being reorganized with a larger budget and tighter controls over the use of its loans). Finally, water and farm infrastructure are effectively subsidized as user costs are set to cover maintenance costs rather than repay capital investment.

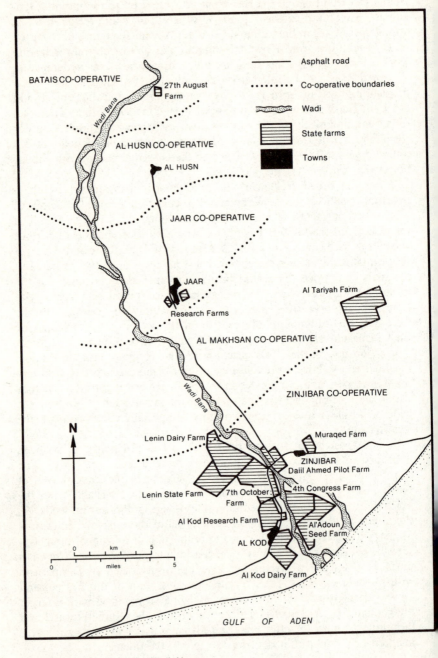

Figure 6.7 Agriculture in Wadi Abyan

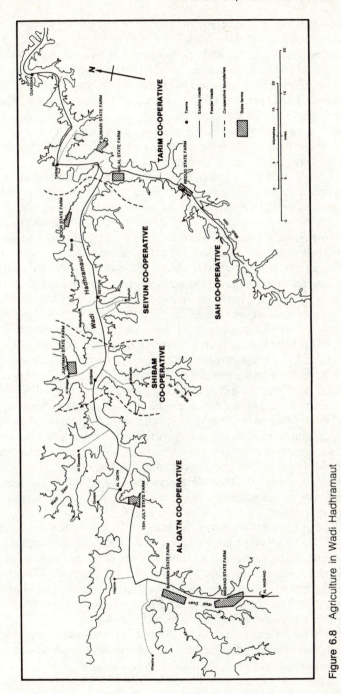

Figure 6.8   Agriculture in Wadi Hadhramaut

Although not paid directly to the Ministry, there are taxes on agricultural production that are intended to meet the cost of its current expenditure. This they currently do but almost entirely because of the tax revenue from *qat* as taxes on other crops products are set at levels that just cover the cost of administering the marketing system and vary little between products. The expansion of agricultural production by these means has always been seen as the most effective way of reducing the enormous income disparity between Aden and the rest of the country that existed prior to independence. However other aspects of rural development have not been neglected, even though it is more difficult to trace specific policies and their effects. The two other spheres of importance in terms of expenditure have been communication and social services' provision, so the aims and achievements in them are summarized in turn below.

Since the country did not exist as such prior to independence there was no national transport network. People and goods moved either by automobile along the few roads (which were unsealed except for some in Aden), by camel or by coastal *dhow* with such difficulty that there was little internal trading. Since the new government regarded the establishment of a natural system of transport as essential for defensive and 'integrative' reasons as well as economic reasons, a road-building programme received top priority in development investment in the first decade. Initially funded from its own resources but later attracting international aid, the programme was based on the provision of an all-weather surface on three main routes – Aden to Ta'izz, Aden to Mukalla and Mukalla to Tarim – and was almost completed in mid-1981 (though subsequent flood damage during the last two rains was widespread). This – and other 'national-priority' investments in airstrips, a telephone system and port improvements – has dominated both investment and attention so that improvements in rural accessibility have been almost entirely achieved as indirect consequences of this programme. The exceptions have been in areas in which 'big projects have operated and paid some attention to feeder roads (see figure 6.8 for the example of Wadi Hadhramaut).

In social services' provision the principal concerns have been the creation of a system of primary education outside the major towns and the availability of basic medical treatment to the majority of the population. Initially investment in both was low due to the general state of the economy but it increased rapidly during the First Five-Year Plan as aid from other Arab countries was directed towards social services rather than production infrastructure. The bulk of this has clearly been in the small towns (*Markaz* centres) of the interior with the result that some education is provided for the vast majority of males and almost half the female children (including the children of Bedouin who have been obliged to attend schools specifically for them since 1973) and rural health standards have improved considerably. In addition to these efforts the government has sought to enhance rural life by encouraging adult literacy, providing courses on domestic management for women and assisting in the formation and activities of cultural groups (including financial help in mosque construction).

Evaluation of the policies

Any assessment of the effects of these policies has to recognize that they have been pursued with differing degrees of enthusiasm, administrative competence and financial support over the years, so that what was once a high priority (such as institutional reform) is almost ignored at present while other issues (such as pricing policies) have come to the fore. There is also a problem of data availability since none of the policies have been effectively monitored (though the IDA and FAO funded projects will eventually be) so the following review will have to be based more on impressions than observations.

Throughout the fluctuations in emphasis and changes in its composition, the government has essentially been concerned with three goals in rural development: (1) improving incomes by raising productivity per worker (part of 'vertical' expansion); (2) increasing the contribution of agriculture to the economy (not least by the 'horizontal' expansion of the productive area); and (3) ensuring a more egalitarian distribution of access to resources. Given the initial state of the rural areas, the fact that there has been any progress at all towards these goals is a mark of the government's commitment to them. The extent of that progress can best be examined under separate headings.

As regards the raising of rural income levels, it is clear then the urban–rural differential has been falling and that there have been improvements in agricultural techniques and the quality of inputs which reflect official investment in irrigation, mechanization and extension, as well as input subsidies. However, agricultural output as a whole has risen more slowly than the size of the labour force so that the productivity gains from this technical modernization would appear to have been very unevenly distributed between types of farm and insufficient in aggregate to counter the effect of a rapid growth in the employable population.

A far more important influence on income levels than rural development policies has been the migration of labour to the Gulf. This has its main effect through remittances, which in one year currently amount to nearly ten times the planned investment in agriculture over the whole of a five-year plan period. Even though not all migrants come from rural areas and by no means all remittances are invested there, the relative importance of this cash flow is obvious and almost every family receives some part of it. However, this means of raising income is not independent of government efforts to raise agricultural productivity – it actually hinders them. Individuals do not invest their remittances in agriculture since this is (more or less) collectively run but rather they import consumer goods, improve their houses, and, occasionally, set up a small scale service or manufacturing establishment. The effect of these investments has been to encourage more male migrants to leave in search of higher incomes to the point where there are now male-labour shortages in some rural areas (exacerbated by the demands of house construction work) and both rural infrastructure projects and State farms cannot attract sufficient labour without increasing their wages beyond the level originally budgeted for.

The second reason for improvements in rural income also owes little to official policy and more to the switch to unregulated agricultural production. This is most readily apparent in the rapid growth of the goat population but can also be detected in the expansion of fodder production, which already accounts for half the cultivated area. The expansion has been such that the increase in total cultivated area associated with the investments of the Wadi Hadhramaut project is less than the increase in area devoted to alfalfa there. Any attempt to deal with this problem will have to tackle the question of the private ownership of livestock – something which involves almost every family, rural *and* urban. There has already been a government reaction to the switch into the other significant unregulated crop, *qat*, as it is now illegal to plant new bushes (except as replacements) and the long-standing restrictions on its use in Aden may gradually reduce domestic demand as beer drinking becomes a more popular daily habit. Finally, the improvement in rural incomes relative to urban ones has not just been a consequence of the changes noted above but also reflects the lack of increase (indeed, between 1967 and 1970, the sharp decline) in average urban incomes. This can be explained by the steady flow of impoverished rural–urban migrants and the unwillingness (or inability) of the government to revive Aden's economy.

Attempts to increase the contribution of agriculture to the economy have also had limited effect. The declining share of agriculture in GDP has already been noted but there has actually been an increase in both the area cultivated and in the value of output. Given the extremely limited extent of agricultural production during the British period, it is not surprising that improvements have been made from that base but unfortunately they have been concentrated in products other than those intended. Increases in food production as a result of agricultural policy have taken place – most notably in both milk and fruit and vegetables from State farms – but most of the change in total area and output is due to unregulated products that do not provide marketed foodstuffs. With export crops the story was similar, as cotton output fell sharply in the mid-1970s due to *qat* planting in its place but it will be drastically cut from 1982 as a result of the flood damage in the Abyan delta. Only the attempt to encourage industrial crops can be regarded as being at all successful and this only so because the production was already sufficient to support the relevant factory.

The reasons for the limited success in increasing production and failure to control its composition seem to be threefold: the emphasis on irrigation in investment planning, the concentration on state farms and the operation of the price control and subsidy system. The former is important because of the long lag between an investment and its effects so that not all the production increases as a result of work started during the First Five-Year Plan have yet been achieved – indeed in the Wadi Abyan project area the cultivated area was actually reduced for eight years while work was in progress. Secondly, the concern with 'horizontal' expansion through the irrigation of an uncultivated land has meant a heavy reliance on State farms as these are the preferred form for colonization. Not only is the bulk of irrigation investment – both national and from the Yemeni–Soviet project – directed towards State farms but few of the farms have yet started to pay user-charges for infrastructure. State farms also receive more favourable treatment than the co-operatives in terms of access to extension workers and equipment from the Machine Renting Stations. Since the farms are not all managed very efficiently, this special treatment has undoubtedly involved a waste of scarce funds, skills and equipment while probably stunting the growth of the more efficient co-operatives. Finally, the price control and subsidy system has been too unwieldy an instrument to effect detailed changes in the composition of output. For ten years prices barely changed although production possibilities for different types of farm reflected rapidly altering conditions of accessibility, water and ability and labour costs or income expectations. Even the recent round of price increases is essentially reactive rather than strongly indicative and it would have been more sensible to reduce the range of crops covered to those that the system could really control than allow market-price trading of any surplus production.

The extent of progress towards equality of access to resources is even more difficult to judge than changes in agricultural productivity and production since only one detailed socio-economic survey has been conducted (SORGEAH, 1981) and that took no account of temporal change. However, broad indications of trends relating to three aspects of equalization can be given. First, the agrarian reform programme greatly reduced inequalities in land ownership by redistributing the largest holdings to either State farms or landless peasants. However, within the fairly generous limits set by the law on ownership there are still inequalities and these are threatening the already fragile structure of the co-operatives for some of the families with large holdings are directing communal resources to their own ends. This affects the second important aspect of equality – cash income. Again, there has been a significant reduction in income disparities since the removal of the sheiks and sultans but the recent trend would seem to be one of increasing divergence. This is due to the differences in landholdings within co-operatives, to growing disparities between the total income of co-operatives and State farms in more favourable areas and those outside them and, most importantly, inequality in access to migrant remittances. The last aspect of equality to be considered here is that of access to social services and the picture is rather better as the amelioration of the initial inequality has been gradual but continuous. Access to basic education and health services is still better in the more densely populated areas, male children still receive priority in education and the quality of the services is often poor but the overall impression is one of slow success.

Socialist transformation?

In terms of the goals which it has set for itself, the PDR Yemen State has had but limited success with its rural development programme. In the light of its initial problems, its limited resources and ambitious targets, this is not entirely surprising and should not be regarded as sufficient reason to abandon its efforts but rather indicates the need to reform and refine the policies. The nature of such reform depends not just on the perceived gaps and weaknesses in past policies but should also be guided by the long-term objective of constructing a socialist State. As the PDR Yemen is heavily influenced by the USSR's conception of socialism – both in terms of international political relationships and the day to day running of the country – the selection of rural-development objectives has been in line with the USSR's own. The sort of rural economy which is being envisaged is one of State farms each of which produces specialized produce for internal (and international?) exchange by using relatively mechanized techniques. This framework is to be co-ordinated by central control through both investment and production plans, according to nationally determined economic and political priorities. The evidence above suggests that such an economy is still a long way off as the mechanized State-farm system has still to prove its value in terms of raising productivity, the central planning system is not that effective in determining the nature and character of production and the construction boom is providing the sort of diversification that this model decries.

It is in light of this record that recent discussion of agricultural policies has become increasingly oriented around questions of market prices, better services for co-operatives and even – partly due to international agencies – greater local participation in agricultural management. Whether this will result in a reorientation of objectives towards at least part of the 'Chinese model' or leads to a growth of capitalism that will threaten the social progress already made is unclear at present. The determining factors will be the composition of the Yemen Socialist Party's leadership and PDR Yemen's international position rather than the dynamics of rural life themselves. What has happened in the past fifteen years has represented the first steps towards a socialist transformation and not the transformation itself. It could appear, with hindsight, to have been no more than sand reform.

Acknowledgements

I am grateful to the Centre for Middle Eastern and Islamic Studies, University of Durham (UK) and the Ministry of Agriculture and Agrarian Reform, Aden (PDRY) for financial assistance provided during a visit to PDR Yemen in 1981. I am also indebted to Yassin Nasser and Salwa Ben-Humam who have taught me so much about their country over the years. The views expressed here, however, are mine alone.

Notes

1. Although this summary is more comprehensive and up to date than any available in English outside the PDR Yemen's Ministry of Agriculture and Agrarian Reform, the World Bank or FAO, it has not been possible to draw upon or reference all the material originally at my disposal as much of this is confidential.
2. In addition to the acknowledgements noted above, I am also indebted to the participants in a conference on 'The Socialist Transition in Agriculture' (held in Cambridge, UK, in April 1982) for stimulating these preliminary reflections.

7
Urban and Regional Development in Zimbabwe

David Drakakis-Smith

Introduction

Zimbabwe has a land area of 391,000 square kilometres and a current population of some 7.7 million. A *per capita* GNP of $US630 makes it a less-developed country by global standards but places Zimbabwe above all other central and southern African states, with the exception of South Africa. Most of this national income is obtained not only from the export of a relatively wide range of primary commodities, both agricultural and mineral, but also from a number of manufactured goods (figure 7.1).

Unlike most of its neighbouring states, therefore, Zimbabwe does not rely solely on a few agricultural or mineral products for its export earnings but this does not imply that a general prosperity, in social or spatial terms, is characteristic of the country. As might be expected, expatriate enterprises continue to dominate the production of the major export commodities, such as gold, asbestos and tobacco, as well as the manufacturing sector. Moreover, in Zimbabwe expatriate domination has a very marked spatial aspect. It derives from the distribution of land between the white and black populations during the colonial period whereby the best farmland was appropriated by white commercial producers and the less-productive areas were allocated to Africans. This geographical distinction in land quality and ownership has been reinforced by trends in almost all other aspects of social and economic development, so that the principal urban centres are almost exclusively located in the former European areas.

This is not to imply that a dual economy exists in Zimbabwe. Although production and reproduction appear to be fairly distinct in socio-spatial terms, there continues to be a strong relationship between black and white, rural and urban, peasant and capitalist sectors of the political economy which is most manifest in the transfer of value through labour exploitation. The situation is further complicated in Zimbabwe by the division and conflict within expatriate capitalism between settler and neo-colonial interests. The nature of this conflict has strongly influenced the spatial dimensions of Rhodesian/Zimbabwean development.

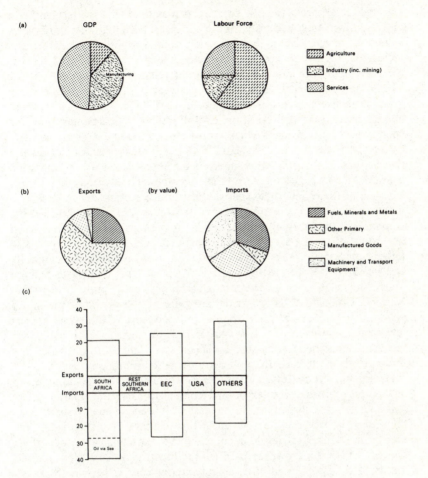

Figure 7.1 Zimbabwe: selected production and trade indicators
(Source: World Bank Development Report 1983.)

The socialist government of Robert Mugabe is pledged to a policy of reconstruction and redistribution through 'growth with equity' – reformist terminology which is an indication of the 'pragmatic socialism' which currently prevails in Zimbabwe. What this means in terms of regional and urban planning will be discussed later in this chapter but, at this point, it is worth noting that beneath the superficial optimism, which international comparisons with other African countries may induce, there are severe development problems facing the country.

Some of these problems stem from the deceleration in economic growth since independence in 1980. In part this is the consequence of cumulative external factors from the last decade, such as the rising scarcity and cost of fuel, which now accounts for 30 per cent of Zimbabwe's imports by value, but internal factors have also contributed to the overall difficulties facing the socialist government. Amongst the most prominent of these is the very rapid population growth, currently standing at 3.6 per cent per annum and expected to rise above 4 per cent by the next decade.

This massive growth, as elsewhere in the Third World, has brought about severe pressure on individual households, particularly in the communal or tribal lands where subsistence production is increasingly falling short of basic requirements. With the gradual relaxation of restrictive legislation this has consequently led to accelerated urban migration in search of cash supplements to family incomes. The socialist government is aware of these new socio-spatial shifts and is attempting to deal with them in its development programme. As in many African countries, the emphasis is on rural development. This is perhaps not surprising given the extent of rural poverty in the tribal areas of Zimbabwe, but it also reflects general Maoist tendencies within African socialism. Ideologically and politically, Robert Mugabe owes a heavy debt to the late Samora Machel of neighbouring Mozambique who protected Zimbabwe African National Union (ZANU) guerrilla bases during the period of the Smith government's unilateral declaration of independence (UDI). Over his last few years, however, Machel was forced by international capitalism towards a modified, pragmatic socialism (see Zafiris, 1982) which has undoubtedly influenced the policies pursued in Zimbabwe.

Given the recency of Zimbabwe's independence, the opening up of new opportunities for international capitalism, and the guarantees to white-settler capitalism that the Lancaster House agreement forced upon the new Zimbabwean government, it is perhaps not surprising that socialist influences in development planning, particularly as they affect regional and urban growth, are difficult to identify. Since 1980 a complex mix of forces from varied internal and external sources has shaped development and it would be counterproductive to examine them merely within the apparent dualistic framework of black, rural peasants and a white, urban industrial economy.

Although the focus of this particular analysis is the nature of urbanization and regional development, such an investigation cannot be separated from an examination of the articulation of three modes of production (monopoly capitalism, settler capitalism and peasant) as this has evolved over the 90 years of Rhodesia/Zimbabwe's existence. Within this general framework, however, the spatial linkages will be emphasized, particularly in their impact on the nature of the urban system and the problems of individual groups within this.

The evolution of socio-spatial inequality

The initial incursions of Europeans into what is now Zimbabwe occurred during the 1890s when the two main African tribal groups, the Ndebele and Shona, were subdued by force of arms by the British South Africa Company (BSAC). The principal goal of BSAC was the acquisition of precious minerals, particularly gold, but the seams in the area were more broken and less remunerative than those further south. During the first two decades of the present century, therefore, BSAC began to encourage and assist in white settlement, of what was by then known as Southern Rhodesia, in order to promote agricultural development for both domestic and external markets. African rights, although recognized, were not respected in this settlement process as whites proceeded to acquire large tracts of the most fertile land in the high veldt country.

Nangati (1982) claims that in these early years the peasant mode of production was quite successfully competing with the white settlers for the domestic market in the small urban and mining centres. Its success was also reducing the 'incentive' for black peasants to enter into wage labour in the mining sector. The result was a gradual spatial restructuring of rural production which began in the 1920s, when Southern Rhodesia was formally annexed by Britain, and reached its early peak in 1930 with the Land Apportionment Act which divided the country between the races. The area reserved for Europeans encompassed much of the best agricultural land and all of the principal mining and industrial sites; a small area of freehold land was allocated to a few small-scale black commercial farmers; the rest was designated Tribal Trust Land. This geographical proportionment was reconfirmed by the Land Tenure Act of 1969 which gave the races some 18.2 million hectares each (see figure 7.2) despite the 19:1 numerical ratio in favour of the black population.

The effect of this redistribution of land was twofold. First, it reserved the most productive primary areas for whites, and second it forced black peasants out of the market economy as producers. It is important to note here that the peasant cultivators were not even permitted to maintain their subsistence economy because the shift to the relatively poor areas of the middle and low veldt made subsistence survival impossible for many households, and forced male labour into the commercial agricultural sector.

This use of spatial planning to reorganize production and reproduction has been examined by Arrighi (1973) and Malaba (1981) and its 'success' can be seen in the rapid decline in the proportion of African earnings coming from the sale of produce (from 70 per cent in 1900 to 20 per cent by 1932). By manipulating space, the State acted on behalf of the white

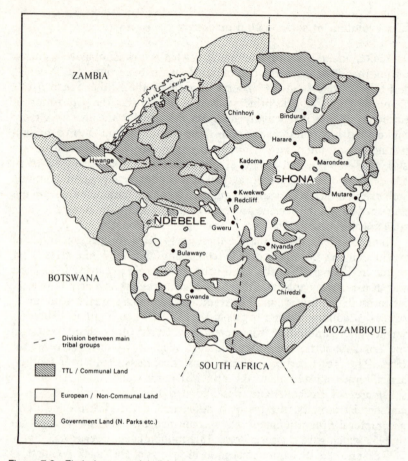

Figure 7.2   Zimbabwe: spatial divisions

primary producers, both to reduce competition in the domestic market and to create a cheap labour force.

This labour force was exploited in the 'absolute' sense by working long hours and by reducing to a minimum the contribution of employers to the means of reproduction. Labour was reproduced primarily in the subsistence agricultural system of the tribal trust lands where the rural household fed and sustained children into an employable condition, and looked after them when sickness or age reduced their labour capacity (Nangati, 1982). In short, white capitalist farms were/are sustained by a migrant labour force, the costs of reproduction of which were externalized.

These black labour reserves have been variously described as semi-feudal or semi-proletarianized, and as interdependent with the modern sector (Arrighi, 1973; Malaba, 1981), but the most convincing conceptualization is that which seeks to identify the situation prior to the Second World War in terms of settler colonialism.

Biermann and Kössler (1981) characterize settler colonialism in terms of what Forbes and Thrift (1984) would describe as a sub-mode of production – incorporating various capitalist relationships into a uniquely national combination. The economy is 'centred around the continuously reproduced exploitation of African labour power kept available at the cheapest possible price, [this] involves a low input of overhead capital' (Biermann and Kössler, 1981, p. 107). As a consequence there is a limited domestic market, little interest or incentive in full proletarianization of African labour and an inclination to spend on consumption rather than reinvest.

There is much truth in this description of the Rhodesian social formation but, as Baylies (1981) points out, the situation was not quite so simple or clear cut. Within the settler mode of production there were differences in outlook between those whose markets were primarily external (tobacco-producers) and those whose markets were internal (maize-producers). Obviously the adherence to the settler 'principles' identified by Biermann and Kössler would be stronger in the former than the latter, for whom the development of an African cash economy was important. Moreover, non-settler capital was always well represented in Rhodesia, particularly in the mining and tertiary sectors of the economy.

The urban system *per se* was a creation of settler colonialism. It functioned primarily in a compradore capacity to facilitate the export of various primary commodities and the import of consumer goods, and has always accommodated the majority of the white population, rising steadily from 52 per cent in 1911 to over 80 per cent today. There was from the beginning a small urban proletariat but for the most part this comprised skilled white workers who, Bermann and Kössler claim, identified themselves primarily with the agrarian-settler class rather than the non-settler capitalists, and certainly not with their black, urban counterparts. Not that there were many black employees in industry or tertiary activities. There were, however, large numbers of black domestic servants in the urban centres whose incorporation into the economy was organized in the same way as farm labour. Their residence in the white towns was tolerated only so long as they had a job (and therefore accommodation) and their families were not permitted to move in from the tribal trust lands. The nature of the relationship was clearly spelled out in the Master and Servants Act of 1901.

The articulation between the settler capitalists and the African peasant mode of production persisted through to independence. Indeed, the basic pressures upon tribal subsistence have increased dramatically since the 1950s as the rural population has expanded and the inherently poor tribal trust lands became further impoverished through over-grazing and over-cropping (Whitlow, 1980, 1982). As a result, by 1969 it was estimated that some 57 per cent of the tribal trust lands were considered to be over-populated. In contrast, production in the commercial areas had been sustained through heavy investment in fertilizers and irrigation, much of which was state subsidized. Not surprisingly, development potential was heavily concentrated in the European zones (figure 7.3).

Changes first began to appear in the settler political economy during the Second World War when the growing demand for semi-processed raw materials, and the curtailing of external supplies of consumer goods began to stimulate domestic manufacturing. This was further encouraged by the creation of increased 'domestic' markets in Southern Africa, particularly during the short-lived Central Africa Federation, and the growing availability of foreign investment funds, particularly from South Africa. This was, of course, an urban-based change – a growth of manu-facturing capital to challenge the domination of the white agrarian bour-geoisie.

The period from 1950 to the declaration of UDI in 1965 is marked by a series of contradictions and conflicts in the role of the State in Zim-babwe. On the one hand, labour movement into the cities was 'encour-aged' by legislation such as the Land Husbandry Act of 1951 which limited the size of agricultural holdings; on the other hand, the growth of 'surplus' labour in the cities was actively discouraged by vigorous enforcement of the vagrancy laws.

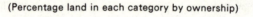
(Percentage land in each category by ownership)

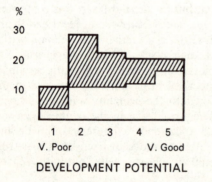

DEVELOPMENT POTENTIAL

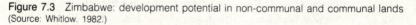
Figure 7.3   Zimbabwe: development potential in non-communal and communal lands (Source: Whitlow, 1982.)

In many ways, however, circumstances were changing irrespective of the wishes of the white agrarian bourgeoisie that formed the 'centre of gravity' of the class structure (Arrighi, 1973). Decades of increasing population pressure on the tribal trust lands had begun to lead to a relative decline in subsistence production so that by the 1960s landlessness had become a feature of the poorest areas. As a result the black migration to the main cities was beginning to expand and, despite the wishes of the administration, squatting and 'lodging' (sub-letting) was appearing in the more densely populated African townships of the capital (Smout and Heath, 1976; Nangati, 1982). However, tight controls ensured that no informal sector appeared, and the growing number of urban unemployed were directly used to keep urban wages low (Davies, 1978).

Nevertheless, manufacturing capital was beginning to challenge some of the basic principles of settler capitalism and class structure by encouraging the development of a black urban proletariat, the recognition of African trade unions and the emergence of a black petty bourgeoisie (Baylies, 1981). In short, it was a drift towards the neo-colonialism which had paralleled the 'independence' process in so many contemporary colonial territories. Settler capitalism thus perceived a threat to its political domination from this alliance of various nascent African urban classes and manufacturing capital (it is one of the lasting ironies that much of this foreign capital was South African). As a result the main settler classes, that is, the agrarian and urban bourgeoisies and the white urban proletariat, closed ranks and declared UDI to preserve their racial superiority in the political economy (see figure 7.4).

The actions by the settlers during UDI were, however, directed more towards internal rather than external measures, towards a firmer repression of Africans rather than a rejection of foreign, manufacturing capital. Thus the Land Tenure Act firmly established the identity of exclusively European lands into which Africans were only allowed on a contract labour basis. This, of course, meant an end to the labour-tenancies of previous years (in which African labour was 'paid' with strips of cultivable land), and as a result increased the number of landless households in the tribal trust lands, in effect creating a full agrarian proletariat compared to the semi-feudalized workforce of the past.

In reality, despite the legislative intent of these and other measures, the UDI period witnessed a continued movement of Africans, as individuals and families, to the urban areas. The process was further intensified by the gradual mechanization of many of the larger commercial farms. This constitutes a virtual admission of the redundancy of settler agrarian principles, the economic bases of which were maximum appropriation of labour surplus and minimum reinvestment. It was not, however, paralleled by an acceptance of the redundancy of racist principles. Cheap African labour was still crucial to the continued dominance of the settler

	Rural space		Urban space	
	African	European	European	African
Principal mode of production	Pre-capitalist subsistence	Settler-capitalism	Settler-capitalism	
			Manufacturing capitalism	
Pre-war classes	Peasant	Settler-bourgeoisie	Settler-bourgeoisie	Proletariat (domestic/tertiary)
		African semi-proletariat	Settler-proletariat	
Post-war classes	Peasant	Settler-bourgeoisie	Settler-bourgeoisie	Proletariat
		African proletariat	Settler proletariat	Bourgeoisie (limited)
			Manufacturing Bourgeoisie	
	Periphery		Internal core	Periphery

[_ _ _] UDI class-alliance

Figure 7.4   Zimbabwe: space, modes of production and classes

agrarian class and the rural wage-labour remains the largest occupational sector in the economy. In 1980, therefore, the 327,000 agricultural workers received only $2150 million in wages compared to the $2400 million received by the 160,000 factory workers ($21.00 = $US0.9).

As the above discussion indicates, the atavism of UDI has been unable to stem the growth of manufacturing capital in the main urban areas in Zimbabwe. Indeed, during UDI the manufacturing contribution to GDP rose from 18 per cent to 24 per cent (Stoneman, 1978). The UDI government thus became increasingly dependent on surplus appropriated through manufacturing and/or foreign-based production units. Given also that major sections of settler capitalism were export or international in orientation, it is not surprising that the settler state did little actively to discourage the growth of neo-colonialist forces within the country. It was 'the blatantly racialist dimension to the local distribution of resources [that] created a commonality of interest of the settler community as against the indigenous population' (Baylies, 1981, p. 121).

In one sense, therefore, UDI can be seen as a temporary halt on the road to neo-colonialism. The black urban population, largely as proletariat, continued to grow to the extent that the ratio of black to white urban residents increased from 3:1 in 1960 to 5:1 in 1980. Significantly, however, this new urban proletariat did not take a significant role in the independence struggle; instead the legislative and racial excesses of UDI stimulated a rural-based guerrilla movement. However, this may well have been ineffectual without the growing internal influence of neo-colonial groups more favourably inclined towards African rule than the hard-core settlers, and able to bring some pressure to bear in the form of international sanctions.

The accelerated phase of the liberation war from 1972 onwards was fought primarily in the rural areas. The various guerrilla groups slowly infiltrated from their external bases in Mozambique and Zambia into the northwestern tribal lands and began to harrass the more vulnerable farms in the settler districts (Martin and Johnson, 1981). Apart from a late and sporadic bombing campaign very little of the war was fought in white-dominated settlements, where the main impact was the growing drain on white salaried staffs through conscription (Cante, 1983). Given the growing importance to the settler economy of the urban industrial economy, this neglect of the urban sector is difficult to understand in retrospect. Internally, therefore, the years leading towards independence are characterized, on the one hand, by an intensifying guerrilla war against a socialist organized rural militia (part proletariat, part landless peasants), and, on the other hand, by growing liberal capitalist pressure for reform and expanded urban industrial growth. This liberal reformism did not extend to wage parity with whites, and the expanded production of the manufacturing capitalism was based as much on cheap labour as was settler agrarian prosperity (at independence, three-quarters of those in manufacturing earned less than $260 per month). What occurred, however, was a spatial shift in the reproduction of labour. Whereas formerly the predominantly male urban labour-force was reproduced through the subsistence activities of the peasant mode of production, it is now reproduced through an urban subsistence role performed by the non-proletarian members of the household. This primarily falls on women who reduce the costs of labour reproduction through their pursuance of peasant activities in the urban setting; for example, by urban gardening on unsettled land or the collection of fuelwood (see Mazambani, 1982).

Occasionally this liberal reformism has been incorporated into basic-needs dialogue in which the role of an expanded 'informal sector' is to be encouraged – ostensibly for the 'humanitarian' reason of letting the poor help themselves but, in reality, creating additional avenues for the transfer of value through an expanded urban labour reserve, an expanded urban consumer market and the evolution of a petty commodity production system.

It is doubtful whether the neo-colonial reformists within the State envisaged the spatial changes in which they were involved as clearly as this. There was certainly no regional and urban planning worthy of the name in the country before independence. Regional planning primarily consisted of 'accepting' that urban growth throughout the European areas would continue to occur and should therefore be 'planned' to avoid excessive concentration in Salisbury and Bulawayo. Most of the planning was physical and designed to cope with demands for accommodation. But the economic basis of such planning is clearly evident in the following statement of 'basic principles:

A regional plan that is oriented towards social objectives would be disregarded by the Government and its agencies, and to the private sector ... would be of academic value only. On the other hand, a regional plan containing proposals for stimulating the economy would be looked at far more seriously ... In short a regional plan ... should be made binding on the public and local authority sectors in order that it may guide and assist private sector development. Foreign investors should find such a plan to be a valuable aid to their own investment programmes. (Horenz, 1979, p. 262)

Despite the ostensible lack of regional or spatial planning in Rhodesia, the reality of 90 years of settler and manufacturing capitalism on the eve of independence was a clear spatial dimension to the class structure and modes of production across the division of lands and the location of the urban system (see figure 7.4). As in most Third World countries the modified pre-capitalist mode of production is predominantly rural, and the imperialist/neo-colonial mode of production is predominantly urban. Unlike most other countries there was a third mode of production, that of the settlers which was predominantly rural and which had evolved and was maintained through the manipulation of space. A further consequence of the control of movement between these spaces and the regulation of urban space was the virtual absence of a sector of petty commodity production.

The following sections will address two issues: First, what is the nature of post-independence changes to the social formation and how have they evolved, and second, what are the prospects for further socialist change in the future?

Post-independence changes in the Zimbabwean space economy

The present decade has witnessed a continuation of the spatial and urbanization processes which emerged during the 1970s under the liberal-reformist banner – accelerated by the rescinding of legislation restricting African movement, residence and property ownership. At present the combined weight of settler- and international-class interests at the Lancaster House negotiations has clearly muted the impact of the socialist government in changing the social relations of production. This is nowhere more evident than in the redistribution of agricultural land (discussed in detail below) which still reflects the European and Tribal Trust patterns formalized half a century ago.

Indeed, rural poverty has increased since independence due to the stagnation of employment opportunities on commercial farms as many of the settler agrarian bourgeoisie emigrate to the 'security' of South Africa. The result has been a notable rise in migration to the urban areas. This process is not clearly represented in the limited aggregated data available and the urban proportion of the rural population has remained fairly constant at around 20 per cent throughout the 1970s (table 7.1).

Given the very rapid population growth rate, however, this apparently static urban proportion represents a doubling of the actual population in the main towns over a period also characterized by a gross national migration loss of some 120,000 people since 1975. Most of the emigrants have been white, urban residents (Davies, 1981), so that the apparently steady urban population growth conceals a more fundamental change in the ethnic and class composition of Zimbabwe's towns (table 7.2). In this context, the replacement of whites by blacks is particularly noticeable in the smaller towns in the former European areas, where the security of the white proletariat and bourgeoisie have been more rapidly eroded.

Since independence this movement into the cities on the part of the black population of Zimbabwe has accelerated for the same reasons as it has in almost all newly independent Third World countries over the last 25 years, that is, the removal of restrictive legislation and the promise of employment, either in industry or in government positions no longer the exclusive prerogative of expatriates. As elsewhere, however, the pace of

Table 7.1   The urban population of Rhodesia-Zimbabwe

Year	Urban population Number (million)	Proportion of total population (%)	Proportion of white population in urban areas (%)	Proportion of non-white population in urban areas (%)
1904	—	—	58	1
1911	—	—	52	2
1921	—	—	56	3
1931	—	—	65	5
1941	—	—	70	7
1951	—	—	73	13
1961	—	17.5	83	17
1969	0.9	17.6	86	16
1974	—	—	88	20
1978	—	—	82	19
1982	1.7	23.0	—	—

Source: Central Statistical Office, Monthly Digest of Statistics, April, 1983

Table 7.2   The white population of Greater Harare

Year	Percentage	No.
1977	19.8	118,300
1978	18.7	113,400
1979	16.6	103,900
1980	15.3	100,000
1981	13.3	91,100

Source: Davies, 1981

the influx has exceeded the capacity of the cities to absorb large numbers of low-income migrants, in terms of jobs, housing and many other aspects.

Although national data are difficult to put together, it would appear that the already dominant cities of Harare, the capital, and Bulawayo have received the main impact of this in-migration (table 7.3). In 1982 the census population of Harare stood at 656,000 which represents an average annual growth over the last decade of less than two per cent. But this is misleading since the population of the capital declined in the mid-1970s and since the end of the war has increased at a rate in excess of five per cent per annum. Moreover, much of the migrational increase during the 1970s, on which the industrial growth was dependent, was deflected to the satellite town of Chitungweza some 20 kilometres to the south-west of the capital. The combined populations of Chitungweza and Harare amount to 830,000 people.

Further evidence of the rapid growth of the capital is the unprecedented appearance of larger squatter settlements on the edge of the city – the biggest of these, at Epworth, grew from 5,000 to 30,000 in 1982 alone. The numbers of households living as 'lodgers' (sub-tenants) in the low-income areas of the city has also increased rapidly since independence.

The growth in the urban population has also brought about changes in the composition of the labour force. In 1969 as a result of employers

Table 7.3   Populations and ratios of three largest cities, 1983

City	Population	Ratio
Harare/Chitungweza	828,600	2:1
Bulawayo	413,800	5:1
Mutare	70,000	

Source: Central Statistical Office, Monthly Digest of Statistics, April, 1983

offering hostel accommodation only to men, the male urban population was almost double that of women. As legislation permitting ownership and family residence was relaxed and removed, so virtual gender parity has been reached. However, bias is still clearly evident in the patterns of employment. Only about 13 per cent of the non-agricultural workforce is female, and this is also true of Harare where the figure is around 15 per cent.

Although this proportion does not appear to have risen much over the last decade, once again the aggregate figures mask considerable change. In absolute terms the number of women entering formal wage employment in the non-agricultural sector has almost doubled during the last five years to more than 100,000. Within this overall growth there has been a major switch in occupational patterns. Most significant has been the massive fall in the importance of domestic service (from one-half to one-fifth of all waged women) – a fall which is likely to continue with the flight of wealthy white families and the introduction of further socialist labour legislation in minimum wages.

The fall in female domestic employment has not been accompanied by a switch to manufacturing. In 1982, of the 76,500 employed in this category in Harare only 6,000 were women. The principal shift appears to be to the distributive trades, which is primarily licensed self-employment in petty commodity trading in food or knitted/crocheted articles. However, as noted above, one of the principal roles of women in the urban context has been to lower the cost of labour by supplementing the domestic costs of reproduction through non-remunerative but economically valuable (and valorizable) activities such as illegal urban gardening. It is clear, therefore, that the increased freedom given to family settlement in the cities has lowered labour costs to manufacturing capital by encouraging the development of urban subsistence and petty commodity sectors, primarily by women.

The position at present is therefore in a state of somewhat unstable stagnation. Settler capitalism is undoubtedly still a major force in the determination of the social relations of production in both rural and urban areas. In the former, it still dominates the commercial export market on which so much of Zimbabwe's foreign earnings depend; in the latter, the settlers are well represented in both the urban bourgeoisie and the urban proletariat where their dominance of trade unions is a major stumbling block to improved wage levels and advancement for African labour (Malaba, 1981). At the same time neo-colonial capitalism continues to grow through foreign investment – not only from British sources but also from South Africa (Stoneman, 1978). Nowhere is this more evident than in the tourist industry where most of the external leakages through repatriation of profits, sources of materials and so on, are towards South Africa, despite the new government's hard-line statements on economic relationships with that country.

There has been positive change for the African population, too, since independence and, despite the publicity given to the redistribution of land, it appears to be in the growth of an African urban proletariat and bourgeoisie that most rapid transformation is occurring. Although this has led to some personal gains in income, and a black bourgeoisie has begun to move into formerly exclusively white middle income suburbs (Harvey, 1982), the greater benefit has accrued to international capitalism which has increased access to both external and internal markets.

Present and future socialist policies towards spatial and urban development

The first question that must be addressed in this section is the extent to which the present government in Zimbabwe can be expected to pursue a socialist programme of redevelopment; the second is to examine the nature of that programme. Several background factors are relevant in answering these questions. One of the most important is that ideologically Mugabe appears to be committed to a rural (Maoist) form of African socialism that is prevalent to varying degrees amongst his socialist Southern African neighbours. Above all, this implies a strong emphasis on (eventual) rural development through collective- or commune-based production units. In real terms this poses a dilemma for Mugabe because he has inherited a national economy which is much more broadly based than that of Tanzania or Mozambique, with a comparatively well-developed industrial sector that even the socialists hope will play an important supply role within the Southern African Development Coordination Council (SADCC) group of 'front-line' Southern African states hostile to South Africa.

There are those, both inside and outside the socialist movement, who argue that uncontrolled enthusiasm for rural investment at the expense of the urban-industrial sector may well undermine this national advantage (see Yates, 1981). Such arguments have been underlined by the 'failures' of the Maoist/African socialism of Nyerere and Machel and their enforced retreat into a 'pragmatic socialism' acceptable to the donors of international aid (Weaver and Kronemer, 1981; Zafiris, 1982).

A second major factor influencing the nature of Zimbabwean socialism is related to this point, and concerns the degree to which the State's prosperity is still dependent on capitalist agriculture and industry – not only in terms of export earnings but also in terms of the national employment structure. Almost 60 per cent of the black workers in Zimbabwe are employed in occupations dominated by private-sector employers (usually white), both domestic and international. In this context, of course, Mugabe is bound by the Lancaster House agreements to guarantee compensation for displaced or nationalized private/white enterprises, so that in constitutional terms the development of socialist programmes has been firmly restricted.

Given such constraints, it is perhaps not surprising that socialist policies directed towards resolving the problems of socio-spatial inequality in Zimbabwe have been limited in both their scale and intensity. In addition, spatial development has been considerably affected by another ethnic factor which has been in existence much longer than the present State, that is the rivalry between the Shona and Ndebele tribal groups. The former comprise some three-quarters of the African population of Zimbabwe, the latter just under one-fifth, and each has a clear geographical distribution within the country (see figure 7.2). More importantly, each has a marked political allegiance – the Shona to Mugabe and the Ndebele to Nkomo. The traditional antagonism between the two groups was suppressed during the independence struggle, but has recently resurfaced in civil disorders in the south-west. Although there is little evidence as yet to indicate that these disturbances are linked to developmental discrimination (and perhaps more to indicate a South African involvement), it is certainly a factor which cannot be discounted within policy formation.

So, within these ideological, constitutional, ethnic and economic constraints: what are the policies of the socialist government with respect to alleviating regional and urban inequality; and what has been achieved since independence?

The principal task recognized by the government has been to improve conditions for Africans living in the former tribal trust lands – now designated communal lands. There are two distinct tasks involved in this respect. The first relates to the *in situ* improvement of amenities, services and returns to agriculture through investment in water provision, rural health facilities, schools, fertilizers and the like. The intention of the government is reflected in the 60–40 split in national investment in favour of the rural areas in sectors such as education and health, but in the immediate post-war years funds have been diverted to the more immediate demands of repairing war-damaged infrastructure.

More serious is the assertion that decades of government neglect and population pressure have affected the basic environment in certain areas to such an extent as to seriously jeopardize any attempts to improve conditions (Whitlow, 1980, 1982). The government has designated such areas 'intensive rural development areas' (IRDA) and intends to concentrate much of its efforts on assistance to their inhabitants. However, increasing criticism is being made by 'reformist' developers that the limited investment funds would be better spent on the 'least marginal' communal lands (see Kay, 1980; Whitlow, 1980, 1982; Cliffe and Munslow, 1981 for some discussion on this).

A more fundamental rural development programme has involved the redistribution of European land to hard-pressed black families. As previously indicated this formed a major negotiating issue at the Lancaster House discussions where the settler agrarian bourgeoisie managed to muster enough international support to insist that landowners must be 'fairly' compensated for any land that changes hands. Agreements in 1981 committed several advanced nations, notably Britain, to providing loan funds to finance this redistribution.

Progress since 1980 has, however, been slow and the government has been forced to split its rural land redevelopment programme into two stages. The long-term manifesto strategy of establishing collectives, co-operatives and State farming as an alternative to 'the present power and monopoly of capitalist agriculture' (Kinsey, 1982, p. 92) has been postponed in favour of the short-term tactics of redistributing land to landless peasants within the framework of the constitutional limits set by the Lancaster House agreement.

Current resettlement policy can be further divided into two programmes (see Kinsey, 1982 for a more complete discussion). The first comprises a three-year programme of intensive resettlement of returned refugees and landless peasants on purchases of former European land. The transaction is on a willing seller–willing buyer basis and as yet the administrative organization is cumbersome and slow, with eight different ministries involved. The second programme is an accelerated, emergency version of intensive resettlement, with minimum planning. A total of 2.4 million hectares is involved up to 1984 and the programme will utilize some 30 per cent of the total aid budget for the benefit of just 4 per cent of the total population of the country, leaving some 185,000 still needing resettlement (Zinyama, 1982, p. 152).

On the positive side, this programme has transferred in two years almost as much land from white to black control as occurred in the last 15 years following Kenya's independence. But, in comparison with the overall needs of the inhabitants of the communal lands and the stated objectives of socialist rural development, it is extremely slow progress, and must be seen as a 'success' for the reactionary forces of settler agrarianism. Although an estimated 60 per cent of the former European farmlands are un(der)used, resettlement from the communal lands has been moderately slow and expensive.

One unforeseen consequence of these problems has been a growing conflict between African rural classes. On the one hand, there are the thousands of returned refugees and landless peasants who have squatted on idle land since independence; on the other hand there are the market-oriented producers who have sought legal title to the land in order to set up small-scale commercial farms. At present the government is favouring the latter group, in conflict with its long-term socialist objectives, although as yet no land-tenure rights have been allocated and all resettled land is simply leased from the government.

As far as the focus of this chapter is concerned, the principal problem is that all contemporary changes are occurring in the absence of a comprehensive (socialist) plan for rural/regional development (Williams, 1982). Indeed, the defects of pre-independent regional planning in Zimbabwe still hold true, so that contemporary goals continue to reflect the reformism of those years, thus undermining some of the basic socialist objectives of the present government.

The stated goals of regional planning, therefore, continue to revolve around the merits and demerits of decentralization from Harare and Bulawayo to alternative urban growth centres. Prior to independence these strategy statements related primarily to growth poles in the European areas of the country (Smout and Heath, 1976; Horenz, 1979; Whitsun Foundation, 1979). The advent of the welfare/liberal reformists of the independence period has, however, extended this debate to the promotion of a series of growth centres throughout the country. This has reached its most sophisticated form in Davies (1981) who distinguishes between economic criteria for growth centres in former European areas and equity criteria for growth centres in communal lands. Even here, however, the main goal appears to be to avoid 'excessive' growth in Harare rather than to promote genuine regional development *per se*. As a result the growth poles in the communal lands, despite the equity concept, are selected on a 'best of a bad lot' basis.

As yet no national urban development strategy has emerged from the socialist government. The main policy document covering urbanization (Riddell, 1981) is essentially an aspatial, sectoral statement but, nevertheless, has considerable spatial implications. Perhaps the most important of these, at least in theory, is the introduction of minimum wage levels which are to go up by a sequence of annual increases. However, although the minimum levels (which have tended to become standard rates) have brought benefit to the farmhands and domestic employees, they have been pitched too low for industrial workers and some multinational firms have been able to *reduce* wages (Nyawo and Rich, 1981, p. 91).

Part of this problem is the result of a weak trade-union movement, still dominated by a settler proletariat, which is finding increasing rapport with the growing African bourgeoisie and skilled proletariat. The present socialist government has never held or sought strong urban socialist support and the present programme of establishing workers committees needs to be rapidly expanded if this is ever to emerge.

In this context, the government seems to place more importance on establishing urban influence and control through a reduction in the independence of the old municipal authorities. Such centralization of what were formerly bastions of white control is understandable and already the central government has achieved some success in reshaping development within the towns themselves. This has been particularly noticeable in terms of housing provision whereby the hitherto generous standards of municipal housing programmes have been reduced in design and costs, through the implementation of a core-housing programme.

The question which must be asked, however, is whether this is a truly socialist housing programme since most of the schemes were initially proposed before independence (Merrington, 1980, 1981). Various types of low-cost self-help housing schemes are now part of the established planning programme of organizations such as the World Bank and United Nations, and can be justifiably subject to the critique of incrementalist policies (see Bromley, 1979), that is, that by temporarily allocating slightly more resources to the urban poor, and thus reducing the likelihood of political protest, they ultimately serve to preserve the *status quo* in class relationships. A more socialist housing programme would surely insist on a more equitable distribution of building resources and in this context the planned introduction of government-organized building brigades seemed to be a more promising and positive measure. Unfortunately, this too has foundered.

Conclusion

The preceding discussion has shown that, prior to independence, development in Zimbabwe occurred within a clear spatial framework, with the articulation of modes of production and the evolution of class relationships taking specific forms within the African and European lands and the urban centres. The examination of the post-independence years unfortunately indicates that not only has the constitutionally hamstrung socialist government been unable to introduce substantial changes to the political economy of Zimbabwe, but also that it lacks a comprehensive socialist development strategy to encourage optimism that any significant redress will occur in the foreseeable future.

The present 'pragmatic socialism' is so strongly based upon reformist development strategies cultivated during the 1970s by liberal capitalist interests that it has permitted the continued entrenchment of capitalist classes in both the urban and commercial farming areas. In the latter, the settler agrarian bourgeoisie has maintained its control of major resources by an alliance with the expanding monopoly capitalism of the cities which in its turn is rapidly co-opting the growing African bourgeoisie and skilled proletariat within its class structure.

Ironically, the major challenge to this capitalist solidarity is coming not from the socialist government itself but from the regional disturbances apparently based on tribal antagonism which have caused the emigration of some agrarian settlers. In this context, it is crucially important for the present government to begin to strengthen its programme of socialization but within a framework that recognizes the fundamental importance of the relationship between capitalism and space. For Zimbabwean socialism must be pushed forward on two quite different fronts – against the agrarian settler rump on the one hand, and against urban industrial capital on the other. As yet the traditional ideological inclination of African socialism towards rural development has left the expansion of urban-industrial capital almost unhindered. It is because of such changes that there is an urgent need for a comprehensive framework for urban and regional development, if socialism in Zimbabwe is not to lose the momentum it still retains.

It is difficult to draw broad 'lessons' from the Zimbabwean experience for urban planning in other socialist states. To a certain extent, this is the consequence of the sheer recency of events and the lack of depth in perspectives on trends and policies. However, it is also a moot point whether such 'lessons' should be drawn from Zimbabwe, not the least because of its own peculiar geopolitics. The present government quite clearly has its ideological heart in the countryside that nurtured the liberation movement and yet Zimbabwe's international clout within Southern Africa, especially SADCC, is undoubtedly related to its urban industrial strength (Wellings and McCarthy, 1983). However, the new government has not yet come to terms with its urban sector. It is ironic in this context that the urban group with whom most *rapprochement* has occurred is the white capitalist, managerial and labour elite. The black proletariat are viewed somewhat suspiciously, while petty commodity production is looked upon with definite disfavour – witness the continuing police campaign against various 'informal sector' activities (see *Sunday Times*, 1983, p.5). It is hardly surprising, in view of these features, that the rapid urban population growth in Zimbabwe appears to have more in common with urbanization in capitalist economies than that allegedly typical of socialist states (Murray and Szelenyi, 1984). Given such circumstances, Zimbabwe appears to have little to offer other socialist states in terms of its (non-existent) urbanization strategies. On the other hand, it must also be admitted that there can be no common socialist strategy – each state must derive its own response to the pressures and problems of urban growth.

8
Guyana: Co-operative Socialism, Planning and Reality

Lesley M. Potter

When considering the nature and effectiveness of territorial policy in a Third World State which claims to be 'in transition to socialism', one must first address a number of fundamental questions:

1. On what criteria may one judge the State to be socialist or even 'protosocialist'? (White, 1983)
2. Given that the State is small, with the typically narrow, raw material-based economy of the dependent periphery, and given that it is riven with communal conflict, what constraints are likely to impede its progress on the non-capitalist path? What options are likely?
3. To what extent has the government's ideological stance changed over time? Does the State appear to be moving steadily toward a permanent socialist outcome or is this claimed 'socialism' merely a temporary phase?

Is Guyana socialist?

In classifications such as Wiles' *New Communist Third World* (1982), Guyana occupies at most a marginal or doubtful position. Clearly, it is not at present thoroughly integrated into the Communist system, at least from the point of view of maintaining close relations with the USSR, or being designated socialist in official Council for Mutual Economic Assistance (CMEA)/Soviet statistics. Aside from the attitude of the USSR, Wiles suggests both economic and political criteria which may be used to distinguish a socialist state. On the economic side, necessary characteristics are seen to be 'thoroughgoing nationalization, agricultural collectivization and the central direction of the economy by a non-market plan'. The political test concerns the nature and effectiveness of control by a Marxist–Leninist party which 'pervades every area of social and economic life and is the origin of all initiatives and where the state apparatus is a close but subordinate parallel of the party apparatus' (Wiles, 1982, p.

18). A *vanguard* policy which persecutes religion and in which the leaders are atheists is important for political control.

Both Wiles and White (1983), who offers a similar list of characteristics, make the proviso that most Third World socialist States have only gone part way along the path to these kinds of transformations, particularly in the economic sphere. What is important is the State's firm intention to proceed, even though complete transformation may take many years.

An examination of the position of Guyana reveals that some of these criteria apply fully, others partly and still others not at all. Nationalization has occurred quite extensively, particularly with regard to 'the commanding heights of the economy' – bauxite and sugar – once foreign-owned. Some local, private firms have also been taken over and the role of foreign banks and insurance companies have been 'miniaturized'. The economy is said to be trisectoral, maintaining a role for the private sector, as well as the co-operative. In practice the public sector is overwhelmingly dominant, although it is claimed that an expanded role for the co-operative may eventually be anticipated.

Collectivization of agriculture, or even milder land reforms, has not been attempted. Newly opened lands are supposed to be settled by co-operatives, but after some early failures, large State farms became characteristic. Worker–management structures have not been altered on the nationalized sugar estates, and peasant agriculture (producing rice for local and export markets and other local food crops) remains largely untouched. About one-fifth of all rice lands is occupied by tenants of large landlords, despite a clause in the 1980 Constitution guaranteeing 'land to the tiller'. Such communal plots as do exist are operated by organizations such as schools.

Central planning of the economy is carried out by the State Planning Commission, established in 1978. This body draws up the annual budget, allocates funds to be administered by appropriate ministries and government corporations, and sets production targets, wage rates and incentive payments for all government workers. External trade is government controlled, internal distribution and retailing increasingly so, although smuggling and blackmarketing are also widespread, forming a 'parallel economy'.[1]

Guyana professes a non-aligned foreign policy. Although its trade and aid relationships with the eastern bloc countries (including China) have recently increased, the West remains more important, especially the USA and the EEC. Since 1977, the country has had a series of dealings with the International Monetary Fund (IMF), the World Bank and the Inter-American Development Bank (IDB). The debt problem has become extremely serious and the IMF's usual severe prescriptions – devaluation, wage freezes, employment cuts and import limitations – have led to increasing internal hardship and discontent. Left-wing opposition critics have claimed that Guyana, under pressure from the World Bank and the IMF, is being forced to dismantle some of its socialist policies. Some joint ventures have been arranged between poorly achieving government-run industries (including bauxite) and foreign firms. The government maintains that it is not abandoning socialism and in November 1983 it suspended formal negotiations with the IMF and requested more help from the eastern bloc. At the time of writing (November 1985) a new agreement with the IMF had not been reached.

On the political side, the People's National Congress party (PNC) was declared paramount over the government, which became merely its 'executive arm' in December 1974 (the Declaration of Sophia). By various electoral manipulations the party has succeeded in remaining in government since 1964. In 1980 it brought in a new constitution which installed former Prime Minister Forbes Burnham as President, with sweeping powers. Although Guyana is not formally a one-party state, the PNC alternately ridicules and represses the two main opposition groups, both of which are left-wing in orientation.

The government has not moved against religion as such, although it has taken control of church-operated schools. In return for their support, it has even encouraged obscure and dangerous groups such as the People's Temple and the House of Israel to settle in Guyana. It has attacked the Catholic Church, one of its leading opponents on human rights' grounds, and restricted access of Catholic clergy to hinterland indigenous (Amerindian) communities. Hindu and Moslem groups have been factionalized, so that the organizations selected to represent those religious bodies are dominated by PNC supporters.

The Guyana government's most difficult problem in carrying out socialist policies (including policies of territorial planning) is that its main power base has been among urban Afro-Guyanese, while the Indo-Guyanese, who largely support Cheddi Jagan's People's Progressive Party (PPP), control the rural areas and produce the export crops of rice and sugar.[2] The urban bauxite workers, formerly government supporters, have been won over by a third political grouping, the Working People's Alliance (WPA), which offers a bridge of worker solidarity across the racial divide. The government has the support of a number of military and paramilitary groups but it now faces a serious decline in popularity among the general population.

In the first few years after the declaration of the Co-operative Republic in 1970, Forbes Burnham was able to appeal to nationalism and anti-colonial feeling as a basis for overcoming communal conflict (Milne, 1981). He did not entirely succeed due to the system of party patronage and pork-barrel politics which had become the norm. The nationalization of the sugar and bauxite industries generally had the approval of the masses and received 'critical support' from the PPP, but Burnham rejected any ideas of power sharing between the African and Indian blocs. Relations deteriorated as sugar prices declined and the country began experiencing economic difficulties, exacerbated by the IMF and its demands in relation to employment and wages. In a country where the State controls so many of the available jobs, the State must also face the consequences of unpopular industrial policies. Inability or unwillingness to implement promised wage increases, strike-breaking activities by the military and politically vindictive retrenchments led to widespread anger and productivity decline. Production has become erratic while emigration continues apace, affecting all levels of the workforce. Recent grudging wage adjustments may not be sufficient to restore worker morale.

The behaviour of the PNC regime would seem to be an almost classic case of the kind of Third World State capitalism noted by Mandle (1982) and earlier discussed by Petras (1977). The failure to alter the relations of production and the position of the working class in the major nationalized industries, the increasing tendency of the military and paramilitary organizations to enforce the power of the State have led to confrontation. 'The masses directly confront the state as their employer and the state confronts the workers as the source of capital expansion' (Petras, 1977, p. 16).

Kalecki's (1972) theory of the 'intermediate regime' (see also Jameson 1980, 1981), is illuminating in its emphasis on the interplay of class forces. A lower-middle-class government manages to retain power by supplanting or buying off the various forces which are likely to oppose it – big business on the one hand and the working class or peasantry on the other. Thus the PNC regime has somehow to try to placate the Indo-Guyanese rice growers and sugar workers, at the same time as maintaining the support of the black urban middle class. Local business interests are weak and largely supplanted, but international big business, in the guise of the World Bank and its agencies, is posing a direct threat to the stability of the regime which is more difficult to overcome. Jameson suggests that poorer groups may be 'bought off' through policies directed towards satisfying their basic needs, while denying them any real participation. Although this may have worked during earlier years, the economic situation has recently forced abandonment of many such policies. Similarly, public service employment, formerly swollen by political and racial patronage, has been cut back savagely to the extent that the Party's middle-class power base has been eroded.

Thomas (1976) and Ohiorhenuan (1980) examined the possibility of Guyana fitting the model of the 'non-capitalist path', whereby the transition to socialism may be achieved in Third World societies in which the industrial proletariat is very weak, through an alliance of workers and peasants, thus by-passing the stage of full capitalist development. Thomas saw the racism in Guyanese society, which is rooted in the social organization of production, as being a severe impediment to working-class unity.

He stressed the need for land reform, diversification in industry and agriculture and workers' control of the production process, before the transition to socialism could begin. Land reform was singled out as the most important test of the extent to which a genuine reorganization was taking place.

There have been two major shifts in ideology in the fourteen years since the declaration of the Co-operative Republic in 1970. The first seven years (1970 to the end of 1976), a time of nationalism and decolonization, saw a concerted attempt by the government to assert sovereignty over the whole territory by breaking out of the coastal strip – the main focus of activity under colonialism – and settling people in the interior. A further expression of sovereignty in those years was the nationalization of sugar and bauxite, prime symbols of colonial domination. The Second Development Plan (1972 to 1976), with its basic needs 'people-oriented' approach was imbued with the philosophy of self-reliance and the building of national self-confidence. The over-ambitious targets of the Plan were not met, but many projects were initiated, thanks largely to the high sugar prices of 1974 and 1975. The co-operative was singled out as the major institution for promoting a more egalitarian and social structure. The second period (from 1977 to 1985), saw a deterioration in the terms of trade and economic recession, a complex relationship with the IMF and an increased commitment to what was termed 'socialism' but was primarily social control. This was implemented through central planning and the new constitution, with its closer attention to local and regional economic organization and, theoretically, grassroots involvement in the development process.

It was also a time of political turmoil and increasing repression, as the PNC sought to retain power at all costs in the face of mounting pressure from opposition groups. The doctrine of 'party paramountcy' was invoked to equate opposition criticism with anti-nationalism, especially in the face of external threat (real or contrived). Threats were perceived to be coming, both from Venezuela (involved in a long-standing border dispute) and from the United States, from whence the Reagan Administration was believed to be working to destabilize the government. The country has been experiencing a very critical time, in which a further shift towards the enunciated socialist goals could take place. The alternative, which is being sought by the World Bank and the IMF, is a reversion to privatized ownership of production and the re-establishment of dependent capitalism. The recent death of President Burnham has raised further questions about Guyana's future direction.

Writers such as Petras (1977),, Wolpin (1981) and Berberoglu (1979) stress the instability of State capitalist regimes, their tendency to make sudden shifts to left or right and the unpredictability of the outcomes. In the case of Guyana (but hardly unique to that State) a further problem which makes prediction hazardous is the failure of rhetoric – particularly the official pronouncements through the controlled media – to match up with reality or even intended policy. In 1981 Wolpin rated the survival chances of the present regime as low. Certainly the Grenada invasion must have induced considerable nervousness regarding US intentions. For the same reason, it is unlikely that the USSR and other eastern bloc states will rush to Guyana's assistance. Planning policies must thus be examined bearing in mind the uncertainties of the State's future course.

Socialism and territorial planning

The main features of socialist policies of territorial planning as enunciated for Eastern Europe and the USSR by Musil (1980) and for Cuba by Slater (1982) and Gugler (1980) include the following:

1. reduction of territorial and regional inequalities by direct aid to backward areas and by redistribution policies;
2. the breaking down of the separation between town and country by 'diffusing the civilizing qualities of town life' to rural communities and concentrating such communities to eliminate the smallest settlements;
3. planning the entire settlement system with establishment of strong regional centres;
4. slowing the growth of larger agglomerations;
5. allowing decentralization of decision-making power away from the centres of the regions.

In order to examine the relevance of these policies to Guyana, one must first analyse the territorial structure inherited from colonialism, and then assess the ways in which socialist-type policies may have been advocated to redress regional disparities. It must also be borne in mind that the policies listed above are not uniquely 'socialist'. Uneven regional development and urban primacy are frequently characteristic of peripheral capitalist states as they achieve independence; no matter what political system they subsequently follow, the desire to achieve national unity is likely to ensure some effort to reduce major inequalities. But ideology, and control by government over the population and its distribution, may well be decisive in the implementation of such policies. This is forcefully illustrated by extreme cases, such as the de-urbanization of Kampuchea. In general terms, if the nature of the capitalist mode of production induces inequality (Soja, 1980), it is only in non-market economies that these inequalities can really be overcome. A further question is the origin of such planning policies, whether top-down, i.e. imposed by the government, or bottom-up, through participation at the grassroots. Top-down policies have generally been the norm, in the USSR as in the west; functioning grassroots policies are few, but are said to exist in Malaysia as well as Cuba and Tanzania (Stöhr and Taylor, 1981; Slater, 1982; Weaver and Kronemer, 1981).

In recent years there has been an attempt to direct bourgeois planning models in Third World countries away from the growth-pole concept (with its underlying aims of economic growth through industrialization and urbanization) towards a more needs-oriented strategy of 'agropolitan development' (Friedmann and Douglas, 1978). The search for an alternative to the industrialization approach has resulted from a realization that urban–rural disparities have been increasing, with continuing serious poverty in rural areas. It is argued that agriculture must become the 'propulsive' sector of the economy, with self-sufficiency in food a major aim, and the agropolis or city-in-the-fields responsible for secondary processing, marketing and transportation functions. The policy is supposed to be bottom-up, addressing local needs and mobilizing local resources. The major problem in its implementation is that it is predicated on a programme of land reform, so that rural surpluses are to be retained in each local district and flows of rents and other resources to the large cities must be stopped. It is in fact a country like Cuba, which has undertaken a programme of agrarian reform as 'part of a package aimed at the transformation of rural economy and society' (Slater, 1982, p. 20) which can serve as a model for agriculture-based planning policies. Even here it appears that the devolution of power to the grassroots has not yet occurred.

Cuba is believed to have provided the inspiration for Guyana's most recent planning strategies. In order to establish a baseline, it is useful to look at the situation in 1970, the year in which the Co-operative Republic was proclaimed and just four years after independence from Britain.

The territorial structure of Guyana in 1970

The settlement pattern which had evolved under colonialism featured the extreme concentration of population and economic activities on a narrow strip of reclaimed swamp land along the Atlantic seaboard, barely 250 kilometres long and 60 kilometres wide. On the rich clay soils of the coast, estates (initially of cotton but quickly replaced by sugar cane) were laid out by African slave labour for their European owners in the eighteenth century. Following emancipation in 1838, peasant villages were established in the same area by the ex-slaves. East Indians[3] who had completed five-year indentures as estate labourers, began setting up their own villages after the ending of automatic re-indenture in 1870. All were still reliant on estate wages, achieving only partial independence from sugar. Rice became established commercially only after the end of the nineteenth century, when a sugar depression resulted in the failure of many estates. Rice-growing became the East Indians' 'safety valve' when sugar work was lacking, and a similar employment alternative in the form of short-term forays into the interior was permitted to the Afro-Guyanese population under strictly controlled conditions. From the interior they extracted gold, diamonds and balata, a kind of rubber. Small interior settlements were formed to regulate and police these activities but most workers retained their homes in the coastal villages. Georgetown meanwhile grew as a trading centre from whence the surplus value extracted from the colonial labour force was remitted to England, in return for imports which included food as well as manufactured items (Potter, 1984). As the major urban focus, Georgetown's primacy was unchallenged. It was the seat of the local legislature, the hub of whatever transport network existed and the home of the black middle classes. New Amsterdam, the only other town, stagnated at the mouth of the Berbice River for most of the nineteenth century. Formerly the capital of a separate colony, its trading activities had languished since Berbice joined Demerara and Essequibo to form British Guiana in 1831.

More recent additions to the settlement pattern included the foreign enclave mining centres, exploiting the sub-coastal bauxite of the Upper Demerara and Berbice rivers and the manganese of the north-west. These settlements came into prominence only during and after the Second World War; the mining was open-cut and the deposits extensive enough to warrant permanent urban development (Potter, 1982). In 1960 an alumina plant was built at Mackenzie (later renamed Linden) 100 kilometres up the Demerara River. This brought an influx of population to that area, and an extensive squatter settlement mushroomed across the river providing services to the company town. By 1970 Linden had become the second largest urban centre and the only moderate-sized population nucleus away from the coast.

Indigenous Amerindian shifting cultivators also inhabited the interior, living in scattered villages. Colonial policy had been aimed at concentration and sedentarization of formerly nomadic groups so that permanent larger settlements could be created and provided with services such as Christian churches and schools. The Amerindians were involved to a varying degree with the money economy: in the north-west district, for example, they provided a convenient agricultural labour force for pioneering coastal settlers, while in the Rupununi savannas of the south, some worked as cowboys on European-owned cattle ranches. Amerindian concentrations in these regions, close to the borders with Venezuela and Brazil, were of strategic political importance.

The census of 1970 defined 29.4 per cent of the population as 'urban'; almost 80 per cent of urban dwellers were located in the city and suburbs

Table 8.1   Percentage distribution by ethnic group and urban rural residence

	Indo-Guyanese	Afro-Guyanese	Mixed	European and other	Chinese	Other (including Amerindian)
Total population (699,848)	52	31	10	0.5	0.5	6
Total urban[a] population (205,774 · 29.4% of total population)	23	54	18	1	1	3
Total rural population (494,074 70.6% of total population)	64	21	7			7

[a] Urban: Georgetown, City and Suburbs, New Amsterdam, Mackenzie (Linden).
Source: Commonwealth Caribbean, 1970

of Georgetown. Table 8.1 indicates the marked racial differences in the rate of urbanization: although East Indians or Indo-Guyanese made up 52 per cent of the total population, they constituted only 23 per cent of the total urban group, while people in the Afro-Guyanese and 'mixed' (usually part-African) categories showed the opposite distribution, forming 41 per cent of the total population but 72 per cent of those designed as urban. It has been pointed out (UNDP, 1977b, no. 4) that the rate of growth of Georgetown and suburbs had in fact been quite low during the decade 1960–70 – a mere 1.01 per cent per annum, compared with 4.4 per cent for Linden. The slow growth of Georgetown (or at least the statistically defined city), was balanced by a very rapid increase in the nearby commuting zones of villages to the east and south, technically in the East Bank and East Coast districts. This partially reflects the 'hemmed in' nature of the capital, surrounded closely by sugar estates which only gradually released land for urban housing. The balance of the city's workforce spilled over into the villages, strung in linear fashion along the Atlantic seaboard and the lower reaches of the Demerara River. Village populations were either Afro- or Indo-Guyanese: serious racial disturbances during 1962–4 had resulted in virtual segregation of the two groups, and much movement of minorities from formerly mixed villages (Potter, 1979).

Because of some persecution of Indo-Guyanese in Georgetown during the disturbances, many of this group preferred to live in their villages outside the city, although taking advantage of urban employment opportunities. Such commuting behaviour has caused an under-estimation by the census of total levels of urbanization and some exaggeration of urban–rural racial differentials. The beginnings of decentralization of industrial activity into these nearby villages, with an accompanying decline in their agricultural production, are further reasons for re-classifying the entire district within about 30 kilometres of the capital as 'peri-urban'.

Concentration in Georgetown of salaried employment in the managerial and professional categories, plus the high wage rates paid to all classes of workers by the Canadian bauxite company at Linden, were partly responsible for clear differentials between urban and rural average incomes. Within the rural districts of the coast, those peripheral areas lacking sugar estates and producing only rice or local food crops tended to be depressed (especially in the western county of Essequibo), when compared with the more central sugar growing areas. Most of the interior was remote and undeveloped, the information about its resources, especially its soils, very patchy and imperfectly understood. It could scarcely be compared with the coast in terms of income, as much production was at subsistence level.

There was ample scope for redressing some of these regional inequalities by the adoption of socialist planning policies. The question was whether the political will existed for the ruling party to achieve much in this direction, given the location of the main racial groups and the urban power base of the PNC.

Planning in the Co-operative Republic, 1970–7

When the Co-operative Republic came into being in February 1970, Guyana was partway through its first post-independence development programme, scheduled to run from 1966–72. The strategies adopted by this programme emphasized employment creation through economic expansion. This was to be achieved through both industrial development and agricultural diversification, mainly by an extension of cropping into new areas. It was anticipated that considerable amounts of foreign capital would be available, both from aid and direct investment. But by 1969 it became clear that foreign industrialists were not keen to invest in Guyana (the new industrial estate was largely empty). The 'easy' projects – big infrastructure works – had been completed, and the flow of external aid was drying up. The basic needs of the population were still only partially met and ethnic mistrust remained strong.

After 'winning' the election of 1968 with a clear majority[4] Forbes Burnham was able to govern without his former right-wing alliance[5] and confidently initiated a major move to the left, announcing that the country would become a Republic, operating under the ideology of 'co-operative socialism'. In an important speech in August 1969, Burnham defined socialism in the context of Guyana as being 'a rearrangement of our economic and social relations ... which will give the worker, the little man, that substantial and preponderant control of the economic structure which he now holds in the political structure'. He identified the co-operative as being 'the means through which the small man can become a real man' (Burnham, 1970a, p. 157). In a further speech Burnham revealed the nationalistic nature of his conception of 'socialism': 'The PNC is a socialist party. Socialist, not in terms of any European or North American definition which others may seek to thrust upon us, but in terms of our own social needs and wants in creating a just society for the people of Guyana' (Burnham, 1970b, p. 10). With the announcement of the Co-operative Republic, a development programme which was more 'people-oriented' was felt to be necessary; the 1966–72 plan was shelved and planners began working on a new set of guidelines under the co-operative socialist ideology.

The Second Development Plan, 1972–6, while noting the same economic difficulties as in the previous programme – the narrow export base and high unemployment – laid more emphasis on social problems. These were identified as racial mistrust, factionalism and lack of national pride, so that imported goods, especially food and clothing were favoured over local products. The strategy was primarily to be one of self-reliance in achieving 'basic needs': the goal was to feed, clothe and house the nation by 1976. Hinterland settlement, through planned regional development, was to receive high priority: such settlement was preferably to be organized in co-operatives. The objectives of the plan emphasized the creation of employment opportunities, equitable income distribution, equitable geographical distribution of economic activities and the establishment of a foundation for self-sustained economic growth (Co-operative Republic of Guyana, Ministry of Economic Development, 1973, p. 81).

To implement the third objective, the country was to be divided into twenty 'production regions', with agricultural targets and suggested crop types drawn from the 1965 Reconnaissance Soil Survey, carried out by FAO. Five of the regions were located on the coastal plain and four along the major rivers (collapsed into one coastal and one 'riverain' region in figure 8.1), while the remaining eleven covered the rest of the interior. It was anticipated that the administration of each region would identify and mobilize local resources, develop farm prototypes and achieve compulsory growth targets by means of increased yields and the bringing of new lands into production. The more even spread of economic activities, which it was hoped would result, was aimed also at 'reducing pressure on the social facilities in the urban areas' (Co-operative Republic of Guyana, Ministry of Economic Development, 1973, p. 126).

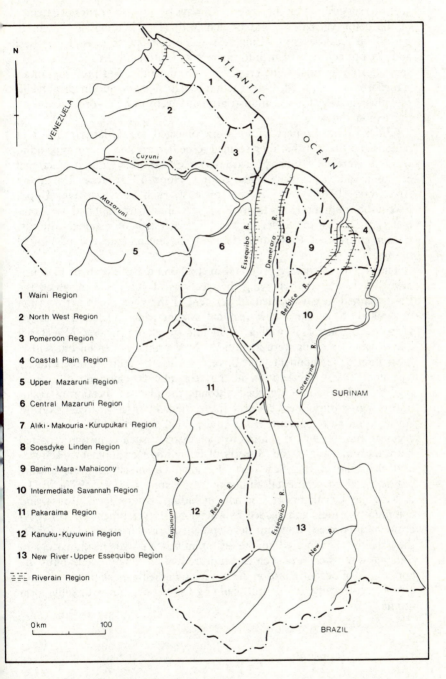

1 Waini Region

2 North West Region

3 Pomeroon Region

4 Coastal Plain Region

5 Upper Mazaruni Region

6 Central Mazaruni Region

7 Aliki - Makouria - Kurupukari Region

8 Soesdyke Linden Region

9 Banim - Mara - Mahaicony

10 Intermediate Savannah Region

11 Pakaraima Region

12 Kanuku - Kuyuwini Region

13 New River - Upper Essequibo Region

Riverain Region

Figure 8.1 Guyana: prospective agricultural regions

It was suggested that there were two ways by which a greater occupancy of the country's lands could be achieved: (1) a gradual movement to peripheral localities still within reasonable distance of coastal facilities; and (2) a direct movement into the remote hinterland. In the latter case, specialization in high-value crops was recommended and provision of air transport was promised. Construction of new housing and medical facilities plus small-scale local industrial plants were to be co-ordinated with the regional system.

Critics of the proposed framework objected to the large number of regions and the fact that they seemed to be demarcated on soil type only, without reference to other physical resources and often without functional economic links to existing urban centres (Hope, 1975; Hope, David and Armstrong, 1976). Some of the 'regions', such as New River–Upper Essequibo (number 13 on figure 8.1) were almost uninhabited and completely undeveloped: to set up the full machinery of regional administration in each of these areas seemed an uneconomic use of limited resources.

The regional structure proposed in the Second Development Plan was never implemented. Following the elections of July 1973 in which the PNC claimed to have gained 70 per cent of the vote, a cabinet decision was taken to set up a new regional system which included elements of political control as well as economic development. Six Ministerial Development regions were identified, three primarily coastal but with a 'near interior' component, and three covering the more remote interior (see figure 8.2). Acceleration in the economic development of the six regions was to be accomplished 'through the planned location of major impact agricultural projects and the settlement of new areas' (TAMS Agricultural Development Group et al., 1976, II, pp. 16, 20). A resident regional minister was appointed to each district: such ministers reported both to cabinet and to the Ministry of Economic Development; each was also chairman of the local Party Regional Organization, thus uniting political and governmental authority in the area (UNDP, 1977a, p. 1). A Regional Development Council, including various local groups and community leaders and headed by a Regional Development Officer (government-appointed) drew up development plans but received no funds, these being all allocated to the sector ministries with headquarters in Georgetown, where any such expenditure had to be approved. Thus the appearance of decentralization did not include the financial power to carry out projects locally: it was still planning from above but with some local inputs.

Figure 8.2   Guyana: administrative regions, 1973–80

Assessment of the Second Development Plan, 1973–7

As the Plan was not drawn up until early 1973, an extra year, 1977, was allowed as a roll-over before the beginning of a new plan period. Performance in two major areas of plan policy will be examined: interior settlement and development and population redistribution, including redistribution away from the central urban area of Georgetown and its surrounds; and satisfaction of basic needs with reduction of inequalities in distribution of income and facilities.

Population redistribution

Although co-operatives had been visualized as the main vehicle for interior development, few seemed to have been successful. In the north-west region, attempts were made at agricultural development through co-operatives in order to stabilize the former manganese mining population of Matthews Ridge–Port Kaituma, following the withdrawal of the American company involved. Most of these co-operatives had collapsed by 1973, the operations being taken over by government corporations which set up large State farms. Lack of leadership experience and problems with viability following a decline in soil fertility were major reasons for failure. Co-operative settlements along the highway connecting Linden with the coast also struggled to maintain production on the sterile white sand which characterized the area. While combinations of poultry and cropping enterprises seemed to be a possible solution (with ample pen manure), many farmers either returned to their coastal villages or lived in Linden and worked their land at weekends. Only one per cent of the projected population was actually resident (Downer, 1982).

It soon became clear that redistribution of people from the coast into the interior was not occurring to any extent. Lack of detailed knowledge about interior districts necessitated slow progress in planning resettlement projects; limited existing infrastructure and unfavourable psychological attitudes on the part of 'coastlanders' constituted other obstacles.

As coastal unemployment (especially among urban Afro-Guyanese) was one of the problems which government had hoped to solve by means of the interior agricultural co-operatives, a different approach was tried. The former Youth Corps, which transferred unemployed youth into a non-coastal location and taught various skills, was expanded into the Guyana National Service, an organization said to be modelled on its Tanzanian counterpart. Beginning in 1974, four National Service centres were established in the interior. Large-scale agricultural production was attempted, especially at Kimbia in the intermediate savannas of Berbice, where cotton was grown as part of the Feed–Clothe–House (FCH) programme.

National Service was made compulsory for all university students and seekers after government employment, thus exposing a wider cross-section of Guyanese to interior conditions, but most had no interest in settling there.

The intermediate savanna district was the scene of a number of agricultural experiments between 1972 and 1980 (see figure 8.1, region 10). It was estimated by the consultants to the Intermediate Savannah Agricultural Development Project (TAMS Agricultural Development Group et al., 1976) that the 'induced population' in the savannas was about 4,000 permanent residents and 900 transients (National Service and Army), as a result of the projects being carried on there. With another 4,000 in the Matthews Ridge–Port Kaituma area and a scattering around other centres, the total population induced to farm in the interior up to 1977 could not have numbered more than 10,000. Figure 8.3 shows the still markedly skewed distribution in 1976. By the census year 1980, those numbers would have declined again with the abandonment of uneconomic projects (but census figures in support of this are not available). The TAMS' report was pessimistic about the viability of almost all existing activities in the intermediate savannas, but it was acknowledged that economic considerations might not always be dominant. The authors stated 'Government must decide how high a priority to assign to the development of the Intermediate Savannah area as part of its programme of decentralization ... whether the financial losses and limited national benefits are too high a price to pay ...' (TAMS Agricultural Development Group et al., 1976, II, p. 185).

The most stable elements of the non-coastal population were the Amerindian communities, who might reasonably be expected to play a larger role in the development of their districts. In 1976, sixty-three Amerindian communities received title to their village lands, vested in their community councils. The boundaries of these lands had been decided upon following an Amerindian Lands Commission in 1967, in which each group was asked to estimate future needs. Invariably their estimates had been reduced in fixing village-land boundaries, the government's aim being to encourage adoption of permanent cropping rather than a continuance of shifting cultivation. The lands could face forfeiture if the minister responsible was satisfied that the community in question 'have shown themselves by act or speech to be disloyal or disaffected towards the State' (Guyana Government, 1976, p. 7). Given the doctrine of the paramountcy of the Party over the State, enunciated in December 1974 (Burnham, 1975), this could be interpreted as disloyalty to the ruling party. Amerindians in the Upper Mazaruni district were not included in the issuing of land titles because of the uncertainty surrounding the fate of their villages if a planned hydropower dam were to be constructed. That project has been shelved but the people remain outside the protection of

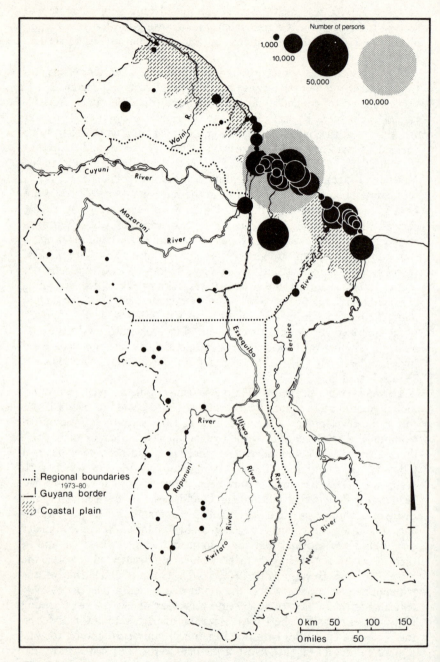

Figure 8.3   Guyana: population distribution, 1976

the Amerinidian Lands legislation and are now under increasing pressure from the uncontrolled encroachment of gold and diamond miners. The Guyana Human Rights Association's 1985 Annual Report has termed the result 'a complete disaster' and 'a pattern of ethnocide' (GHRA, 1985, p. 18). On the other hand, those Amerindian groups who received their land titles have the power to restrict entry of outsiders.

Integration of Amerindian groups into the regional system has been encouraged, increasing their dependency on the coast for certain processed food items and drawing them further into the cash economy to solidify political control. Amerindians were exhorted to produce cash crops for sale on the coast or in interior centres, crops which sometimes rotted on airstrips waiting for promised transport. However, the amount of surplus produce shipped from the interior to the coast has remained small; according to 1977 data, the volume, largely beef and balata products, was only about one-third of the movement of food, fuel and other produce in the opposite direction (Guyana Airways Corporation, 1977).

While the 'thrust into the interior' may have brought about some minimal population redistribution, there is no evidence of any major shift along the coastal strip itself, away from the primary city. The nationalization of various activities and the setting up of a number of public corporations in their place resulted in 80 per cent of the economy being under government control by 1976. If anything, the centralization of management and professional activities in the city increased and there was initially a burgeoning of the public service. However, the dominant role of the public sector in providing employment meant considerable government power over the workforce, a power which became apparent when retrenchments or 'redeployment' became effective in 1977 at the insistence of the IMF.

Comparison of the results of a Labour Force Survey held in 1977 (Caesar and Standing, 1978) with employment figures from the 1970 Census indicates a decline in that section of the labour force designated 'urban'[6] from 31.8 per cent to 27.9 per cent, and an overall drop in numbers of people employed. The latter decline was confined to males, the number of employed females in fact rising, although participation rates were still low. It is difficult to draw conclusions from these figures without the 1980 Census as a check, but that information is still classified. Nevertheless, the figures which have been 'leaked' indicate a very small decrease in the overall total population between 1970 and 1980, most of the natural increase being removed by emigration.[7] Given that such migration may be expected to occur disproportionately among the more educated sections of the population, a higher rate of movement from the urban districts, especially Georgetown, might be anticipated (UNDP, 1977b, no. 4). This heavy emigration, coupled with differential fertility levels between urban and coastal rural areas (Singh, 1979), and the likelihood of high fertility among Amerindians in the interior,[8] might in fact be changing the proportions of the population resident in the different districts and taking some of the pressure off the primary city, though not for the reasons anticipated by the planners. Rural–urban wage differentials were still considerable in 1977, with Linden continuing to show much higher rates than elsewhere, despite nationalization of the bauxite industry (see figure 8.4). Male–female disparities in mean wages were also marked (table 8.2). Differences between the city and suburbs of Georgetown result from construction of elite housing areas on former sugar-estate land both within and just beyond the official city boundaries. Although such housing is hardly consistent with socialist ideology, it must be admitted that on the whole, economic differentials are lower in Guyana than in many Third World States. The continuing flight of the middle class and government control over employment, together with accompanying wage freezes, has tended to reduce inequities. The major gap lies between the bulk of the workforce and the bureaucratic party elite.

Basic-needs fulfilment

A recurring theme in the literature on the socialist Third World is that such societies tend to perform very well in satisfying the needs of their populations for basic food, housing, pure water, medical and educational facilities, even though their economic achievement may fall below that of their capitalist counterparts. Thus Jameson (1980, 1981) rates Guyana quite high on a basic needs' scale, though not as high as Cuba. He uses the Physical Quality of Life Index (PQLI) based on infant mortality, life expectancy and literacy. It must be noted, however, that some Guyanese achievements, such as high literacy levels, occurred well before the period in question and cannot be attributed to government policies. Further, the figures are averaged and may mask important regional or group inadequacies.

Guy Standing (1979), while using somewhat similar data, is more cautious about Guyana's situation and highlights areas of severe shortfall. Some improvements were noted in the nutritional status of young children between 1971 and 1976 – an overall drop in nutritional deficiencies from 18.2 per cent to 10.7 per cent – but regional disparities were marked. About 25 per cent of the two- to five-year-olds on the Essequibo coasts were found to have second- or third-degree malnutrition in 1976, with West Demerara not far behind (Standing, 1979; Omawale and Rodrigues, 1979). Roopchand (1979) also presented evidence of severe malnutrition among the children of settlers along the Linden Highway – 22.7 per cent with second-or third-degree malnutrition. He blamed low agricultural output, low income and living standards. Malnutrition has been found

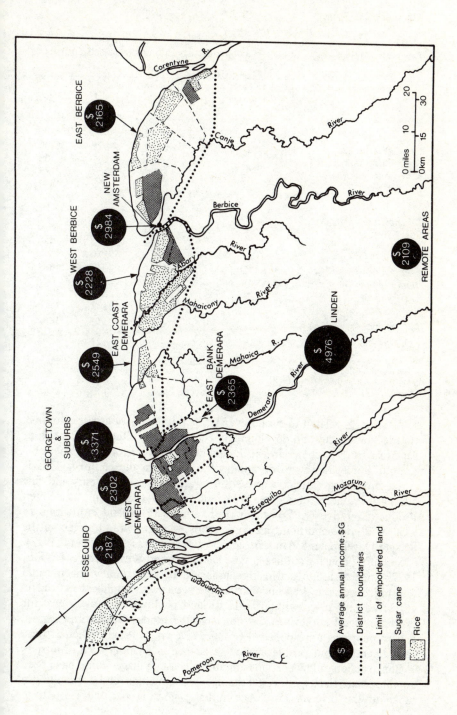

Table 8.2   Annual mean net earned income by rural and urban district[a] and sex, 1977

District	Total G$	Male G$	Female G$
Urban areas			
Georgetown (city)	3087	3351	2739
Georgetown (suburbs)	3655	4384	2408
New Amsterdam	2984	3368	2138
Linden	4976	5055	4678
Urban mean	3628	4192	2633
Coastal–rural districts			
East Coast Demerara	2549	2818	1672
East Bank Demerara	2365	2773	1331
West Demerara	2302	2578	1356
Essequibo	2187	2272	1688
West Berbice	2228	2289	1807
East Berbice	2165	2373	1259
Remote Areas	2109	2164	1875
Rural mean	2306	2518	1482
Guyana mean	2714	2959	1989

[a] See figure 8.4 for district boundaries.
Source: Caesar and Standing, 1978

to be strongly related to rural poverty 'including lack of access to opportunities for agricultural development' (Omawale and Rodrigues, 1979, p. 2). Mean rural incomes in the Essequibo area were found by the 1977 Labour Force Survey to be among the lowest in the country. This is a rice-producing district with serious water control problems and few alternative sources of income (see figure 8.4). Despite an Accelerated Production Drive (APD) which paid farmers to expand cultivation of local food crops and non-traditional crops, the agricultural targets of the Second Development Plan were achieved in few areas (Roopchand, 1977).

Similarly, housing studies in East Demerara and West Berbice (UNDP, 1977b, no. 5) found a direct relationship between family income and housing condition. Low-income people seasonally employed or underemployed and those without title to land could not get assistance for housing. Local authorities did not have the funds to assist and central government was unresponsive. Co-operative and self-help schemes were recommended, but people needed assistance in forming such groups.

Construction of low-cost housing in the coastal villages could have been an important strategy for stabilizing the population and forestalling a move to the primary city. An immense housing backlog existed, to overcome which optimistic targets were set out in the Plan: the government and co-operative sectors were expected to complete half the required 65,000 units and private individuals the rest. However, a 1977 study estimated performance by the end of 1975 as at best 18,000 units (UNDP, 1977b, no. 5). Most construction was concentrated in region 3 (which included Georgetown and Linden) or in other urban centres such as New Amsterdam. A 'co-operative village' set up 30 kilometres east of Georgetown as a model for rural development, housed government supporters from the police and the army, and had cost $G10 million by 1979, including a large shopping complex, a garment factory and piped water to all houses. Members repaid the housing co-operative according to their income rather than the value of the property, which was said by a commentator (Payne, 1982) to adhere to the principles of socialism. It no doubt heightened the suspicion of the impoverished Indo-Guyanese farmers of the district that 'Co-operatives are ... mainly for the benefit of the Africans' (Milne, 1975, p. 364).

There has been considerable discussion by observers concerning the validity of the co-operative as an institution for building socialism (e.g. Hope, 1979; Mandle, 1982; Milne, 1975, 1977; Sackey, 1979; Singh, 1972; Standing, 1979). Critics have focused on two issues: the lack of information about the role actually being played by co-operatives in the economy (and their lack of accountability) and the social relations of production which characterized them. On the first point it was agreed that despite an increase in the number of registered groups and members – from 811 in 1969 to 1,327 by 1975 with over 120,000 members in the latter year (Sackey, 1979) – their role was quite limited. Many were simply small savings and credit societies, such as schools, while others were registered as co-operatives 'for secondary objectives'. Very little was known about their financial status, although many were believed to be defunct. The 1982 report of the International Fund for Agricultural Development (IFAD) stated: 'Out of 120,000 acres controlled by co-operatives a large proportion remains idle' (quoted in the *Mirror*, 30 January 1983). On the second point, it was noted that many housing and construction co-operatives were simply another form of private company, being permitted to hire labour, while some which Standing (1979) calls 'partial co-operatives' appeared to assist larger operators by channelling credit to them. They thus represented an extension of capitalism, or as Singh (1972, p. 19) had feared, 'a fascist alternative to genuine socialism' and could be used as a tool for political power and corruption.

Central planning, 1978–81

Following the 1976 report of a committee, set up to study 'ways of restructuring the economy along socialist lines', the government took its first step towards the institution of co-ordinated central planning by announcing the establishment of the State Planning Commission. Finance Minister Hope described the new body as 'the nerve centre of co-operative and co-ordinated planning and monitoring activity for the whole economy' (Co-operative Republic of Guyana, 1978, p. 6). Not only was the Commission to take over responsibility for the preparation of future development plans but it was to examine the performance of the entire public sector including the state enterprises, 'and indicating by implication to the co-operative and private sectors the roles which they need to play in the whole scenario' (Co-operative Republic of Guyana, 1978, p. 7). The description of the Commission's functions in the enabling Act (no. 24 of 1977) included the following:

to advise the Government on the planning of (i) the orderly, balanced economic and social development of Guyana; (ii) the most effective, efficient and rational utilisation of the human, material, and financial resources of Guyana in order to achieve the most rapid economic growth consistent with the continuous improvement in the standard of living, the quality of life and the general material and cultural well-being of the nation. (Guyana Government, 1977, p. 3)

The point about rational utilization of human resources was taken up by Ivelaw Griffith (1978) in his review of the establishment of the Commission in the government-controlled *Sunday Chronicle* newspaper. Referring to the 'redeployment' exercise carried out in the public sector during 1977 in the interests of 'Proper Labour Placement', Griffith stated: 'Within the whole arena of central economic organization, the rationalization of labour, as well as skill and labour flows, would become vital. Proper Labour Placement has been set in train, but I see the need for more clear-cut understanding arrangements and inter-related labour adjustments.' The Trades Union Council, consumer organizations and producer groupings were all to be involved in 'truly democratic co-ordinated planning'. Production targets were to be instituted and incentive schemes introduced, the emphasis to be placed firmly on productivity and accountability (Griffith, 1978, p. 4).

The beginning of centralized planning must be seen against the backdrop of economic difficulties which were becoming severe by 1977. After three reasonably good years the country had been plunged into economic crisis. This resulted from the combined effects of a drop in international sugar prices and higher fuel import bills, with Guyana finally feeling the impact of the OPEC oil-price increases which had been cushioned by high sugar earnings in 1974–5. Conditions were exacerbated by a prolonged strike in the sugar industry in late 1977 with consequent loss of export earnings, making more difficult the repayment of agreed compensation to the nationalized sugar and bauxite companies. Political tensions increased with the renewed atmosphere of confrontation between Burnham and Jagan, following the latter's abandonment of 'critical support' after Burnham rejected his proposals for a permanent coalition government. The rise of the Working People's Alliance (WPA), with its ability to erode Afro-Guyanese support for the PNC, was also causing concern in party circles.

Hence the attempted restructuring of the economy towards more explicitly socialist objectives was taking place at a time of growing domestic difficulty and, paradoxically, increasing involvement by the IMF which was to exert strong counter-pressures and greatly restrict the government's freedom to manoeuvre.

The 1978–81 Four-Year Development Programme, introduced in February 1978, placed heavy emphasis on the productive sector and improvement of export earnings from bauxite, sugar and rice. Three extensive drainage and irrigation schemes were designated to increase coastal rice production while a large hydropower development projected for the remote Upper Mazaruni was to have the dual aim of reducing fuel imports and making possible the establishment of aluminium processing and associated heavy industry in the Linden area. The bulk of the development expenditure was to be carried out by the public sector, with co-operatives and private firms playing only a minor role; it was expected that a fair proportion of the total costs would have to be met by external aid (Co-operative Republic of Guyana, 1978). The Programme was thus markedly different from the previous plan in its allocation of investments: only seven per cent of the total outlay was to be directed to social projects (table 8.3).

Table 8.3 Comparison of planned investment allocations, Second and Third Development Plans

	Plan	
	1972–6	1978–81
Sector	(%)	(%)
Agriculture	15.5	33.5
Forestry and fishing	7.9	7.8
Mining and quarrying	5.3	12.9
Manufacturing	4.1	3.9
Power	7.1	6.5
Education and social development	21.7	3.9
Health and housing, roads	18.1	6.8
Sea defence	3.5	1.9
General administrative and other services	12.2	19.9
Engineering and construction	0.8	
Distribution	3.2	
Total programmed public investment	$G 1,018.3 m	$G 1,122 m
Private investment	$G 132.7 m	$G 160 m
Total investment	$G 1,152.9 m	$G 1,282 m

Source: AID Project Paper, 'Small Farm Development – Black Bush Region', May 1978, published in Rhodes-Checchi, 1979

Table 8.4 Proposed capital investment programme, 1979, percentage share by sector, region and urban area

Sector	N.W.	Essequibo and West Demerara	Central Demerara and West Berbice	East Berbice	Mazaruni Potato	Rupununi	George-town	Linden	New Amsterdam	Total	Sectoral % of total allocation
Agriculture, forestry, fishing	0.9	23.5	32.3	7.2	0.2	0.2	23.1	2.0	7.7	100.0	31.0
Mining	0.0	0.4	1.6	28.7	3.5	0.5	0.3	5.0	0.0	100.0	11.2
Manufacturing	0.0	0.0	37.6	0.0	0.0	0.0	45.0	0.0	17.5	100.0	9.0
Education and social development	1.3	6.7	9.9	8.4	1.0	1.0	62.6	7.3	1.5	100.0	9.9
Health and housing	0.4	7.3	15.7	6.4	0.6	2.0	21.4	38.2	7.9	100.0	6.6
Roads, construction, transport, communication	2.8	20.8	36.3	7.9	2.3	0.5	25.0	3.6	0.7	100.0	19.5
Defence, administration, finance, utilities and other government services	8.7	3.6	5.9	10.1	15.4	4.2	39.5	12.6	0.1	100.0	12.8
Regional % of total allocation	2.1	13.0	24.3	9.6	3.0	1.0	28.8	4.8	13.5	—	100.0

Source: State Planning Commission, reworked from Bernard, 1979

Although it was stated that production targets were to be applied at regional as well as national levels, the mechanisms for this were not stated clearly. In the 1979 budget the public sector capital spending was broken down for the first time on a regional basis (table 8.4). It is clear that the bulk of such spending was to be in the central region (region 3), which included both Georgetown and Linden. In spite of the overall emphasis on agriculture in the rhetoric of the plan, 45 per cent of total capital investment in 1979 was to be on expansion of the bauxite works and urban road construction. Even social expenditures such as education, housing and health were concentrated heavily in the urban areas. The other point which emerged from these figures was the retreat from interior projects, regions 1, 5 and 6 being allocated only six per cent of the total expenditure. It may be objected that as most of the people lived on the coast, expenditures should be concentrated there for the sake of efficiency. However, when the figures are examined on a *per capita* basis[9] some interesting variations may be noted. Spending in the urban areas was heavy, particularly in Linden, but two of the three interior regions showed up quite well: it was in fact the 'peripheral' coastal areas (regions 2 and 4), the heartland of the Indo-Guyanese population, which were not receiving their fair share: the big rice projects which formed the bulk of their agricultural investment only affected a small part of each of these districts. When explaining the skewed distribution of capital spending between the regions, Economic Development Minister Hoyte commented 'this type of distribution reflects the state of development of the regions ... one of the prime objectives of regional planning would be to devise plans for the balanced development of the regions' (Co-operative Republic of Guyana, 1979, p. 49).

Internal political problems and external pressures from the IMF and also from Venezuela[10] made it unlikely that questions of regional imbalance would be accorded a high priority between 1979 and 1981. As the foreign-exchange crisis worsened, keeping the export industries functioning became the major objective. The government's refusal to honour an agreement to lift the minimum wage from $G11 to $G14 per day in 1979,[11] led to a strike in the bauxite industry which then extended to sugar and the employees of urban-based corporations (the Clerical and Commercial Workers Union – CCWU). Both CCWU and the mining union (Guyana Mine Workers Union – GMWU) had been part of the PNC's traditional urban–industrial power base; they now largely transferred their allegiance to the WPA.

Increased political activity among the opposition forces was aimed particularly at the government's actions in relation to the proposed new constitution. A referendum in July 1978[12] was supposed to have given a mandate to government to prepare a new 'socialist constitution'. Elections due at that time were postponed as government appointed itself a Constituent Assembly and began canvassing submissions for the constitution.

Despite consultations with many groups, the final version of that document, adopted in February 1980, largely reflected the original PNC draft.

Export earnings in 1979 decreased from \$G1,844 million to \$G737 million and negative GDP growth was recorded (see table 8.5). The 1980 budget continued to reflect the low priority accorded to social expenditures – 4.8 per cent to health and 7 per cent to housing – while increasing attention was given to security, with 10.4 per cent devoted to the police, army and paramilitary organizations. Much of the security activity appeared to be directed against opposition groups, especially the WPA. An observer in mid-1980 (Gittens, 1980) analysed the social situation: unemployment was high, essential consumer items were either banned or in short supply and basic services were running down. In Georgetown electricity, water and public transport systems were all collapsing. On the subject of health, Gittens observed:

more than 50% of all medical and health personnel are believed to be based in Georgetown, despite the expenditure of milions of dollars since 1974 on the establishment of a 'regional development' administration to decentralise social services and improve the living standards of non-urban Guyanese. (Gittens, 1980, p. 7)

A major loan agreement for a three-year period was granted by the IMF in 1980 in return for an economic austerity programme. Burnham's acceptance of this was seen by the US as significant in their attempts to check the spread of 'leftist influence' in the Caribbean. He had apparently been informed that the financing of the Upper Mazaruni hydropower project was contingent on his accepting the programme (\$G40 million had already been spent on access works) but no such financing was forthcoming. Economic performance was again poor in both 1980 and 1981. Prices received for exports remained at low levels as a result of the world recession and balance of payments' problems continued, despite strict limitations on imports. With inflationary increases in living costs and wages pegged, the morale of workers was low. Exhortations for higher production went ignored, compounding the shortfall in export income. In February 1982 Burnham identified the major problem as *misplacement of labour*. More people were said to be needed in agriculture and forestry: 'there are too many comrades doing routine jobs in the public sector while our national assets lie wasting and unexploited' (Burnham, 1982, p. 6). Such a speech was the inevitable prelude to more large-scale retrenchments as the news came that the country was officially bankrupt, and could not pay for its imports or debt instalments. Reports of a 'deal' with the World Bank to allow equity participation by foreign companies in the bauxite industry and majority ownership by private capital in money-losing government corporations seemed to spell the end

Table 8.5a   Guyana: percentage value added to GDP by sector, selected years

Sector	1970	1974	1977	1980	1983
Agriculture	19.1	30.5	24.5	17.1	20.6
Mining	20.3	13.4	16.3	11.0	5.1
Manufacturing	12.1	13.6	8.6	17.9	18.2
Construction	7.8	6.4	7.6	8.0	8.3
Wholesale and retail trade	11.4	8.9	9.5	11.3	8.6
Transport and communication	5.9	5.1	5.8	7.9	9.6
Financial and other services	9.6	7.3	8.5	6.8	7.2
Government	13.7	14.7	19.1	20.1	22.4

Source: IDB, 1984; World Bank, 1976, 1980 (World Tables)

Table 8.5b   Total GDP and GDP per capita

	1960	1970	1980	1983
Total GDP in 1982 $US	435.5	609.3	912.4	576.3
GDP per capita (selected years)	721.0	851.0	905.2	711.5

Source: IDB, 1984, p. 420

Table 8.5c   Growth rates of GDP total and by selected sectors, 1979–84

Sector	1979	1980	1981	1982	1983	1984
Total GDP (market prices)	−3.8	2.1	1.2	−10.6	−10.6	2.0 (est.)
Agriculture	−9.6	0.4	2.3	−1.3	−2.8	na
Mining	−8.9	6.1	−11.5	−31.4	−38.0	na
Manufacturing	4.0	0.7	6.0	−12.8	−10.5	na
Construction	2.9	2.8	−4.0	−7.7	−4.8	na

Sources: IDB, 1982, 1983, 1984

of Guyana's socialist experiment. The government hesitated, the 'Action programme' was delayed and demands by the IMF at the end of the year for further severe devaluation (the Guyana dollar had been cut 18 per cent in 1981) were resisted. A stalemate resulted. Guyana was not eligible for further loans and had to resort to barter-trade for certain commodities. Meanwhile, with the adoption of the new constitution in October 1980 a new regional system was installed which attempted further measures of party control over the population. We will now examine the rationale for that system and try to assess its results.

The new regional system – local democratic organs

While in the mid-1970s Tanzania was a model for several of Guyana's experiments in national mobilization, by the 1980s this had largely been replaced by Cuba.[13] The Cuban model was particularly apparent in the new regional framework. According to the constitution, the prime objective of the political system was to 'provide increasing opportunities for *participation* of citizens in the management and decision-making processes of the state' (Co-operative Republic of Guyana, 1980, p. 22). Likewise, national economic planning was expected to provide for participation at every level – 'enterprise, community, regional and national' (Co-operative Republic of Guyana, 1980a, p. 23). The structures necessary for such participation involved a major reorganization of the regional system. There were to be ten regions (replacing the former six) subdivided into a varying number of sub-regions, then broken down further into smaller districts 'for the purpose of organising local democratic organs'. Each region was to contain six sectors of Local Democratic Power – region, sub-region, district, community, neighbourhood and 'People's co-operative unit'. Regional boundaries were designed to take account of

the population, physical size, geographical characteristics, economic resources and the existing and planned infrastructure of the area, as well as the possibilities of facilitating the most rational management and use of such resources and infrastructure, with a view to ensuring that the area is or has the potential for becoming economically viable. (Guyana Government, 1980a, p. 7)

The boundaries were described as 'natural boundaries, following rivers or watersheds' (figure 8.5). A comparison with the 1973 subdivisions reveals that while two of the former six entities remain regions, the others have been subdivided or the boundaries re-drawn to form compact and logical areas. Region 4, Demerara–Mahaica, includes Georgetown and its main commuting and peri-urban industrial zones on the east and south. This region, which must still contain about 40 per cent of Guyana's population, is a more carefully defined 'central area' than was the old region 3. Another important change made by the new system was the incorporation of all existing local government areas (previously confined to the settled coast) into a country-wide network of 'local democratic organs'.

According to Economic Development Minister Hoyte, in his speech on the Local Democratic Organs (LDO) Bill (18 August, 1980), local

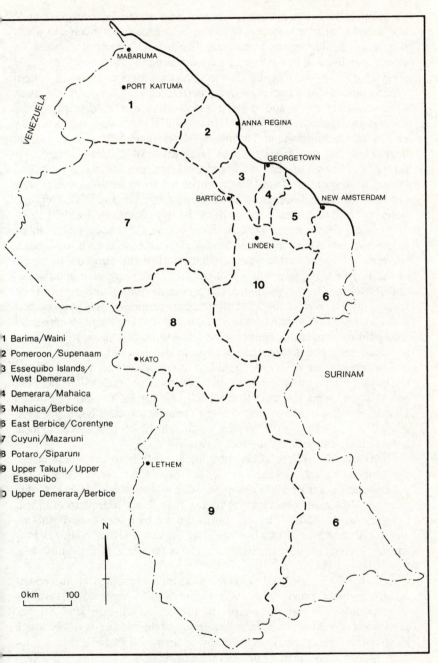

1 Barima/Waini
2 Pomeroon/Supenaam
3 Essequibo Islands/
   West Demerara
4 Demerara/Mahaica
5 Mahaica/Berbice
6 East Berbice/Corentyne
7 Cuyuni/Mazaruni
8 Potaro/Siparuni
9 Upper Takutu/Upper
   Essequibo
10 Upper Demerara/Berbice

Figure 8.5   Guyana: administrative divisions, post-1980

government was thus assigned 'a pivotal role' in the political, economic and social life of the country. It was to be organized to involve as many people as possible 'in the management and development of the areas in which they live and in the various decision-making processes of the State' (Hoyte, 1980, p. 4). Arguing that the previous local government system which evolved during colonial times performed only a minor role, such as maintaining roads and digging canals, and had to depend on the Central Government for major initiatives, Hoyte stated that the new system of local government would be 'development oriented, with real devolution and decentralization of large areas of Central Government activities'. He also foreshadowed personnel transfers, indicating the possibilities of movement of staff from central government to LDO's or vice versa. A National Council of Local Democratic Organs (NCLDO) has been set up to co-ordinate the work of the local authorities, to be known as Regional Democratic Councils. Each region elects a councillor to sit in the National Assembly, while the full NCLDO meets with the Central Government about once a year in what is called the Supreme Congress of the People. Among the duties of the LDO is to 'organise popular co-operation in respect of the political, economic, cultural and social life of their areas ... to raise the level of civic consciousness, preserve law and order, consolidate socialist legality and safeguard the rights of citizens'. The political crunch, as enunciated by Hoyte, is that the system cannot tolerate negative and disruptive activities (Hoyte, 1980, p. 10).

The 'organs' are devices for political mobilization and control. In the elections of December 1980, of the 205 seats on regional democratic councils, 169 were allocated to the PNC, 35 to the PPP and one to the United Force, an almost non-existent remnant of the PNC's pre-1968 ally. PNC members were elected to fill all 12 of the allocated local government seats in the National Assembly.

Hoyte saw the system as contributing powerfully to the real development of each area. Political opposition played no part in this – the paramountcy doctrine of the one-party state was obviously at work. Prime Minister Reid made that clear in August 1983 in his address to the Fifth Congress of the PNC: 'The Party must be the base for all the economic and social activities of our districts, our regions and our country. It is the Party which must be the motivating lever for all the national mobilisation' (Reid, 1983, p. 1).

As well as being allocated funds from Central Government to implement its development programme, each region is now expected to raise its own revenue. They are also expected to become self-sufficient in food, producing at least up to a minimum level, while housing is to become a local, rather than central responsibility (Parris, 1983).

In August 1984 the first attempts were made to mobilize the population into People's Co-operative Units, the lowest level of local organization in the regional system. PNC activists went from house to house collecting information about the inhabitants, and a leaflet was issued from the Ministry of Regional Development outlining the new system. The Units were scheduled to become channels for distribution of scarce foodstuffs at designated shops as well as providing 'practical opportunities for involvement and participation in vital decision-making' (leaflet quoted by Morrison, editor *Catholic Standard*, 2 September 1984). Such decision-making largely concerned community improvement, schools' fund raising, and the organizing of residents in civil defence activities. Morrison commented that people were apprehensive, that the Units were 'yet another means of the Ruling Party exercising control over their lives'.

The kind of mobilization which is intended here sounds rather like the system of 'Popular Power', established nationally in Cuba in 1976. In that system, local responsibility is taken for the organization of a range of productive activities and management of such areas as sports and housing, while the local committees oversee the operation of programmes in education and health which are organized centrally. Eckstein suggests that 'Popular Power' works well at grassroots level because delegates are accountable to their communities and may be replaced by democratic process. In her description of the system, she notes, however, that the masses do not control the work process and have limited means of influencing national policies. Discontent and dissidence can be controlled at the grassroots level: 'Grassroots participation ... a mechanism for consensus building' (Eckstein, 1982, p. 213).

There is evidence that the PNC Government has been anxious to push through such decentralization policies, first in the hope of breaking through the widespread disincentive to produce and the alienation of workers which are creating havoc with economic targets; secondly in order to 'redeploy' a large proportion of the workforce presently employed in central government or public enterprises around Georgetown. Everywhere people are exhorted to take up more productive activities especially in agriculture. 'Government has taken a decision to relocate personnel to various regions with the intentions of correcting the regional imbalance in the distribution of human resources' (Green, 1983).

However, unless the government is also prepared to allocate money for creation of a range of employment opportunities and the upgrading of general facilities in the 'lagging' regions, it is unlikely that former urban workers will *voluntarily* be persuaded to move there seeking agricultural opportunities which might well prove chimerical. P.E. Williams (1983, p. 19) has asserted that Guyana 'cannot afford the luxury of spatial equity in planning': regional self-reliance may be a euphemism for central neglect. East Berbice (region 6), a depressed area with considerable potential for agriculture, has seen a large exodus of its men to work in rice, timber or sugar across the border in neighbouring Surinam, where jobs are available and wages higher, even though Guyanese are regarded there as cheap labour and often exploited (Menke, 1983).

Economic realities, 1981–4

Despite the bold showing of independence towards the IMF, government planners cannot afford to ignore the demands of that institution. Every year since 1981 the servicing of Guyana's external debt has absorbed much government revenue. The country's creditors will not reschedule these debts until an agreement with the IMF is reached. Aid from western countries is increasingly being held up, while a search for eastern bloc help has produced little direct financial assistance. The government is now moving quietly to meet some of the IMF's conditions: two further devaluations of the Guyana dollar took place during 1984 and more private joint-venture involvement is being arranged in money-losing government corporations. Table 8.5 summarizes the downward trend in the economy between 1979 and 1984, particularly the collapse of the bauxite industry, production of which fell by 40 per cent during 1983 (IDB, 1984). A joint-venture agreement with Reynolds Metals[14] for rehabilitation of the industry has been announced and more private involvement is to be permitted in the milling and marketing of rice. Production of that commodity remains low, as farmers refuse to plant, blaming low prices, bureaucratic bungling and 'rampant corruption' (GHRA, 1985, p. 12).

The local subsidy on sugar has been removed, while restrictions continue on imports of essential foods and other consumer goods. Government policy is that Guyanese rely entirely on locally produced foods. Where these are not available, items may be obtained on the parallel market but at grossly inflated prices. The human consequences are summed up in the following paragraph:

A 1981 survey of expenditure which has been accepted by Cabinet placed the monthly average household expenditure on food at $G62 per month. Since this item alone is higher than the total reported incomes of over 50% of the working population, it suggests that a large section of the population is eating inadequately, earning clandestinely or turning to crime to meet its needs. (GHRA, 1985, p. 13)

The Labour Amendment Act (1984), which removed the power of trade unions to negotiate over wages, leaving this power solely in the hands of the President, might also have been prompted by IMF pressure. That measure led to a revolt of the Trades Union Congress, which finally threw off PNC control and brought a grudging approval from Burnham for a wage increase. The rate decided, $G15.10 per day, is still well below the 'survival minimum' of $G25 which was requested by the TUC, and is the lowest in the English-speaking Caribbean.

Conclusions: co-operative socialism, planning and reality

When one examines the actual performance of the Guyana government over the years between 1970 and 1985, setting aside the rhetoric of official pronouncements, one must concur with other analysts (e.g. Mandle, Sackey, Thomas) that a socialist transformation has not occurred. Following an initial period of decolonization the government attempted to steer a middle course which it called co-operative socialism, 'a sort of halfway house between capitalism and communism' (*Guyana Graphic*, 18 October 1969). This rapidly became State capitalism and the pursuit of power by a bureaucratic elite. Military and paramilitary organizations became more important than co-operatives, many of which were indistinguishable from private firms and which strongly reflected party patronage. Neither have policies of socialist planning been carried through, despite experimentation with systems of regional subdivision and exhortation to local self-reliance. Political power remains centralized, the economy remains export-oriented and the exigencies of the present recession indicate that polarized development planning will also continue. The opportunity for re-structuring along different lines following the economic collapse would seem to have been diverted by external economic pressures. Thomas asks: 'How much of an "alternative development" could co-operative socialism entail if it is being built on IMF/World Bank Group credit?' (Thomas, 1984, p. 13). That would seem to be the fundamental problem unless the country can find alternative ways out of its debt crisis.

The interesting question is – 'where to now?' In his 1982 analysis, Mandle singled out the rural–urban split which parallels the Indian–African racial split as the major problem impeding Guyana's development. Thomas believes that the Afro-Guyanese PNC regime has continued to be propped up by international agencies simply because it is still seen as ideologically to the right of the doctrinaire, Moscow-oriented PPP of the Indo-Guyanese. Early in 1985 Burnham was reported to be moving to invite the PPP's Cheddi Jagan to some form of coalition (LARR *Caribbean report*, 29 March 1985). Such initiatives were suspended following Burnham's death on 6 August. Desmond Hoyte, the new president, has been treading cautiously, anxious to maintain the impression of continuity in PNC policy and to restate the commitment to co-operative socialism. A few concessions have recently been made to the PPP and the resumption of talks in October seemed to offer some hope of a power-sharing formula being evolved among the three socialist-oriented parties – the PNC, PPP and WPA. However the announcement of arrangements for the impending election (in particular Hoyte's refusal to allow a team of Commonwealth observers) resulted in a re-polarization of positions (LARR *Caribbean report*, November 1985). Unfortunately elections in Guyana always heighten the racial divisions in the society. It remains to be seen whether progress toward a coalition can be made after the tensions of the election have subsided and if so, what the reaction of Washington and the international financial agencies might be to such an alliance.

Notes

Note on sources
In addition to the detailed sources listed in the Bibliography, a range of newspaper sources was consulted, particularly for the 1980–5 period. These included the regional *Caribbean Contact*, monthly organ of the Caribbean Council of Churches and a useful source of political and economic comment, and Guyanese opposition papers, the weekly *Catholic Standard, Mirror, Day Clean* and *Open Word*. The government-controlled press was represented by the *Daily* and *Sunday Chronicle* and (occasionally) *New Nation* (the PNC organ).

1. Despite periodic 'crackdowns' and claims that smuggling and the 'parallel economy' have been brought under effective government control, some Guyanese observers contend that if anything, it is the parallel rather than the official economy which is functioning and relevant to their daily lives.
2. The two main population groups in Guyana, Indo-Guyanese (51 per cent) and Afro-Guyanese (31 per cent) are descendants respectively of indentured workers from the Indian sub-continent and slaves from west and central Africa. There is also a mixed group (part-African) which accounts for 10 per cent of the total and is often also considered to be Afro-Guyanese. Indigenous Amerindians (5 per cent) and 'other' (2 per cent) – Chinese, Portuguese, other Europeans, etc. – complete the population mix. These figures are quoted in the *Guyana Human Rights report* (GHRA, 1985, p. 3) and are probably from the 1980 census, although this is not explicitly stated.
3. 'Indo-Guyanese' is a more recent term for the East Indian population. They had been called East Indians in the Caribbean to distinguish them from indigenous Amerindians, groups of whom still survive in the mainland territories, although almost extinct on the islands.
4. This was the first of three elections (1986, 1973, 1980) which the PNC has been accused of 'rigging', by opposition groups and outside observers. The technique of padding the voters' rolls with expatriate Guyanese – at that stage mainly of African origin, therefore pro-PNC – was adopted for the first time in 1968. Many names on these lists were considered fictitious or erroneous.
5. The United Force, a grouping mainly of business interests. In 1964 Burnham was considered by the UK and US to be less radical than Cheddi Jagan of the PPP; Burnham's PNC–UF alliance was helped to power through the

change in the voting system to proportional representation and CIA destabiliz-
ation of Jagan's government (see Premdas, 1977, 1978, 1979).

6. The 'urban' population included the population of Georgetown, suburbs of
Georgetown, Linden and New Amsterdam.

7. Thomas gives a figure of total population increase between 1970 and 1980 of
only 7,000 (*Guardian*, 15 May 1984). The Inter-American Development Bank
estimates a one per cent increase between 1970 and 1981, which appears
more reasonable.

8. The large increase in the enumerated numbers of Amerindians between 1960
and 1970 has been explained partly by inadequate coverage at the 1960
Census. Nevertheless, birth-rates are high in most Amerindian areas; the
true rate of natural increase should be revealed when the 1980 figures are
released.

9. Presuming the proportional distribution of population among the regions to
have remained about the same as 1970, and allocating numbers according to
the IDB estimates of a one per cent increase per year: this may well be
inaccurate but there are no better data as outmigration rates by region are
unknown.

10. Venezuela claims all of the western county of Essequibo. It is believed that
pressure from Venezuela has caused the World Bank's failure to fund the
large Upper Mazaruni Hydro Scheme, located within Essequibo.

11. In 1979, $US1.00 = $G2.50.

12. As usual, widespread 'rigging' of the results was claimed by the opposition.

13. The 1980 Constitution however, drew upon many sources for its formulation,
including Cuba, North Korea, China and East Germany but also Zambia
(James, 1982; Lutchman, 1982).

14. This move is ironical as Reynolds was one of the two bauxite companies to
be nationalized during the 1970s, though admittedly the smaller of the two.

9
Spatial Equality and Socialist Transformation in Cuba

Paul Susman

Introduction

Statement of the Problem

Increasingly, Third World leaders are paying attention to the relationship between development objectives and the spatial organization of economic, political, and social activities. Their goals often explicitly include promoting spatial integration by more closely linking all sectors and parts of their societies together. Achieving spatial integration depends not only upon existing economic, political, and social conditions, but also upon the development path selected, whether capitalist, socialist, or other. Each path implies (but does not determine) a set of long-term economic and social equity objectives. This chapter examines the case of spatial development in Cuba in order (1) to assess the relationship of changes reflected in spatial organization and development to government equity objectives, and (2) to discern the particularly socialist aspects of territorial changes in this Third World country's development process.

In Cuba one finds efforts to achieve socialist equity objectives in all aspects of development planning including in spatial organization. This is true at the local level as well as at the provincial and national levels. For example, from early in the post-1959 revolutionary period, programmes designed to overcome urban–rural differences at the local level define one important policy focus. At a higher level of spatial organization, balancing economic activity over the provinces defines another goal and, to a considerable degree, reducing Havana's urban primacy in the country as a whole constitutes another thrust of Cuban planning. Many Cuban programmes are directed at achieving equity goals while creating an economic structure suitable to meet growth requirements. It should be understood that Cuban officials, drawing on socialist thought, see economic change as part of a unity with political processes. Thus, Cuban development planning includes creating political institutions to provide for greater input from across the country and to devolve some decision-making to municipalities and provincial governments in political as well as in economic activities. This paper discusses each of these in turn.

Unlike many other countries limiting their goals to increasing *consumption per capita* via the distribution of goods and services, Cuban objectives include increasing equality in participation in decisions concerning *production* activities over the population and country as a whole. This thematic ties in to a socialist vision of development. By defining its equity objectives as achieving equality, in the long run Cuba faces difficulties inherent in any effort to increase simultaneously economic growth and achieve social and spatial equality. Ward (1967, p. 14) refers to this dilemma as the 'socialist controversy'. Traditionally, planners view economic growth and equality objectives as necessary trade-offs, at least in the short run. Simultaneously maximizing two variables, growth and equality, is not possible in a mathematical sense but may be in a political economic context in which consciousness, individual and group effort, and many other factors affect economic outcomes (Mihailovic, 1975, p. 48). Related to this trade-off between growth and equality are choices in spatial development strategies, most starkly summarized as polarized versus equal growth. While both strategies are usually seen as attempts to attain economic equilibrium expressed over space, neither will necessarily lead to economic equality. Each gives rise to choices in respect of sectoral investment decisions, the location of new investment in selected areas and whether urban or rural, and the degree of economic and spatial concentration or dispersal. Often investment in industry is seen as an alternative to agricultural investment and, similarly, an urban focus is counterposed to rural as the appropriate site of economic activity.

Each strategic alternative assumes some combination of market forces and institutional adjustments will lead to an optimal allocation of resources in terms of some equity objectives, but equality is not the goal. Since the revolution, the Cubans have violated the traditional dichotomy between polarized and balanced development strategies by defining equality as the goal and by embarking upon equal spatial development and economic growth paths in which economic growth and equality over the population and space are viewed as complementary rather than contradictory. At different periods in the post-1959 history, the Cuban leadership emphasized one facet more than another, but the overall trend appears to be towards balancing both growth and equality objectives.

Equality, democracy and socialism

There are important differences in the conception of equality and democracy that prevail in capitalist and socialist societies. Among capitalist countries, there exists a spectrum of political systems ranging from democracy to dictatorship. The political system does not define the fundamental economic organization of these countries. Within capitalist democracies, the idea of equality is most often expressed in purely political terms. The right to vote, however, is not synonymous with a right to an equal share of economic or other benefits within the society. Allocating benefits or social resources for the most part remains the right of those who own or control them and is pursued, following the logic of capitalist accumulation, to enhance the economic position of the individual or firm. In capitalist societies, distribution of benefits, such as housing, education and health care, to the population as a whole may be politically mandated, but this does not bring the entire population into the decision-making process about what will be produced and under what conditions.

In fact, because real economic power resides in the firm or individual, and not society as a whole, a level of benefits established for the population at one time may be lost as a consequence of later political developments. For example, the election of Reagan in the United States led to a reduction of benefits to low-income people in accord with changing political attitudes. Thus it may be said that the concept of equality in capitalist democracies extends to the political but not the economic realm. Furthermore, to the extent that a capitalist society makes a political choice for redistribution of income or wealth, it is expressed invariably in terms of consumption such as the level of disposable income and access to goods and services, but not in terms of shifting control over a nation's resources to the population as a whole. Piven and Cloward (1982) describe this separation between political and economic spheres as a relatively recent ideological development accompanying capitalism. Prior to the separate-spheres perspective taking hold, the 'moral economy' view dominated throughout the feudal period in which each individual's access to the means of subsistence was the rule, one beyond the general discretion of those in power.

Democracy, in socialist terms, refers to popular control over both political and economic conditions: 'Indeed, in a general way, Marx's socialism (communism) as a political program may be most quickly defined, from the Marxist standpoint, as *the complete democratization of society*, not merely of political forms' (Draper, 1977, p. 282). Socialist theory emphasizes control over the means of production by the population. From control over the means and organization of production follows the creation of socially desirable amenities for the population and a particular distribution of benefits including the ability to participate in consumption activities (Marx, 1970a, pp. 195–204). Allocation of a society's resources, including its labour power, is, in the socialist ideal, a reflection of collective ownership and commitment to particular social goals. Democracy, according to Marx, 'requires a new social content – socialism' for it to be complete (Draper, 1977, p. 283).

In this chapter, the fundamental question remains to assess Cuban development, and especially spatial restructuring, in terms of both the overall goals of the revolution and a definition of socialism as a transition to an equal and, therefore, democratic society.

Although analysis of capitalism's spatial impact on Cuba and other Third World societies is plentiful, there is relatively little explanation of the relationship between socialist forms of organization and consequent spatial patterns (Slater, 1982, p. 3). Also, it is difficult to generalize from observations of a transition in process in Cuba, a particular historical case, rather than from an accomplished political–economic system (Dupuy and Yrchik, 1978, p. 50). Yet in Cuba, to provide a conclusion in advance, evidence of changes in the social and spatial organization generally support government declarations of trying to promote goals of socialist equality. Importantly, many of the changes in Cuban spatial organization may be linked to the goal of overcoming urban–rural disparities in both consumption and production spheres.

Spatial organization and Cuban socialist vision

From the earliest days of their struggle, the Cuban revolutionaries saw the exploitive role of urban-based elites in perpetuating inequality as antagonistic to its goals. This view of urban–rural-based relations is found in classical works by Marx and Engels. For example, commenting upon the separation between town and country, Marx and Engels (1970, p. 69) emphasize the relationship between the development of the division of labour in society and the opposition of interests in town and country populations. The antagonism that exists:

is the most crass expression of the subjection of the individual under the division of labour, under a definite activity forced upon him – a subjection which makes one man a restricted town animal, the other into a restricted country animal, and daily creates anew the conflict between their interests. (Marx and Engels, 1970, p. 69)

However, with development of industrial capitalism, the pattern of exploitation loses some of its spatial specificity and is recast more strongly in class terms and in the context of the capitalist division of labour. Ironically,

abolition of the antithesis between town and country is not merely possible. It has become a direct necessity of industrial production itself, just as it has become a necessity of agricultural production and, besides, of public health. (Engels, 1978, p. 723)

But, capitalist exploitation continues, with fewer encumbrances as urban- and rural-based capitals are better co-ordinated. The urban-based bourgeoisie comes to dominate both urban and rural workers.

Engels' goal is the elimination of the capitalist division of labour and creation of a classless society in which a 'fusion of town and country' is part of the process of reducing and ending alienation (Engels, 1978, p. 723). In this sweeping phrase, Engels suggests that traditional categories of 'urban' or 'rural' will lose their meaning as integrated socialist landscapes evolve, replacing capitalist landscapes of spatial and class inequality. If elimination of capitalism in both class and spatial terms is, indeed, the goal of Cuba, the outcome should be evident on the landscape.

Creating a socialist consciousness, considered essential to the success of building a socialist society in Cuba by the leadership, requires that class differences and patterns and attitudes of domination be overcome. In another of many references to goals of the revolution, Castro again draws from Marx's analysis of the causes of alienation in capitalism including a spatial expression in town–country differences:

Every citizen will become used to viewing a fellow human being not as an enemy, not as a beast against which he has to protect himself, but as a truly human person, as a brother or sister who can help him in time of trouble; he will not see his fellow as a superior or inferior being, but as an equal; merit will take the place of privilege, for merit will be what distinguishes one citizen from another since merit will be the only rule by which a citizen is judged. (Castro, 1982, p. 70)

The pre-1959 spatial division of labour and power reflected patterns of exploitation and accompanying attitudes of domination, further separating people in towns from those in the countryside. Hence, so much of Cuban development may be interpreted in the context of efforts to integrate town and country, economically, politically and at different geographic scales – local, provincial, and national.

The concentration in the cities of wealth created in the countryside has been a historical constant in the development of Cuban society. For the most part such wealth was channeled into Havana, creating disequilibria that, since 1959, have constituted one of the principal concerns of a country pursuing its own integral development. To bring the economy into regional balance, therefore, priority was given to elevating the rural standard of living. (Acosta and Hardoy, 1973, p. 10)

The long-term nature of the commitment to improving the lot of the rural population and eliminating urban–rural disparities is evident in Castro's famous 'History Will Absolve Me', in which he details the hardship of rural life and calls for improving it:

If Cuba is an eminently agricultural country; if farmers make up a large proportion of the population; if the cities depend on the rural sector; . . . if the greatness and the prosperity of our nation depends on a healthy and vigorous farmer who loves his land and knows how to cultivate it and on a state that would protect and guide him, how can this state of affairs go on forever? (Castro, 1972, p. 1986)

Thus, since the revolution, a goal of the Cuban government is to promote a spatial distribution of population, production and consumption activities, and services in a pattern reflecting the leading role of agriculture in the economy.

Although this paper addresses spatial aspects of Cuban development, it is not intended to assign space the central role in development, but rather to emphasize it as one of many factors undergoing transformation since 1959. The changes observed cannot be understood separately from the overall changes in economic activity, political processes, and the identified needs and wants of the population. Slater's (1982) and Anderson's (1979) admonitions to avoid fetishizing space are well-taken.

Understanding the Cuban revolution necessitates examining goals and achievements since 1959. Thus, observing the transformation of the economic landscape from pre-revolution patterns established during 400 years of Spanish rule and 60 years of United States domination provides insights into the degree of equality attained in economic activity and service provision. From the perspective of the ability to consume (income factors), improvements in any place may be readily measured by general indicators, such as access to health-care facilities, educational institutions, employment opportunities, housing and changes in income levels, both absolute and relative to the country as a whole. Further inquiry into spatial aspects of development may shed light on the structure of control in production and the degree of popular participation in the political process and issues of socialist democracy, in general.

Overcoming inequalities

Broad indicators of income distribution and access to goods and services over the population and space suggest that Cuba is moving towards income equality. Even critics of Cuban development programmes, such as Mesa-Lago (1981), say that Cuba's income distribution is the most egalitarian in Latin America. Cuban incomes have varied, on average, by ratios of 4.3:1 to 5.3:1 (Brundenius, 1984b; Mesa-Lago, 1981). By comparison, in Peru and Brazil, incomes differ by 24:1.

A brief review of some of the programmes and outcomes follows. In Cuba, programmes addressing the spatial aspect of inequality include those affecting Havana's primacy and the long-standing urban–rural dichotomy. These are tied to efforts to achieve some level of equality in access to goods and services, what has been called the consumption sphere, over the population and the country as a whole. After summarizing some of the indicators of consumption equality, the discussion turns to indicators of production equality and democratization of Cuban political-economy with particular emphasis on spatial aspects.

National spatial scale: reducing Havana's urban primacy

In 1959 the new government inherited a spatially distorted economic landscape (Susman, 1974). Urban primacy was well established and the government set out to reduce Havana's disproportionate share of population, new construction and industrial activity, as well as its overwhelming dominance in almost every other aspect of Cuban society and economy. Examples demonstrate the difference in Havana's standing in 1959 and after a few decades of the revolution. For instance in 1959 Havana, the capital city, contained 1.5 million people or about 25 per cent of the nation's population and absorbed 26 per cent of the total annual increase in Cuba's population. Yet by 1980 Havana's population had grown to only 1.9 million, accounting for under 20 per cent of the national population, a smaller share than in 1959. This represents a total increase of 26 per cent from 1959. In comparison, in 1959, fourteen other principal cities together barely contained one million people. By 1980, they increased about 80 per cent, from about one million to 1.8 million, as a result of government policies to encourage growth away from Havana.

The change in urban primacy is also reflected, for example, in construction data. In 1958, 70 per cent of Cuba's construction activity was in Havana. By 1982, Havana's share of construction had dropped to 27 per cent (Comite Estatal de Estadisticas, 1982). Other examples may be drawn from almost any sector. In each, the concentration of wealth, industry, government services, etc. in the primary city may be counterposed to inadequate income levels, services, etc. in other parts of the country. Thus, primacy in pre-revolutionary Cuba was clearly associated with inequality.

Inequalities between Havana, or at least some residents of that city, and the rest of the country are suggested by data from a 1953 survey. It indicates that family incomes in Havana were 14.7 per cent above the national average, while family incomes in the rest of the country were 10.6 per cent below the national average, or a gap of 25.3 per cent. By 1972 that gap had been reduced to 19.2 per cent, with Havana's family income 13.2 per cent above the national average and the rest of the country 6 per cent below it (Mesa-Lago, 1981, pp. 144–5). By 1978–9 the average income in Havana was down to 12 per cent above the national average, although the gap may be larger than the figures suggest if food consumption, rather than cash income figures are used. For example, using 1978–9 data, Brundenius (1981, p. 159) shows that food consumption in Havana was one-third higher than the national average, considerably more than the cash-income level. However, as Eckstein (1980, p. 95) points out, the contribution of all non-cash income, especially to the poorest workers, in the form of 'free health care, education, and social security' may not show up in traditional measures of income but may mean that the degree of consumption equality is even greater than these general statistics reveal. Studies of social welfare programmes in other countries show a similar effect in reducing regional disparities beyond what income data alone might suggest (Richardson, 1979, p. 230). Such indicators of reduced primacy in Cuba reflect the outcome of policies to affect these changes. Because efforts to reduce Havana's primacy are discussed in considerable detail elsewhere (for example, Eckstein, 1980; Garnier, 1973; Gugler, 1980; Hamberg, 1986) only a few major points will be made here.

During the early years of the revolution, from 1959 to about 1965, government plans called for Havana to grow as slowly as possible. Responding to both ideological preferences for reduced urban dominance and to difficulties of meeting labour needs in rural areas and maintaining basic economic functions, investment and infrastructural development policies were designed to diminish Havana's primacy. Thus, rural areas accounted for about half of new housing built. In contrast, Havana's housing availability in the 1960s reflected outmigration to the United States and elsewhere, rather than new construction. Nevertheless, cities proved quite attractive to rural labourers.

The labor scarcity in the countryside was due to the 'movement of a considerable number of agricultural workers to the cities where they found less difficult work and better returns at the beginning of the first phase of the Revolution, to the necessary mobilization of human resources for the defense of the nation, to the disappearance of seasonal unemployment which facilitated a labor surplus ... to the disappearance of hidden unemployment in peasant families ... and to the rapid increase in rural constructions' (Cuba, 'El Desarrollo Industrial de Cuba', p. 161, quoted in Acosta and Hardoy, 1973, p. 18)

New investment and growth were to occur in other large- and medium-sized cities as a means of reducing Havana's importance and as a means of distributing both consumption and production activities more evenly across the country. Success of these programmes is reflected in a steadily decreasing contribution of *migration* to the annual rate of urban population growth from about 1.4 per cent before the revolution to 0.6 per cent per year for the period 1959–70 (Estevez, 1982; Gugler, 1980). By 1982, Havana's net gain from migration was at a lower rate (0.4 per cent) than the national urban rate (Comite Estatal de Estadisticas, 1982, pp. 84–5).

However, Havana's industrial and other infrastructure, the legacy of an 'overurbanized' past, continued to haunt the new regime. It was too costly to duplicate existing facilities and it was not possible to create sufficient new housing in other places to attract potential migrants away from Havana. Thus, beginning in 1965, the Master Plan for Havana recognized some merit in moderate growth and allowed for the location of new economic activity in Havana if it fulfilled certain criteria. By 1968 the decision to permit controlled industrial location was evident in the new activities being established in Havana (Garnier, 1973, pp. 235–6). In a reversal of earlier policy:

In 1969 Fidel announced that new industries would be located in the capital because the city has good port facilities, infrastructure to support the new industries, a concentrated consumer market, an abundant supply of labor (especially of skilled domestic labor and foreign technicians), and a disciplined and experienced labor force. Supposedly Havana was no longer parasitic. (Eckstein, 1981, p. 126)

Since that time, Havana's population has continued to grow, very slowly, at a compound rate of 0.71 per cent per year (1970–82). (This experience is in stark contrast to other Latin American cities growing much more rapidly. For example, Bogota, 7.3 per cent per year; Sao Paulo, 6.3 per cent; Mexico City, 5.8 per cent; Lima, 5.1 per cent.) By controlling the location of employment opportunities and housing availability, the government has limited Havana's growth while encouraging growth elsewhere (Eckstein, 1980). 'Indeed, officials consider housing nearly as important as wage incentives in assuring labor force stability in developing areas' (Hamberg, 1985, p. 39).

In the *Socioeconomic Guidelines for the 1981–85 Period*, there is an explicit policy of limiting the location of new industry in Havana, as part of the ongoing concern about disproportionate growth. The plan is to limit

the location in Havana of new, enlarged or remodeled industries to those that have advanced technology and high productivity per worker, use relatively little water, occupy a limited area and do not pollute the atmosphere or overburden the transportation system. (Second Congress of the Communist Party of Cuba, n.d., p. 125)

Even while limiting industrial growth, new housing investment and infrastructural improvement is now being poured into Havana to compensate for years of neglect. For example, there are now efforts to repair much of the housing stock in the Vedado section, which was allowed to deteriorate for a decade before the revolution, as well as during the post-1959 period as a consequence of plans to de-emphasize Havana and use scarce resources for building factories, schools and hospitals. Also, because of its population density and land-use pattern, the increasingly popular solution to the housing shortage in the rest of the country of self-built housing is not feasible. Thus, in 1982, the less than 20 per cent of the national population in Havana received 27.6 per cent of national expenditure on housing construction (built by State construction companies) (Comite Estatal de Estadisticas, 1982, p. 191). In that year Havana gained about 23 per cent of housing completed by construction companies in Cuba in 1982. When adjusted for population (see table 9.1), Havana's share of investment and units completed is in line with national trends. In fact, between 1972 and 1982, investment in construction in Havana is very closely correlated with national investment ($r = 0.98$), again in keeping with government policy not to favour Havana disproportionately.

Table 9.1 Government housing investment, urbanization, and housing completion by construction companies compared to population by province (1982)

Province	% investment by % population	Urban index	% units completed by % population
Pinar del Rio	0.9	71.3	1.3
Le Habana	1.3	105.5	1.7
Havana City	1.4	143.5	1.2
Matanzas	1.1	109.2	1.0
Villa Clara	1.0	99.6	0.6
Cienfuegos	1.6	104.3	1.5
Sancti Spiritus	0.7	91.5	0.9
Ciego de Avila	0.7	96.7	1.7
Camaguey	0.9	104.6	0.7
Las Tunas	0.6	72.9	0.8
Holguin	0.8	73.0	0.8
Granma	0.6	74.5	0.5
Santiago de Cuba	1.0	90.2	1.0
Guantanamo	0.4	78.5	0.4
Cuba	1.0	100.0	1.0
Mean Index	0.9	—	0.9
Stand. Deviation	0.4	—	0.5

Source: Derived from Comite Estatal de Estadisticas, 1982, pp. 64, 72, 191, 193

Provincial scale: equality in consumption

Efforts to de-emphasize Havana are complemented by the commitment to improve economic conditions and access to goods and services in the rest of the country. By the end of the first decade of the revolution considerable progress has been made in bringing into balance the population, area and economic activity of the provinces (Susman, 1974).

However, remaining inequalities, problems in economic organization and production, and a need (MacEwan, 1981, p. 128) or desire to incorporate the population more fully into decision-making processes as a prerequisite for improving both economic and political performance led, in the early 1970s, to changes in the political and economic structure of Cuba.

These changes followed the national mobilization to achieve an unprecedented ten-million-tonne goal in the 1970 sugar harvest. However, although resulting in the largest sugar harvest ever (8.5 million tons), the 1970 *zafra* proved a failure in the terms of the high economic costs, resource commitment and dislocation, as well as the national pride and psychic investment tied to the ten-million-tonne goal. Castro, in a noted speech, blamed himself for the failure and initiated a process of sweeping changes in the economic and political institutions of Cuba that came to be known as the 'institutionalization' of the revolution.

The reorganization included a heightened emphasis on spatial organization as central to the planning process. In Cuba, 'physical planning has always been, to a greater or lesser extent, of one sort or the other, *also* economic planning' (Menendez and Baroni, 1980, p. 159; my translation). However, in the 1970s, the government undertook a comprehensive study of the provincial urban systems as a basis for integrating the economy and population over the territory of each province. Thus, the First Congress of the Communist Party of Cuba in 1975 redrew provincial and municipal boundaries for both political and economic planning purposes.

Broadly speaking the new structure has two spatial components. First, the number of provinces was more than doubled from 6 to 14, to create more homogeneous and rational political and economic units (see figures 9.1 and 9.2). Also, 407 municipalities (similar to counties in the US) were restructured into an aggregated 169 with an average urban population of 20,290, excluding Havana City. Second, each provincial capital was to serve as the highest order centre of a local central place hierarchy in an increasingly integrated urban–rural landscape. The population-based hierarchy formed by the reorganization is extremely regular, generating a ratio between levels of about 3:1. On average, the ratio of population in provincial capitals to the next level of the provincial urban hierarchies is 3.3:1, and from the second to third levels, the average ratio is 2.9:1 (derived from Comite Estatal de Estadisticas, 1982, pp. 66–8). The explicit intent in designing a regular hierarchy was to provide a spatial context conducive to achieving equality goals in both economic and political terms. Potentially, it maximizes the entire population's access to services, economic infrastructure and political bodies at all levels.

The provincial capitals were also designated as the location for new investment in industries with an emphasis on those strongly linked to local agricultural production. Thus, each provincial capital was to grow as part of an integrated agro-industrial economy. Coming out of the Second Congress of the Communist Party is the goal to

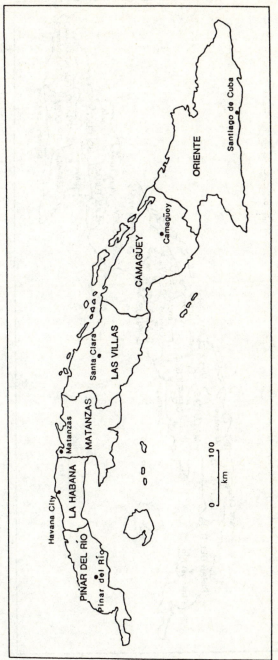

Figure 9.1 Cuba: pre-1975 boundaries

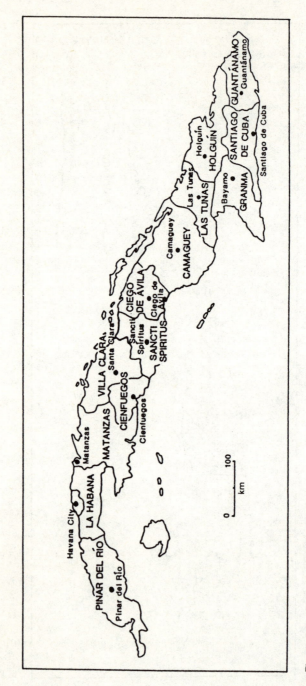

Figure 9.2 Cuba: post-1975 boundaries

Build up the provincial capitals, improving their occupational structure, developing them as industrial centers and consolidate their possibilities to offer services
to the rest of the province. Promote population growth in the municipal centers
– particularly the smaller ones – by developing industrial investments compatible
with their characteristics, and increase their role in offering intermediate services
to the population. (Second Congress of the Communist Party of Cuba, n.d., p.
130)

Investments are directed primarily to agro-industries and other resource-
oriented industries in the provincial capitals in order to establish strong
links between the primary and secondary sectors. There is also the
expectation that induced demand on industries in smaller centres will
generate greater overall provincial economic growth, thus giving the
provincial population an equitable share of national growth and of national
production. In Cuba, links may remain within a spatially contiguous
region such as a province as a consequence of central planning. Agglomeration and urbanization economies may be realized, thus rewarding local
enterprises with surpluses which may be redirected to any number of
options permitted under the System of Economic Direction and Planning
(beginning 1976). Under this system, JUCEPLAN's (central planning
agency) decisions control growth in the provincial economies to the extent
that national policy determines the allocation of resources to particular
sectors in set locations, thus allowing the national government to address
inter-provincial inequalities and other problems.

Several broad indicators of the distribution of population and activity
at the provincial level further illustrate the degree of equality achieved
since 1959. For instance, extending the housing investment example
(presented in the discussion of Havana above), the allocation of national
housing investment in each province is almost proportional to each province's share of national population in 1982 (and the two are highly correlated
$(r = 0.92)$) showing considerable equality over the population and space.
However, this example does not include self-built housing and micro-
brigade construction, both of which would reduce Havana's share in the
national total of new housing activity, reflecting greater relative activity
outside of Havana.

Within each province, there is evidence of a government bias towards
urban housing construction to increase and control urban population
densities as well as to promote the overall goal of urbanizing the countryside. Self-built housing, in both urban and rural areas, constructed outside
the planning process, is the population's response to a housing shortage.
About one-third of all households occupy self-built housing (Hamberg,
1985). In urban areas, self-built housing is almost entirely single-family
detached dwellings, exacerbating problems of low-density sprawl. While
urban self-built housing averaged 30,000 units a year between 1981 and

1983, only 6,000 building permits were issued nationally (Estevez, 1984). Since 1959, Cuba's urban population doubled, but urban land-use tripled. This rapid and unplanned growth exceeded and hindered infrastructural expansion (CTVU, 1984). While rural areas contain one and a half times more self-built housing than their share of households, encouraging dispersed rural housing would generate additional infrastructural costs and make delivery of social services more difficult. Thus, the government directed housing funds to urban areas within the provinces to begin meeting housing and infrastructural needs while preserving urban density standards and also to encourage agglomerated settlements. There is a clear relationship between housing investment *per capita* and the relative rate of urbanization in each province in 1982 (see table 9.1), yielding a correlation coefficient of 0.71.

Another indicator of a high degree of equality throughout the nation is drawn from an examination of median salaries in each province (see table 9.2 and figure 9.3). Based on a national value of 100 in 1982, the median salary in each province ranges from a high of 106.2 in Cienfuegos to a low of 9.16 in Guantanamo, a ratio of median salaries of 1.16:1. This very low differential in median salaries across the country is the product of the government's explicit commitment to promote economic activity in each province.

This commitment to generate equality over the landscape prevailed even when individual wage differentials increased from 4.3:1 (1963) to

Table 9.2   Median annual salary by province, 1977, 1980, 1982 (Cuba = 100)

Province	1977	1980	1982	% increase 1977–82
Cuba	100.0	100.0	100.0	27.6
Pinar del Rio	94.0	94.6	92.4	25.4
La Habana	97.2	101.5	100.8	32.2
Havana City	112.7	108.8	105.2	19.1
Matanzas	101.6	99.2	100.5	26.1
Villa Clara	101.8	100.9	100.3	25.8
Cienfuegos	105.1	102.8	106.2	28.9
Sancti Spiritus	94.5	96.8	99.8	34.7
Ciego de Avila	98.0	98.0	101.4	32.0
Camaguey	99.2	98.1	98.6	26.8
Las Tunas	90.9	94.2	98.9	38.8
Holguin	91.1	96.7	99.2	39.0
Granma	88.6	90.9	94.5	36.2
Santiago de Cuba	94.7	94.6	94.5	27.3
Guantanamo	86.5	89.9	91.6	35.1

Source: Data derived from Comite Estatal de Estadisticas, 1982, p. 118

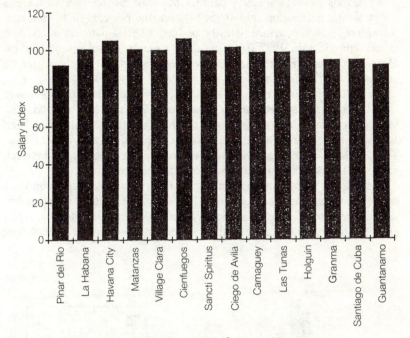

Figure 9.3   Median salary by province, 1982 (Cuba = 100)
(Source: Comite Estatal de Estadisticas, 1982, p. 118.)

5.3:1 (1981) as a result of the 1980 wage reform. The wage reform, implemented to enhance the material incentives of the new Economic System, generated a 15 per cent average salary increase. Also productivity jumped 35 per cent (1977–81), perhaps as a consequence of more vigorous material incentives, namely the 15–25 per cent of salaries reserved for individuals or enterprises exceeding norms (Perez-Stable, 1985, p. 298).

Continually increasing equality in median salary levels in each province is evident not only from the beginning of the revolutionary period to the present, but also since the political and economic reorganization of the mid-1970s. For example, in 1977 the ratio of highest to lowest median salaries by province was 1.3:1. By 1980 it had dropped to 1.21:1, and then reached the lower figures (1.16:1) cited in 1982. While the median salary for Cuba increased by 27.6 per cent (1977–82), it grew by only 19.1 per cent in Havana, thus accounting for less of a gap between Havana and the rest of the country. Havana's shift from having a median salary level index of 112.7 in 1977 to 105.2 in 1982 illustrates the outcome of conscious policy to encourage settlement and economic activity elsewhere, and to reduce the persistent disparities in income levels dating from before the revolution.

When looking at total salary paid (rather than median levels), there is a very strong correlation (r = 0.95) between the per cent of the national workforce in each province with the per cent of total national salary paid in each province in 1982 (see figure 9.4). Again, these figures confirm Mesa-Lago's and Brundenius's conclusions about the high degree of income equality in Cuba.

A careful examination of the data generally supports the notion that since 1959 there has been an increasing equality in access to goods and services across the country. This is not to argue that equality has been achieved, either across all segments of the population or in all locations. Thus, a large-scale map would reveal remaining inequalities between urban and rural areas within provinces and between them.

Yet, simply identifying pockets of inequality by map inspection does not reveal the extent of changes in and the degree of equality or inequality with respect to particular services. Roca (1983, p. 76), for example, identifies the extremes in the medical system, as one example of persistent

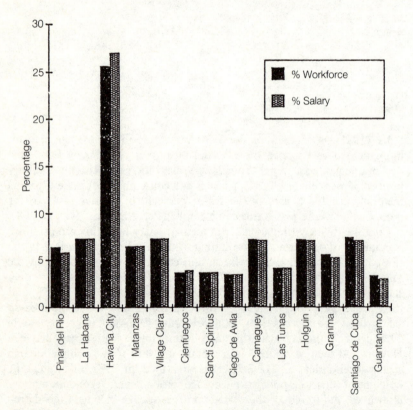

Figure 9.4   Cuba: percentage workforce – percentage salary by province, 1982

service inequality. In the city of Matanzas the population per physician ratio was 263:1 (1980) compared to rural Granma province with a ratio of 1,750:1. He describes the government commitment to equalizing health care by allotting about 60 per cent of new hospital beds '(i)n recent years ... to the most needy rural areas' (Roca, 1983, p. 76). (By 1982, the population per physician ratio in Granma Province had dropped to 1,383:1.) Roca, however, omits the long-standing differences in infrastructure and organization between one of the major Cuban cities and one of the traditionally most isolated and poorest provinces. Rather than examine progress or changes in each, he focuses upon extremes (including in scale between a city and province). Roca also fails to discuss the structure of the medical system in which primary care facilities are almost ubiquitous, located in rural areas as well as in urban areas of all sizes, while secondary and tertiary health facilities, free and accessible to all, are in towns and cities throughout Cuba. It is no surprise that an urban area has a much more favourable ratio. Also, it is not clear if Roca distinguishes between rural and urban as defined in Cuba. The Cuban census considers any agglomeration urban if (Comite Estatal de Estadisticas, 1982, p. 59): (1) it has 2000 or more people; or (2) at least 500 people and four of the following: paved streets; piped water system; sewer system; medical services; educational centre; or (3) at least 200 people and all of the five infrastructural characteristics listed in (2) plus electric street lights. By definition, a rural, rather than urban agglomeration may lack medical services and thus rely upon nearby urban facilities.

Another indicator of equality in the health care system is the provincial figures of hospital beds per 1000 inhabitants (1982) (see figure 9.5). If the scale of analysis is consistent at the provincial level, there is less inequality than Roca suggests. For example, while Matanzas Province has 4.0 beds per 1000 population, predominantly rural Granma Province has 3.6, a lower but close number to Matanzas. The city of Havana, with the most hospitals and highest order facilities serving both the rest of Havana Province, of which it was a part until 1976, and the national population, has the highest ratio of 9.2 per 1000 inhabitants. Outside Havana City, the ratios are quite close overall.

These data suggest a high level of equality in terms of population distribution and access to goods and services at the provincial scale of spatial aggregation. The commitment to equality is made explicit in the five-year plan guidelines formulated at the Second Congress of the Communist Party in 1980. One goal is to promote growth in the least developed provinces by 'increasing participation by the least developed provinces and cities in production and industrial employment' (Second Congress of the Communist Party of Cuba, n.d., p. 124). In order to promote equality in consumption, Cuban development programmes also focus on the sub-provincial scale.

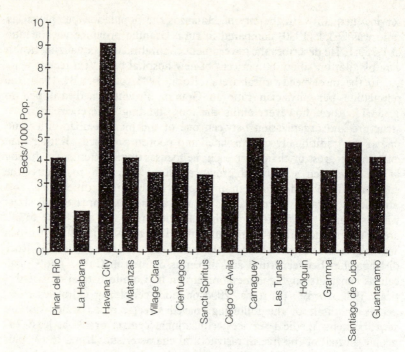

Figure 9.5   Cuba: hospital beds per 1,000 population by province, 1982
(Source: Comite Estatal de Estadisticas, 1982, p. 473.)

Local spatial scale: overcoming urban–rural disparities – income

Programmes aimed at improving living conditions in rural and traditionally underserved areas (including low income areas within cities) complemented the de-emphasis on Havana. Before the revolution, disparities
in living conditions between urban and rural areas were considerable.
Manitzas (1973, p. 5) details gross inequalities in service provision,
health, and infrastructure. Mesa-Lago (1981, p. 142), examining surveys
conducted in 1953 and 1957, concludes that the '(f)igures suggest that
the ratio between urban and rural average incomes was about 17:1.'

According to a private survey undertaken in 1957 in selected rural areas, more
than 50 per cent of rural families in those areas had annual incomes below 500
pesos a year and only 7 per cent earned more than 1,000 pesos a year. The
average family of six had an annual income of 549 pesos, a little more than 91
pesos per capita. (Mesa-Lago, 1981, p. 142)

Mesa-Lago concludes his survey of the data: 'Again, a true comparison
among all these figures is not possible, but they suggest extreme inequalit-

ies in income in Cuba, particularly between the urban and rural sectors.' The commitment of the revolution to improve rural conditions and overcome urban–rural disparities led to a number of policies resulting in a significant increase in rural incomes (Brundenius, 1981, p. 145).

Almost immediately upon taking power, the revolutionary government began to address problems of land tenure and working conditions for agricultural labourers. As a consequence of the Agrarian Reform Law of May 1959 and the guarantee of stable employment to 350,000 cane cutters, their wages were almost doubled. Dumont (1970, p. 33) claims a 20 per cent increase of wages in rural areas during the first two years of the revolution. He further argues that '(i)f account is taken of the increase in the number of work days, which often went from 160 to 200–240 days per year, salaries in general were perhaps raised more than 60 per cent in two years.'

Similarly, in urban areas, the Rent Reduction Law (Ley no. 135) of 1959 contributed to a massive income redistribution to residential tenants amounting to an estimated 80 million pesos. Hamberg (1985, p.5) points out that the only measures with greater income redistribution effects were wage increases (150 million pesos) and elimination of most forms of gambling (100 million pesos). In addition, purchasing power of the poor in Cuba increased 15 million pesos due to lower prices for medicine and another 15 million as a result of a reduction in electricity rates. Brundenius (1981, p. 146) estimates a total increase in purchasing power of 382.5 million pesos as a result of all these programmes. When compared to income changes in other Latin American countries, increases in Cuban income equality are more striking. Brundenius (1984a, p. 39) compares Cuba to Brazil and Peru. Whereas in Cuba the share of national income of the poorest 40 per cent increased almost four times (1958–78), in both Brazil and Peru the share of the poorest 40 per cent decreased while the share of the richest 5 per cent increased (see table 9.3). Cuba's GDP *per capita* grew at an annual average of 2.7 per cent (1980–82), a performance better than the rest of the Caribbean, Central America and most of South America, enabling it to equal the Latin American average (Brundenius, 1984a, p. 35).

Mesa-Lago (1981, p. 144) argues that the greatest benefits to the poorest segments of the population occurred in the early years, and thereafter benefits accrued more to the middle-income groups. But Brundenius (1984b, pp. 109–10) presents a picture of steadily increasing income equality as the gap between agricultural and industrial workers diminished. He also finds that between 1962 and 1973, agricultural workers' average income increased from 954 pesos to 1,416 while industrial workers' average income decreased from 1,941 to 1,603 pesos. Furthermore, relying on Mesa-Lago's data but assuming a smaller per cent of workers in the highest-pay category, Brundenius estimates slightly

Table 9.3 Income shares: Cuba, Brazil, Peru

Country	% national income lowest 40%	% national income upper 5%
Cuba		
1958	6.5	26.5
1978	24.8	11.0
Brazil		
1960	11.5	27.7
1980	9.9	34.2
Peru		
1961	10.0	26.0
1979	8.2	37.3

Source: After Brundenius, 1984a, *Cuadro*, no. 13, p. 39

smaller income differentials between agricultural and industrial workers for 1978 than for 1973, showing a continuation of the equalization trend.

Throughout the revolutionary period, the standard of living seems to have been rising for the vast majority of Cubans. In addition to direct cash income, examination of the broad category of basic needs leads to the same conclusions (and thus supports Eckstein's point made above). Brundenius (1981) documents achievements in all areas of basic needs' provision. Based on both Cuban and Canadian prices, he constructed indices of basic needs and observes (1981, p. 132) 'A comparison of the two indices shows that per capita expenditures on basic needs were 52 per cent higher in 1980 than in 1958 if measured in Cuban prices, but 88 per cent higher if Canadian prices are used'.

Thus, over the population as a whole, income measured in both cash and basic needs' terms has increased and become more equal, whether looking at the national scale or at differences between urban and rural areas.

Local spatial scale: education in the countryside

One of the emphases of the Cuban revolution, from its inception, has been on education. Two aspects of the educational system concern us here. One is the commitment to providing educational services to the underserved rural population. A second is the deliberate use of the educational system to bring urban and rural populations together as a means of overcoming attitudes of domination (and subjugation). This includes integrating all students into productive activity so they come to appreciate the effort and skill involved in providing their subsistence while, at the same time, contributing to their own upkeep and to society.

In 1961 the government launched the famous 'Literacy Campaign'. About 100,000 student volunteers and another 140,000 adults served as teachers to the estimated million illiterate adults, mostly in rural areas. The texts and techniques used in the programme directly referred to and relied on identifying and discussing important socio-economic issues affecting the population.

In addition to teaching people how to read, the literacy campaign taught people how to teach and how to teach people how to teach. And perhaps of greatest importance, the literacy campaign was a training ground for organisational and leadership abilities. In a process of learning by doing, the participants in the literacy campaign learned how to mobilise themselves. (MacEwan, 1981, p. 77)

Also, because the (urban-originating) volunteer teachers went to live with rural families and worked in the fields during the day and taught in the evenings, they came to know rural life. Kozol quotes Dr. Mier Febles, 'one of Cuba's oldest Marxist educators':

The goal of the campaign was always greater than to teach poor people how to read. The dream was to enable those two portions of the population who had been most instrumental in the process of revolution from the first to find a common bond, a common spirit, and a common goal. The students discovered the poor. Together, they all discovered their own *patria*. (Kozol, 1978, p. 22)

The programme that began in 1971 to locate secondary schools in the countryside is another example of the commitment to providing educational opportunities to the rural population while integrating urban and rural populations. These schools would ensure that city children would have rural experiences. In addition to regular classes the students spend part of each day in agricultural labour. Aside from the ideological commitment to bringing urban and rural people together, students help alleviate the agricultural labour shortage. When the programme began in 1971/2, only seven (or 1.5 per cent) of 478 secondary schools were located in rural areas. By 1979–80, 533 (or 40.4 per cent) of 1318 secondary schools were in rural areas (Comite Estatal de Estadisticas, 1980, p. 220). About 37 per cent of junior-high students and 47 per cent of high-school students are now enrolled in these schools (Leiner, 1985, p. 35).

The educational system involves temporarily shifting urban populations to rural areas. By offering easy access to education for rural populations, it constitutes part of the overall effort to improve conditions in the formerly under-served rural areas. Also it is designed with the ideological intent of eradicating elitest attitudes in part by the focus on agricultural labour. Thus, the educational system may be viewed as having the role of 'ruralizing' the urban population's outlook with the effect of further integrating the entire national population.

Local spatial scale: new towns programme: 'urbanizing' the countryside

From the beginning of the revolutionary regime, the importance of providing services to the rural population was central to planning concerns. In 1959, the government instituted a programme to construct new towns to accomplish various objectives. One was to locate population in concentrated communities as part of the organization of the State-farm system. By ceding their land to the State, and thereby facilitating large-scale agricultural production, farmers gained life-time pensions for the land they sold as well as access to housing in the new communities when available. A second goal of the new towns programme was to create a spatially efficient setting for providing modern infrastructure and social services to the population. Education and health services, water and power for example, could be provided much more cheaply to clustered settlements than to a dispersed population. Constructing housing in the new communities to replace some of the dispersed rural substandard dwellings became the expression of government rural-housing policy as well as a means of inducing rural dwellers to particpate in the State farm and later producer-co-operative agricultural system. Under the 1985 Housing Reform, co-operative members, along with the rest of the population, now build their own homes with loans from the State bank.

By 1964, 150 new towns with 26,000 housing units had been constructed. These were generally quite small, with some containing only about twenty houses, although, there were larger settlements with a few thousand units each (Segre, 1980). By 1982, there were 360 new towns housing about 4.5 per cent of the rural population.

Initially, the new towns were associated with the creation of State farms. In 1977 Castro announced a new programme to encourage small private farmers to form voluntarily producer co-operatives (still owned by members) centred in new communities (or in existing communities) constructed by residents. In a speech to the Fifth Congress of the Association of Small Farmers (ANAP), Castro (1977) estimated that it would take thirty years for the government to be able to provide basic infrastructure and services to private farmers if based on the new towns programme alone – hence, the emphasis on self-built housing by members of co-operatives. A second aspect of the co-operative movement reflects the ideological orientation of the revolution:

The construction of socialism requires that the socialist relations of production be developed and extended until they are the only kind of relations in the country.

In Cuba's present condition of the socialization of the basic means of production, the grouping of small farmers' holdings in agricultural production cooperatives is set forth as an absolutely necessary goal. This should be achieved completely voluntarily, through a gradual process of integration in cooperatives. (Second Congress of the Communist Party, 1981, p. 172)

Furthermore, the co-operatives are viewed as potential urban agglomerations and the goal, as elaborated at the Second Congress of the Communist Party, is to promote settlement by 'agricultural workers in rural areas, simultaneously initiating the process of urbanizing cooperatives, in line with their size and level of development' (Second Congress of the Communist Party, 1981, pp. 204–5).

The Cubans, along with foreign commentators such as Gugler (1980) and Hamaberg (1985), among others, point to the new towns programme as a means of 'urbanizing' the rural population. However, in Cuba, 'urban' may have little relationship to the definition prevailing in other countries. As defined above, 'urban', as used in Cuba, is based on both population size and the attributes of particular forms of infrastructure and services. The goal of 'urbanizing' the countryside may be expressed more clearly, perhaps, by the early notion of 'city' derived from *civitas* referring to 'the body of citizens rather than a particular settlement or type of settlement' (Williams, R., 1983, p. 40). Viewing the 'city' as that space and population fully integrated into a State with its laws and systems of production also allows the historical specificity of 'urban' to be examined in a particular context. Castells (1977, p. 17) chooses, in this light, to speak of 'the theme of the social production of spatial forms rather than speak of urbanization.' Holton (1984, p. 15), drawing on Pahl (1968), further argues that the terms 'urban' and 'rural' do not have a specific 'generic content' and that the traditional sociological view of a continuum between urban (modern) and rural (traditional) cannot be linked 'to any consistently definable contrasts in social organization, economic function, or a set of cultural values'. Thus, the terms are better understood in the context of the broader social relations of production.

In Cuba the goal of urbanizing the countryside really takes on the meaning of providing access to at least the minimal level of goods and services consonant with the social goals of improved service provision and enhanced equality. Thus, it is no surprise to find a high rate of urbanization in Cuba, with the urban share of population increasing from 60.5 per cent in 1970 to 69.7 per cent in 1982 (see table 9.4).

A question that remains to be answered is the extent to which increased equality in the consumption sphere over the population and space has been accompanied by increased equality in the production sphere, a crucial component of the socialist vision.

Increased democratization post-1970: production equality

The sum of the changes, thus far, suggests a commitment to increasing popular participation in decision making in both civic and economic matters. Although observers differ in their assessments of the extent of democratization and participation in one or another facets of the new

Table 9.4   Urban population proportion, 1970–82 (nation and provinces)

Provinces	1970	1982	Increased urban proportion
Cuba	60.5	69.7	9.2
Pinar del Rio	38.1	49.7	11.6
La Habana	68.1	73.5	5.4
Havana City	100.0	100.0	0.0
Matanzas	62.2	76.1	13.9
Villa Clara	54.0	69.4	15.4
Cienfuegos	59.2	72.7	13.5
Sancti Spiritus	50.8	63.8	13.0
Ciego de Avila	54.7	67.4	12.7
Camaguey	64.9	72.9	8.0
Las Tunas	34.2	50.8	16.6
Holguin	39.0	50.9	11.9
Granma	39.9	51.9	12.0
Santiago de Cuba	51.3	62.9	11.6
Guantanamo	40.8	54.7	13.9

Source: Comite Estatal de Estadisticas, 1981, p. 41; 1982, p. 64

programmes, the conclusion drawn here is that the overall direction is, indeed, towards greater participation. Even if participation is less than a textbook image of socialist democracy, the question remains of why would the government initiate and propagandize the new system? Would a government not wanting much more participation raise expectations only to dash them?

People's power – political participation

As discussed above, the failure to achieve a ten-million-tonne harvest in 1970 prompted the institutionalization of the revolution. After several years of preparation, the First Congress of the Communist Party adopted a plan for new economic and political structures in 1975 the implementation of which began in July 1976. Also, a new System of Economic Planning and Direction was instituted, giving local enterprises more responsibility and accountability for efficiency, profits and losses. Politically, the system of People's Power regularized popular participation with new electoral and voting processes, including the right to recall elected delegates. ('By February 1979, 108 out of 10,725 municipal-level delegates had been recalled' (Ritter, 1985, p. 280)).

The new system was first implemented experimentally in Matanzas Province. Decentralization of decision-making in locally oriented production and service activities became a central part of the new economic system. Speaking of the changes to occur in that province, Castro said in his 26 July 1974 speech:

A total of 5,597 production and service units throughout the province will be handed over to the organs of People's Power. This is the basic criterion: all production and service units that serve the community – that is the grass roots – must be controlled at the grass-roots level. (Castro, 1983, p. 200)

Based on the outcome in Matanzas Province, the First Congress of the Communist Party of Cuba adopted People's Power and the new economic system for the entire country. As part of the programme, government functions and lines of authority were streamlined and five-year economic and physical planning processes were initiated across the country.

People's Power includes a representative electoral system based on direct elections at the local (municipal) level and indirect elections at the provincial and national levels. Indirect elections are supposed to ensure that particular interest groups are represented along with popular representation in general. In addition to economic planning and issues being part of the legitimate concern of political bodies in Cuba, the new system of People's Power gives the local level of government direct control over locally oriented economic production and service sectors. In short, the new system of economic and political organization was designed to increase significantly direct popular participation in decision making with respect to production activities, at least those affecting the local community.

However, priorities are still established at the national level. When viewing the political goals in the context of socialist democracy, conflicting objectives arise. On the one hand, a goal may be to integrate increasingly the entire population, wherever they live, in decision-making processes affecting not only the political sphere, including the distribution of benefits, but also the realm of economic production. Only when the population is fully engaged in both political and economic spheres together can socialist democracy be realized. On the other hand, total decentralization of decision-making may prove antithetical to spatial-equality objectives such as promoting even development. Even development over the entire country requires, as the Cubans see it, centralized co-ordination in at least the economic sphere. People's Power, in its democratic centralist structure, seems to be an effort to increase popular participation within a context of nationally determined goals.

Establishing the Organs of People's Power (Organos del Poder Popular) was a natural continuation of the 'institutionalization' of the revolution that had begun in the early 1970s. Eckstein (1981, p. 132) cites two studies conducted in the late 1970s indicating high rates of political participation in the new system. However, she argues that the true extent of democratization is still quite limited and the new system created a situation in which 'skilled, productive, and politically well-connected workers came to monopolize a disproportionate share of available scarce, valued resources' (Eckstein, 1981, p. 133). Ritter (1985, p. 280) shares Eckstein's scepticism. On the other hand, Ritter refers to a 1984 report based on a 1979 study. In it, B. Jorgensen (1983) concludes:

OPP [Organs of People's Power] has been an important instrument in improving local service systems ... the OPP seems to have functioned effectively in leading People's energies towards making solutions to their own problems. At this level, the OPP decentralize decision-making and place the discussion of problems close to the problems themselves. (in Ritter, 1985, p. 280)

The available evidence suggests the population as a whole is more involved in selecting leaders, discussing problems and asserting their political and economic interests at the municipal (local) level of People's Power. At the provincial and national assembly levels there is less participation evident than at the municipal level, both because of indirect election to these assemblies and because national leadership still establishes the political and economic agenda at the provincial and national levels rather than officials at each constituent level of the system. Furthermore, final approval of the agenda rests with the top levels of the Communist Party. Rabkin (1985, p. 258), for example, citing both *Granma* and William LeoGrande, says '(T)he Politburo customarily reviews draft legislation before it is submitted to the (National) Assembly and decides certain policy matters on its own authority' (parentheses added). At all levels, enterprises now under the jurisdiction of local People's Power authorities still are constrained by national economic plans, pricing policies and resource-allocation decisions. Nevertheless, local control over some services and locally oriented production occurs at both the level of political institutions and within workplaces themselves.

System of Economic Planning and Direction – economic participation

Under the new System of Economic Planning and Direction and People's Power in the workplace, workers are more involved in decision making with respect to production plans as well as allocation of housing and luxury goods. Zimbalist (1985, p. 221) cites evidence suggesting an increase in worker participation from the early 1970s, with the beginning of the institutionalization process. Surveys conducted in 1975 by Perez-Stable and in 1979 by Herrera and Rosenkranz suggest that workers believed they had a significant voice in issues concerning their particular workplace (Zimbalist, 1985, pp. 220–1):

According to Castro's reports to the First Party Congress (1975) and Second Party Congress (1980), the number of workers participating in the discussion of the annual economic plan in their enterprises rose from 1.26 million in 1975 to 1.45 million in 1980 (or by 15 per cent). (Zimbalist, 1985, p. 221)

Also, efforts to revitalize unions as more independent organizations led to new union elections, in many areas resulting in new leadership replacing incumbents: 'A 1984 study of Cuban unions by Linda Fuller finds that workers have more power today than in the past to dismiss bad management' (Zimbalist, 1985, p. 221). According to Eckstein (1981, p. 131), at the grass-roots level there is much more participation in both political and economic decision making.

Data on local institutional control of economic activities before and after the creation of People's Power are scarce. Nevertheless, it is clear that the local level of the State sector (as against self-employed and agricultural sectors) has maintained a stable share of a growing economy during the period following reorganization. For example, between 1978 and 1982, the local State sector (municipality and provincial levels) maintained control of a fairly constant 8.5 per cent to 9 per cent of the total State sector (see table 9.5).

The State sectoral component of the 32.3 per cent (1980–2) increase in global social product may be disaggregated to local State sector growth of 25.8 per cent and national State sector growth of 28.1 per cent, excluding commerce from the calculations (Comite Estatal de Estadisticas, 1980, 1981, 1982). If commerce were included, it is likely that the local component would exceed the national because under People's Power commerce is now controlled mostly by provincial and municipal authorities. Similar cases of locally or provincially controlled sectors growing more rapidly than the global social product are becoming more common. For example, between 1980 and 1982, locally controlled non-ferrous mining and metals jumped 157.1 per cent; wood and forest industries

Table 9.5 Share of local State sector in total State sector, 1978–82

	National State sector (millions of pesos)	Local State sector	% Local
1978	9403.4	931.8	9.0
1979	9689.6	931.8	8.8
1980	11794.2	1158.3	8.9
1981	14615.6	1363.1	8.5
1982	15109.7	1456.7	8.8

Source: Comite Estatal de Estadisticas, 1979, 1980, 1981, 1982

increased 139.9 per cent; construction grew by 51.2 per cent; and transportation by 87 per cent. In each of these sectors, there has been an effort to shift responsibility to the local levels as part of the economic reorganization.

Additional decentralization of decision making in the economic system is also a result of recently permitted freer trade among different enterprises without going through the central government. Thus, local economic links are stronger while bureaucratic hindrances diminish. Whether worker participation in decision-making increases and whether the Cuban economy is able to continue to grow while promoting social equity goals has yet to be seen. However:

(a)ccording to an August 1982 study by the Cuban Central Bank, Gross Social Product in Cuba grew at a real annual rate of 7.5 per cent from 1970 to 1975 and 4.0 per cent from 1976 to 1980; real growth in 1981, 1982, and 1983 was reported at 12 per cent, 2.7 per cent, and 5.2 per cent respectively. (Zimbalist, 1985, p. 226)

In 1984 the Cuban economy grew 7.4 per cent, and in 1985 at 4.8 per cent, leading to an average annual 7 per cent growth rate in the most recent five-year period, 1981–5 (*Cuba Update*, Winter–Spring 1985, p. 15; *Granma*, 31 December 1985, p. 2). As Zimbalist says (1985, p. 226): 'These figures represent a very creditable growth rate, particularly given the severe international recession and low commodity prices over most of the period'.

Transferring control over some locally oriented economic activities to local authorities does not mean, at this time, a decentralization of the fundamental planning decisions for Cuba. Yet it is a means of being more responsive to popular will. At the same time, Cuban aggregate economic performance appears to be satisfactory and improving. Whether this reflects the contribution of enhanced popular participation cannot be determined, yet it is highly suggestive of the importance of changes in political–economic relationships in Cuba to quantifiable outcomes.

Conclusions

In this chapter, issues of equality and, more specifically, socialist equality, have been examined in the context of Cuba. It has been argued that socialist equality, in principle, encompasses both the spheres of consumption and production. Furthermore, socialism as a system of equality, necessarily brings the consumption and production spheres together with the political system in a direct manner. Without popular participation in decisions regarding the allocation of social resources and the means of production, equality cannot be achieved. Consequently, socialism implies democracy.

Equality and full participation in decisions concerning resource allocation and distribution of benefits further leads towards reducing or, perhaps, eliminating alienation as understood in the Marxist sense. This necessitates eliminating class differences that prevail in capitalist societies. It also may necessitate restructuring the spatial organization of society to minimize inequalities between places and areas and to integrate each part of the country into socialism. Thus, efforts to construct socialism entail promoting equality over consumption and production spheres, necessitating increasing democratic participation in all aspects of decision-making, and creating a landscape of socialist integration, enhancing equality over the population in all spheres of activity.

This chapter examined a variety of policies and programmes designed to correct inequities of the past and promote equality in contemporary Cuba with special attention to their spatial aspects. Many indicators suggest a considerable degree of equality over the population and over space, at different scales. Havana's overwhelming dominance in the national space-economy has been diminished, although it continues. The advantages of agglomeration economies and existing infrastructure in Havana contribute to its persistence as a dominant centre. Efforts to overcome disparities at the provincial level include the deliberate spatial allocation of investments, services, housing and employment opportunities. Focusing investment in provincial capitals and linked urban systems ensures a greater distribution of economic activity across the country and out of Havana.

Provincial capitals (including Havana) offer the clearest example of large-scale urbanization in Cuba. These cities provide the greatest contrast with the smallest urban settlements as well as with rural areas. At the level of the provincial capital 'urbanization' may continue to have a traditional meaning. However, at the smallest scale, notions of urbanization may be better expressed in the context of socialist spatial integration.

Socialist spatial integration in Cuba is the goal of steadily increasing urbanization. This has and will continue to have the effect of eliminating population dispersal and pockets of low participation in the national political and economic life. While bringing the population together in clustered settlements offering better access to services, urbanization is also part of the process of integrating the population into political and collective economic participation. As relatively high degree of equality in services and consumption has been achieved, the Communist Party and the government introduced structures for much more participation in production sphere decisions and in the political and economic life at the local level. An interesting question that emerges from this study is whether increased participation at the local level in both consumption and production decisions is a consequence of urbanization and changes in consciousness. As settlement density increases, people may better recognize common interests and voice their concerns. (This is, of course, similar to the contradiction in capitalism of bringing workers together in the workplace for efficiency and economies of scale; such amassing also allows them to identify common interests to organize around.) Given encouragement by the State to individuals to participate in mass organizations and local organs of People's Power (among other groups), it would seem reasonable to conclude that there is a definite relationship between government programmes to move the country towards socialism and a clear need to include the population more actively in the process. It is likely that, in addition to government prompting, demand for more participation comes from the grass roots.

The process of creating a landscape of socialist integration is one of contradiction between tendencies towards centralization of economic activity and political power, on the one hand, and decentralization, on the other. Generating increased equality in the consumption sphere over the population and space has the effect of reducing differences between people and places, albeit in a positive way. Guidelines established by a central authority for delivery of equal local services may favour linking settlements together to fit better into centralized delivery networks. Even with more decentralization of service provision under People's Power, as part of an effort to reduce waste in delivery and to make services more responsive to local need, homogenizing the landscape seems the trend in the consumption sphere. Also, to the degree that national government planning determines investments and production issues, centralization of economic and political power prevails. On the other hand, the Cubans have, in recent years, decentralized decisions in both economic and political spheres. Locally oriented economic activity may fall under the jurisdiction of the local organs of People's Power. Greater participation within the workplace may give the labour force much more input into the organization and intensity of production and related decisions. Also, the institutionalized political system is designed to promote local political mobilization. Although the political agenda is established at the national level (centralization), again mobilization may lead to expressions of popular will not before heard, including calls that would lead to greater local identity.

In sum, Cuban efforts in the consumption sphere seem relatively successful. Since 1975 and the introduction of People's Power and the System of Economic Planning and Direction, a commitment to greater popular participation in production sphere and political decisions, in general, is evident. An important component of the post-1975 changes is the spatial reorganization leading to more homogeneous and economically rational provincial and municipal structures. The more even distribution across Cuba of the new provincial capitals, compared to the old capitals, facilitates more even investment patterns and growth over the country. Linking larger cities to each other, on the one hand, and to local urban systems, on the other, further contributes to possibilities for achieving equality in infrastructure and economic activity.

To the extent that the population is brought into fuller and more active participation in all locations in both direct political decisions as well as in issues emanating from the workplace, it would appear to be part of a movement towards socialism, including full equality. Cuba's record in consumption supports this conclusion. In the production sphere and in direct political participation, the promise of democracy has been made and some clear steps have been taken. The outcome has yet to be seen.

10
Socialism, Democracy and the Territorial Imperative: A Comparison of the Cuban and Nicaraguan Experiences

David Slater

> Without general elections, without unrestricted freedom of press and assembly, without a free struggle of opinion, life dies out in every public institution, becomes a mere semblance of life, in which only the bureaucracy remains as the active element.
> R. Luxemburg, *The Russian Revolution*, Ann Arbor, 1961, p. 71, quoted in Poulantzas, 1978, p. 253

> the political behaviour of revolutionary leadership must be conceptualized as the development of a popular hegemonic practice that is differentiated from the models usually associated with the 'dictatorship of the proletariat'.
> Coraggio, 1984, p. 5

The diverse faces of socialism

Haunting many of the discussions on post-revolutionary societies and territorial re-organization is the central problem of defining what is or can be meant by socialism. In my view, this problem, intricate and involved of itself, has been made even more tortuous through the proliferation of a literature that tends to evade the pivotal difficulties of setting out the contrasting meanings and definitions of socialism. Nowhere is this more dramatically evident than in the context of the general discussion of Third World development[1] For within this broad sphere there exists a veritable plethora of terms ranging from, for example, the notion of 'African socialism' to those more geographically constrained concepts of Algerian, Chilean or Tanzanian 'socialisms'. Moreover, a series of longer-established and conflicting categories of 'State socialism', 'State capitalism', 'workers' States', 'State bureaucracies' and so on has often been applied to a wide gamut of African, Asian or Latin American situations.

In his consideration of this issue, White (1983, p. 2) suggests that the term 'socialist societies' is too uncritical, whilst 'state capitalist' remains 'unconvincingly damning'; and again 'non-capitalist' is too vague just as 'transitional societies' by-passes the political nature of the route and destination of the 'transition'. White prefers the term 'proto-socialist', leaning thus in Bahro's (1978) direction;[2] or 'State socialist' denoting the highly *étatised* character of this form of socialism. He further contends that countries such as Cuba, Mozambique, Vietnam, and PDR Yemen share at least two basic structural commonalities; first, that they have broken the autonomous power of private capital in the political sphere, and second that they have carried through (in varying degrees) nationalization of industry, socialization of agriculture, limitation of markets, the setting up of centralized planning systems, and the establishment of a 'politico-ideological system bent on the transition to an ultimate communist society'. At the same time White posits a distinction between actual 'proto-socialist' societies and a future 'fully socialist' ideal which would conform to a society bereft of classes and the State, exemplifying thus political democracy and 'conscious control of the social economy by the associated producers'. Accordingly, 'proto-socialist societies could be said to be genuinely engaged in 'the transition to socialism' to the extent that efforts are made and institutions designed in such a way as to pre-figure or increasingly to embody the eventual forms of 'full socialism'. And, 'without this dynamic, "socialist society" calcifies into a static "mode of production" which imposes structural and institutional constraints on progressive change' (White, 1983, p. 2).

There are a number of questions which flow from White's situating remarks but here I shall just mention two. Since 'political democracy' is taken to be a key constitutive element of 'full socialism' does this imply that in the 'proto-socialist' phase the seeds of democracy are already present, and what precisely do we mean by 'political democracy'? Second, are there any tensions between the 'Marxist–Leninist' vision of socialism, often invoked and defended in the post-revolutionary societies of the periphery, and the essential substance of democracy – and if so, how are these tensions to be dissipated? Before examining these questions in some more detail I want to outline four perspectives or approaches to 'socialism', as a way of attempting some clarification of the analytical terrain. Next, I intend to contrast the Cuban and Nicaraguan experiences, taking the debate on Leninism and popular democracy as an organizing theme. Subsequently, I shall consider the issue of decentralization and the territorial dimensions of political strategy, again drawing out some differences between the Cuban and Nicaraguan examples. Finally, these differences will be located in a concluding section which sets out some general implications for revolutionary change in peripheral societies.

Social democracy as socialism

In the capitalist democracies of Western Europe there exists a long-established political tradition that equates socialism with using the State to develop social welfare, redistribute income, nationalize certain sectors of the economy and promote social justice. Also, within this tradition, it is contended that through winning power in parliament the interests of the working class can be thus represented. This representation and implied or stated belief in the protection of the material well-being of the working class is not linked to any direct attack on the political dominance of capital. In fact, in this sense, socialism is assumed to be attainable without superceding capitalist relations of production and overall bourgeois hegemony. Hence, and expressed in class terms, at the root of this conception of socialism lies a debilitating and contradictory belief – namely, that the interests of the class which produces surplus value for capital can be safeguarded without challenging the dialectically opposed interests of capital whose reproduction depends on the continuing exploitation of that same working class. In time of crisis this contradiction becomes more transparent, giving rise to a series of damaging political effects for the social-democratic and labour parties. One can reasonably argue therefore that this vision of socialism, is more appropriately captured by the term 'social democracy' where the essential intent is to reform rather than replace capitalism.[3]

Radical nationalism as Third World socialism

Under this second rubric we can place a considerable number of peripheral capitalist societies, ranging from Tanzania and Zimbabwe to Jamaica under Manley, Chile under Allende, Egypt under Nasser and present-day Algeria. Notwithstanding obvious variations in historical circumstance and ideological content three common factors can be discerned.

In all cases, and in contrast to Cuba and Nicaragua for example, no structural break has been made from the internalized dominance of international capital. These societies remain, or in previous periods remained, integrated into the world capitalist economy in ways which prevent the nation-State from assuming domestic control over the development of productive forces and the planned re-investment of the social surplus product based on a national strategy of social and economic transformation. In this sense therefore they remain peripheral capitalist societies.

Second, these societies are characterized by the presence of political regimes which, in differing contexts and to varying degrees, enunciate programmes of social reform, redistribution of income, promotion of social justice and equality and the nationalization of private property, albeit at contrasting levels and with diverse degrees of intensity. However, as with the example of social democracy under advanced capitalism, there is no abolition of the capital relation and no programme for the super-session of capitalist production – instead one finds that distribution becomes the cynosure of political discourse.[4]

Last, and perhaps in relation to the first feature somewhat paradox-ically, the protagonists of Third World socialism express anti-imperialist positions, which at least in terms of political idiom find a close parallel with the post-revolutionary societies of the periphery. The underlying reality, nevertheless, is quite distinct since eloquent denunciations of imperialism have not been concurrent with a revolutionary mobilization of popular forces and the establishment of a genuinely independent political base. Conversely, the genuine vehemence of an anti-imperialist discourse has evoked a certain dissonance in relation to the norms and values of West European social democracy. Consequently, it should not be assumed that 'Third World' and 'First World' social reformist dis-courses are synonymous.

The Leninist model and its enduring authoritarianism

With this third approach which has its origin in the Bolshevik Revolution, production, the party and the 'dictatorship of the proletariat' are vested with a cardinal importance. State ownership and control of the means of production, consequent upon the revolutionary seizure of power, is viewed as a fundamental condition for the development of socialism. In the Soviet Union and elsewhere – Cuba, North Korea, Vietnam – the State, within the Leninist perspective, is assumed to embody the interests of the proletariat in a society freed from class contradictions. Predomi-nantly, the problems of social–economic development are situated in a technical discourse that prioritizes the significance of the productive forces and locates the terrain of politics within a pre-given sphere of 'scientifically' proven theses.[5]

Moreover, within the Leninist model, it is already preconceived that the workers are represented by a State which protects and carries forward their interests. But the reality has been different. Claudin-Urondo (1977), for example, in a well-argued and perceptive analysis of Lenin's views on culture and the organization of the labour process in the Soviet Union convincingly shows that it was Lenin's own approach to labour organization which took away from the proletariat the possibilities for popular control at the point of production. As she expresses it: 'the dialectic of the disintegration of popular control ... is ... *both cause and effect* of a situation in which the masses are more 'acted upon' than 'acting' (emphasis in original) (Claudin-Urondo, 1977, p. 99). In the context of political control, the proletariat not only delegated its power to the party in the sphere of societal leadership, but also in the organiz-ation of the factory and the production process, it did likewise. For

Lenin, management had to be 'entrusted by the Soviet power to capitalists not as capitalists but as technicians and organizers, for higher salaries' since 'it is precisely these people whom we, the proletarian party, must appoint to "manage" the labour process and the organization of production, for there are no other people who have practical experience in this matter' (quoted in Claudin-Urondo, 1977, pp. 96–7). However, blind obedience does not engender a spirit of initiative and as Claudin-Urondo observes the fact that the 'man in charge is "Soviet"' does not alter the process by which functional subordination leads to social subjection (Claudin-Urondo, 1977, pp. 96–7).

This social subjection and its concomitant party-directed authoritarianism provided one area of critical concern for Rowbotham (1980) in her influential examination of left-wing political organization and the women's movement. In this example, the political setting was contemproary Britain rather than the Soviet bloc countries or related post-revolutionary peripheral societies. Nevertheless her critique of the values and centralist forms of political organization associated with Leninism, and, in the UK example, Trotskyism, possesses a much wider relevance.

Similarly, and more recently, Ryan (1984) in a broad treatment of deconstruction and Marxism takes a critical look at Leninism, relying for a good part of his argument on Rowbotham's earlier text. He advances the view that the revolution is not simply the embodiment of the party and the question of revolution, its inner meaning, changing course etc., must be kept open. However, the Leninist party closes the question of revolution, 'putting the proletariat back to work, and declaring the revolution over after a transfer of power which retains domination'. According to Ryan this attitude 'arises from a fear and distrust similar to that which motivates the metaphysical desire for a seamless world, one without fissure, rupture, heterogeneity, or crisis'. In a related fashion, 'Leninist metaphysics is founded on an overcoming of crisis (the potential proliferation of movements and the dissemination of workers' power) through the abstract and formal disciplinary party form' (Ryan, 1984, p. 194).[6]

As is well-known, the party holds a pivotal position in the Leninist model of socialist development, and it is the nature of the political tenets on which the revolutionary party is founded and maintained that normally gives rise to the critique of authoritarianism. To a certain extent, Polan (1984) follows this line too, drawing attention to the intolerance, exclusivity, hierarchical structure and pyramidal concentration of effective power of the Bolshevik party. For Polan the Leninist party has two ominous qualities: an extremely rigid and centralist character, summed up in the term 'democratic-centralism'; and, in its relations with the outside world, it arrogates to itself the privilege of possessing, in contrast to other political tendencies, a 'scientific', almost omniscient status. Despite the existence of these authoritarian characteristics, Polan argues that the predominance of 'democratic centralism' is not sufficient to account for the decline of democracy in the Soviet Union. Instead this decline has to be rooted in Lenin's vision of the State form in general.

Hence, Polan locates the problem of authoritarianism not simply at the level of the party but rather in terms of the constitution of the state itself: 'Lenin's state form is one-dimensional' and 'it allows for no distances, no spaces, no appeals, no checks, no balances, no processes, no delays, no interrogations and, above all, no distribution of power'. In this way, administration or bureaucracy and the party become conflated and the institutions of the Soviet State – the factory committees, the party cells – 'take on the culture of administrative apparatuses, forced to accept the limited powers and rights of knowledge and discussion more appropriate to the administration' (Polan, 1984, pp. 128–9). There are no real divisions of power and the State form becomes unilinear.

Given this extreme concentration of political power, legitimation becomes crucial, and it is here that the Leninist or Marxist–Leninist discourse comes to play the role of ideological cement and monolithic sanctifier of a world view. In Polan's (1984, p. 30) words, 'it colonizes the whole planet of thought and leaves no enclaves from which resistance may be mounted'. The field of democratic politics is purified of difference and thus rendered obsolete.

It is not my intention to attempt any tracing out of the Leninist model of socialism,[7] as it has been adopted in a number of post-revolutionary Third World societies, but, certainly, if we survey many of these societies (Cuba, Ethiopia, North Korea, PDR Yemen, Vietnam) it is possible to detect many connections with the Leninist vision, even if there are, quite naturally, other *sui generis* features of political ideology and organization. I *do* intend, none the less, to relate the Cuban experience to the Leninist model, but before doing so, it is necessary to sketch out a fourth vision of socialism, which, I believe, can furnish some elements for a consideration of the Nicaraguan case, to be taken up along with Cuba in the following sections of the paper.

Towards a democratic socialism through popular hegemony

In order to understand more clearly what can be meant by democratic socialism and popular hegemony let us first recall that the classical Marxist approach to socialism expressed the view that social differences could be assimilated within an over-determining and unified class perspective.[8] Consequently, the national question, the problem of the intellectuals and the historical challenge presented by the peasant question could be embedded and explained within a class perspective that reduced these various differences to a deterministic unity based on a unique class contradiction between capital and wage-labour. Equally, within this perspective, it was and continues to be presupposed that an essential unity of interests can be predicated on a shared position in the economic structure of capitalist society. The relations of production as a site of political struggle is granted a determining position, and the formation of the socialist consciousness of social subjects can only be effected through class struggle. This approach, certainly in the Leninist variant, leads to the related viewpoint that the abolition of private property in the means of production is by itself a sufficient condition for the abolition of all forms of alienation.

In contrast to this classical tradition, there exists another approach, emanating largely from Gramsci, which emphasizes the absence of any pre-given automatic socialist consciousness rooted in class determination. In this perspective the development of popular hegemony involves struggles around the political construction of the subjectivity of the masses, without any prioritization being granted to a pre-constituted site of political struggle, such as the point of production. Therefore, as Mouffe (1984, p. 143) argues,

what is really at stake in the articulation of the multiplicity of struggles against all social relations of subordination into a new radical project of transformation of society is, in fact, *a redefinition of socialism as the extension of democracy to all fields of social existence*. (Emphasis added.)

In a post-revolutionary situation such an approach to socialism would necessarily imply the guaranteeing of plurality and the continuing struggle for an effective abolition of all forms of alienation. In this sense the seizure of State power would not be taken as constitutive of an end of politics but rather one step, albeit an essential one, in the struggle to develop and securely root new forms of popular control and organization.

Although written in relation to different political and historical circumstances Poulantzas (1978) in the final chapter of his last book introduces a number of issues which are highly germane at this point. In particular, in discussing the democratic road to socialism he asks:

how is it possible radically to transform the State in such a manner that the extension and deepening of political freedoms and the institutions of representative democracy (which were also a conquest of the popular masses) are combined with the unfurling of forms of direct democracy and the mushrooming of self-management bodies? (Poulantzas, 1978, p. 256)

For Poulantzas this question was only obscured by the notion of the dictatorship of the proletariat: the fundamental problem lies in finding ways of 'combining a transformed representative democracy with direct, rank-and-file democracy' (Poulantzas, 1978, p. 256). It is, in fact, the contemporary political project of the Nicaraguan government to find ways of combining these two 'forms' of democracy, in a society bereft of any kind of democratic tradition – and in this context Nicaragua can be clearly demarcated from other post-revolutionary peripheral societies such as Cuba. Let us now turn therefore to a consideration of some key differences between these two societies.

On the historical specificity of revolutionary change – or why Nicaragua is not a second Cuba

Before outlining the reasons why Nicaragua is not a second Cuba it is important to note the existence of two overall similarities. First, it is evident that in both cases a revolutionary break has been made with the internalized dominance of international, and especially United States, capital and a new State has emerged to take on the tasks of formulating and sustaining a genuinely national development strategy. Second, in terms of the geo-political perception of the Reagan administration, both countries constitute a direct threat to United States' hegemony in the Central American and Caribbean region.[9] I shall return to these two basic similarities below.

Now, with respect to the divergences, and keeping in mind the previous suggestions on the diverse faces of socialism, especially the notes on the third and fourth perspectives, I want to begin with three preliminary propositions.

1. As regards Cuba the ideological foundation of the State has more in common with the Leninist model of post-revolutionary society than is the case in Nicaragua, where the concepts of the 'national', the 'popular', and the 'democratic' figure more prominently than 'Marxist–Leninist' or 'communist'.
2. Civil society in Cuba is far more bureaucratized than in Nicaragua where a greater degree of autonomy exists for groups and popular organizations outside the State structure. In Cuba, plurality, even viewed favourably, is much more centrally circumscribed.
3. In contrast to the Leninist model, in Nicaragua the seizure of State power is seen as the beginning of a process of *political* struggle for popular hegemony. In other words, the terrain of politics is not closed off by an all-pervasive 'democratic centralism'. On the contrary, political space is viewed far more as a heterogeneous entity where social differences must be recognized and allowed ambits for autonomous expression.

It is important to add here that these three propositions do not take any account of the fact that the duration of the two post-revolutionary processes differs considerably and it might well be speculated that Nicaragua could in the future, as a consequence of continuing external aggression, come to approximate the Cuban model of political organization. Furthermore, some observers have pointed to the presence of Leninist conceptualizations within some Sandinista policy discussions. However, these orientations certainly do not constitute a dominant feature of the Sandinista discourse as a whole and as I shall argue below it can be seen that acute internal problems rather than a shift to more centralist control have generated the development of more socio-political autonomy. It is also instructive to bear in mind that the Nicaraguan political model continues to exert an influence on the Cuban leadership although as of now there are no tangible signs of this influence being reflected in an internal re-structuring of the Cuban political system.

Examining Nicaragua in some detail, Coraggio (1985) makes a series of observations that are relevant to any attempted comparison of the Cuban and Nicaraguan roads to revolutionary transformation. As one instance, he highlights the importance of the form of articulation between political party and social movements, indicating the existence of a 'verticalist' option whereby mass organizations are located in a subordinate position *vis-à-vis* the party, and an option wherein the party articulates various social identities and their corresponding popular organizations 'horizontally'. The former option can take us along the Cuban road, with all the organizational and ideological implications of an adapted Leninist model. However, the so-called 'horizontal' option comes much closer to the Nicaraguan experience and recalls an argument recently advanced by Laclau and Mouffe (1985) on the role of the party.

They note, for example, that the party as a political institution can be one of two things: either 'an instance of bureaucratic crystallization which acts as a brake upon mass movements', or 'the organizer of dispersed and politically virgin masses', whereby it can serve as an 'instrument for the expansion and deepening of democratic struggles' (Laclau and Mouffe, 1985, p. 180). The second instance relates more closely to the Nicaraguan case, although the term 'politically virgin masses' is not entirely pertinent since the masses in Nicaragua have a longer history of struggle and political consciousness than is often assumed (Gianotten and de Wit, 1986).

In Nicaragua all social sectors have been given the opportunity to participate in the Sandinista project of democratic transformation. Developing this idea further, Coraggio and Irvin write as follows:

Nicaragua is a unique example of a social revolution in which the old ruling class ... has been called upon to cooperate in the process of national reconstruction. Indeed to maintain national unity the FSLN has had to restrain some of its own followers and, more generally, to exercise a mediating role between contending social and economic demands in the name of the state. The first months of the revolutionary government involved, *inter alia*, restraining those elements who wished to dispossess the bourgeoisie regardless of whether or not they continued to produce. Private property was given full legal protection on the condition that it continued to fulfil its social role in production. The government was engaged in almost continuous dialogue and mediation between different social groups; it encouraged reasoned argument, sought consensus whenever possible, and used minimal coercion. (Corragio and Irvin, 1985, p. 25)

It might then be asked: to what extent is this kind of pluralism compatible with popular hegemony? Obviously pluralism in Nicaragua does not embrace those (mainly ex-Somocista) leaders who are actively and continually engaged in armed counter-revolutionary struggle, but no class or social group per se is excluded. According to Coraggio and Irvin (1985, p. 26) 'different classes, social groupings and political parties are represented at every level, be it in the *Junta de Gobierno*, the Cabinet, the Council of State, the judiciary system, the mass media, popular organizations, churches, the school system, and so forth'. Also, although the FSLN maintains close control of the Revolutionary Popular Army, the Popular Militia, which is considerably larger than the army, is open to all.

It is also important to remember that the Sandinistas acquired power in a situation where the institutions of civil society were very undeveloped, and the State, after forty years of Somocista rule, had become an extremely crude apparatus of repression based on personal authority and patronage. Consequently, for the Sandinistas a central aim of revolutionary development has been the creation of bases for the construction of a new civil society in which the majority, through its own autonomous social and political organizations, is able to exercise decisive influence over the State. The literacy campaign, the fundamental changes carried through in the rural sector through agrarian reform, and the very considerable extension of basic social services are part of this construction, as also is the growth of popular organizations such as the revolutionary defence committees (CDSs), women's organizations, youth organizations, popular militias and organizations for ethnic groupings.[10]

The first set of changes – literacy, social welfare and agrarian reform – are also, of course, a positive feature of Cuban policies, post-revolution, whereas with regard to the growth of popular organizations, we encounter a series of more striking contrasts. This does not mean to imply that, in terms of the organization of social welfare or agrarian reform, there are no differences, but rather that in trying to specify the determining variations between the two post-revolutionary societies the relations between State and the institutions of civil society constitute a much more indicative and symptomatic terrain on which to mount such a specification.[11]

As a way of making the discussion more concrete I now want to consider some aspects of those relations, and in particular, the issue of 'popular power' and state control, giving especial attention to the territorial dimensions of the contrasting political approaches.

From the territory of revolution to the revolution of territory

In a previous paper on State and territory in Cuba, I distinguished four forms of spatial policy: (1) the reduction of territorial disparities in social and economic development; (2) the promotion of regional interdependence and integration; (3) the breaking down of the separation of town and country; and (4) the administrative decentralization of decision-making power (Slater, 1982, p. 14). The first three forms are predominantly related to the territorial re-organization of social-economy, although the direction, content and discursive formulation of their realization cannot be seen as outside the political realm. The fourth type however connects more directly with our present theme, providing a counterpoint to the comparable spatial-policy form in Nicaragua.

Furthermore, in the case of Nicaragua, the vexed question of regionalist autonomy, linked to the ethnic heterogeneity of the society in general, can be seen as an object of spatial policy, which, although being closely related to the administrative decentralization of decision-making power, does possess its own identity. In Cuba, on the other hand, no comparable 'ethnic-regionalist' question presents itself. However, at another level, the way in which the Sandinista government has and is approaching the socio-political problems which have emerged in the Atlantic-coast region (Mosquitia) is highly illustrative of its overall political conception of territorial strategy. Therefore, although there are not strict grounds for comparability I shall look at this question below.

A sixth possible form or component of spatial policy relates to the defence and protection of the integrity of national territory. In this example, despite the obvious geographical differences in size and location – Nicaragua having to protect the land borders of its territory on two fronts – there is, in principle at least, a basis for a comparison. There have, however, been few if any studies of this domain of territorial–military strategy; one exception is the work of the French geographer Foucher (1979, 1982 and 1985) who has examined the territorial dimensions of the guerrilla war in Nicaragua, and the contemporary situation along the Nicaragua–Honduras frontier. Similar work on Cuba is conspicuous by its absence, although given the highly sensitive nature of the issues involved we should not find this surprising. Cuba has the advantage, in comparison to Nicaragua, of being an island, and therefore its potential territorial permeability is more easily guarded. In the Nicaraguan case territorial permeabiity has presented and continues to present many difficulties, not least the danger that a particular zone could be taken over by enemy forces and used to establish an 'alternative government'. The manner in which the Sandinistas have faced this danger, by encouraging regional groupings to arm themselves in order to protect their interests and their territory, together, of course, with military support, is indicative of a more open and flexible approach than one might expect to find in other post-revolutionary societies.

In order to specify more concretely some of the salient contrasts between Cuban and Nicaraguan spatial policy, especially bearing in mind our previous discussion of socialism and democracy, I shall now examine two related issues; first, the question of political decentralization in its territorial setting, and including therein a treatment of regional autonomy (in other words, a combination of the fourth and fifth forms of spatial policy), and second, the problems associated with territorial permeabiity and the strategic interpretation of the territorial imperative (the sixth form of spatial policy).

State and popular power in Cuba

The main organ of State power in Cuba is the National Assembly of People's Power (*Asamblea Nacional del Poder Popular*) in which constitutional and legislative power is concentrated. The Assembly elects a standing committee called the Council of State (*Consejo de Estado*) which is vested with the supervision of the Council of Ministers, the judiciary and local government. The Council of Ministers is defined, according to the Constitution,[12] as the highest executive and administrative organ of State power, and is composed of the president, vice-presidents and secretary of the Council of State, the ministers, and the president of the Central Planning Board (JUCEPLAN). The secretary general of the Confederation of Cuban Workers (*Central de Trabajadores de Cuba*) has the right to participate in the Council of Ministers, which is accountable and responsible to the National Assembly and also in the Council's executive committee. Provinces and municipalities are ruled by the People's Power Assemblies which are elected every two and a half years, whereas the National Assembly is elected every five years.

In respect of the political management of the economy, in which, unlike Nicaragua, there is effectively no private sector, the organs of popular power (OPP), at provincial and municipal levels, have a quite limited role. Economic units are divided into two categories; those of high importance nationally (for example, the sugar industry) and those with a more localized significance (small industry, basic food production). As far as the first category is concerned, the OPP have little if any potential for control. Financial power remains at the level of central government as well as other planning functions such as the setting up of norms and procedures, the deployment of technical experts, the training of specialist cadres and research and development. Thus at the local and provincial levels the tasks are essentially those of the *implementation* of centrally formulated planning objectives and directions. Furthermore, with the principle of dual subordination not only are the planning and administrative departments at provincial and municipal levels subordinate to their respective assemblies and executive committees *but also to central planning and administrative agencies at the national level. In fact, according to an official party resolution, these agencies are meant to enhance their control over the activities and enterprises administered by the local organs of People's Power – thus, 'the process of decentralization ... shall be accompanied by a strengthening of the role of planning and of systematic control on the part of central state agencies'* (emphasis added).[13]

In addition, it has to be remembered that while the central administration retains planning control over major industries, agriculture, mining and finance, the OPP, in contrast, predominantly manage services. Furthermore, decisions taken by the OPP can be changed or revoked by other organs of the State system and at the same time the party has the final authority to decide who is able to attend and chair the OPP executive committees which are the key decision-making and managerial bodies of *poder popular*.[14]

When we look at the political composition of the assemblies of popular power, it is evident that the higher the level the greater is the preponderance of party members, so that while in 1979 65 per cent of the municipal assembly representatives were members of the party or candidates for membership, this same category rose to 92 per cent for the representatives in the national assembly. Correspondingly, while 91 per cent of the executive committees of the municipal assemblies were thus categorized, the Council of State was entirely composed of members of or candidates for the party (Jorgenson, 1983, p. 39). With respect to differentiation by the category of gender, only 7 per cent of municipal assembly representatives were women, although this figure did rise to 22 per cent for the national assembly.

It would also seem that those representatives who are not attached to the party do not articulate political ideas that run counter to the dominant orientation, certainly not in an organized fashion, and therefore the grounds for genuine debate based on the unhindered expression of real differences are extremely tenuous. This does not mean that there are no individual variations of opinion, but rather that a sustained plurality is hardly visible.[15] The lack of such plurality, in contrast to Nicaragua, can be situated in the Cuban leadership's vision of democracy. Raul Castro, for instance, argued in the 1970s that when a State like the Cuban State represents the interests of the workers, it is a much more democratic state than any other kind which has ever existed in history, because the state of the workers, the state which has undertaken the construction of socialism is, in any form, a majority state of the majority while all other previous states have been states of exploiting minorities. (Quoted in Rabkin, 1985, p. 256)

In Cuba not only is power highly centralized at the apex of the political system – in the Council of State, Council of Ministers and the Central Committee of the Party[16] – but also the structuring of power at the local and provincial levels does not represent the experiment in decentralized self-government but more the institutionalized working-out of an efficient and regulated apparatus of local government, closely fenced in by the circumscribing power of the centralized organs of the State. In his brief comparison of Cuba and Nicaragua, Weber (1981) remarks that the masses are asked to choose between several candidates not on the basis of varying political position, but according to differentiated personal background, and since everyone appears to agree fully with the party programme, the 'only task is to select the most devoted, hard-working and efficient candidate'. For Weber (1981, p. 131) the right to vote can only be real 'if the election system allows a genuine clash between distinct political projects and between the parties that give them expression'; he goes on, 'and that presupposes full exercise of the various political rights: freedom of expression and association, freedom to demonstrate, the right to strike'. These types of freedom, which are rooted in an acceptance of social difference and a heterogeneity of identities, cannot flourish when the dominant vision emanates from Leninist precepts. Such a vision has been clearly expressed by Fidel Castro in his extensive report to the First Congress of the Cuban Communist Party in 1975. On the question of the party and political organization, he asserted that 'the Party is the synthesis of everything. Within it, the visions of all the revolutionaries in our history are synthesized; within it, the ideas, principles and strength of the Revolution assume concrete form' ... and further 'the Party is today the soul of the Cuban Revolution' [17]

Equally, and in direct connection to our previous consideration of the Leninist model of politics, it can be added that in official discussions of the relationship between the party and the State one can find a confusion which reminds us of Polan's general treatment of the Leninist view of this relation. For example, in 1981, *Granma*, the official organ of the Cuban Communist Party, suggested that the 'Party is the ruling organ of the entire society', while the organs of People's Power 'administer the state'. However, in other documents including the 1976 Cuban Constitution, the organs of People's Power are viewed not merely as administrative bodies, but as 'decision-making structures that embody the popular will' (Rabkin, 1985, p. 256). Although it is necessary to draw links with our previous examination of the Leninist model of politics, thus revealing the influence within Cuba of many traditional Marxist–Leninist precepts of political organization and theory, I also want to point out that I am not subscribing to the simplistic 'Sovietization of Cuba thesis'. As Azicri (1985) argues in his brief analysis of socialist legality and practice, Cuba maintains autochthonous elements which blend together with Leninist ideology, so that, for instance, its constitution combines previous national constitutional experience with the influence of Soviet law *and* its own original 'political pragmatism' (Azicri, 1985, p. 323).

In the above-quoted speech, Castro includes in his discussion of the Party a reference to the fact that 'Cuba is building socialism just a few steps away from the most aggressive and criminal imperialist country, whose government has not ceased attacking and threatening our Homeland, with the use of all its means' (Azicri, 1985, p. 323). This threat has not receded and party calls for more production, discipline and continual vigilance remain intimately associated with the leadership's conception of defence. Similarly, the organization of popular power, the institutionalization of the revolution over the territory of Cuba constitutes an essential arm in the protection of the integrity of that territory. It is in this sense that we can detect an intimate connection between our posited fourth and sixth forms of spatial policy in Cuba. In 1984 the Cuban Politburo set up a National Council for Defence which has subsequently begun to implement an adaption of the original Vietnamese approach to the war against United States' imperialism – in the Cuban case this consists of aiming to provide every citizen with a shelter and a weapon against the possibility of external aggression. The French correspondent Ramonet (1985, p. 3) describes the new Cuban approach as the 'demilitarization of defence and the re-militarization of the people', since the Cuban leadership is increasingly strengthening the military role of the MTT (militias of territorial troups) which are estimated to be almost five times greater in number than the regular army. Confronted thus with the perceived threat of a possible future violation of Cuban territory the response of the Party is to convert, on a comprehensive territorial basis, the population into an omnipresent armed force able to resist and repel. As we shall see Cuba's management of the imperative of territorial defence contrasts with a less centralized approach in the case of Nicaragua.

Popular hegemony and the territorial imperative in Nicaragua

For the Sandinista government as well for Cuba, territorial control is a cardinal element of overall policy – its fulcrum in fact. From the re-organization of local government,[18] through the development of grassroots' organizations to the recent draft plan to grant a certain regional autonomy to the Atlantic-coast zone, territorial control, and above all the central imperative of trying to prevent or reduce the degree of violation of the integrity of Nicaragua's territory, forms a continuing guiding thread. Faced with a concerted imperialist strategy to undermine and remove its presence from the Central American region the Nicaraguan government might well have veered in a much more authoritarian and inflexible direction. Such tendencies may well be present but so far they have not become dominant. In fact, in contrast to the centralized organizational form of militias of territorial troups, the Nicaraguan government has tended to seek a more flexible and open military policy whereby peasants in the war zones are encouraged to organize their own defence in a collective and locally directed fashion. At the same time, rather than accelerate a process of state control over land, the recent tendency has been to distribute land to the peasants as a way of giving them a direct stake in their own production (see Instituto Historico Centroamericano, 1985, pp. 1c–19c).

Nowhere has the challenge to the integrity and potential permeability of Nicaragua's territory been more evident than in the Atlantic-coast region. Here, a combination of *contra*-activities and the existence of ethnic groupings (Miskitus, Ramas, Sumos, Creoles and Garifunas), who have not readily identified with the Sandinista government, has generated a series of apparently intractable political problems – the Sandinistas have referred to it as their 'number one domestic problem'.[19] According to Coraggio (1985) the difficulties that have emerged in this region are the result of the contradiction between the need to defend territorial integrity against external aggression and the desire to allow self-determination and a gradual re-articulation of the indigenous communities to the society-in-revolution.

A basis for a possible solution to this contradiction was recently formulated through the establishment of a statute on autonomous rights for the Atlantic-coast region. This will be included in the Political Constitution to be decreed by the National Assembly in 1986. In order to co-ordinate the project a National Commission for Autonomy has been set up, consisting of representatives from the different communities, social scientists and political cadres; the Commission's President is Tomás Borge, the Minister of the Interior. The Commission has recently produced a draft document entitled 'Principles and Objectives of Autonomy for the Atlantic Coast'. Some of the most salient features of the document are as follows:

1. The area traditionally occupied by the indigenous peoples and other communities of the coast will be preserved by the central government as an indispensable material base for the planning of development. National resources will be controlled at the regional level. It is also

stated that of the profits to be generated through the utilization of the area's national resources, a certain percentage will be taken to strengthen the national economy, while the other proportion will be re-invested in the region, according to criteria decided at that level. However, the balance between regional re-investment and national appropriation is not specified.[20]

A relevant point here is the fact that until the end of 1984 investment projects falling under the direction of the Ministry of Agriculture and Agrarian Reform were predominantly located in the Pacific and North–Central agro-ecological zones: in contrast only 13 per cent of such projects were found in the coastal humid–tropical zone (figures calculated from Wheelock, 1985, pp. 128–30).

2. The draft statute proposes to establish two autonomous regions (North and South Zelaya) in which there will be regional governments composed of an assembly and an executive. The regional assembly will consist of delegates from each of the communities elected by direct vote and in a number proportional to the size of the community. The assembly will have the functions of defining the socio-cultural and economic policies of the region, adapting national laws to the characteristics of the zone, proposing laws to the National Assembly, appointing local government officials and defining the most appropriate form of integration of the coastal population into the defence of the nation.

   Nevertheless, the major responsibility of defence of the Atlantic-coast zone will remain in the hands of the central government, albeit in co-ordination with the regional governments.

3. There will be bilingual education programmes, which, to an important extent, have been motivated by the high drop-out rate and low academic level of the students of the zone (see Yih and Slate, 1985, pp. 23–7).

   At this stage, it is difficult to envisage how successful the proposal will be – the document talks of the fulfilment of an historical compromise – but certainly, in comparison to earlier Sandinista approaches to the ethnic–regional question,[21] there would appear to be a more favourable potential embedded in the draft statute proposal, especially since discussion and preparation of the document was broadly conceived and took into account the views of representatives of the six groups of the coastal zone.

Political democracy at 'grassroots' level' is also reflected through the popular organizations – specifically, the Sandinista Defence Committees, the Sandinista Workers' Association, the National Union of Farmers and Ranchers, the Luisa Amada Espinosa Nicaraguan Women's Association, and the Sandinista Youth (Serra, 1985, pp. 65–89). By 1984 these organizations had become the main channels of popular participation, promoting collective decision-making in workplaces, schools, neighbourhoods, universities, the militia, co-operatives and rural communities across the country. In this way, they have become 'permanent schools of political socialization and democratic consciousness-raising' (Serra, 1985, p. 88). Their ideological orientation can be characterized in terms of the objective of defending and developing popular hegemony but they do not carry the same party-political imprimatur as do comparable organizations in Cuba. Moreover, the guaranteeing of political plurality in the form of elections also clearly differentiates Nicaragua from Cuba, and provides an important contextual factor for any consideration of the working of the grassroots' organizations – that is to say, they do not operate in a space where political heterogeneity has been monolithically synthesized by one party.[22]

The existence of political plurality in Nicaragua is crucial for any comparison of Cuba and Nicaragua in relation to our previously suggested fourth and fifth forms of spatial policy. Thus, in terms of political decentralization the Nicaraguan Council of State, which was established in 1980, is comprised of representatives from the popular organizations, the various labour unions, commercial, professional and other social associations, and the political parties. Unlike the Cuban Council of State it cannot be interpreted in the context of the ideological orientation of one dominant party. At the local level, too, plurality is the guiding orientation in Nicaragua, and as we have seen in relation to the Atlantic-coast region the Sandinistas are now attempting to develop new forms of regional autonomy. Similarly with our sixth form of spatial policy, concerning questions of territorial permeability and defence Nicaragua is characterized by a greater degree of political openness as regards the collective organization of territorial control and the lower profile given to centrally directed military strategy.

Concluding implications for revolutionary change in the Third World

It ought to be clear from the structuring of this chapter that I strongly subscribe to the view that we cannot begin to understand the various spatial dimensions of revolutionary change unless we first begin with a treatment of determinant social and political processes. In the case of discussions on socialism and spatial reorganization it is important to tackle the difficult but central issue of what can be meant by 'socialism'. In debates on the Third World one often finds an unfortunate tendency to group together, as socialist, societies which have quite fundamental political asymmetries – for example, classifying Cuba and Tanzania as both being socialist without any consideration of their historical and political specificity.

In my previous delineation of the Leninist variant of socialism and the discussion of its application in the Cuban case, I emphasized the underlying contrasts with the Nicaraguan approach to post-revolutionary construction and development. In particular, the notions of popular hegemony, and political plurality were introduced to capture the key difference with the Cuban system within which the political terrain remains essentially homogeneous and closed. This key difference has a wider relevance as can be seen in the unfolding of events in Grenada where an internal conflict between an orthodox Leninist orientation and a much more open pluralist approach to popular democracy, centred around the ideas of Maurice Bishop, led to an erosion of unity and a final and fatal facilitation of external aggression and imperialist revanchism (see Henfrey, 1984, pp. 15–36). Paradoxically, Grenada demonstrates *not* the strength of US imperialism but its relative weakness, since it was only because of a process of growing internal political paralysis that military intervention became a realistic political option for the Reagan administration.

Today, one central component of Nicaragua's political significance is that it still offers an alternative model of post-revolutionary development, which cannot, within any thoughtful approach, be dismissed as an imitation of the Soviet-model. And it is precisely because of its political novelty, its creative and autochthonous specificity, and its potential effect on other Third World societies and movements that it continues to be subjected to a battery of measures designed to erase it from the geopolitical map of Central America. Its survival is thus vital for the future development of Third World revolutions, as well as, of course, for the future of its people.

Notes

This chapter was originally prepared for the ISA/CUSUP Conference on 'The Urban and Regional Impact of the New International Division of Labour', held at the Centre of Urban Studies and Urban Planning, University of Hong Kong, 14–21 August 1985. A slightly extended version of the paper has been published in *Antipode*, September 1986.

1. An example here would be Petras' (1978) paper on 'socialist revolutions'. A more recent and in many ways more useful review is provided by White (1983) who does tackle this problem of definition. I shall return to this paper below.
2. Bahro, 1978. Bahro's term 'actually existing socialism' is also invoked.

3. See Hall (1984) for a good discussion.

4. It is perhaps worth noting here that in strong opposition to the social democrats of his day Marx continually emphasized that the development of socialism was not equivalent to some moralistic vision of what might constitute a 'fairer', 'more equal' distribution of wealth and income. As he observed in the *Critique of the Gotha Programma*, 'vulgar socialism . . . has taken over from the bourgeois economists the consideration and treatment of distribution as independent of the mode of production and hence the presentation of socialism as turning principally on distribution' (Marx, 1970b, p. 20).

5. The belief, still current today, that the solution of the difficulties faced in securing socialism lies through the development of production goes back to the beginning of the 1920s. As Corrigan et al. (1978, p. 107) remark: 'this stark theory of productive forces was 'clearly conceptualized by Bukharin and Preobrazhensky in 1920 – they wrote that: "the foundation of our whole policy must be the widest possible development of productivity".'

Similarly, of course, Lenin gave a related emphasis, noting in a well-known passage that an immediate transition to socialism is to a certain degree conceivable provided that, 'we construct scores of district electric power stations . . . and transmit electric power to every village'; then, 'if we obtain a sufficient number of electric motors and other machinery, we shall not need, or shall hardly need, any transition stages or intermediary links between patriarchalism and socialism' (Lenin, 1975, p. 350).

6. It is perhaps worthwhile to indicate that Ryan uses a definition of metaphysics that follows Derrida: 'metaphysics, Derrida shows, is not a historically periodizable school of thought; it is, rather, a permanent function of a kind of thinking which overlooks . . . its own historicity, differentiality and materiality' (Ryan, 1984, p. 117).

7. There is no space in this chapter to deal with the long and disputatious discussion of the so-called transition from socialism to communism – for some useful analysis see Buick, 1975, 1978.

8. The following paragraphs are largely based on Laclau, 1983, and also Laclau and Mouffe, 1985.

9. This does not imply that the policies of the Reagan administration are identical in relation to Cuba and Nicaragua, but both countries, and Grenada previously, are perceived as antagonistic to US interests. The extremist reaction of the US administration to the Sandinista government can also be linked with Nicaragua's potential as an alternative political model for other peripheral societies.

10. For a detailed discussion, see Coraggio, 1985, pp. 210–20.

11. As far as I am aware there have not been very many comparative studies of Cuba and Nicaragua, but the articles of Azicri (1982) and Fitzgerald (1985) do provide some guidelines. Azicri tends to concentrate more on trends in political organization and ideological background, whereas Fitzgerald focuses on divergences in economic strategy.

12. For an analysis of the Cuban constitution, and its Soviet counterpart, see De la Cuesta, 1976.

13. Party Resolution on the Organs of People's Power, *Granma*, 11 January 1976, p. 9, quoted in Gonzalez, 1976, p. 10.

14. See Slater, 1982, for some further details.
15. For a more optimistic outlook, see Green, 1985, pp. 37–43. Also visiting the same province as Green, namely Cienfuegos, in 1985, together with a group of planners and architects from London, it did not appear to me that the concept of conflict and alternative options came into play; not at least as far as the representatives of popular power we interviewed were concerned – this was especially the case in terms of the decision to develop nuclear power facilities in the Cienfuegos area.
16. At the beginning of 1980 the State structure was reorganized in terms of the centralization of command of the entire administrative apparatus. It was clearly established that the top level of government would be the executive committee of the Council of Ministers. Fidel Castro's position as first secretary of the CP, and president of the Council of Ministers, was combined with his position as commander-in-chief of the armed forces, and his brother Raul Castro was appointed as his deputy in these positions. Together with Carlos Rafael Rodriguez these three have been the only people to be members of the party's Central Committee, Council of Ministers and Council of State.
17. Castro, 1976, pp. 231–2. There are also in this same passage allusions to military discipline which recall Lenin's views on the party, its tasks and proletarian order.
18. Downs and Kusnetzoff (1982, p. 547) in their discussion of the Nicaraguan experience conclude by writing that in their opinion Nicaragua is an 'auspicious example of a revolutionary process which has developed concrete organs of popular participation and management while thus far avoiding some of the negative effects – such as excessive centralization and bureaucratization – often considered as inevitable results of the consolidation of such a process'.
19. There are an estimated 90,000 Miskitus, 10,000 Sumos, 840 Ramas and a smaller group of Garifuna. The region – 56 per cent of national territory – also has about 60,000 'Creoles'; English-speaking blacks who largely reside in the south. For an analysis see Bourgois, 1981, 1982 and Brown, 1985.
20. For a summary, see *Barricada Internacional*, 4 July 1985, Managua, pp. 8–9 – also *Latin America Regional Reports*, Mexico and Central America, RM 8506, 12 July 1985.
21. In an interview with *Pensamiento Propio*, the vice-minister of the interior, Luis Carrion, gives some background on the reasons for changes in Sandinista policy – *Pensamiento Propio*, Año III, no. 20, enero-febrero, 1985, pp. 27–31. *Dirigista* perspectives seem to have been modified in the direction of accepting more flexibility and democratic pluralism.
22. For a discussion of the Nicaraguan elections see Robinson and Norsworthy, 1985, pp. 83–110.

Bibliography

Abdi, N. 1975: Réfórme agraire en Algérie. *Maghreb-Machrek*, *69*, 34–40.

Acosta, Maruja and Hardoy, Jorge E. 1973: *Urban Reform in Revolutionary Cuba.* New Haven: Yale University Antilles Research Program.

Almanac of China's Economy 1983, Beijing.

Anderson, James 1979: *Engels' Manchester: Industrialisation, Workers' Housing, and Urban Ideologies.* London: Planning School, Architectural Association.

Anon. 1983: Notes on the national seminar on development of production in less-developed regions'. *Jingji Dili*, 2, 154–6.

Arabia 1983: Algeria takes a planned step – but to where? *Arabia*, *19*, March 1983, 54–5.

Arrighi, G. 1973: Labour supplies in historical perspective: a study of the proletarianization of the African peasantry in Rhodesia. In G. Arrighi and J. S. Saul (eds), *Essays in the Political Economy of Africa*, New York: Monthly Review Press, 180–234.

Azicri, M. 1982: A Cuban perspective on the Nicaraguan revolution. In T. W. Walker (ed.), *Nicaragua in Revolution*, New York: Praeger, 345–73.

—— 1985: Socialist legality and practice: the Cuban experience. In S. Halebsky and J. M. Kirk (eds), *Cuba: Twenty-Five Years of Revolution, 1959–1984*, New York: Praeger.

Aziz, S. 1978: *Rural Development: Learning from China.* London: Macmillan.

BBC (British Broadcasting Corporation) various dates, *Summary of World Broadcasts: The Far East.* London.

Bahro, R. 1978: *The Alternative in Eastern Europe.* London: New Left Books.

Barkin, D. 1980: Confronting the separation of town and country in Cuba. *Antipode*, 12 (3), 31–40.

Basso Farina, M. 1980: Urbanization, deurbanization and class struggle in China 1949–69. *International Journal of Urban and Regional Research*, 4, 485–502.

Baylies, C. 1981: Imperialism and settler capitalism: friends or foes. *Review of African Political Economy*, 18, 116–26.

Berberoglu, B. 1979: The nature and contradictions of state capitalism in the Third World. *Social and Economic Studies*, 28 (2), 341–63.

Bernard, D. 1979: Regional development planning in Guyana. Unpublished conference paper, Geographical Association of Trinidad and Tobago, Seminar New Directions in the Caribbean, St Augustine, Trinidad, 8–10 June 1979.

Bernstein, T. P. 1977: *Up to the Mountains and Down to the Villages: The Transfer of Youth from Urban to Rural China.* New Haven: Yale University Press.

Bettelheim, C. 1975: *The Transition to Socialist Economy*. Brighton: Harvester.
—— 1976: *Economic Calculation and Forms of Property*. London: New Left Books.
Biermann, W. and Kössler, R. 1981: The settler mode of production: the Rhodesian case. *Review of African Political Economy*, *18*, 106–16.
Birks, J. S., Sinclair, C. A. 1980: *International Migration and Development in the Arab Region*. Geneva: International Labour Organisation.
Bourgois, P. 1981: Class, ethnicity and the state among the Miskitu Amerindians of north-eastern Nicaragua. *Latin American Perspectives*, *8*, (2), 22–39.
—— 1982: The problematic of Nicaragua's indigenous minorities. In T. W. Walker (ed.), *Nicaragua in Revolution*, New York: Praeger.
Bromley, R. (ed.) 1979: *The Urban Informal Sector*. Oxford: Pergamon.
Brown, G. 1985: Miskito revindication: between revolution and resistance. In R. Harris and C. M. Villas (eds), *Nicaragua: A Revolution Under Siege*, London: Zed Books, 175–201.
Brûlé, J. C. and Mutin, G. 1982: Industrialisation et urbanisation en Algérie. *Maghreb-Machrek*, *96*, 41–66.
Brundenius, Claes 1981: *Economic Growth, Basic Needs and Income Distribution in Revolutionary Cuba*. University of Lund, Sweden: Research Policy Institute.
—— 1984a: Crecimiento con equidad: Cuba 1959–1984. Instituto de Investigaciones Economicas y Sociales, Managua, 1–43.
—— 1984b: *Revolutionary Cuba: The Challenge of Economic Growth with Equity*. Boulder: Westview Press.
Buchanan, K. 1967: *The Southeast Asian World*. London: Bell.
Buick, A. 1975: The myth of the transitional society. *Critique*, *5*, 59–70.
—— 1978: Ollman's vision of communism. *Critique*, *9*, 152.
Burnham, L. F. S. 1970a: *A Destiny to Mould*. Compiled by C. A. Nascimento and K. A. Burrowes, New York: Africana Publishing Company.
—— 1970b: A vision of the co-operative republic. In L. Searwar (ed.), *Co-op Republic – Guyana*, Guyana: Georgetown, 9–19.
—— 1975: *The Sophia Declaration*. Guyana: Ministry of Information and Culture, Georgetown.
—— 1982: Republic anniversary speech, reported in *Daily Chronicle*, 24 February 1982.
CTVU (Centro Tecnico de la Vivienda y el Urbanismo) 1984: XI Seminario de la Vivienda y el Urbanismo: La construccion de viviendas por esfuerzo propio. Havana: Memoria Editorial del Centro de Informacion de la Construccion.
Cao, A. D. 1978: Development planning in Vietnam: a problem of postwar transition. *Asia Quarterly*, *4*, 263–76.
Caesar, N. and Standing, G. 1978: *The Labour Force in Guyana in 1977 – A Preliminary Report*, Geneva: WEP Research and Working Papers – Population and Employment, ILO.
Cante, D. 1983: *Under the Skin: The Death of White Rhodesia*. London: Allen Lane.
Castells, Manuel 1977: *The Urban Question*. London: Edward Arnold.
Castro, Fidel 1972: History will absolve me. In R. E. Bonachea and N. P. Valdes (eds), *Revolutionary Struggle 1947–1958: Vol 1 of the Selected Work of Fidel Castro*, Cambridge: MIT Press.

—— 1976: *First Congress of the Communist Party of Cuba, Havana* Moscow: Progress Publishers.

—— 1977: En la clausura del V Congreso de la Asociacion Nacional de Agricultores Pequenos (ANAP). Reprinted in *Discursos Fidel Castro* Havana: Edicion Politica, 1979.

—— 1982: Speech to the 4th Congress of the Young Communist League. In *Speeches at Three Congresses*, Havana: Editoria Politica.

—— 1983 *Fidel Castro Speaks, Volume II: Our Power is that of the Working People* New York, Pathfinder Press, [Michael Taber, editor].

Central Committee, Communist Party of Vietnam 1977: *4th National Congress. Documents*, Hanoi: Foreign Languages Publishing House.

Central Statistical Office 1983: *Monthly Digest of Statistics (April)*, Harare: CSO.

Central Statistical Organization 1980: *Statistical Yearbook* Aden: CSO.

Chaliand, G. and Minces, J. 1972 *L'Algérie Indépendante (Bilan d'une Révolution Nationale)*. Paris: Maspéro.

Chanda, N. 1981a: A last minute rescue. *Far Eastern Economic Review, 111* (10), 28–34.

—— 1981b: An idealogue in charge. *Far Eastern Economic Review, 113*, (29), 14–16.

—— 1981c: Cracks in the edifice. *Far Eastern Economic Review, 114*, (5), 84–5.

—— 1984: Toeing a liberal line. *Far Eastern Economic Review, 123*, (2), 32.

Chase-Dunn, C. (ed.), 1982: *Socialist States in the World System*. Beverly Hills: Sage.

Chen, Baosheng 1981: On the theory and practice of public finance, lending and materials. In Zhongguo Caizheng Bu (ed.), *Zhongguo Caizheng Wenti*, Tianjin Kexue Chubanshe, 242–80.

Chen, Pi-chao and Kols, A. 1982: Population and birth planning in the People's Republic of China. *Population Report*, series J, *25*, 577–618.

Claudin-Urondo, C. 1977: *Lenin and the Cultural Revolution*. Brighton: Harvester Press.

Cliffe, L. and Munslow, B. 1981: The prospects for Zimbabwe. *Review of African Political Economy, 18*, 1–6.

Comite Estatal de Estadisticas 1979: *Anuario Estadistico de Cuba*. Various dates. Havana: Comite Estatal de Estadisticas.

Commonwealth Caribbean 1970: *1970 Population Census of the Commonwealth Caribbean*. University of the West Indies: Census Research Programme.

Co-operative Republic of Guyana, Ministry of Economic Development 1973: *Draft Second Development Plan 1972–1976*. Guyana, Georgetown: Government Printery.

Co-operative Republic of Guyana 1978: *Budget 1978*. Budget Speech, National Assembly, 27 February 1978.

—— 1979 *Budget 1979*. Budget Speech, National Assembly, 12 March 1979. Sessional Paper, no. 1, 1979.

—— 1980: *Constitution of the Co-operative Republic of Guyana*. Act no. 2 of 1980, Georgetown: Government Printery.

Coraggio, J.-L. 1984: Revolucion y democracia en Nicaragua. *Cuadernos de Pensamiento Propio*, Managua.

—— 1985 Social movements and revolution: the case of Nicaragua. In D. Slater (ed.), *New Social Movements and the State in Latin America*, Holland: Foris Publications.

—— and Irvin, G. 1985: Revolution and democracy in Nicaragua. *Latin American Perspectives*, *12* (2).

Corrigan, P., Ramsey, H. and Sayer, D. 1978: *Socialist Construction and Marxist Theory – Bolshevism and its Critique*. New York: Monthly Review Press.

Dao Van Tap 1980: On the transformation and new distribution of population centres in the Socialist Republic of Vietnam. *International Journal of Urban and Regional Research*, *4* (4) 503–15.

Davies, D. 1981: Caught in history's vice. *Far Eastern Economic Review*, *114*, (53), 17–21.

Davies, D. H. 1981: Towards an urbanization strategy for Zimbabwe. *Geojournal* (supplementary issue 2), 73–84.

Davies, R. 1978: *The Informal Sector: a Solution to Unemployment? From Rhodesia to Zimbabwe, no. 5*. Salisbury: Mambo Press.

De La Cuesta, L.-A. 1976: The Cuban socialist constitution: its originality and role in institutionalization. *Cuban Studies*, *6* (2), 15–30.

Diamond, D. R., Hottes, K. and Wu, C. (eds) 1984: *Regional Planning in Different Political Systems – The Chinese Setting*. Germany: Bochum.

Dobb, M. 1951: *Studies in the Development of Capitalism*. London: Routledge and Kegan Paul.

—— (ed.) 1954: *The Transition from Feudalism to Capitalism*. New York: Science and Society.

—— 1969: *Welfare Economics and the Economics of Socialism*. Cambridge: Cambridge University Press.

Domes, J. 1980: *Socialism in the Chinese Countryside: Rural Societal Policies in the Peoples Republic of China 1949–1979*. London: Hurst.

Dong Thao 1983: The district and large-scale agricultural production along socialist lines. *Vietnam Courier*, *19* (1), 6–7.

Downer, A. V. 1982: Settlements on the white sands. In M. P. Thomas (ed.), *The Caribbean Peoples and their Environment*, Commonwealth Human Ecology Council, Publication no. 18, 193–205.

Downs, C. and Kuznetzoff, F. 1982: The changing role of local government in the Nicaraguan revolution. *International Journal of Urban and Regional Research*, *6* (4), 533–48.

Draper, Hal 1977: *Karl Marx's Theory of Revolution: State and Bureaucracy*. Vol. 1, New York: Monthly Review Press.

Dumont, Rene 1970: *Cuba: Socialism and Development*. New York: Grove Press Inc.

Dupuy, Alex and Yrchick, John 1978: Socialist planning and social transformation in Cuba: a contribution to the debate. *Review of Radical Political Economics (4)*, 48–60.

Eckstein, S. 1980: Income distribution and consumption. *Cuban Studies*, January.

—— 1981: Cuba and the Transition to Socialism. *Socialist Review*, *60*.

—— 1982: Cuba and the capitalist world economy. In C. Chase-Dunn (ed.), *Socialist States in the World System*. Beverley Hills: Sage Publications, 203–17.

Economist Intelligence Unit 1977: *Quarterly Economic Review of Algeria*. Annual Supplement, 1–28.

—— 1981: *Quarterly Economic Review of Indochina: Vietnam, Laos, Cambodia. Annual Supplement.* London: The Economist Intelligence Unit Ltd.

—— 1982: *Algeria: the giant market of North Africa.*

Edwards, R., Gordon, D. and Reich, M. 1982: *Segmented Work, Divided Workers.* Cambridge: Cambridge University Press.

Elliott, D. 1975: Political integration in North Vietnam: the cooperativization period. In J. J. Zasloff and M. Brown (eds), *Communism in Indochina*, Lexington: Lexington Books, 165–93.

—— 1979: Vietnam: institutional development in a time of crisis. In *Southeast Asian Affairs*, Singapore: Institute of Southeast Asian Studies, 348–63.

Elsenhans, H. 1982: Contradictions in the Algerian development process: the reform of the public sector and the new approach to the private sector in industry. *The Maghreb Review*, 7 (3–4), 62–72.

Engels, Friedrich 1978: On the division of labour in production. In Robert C. Tucker (ed.), *The Marx Engels Reader*, New York: W. W. Norton and Company, 718–24.

Epstein, D. G. 1973: *Brasilia, Plan and Reality*. Berkeley: University of California Press.

Estevez Curbelo, Reynaldo 1982: *La Vivienda y El Urbanismo En Cuba*. Havana: Ministerio de Construccion.

—— 1984: *Analisis de las realizaciones de viviendas en Cuba y en otros paises socialistes (presented at the XI Seminario de Viviendas y Urbanismo, Havana, March).* Havana: Centro Technico de la Vivienda y el Urbanismo.

FLN (Front de Libération Nationale) 1980: Resolutions of the extraordinary Congress–June 1980. *Révolution Africaine*, 4–10.7.1980, *14*.

Fall, B. B. 1967: *The Two Vietnams: A Political and Military Analysis*. London: Pall Mall Press, 2nd edn.

Far Eastern Economic Review Asia Yearbook, various issues: Hong Kong: *Far Eastern Economic Review*.

Feng, Haohua 1983: A study of the history of immigration into and cultivation of Qinghai Province. *Jingji Yanjiu*, 5, 52–7.

Field, R. M., Lardy, N. R. and Emerson, J. 1975: *A Reconstruction of the Gross Value of Industrial Output by Province in the People's Republic of China, 1949–73*. Washington, DC: US Department of Commerce.

Firebrace, J. 1984: *Never Kneel Down. Drought, Development and Liberation in Eritrea*. Nottingham: Spokesman Books.

Fitzgerald, E. V. K. 1985: The problem of balance in the peripheral socialist economy: a conceptual note. *World Development*, *13* (1), 5–14.

Forbes, D. K. 1981: Production, reproduction and underdevelopment: petty commodity producers in Ujung Pandang, Indonesia. *Environment and Planning A*, *13*, 841–56.

—— and Thrift, N. 1984: Determination and abstraction in theories of the articulation of modes of production. In D. K. Forbes and P. J. Rimmer (eds), *Uneven Development and the Geographical Transfer of Value, Human Geography Monograph 16*, Canberra: Australian National University, 111–34.

—— and Wilmoth, D. 1986: *Urban Development Problems and Opportunities in Tianjin, China*: Report prepared as part of the exchange programme between Tianjin and the Australian Institute of Urban Studies, Canberra.

Foucher, M. 1979: Enquête au Nicaragua (I). *Herodote*, *16*, 5–35.

—— 1982: Le bassin Méditerranen d'Amérique: approches geopolitiques. *Herodote*, *27*, 16–40.

—— 1985: Problémes stratégiques et politiques de la frontière nord du Nicaragua. *Cahiers des Amériques Latines*, *1*, 69–89.

Fraser, S. E. 1983: Vietnam: population statistics. *Vietnam Today*, *24*, 8–10.

—— 1984: Vietnam's population: current notes. *Contemporary Southeast Asia*, 6 (1), 70–88.

—— 1985: Vietnam struggles with exploding population: *Indochina Issues*, *57*, 1–7.

French, R. A. and Hamilton, F. E. I. (eds), 1979: *The Socialist City: Spatial Structure and Urban Policy*. Chichester: Wiley.

Friedmann, J. and Douglas, M. 1978: Agropolitan development: towards a new strategy for regional planning in Asia. In F. C. Lo and K. Salih (eds) *Growth Pole Strategy and Regional Development Policy*, Kuala Lumpur: Pergamon, 163–92.

Fuchs, R. J. and Demko, G. J. 1981: Population distribution measures and the redistribution mechanism. In United Nations, *Population Distribution Policies in Development Planning*, New York: United Nations, Population Studies no. 75, 70–84.

GHRA (Guyana Human Rights Association) 1983, 1984, 1985: *Guyana Human Rights, Report*.

Garnier, Jean-Pierre 1973: *Une Ville, une révolution: la Havana de l'urbain au Politique*. Paris: Editions Anthropos.

Ge, Zhida 1981: Increase the efficiency of utilizing financial subsidies in agriculture. In Zhongguo Caizheng Bu (eds), *Zhongguo Caizheng Wenti*, Tianjin Kexue Chuban-she, 626–39.

Ghiles, F. 1983: Algerian entrepreneurs come in from the cold. *Financial Times*, 19 April 1983, 4.

Gianotten, V. and de Wit, T. 1986: Organización y Reforma Agraria en Nicaragua: Algunas Lecciones de la Historia. Mimeo, Amsterdam: Centre for Latin American Research and Documentation.

Giddens, A. 1981: *A Contemporary Critique of Historical Materialism*. London: Macmillan.

—— 1985: *The Nation State and the Means of Violence*. Cambridge: Polity Press.

Gilbert, A. and Gugler, G. 1982: *Cities, Poverty and Development. Urbanization in the Third World*. Oxford: Oxford University Press.

Ginsburgs, G. 1963: Local government and administration in the Democratic Republic of Vietnam since 1954 (part 2). *The China Quarterly*, *14*, 195–211.

Gittens, P. 1980: Guyana on its 14th freedom year. *Caribbean Contact*, May.

Glass, R. 1955: Urban sociology in Great Britain. *Current Sociology*, *4*, 5–19.

Gonzalez, E. 1976: The party congress and *Poder Popular*: orthodoxy, democratization and the leader's dominance. *Cuban Studies*, 6 (2).

Gordon, A. 1981: North Vietnam's collectivisation campaigns: class struggle, production, and the middle-peasant problem. *Journal of Contemporary Asia*, *11* (1), 19–43.

Green, G. 1985: *Cuba ... The Continuing Revolution*. New York: International Publishers.

Green, H. 1983: Speech to management seminar of Mahaica–Mahaicony–Abary Agriculture Development Authority, 4 February 1983, reported in *Daily Chronicle* 5 February 1983, 1.

Griffith, I. 1978: Central planning and the economics of socialism. *Sunday Chronicle*, Georgetown, March 3.

Gruchman, B. 1982: Relevance of regional economics to centrally planned and socialist countries. In R. P. Misra (ed.), *Regional Development*, Singapore: Maruzen Asia, 31–42.

Gugler, J. 1980: A minimum of urbanism and a maximum of ruralism: the Cuban experience. *International Journal of Urban and Regional Research, 4* (4), 516–35.

—— 1982: The urban character of contemporary revolutions. *Studies in Comparative International Development*, XVII (2), 60–73.

Guyana Airways Corporation 1977: *Report on total air freight traffic to and from the interior for the year 1977.*

Guyana Government 1976: Amerindian Lands Amendment Bill.

—— 1977: Act no. 24, 1977 State Planning Commission Act.

—— 1980a: Act no. 12 1980a Local Democratic Organs Act.

—— 1980b: Act no. 69 1980b Order made under Local Democratic Organs Act.

Hall, S. 1984: The State – socialism's old caretaker. *Marxism Today, 28* (11), 24–30.

Halliday, F. 1974: *Arabia without Sultans*. Harmondsworth: Penguin.

—— 1979: *Yemen's Unfinished Revolution*. MERIP Report 81.

Hamberg, Jill 1985: The dynamics of Cuban housing policy. Unpublished manuscript.

—— 1986 Under construction: housing policy in revolutionary Cuba. In R. Bratt, C. Hartman and A. Meyerson (eds), *Critical Perspectives on Housing*, Philadelphia: Temple University Press.

Harvey, S.D. 1982: Black residential mobility in a part-independent Zimbabwean city. Mimeo, South African Conference of the Commonwealth Geographical Bureau, University of Zambia, Lusaka.

Haviland, H. F., Fabian, L. L., Mathiasen, K. and Cox, A. M. 1968: *Vietnam After the War: Peacekeeping and Rehabilitation*. Washington DC: The Brookings Institution.

Henfrey, C. 1984: Between populism and Leninism – the Grenadian experience. *Latin American Perspectives, 11* (3), 15–36.

Hill, R. D. 1983: Aspects of land development in Vietnam. Unpublished manuscript, University of Hong Kong.

Hodgkin, T. 1981: *Vietnam. The Revolutionary Path*. London: Macmillan.

Holton, R. J. 1984: Cities and the transition to capitalism and socialism. *International Journal of Urban and Regional Research, 8* (1), 13–37.

Honjo, M. (ed.), 1981: *Urbanization and Regional Development*. Singapore: Maruzen Asia.

Hope, K. 1975: Regional community development planning in Guyana. *Social and Economic Studies, 24* (4), 504–8.

—— 1979: *Development Policy in Guyana: Planning, Finance and Administration*. Boulder: Westview Press.

—— David, W. and Armstrong, A. 1976: Guyana's Second Development Plan,

1972–76: a macro-economic assessment. *World Development, 4* (2), 131–41.

Horenz, W. G. 1979: An approach to regional planning. *Zimbabwe Rhodesia Science News, 13* (11), 261–64.

Hoyte, D. 1980: *Local Democratic Organs*. Speech made by Comrade Desmond Hoyte, Minister of Economic Development and Co-operatives during debate on the Local Democratic Organs Bill in the National Assembly, 18 August 1980, Guyana: Publications Division, Ministry of Information.

Hua, Juxian 1981: Nationalities/minorities in China. In *Zhongguo Jingji Nianjian*, Beijing: Jingji Guanli Chubanshe, I–32–34.

Instituto Histórico Centroamericano 1985: Special issue. *Envio, 4* (51), 1c–19c.

IDB (Inter-American Development Bank) 1982, 1983, 1984 *Economic and Social Progress in Latin America 1981–82, 1982–83, 1983–4.*

INEAP (Institut National D'Etudes et D'Analyses Pour La Planification) 1979: *Les Plans Communaux de Développement (PCD)*, Algiers: INEAP, 2 vols.

Indochina Chronology 1984: Statistics from Vietnam: 1984. *Indochina Chronology, 3* (3), 28.

—— 1985: Vietnam chronology. *Indochina Chronology, 4* (1), 3–10.

Indochina Report 1984: The Vietnamisation of Kampuchea. *Indochina Report*, pre-publication issue, 2–19.

Institute for Strategic Studies 1984: *The Military Balance 1984/85.* London.

Ip, D. F. and Wu, Chung-Tong 1983: From subsistence to growth: Chinese strategies of rural transformation. In D. A. M. Lea and D. P. Chaudhri (eds), *Rural Development and the State*, London: Methuen, 301–32.

James, R. 1982: National goals and basic social rights and obligations in the Guyana constitution. *Transition, 5*, 56–70.

Jameson, K. 1980: An intermediate regime in historical context: the case of Guyana. *Development and Change, 11* (1), 77–95.

—— 1981: Socialist Cuba and the intermediate regimes of Jamaica and Guyana. *World Development, 9* (9–10), 871–88.

Japan–China Economic Association 1982: *Shanghai Economic Study Report.* (In Japanese). March.

Jiang, Weizhuang 1981: A treatise on the financial relationships between the national government and the collective agriculture sector. In Zhongguo Caizheng Bu (eds), *Zhongguo Caizheng Wenti*, Tianjin Kexue Chubanshe, 647–76.

Jones, G. W. and Fraser, S. E. 1982: Population resettlement policies in Vietnam. In G. W. Jones and H. V. Richter (eds), *Government Resettlement Programmes in Southeast Asia*. Australian National University: Development Studies Centre, Monograph no. 30, 113–33.

Jorgenson, B. 1983: The interrelationship between base and superstructure in Cuba. *Ibero–Americana–Nordic Journal of Latin American Studies, 13*, 1.

Kalecki, M. 1969: *Introduction to the Theory of Growth in a Socialist Economy.* Oxford: Blackwell.

—— 1972: *Selected Essays on the Economic Growth of the Socialist and Mixed Economy.* Cambridge: Cambridge University Press.

Kay, G. 1980: Towards a population policy for Zimbabwe–Rhodesia. *African Affairs, 79* (314), 95–114.

Khong Dien 1983: Population distribution in urban and rural areas of Vietnam. *Vietnam Courier, 19* (10), 26–7.

Kinsey, B. H. 1982: Forever gained: resettlement and land policy in the context of national development in Zimbabwe. *Africa*, *52* (3), 78–112.

Kirkby, R. J. R. 1985: *Urbanisation in China: Town and Country in a Developing Economy 1949–2000 A.D.* London: Croom Helm.

Konrad, G. and Szelenyi, I. 1977: *Social conflicts of underurbanization*. In M. Harloe (ed.), *Captive Cities*, New York: John Wiley, 73–92.

Kozol, Jonathan 1978: *Children of the Revolution*. New York: Dell Publishing Co.

LARR (Latin American Regional Report), various issues, *Caribbean Report*.

Laclau, E. 1983: Socialisme et transformation des logiques hegemoniques. In C. Buci-Glucksmann (ed.), *La Gauche, le Pouvoir, le Socialisme*, Paris: PUF, 331–8.

—— and Mouffe, C. 1985: *Hegemony and Socialist Strategy*. London: Verso.

Lang, M. H. and Kolb, B. 1980: Locational components of urban and regional public policy in postwar Vietnam: the case of Ho Chi Minh City (Saigon). *GeoJournal*, *4* (1), 13–18.

Lange, O. 1966: *Socialism and Socialist Economy*, Warsaw.

Le Dan 1983: Ho Chi Minh City: industry in the service of agriculture. *Vietnam Courier*, *19* (7), 4–6.

Le Hong Tam 1980: Development policy: providing the basics. *Southeast Asia Chronicle*, *76*, 22–5.

Leiner, Marvin 1985: Cuba's schools: 25 years later. In Sandor Halebsky and John M. Kirk (eds), *Cuba: Twenty-Five Years of Revolution, 1959–1984*, New York: Praeger Publishers, 27–44.

Lele, U. 1975: *The Design of Rural Development*, Baltimore: Johns Hopkins University Press.

Lenin, V. I. 1975: *Collected Works, Volume 32*. Moscow: Progress Publishers.

Li, Zhe 1981: On the questions of making use of the advantages of regional economies. *Jingji Yangjiu*, *12*, 49–51.

Li, Zhankui, Yunhang, Ai and Hangsheng, Zhou 1982: Speed up the construction of mountainous districts: an important strategy of realizing agricultural modernization. *Nongye Jingji Wenti*, *12*, 36–40.

Liang, Wensun 1981: Issues of balance between industries and agriculture in the development of the Chinese economy. In Xu Dixin, et al. *Zhongguo Guomen Jingji Fazhan Zhong Di Wenti*, Beijing: Zhongguo Shehui Kexue Chubanshe.

Liang, Wensun and Jianghai, Tian 1983: Some problems regarding accumulation. *Zhongguo Shehui Kexue*, *1*, 95–112.

Lilienthal, D. E. 1969: Postwar development in Vietnam. *Foreign Affairs*, *47* (2), 321–33.

Liu, Guoguang (ed.), 1981: *Issues in the Theory of Balance in National Economic Development*. Beijing: Zhongguo Shehui Kexue Chubanshe.

Liu, Guoguang and Xiangming, Wang 1980: A study of the speed and balance of China's economic development. *Zhongguo Shehui Kexue*, *4* (July), 3–22.

Lo, F. C. and Salih, K. (eds), 1978: *Growth Pole Strategy and Regional Development Policy: Asian Experiences and Alternative Approaches*. Oxford: Pergamon.

Lu, Zhuoming 1982: The geographical structure and the regional economic advantages. *Jinqj Kexue*, *3*, 10–16.

Lundqvist, J. 1981: Tanzania: socialist ideology, bureaucratic reality, and development from below. In W. B. Stöhr and D. R. F. Taylor (eds), *Development from Above or Below?* Chichester: Wiley, 329–50.

Lutchman, H. 1982: The presidency in the constitution of the Co-operative Republic of Guyana. *Transition, 6*, 27–66.

MPAT (Ministere de la Planification et de L'Aménagement du Territoire) 1979: *Bilan Aménagement du Territoire et Développement Régional.* Algiers.

—— 1980a: *Synthèse de bilan économique et sociale de la décennie 1967–1978.* Algiers.

—— 1980b: *Rapport Général, 1er. Plan Quinquennial.* Algiers.

MPAT/DSCN (Direction des Statistiques et de la Comptabilite Nationale) 1979: *Bilan régional 1967–1978. Instrument d'analyse spatiale et identification des inégalits régionales.* Algiers: Internal document, DSCN.

Ma, Hung 1982: Fully develop the use of coastal cities in the four modernization processes. *Jingji Guanli, 1* (January), 5–8.

Ma, L. C. J. and Hanten, E. W. (eds), 1981: *Urban Development in Modern China.* Boulder: Westview Press.

Ma, Quingyu 1983: A preliminary analysis of the development trends and characteristics of our nation's cities and towns. *Jingji Dili, 2*, 126–31.

Mao, Zedong 1969: *Mao Zedong Sixinq Wanshi.*

MacEwan, Arthur 1981: *Revolution and Economic Development in Cuba.* New York: St Martin's Press.

Malaba, L. 1981: Supply, control and organisation of African labour in Rhodesia. *Review of African Political Economy, 18*, 7–28.

Mandle, J. 1982: *Patterns of Caribbean Development.* New York: Gordon and Breech.

Manitzas, Nita R. 1973: The setting of the Cuban Revolution. Andover: Warner Modular Publications, Module 260.

Martin, D. and Johnson, P. 1981: *The Struggle for Zimbabwe.* London: Faber and Faber.

Marx, Karl 1970a: Introduction. *A Contribution to the Critique of Political Economy*, Moscow: Progress Publishers, 195–204.

—— 1970b: *Selected Works.* Moscow: Progress Publishers.

—— and Engels, Frederick 1970: *The German Ideology.* New York: International Publishers.

Maxwell, N. 1975: Learning from Taichai. *World Development, 3*, (7/8), 473–95.

Mazambani, D. 1982: Peri-urban cultivation around Harare's low-income residential areas, 1955–1980. *Zimbabwe Science News, 16* (6), 134–8.

Menendez, Cecilia and Baroni, Sergio 1980: Esquemas de Desarrollo Y Distribucion de las Fuerzas Productivas Enfoque Metodologico. *Cuestiones de la Economia Planificada, 3* (4), 157–92.

Menke, J. (ed.), 1983: *The Impact of Migration on the Social and Economic Transformation of Suriname*, University of Suriname: Institute of Economic and Social Research.

Merrington, G. J. 1980: Some aspects of the low-cost housing programme in Zimbabwe. Mimeo, Windhoek: Building Industry Action Committee.

—— 1981: A new look at the low-income housing plan. Mimeo, Victoria Falls: Local Government Association Annual Conference.

Mesa-Lago, Carmelo 1981: *The Economy of Socialist Cuba: A Two Decade*

Appraisal. Albuquerque: University of New Mexico Press.

Mihailovic, Kosta 1975: *Regional Development Experience and Prospects in Eastern Europe.* Geneva: UN Research Institute for Social Development.

Milne, R. S. 1975: Guyana's co-operative republic. *Parliamentary Affairs, 28,* 352–67.

—— 1977: Politics, ethnicity and class in Guyana and Malaysia. *Social and Economic Studies, 26,* 18–37.

—— 1981: *Politics in Ethnicity Bipolar States.* Vancouver: University of British Columbia Press.

Mingione, E. 1977a: *L'uso del territorio in Cina.* Milan, Mazzotta.

—— 1977b: 'Territorial social problems in socialist China' in J. Abu Lughod and R. Hay (eds), *Third World Urbanization,* Chicago: Maaroufa Press, 370–84.

—— 1980: Land use and class conflict: problems of the transition to socialism. (Translation into Polish published in *Planowanie spolezneqo rozowjy miast i spoteczosci terytorialnycm a badania socjoloqiczne* Ossolineum, Warsaw.)

—— 1981: *Social Conflict and the City.* Oxford: Blackwell.

—— 1984: The informal sector and the development of Third World cities: preliminary answers to some basic questions. *Regional Development Dialogue,* 5 (2), 63–74.

Ministère des Industries Légères 1977: *Emplois et Implantations des Projects Industriels en 1976 et en 1982.* Algiers.

Mittelman, J. 1981: *Underdevelopment and the Transition to Socialism: Mozambique and Tanzania.* London: Academic Press.

Monnier, A. 1981: Données récentes sur la population du Vietnam. *Population, 36* (3), 610–19.

Morrow, M. 1982: Ready for a rebound. *Far Eastern Economic Review, 115* (5), 48–9.

Mouffe, C. 1984: Towards a theoretical interpretation of 'new social movements'. In S. Hännen and L. Paldan (eds), *Rethinking Marx,* Berlin: International Socialism, discussion 5.

Mumford, L. 1966: *The City in History.* Harmondsworth: Penguin.

Murray, P. and Szelenyi, I. 1984: The city in the transition to socialism. *International Journal of Urban and Regional Research, 8* (1), 90–107.

Musil, J. 1980: *Urbanisation in Socialist Countries.* London: Croom Helm.

Nacer, M. 1979: *Regional Disparities and Development in Algeria,* University of Sheffield: MA Thesis, Department of Town and Regional Planning.

—— 1983: Processus de Planification et Cadre Administratif à Alger. *ONRS Informations, 13,* Algiers: Organisme Nationale de la Recherche Scientifique, 51–64.

Nangati, F. 1982: Harare Musika: an urban squatter settlement. Mimeo, Harare: Department of Social Services.

Nellis, J. R. 1977: Socialist management in Algeria. *The Journal of Modern African Studies, 15,* 529–54.

Nelson, J. M. 1979: *Access to Power: Politics and the Urban Poor in Developing Nations.* Princeton: Princeton University Press.

Nghiem, Dang 1966: *Vietnam. Politics and Public Administration.* Honolulu: East–West Center Press.

Nguyen Cong Thang and Nguyen Thi Xiem 1983: Population growth and family planning. *Vietnam Courier, 19* (10), 22–4.

Nguyen Duc Nhuan 1978: *Désurbanisation et Développement Régional au Vietnam (1954–1977)*. Paris: Centre de Sociologie Urbaine.

—— 1984a: Do the urban and regional policies of socialist Vietnam reflect the patterns of the ancient Mandarin bureaucracy? *International Journal of Urban and Regional Research, 8* (1), 78–139.

—— 1984b: Constraintes démographiques et politiques de développement au Vietnam, 1975–1980. *Population, 2*, 313–38.

Nguyen Khac Vien 1978: Letter from Ho Chi Minh City. *Vietnam Courier, 70*, 18–20.

—— 1980: Historical research in Vietnam since independence. *Journal of Contemporary Asia, 10* (3), 241–49.

—— 1982: Ho Chi Minh City – 1982: the releasing process. *Vietnam Courier, 18* (4), 20–3.

Nguyen Lam 1982: Government report at national assembly. *Vietnam Courier, 18* (1), 3–5.

Nguyen Tien Loc 1983: Studies on the district by the college of economics and planning. *Vietnam Courier, 19* (1), 8–9.

Nguyen Vinh Vien 1974: Hanoi, one year after. *Vietnam Courier, 20*, 4–7.

Nguyen Xuan Lai 1983: Economic development 1976–85. *Vietnamese Studies, 1* (71), 22–64.

Nhan Dang 1984: Hanoi: from consumer city to industrial city. *Vietnam Courier, 20* (11), 12–15.

Nongye Bu, Zhengce Yanjiushi (ed.) 1982: *Zhongguo Nongye Jingji Caiyao*. Beijing: Nongye Chubanshe.

Nyawo, C. and Rich, A. 1981: Zimbabwe after independence. *Review of African Economy, 18*, 89–93.

Nyland, C. 1981: Vietnam, the plan/market contradiction and the transition to socialism. *Journal of Contemporary Asia, 11* (4), 426–48.

O'Connor, A. 1984: *The African City*. London: Hutchinson.

Ohiorhenuan, J. 1980: Dependence and non-capitalist development in the Caribbean: historical necessity and degrees of freedom. *Science and Society*, XLIII (4), 386–408.

Omawale and Rodrigues, A. M. 1979: Agricultural credit related to nutrition and national development in the Caribbean: a study of the Guyana Agricultural Co-operative Development Bank. *Tropical Agriculture. Trinidad, 56* (1), 1–9.

Ottaway, D. and Ottaway, M. (eds) 1981: *Afrocommunism*. London: Holmes and Meier.

Pahl, R. 1968: The rural–urban continuum. In R. Pahl (ed.), *Readings in Urban Sociology*, Oxford: Pergamon Press.

Paine, S. 1978: Some reflections on the presence of 'rural' or of 'urban bias' in China's development policies, 1949–1976. *World Development, 6* (5), 693–708.

Pannell, C. W. 1981: Recent growth and change in China's urban system. In L. C. J. Ma and E. W. Hansen (eds), *Urban Development in Modern China*, Boulder: Westview Press, 91–113.

Parris, H. 1983: Report of speech to regional executive officers from Chairman, State Planning Commission. *Daily Chronicle, 18*, January 1983.

Payne, G. 1982: Analysis of a settlement – the Guyana situation – Melanie

Damishana. In M. P. Thomas (ed.), *The Caribbean Peoples and Their Environment*, Commonwealth Human Ecology Council, publication no. 18, 206–22.

Perez-Stable, Marifeli 1985: Class, organization, and *Conciencia*: the Cuban working class after 1970. In Sandor Halebsky and John M. Kirk (eds), *Cuba: Twenty-Five Years of Revolution 1959 to 1984*, New York: Praeger, 291–306.

Petras, J. 1977: State capitalism and the third world. *Development and Change*, 8, 1–17.

—— 1978: Socialist revolutions and their class components. *New Left Review*, 111, 37–66.

Pham Tri Minh 1978: Planning of district capitals. *Vietnam Courier*. 68, 9–13.

Piven, Frances Fox and Cloward, Richard 1982: *The New Class War*. New York: Pantheon Books.

Pike, D. C. 1966: *The Vietcong: Organization and Technique of the NLF of South Vietnam*. Cambridge: MIT Press.

—— 1982: Vietnam in 1981: biting the bullet. *Asian Survey*, 22 (1), 69–77.

Polan, A. J. 1984: *Lenin and the End of Politics*. London: Methuen.

Ponchaud, F. 1978: *Cambodia Year Zero*. Harmondsworth: Penguin.

Potter, L. 1979: Georgetown, parasite on the villages? A case study of the interrelationship between Georgetown and East Demerara, 1946–75. *Rural–Urban Transition and the Environment*, Georgetown: National Science Research Council.

—— 1982: Coastal and interior settlements in Guyana: development, problems and prospects. In M. P. Thomas (ed.), *The Caribbean Peoples and their Environment*, Commonwealth Human Ecology Council, publication no. 18, 34–9.

—— 1984: Transfer of value and uneven development in peripheral economies: the case of Guyana. In D. K. Forbes and P. J. Rimmer (eds), *Uneven Development and the Geographical Transfer of Value*. Canberra: Australian National University, Human Geography Monograph 16, 239–66.

Poulantzas, N. 1978: *State, Power, Socialism*. London: New Left Books.

Premdas, R. 1977: Guyana: communal conflict, socialism and political reconciliation. *Inter-American Economic Affairs*, 30, 63–83.

—— 1978: Guyana: socialist reconstruction or political opportunism? *Journal of Inter-American Studies and World Affairs*, 20 (2), 133–64.

—— 1979: Guyana: socialism and destabilisation in the western hemisphere. *Caribbean Quarterly*, 25 (3), 25–41.

Prenant, A. 1978: Centralisation de la décision à Alger. Decentralisation de l'exécution en Algérie. La mutation des fonctions capitales d'Alger. *Revue française d'études politiques méditerranéennes*, 3 (2–3), 30–31 and 128–50.

Quinn-Judge, P. 1982: A Vietnamese cassandra. *Far Eastern Economic Review*, 115 (9), 14–16.

—— 1983a: Contracts for change. *Far Eastern Economic Review*, 119 (6), 32–3.

—— 1983b: Ideological backtracking. *Far Eastern Economic Review*, 121 (29), 23.

—— 1984a: A non-industrial revolution. *Far Eastern Economic Review*, 123 (5), 46–7.

—— 1984b: Vietnam. Pragmatists win the day. *Far Eastern Economic Review*, 125 (32), 14–16.

—— 1985a: Hanoi's bitter victory. *Far Eastern Economic Review*. *128* (17), 30–6.

—— 1985b: Vietnam. No more free lunch. *Far Eastern Economic Review*, *129* (29), 36–7.

RADP (République Algérienne Démocratique et Populaire) 1974: *Ile Plan Quadriennal 1974–77. Rapport Général*. Algiers: Imprimerie Officielle.

Rabkin, R. P. 1985: Cuban political structure: vanguard party and the masses. In S. Halebsky and J. M. Kirk (eds), *Cuba: Twenty-Five Years of Revolution 1959–1984*, New York: Praeger, 251–69.

Race, J. 1972: *War Comes to Long An*. Berkeley: University of California Press.

Ramonet, I. 1985: Cuba-renovation dans la revolution? *Le Monde Diplomatique*, September.

Reid, P. 1983: Address to 5th Biennial Congress of the P.N.C. Reported in *New Nation*, 21 August 1983.

Renaud, B. 1981: *National Urbanization Policy in Developing Countries*. New York: Oxford University Press.

Revue de Presse 1983: Petite et Moyenne Industrie, la solution d'avenir. *Révolution Africaine*, no. 1000, reprinted in *Revue de Presse* (Algiers), no. 274, May.

Rhodes-Checchi Co. Ltd. 1979: *Rice II – Second Rice Development Project in Guyana*.

Richardson, Harry W. 1979: *Regional Economics*. Chicago: University of Illinois Press.

Riddell, R. 1981: *Report of the Commission of Inquiry into Prices, Incomes and Conditions of Service*. Salisbury: Government of Zimbabwe.

Ritter, Archibald R. M. 1985: The organs of people's power and the Communist Party: the nature of Cuban democracy. In Sandor Halebsky and John M. Kirk (eds), *Cuba: Twenty-Five Years of Revolution 1959 to 1984*, New York: Praeger, 270–90.

Roberts, B. 1978: *Cities of Peasants*. London: Edward Arnold.

Robinson, W. I. and Norsworthy, K. 1985: Elections and U.S. intervention in Nicaragua. *Latin American Perspectives*, *45* (2), 83–110.

Roca, Sergio 1983: Cuba confronts the 1980s. *Current History*, February 1983, 74–8.

Rondinelli, D. A. 1971: Community development and American pacification policy in Vietnam. *Philippine Journal of Public Administration*, *15* (2), 162–74.

—— 1973: Postwar reconstruction in Vietnam: the case for an urbanisation policy. *Asian Forum*, *5*, 1–15.

Roopchand, C. 1977: Targets for agricultural production. Unpublished paper presented at seminar 'Rationalisation of Research in Agriculture and Related Fields' Guyana, 15 March 1977.

—— 1979: Environmental problems and considerations in the growth of rural settlements: the case of Kuru Kururu, Linden/Soesdyke Highway. *Rural–Urban Transition and the Environment*, Guyana: National Science Research Council.

Rowbotham, S. 1980: The women's movement and organizing for socialism. In S. Rowbotham, L. Segal and H. Wainwright, *Beyond the Fragments – Feminism and the Making of Socialism*, London: Merlin Press, 21–156.

Ryan, M. 1984: *Marxism and Deconstruction – A Critical Articulation*. Baltimore: Johns Hopkins University Press.

SEP (Secretariat d'état au plan) 1980: *Annuaire Statistique de l'Algérie*. Algiers.
—— 1981: *Les résultats de l'enquête emploi et salaries de 1979*. Algiers.
—— 1982: *Annuaire Statisque de l'Algérie, 1980*. Algiers.
—— 1983: *L'Algérie en Quelques Chiffres 1982*. Algiers.
Sackey, J. A. 1979: Dependence, underdevelopment and socialist oriented transformation in Guyana. *Inter-American Economic Affairs, 33*, 29–50.
Sarin, M. 1982: *Urban Planning in the Third World. The Chandigarh Experience*. London: Mansell.
Sayigh, Y. A. 1982: *The Arab Economy*. Oxford: Oxford University Press.
Schnetzler, J. 1981: *Le développement Algérien*. Paris: Masson.
Schram, S. R. 1963: *The Political Thought of Mao Tse-tung*. London: Pall Mall Press.
Second Congress of the Communist Party of Cuba, n.d.: *Socio-economic Guidelines for the 1981–1985 Period*. Havana: Political Publishers for the Second Congress of the Communist Party of Cuba.
—— 1981: *Documents and Speeches*. Havana: Political Publishers.
Segre, Roberto 1980: *La Vivienda en Cuba en el Siglo XX: Republica y Revolucion*. Mexico: Editorial Concepto.
Serra, L. H. 1985: The grass-roots organizations. In T. W. Walker (ed.), *Nicaragua: The First Five Years*, New York: Praeger, 65–89.
Shaw, M. (ed.) 1984: *War, State and Society*. London: Macmillan.
Sid Ahmed, A. 1982: Petrole et développement: le cas algrien. *The Maghreb Review*, 7 (3–4), 49–61.
Singer, P. 1984: Capital and the national state: a historical interpretation. In J. Walton (ed.), *Capital and Labour in the Urbanized World*, Beverly Hills, Sage, 17–42.
Singh, J. 1979: Selected aspects of Guyanese fertility: education, mating and race. *A.N.U. Development Studies Centre Occasional Paper No. 21*, Canberra.
Singh, P. 1972: *Guyana: Socialism in a Plural Society*. London: Fabian Society, Fabian Research Series 307.
Slater, D. 1982: State and territory in post revolutionary Cuba: some critical reflections on the development of spatial policy. *International Journal of Urban and Regional Research*, 6 (1), 1–34.
Soja, E. 1980: The socio-spatial dialectic. *Annals of the Association of American Geographers, 70*, 207–25.
SORGEAH 1981: *Wadi Hadhramant Feasibility Study, Appendix 3*, SORGEAH.
Smout, M. and Heath, R. 1976: Urbanization and regional planning in Rhodesia. Mimeo. Salisbury: Whitsun Foundation.
Spragens, J. 1980: Looking ahead. *Southeast Asia Chronicle*, 76, 9–18.
Standing, G. 1979: Socialism and basic needs in Guyana. In G. Standing and R. Szal, *Poverty and Basic Needs: Evidence from Guyana and the Philippines*, ILO, World Development Programme, 17–79.
Statistical Yearbook of China, various years: Beijing: State Statistical Bureau.
Stern, L. M. 1981: District development, the new economic zones, co-operativization and South Vietnam's new economic policies of 1979. *Asian Profile*, 9 (4), 363–70.
Stöhr, W. B. and Taylor, D. R. .F. (eds) 1981: *Development from Above or Below? The Dialectics of Regional Planning in Developing Countries*. Chichester: Wiley.

Stoneman, C. 1978: Foreign capital and the reconstruction of Zimbabwe. *Review of African Political Economy*, *11*, 62–83.

Stretton, H. 1978: *Urban Planning in Rich and Poor Countries*. Oxford: Oxford University Press.

Sun, Hungzhi 1983: The advantages of core-cities leading villages. *Jingji Guanli*, 2, 37–9.

Sun, Jian 1980: *An Economic History of the People's Republic of China 1949–1957*. Jilin: Rehmin Chubanshe.

Sun, Shangqing (ed.) 1984: *Lun Jingji-jieqou Duice (Policies towards economic structure)*. Beijing: Zhongguo Shehui Kexue Chubanshe.

Susman, Paul 1974: Cuban development: from dualism to integration. *Antipode*, 6 (3), 10–29.

Sutton, K. 1976: Industrialisation and regional development in a centrally-planned economy – the case of Algeria. *Tijdschrift für Economische en Sociale Geografie*, 67 (2), 83–94.

—— 1981: Algeria: centre-down development, state capitalism, and emergent decentralization. In W. B. Stöhr and D. R. F. Taylor (eds), *Development from Above or Below?* Chichester: Wiley, 351–78.

—— 1982: Agrarian Reform in Algeria – Progress in the face of disappointment, dilution, and diversion. In S. Jones, P. C. Joshi and M. Murmis (eds), *Rural Poverty and Agrarian Reform*. New Delhi: Allied Publishers, 356–75.

Szelenyi, I. 1981: Structural changes of and alternatives to capitalist development in the contemporary urban and regional system. *International Journal of Urban and Regional Research*, 5 (1), 1–14.

—— 1983: *Urban Social Inequalities under State Socialism*. Oxford: Pergamon.

TAMS Agricultural Development Group, Aubrey Barker Associates, Intermediate Savannah Agricultural Development Project, Ministry of Agriculture, Co-operative Republic of Guyana 1976: *The Intermediate Savannahs Report – Volume 1, Resources: Volume II, Development Planning*.

Tao, Weichun 1983: Fully develop water transport enterprises – let Zongqing develop its role as a core city. *Jinqji Guanli*, 3, 12–15.

The Nguyen 1978: The commercial growth in the southern provinces: from capitalism to socialism. *Vietnam Courier*, 72, 11–18.

Thien Anh 1983: An Ha: the shaping of a New Economic Zone. *Vietnam Courier*, 19, (12), 17–18.

Thiery, S. P. 1980: Emploi, formation et productivité dans l'industrie algérienné. *Annuaire de l'Afrique du Nord*, XIX, 181–93.

—— 1981: Quelques réflexions sur l'emploi non agricole et l'évolution du rapport salarial en Algérie. *Mondes en Développement*, 9 (36), 45–61.

Thomas, C. Y. 1976: Bread and justice: the struggle for socialism in Guyana. *Monthly Review*, September, 23–35.

—— 1984: Developing nations' growth slows: I.M.F. reason? Guyana: a case study. *Caribbean Contact*, 4 November.

Thrift, N. J. and Forbes, D. K. 1985: Cities, socialism and war: Hanoi, Saigon and the Vietnamese experience of urbanisation. *Society and Space*, 3 (3), 279–308.

—— 1986: *The Price of War: Urbanisation in Vietnam 1954–1985*. London: Allen and Unwin.

Tian, Jianghai and Guangan, Li 1981: A summary of discussion on the theory of overall balance of the national economy. *Zhongguo Shehui Kexue, 3,* 97–102.

To Huu 1984: Building up the district echelon: all round development for 400 districts. *Vietnam Courier, 20* (12), 16–18.

Ton That Thien 1967: Vietnam: a case of social alienation. *International Affairs, 43* (3), 455–67.

Tran Dang Van 1978: Re-deployment of the labour force in Vietnam. *Vietnam Courier, 69,* 10–12, 30.

Turley, W. S. 1975: Urbanization in war: Hanoi, 1946–1973. *Pacific Affairs, 48,* 370–95.

UNDP and Government of the Co-operative Republic of Guyana 1977a *Urban and Regional Planning Project: The Administrative System in Guyana at Regional, Sub-Regional and Local Level* Studies in Region 3, Sub-regions East Coast Demerara and West Coast Berbice.

—— 1977b: *Urban and Regional Planning Project Guy/74/005 National Physical Development Strategy Background Studies: No. 2 – Population No. 4 Settlement: No. 5 – Housing.*

United Nations 1980: *Patterns of Urban and Rural Population Growth.* New York: United Nations, Population Studies no. 68.

—— 1985a: *1982 Statistical Yearbook.* New York.

—— 1985b: *1983 Demographic Yearbook.* New York.

Vietnam Courier 1982a: Economic achievements in five years (1976–1980). *Vietnam Courier, 18* (2), 6–8.

—— 1982b: Facts and figures. *Vietnam Courier, 18* (10), 25.

—— 1982c: Housing – a real problem. *Vietnam Courier, 18* (4), 24.

Vietnamese Studies 1983: Vietnam's population. *Vietnamese Studies, 1* (71), 179–82.

Viviani, N., Thayer, C., Van Der Kraan, A. and McCoy, A. 1981: Report on discussions held with the Social Sciences Committee of Vietnam and others during a visit to Vietnam and Kampuchea by an academic delegation of the Australian Committee for Scientific Cooperation with Vietnam, 19 August–9 September 1981. Unpublished typescript.

Vo Van Kiet 1983: Orientation, tasks and essential objectives of the socio-economic plan for 1983 and the objectives to be attained by 1985. *Vietnam Courier, 19* (2), 2–6.

Ward, Benjamin 1967: *The Socialist Economy.* New York: Random House.

Weaver, J. and Kronemer, A. 1981: Tanzanian and African Socialism. *World Development, 9* (9/10), 839–49.

Weber, H. 1981: *Nicaragua – The Sandinista Revolution.* London: New Left Books.

Weinand, H. C. 1972: A spatio-temporal model of economic development. *Australian Geographical Studies,* X (1), 95–100.

Wellings, P. A. and McCarthy, J. J. 1983: Whither southern Africa human geography. *Area, 15* (4), 337–45.

Wheelock, J. 1985: *Entre la Crisis y la Agresión – La Reforma Agraria Sandinista.* Managua: Editorial Neuva Nicaragua.

White, C. 1982: *Debates in Vietnamese Development Policy.* University of Sussex, Institute of Development Studies, discussion paper 171.

White, G. 1983: Revolutionary socialist development in the third world: an overview. In G. White, R. Murray and C. White (eds), *Revolutionary Socialist Development in the Third World*. Brighton: Wheatsheaf Books, 1–34.

—— 1984: Developmental states and socialist industrialisation in the third world. *Journal of Development Studies*, 21, 97–120.

—— Murray, R. and White, C. (eds) 1983: *Revolutionary Socialist Development in the Third World*. Brighton: Wheatsheaf Books.

Whitlow, J. R. 1980: Environmental constraints and population pressures on the tribal areas of Zimbabwe. *Zimbabwe Agricultural Journal*, 77 (4), 173–81.

—— 1982: Marginality and remoteness in Zimbabwe. *Zimbabwe Agricultural Journal*, 79 (4), 139–47.

Whitsun Foundation 1979: A changing emphasis in urban planning. Mimeo, Salisbury.

Wiegersma, N. 1983: Regional differences in socialist transformation in Vietnam. *Economic Forum*, 14 (1), 95–109.

Wilber, C. K. and Jameson, P. (eds) 1982: *Socialist Models of Development*. Oxford: Pergamon Press.

Wiles, P. D. (ed.) 1982: *The New Communist Third World*. London: Croom Helm.

Williams, P. E. 1983: An analysis of regional inequality and public investment in Guyana. Unpublished paper presented to Conference on the Spatial Dimension in Development and Planning, University of Guyana, 14–17 April 1983.

Williams, Raymond 1983: *Key Words: A Vocabulary of Culture and Society*. New York: Oxford University Press.

Williamson, J. B. 1965: Regional inequality and the process of national development: a description of patterns. *Economic Development and Cultural Change*, 13, 3–43.

Wolpin, M. 1981: Contemporary radical third world regimes: prospects for their survival. *Caribbean Quarterly*. 27 (1), 63–82.

Woodside, A. B. 1971: *Vietnam and the Chinese Model. A Comparative Study of Nguyen and Ch'ing Civil Government in the First Half of the Nineteenth Century*. Cambridge: Harvard University Press.

World Bank 1975: *Rural Development*. Washington, DC: World Bank.

—— 1976: *Economic Memorandum on Guyana*. 21 June 1976, report no. 1005 GUA.

—— 1979: *PDR Yemen Review*. Baltimore: Johns Hopkins University Press.

—— 1981a: *China: Socialist Economic Development*. (9 volumes) Washington, DC: World Bank.

—— 1981b: *World Development Report 1981*. New York: Oxford University Press.

—— 1983: *World Development Report 1983*. New York: Oxford University Press.

—— 1984: *World Development Report 1984*. New York: Oxford University Press.

—— 1985: *China: Long-Term Development Issues and Options*. Washington, DC: World Bank.

Wu, Chung Tong 1979: Development strategies and spatial inequality in the People's Republic of China. WP79–06, United Nations Centre for Regional Development. Japan: Nagoya.

—— 1980: Transforming rural development strategies: a preliminary report. *Regional Development Dialogue*, 1 (2), 102–23.

—— and Ip, D. F. 1980: Structural transformation and spatial equity: lessons from China. In C. K. Leung and N. Ginsburg (eds), *China: Urbanization and National Development*, Chicago: Department of Geography Research Paper, no. 196, 56–88.

—— 1982: *Regional Autonomy and Rural Development in China: Recent Directions in Shangdong and Guangdong*. Report presented to the United Nations Centre for Regional Development, Nagoya, Japan.

Wu, Youren 1981: The question of socialist urbanization in China. In Institute of Population Economics, Beijing College of Economics (ed.) *Zhongguo Renkou Kexue Lunji*, Beijing: Zhongguo Xuexue Chubanshe, 96–104.

Xu, Yi 1981: On the relationship between the structure of the distribution of public finance and the structure of the national economy. In Zhongguo Caizheng Bu, *Zhongguo Caizheng Wenti*, Tianjin: Tianjin Kexue Chubanshe, 202–20.

Xu, Dixin 1982: *Zhongguo Shehui Zhuyi Jingji Fazhen Zhong di Wenti*. Beijing: Zhongguo Shehui Kexue Chubanshe.

—— et al. 1981: *Zhongguo quomen jingji fazhen zhong di wenti (Problems of China's National Economic Development)*. Beijing: Zhongguo Shehui Kexue Chubanshe.

Yang, Jianbai and Li Xuezeng 1980: China's historial experience in handling the relations between agriculture, light industry and heavy industry. *Zhongguo Shehui Kexue*, 3, 19–40.

Yates, P. 1981: The prospects for socialist transition in Zimbabwe. *Review of African Political Economy*, 18, 68–88.

Yearbook of Chinese Agriculture various years: *Zhongguo nonye nianjian*. Beijing: Nongye Chubanshe.

Yih, K. and Slate, A. 1985: Bilingualism on the Atlantic Coast – where did it come from and where is it going? *Wani*, 2–3, May–December, 23–7.

Yu, Guangyuan 1982: *Strategy of Economic and Social Development*. Beijing: Zhongguo Shehui Kexue Chubanshe.

Zafiris, N. 1982: The People's Republic of Mozambique: pragmatic socialism. In P. Wiles (ed.), *The New Communist Third World*, London: Croom Helm, 114–64.

Zasloff, J. J. c.1963: Rural resettlement in Vietnam. An agroville in development. Michigan State University Vietnam Advisory Group, Agency for International Development.

Zhang, Chunyuan 1982: On some problems of our per capita GNP. *Jingji Kexue*, 2, 8–15.

Zhang, Shuguang 1981: Economic structure and economic results. *Zhongguo Shehui Kexue*, 6, 41–58.

Zhang, Tianlu 1981: Of family planning and national prosperity. In Institute of Population Economics, Beijing College of Economics (ed.), *Zhongguo Renkou Kexue Lunji*, Beijing: Zhongguo Xuexue Chubanshe, 113–20.

Zhao, Ziyang 1983: Report to the 6th National People's Congress, first session. *Nangfang Ribao*, 24 June.

Zhonggong Wehai Shiwei 1983: Towards the path of uniting city and rural economic development. *Jingji Yanjiu*, 4, 66–8.

Zhonggong Zhongyang 1979: *Decisions Concerning the Speedy Development of Agriculture*. 4th Plenary Session of the Central Committee of the Chinese Communist Party 11th Congress.

—— 1986: *The 7th Five Year Plan*. In *Wenhuibao* 15 April.

Zhongguo Baike Nianjian (Encyclopedia of China) 1982: Beijing: Baike Nianjian Chubanshe.

Zhonghua Renmen Gongwoguo 1983: *Guomen Jingji wo Shehui Fazhen: di Luigu Wunian Jihua*, 1983–1985. Beijing: The Sixth Five-Year Plan 1981–1985: National Economic and Social Development Plan, Renmen Chubanshe.

Zhou, Xiaohan 1982: A macroeconomic analysis of the relationship between the growth rate and economic results. *Zhongguo Shehui Kexue*, 4, July, 133–56.

Zimbalist, Andrew 1985: Cuban economic planning: organization and performance. In Sandor Halebsky and John M. Kirk (eds), *Cuba: Twenty Five Years of Revolution, 1959–1984*, New York: Praeger Publications, 213–30.

Zinyama, L. 1982: Post-independence land resettlement in Zimbabwe. *Geography*, 67, 149–52.

Notes on Contributors

David Drakakis-Smith is Senior Lecturer in Geography at the University of Keele. His main research interests lie in the problems posed for low income groups by rapid urbanization. He has researched on these problems in Australia, Southeast Asia, Southern Africa and the Middle East. He is the author or editor of several books including *Urbanisation, Housing and the Development Process*, *Urbanisation in the Developing World*, and *Multinationals and the Third World*. He is currently writing a book about urban employment in developing countries.

Dean Forbes works in the Appraisals, Evaluation and Sectoral Studies Branch of the Australian Development Assistance Bureau, and is also a Senior Research Fellow in the Research School of Pacific Studies in the Australian National University. He has undertaken research and worked on development projects in Indonesia, Papua New Guinea, Vietnam, China and the Philippines. He is the co-editor of *Australian Overseas Aid* and joint author with Nigel Thrift of *The Price of War*. He is on the Editorial Boards of *Society and Space* and *Antipode* and an editorial consultant to *Inside Indonesia*. His current work is on China.

Jim Lewis' main interests are in Southern Europe and Southeast Asia. He is a lecturer in the Department of Geography at the University of Durham. His publications include *Uneven Development in Southern Europe* (co-edited with R. Hudson).

Enzo Mingione is Professor of Sociology at the University of Milan and the University of Messina. His previous publications have included *Social Conflict and the City*.

M'hamed Nacer is an Algerian geographer-planner attached to the Unite de Recherches en Amenagement du Territoire (URAT) within the National Agency for Planning (ANAT) in Algiers, having previously lectured in geography at the Houari Boumedienne University of Algiers. He studied at the University of Algiers and then in the Department of Town and Regional Planning at the University of Sheffield. He is currently research-ing into regional development in Algeria with particular reference to the Greater Algiers region.

Lesley Potter lectured at the University of Guyana from 1971 to 1979, and was head of department from 1976. She is now Senior Lecturer in Geography at the University of Adelaide, but maintains an interest in Guyana. Her current research activities are focused on Indonesia, specifi-cally Kalimantan, where she is studying agricultural change and the evolution of environmental policy. Her recent publications include a chapter on Guyana in *Uneven Development and the Geographical Transfer of Value* and a chapter on Indonesia in *Land Degradation and Society*.

David Slater is Associate Professor in Social Geography in the Centre for Latin American Research and Documentation in Amsterdam. His research interests include problems of urban and regional development in peripheral capitalist societies, and theories of the peripheral state, and revolution and territorial re-organization. His recent publications include books on *State and Issues of Regional Analysis in Latin America*, *New Social Movements and the State in Latin America* (ed.) and *Capitalism and Urbanization at the Periphery*.

Paul Susman is an Associate Professor in the Geography Department of Bucknell University. His research focuses include the political economy of development in both first and third world contexts. His current inter-ests include socialism and spatial development in Cuba, technological change and industrialization in first and third world countries, especially in the Caribbean Basin, and energy and economic development in the third world.

Keith Sutton is a Senior Lecturer at the University of Manchester. After post-graduate research at the University of London on the historical geography of the Sologne region in France he developed an interest in the economic development problems of North Africa which he has regu-larly visited since 1967. He is the author of several chapters and articles on nineteenth-century land clearance in France, on population change, agrarian reform, rural settlement, and socialist villages in Algeria, and on rural resettlement in Africa and Malaysia. Several of these are listed in his article on 'Progress in the Human Geography of the Maghreb', published in *Progress in Human Geography*.

Nigel Thrift's main interests are in the City of London, multinational corporations, time geography, social theory and the socialist third world.

He is currently a lecturer in the Department of Geography at the Universiy of Bristol. He has also been University of Wales Reader in the Geography Department at Saint David's University College, Lampeter, and prior to that held appointments at The Australian National University, University of Leeds and Cambridge University. He is Co-editor and Reviews Editor of *Environment and Planning A*, Reviews Editor of *Environment and Planning B*, and a member of the Editorial Board of *Society and Space*. His recent publications include *Time, Spaces and Places* (with D. Parkes) and *The Price of War* (with Dean Forbes).

Chung-Tong Wu is Associate Professor of Town and Country Planning at the University of Sydney. He has published a number of articles and chapters in books on rural development in China. His recent papers and research deal with rural industrialization and the development of small towns in China. He is also co-ordinator of an international project on export processing zones in Asia.

Index

References in italics indicate figures and tables. The letter 'n' indicates a reference to a note to the main text.

accumulation, socialist, 38–9, 43, 44
  China, 61, *63*
Aden, 171, 171–2, 188
administrative structure
  Algeria, *132*, 148
  Guyana, 226, *227*
  Vietnam, 99–100, 104–5
Afro-Guyanes, 216, 222, *222*
Agrarian Reform Law
  Cuba (1959), 269
  PDR Yemen (1968), 177–9
agricultural classes, 39, 41
Agricultural Development Fund,
    PDR Yemen, 185
agriculture, 34, 37–8
  Algeria, 129, 132, 135, 156
  China, 46, 47, *66*, 69, *70–1*, 91;
      Five-Year Plans, 55, *56*, 57,
      58, 59, 60; state investment,
      61, *62*, *64*, 64
  Cuba, 15, 255, 271
  Guyana, 220, 224, *225*
  PDR Yemen, 170, 175–91
  Vietnam, 109–11, *110*, 112
  Zimbabwe, *195*, 198
  see also brigades; communes;
      cooperatives; state farms;
      workteams
agrovilles, 102
Algeria, *18*, 19, 21–2, 129–31, 166–8
  development plans, 146–58
  industrial development, 158–66
  regional disparities, 132–46
Algiers, 137, 146, 150, 153, 159
Amerindians, Guyana, 221, 229–31,
    232, 249n

Annaba, Algeria, 137, 146, 150, 153,
    159
Atlantic coast, Nicaragua, 297, 298
authoritarianism, and Leninist model,
    286–7
autonomy, Nicaragua, 293, 297–9

backward areas
  Algeria, *140*, 142–5, *144*
  China, 85–6, *86*, 87, 93
Batna, Algeria, 142
bauxite industry, Guyana, 216, 237,
    239, 246, 249n
Bechar, Algeria, 137, 142
Beijing, China, 20, 47, 68–9, 72
birth control in China, 56, 59, 60–1,
    61
Blida, Algeria, 137, 150, 153
Bouira, Algeria, 142
brigades, Chinese, 58, 59, 95n
  diversification, 72–3, 75–8, *76–7*
  poverty, 78, *82*, 82, *84*
British occupation of Aden, 171, 172
British South Africa Company, 197
Brûlé, J. C., 159
Brundenius, C., 269–70
building sector and urbanization,
    34–5
Bulawayo, Zimbabwe, 206, *206*, 211
Burnham, F., 216–17, 223–4, 240,
    248n

capitalism and socialism, 4, 5, 284–5
  see also private enterprise
Castro, F., 254–5, 260, 272, 275, 296
Castro, R., 294–5

cement industry, Algeria, 130
centralization
  Algeria, 130, 131, 132, 165
  planning in Cuba, 25, 294, 295
  Vietnam, 100, 126–7
China, 14–15, *18*, 19, 20–1, 53–4,
  91–4
  backward regions, 85–6, 93
  Five-Year Plans, 55–61, 86, 88–9,
  93
  regional development, 66–85, 86–91
  rural strategy, 170–1
  urbanization, 10–11, 44–50, 52n
Chitungweza, Zimbabwe, 206
cities, 10–12
  China, 45, 47–8, 56, 58, 59, 94n;
  core, 90, 91, 92
  Cuba, 256, 257, 273; *see also*
  Havana
  Guyana *see* Georgetown; Linden
  pre-industrial, 28, 50n
  Vietnam, 104, 113–14, 121–6;
  agrovilles, 102
  Zimbabwe, 205–6, *206*; *see also*
  urban primacy
class and socialist development, 4, 6,
  12–13, 254, 287–8
Claudin-Urondo, C., 285–6
coastal regions, China, 55, 60, 69,
  73, *74*, 92
  migration from, 56, 59
  poverty, 83, 84
collective farming, 170
  see also brigades; cooperatives; state
  farms, workteams
collective income, China, 78–81, *79*,
  80, *82*, *83*, *84*, 95–6n
colonialism, Guyana, 220
commercial farms, Zimbabwe, 210
Communal Development Plans,
  Algeria, *see Plans Communaux*
communes
  Algeria, 148, 153, 154, 161; *see*
  also Plans Communaux
  China, 57, 58, 94–5n;
  diversification, 48, 72–3, 75–8,
  76–7
  Vietnam, 100–1
communications
  Algeria, 151, *152*

PDR Yemen, 188
Communist Party, Cuba, Congresses,
  260, 267, 274, 277
community development programmes,
  Vietnam, 101–4
Congresses
  Communist Party, Cuba, 260, 267,
  274, 277
  FLN, Algeria, 148–9, 153
Constantine, Algeria, 137, 146, 150,
  153
construction industry
  Algeria, 158, *158*
  Cuba, 256
consumer cities, 10
consumption equality, Cuba, 255–73
cooperative socialism, 247
cooperatives
  China, 55
  Cuba, 272, 273
  Guyana, 218, 224, 228
  Vietnam, 105–6, 108, 110
  Yemen, 177, *179*, 181, 185, 191,
  192
Coraggio, J. L., 282, 290, 290–1, 297
core cities, China, 90, 91, 92
cotton production, Yemen, 177
Council of Ministers, Cuba, 293
Cuba, 15, *18*, 25–6, 250–5, 278–81
  consumption equality, 255–73
  model for Guyana, 220, 242, 245
  Nicaragua compared, 289–96, 300
  production equality, 273–8
culture and urbanization, 28–9

daïrate, *132*, 161
decentralization
  Algeria, 131, 136, 149, 155–6; and
  deconcentration, 153–5, 164,
  167–8
  China, 55, 57, 92
  Cuba, 26, 274, 278, 292, 294, 295
  Nicaragua, 26, 292, 299
  Vietnam, 103, 126–7
deconcentration, Algeria, 153–4, 164,
  167–8
defence, Nicaragua, 292–3, 298, 299
democracy, 283, 288–9
  capitalist, 251–3
  social, 284

socialist, 252–3
Democratic Republic of Vietnam, 104–8
democratic socialism, 287–9
democratization, Cuba, 273–8, 281
Development Plan, Guyana, 24, 218, 224, 226, 228–35
district plan, Vietnam, 119–21, 127
district towns, Vietnam, 120, 121

East Indians, *see* Indo-Guyanes
Eckstein, S., 275–6
economic regions, China, 88–9, 96n
economic zones
   Algeria, 132, *134*, 160, *160*
   China, 90–1
   Vietnam, 106–8, 116, 117–19
education
   Cuba, 270–1
   Nicaragua, 298
El Asnam, Algeria, 142
employment, Algeria, 158, *158*, *160*, *162*
energy industries, China, 55
Engels, F., 9, 10, 253, 254
England, industrialization, 31–2
equality, Cuba, 251–3
   consumption, 255–73
   production, 273–8
Eritrea, Ethiopia, 17, *17*
European lands, Zimbabwe, 197, *198*, 200, 201, 210
exports, Zimbabwe, 194, *195*

farms, *see* commercial; cooperatives; State farms
feudal system in Vietnam, 98–9
fiscal stability, China, 64–5, 67, 88
fishing, Yemen, 22, 175–6, *176*
Five-Year Plans
   Algeria, 22, 148–53
   China, 55–61, 86, 88–9, 93
   Vietnam, 106, 108–9, 110, 112, 116, 117
   Yemen, 175, 181, 191
foreign debt and urbanization, 34, 35
Four-Year Plan, Guyana, 237
French rule in Vietnam, 100, 103–4
Front de Libération Nationale, Algeria, Congress, 148–9, 153
Fujian province, China, 73

GDP growth, Guyana, 240, *241*
Geizhou province, China, 86
geography and socialist development, 4, 6
Georgetown, Guyana, 221, 222, 232, 235, 240
governorates, PDR Yemen, 171, *172*, *173*
'Great Cultural Revolution', China, 58, 65, 87–8
'Great Leap Forward', China, 47, 57, 65
Grenada, 300
Griffith, I., 236
Guyana, *18*, 19, 24, 214–19, 246–8
   planning, 219–20, 223–42
   regional structure, 220–3, 242–6

hamlet programme, Vietnam, 102–3
handicraft industries in China, 56
Hanoi, Vietnam, 122, *123*, 124, 125–6
Harare, Zimbabwe, *206*, 206, 207, 211
Havana, Cuba, 11, 15, 25, 250, 256–9, 279
heavy industries, 15
   China, 57, 60, *66*, 91; state investment, 55, *56*, 58, 61, *62*, *64*
   Vietnam, 106
hegemony, popular, 25, 287–9, 291, 296–8
history and socialist development, 4, 6
Ho Chi Minh City, *formerly* Saigon, 10, 122–6, *123*
horizontal expansion, PDR Yemen, 181, 189, 191
hospital bed provision, Cuba, 267, *268*
housing, 39–41
   China, 45–6
   Cuba, 257, 258–9, *259*, 263–4, 272
   Guyana, 232, 234–5
   Hanoi, Vietnam, 122, 126
   Zimbabwe, 212
Hoyte, D., 247

IMF (International Monetary Fund), 215–16, 218, 240, 241, 246

imports, Zimbabwe, *195*, 195–6
income distribution
  China, 78–81, *79*, *80*, *82*, *83*, *84*,
    95–6n
  Cuba, 255, 256, 264–6, *264*, *265*,
    266, 268–70
  Guyana, *233*, *234*, 234
  PDR Yemen, 189–90, 192
  Indo-Guyanes (East Indians), 216,
    220–1, 222, 235, 248n
industrialization, 14–15, 19–20
  Algeria, 22, 129, 130–1, 146,
    158–66, 167
  and urbanization, 19–20, 27–8;
    advanced countries, 30–5;
    Chinese example, 44–50;
    socialist countries, 35–44
  China, 54, *56*, 59, 68–9, *70–1*, 72
  Cuba, 258, 263
  PDR Yemen, *176*, 176–7
  Vietnam, 106, 109, 110, *110*
  Zimbabwe, 195
  see also heavy industries; light
    industries; rural
    industrialization
informal sector, 34, 35, 212
inland region, China, 55, 60, 69, *73*,
  74, 84
  migration to, 56, 59
Inter-American Development Bank
  (IDB), 215
intermediate region, China, 69, *73*,
  74, 84
International Monetary Fund, *see*
  IMF
international relations, 4, 6, 44
  and industrialization, 32, 36–7
  China, 46–50
investment, *see* state investment
iron and steel industry, China, 55,
  57, 58
irrigation, PDR Yemen, 171, 179,
  180, 181, 185, 191
Irvin, G., 290–1
Italy, urban data, 29

Jagan, C., 216, 236, 247, 248n
Jiangsu province, China, 75–6

Kampuchea, 10

Labour Amendment Act, Guyana,
  246
labour
  and industrialization, 30–1, 32, 33
  Zimbabwe, 198–9, 200
  see also employment; trade unions
Laclau, E., 290
Laghouat, Algeria, 137, 142, 146
Lancaster House agreement, 196,
  208–9, 210
Land Apportionment Act, Zimbabwe,
  23, 197
land distribution, Zimbabwe, 197,
  198, 200, 201, 204, 210
Land Husbandry Act, Zimbabwe,
  200
land reform
  Guyana, 218
  PDR Yemen, 177, 191
  Vietnam, 105
Land Tenure Act, Zimbabwe, 197,
  201
Leninist model of socialism, 4, 6, 25,
  285–7, 289, 300, 301n
  Cuba, 289, 295–6
Liaoning, China, 20, 68–9, *72*
light industries
  Algeria, 161
  China, *56*, 60, 61, *62*, 66
  Vietnam, 112
Linden, Guyana, 221, 222, 232, 239
Linyi prefecture, China, 86
'lodging', Zimbabwe, 201, 206
Literacy Campaign, Cuba, 271
Local Democratic Organs, Guyana,
  242, 244
Luxemburg, R., 282

Machel, S., 196
Machine Renting Stations, PDR
  Yemen, 181, 185
mandarins, Vietnamese, 99, 100
Mao Tse-Tung (Zedong), 10, 53, 55,
  56
Maoist tendencies, Zimbabwe, 196,
  208
market socialism, 111–12, 127
Master and Servants Act (1901),
  Zimbabwe, 199
Matanzas province, Cuba, 274, 275
Marx, K., 9, 10, 252, 253, 287·

medical system, Cuba, 266–7
Mekong Delta, Vietnam, 113, 116
Mesa-Lago, C., 268–9
migration
    Algeria, 159
    and urbanization, 33–4, 42
    China, 46, 47, 56, 58, 59, 97n
    Cuba, 257
    Guyana, 231
    PDR Yemen, 189–90
    Vietnam, 117
    Zimbabwe, 196, 201, 205–6
Milan, population data, 29
military participation, 13–14, *14*
    Guyana, 216, 217
mining, Guyana, 221
    see also bauxite
Ministry of Heavy Industries,
    Algeria, 163
Ministry of Light Industry, Algeria,
    160–1, 167
minorities in China, 85, *87*, *88*, 93,
    96n
Mouffe, C., 290
Mugabe, R., 23, 194–5, 196, 208
Murray-Szelenyi model of
    urbanization, *12*, 12–13, 41–2
Mutare, Zimbabwe, *206*
Mutin, G., 159

National Assembly of People's Power,
    Cuba, 293, 294
National Commission for Autonomy,
    Nicaragua, 297
National Liberation Front
    PDR Yemen, 171, 174
    Vietnam, *16*, 17
National Plans, Algeria, 129, 132,
    135–6, 146–53, *150*, 157
National Service, Guyana, 228–9
nationalities in China, 85, *88*, *93*, 96n
nationalization, 3, 283
    Guyana, 215, 218
Ndebele tribe, Zimbabwe, 197, *198*,
    209
Nellis, J. R., 129–30
neo-colonialism, 202, 203, 207
New Amsterdam, Guyana, 221, 235
New Economic Policies, Vietnam, 21,
    109, 111, 124

New Economic Zones, Vietnam, 21,
    106–8, 116, 117–19
'new life hamlets', Vietnam, 103
new towns, 35, 50n
    Cuba, 272–3
Nguyen dynasty, Vietnam, 99, 101
Nicaragua, *18*, 19, 25–6, 289–92,
    292–3, 300
    popular hegemony, 296–9
North Vietnam, 99, 104–8, 111, 113,
    116–17

Options Hauts-Plateaux, 150–1, *152*
Oran, Algeria, 137, 146, 150, 159,
    164–5
Organs of People's Power, Cuba,
    275, 276, 293, 294
Ouargla, Algeria, 137, 142, 146
overproduction, industrial, 32, 33,
    34, 39

'parallel economy', Guyana, 215,
    248n
PDR Yemen, *see* Yeman, PDR
peasant mode of production,
    Zimbabwe, 196, 197, 199–200
peasants and revolution, 41, 45
People's Cooperative Units, Guyana,
    244–5
People's Liberation Front, Eritrea, 17
People's National Congress Party
    (PNC), Guyana, 216, 217, 218,
    244, 245, 247
People's Power, Cuba, 274–6
People's Progressive Party, Guyana,
    216
Petites et Moyennes Industries
    programme, Algeria, 164
Phnom Penh, Kampuchea, 10
Plans, national, *see* Development
    Plan; Four-Year Plan; Five-Year
    Plans; National Plans; Three-
    Year Plans
Plans Communaux de Developpement
    (Communal Development Plans),
    Algeria, 22, 136–7, 146–8,
    155–8, *156*, *157*, 163
Plans de Modernisation Urbaine,
    Algeria, 148

pluralism, Nicaragua, 25, 289, 291, 299
PNC, see People's National Congress
Polan, A. J., 286–7
polarization reversal, 6–7, 12
popular hegemony, 25, 287–9, 296–8, 300
population distribution
　Algeria, 142, 151, 153
　China, 54, 56, 59, 94n; see also birth control
　Cuba, 260, 263
　Guyana, 24, 221–2, 222, 228–32, 230
　Vietnam, 101, 106–8, 114–19, 115, 117, 118; cities, 122, 123, 125
　Zimbabwe, 196, 210
　see also migration
Poulantzas, N., 288
poverty
　China, 78–84, 85
　PDR Yemen, 171, 174
　Zimbabwe, 196, 205
'pragmatic socialism', 195, 196, 212
prefectures, Chinese, 90, 96n
pre-industrial inheritance and urbanization, 28–9, 50n
price control, PDR Yemen, 185, 191
prices, agricultural, China, 65, 66–7, 68, 78, 95n
primacy, 6, 8, 9, 250, 256, 257
private enterprise
　Algeria, 131, 167
　Guyana, 215, 218
　Vietnam, 111
　Zimbabwe, 208
private plots in China, 58, 59, 95n
production cities, 10
production equality, Cuba, 273–8
production incentives, Vietnam, 111–12
'proto-Socialist', 283
provinces
　Chinese, 73
　Cuba, 259–60, 260, 261, 262; capitals, 260–3, 279; consumption equality, 259–68
　Vietnam, 107, 118, 119
Provincial Development Fund, Vietnam, 103

qat, 185, 188, 190, 191

racism
　Guyana, 217–18
　Zimbabwe, 201–2
radical nationalism, 25, 284–5
railways, Algeria, 151, 152
Rapport Général Plan, Algeria, 149
Regional Development Council, Guyana, 226
religion, Guyana, 216
Rent Reduction Law, Cuba, 269
research data
　comparison problems, 29–30
　sources for China, 44–5, 51–2n
Resolution 6, New Economic Policies, Vietnam, 111, 124
resource distribution, 36, 37–9, 43, 44
　China, 48, 49
　PDR Yemen, 189, 191
　Vietnam, see equality
rice growing, Guyana, 221, 237
Roca, S., 266–7
Rome, population data, 29
rural cities, Vietnam, 102
rural development, socialist defined, 169–70
rural industrialization, China, 72–3, 75–8, 76–7, 92
Ryan, M., 286, 301n

Saigon, see Ho Chi Minh City
salaries, see income
Sandinista policy, Nicaragua, 290, 291, 292, 296–7, 298–9
savannah district, Guyana, 229
scale economies and urbanization, 30, 31
scientific socialism, 174
security, Nicaragua, 292–3, 298, 299
segmentation and industrialization, 33
self-built housing, Cuba, 263–4, 272
Sétif, Algeria, 142
settler capitalism, 196, 199–200, 201, 202, 204, 207
Shandong province, China, 79–81, 86
Shanghai-Changjiang Delta, China, 89
Shanghai, China, 20, 68–9, 72, 78
Shanxi region, China, 89

Shona tribe, Zimbabwe, 197, *198*, 209
Sidi-bel-Abbès, Algeria, 137, 153
Skikda, Algeria, 153, 159
social democracy, 25, 284
social services, PDR Yemen, 188–9
socialism defined, 1–5, 27, 36, 50–1n, 282–9, 299–300
socialist developing countries
  defined, 1, 3
  heterogeneity, 4–5
  territorial organization, 5–18
  transitional, 214–15
Socialist Republic of Vietnam, 108–28
Sonatrach, 131, 165–6, *166*, 167
South Vietnam, 99, 101–4, 113, 116–17
Soviet Union, rural development strategy, 170–1, 192
Special Programmes, Algeria, 132–6, *133*, *135*, *136*, 154
squatting, Zimbabwe, 201, 206
Stalin, J., 11
state companies, Algeria, 131, 165, *165*
  see also Sonatrach
state farms
  Cuba, 272
  Guyana, 215
  PDR Yemen, 179, *179*, *180*, 185, 191, 192
state investment
  Algeria, 130, 146, *147*, *156*, *157*
  China, 55, *56*, 57, 61, *62*
  Guyana, 237–42, *237*, *238*
  PDR Yemen, 175, *180*, 181
  Vietnam, 106
  Zimbabwe, 209
State Planning Commissions
  Guyana, 215, 236
  Vietnam, 104–5, 108
'State socialist', 283
statistics
  Algeria, 137, 159
  comparison problems, 29–30
  sources for China, 44–5, 51–2n
subsidies, PDR Yemen, 185, 191
sugar production, Guyana, 221, 237, 246
survival levels and industrialization, 40–1, 42

System of Economic Planning and Direction, Cuba, 276
Szelenyi, I., *12*, 12–13, 27, 41–2

TAMS report, Guyana, 229
taxes, PDR Yemen, 188
Tay Son movement, China, 99
Ten-Year Economic Development Plan, China, 59, 60
Thiery, S. P., 158–9, *159*–60, 166–7
Thomas, C. Y., 217–18
Three-Year Plans, PDR Yemen, 175, 181
Tianjin, China, 20, 68–9, *72*
Tiaret, Algeria, 142, 163
Tlemcen, Algeria, 137, 146
towns
  district, Vietnam, 120–1
  new, 35, 50n, 272–3
  PDR Yemen, *172*
trade unions
  Cuba, 277
  Guyana, 239, 246–7
  Zimbabwe, 211
transport
  Algeria, 151, *152*
  PDR Yemen, 188
Tribal Trust lands, Zimbabwe, 197, *198*, *198*, 200, 201, 209

UDI, Zimbabwe, 201, 202–3
underdevelopment and urbanization, 34–5
United States of America, 289, 301n
urban primacy, 6, *8*, *9*, 250, 256, 257
urban–rural differences, Cuba, 250, 253, 254–5
  income, 268–70, *270*–1, 272–3
urbanization, 6–13, *7*, *8*, *9*, *12*, 15
  Algeria, 151–3
  China, 54
  Cuba, 273, 274, 279, 280
  Guyana, 221–3, *222*
  Vietnam, 127
  Zimbabwe, 203, *205*–7, 205, 211, 213
  see also cities; industrialization; towns
USSR, rural development strategy, 170–1, 192

vertical expansion, PDR Yemen, 181,
    189
Vietnam, 10, 13, *18*, 19, 21, 98–128
    cities, 121–6
    National Liberation Front, *16*, 17
    planning, 108–13
    population, 114–19
    territorial organization, 98–108,
        113–14, 119–21
Village, Self-Development
    Programme, Vietnam, 103
villages, *see* communes

wadis, PDR Yemen, 181
    Abyan, 127, 185, *186*, 191
    Hadhramaut, 177, 185, *187*, 190
    Tuban, 177, 181–5, *184*
war, 13–14, *14*
    Vietnam, 106, 122
    Zimbabwe, 203
wasteland reclamation, China, 57
wilayate, Algeria, 132, *132*, *141*, 148,
    153
    comparisons, 137–45, *138–9*, *140*,
        143, *144*, *145*
    state investment, 146, *147*, 157,
        157
White, G., 1–3, 283
Wiles, P. D., 2, 3, 214–15

women
    Guyana, 231, 232, 234
    Zimbabwe, 203, 207
work teams, China, 58, 59, 78, *82*,
    95n
Working People's Alliance, PDR
    Yemen, 216, 237
World Bank, 169–70, 215, 216, 217,
    218, 240
world market exposure, *see*
    international relations

xian, China, 75, 76, 82, *82*, *83*
xiang, China, 76
Xinjiang province, China, 78–9
Xizang province, China, 78–9

Yemen, PDR, *18*, 19, 22–3, 169,
    171–5, 192–3
    rural development, 175–92
Yemeni–Soviet project, 179, 181
youth rustication, programme, China,
    11, 57, 59

Zimbabwe, *18*, 19, 23–4, 194–7,
    212–13
    spatial inequalities, 197–208
    spatial policies, 208–12

Index by A. R. Crook